COMMUNICATIONS SATELLITE HANDBOOK

COMMUNICATIONS
SATELLITE HANDBOOK

WALTER L. MORGAN AND GARY D. GORDON

A Wiley-Interscience Publication

John Wiley & Sons

New York • Chichester • Brisbane • Toronto • Singapore

Copyright © **1989** by John Wiley & Sons, Inc.

All rights reserved. Published simultaneously in Canada.

Reproduction or translation of any part of this work
beyond that permitted by Section 107 or 108 of the
1976 United States Copyright Act without the permission
of the copyright owner is unlawful. Requests for
permission or further information should be addressed to
the Permissions Department, John Wiley & Sons, Inc.

Library of Congress Cataloging in Publication Data:

Morgan, Walter L.
    Communications satellite handbook/Walter L. Morgan and Gary D.
  Gordon.
       p.   cm.
   ISBN 0-471-31603-2
   1. Artificial satellites in telecommunication—Handbooks, manuals,
  etc. I. Gordon, Gary D. II. Title.
  TK5104.M67 1989
  621.38'0422—dc19   88-6077
                                                 CIP

Printed in the United States of America

10 9 8 7 6 5 4 3

PREFACE

We have worked with communications satellites since they were first launched in the early 1960s. We first worked at the RCA (now GE) Astro Space Center (Princeton, New Jersey) and then at the prestigious COMSAT Laboratories (Clarksburg, Maryland), a leader in research on communications satellites. Now we are each consultants, sharing our aerospace expertise with many.

We have accumulated many years of experience and have many friends who are the leading technical experts. We have presented technical papers, given lectures, taught industrial and graduate level courses, and answered thousands of questions. We not only know the answers, but know the questions that many ask. Many of the answers have been incorporated into this book, so that they can be read by readers who have the same, or similar, questions. We have combined our expertise in this, the first full-scope sourcebook of satellite telecommunications.

This book will appeal to advanced students, technicians, engineers, professors, planners, regulators, and other professionals. We wrote it both for the newcomer just starting work in communications satellites and for the expert in one area that needs or wants to know more in another related area. The system planner will find the reasons, limitations, and trade-offs needed in using communications satellites. The earth station operator can learn the reasons the satellites move as they do, and the astronomical limitations they face. The student will find this handbook a useful companion to a textbook, to provide alternate treatment, depth, and breadth to the subject. All can benefit from a better understanding of the subsystems on the spacecraft, and how it operates.

We have written self-contained topics, so the reader can go directly to a point of interest. Most topics can be found through the Contents or the Index. Extensive use of figures and tables pack a maximum of information. Graphs provide sufficient accuracy for both understanding and for many applications. When high precision is needed, the equations provided can be used with an engineering pocket calculator. We have tried to include material of lasting value; where there is change, we have tried to document the trends.

This book is about communications satellites: the traffic, multiple-access techniques, link budgets, the spacecraft bus, and the geostationary orbit used. Knowledgeable individuals will find many items not available in other books. As we have tried to fill a need, our book tends to complement other books on communications satellites. If an answer cannot be found in other books, we hope the answer will be here.

A communications satellite system has many facets. A system starts with the traffic to be handled (Chapter 2), the way this traffic flows through the system (Chapters 3 and 4), the spacecraft that supports the traffic (Chapter 5) and the orbital environment in which it is placed (Chapter 6). More on how the book is organized can be seen in the Introduction.

WALTER L. MORGAN
GARY D. GORDON

Clarksburg, Maryland
Washington Grove, Maryland
November 1988

ACKNOWLEDGMENTS

This book is a step in a continuing education program, and is the combined effort of hundreds. We appreciate all who provided inspiration, insight, answers, or questions—at the RCA Space Center (now GE), COMSAT Laboratories, and our clients.

Burt I. Edelson and Fred H. Esch provided special approval and encouragement. Geoffrey Hyde and Leonard Golding helped in the initial formulation of the book. Blanche Reid, Cynthia Miller, Cathy Scott-Cahill, and many others typed and retyped the manuscript. George Telecki and Sam Sieleman provided a final push towards completion. The professional and patient staff at John Wiley & Sons turned our sketches and rough manuscripts into a classic book.

The chapters on telecommunications traffic and technology are the result of training under masters such as Jack Keigler, John Keisling, Bill Garner, and Louis Pollack. Our clients keep reminding us what engineering is all about: practical applications of technology.

Chapter 5 on spacecraft technology is a composite of many technical areas, and many at COMSAT Laboratories helped directly and indirectly. We particularly thank Brij N. Agrawal, Wilfred J. Billerbeck, Richard S. Cooperman, Denis J. Curtin, George R. Huson, Nelson M. Jacobus, Jr., James R. Owens, Alberto Ramos, and Joseph F. Stockel. Encouragement to produce a set of video tapes on spacecraft technology came from Wil F. Zarecor.

For Chapter 6 on Satellite orbits we owe a debt to the COMSAT Orbital Mechanics Group, to William D. Kinney who led the group, and to Victor J. Slabinski, the orbits master par excellence.

Our wives, Emily Morgan and Doris Gordon, have been patient about lost vacations, evenings, and weekends, while this book was written, revised, and redrawn. Without the encouragements of many, this book would not have been possible.

CONTENTS

LIST OF ACRONYMS   xvi

LIST OF CONSTANTS   xxii

LIST OF SYMBOLS   xxiv

1. INTRODUCTION   1

   1.1 Purpose, 1
   1.2 Why Use a Satellite for Communications?, 3
   1.3 What Roles Do Communications Satellites Perform?, 5
       1.3.1 Traditional Markets, 5
       1.3.2 New Markets, 6

   1.4 Basic Elements of Communications Satellite Service, 7
       1.4.1 Space Segment, 7
       1.4.2 Earth Segment, 8

   1.5 Obtaining Access to Satellite, 8
   1.6 Where Is Satellite?, 8
   1.7 How Is Success Measured in Satellite Systems?, 9

2. TELETRAFFIC   11

       Introduction, 11
   2.1 Types of Traffic, 12

2.2   Interfaces between Terrestrial and Satellite Systems, 23
2.3   Sources of Satellite Traffic, 28
2.4   Teletraffic, 42
2.5   Telegrowth, 64
2.6   Telecommunications Networks, 85
      2.6.1   Introduction to Networking, 85
      2.6.2   Networking Forms, 95
      2.6.3   Equipment, 101
      2.6.4   Performance, 101

3. COMMUNICATIONS SATELLITE SYSTEMS   124

      Introduction, 124
3.1   Basic Resources, 124
      3.1.1   Frequencies, 134
      3.1.2   Band and Emission Designations, 180
      3.1.3   Illumination Level and Power Flux Density, 183
      3.1.4   Frequency Reuse, 242
      3.1.5   Utilization of Various Frequency Assignments, 252
      3.1.6   Space Resource, 258
      3.1.7   Satellite Link Delays, 270
      3.1.8   Earth Segment Resource, 273

3.2   System Performance, 275
      3.2.1   Antennas, 276
      3.2.2   Receiving Equipment Parameters, 291
      3.2.3   Transmitting Equipment Parameters, 324
      3.2.4   Link and Link Losses, 326
      3.2.5   Carrier and Noise Levels, 328
      3.2.6   System Performance, 331

3.3   System Modeling, 337
      3.3.1   Space Segment Modeling, 343
      3.3.2   Examples of Modeling, 353

3.4   Overall System Calculations, 354
      3.4.1   Up-Link Budgets, 355
      3.4.2   Down-Link Budgets, 361
      3.4.3   Intersatellite Link Budgets, 369
      3.4.4   Overall Link Performance, 371
      3.4.5   Signal-to-Noise Ratios, 383
      3.4.6   Digital Link Performance, 386
      3.4.7   Summary of Useful Up-Link Equations, 388

3.4.8   Summary of Useful Cross-Link Equations, 391
3.4.9   Summary of Useful Down-Link Equations, 395
References, 399

4. MULTIPLE-ACCESS TECHNIQUES   401

4.1   Overview of Multiple Access, 401
      4.1.1   Access, 401
      4.1.2   Up-Link Signals, 408
      4.1.3   Digital Traffic, 412
      4.1.4   On-Board Regeneration and Remodulation, 416
      4.1.5   Spread Spectrum, 417
      4.1.6   Interservice Links, 418

4.2   Frequency Domain Multiple Access (FDMA), 418
      4.2.1   Basic FDMA Characteristics, 418
      4.2.2   FDMA Impairments, 432
      4.2.3   Interference and Noise Sources, 434
      4.2.4   Single Channel per Carrier, 445
      4.2.5   TV/FM/FDMA and Narrow-Band TV/FM/FDMA, 451

4.3   Time Domain Multiple Access, 458
      4.3.1   Basic TDMA Characteristics, 458
      4.3.2   Types of TDMA, 465
      4.3.3   TDMA Operations, 468
      4.3.4   TDMA Impairments, 484

4.4   Space Domain Multiple Access (SDMA), 490
      4.4.1   Basic SDMA Characteristics, 490
      4.4.2   SDMA with FDMA, 492
      4.4.3   SDMA with TDMA, 498
      4.4.4   SDMA with CDMA, 506
      4.4.5   Other Forms of SDMA, 506

4.5   Code Domain Multiple Access (CDMA), 515
      4.5.1   CDMA Basics, 515
      4.5.2   Spread Spectrum, 517
      4.5.3   CDMA Operations, 522

4.6   Random Multiple Access (RMA), 528
      4.6.1   Introduction to Random Multiple-Access Systems, 528
      4.6.2   RMA Operations, 530
      4.6.3   ALOHA Forms of RMA, 533
      References, 539

5. SPACECRAFT TECHNOLOGY   541

5.1   Environment of Space, 541
       5.1.1   Zero Gravity, 541
       5.1.2   Lack of Atmosphere, 541
       5.1.3   Intensity of Solar Radiation, 542
       5.1.4   Temperature of Space, 543
       5.1.5   Particles in Space, 544
       5.1.6   Earth's Magnetic Field, 546

5.2   Space Configuration and Subsystems, 547
       5.2.1   Communications Subsystem, 550
       5.2.2   Telemetry, Tracking, and Command, 551
       5.2.3   Electric Power, 552
       5.2.4   Attitude Control, 554
       5.2.5   Spacecraft Structure, 555
       5.2.6   Thermal Control, 557
       5.2.7   Reliability, 557

5.3   Communications Subsystem, 558
       5.3.1   Introduction, 558
       5.3.2   Receiver, 561
       5.3.3   Traveling-Wave Tube, 564
       5.3.4   Filters and Multiplexers, 566
       5.3.5   Antennas, 568

5.4   Telemetry, Tracking, and Command, 573
       5.4.1   Introduction, 573
       5.4.2   Telemetry, 578
       5.4.3   Command, 592
       5.4.4   Tracking and Ranging, 597

5.5   Solar Arrays, 602
       5.5.1   Introduction, 602
       5.5.2   Solar Radiation, 603
       5.5.3   Solar Cells, 605
       5.5.4   Reliability and Lifetime, 609
       5.5.5   Mass, 611

5.6   Electric Power, 612
       5.6.1   Introduction, 612
       5.6.2   Difference on Spinner and Body-Stabilized
               Satellites, 613
       5.6.3   Power Budget, 614
       5.6.4   Bus and Bus Regulation, 615

5.6.5 Rechargeable Batteries, 619
5.6.6 Distribution, 628
5.6.7 Conclusion, 629

5.7 Spacecraft Attitude, 630
5.7.1 Introduction, 630
5.7.2 Body-Stabilized Satellites, 636
5.7.3 Simple Spinner Satellites, 637
5.7.4 Dual-Spinner Satellites, 640
5.7.5 External Torques, 642

5.8 Attitude Control, 646
5.8.1 Introduction, 646
5.8.2 Attitude Sensors, 646
5.8.3 Gyroscopes and Wheels, 650
5.8.4 Propulsion System, 651
5.8.5 Control System, 657
5.8.6 Errors and Accuracy, 659

5.9 Structure, 661
5.9.1 Introduction, 661
5.9.2 Basic Elements in Structure Design, 663
5.9.3 Resonance Effects, 671
5.9.4 Material Properties, 677

5.10 Thermal Control, 682
5.10.1 Introduction, 682
5.10.2 Transients, 688
5.10.3 Radiative Coupling in Simple Two-Body Cases, 692
5.10.4 Configuration Factors and Radiation Coupling Factors, 698
5.10.5 Examples of Radiatively Coupled Spacecraft, 706
5.10.6 Heat Transfer by Conduction and Radiation, 712

5.11 Spacecraft Testing, 716
5.11.1 Introduction, 716
5.11.2 Solar Cell Testing, 718
5.11.3 Vibration Testing, 720
5.11.4 Thermal Vacuum Tests, 732
5.11.5 In-Orbit Tests, 736

5.12 Reliability, 737
5.12.1 Introduction, 737
5.12.2 Combining Probabilities, 741
5.12.3 Constant Failure Rates, 742

5.12.4   Wear-Out Distribution, 745
5.12.5   Redundancy, 746
5.12.6   Reliability Model, 752
5.12.7   Improving Reliability, 759
5.12.8   System Trade-Offs, 761
5.12.9   Monte Carlo Method, 766
References, 772

6. SATELLITE ORBITS   **774**

6.1   Fundamentals of Circular Orbits, 774
      6.1.1   Geocentric Equatorial Coordinate Systems, 774
      6.1.2   Planar Projection, 776
      6.1.3   Polar Projection, 778
      6.1.4   Spherical Trigonometry, 779
      6.1.5   Circular Orbit Parameters, 781
      6.1.6   Orbits of Different Radii, 785
      6.1.7   Near-Geostationary Orbits, 786

6.2   Direction of Orbit Normals and of Sun, 790
      6.2.1   Orientation of Orbit Plane, 790
      6.2.2   Solar Eclipses, 794
      6.2.3   Direction of Sun, 797

6.3   Elliptical Orbits in Plane, 802
      6.3.1   Properties of Ellipses, 802
      6.3.2   Kepler's Equation, 805
      6.3.3   Summary of Equations for Elliptical Orbits, 806
      6.3.4   East–West Oscillations due to Eccentricity, 809

6.4   Position from Orbital Elements, 810
      6.4.1   Orientation of Elliptical Orbit, 811
      6.4.2   Satellite Position and Velocity, 812

6.5   Orbital Elements from Position and Velocity, 816
      6.5.1   Mathematical Formulation of Equations, 816
      6.5.2   Classical Orbital Elements, 818

6.6   Earth Station—Azimuth, Elevation, and Range, 820
      6.6.1   Longitude of Satellite, 820
      6.6.2   Azimuth and Elevation, 825
      6.6.3   Range and Range Rate, 828

6.7   Lunar and Solar Perturbations, 831
       6.7.1   Effect of Gravity Gradient on Orbit Normal, 832
       6.7.2   Perturbations due to Moon, 834
       6.7.3   Perturbations due to the Sun and Moon, 838

6.8   Perturbations from Nonspherical Earth, 840
       6.8.1   Effect on Orbit Normal, 840
       6.8.2   Effect on Satellite Longitude, 845

6.9   Stationkeeping in Geostationary Orbit, 849
       6.9.1   North–South Stationkeeping: Inclination, 850
       6.9.2   East–West Stationkeeping: Longitude, 854
       6.9.3   Control of Eccentricity, 858

6.10   Launching into Geostationary Orbit, 859
       6.10.1   Steps to Geostationary Orbit, 863
       6.10.2   Repositioning A Satellite, 865
       6.10.3   Rockets and Launch Vehicles, 867
       Further Reading, 869
       References, 870

INDEX   871

LIST OF ACRONYMS

This is a list of terms used in this and other books. See also the List of Symbols for additional reference. Some acronyms (e.g., SS/TDMA) are a composite of several terms.

8PSK	Eight phase shift keying
ACK	Acknowledgment
A/D	Analog to digital conversion
ADMA	Amplitude domain (or division) multiple access
ADPCM	Adaptive differential pulse-code modulation
AFC	Automatic frequency control
AGC	Automatic gain control
AKM	Apogee kick motor
ALC	Automatic level control
AM	Amplitude modulation
AOR	Atlantic Ocean Region (Intelsat)
AOS	Attitude and orbit control subsystem
ARQ	Automatic repeat request
ASK	Amplitude shift keying
AT&T	AT&T Communications
BAPTA	Bearing and power takeoff assembly
BBC	British Broadcasting Corporation
BER	Bit error rate
B-MAC	Multiplexed analog component TV (type B)
BO	Backoff
BOL	Beginning of life
BPF	Bandpass filter
BPSK	Binary phase shift keying
BSS	Broadcasting satellite service
BT	Bandwidth time product
BW	Bandwidth

CAD	Computer-aided design (or drafting)
CAE	Computer-aided engineering
CAM	Computer-aided manufacturing
CATV	Cable television (community antenna television)
CC	Common carrier
CCIR	Comite Consultatif International des Radiocommunications (International Radio Consultative Committee)
CCITT	Comite Consultatif International Telephonique et Telegraphique (International Telephone and Telegraph Consultative Committee)
CCS	Hundreds of call seconds
CDMA	Code domain (or division) multiple access
CEPT	Conference of European PTTs
C-MAC	Multiplexed analog component TV (type C)
CMSA	Consolidated metropolitan statistical area
CO	Central office
CODEC	Coder–decoder
CONUS	Contiguous 48 states (USA)
CPFSK	Continuous phase frequency shift keying
CPSK	Coherent phase-shift keying
CSC	Common signalling channel
CSSB	Companded single sideband AM
CT	Centre du Transit
CVSDM	Continuously variable-slope delta modulation
CW	Continuous wave
DA	Demand assignment
D/A	Digital to analog
DAMA	Demand assignment multiple access
dB	Decibel
dc	Direct current
DCPSK	Differentially coherent PSK
DEMUX	Demultiplexer
DES	Digital encryption standard
DM	Delta modulation
D-MAC	Multiplexed analog component TV (type D)
D2-MAC	Multiplexed analog component TV (type D2)
DOPSK	Differential offset phase shift keying
DPCM	Differential pulse code modulation
DPSK	Differential phase-shift keying
DQPSK	Differential four phase-shift keying
DS	Direct sequence
DS1, etc.	Digital services
DSB	Double sideband

DSI	Digital speech interpolation
DUV	Digital under voice
EBHC	Equated busy hour call
EDW	Energy dispersion waveform
EFT	Electronic funds transfer
EHF	Extremely high frequency
EIRP	Equivalent isotropically radiated power
EOL	End of life
EPC	Electronic power conditioner
ET	Ephemeris time
E/W	East/west
FBSS	Full band spread spectrum
FCC	Federal Communications Commission
FDM	Frequency division (or domain) multiplex
FDMA	Frequency domain (or division) multiple access
FEC	Forward error correction (or control)
FFT	Fast Fourier transform
FH	Frequency hopping
FITS	Failures in 10^9 hours (see page 740)
FM	Frequency modulation
FSK	Frequency shift keying
FSS	Fixed satellite service
FT	Fiber optics transmission standards
FVDA	Fully variable demand access
GDP	Gross domestic product
GEO	Geostationary (or geosynchronous) orbit
GHA	Greenwich hour angle
GM	Gravitational parameter
GMT	Greenwich mean time
GNP	Gross national product
GRARR	Goddard range and range rate
G/T	Antenna gain to system noise temperature ratio
GTO	Geostationary (or geosynchronous) transfer orbit
H	Horizontal (polarization)
HDTV	High definition television
HF	High frequency
HPA	High power amplifier
IF	Intermediate frequency
IFRB	International Frequency Registration Board

IM	Intermodulation noise or distortion
IMP	Interface message processor
IOR	Indian Ocean Region (Intelsat)
IRIG	Inter Range Instrumentation Group
ISDN	Integrated services digital network
ISL	Intersatellite link
ITU	International Telecommunications Union
IXC	Interexchange carrier
LAN	Local area network
LATA	Local area transport arrangement
LEC	Local exchange carrier
LEO	Low earth orbit
LNA	Low noise amplifier
LNB	Low noise block-down converter
LNC	Low noise converter
LPF	Low pass filter
MAC	Multiplexed analog component TV
MAMSK	Multitone AMSK
MBPC	Multiple burst per carrier
MBPT	Multiple burst per transmitter (or transponder)
MCPC	Multiple channels per carrier
MCPT	Multiple channels per transmitter (or transponder)
MECO	Main engine cutoff
MODEM	Modulator–demodulator
MPSK	M-ary phase shift keying
MSA	Metropolitan statistical area
MSK	Minimum shift keying
MSS	Mobile satellite service
MTS	Message telecommunications service
MUX	Multiplexer
N/S	North/south
NACK	No acknowledgment
Ni-Cd	Nickel cadmium
Ni-H_2	Nickel hydrogen
NOC	Network Operations Center
NOR	Not OR function
NRZ	Non return to zero
NRZ-L	No return to zero—low
NRZ-M	No return to zero—mark
NRZ-S	No return to zero—space
NTSC	National Television Standards Committee

OCC	Other common carrier
OQSPK	Offset four phase shift keying
P.	Grade of service
PA	Power amplifier
PAL	Phase alternate line
PAM	Pulse amplitude modulation
PAMA	Preassigned multiple access
PBS	Public Broadcasting System
PBX or PABX	Private (or private automatic) branch exchange
PCM	Pulse code modulation
PLL	Phase lock loop
PM	Phase modulation
PN	Pseudorandom noise
POR	Pacific Ocean Region (Intelsat)
PSK	Phase shift keying
PTT	Postal, Telegraph & Telephone authority
PWM	Pulse width (or duration) modulation
QAM	Quadrature amplitude modulation
QASK	Quadrature amplitude shift keying
QPSK	Quadrature (or quadriphase) phase shift keying
RARC	Regional Administrative Radio Conference
RBOC	Regional Bell operating company
RCS	Reaction control subsystem
RDSS	Radio determination satellite service
RF	Radio frequency
RIP	Ring interface processor
RMA	Random multiple access
RSS	Root of the sum of the squares
RX	Receiver
RZ	Return to zero
SBPC	Single burst per carrier
SBPT	Single burst per transmitter (or transponder)
SC	Supressed carrier
SCC	Specialized common carrier
SCPC	Single channel per carrier
SDMA	Space domain (or division) multiple access
SECAM	Sequential Couleur A Memorie
SECO	Secondary (sustainer) engine cutoff
SHF	Super high frequency
SLOHA	Slotted ALOHA

SPADE	Single channel per carrier, pulse code modulation, multiple access, demand assignment equipment
SS	Satellite switched
SS	Spread spectrum
SSB	Single sideband AM
SSMA	Spread spectrum multiple access
SSPA	Solid state power amplifier
STS	Space Transportation System (the Shuttle)
T1, etc.	Digital transmission system
TASI	Time-assigned speech interpolation
TASO	Television Advisory Standards Organization
TAT	Transatlantic telephone cable
TDM	Time division (or domain) multiplex
TDMA	Time domain (or division) multiple access
TDRS	*Tracking and Data Relay Satellite*
TH	Transponder (or time) hopping
TIM	Terrestrial interface module
TT&C	Tracking, telemetry and control (or command)
TTY	Teletype
TV	Television
TVRO	Television receive only
TWT	Travelling wave tube
TWTA	Travelling wave tube amplifier
TX	Transmitter
UC	Unit call
UHF	Ultra high frequency
USA	United States of America
UW	Unique word
V	Vertical (polarization)
VCO	Voltage controlled oscillator
VHF	Very high frequency
VOX	Voice activation
VSAT	Very small aperture terminal
VSB	Vertigal sideband
WARC	World Administrative Radio Conference
XMIT	Transmit
XPDR	Transponder

LIST OF CONSTANTS

ELEMENTARY CONSTANTS

Value (dB)	Units	Symbol	Absolute	Quantity
−228.60	dBW/K Hz	k	1.3803×10^{-23}	Boltzmann constant
−35.23	dB km GHz	c_k	0.000299792	Speed of light (km/ns)
−5.23	dB m GHz	c_m	0.299792458	Speed of light (m/ns)
−2.22	dB	η	0.6	Antenna efficiency
4.97	dB	π	3.1415926536	Pi
46.20	dB/km	S	41679	Maximum range
46.25	dB/km	r	42164.570	Geostationary radius

DERIVED CONSTANTS

Value (dB)	Units	Equation	Page Used
20.41	dB/m² GHz²	$(\pi c_m)^2$	187
−21.46	dB m² GHz²	$c_m^2/4\pi$	328
−43.75	dBW/Hz K GHz²	$k(4\pi S/c_k)^2$	389
−64.16	dBW m²/Hz K	$k(4000\ S)^2$	389
70.99	dB/m²/km²	$\pi(2000)^2$	189
72.04	dB/m²/km²	$(4000)^2$	227
92.45	dB/km² GHz²	$(4\pi/c_k)^2$	187
−136.15	dBW/Hz K km² GHz²	$k(4\pi/c_k)^2$	392
−156.56	dBW m²/Hz K km²	$k(4000)^2$	392
163.39	dB/m²	$\pi(2000\ S)^2$	190

Symbol	Units	Equation or Table	Meaning of the Symbol	
164.44	dB/m^2		$(4000\ S)^2$	388
166.66	dB/m^2		$(4000\ S)^2/\eta$	227
169.51	dB/m^2		$\pi(4000\ r)^2$	371
−176.97	dB/m^4 GHz^2 W/Hz K km^2		$k(4000\ c_{\mathrm{m}}/\pi)^2$	393
184.85	dB/GHz^2		$(4\pi S/c_{\mathrm{k}})^2$	388
190.97	dB/GHz^2		$(8\pi r/c_{\mathrm{k}})^2$	369
−207.14	dBW/Hz K m^2 GHz^2		$4\pi k/c_{\mathrm{m}}^2$	476
−227.55	dBW/Hz K		$4k/\pi$	399
−228.60	dBW/Hz K		k	329

LIST OF SYMBOLS

This is a list of symbols to assist the readers in using the equations and text of this handbook. The number of the first equation (or Table) where the symbol is used will lead the reader to the start of the chain of equations. For further assistance, consult the index using key words from the meaning of the term. In some cases the same symbol may have several meanings so be careful to look at all variations. Generally, the one you need is near the equation you are trying to solve.

Symbol	Units	Equation or Table	Meaning of the Symbol
B_{fax}	Hz	2.1	Bandwidth needed for facsimile transmission
f_c	Hz	2.1	Frequency, carrier
H	in	2.1	Facsimile page height
W	in	2.1	Facsimile page width
d	1/in	2.1	Facsimile scanning density
t	min	2.1	Time to transmit a facsimile page
M	%	2.2	Market share
R	#	2.2	Rank
T_{12}	chan	2.3	Traffic between places 1 and 2 in voice channels
s	chan	2.3	Scaling factor
P_1	#	2.3	Population of place 1
P_2	#	2.3	Population of place 2
d_{12}	km	2.3	Distance between places 1 and 2
LAT_1	deg	2.4	Latitude of place 1
LAT_2	deg	2.4	Latitude of place 2
LNG_1	deg	2.4	Longitude of place 1
LNG_2	deg	2.4	Longitude of place 2
k_d	†	2.4	Constant for distance (Table 2.14)
V	#	2.5	V coordinate for distance measuring
H	#	2.5	H coordinate for distance measuring
C	†	2.7	Calling rate (average calls per hour)

Symbol	Units	Equation or Table	Meaning of the Symbol
H	hr	2.7	Holding time or average call duration
CCS	†	2.8	Traffic, hundreds of call-seconds per hour
C_s	†	2.8	Calling rate (average calls per hour)
H_s	†	2.8	Holding time or average call duration
$P.$	ratio	2.9	Grade of service
C_b	Calls	2.9	Calls busied out (blocked)
C_m	Calls	2.9	Call attempts made
E	†	2.10	Traffic level in Erlangs
n	†	2.11	Number of circuits in a trunk
n_0	†	2.16	Traffic, current
g	ratio	2.16	Growth rate of traffic
y	yr	2.16	Years
a	ratio	2.19	Availability
t_a	hr	2.19	Time available for use
t	hr	2.19	Time, total
MTBF	hr	2.20	Mean time between failures
MTTR	hr	2.20	Mean time to repair
a_s	hr	2.21	Availability of entire system
a_1	hr	2.21	Availability of subsystem 1
a_n	hr	2.21	Availability of subsystem n
f_{13}	GHz	3.1	Center frequency for BSS channel in regions 1 and 3
N_{c13}	#	3.1	BSS channel number in regions 1 and 3
f_2	GHz	3.2	Center frequency for BSS channel in region 2
N_{c2}	#	3.2	BSS channel number in region 2
f	†	3.3	Frequency (may be in Hz, kHz, MHz or GHz)
c	†	3.3	Speed of light (matching units to f)
λ	†	3.3	Wavelength (metric units to match c and f)
dBW	dBW	3.4	Decibels (referenced to 1 watt)
P_1	W	3.4	Power level being measured or compared
P_2	W	3.4	Power level being used as a reference (1 W)
$\log_{10}$		3.4	Base 10 logarithm
W	dBW/m^2	3.5	Illumination level
L	dB	3.5	Loss
G	dBi	3.5	Gain of an antenna, isotropic reference
EIRP	dBW	3.5	Equivalent isotropically radiated power
S	km	3.6	Distance
f	GHz	3.6	Frequency
k_w	dB	3.11	See Figure 3.22
S_1	km	3.12	Distance from earth station to satellite 1
S_2	km	3.12	Distance from earth station to satellite 2
θ_g	deg	3.12	Topocentric angle (see Figure 3.31)

Symbol	Units	Equation or Table	Meaning of the Symbol
θ_t	deg	3.12	Geocentric angle (see Figure 3.31)
P	dBW	3.13	Power from the receiving antenna
D	m	3.14	Diameter of an antenna
η		3.14	Efficiency (a value between 0 and 1)
PFD_B	dBW/m^2	3.17	Power flux density at edge of earth in B_{CCIR}
B_{CCIR}	Hz	3.17	Bandwidth specified by CCIR (Table 3.22)
B_t	Hz	3.17	Transponder bandwidth actually in use
PI_t	ratio	3.18	Polarization isolation for the total system
PI_u	ratio	3.18	Polarization isolation for the up link
PI_d	ratio	3.18	Polarization isolation for the down link
PI_x	ratio	3.18	Polarization isolation for the cross-link
L_{range}	ratio	3.21	Loss due to the slant range distance
k_r	†	3.22	Slant range constant (Table 3.25)
R	km	3.23	Radius of the earth (6378.14 km)
r	km	3.23	Center of the earth-to-satellite distance
h	km	3.23	Satellite altitude
β	deg	3.23	Coverage angle (see Figure 3.19)
t_{terr}	msec	3.24	Time delay in the terrestrial plant
Q	dB	3.25	Quality factor for a link
G/T_s	dBi/K	3.25	Figure of earth station merit
G	ratio	3.29	Antenna gain ratio (with respect to isotropic)
A	cm^2	3.29	Antenna reflector area
λ	cm	3.29	Wavelength
r	cm	3.30	Antenna reflector radius
D	cm	3.30	Antenna reflector diameter
D_1	cm	3.35	Antenna reflector minor axis
D_2	cm	3.35	Antenna reflector major axis
k_b	†	3.37	Constant (see Table 3.31)
θ_3	deg	3.38	Antenna beamwidth (between −3 dB points)
π		3.40	3.1416
A_b	deg^2	3.40	Antenna beam cross-section area
θ_{31}	deg	3.40	Antenna beam minor axis at −3 dB
θ_{32}	deg	3.40	Antenna beam major axis at −3 dB
e	cm	3.41	Antenna surface tolerance (rms)
η_a		3.42	Efficiency (a value between 0 and 1)
G_{1m^2}	ratio	3.44	Antenna gain (ideal case for reference)
T_s	K	3.45	Noise temperature of the entire system
T_{atten}	K	3.46	Noise temperature due to attenuation
a_{atten}	ratio	3.46	Loss mechanism attenuation (range: 0 to 1)
T_m	K	3.46	Temperature, ambient (generally 290 K)
L_{atten}	dB	3.47	Loss due to the attenuation mechanism
T	K	3.48	Temperature

Symbol	Units	Equation or Table	Meaning of the Symbol
T_{ant}	K	3.48	Noise temperature of the antenna (and sky)
a_r	ratio	3.48	Line loss mechanism attenuation (range: 0 to 1)
T_R	K	3.48	Reference temperature (generally 290 K)
T_{rx}	K	3.49	Receiver noise temperature
T_{A1}	K	3.49	Noise temperature of receiver stage 1
T_{An}	K	3.49	Noise temperature of receiver stage n
T_{final}	K	3.49	Noise temperature of the final receiver
G_1	ratio	3.49	Receiver stage 1 gain
G_3	ratio	3.49	Receiver stage 3 gain
F	ratio	3.51	Noise figure
dBNF	dB	3.51	Noise figure
P	W	3.54	Power into the antenna
α	deg	3.57	Longitudinal separation (see Figure 3.158)
k	dBW/KHz	3.59	Boltzmann constant (-228.6)
C/kT	dBHz	3.59	Carrier to thermal noise with k
C/T	dBW/K	3.58	Carrier to thermal noise
C/kTB	dB	3.60	Carrier to noise in bandwidth B
C/N	dB	3.60	Carrier to noise in bandwidth B
B	Hz	3.60	Bandwidth (usually IF or detector)
C/N_0	dB	3.60	Carrier to noise density ($B = 1$ Hz)
B_{if}	Hz	3.61	Bandwidth for FM using Carson's rule
B_{high}	Hz	3.61	Highest baseband frequency
d_{peak}	Hz	3.61	Peak FM deviation
$()^*$	ratio	3.62	Ratio values of (C/N)
C_t/N_t	†	3.62	Total carrier to noise (ratio if accompanied by *)
C_u/N_u	†	3.62	Up-link carrier to noise (ratio if accompanied by *)
C_d/N_d	†	3.62	Down-link carrier to noise (ratio if accompanied by *)
R	b/s	3.64	Digital transmission capacity in bits/second
E_{b*}	W	3.64	Energy per bit
N_{0Th*}	W	3.64	Noise density at the threshold for a given bit error rate
E_b/N_{0Th}	dB	3.64	Ratio of E_b and N_{0Th}
M_i	dB	3.65	Implementation margins total
P_E	W	3.66	Eclipse power requirement
K	ratio	3.66	Eclipse factor (see text)
P_T	W	3.66	Noneclipse transponder dc power load
P_H	W	3.66	Housekeeping power load
η_b	#	3.66	Boost regulator efficiency
η_r	#	3.66	Storage regulator efficiency

Symbol	Units	Equation or Table	Meaning of the Symbol
P_{ES}	W	3.67	Charging array requirement
η_e	#	3.67	Charging efficiency
η_c	#	3.67	Charge controller efficiency
P_{AR}	W	3.68	Solar array power requirement
η_m	#	3.68	Main regulator efficiency
M_{AR}	kg	3.69	Solar array mass
γ	W/kg	3.69	Solar array specific power
σ	W/kg	3.70	Energy storage element specific power
M_{ES}	kg	3.70	Energy storage mass
P_{tx}	W	3.71	Transmitter dc power requirement
P_{t0}	W	3.71	Transmitter dc input power when there is zero rf output power
β_p	#	3.71	dc to rf conversion efficiency
P_0	W	3.71	Saturated power amplifier power rf output
M_{tx}	kg	3.72	Transmitter mass (per power amplifier)
M_{t0}	kg	3.72	Transmitter mass in the fixed portion
β_w	†	3.72	Transmitter specific mass in W(rf)/kg
X	#	3.73	Up-link (receive) beams
Y	#	3.73	Down-link (transmit) beams
M_T	kg	3.73	Transponder subsystem mass
S_r	ratio	3.73	Reception equipment spares ratio
M_{rx}	kg	3.73	Receiver mass (each)
M_{dm}	kg	3.73	Demodulator mass (each)
M_{dc}	kg	3.73	Decoder mass (each)
M_m	kg	3.73	Cross-connect matrix element mass (each)
S_t	ratio	3.73	Transmission equipment spares ratio
M_{rc}	kg	3.73	Mass of each recoder (if present)
M_{rm}	kg	3.73	Mass of each remodulator (if present)
T	#	3.73	Transmitters per beam
P_T	W	3.74	Communications total power (dc)
P_{rx}	W	3.74	Receiver dc power (each)
P_{dm}	W	3.74	Demodulator dc power (each)
P_{dc}	W	3.74	Decoder dc power (each)
P_m	W	3.74	Cross-connect matrix element dc power (each)
P_{rc}	W	3.74	Recoder dc power (each, if present)
P_{rm}	W	3.74	Remodulator dc power (each, if present)
L	m	3.75	Physical reflector diameter
D	m	3.75	Effective reflector diameter
M_R	kg	3.76	Reflector mass
A	m^2	3.76	Reflector area
ρ (rho)	kg/m^2	3.76	Reflector density
M_A	kg	3.77	Antenna subsystem mass

Symbol	Units	Equation or Table	Meaning of the Symbol
M_f	kg	3.77	Feed mass per beam
M_{sup}	kg	3.77	Mass of supports and mounting hardware
G_t	ratio	3.79	Antenna gain
η_a	#	3.79	Antenna efficiency
θ_3	deg	3.80	Half-power (-3dB) beamwidth of an antenna
$EIRP_c$	W	3.81	Equivalent isotropically radiated power at beam center
a_t	#	3.81	Transmission line factor (see text)
P_{0B}	W	3.82	Saturated power amplifier requirement
BO_0	dB	3.82	Output backoff
M_P	kg	3.83	Payload mass (total)
M_{reg}	kg	3.83	Power regulators mass (total)
M_{dist}	kg	3.83	Power distribution mass (total)
C	deg	3.85	Hypotenuse of an error rectangle (Figure 3.147)
A	deg	3.85	Error rectangle (latitude) (Figure 3.147)
B	deg	3.85	Error rectangle (longitude) (Figure 3.147)
Q_d	dB/carr	3.86	Figure of quality for the down-link per carrier
L_{range}	dB	3.87	Range loss
α	deg	3.87	Orbital separation between two satellites
E	dBμV/m	3.89	Electric field strength
S_t/N_t	dB	3.95	Single-to-noise ratio
d	ratio	3.95	Peak-to-peak modulation index (f/f_m)
f_m	Hz	3.95	Modulation bandwidth
Q_p	dB	3.95	Pre-emphasis and weighting factors
M	dB	3.95	Margins
$()_{req}$	†	3.96	Required value (e.g., decibels)
E_b/N_0	dB	3.98	Energy per bit/noise density
R	dBb/s	3.98	Information rate
L_{add}	dB	3.99	Total additional losses
D_u	m	3.99	Up-link earth station antenna diameter
$EIRP_u$	dBW	3.100	Up-link EIRP
P_u	W	3.99	Up-link power
f_u	GHz	3.101	Up-link frequency
W_u	dBW/m^2	3.106	Up-link illumination level
B_u	Hz	3.110	Up-link bandwidth
M_{du}	dB	3.113	Up-link margin with degradations
R_u	b/s	3.119	Up-link bit rate capacity
$()_x$		3.121	Cross-link subscript
$()_u$		3.99	Up-link subscript
$()_d$		3.141	Down-link subscript
$()_t$		3.98	Total subscript
D	m	3.127	Cross-link antenna diameter

Symbol	Units	Equation or Table	Meaning of the Symbol
$EIRP_x$	dBW	3.131	Cross-link EIRP
P_x	W	3.130	Cross-link power
f_x	GHz	3.121	Cross-link frequency
W_x	dBW/m^2	3.133	Cross-link illumination level
B_x	Hz	3.139	Cross-link bandwidth
R_x	b/s	3.140	Cross-link bit rate capacity
β	†	3.126	See Equation 3.126
M_x	dB	3.113	Cross-link margins
T_{sx}	K	3.132	Cross-link system noise temperature
η_x	#	3.127	Cross-link antenna efficiencies
c	km/sec	3.125	Speed of light (2.998×10^3 km/sec)
t_x	msec	3.124	Cross-link time delay
D_{esd}	m	3.157	Down-link earth station antenna diameter
D_{sad}	m	3.153	Down-link satellite antenna diameter
$EIRP_d$	dBW	3.142	Down-link EIRP
P_d	W	3.141	Down-link power
f_d	GHz	3.141	Down-link frequency
W_d	dBW/m^2	3.145	Down-link illumination level
B_d	Hz	3.153	Down-link bandwidth
R_d	b/s	3.155	Down-link bit rate capacity
M_{dd}	dB	3.113	Down-link margins with degradations
T_{es}	K	3.154	Down-link system noise temperature
η_{ad}	#	3.153	Down-link earth station antenna efficiency
D_{sa}	#	3.153	Down-link satellite antenna efficiency
f_p	#	4.1	FEC bits
d_p	#	4.1	Data bits
R	#	4.1	Coding rate (e.g., 3/4)
C/I	dB	4.2	Carrier-to-interference
$(C/I)*$	ratio	4.2	Carrier-to-interference ratio
N_w	dBW	4.3	Noise power
N_w*	pW	4.3	Noise power
Q_d	dB	4.4	Down-link figure of quality
$a_i(t)$	†	4.5	Carrier code for spread spectrum multiple access
$S_i(t)$	†	4.5	Signal (information) code for spread spectrum multiple access
N	#	4.5	Number of spread spectrum accesses
j	†	4.5	Interfering signal(s)
g	ratio	4.6	Spread spectrum processing gain
P	b/s	4.6	Spread spectrum PN ("chip") rate
R_b	b/s	4.6	Spread spectrum information rate
t	sec	4.5	Time

Symbol	Units	Equation or Table	Meaning of the Symbol
B_S	Hz	F4.72	Spread spectrum bandwidth
IM	W	4.7	Intermodulation noise power
B_{xpdr}	Hz	4.7	Transponder bandwidth
$()_s$		4.8	Spread spectrum subscript
S_s/N_s	dB	4.8	Spread spectrum signal-to-noise ratio at the receiver input
t_{CDMA}	sec	4.10	Time to achieve CDMA lock
C_{STA}	sec	4.10	Rise time of the CDMA receiver circuitry
B_d	Hz	4.10	Postcorrelation receiver bandwidth
R_s	†	4.11	Searches per second by a CDMA receiver
m	†	T4.8	Channel capacity (half-circuits)
$EIRP_d$	dBW	T4.8	Down-link EIRP
L_d	dB	T4.8	Down-link free-space path loss
k	†	T4.8	Boltzmann constant (-228.6 dBW/K/Hz)
VA	dB	T4.8	Voice activity factor (typically 4 dB)
$(C/N_0)_R$		T4.8	Required carrier-to-noise density ratio (dB)
M	dB	T4.8	Miscellaneous loss margins (including rain)
G_{1m^2}	dBi	T4.11	Antenna gain for a theoretically perfect 1 m^2 antenna
BO_i	dB	T4.11	Input backoff
W_u	dBW/m^2	T4.11	Up-link illumination level
W_s	dBW/m^2	T4.12	Saturated down-link illumination level
$C/(N+I)$	dB	T4.13	Carrier to (noise + interference)
$(q_w)^*$	ratio	T4.14	Psometric + pre-emphasis weighting factors
f_u	Hz	T4.14	Test tone deviation at 0 dBm0
f_{max}	Hz	T4.14	Top FDMA baseband frequency
B	Hz	T4.17	Carson's rule bandwidth
b	Hz	T4.14	Channel bandwidth
l	ratio	T4.17	CCIR multichannel load factor
g	ratio	T4.17	Peak factor
XTR	dB	T4.18	Crosstalk ratio for backed-off multicarrier operation
K_p	deg/dB	T4.18	Average AM/PM coefficient
S	ratio	T4.18	Linear gain slope
f_w	Hz	T4.18	Peak deviation of the wanted carrier
f_i	Hz	T4.18	Peak deviation of the interfering carrier
P_t	W	T4.18	Total power
P_i	W	T4.18	Interfering power
M_t	dB	T4.23	Sum of all margins
N	#	T4.31	Number of earth stations or time slots
P	#	T4.31	Number of preamble bits
C	b/s	T4.31	Information bit rate

Symbol	Units	Equation or Table	Meaning of the Symbol
F	sec	T4.31	Frame period(s)
G	sec	T4.31	Guard time(s)
A	ratio	T4.33	Information bits per symbol
Γ	†	5.1	Magnetic field intensity (10 microgauss)
Ω	deg	6.8	Right ascension of ascending node
α	deg	6.1	Right ascension
α_0	deg	F5.84	Angular displacement, seen by satellite
α	#	5.90	Absorptivity to sunlight
α	deg	5.51	Angle of coil plane
α	deg	5.68	Semivertex angle of cone
α	10^{-6}/K	T5.22	Thermal expansion
α_e	#	5.113	Absorptance, effective
α_m	deg	6.79	Right ascension of moon's orbit normal
α_w	deg	6.16	Right ascension of orbit normal
β	#	5.69	Function of plate aspect ratio
β_0	deg	F5.84	Central angle, earth station to subsatellite
δ	deg	6.1	Declination
δ_e	deg	6.26	Angle between sun's rays and orbit plane
δ_m	deg	6.79	Declination of moon's orbit normal
δ_w	deg	6.16	Declination of orbit normal
δ_{ij}	#	5.128	Identity matric
ϵ	#	5.89	Emissivity, thermal
ϵ	#	T5.19	Strain, mechanical
ϵ	deg	6.31	Obliquity of ecliptic (ecliptic–equation angle)
ϵ_e	#	5.108	Emittance, effective
η	#	5.9	Efficiency, antenna
η	#	6.24	Fractional time in eclipse
κ	W/mK	5.140	Thermal conductivity
λ	deg	6.49	Longitude
λ	μm	5.22	Wavelength
λ	rad/sec	5.44	Precession frequency (often called nutation)
λ	FITS	5.161	Failure rate (failures per 10g hr)
l_e	m	5.149	Effective length, thermal
μ	W/m^2	5.110	Thermal radiation from space (defined?)
μ	#	T5.19	Poisson's ratio
μ	m^3/sec^2	T6.1	Gravitational parameter GM
μ_m	km^3/s^2	T6.1	Lunar gravitational constant
μ_s	km^3/sec^2	6.77	Sun's gravitational parameter
ω	deg	6.8	Argument of perigee
ω	rad/sec	T5.14	Angular speed
ϕ	rad	5.76	Phase
ϕ	km^2/sec^3	6.81	Earth's gravitational potential

Symbol	Units	Equation or Table	Meaning of the Symbol
ψ	#	5.136	Fractional suntime
ρ	m	5.65	Radius of gyration of cross section
ρ	kg/m^2	5.72	Mass per area
ρ	km	6.70	Range. Distance from satellite to point on earth
σ	W/m^2K^4	5.88	Stefan-Boltzmann constant ($5.6703 \times 10_{-8}$ W/m_2K_4)
σ	#	5.45	Ratio of moment of inertia, spin to transvers
σ	N/mm^2	&5.19	Stress, mechanical
σ	J/kg K	T5.22	Specific heat
τ	sec	5.46	Time constant
τ_d	sec	5.60	Ratio of two gains, control
θ	rad	T.514	Angle
θ	deg	5.46	Coning angle during nutation
θ	deg	F5.91	Misalignment
θ	deg	6.3	Central angle
θ	deg	6.65	Right ascension of earth station
θ_p	deg	5.43	Pitch angle
θ_r	deg	5.43	Roll angle
θ_s	deg	5.136	Angle to the normal
Ξ	#	5.88	Area, total surface
A	m^2	5.88	Area, total surface
A	m^2	5.140	Area, cross sectional
A	deg	6.67	Azimuth
A,B,C	deg	6.4	Angles of a triangle
C	#	5.102	Constant of integration
C	W	T5.8	Carrier-received power
C	km	6.64	Constant used in earth's geoid
Cd	#	F5.68	Cadmium
D	#	5.32	Depth in discharge of battery
D	N	5.73	Stiffness
D	sec/km	6.54	Parabolic anomaly (similar to)
D	deg/day	6.76	Drift rate of mean set longitude
E	eV	5.22	Energy
E	N/mm^2	T5.19	Modulus of elasticity
E	deg	6.38	Eccentric anomaly
F	N	6.10	Force
F_{ij}	#	5.121	Form fractor, thermal
G	N/mm^2	T5.19	Modulus of rigidity
G	#	5.6	Gain, antenna
G	Nm^2/kg^2	T6.1	Fundamental gravitational constant
GHA	deg	6.58	Greenwich hour angle
H	gauss	5.1	Magnetic field

Symbol	Units	Equation or Table	Meaning of the Symbol
H	#	F5.69	Hydrogen
I	A	5.23	Current, electric
I	Nmsec^2	T5.14	Moment of inertia
I	W/m^2	5.116	Intensity of radiation
I_{mp}	mA	5.26	Current, maximum power
I_s	sec	5.57	Specific impulse, thruster
I_{sc}	mA	5.26	Current, short circuit
I_{sp}	sec	6.84	Specific impluse (thrust/mass flow rate)
I_t	kgm/sec	5.58	Total impluse
J2	#	6.81	Coefficient measuring oblateness of earth
JDO	day	6.58	Julian day at midnight
K	mohm/yr^2	5.34	Constant, aging battery
K	Nm/deg	5.60	Gain, control
K	oct/min	5.151	Sweep rate
K_c	#	5.66	Buckling constant
L	deg	5.1	Latitude, magnetic
L_{mean}	deg	6.30	Sun's mean longitude along ecliptic
L_{sun}	deg	6.30	Longitude of sun along ecliptic
M	kg	5.75	Mass
M	kg	6.1	Mass of earth
M	deg	F6.23	Mean anomaly. Location of satellite in orbit
M	kg	6.83	Mass of spacecraft
M_E	deg	6.30	Earth's mean anomaly
N	Nm	T5.14	Torque (force times lever arm)
N	#	5.59	Nitrogen
N	#	5.29	Number of components
Ni	#	F5.68	Nickel
N_0	W/Hz	T5.8	Noise density
O	#	F5.69	Oxygen
P	N	T5.19	Force
P	W	5.7	Power, electric
P	#	5.163	Probability
P	sec	F6.23	Period of orbit
P,Q,W	#	6.50	Unit vectors defining orbit plane
P_e	#	5.11	Probability of error
Q	W	5.90	Internal dissipation (electric to thermal)
R	mohm	5.34	Resistance, electrical
R	#	5.3	Ratio
R	#	5.158	Reliability
R_e	km	T6.2	Earth's equatorial radius
R_{ki}	m^2	5.127	Radiation coupling factor
R_s	ohm	5.28	Resistance, series

Symbol	Units	Equation or Table	Meaning of the Symbol
R_{sh}	ohm	5.28	Resistance, shunt
S	W/m^2	5.12	Solar constant (power from sun)
S	km	F5.84	Range, station to satellite
S	#	T5.6	Status
S	km	6.64	Constant used in earth's geoid
T	K	F5.2	Temperature
T	day	6.85	Time intervals between maneuvers
UT	hr	6.58	Universal time (similar to GMT)
V	V	5.23	Voltage
V	km/sec	6.23	Velocity of satellite
V_{mp}	mV	5.26	Voltage, maximum power
V_{oc}	mV	5.26	Voltage, open circuit
W	m	F5.106	Width
W	#	6.16	Orbit normal
X	#	5.59	Fraction of ammonia disassociated
a	m^2	5.90	90 Area, projected
a	#	5.23	Constant
a	#	5.7	Ratio of frequencies
a	km	6.12	Semimajor axis of orbit
a,b,c	deg	6.5	Sides of a spherical triangle
b	km	F6.23	Semiminor axis of orbit
b_{ij}	1/m^2	5.134	Internal dissipation term
c	Ws/kgK	5.92	Specific heat
c	km/sec	5.154	Molecular speed, rms
c_{iw}	#	5.135	Coefficient for external input
d	m	5.137	Cylinder height
d	days	6.29	Time in days from 2000 January 0
dV	km/sec	6.83	Velocity change
e	#	6.36	Eccentricity
f	Hz	5.151	Frequency
f	1/yr	5.160	Failure density
g	km/s^2	T6.2	Acceleration of gravity, standard
h	deg	F5.84	Elevation, angle from horizon
h	mm	5.71	Distance between two surfaces
h	sec	5.157	Time interval
h	km^2/sec	F6.23	Angular momentum per unit mass
h	deg	6.68	Elevation above horizon
h_{inf}	deg	6.66	Elevation of star at infinity
i	deg	5.49	Angle of incidence
i	W/m^2	5.104	Irradiance, thermal
i	deg	6.8	Inclination of orbit
i_m	deg	6.79	Inclination of moon's orbit (from ecliptic)

Symbol	Units	Equation or Table	Meaning of the Symbol
k	Ws/K	5.23	Boltzmann constant
l	m	F5.106	Length
m	yr	5/169	Mean time between failures
m	kg	6.10	Mass of satellite
m	kg	6.103	Mass of rocket
$\dot{m}$	kg/sec	5.56	Mass flow rate, thruster
n	rad/sec	F6.23	Angular velocity
n	rad/s	T6.2	Mean angular velocity of earth
p	N/m^2	5.154	Pressure, gas
p	km	F6.23	Parameter (constant of elliptical orbit)
q	C	5.23	Charge, electric (of electron)
q	W	5.88	Heat radiated
q	W	5.119	Total thermal power
r	m	5.137	Cylinder radius
r	km	6.1	Distance from origin
r	km	T6.2	Radius of geostationary orbit
r_A	km	F6.23	Apogee. Satellite farthest from earth
r_g	m	T5.14	Radius of gyration
r_m	km	6.13	Moon–earth distance, average
r_e	km	5.1	Radius of earth
r_p	km	F6.23	Perigee. Satellite closest to earth
r_s	km	6.33	Distance from earth to the sun
$\dot{s}$	km/sec	6.12	Velocity
$\dot{s}_A$	km/sec	F6.23	Satellite velocity at apogee
$\dot{s}_c$	km/sec	F6.23	Characteristic satellite velocity
$\dot{s}_p$	km/sec	F6.23	Satellite velocity at perigee
t	mm	5.70	Thickness
t	sec	6.21	Time
t_i	sec	6.72	Time from passage of ascending node
t_p	sec	6.72	Time from passage of perigee
v	deg	6.8	True anomaly (angle from perigee to satellite)
v_e	km/sec	5.56	Effective exhaust velocity of thruster
w	kg/m^2	5.69	Load per area, uniform
$\dot{w}$	kg/sec	5.56	Weight flow rate, thruster
x	m	5.76	Displacement
x,y,z	km	6.1	Geocentric equatorial coordinates
x_ω	km	6.34	Orbit plane coordinate towards perigee
y_ω	km	6.34	Orbit plane coordinate perpendicular to perigee

NOTE: † means "See meaning for units"; # means "A quantity or number"; * means "Absolute quantity (a ratio, *not in dB*)".

COMMUNICATIONS SATELLITE HANDBOOK

INTRODUCTION

1.1 PURPOSE

This handbook has been compiled to assist the reader in the design of satellite communications systems, including the spacecraft, the earth stations, services, and ancillary equipment.

It is the result of many years of experience in this field by the authors. We hope that by sharing this information with you, you may learn and the industry may profit from our experience.

To assist you in using this handbook, we have provided a detailed table of contents and index. Because of the breadth of the book, some topics may be covered from different aspects in several chapters.

The preparation of a handbook is not a trivial task if it is to have any longevity. One basic objective throughout the writing was to envision the needs of the reader 10 years hence, and because it was impossible to guess the exact question a reader may pose, the second basic objective is to provide sufficient information on how to approach the solution so that a knowledgeable reader could then fashion his or her own technique and eventually the answer. Because of the rapid advances in this business, we have eliminated many examples (which get obsolete at an incredible rate) and have tried to stick to the basics.

The purpose of this handbook is to provide the tools by which solutions to current and future problems may be found. The emphasis is on the fundamental principles in each technology area. Equations and graphs provide tools to make calculations of many important parameters. In general, we have avoided sections on the historical development, latest state-of-the-art developments, detailed descriptions of current satellite programs, or future trends.

The handbook is organized in chapters for convenience. This chapter contains generalized material covering the rationale for using communications satellite systems.

Chapter 2 sets the background identifying the types and sources of satellite traffic and providing tools for the scaling of this traffic with time. It also identifies the particular requirements on the communications system for each of the services provided.

Chapter 3 discusses the various technical disciplines and resources required for design, operation, and evaluation of communications satellite systems.

Chapter 4 describes the various methods by which the resources of a communications satellite system are shared among many users by utilization of various multiple-access techniques. In addition, the chapter deals with the modulation and transmission techniques for impressing information on an rf carrier and transmitting the resulting signals to and from the communication satellites.

Background information for the design of spacecraft is given in Chapter 5.

The launching and operation of a satellite requires understanding of the various orbits and, in particular, the capabilities of the geosynchronous/geostationary orbit. This subject is covered in Chapter 6.

Communications satellites are, of course, only one means of telecommunications transmission. The traditional means include copper wire and microwave point-to-point links. Newer techniques involve the use of optics—either point-to-point infrared or fiber optics. Point-to-point radio systems such as short-wave radio may also be used. Although this handbook concentrates on satellite communications, many of the basic principles apply to all of these forms. Some links between end users may require the use of all of these forms in a chain.

Communications satellite services can be broken down into several categories: point to point, point to multipoint, multipoint to point, and multipoint to multipoint. Satellites tend to provide services that are distance insensitive and, in general, that are not affected by the location of the points being connected.

The satellite services are particularly unique in their capabilities with respect to mobile terminals (such as ships, aircraft, trucks and cars, etc.), to various modes of broadcast operations, and to the spanning of jungles, deserts, oceans, and so on.

The basic elements of a communications satellite service are divided between the space segment and the earth segment. The space segment consists of the spacecraft and the launch mechanism. The earth segment comprises the earth stations and the network control center of the entire satellite system.

Access to a satellite is limited to authorized users. The satellite and its transponders are property of one or more organizations, and unauthorized use constitutes theft of service and is a violation of Federal Communications Commission (FCC) rules. Since the satellite and its transponders are valuable assets, there is usually a charge for their use. The owners, lessors, or sublessors of the satellite should be contacted to obtain access. Many transponders are in use 24 h/day, but others are available for lease or sublease on an hourly basis.

This handbook will concentrate on geostationary satellites. These are satellites that are placed in an equatorial orbit and hover over a particular spot on the earth's equator. On-board controls keep the satellite over that spot. When viewed by an earth station antenna, such satellites appear to be stationary in the sky. Therefore, the earth stations may use antennas that are fixed; this results in a less expensive station with less maintenance requirements than if the satellite were moving.

Satellites in nongeostationary orbits will appear to move with respect to an earth station; therefore, tracking antennas are necessary at the earth station.

1.2 WHY USE A SATELLITE FOR COMMUNICATIONS?

Because of their inherent capabilities, satellites have found uses in the communications community for the following reasons:

1. Distance is no problem for satellites. In general, the cost of sending a message via satellite is the same for a link between two points 2000 km apart as for a link between two points only 10 km apart.

2. Satellites are inherently wideband devices. Tens of megahertz are available in each transponder channel, and each channel may be utilized between any two or more points within the coverage of the satellite. Terrestrial radio links are generally limited to a few low-capacity trunks between major markets.

3. The satellite may technically operate into any station within its antenna's view. As illustrated in Figure 1.1, approximately 42% of the earth's surface is within view of a geostationary satellite. Regulatory restrictions of the International Telecommunications Union (ITU) and national regulatory agencies (e.g., the FCC in the United States) generally limit the actual coverage to a much smaller region.

4. Satellites do not respect heretofore natural limitations such as mountains, cities, deserts, oceans, and so on. Thus, a nation that comprises separated elements (e.g., offshore) may utilize satellites to provide a unified national service. An example of this is the United States with Hawaii and Alaska.

5. A satellite system may serve large and small cities with identical forms of service. Traditional terrestrial telecommunications networks have favored the large, dense cities with the most modern equipment in the system, whereas the rural areas have been handled by older equipment (often retired from the cities). Thus, the flow of industry and its data-processing facilities from the cities to the more rural areas could be accelerated by the availability of satellite communications. This ability of the satellite, to serve equally both the developed and the undeveloped areas of a nation (or the world), would permit diversification of industry that heretofore has been considered impossible because of problems associated with labor, defense, and manufacturing reasons.

6. The wideband go-anywhere service provided by the satellites plus the competition from the terrestrial alternatives has led to the need to investigate the possibilities for satellite systems to serve either new markets, previously underserved markets, or underutilized communications links. This need and desire has resulted in the emergence of specialized satellite television networks for cable television (pay-TV), ethnic and language groups, religious groups, sports, and so on.

7. The availability of satellites has led to the emergence of new common carriers

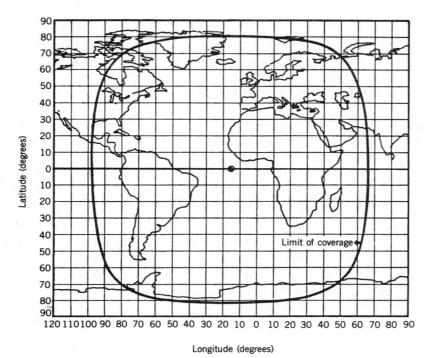

FIGURE 1.1 A geosynchronous satellite has an orbital period equal to the earth's rotation period and any orbital inclination to the earth's equatorial plane. A geostationary satellite is merely a geosynchronous satellite whose orbital plane is in the earth's equatorial plane and therefore hovers in a fixed position at orbital altitude over some equatorial point.

in the United States and Canada. Some of these are owners of satellite systems (e.g., Contel-ASC and Telesat Canada) who are simultaneously in the bulk transponder business at the wholesale level and who market individual circuits at the retail level. Other common carriers take the wholesale leases and sublease individual circuits or television channels to others. The resulting competition produces opportunities for further expansion of the markets for satellite services.

8. The unique capabilities of satellites have resulted in totally new concepts in communications. Traditional methods of midocean communications (high-frequency radio) have been stretched to the limits of technology and still were not providing adequate service under many conditions, particularly in an emergency: Inmarsat is a particular example of a satellite service. It provides mobile voice, data, and low-speed video to ships via satellites. Collection of data from a regional or global sensor platform may be done by using communications satellites to pick up the data and relay it to one or several data reduction points. The same satellite may then be used to disseminate the results (e.g., earthquake warnings) back to the source areas.

9. Very small aperture earth station terminals (VSATs) use satellites to provide immediate access to data bases, branch offices, and management information systems. These terminals are usually placed directly on the customer's premises. For further information on these terminals see Morgan and Rouffet (1988). Morgan, W. L. and Rouffet, D. D. (1988) *Business Earth Stations for Telecommunications*, Wiley, New York.

1.3 WHAT ROLES DO COMMUNICATIONS SATELLITES PERFORM?

Communications satellites serve two basic markets:

1. They serve as a replacement for existing terrestrial service.
2. They provide services to previously unserved or underserved markets.

The role of communications satellites in these markets will be discussed in the following sections.

1.3.1 Traditional Markets

Terrestrial carriers of the 1960s and 1970s used wire cables (above or below the ground and underseas) and chains of microwave towers to provide the traditional analog services. In the late 1970s digital services were added by overbuilding existing microwave links.

Fiber-optics networks were installed in the 1980s and have drawn off much of the satellite traffic that handled bulk telephone and data circuits.

Satellites entered the international communications market in the 1960s in competition with underseas cables in the telephony and telegraphy markets. The satellites in the sixties and early seventies served primarily as alternative paths. The markets for international data exchange had not emerged. The mid-1970s saw the emergence of domestic satellites. Whereas the international satellites were carrying public traffic between PTTs (postal, telephone, and telegraph authorities), most of the U.S. domestic satellite systems concentrated on private (business) telephone and data lines. These services provide point-to-point (and sometimes multipoint) links between an office and a manufacturing plant, store, depot, and so on.

Low-speed data (up to 9600 bits/sec) can be handled on conditioned telephone lines. The satellites entered this market by providing wide-area multipoint networks that had very low bit error rates.

Television networks (commercial and public) had long depended on the long-distance carriers (primarily AT&T) for their interstation connections. Domestic satellites first entered this market by providing a means for extending the real time length of these chains (to Hawaii, Alaska, and Puerto Rico). Later the satellites provided transcontinental hops for feeds from remote stations into the national switching centers from which the programs then entered the traditional terrestrial distribution chain. Still later, satellites were used to replace the land-based links between stations. This permitted programming to enter the network (via satellite) from any of the stations. For the first time it became economically feasible to draw upon the best resources of each of the stations in the network. Further evolution led to transportable earth stations that could go directly to the scene of a news event. At this time the network could be entered from this remote pick-up location via the satellite. This service is now called electronic news gathering.

1.3.2 New Markets

International television was one of the earliest of the new markets developed by satellites. Prior to communications satellites a few primitive attempts were made at using underseas telephone cables for television. The frame rate was reduced until what resulted was more like facsimile than television. An event of a few minutes duration took hours to transmit, even if many successive frames were omitted. The results were recorded and then played into the television networks at a speeded-up rate. The results were as crude as some early silent-movies.

The international communications satellite changed that situation rapidly. High-quality, full-standard color television became immediately available. The term *live, via satellite* became commonplace. So successful was this new service that it was considered most likely that future satellites would mainly carry television, plus a smaller amount of telephone/telegraph traffic. However, this was not to be the case: International television was to be mainly limited to sports, news, and special events. Worldwide dissemination of the entire schedule of the BBC, French and U.S. networks, and so on, was not to take place (for a variety of reasons) for many years. By 1988 television represented only a few percent of the international satellite revenue.

National television by satellite has proceeded on two fronts: In countries where an extensive network was in place (e.g., Canada and the United States), satellites were used to extend the chains to more remote locations; in nations with more formidable terrain problems and lesser developed video facilities (e.g., Brazil, Algeria, and Indonesia), extensive networks could now be deployed via satellite.

Satellites also led to the development of alternative television programming. In the United States this took two forms: Cable television systems were provided with a choice of sponsored and pay-TV programs via domestic satellites; independent (nonnetwork) television stations joined in a loose federation to exchange video news material, sports, and special programming. Thus, the satellite became a tool for pooling and tapping previously remote talents and facilities.

International television involves links terminating in a network control center in New York, Paris, Tokyo, and so on. Domestic television services are often intended to terminate in a local broadcast or cable station. Over the years a large number of home stations emerged that could view these signals. Scrambling was instituted on some channels to control who could receive useful pictures.

High-speed data, particularly at the "natural rates" found inside a computer and desired for computer-to-computer telecommunications, are handled especially well as a new service by satellites. Satellite transponders are inherently wideband devices capable of multimegabit-per-second communications. Very small aperture terminals (VSATs) technically subdivide a transponder's resources into small pieces to provide low data rate services to the commercial mass market. In these cases low costs are the primary concern.

An obvious advantage of the satellite over terrestrial services that are bound to the use of wire, cable, optical fibers, and so on, is in applications where the user is mobile (e.g., ships, planes, automobiles, trucks, trains, and on foot). For these mobile users communications via satellite are now possible.

A businessman can sit in his London office and redirect the ships in his fleet from one destination (where his product is no longer in profitable demand) to a alternative port. In such a case the high-frequency radio of marine communications often took a day for the change in sailing instructions to be received on the ship. Satellites accomplish more in a fraction of a second. The savings realizable in one such instance alone were about equal to the first cost of the ship-board terminal for marine satellite service.

A passenger on board a long-distance airplane can now call his office, delay an appointment if the plane is late, and so on, using an on-board satellite telephone or data terminal.

Speed is of particular importance in the case of mobile emergencies. Help for a damaged ship, plane, or vehicle is often delayed by the lack of convenient, reliable, and organized communications. Many studies have shown the relationship between emergency response time and the odds of survival. Mobile users carry equipment that automatically reports their location to a central facility. This may be used to prevent congestion and collisions and to make better use of traffic lanes and terminal facilities. By knowing their locations, more users can be safely accommodated.

1.4 BASIC ELEMENTS OF COMMUNICATIONS SATELLITE SERVICE

Communications satellite systems may be subdivided into two segments: space and earth.

1.4.1 Space Segment

The space segment includes the satellites (with all of their on-board equipments), the launch vehicles, and the facilities required to place, maintain, and replace these satellites. This includes the telemetry, tracking, and control (TT&C) functions.

1.4.2 Earth Segment

The most obvious equipment in the earth segment are the earth stations with their antennas, rf amplifiers, modulation/demodulation, and baseband equipments. Equally important are the links to and from the terrestrial networks or users. Another portion is the organization for marketing of the unique satellite services: This is done within the existing organization (e.g., a satellite common carrier or a retailer of telecommunications services).

1.5 OBTAINING ACCESS TO SATELLITE

Chapter 4 discusses the various methods of multiple access to a satellite. These may be summarized as methods that permit one user at a time to use a given facility (e.g., a satellite transponder) and methods that permit many simultaneous users. Time domain (or time division) multiple access (TDMA) is an example of the first case and frequency domain multiple access (FDMA) of the latter.

These two basically different forms of accessing have subforms that combine with the various modulation methods (AM, FM, and PM) and among themselves to form still other methods of communications. Out of this bewildering set of choices emerge the few in current use.

Access to a satellite and its system may be done on a preassigned basis. Each earth station has an assigned time or frequency slot that is infrequently (e.g., perhaps once a year) changed. As such, it is like a dedicated leased line.

Demand assignments are those made at the time of and for the duration of each request (e.g., for the period of a single telephone call or a burst of data bits). These assignments are made within a fraction of a second from the time of the request. When the user has completed his requirement the circuit is "torn down" and returned to the pool for reassignment. On-demand services are akin to the dial-up telephone facilities. Unlike the preassigned dedicated services, the on-demand service may be configured to permit a user to establish contact with any other properly equipped subscriber.

1.6 WHERE IS SATELLITE?

There is a variety of orbits for communications satellites. Out of all these choices a few obvious and standardized orbits appear. The earliest communications satellites (*Score*, *Echo*, *Courier*, *Telstar*, and *Relay*) used low orbits (altitudes of a few hundred to ten thousand kilometers) because of the limitations imposed by their launch vehicles: The *Syncom* satellite demonstrated the possibilities of the geosynchronous and geostationary orbits. The geostationary orbit† has become the world

---

†Defined as an orbit having an inclination of near zero with respect to the earth's equatorial plane and a period equal to the earth's rotation period (see Chapter 6).

standard for most communications satellites. If all or part of this important resource (i.e., the geostationary orbit) were to become unavailable, communications satellite programs would suffer seriously. Such unavailability might result from debris in or intersecting the orbit (i.e., from "space junk" or man-caused destruction of satellites) or from overcrowding, resulting in excessive mutual interference or destruction.

The high-latitude geography of the Soviet Union, may be served by either geostationary satellites, or by an elliptical orbit as exemplified by the *Molniya* series of satellites. In this program a 12-h orbit is inclined about 65° (where the tendency for a change in orbital inclination is zero) and one of the apogees (the highest point in the orbit) is made to occur over the USSR in the Northern Hemisphere. Multiple satellites that are phased along the orbit track and tracking earth stations are required.

1.7 HOW IS SUCCESS MEASURED IN SATELLITE SYSTEMS?

The most common criteria for determining the success of a satellite communications system are the amount of traffic it handles and its economic return on the funds invested. Certainly, the International Satellite Organization (Intelsat) has met these criteria. After an initially explosive beginning (typical of many new services) the traffic growth rate has settled at a still impressive 15–20% per year in 1988. (See Figure 1.2). It has returned 17% per year on the funds invested by its members.

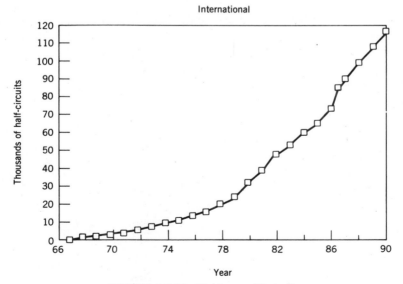

FIGURE 1.2 Total Intelsat satellite traffic.

TABLE 1.1 Advantages of Satellites

1. Wideband capability
2. Wide area coverage readily possible
3. Distance-insensitive costs
4. Counterinflationary cost history
5. All users (large, small, rural, and urban) have same access possibilities
6. Point to point, point to multipoint (broadcast), and multipoint to point (data collection) are all possible
7. Inherently suited for mobile applications
8. Readily compatible with new technology (e.g., with both digital computers and analog high definition TV)
9. Capabilities sometimes greater than immediate needs
10. New and fresh concepts and opportunities
11. Services directly to the users premises.

TABLE 1.2 Limitations of Satellites

1. High initial investment in space segment normally required
2. New investment may be required in earth stations
3. Short satellite lifetimes (7–10 y)
4. Orbit/spectrum crowding, frequency sharing, power flux density limits
5. Access to end user may be difficult for engineering or regulatory reasons
6. Institutional/legal/regulatory aspects
7. Possible difficult, awkward, or expensive maintenance
8. Launch vehicle reliability

Another form of success may be seen in the new forms of communications that the satellite has made possible (see Section 1.3.2). Growth in satellite traffic must come from the generation of new forms of communications.

Tables 1.1 and 1.2 indicate the general advantages and limitations of communications satellites.

TELETRAFFIC

Introduction

The purpose of this chapter is to provide the reader with the tools for determining how much traffic a telecommunications network should be expected to carry, the types of traffic, the interfaces between the terrestrial and space facilities, the traffic sources, and how the traffic may grow. Sections are provided on the types of networks, the equipment used, and its performance in a network configuration.

Since satellite communications may result in the creation of new forms of traffic (just as it did in the United States in the late 1970s when satellite cable television distribution was introduced), this chapter can provide only the historic traffic growth trends and the underlying mechanisms of expansion. Additional information may be found in the index under terms such as *economy of scale*, *growth*, *new services*, and so on.

Two limitations have been historically present: (1) the frequent underestimation of the amount of future traffic and (2) the lack of facilities to allow new forms of traffic to become significant. The two eventually become intertwined. The lack of foresight leads to an inadequate capacity, which in turn discourages new and innovative communications, which in turn discourages or inhibits adequate foresight, thus continuing the degrading cycle.

The advent of satellites and fiber optics offers the hope, at least initially, of breaking this cycle. Many systems have a beginning-of-life capacity much larger than the demand. This leads to an innovative marketing plan to sell the unused capacity. New services emerge, and the systems' capacities become more and more fully utilized. Generally a new service method (e.g., fiber optics) initially tries to be a replacement medium for existing services. This causes a change in the existing market conditions. Eventually both the new and old media develop specialized services to consume their capacity. In satellites, the alternative may be VSAT (very small aperture terminals) services which are cost effective and may consume large amounts of satellite capacity per channel.

2.1 TYPES OF TRAFFIC

The types of telecommunications traffic may be listed in a variety of manners:

1. By direction:
    a. one-way (such as inbound only or outbound only) and
    b. two-way (typical of voice services).
2. By bandwidth:
    a. narrow-band (on–off states, telegraph and low-speed data);
    b. voice frequency band (to 3 or 4 kHz); and
    c. wideband (56 kHz to hundreds of megahertz).
3. By timeliness:
    a. nonreal time (days);
    b. near-real time (hours); and
    c. real time (seconds or less).
4. By type of network:
    a. switched (public telephone, telegraph, and some complex private networks);
    b. dedicated (point-to-point services such as private lines);
    c. broadcast (radio, television, time, weather warnings);
    d. demand access (VSATs).
5. By type of signaling:
    a. analog (voice, video, and raw measurements) and
    b. digital (data, digitized measurements, digitized voice or video, telegraphy, on–off states).

The various classes of telecommunications are identified in Table 2.1. Some originate at many (n) locations and terminate in one place (n to 1), others are point to point (1 to 1), and some are in a "broadcast" (1 to n). Some are switched services (n to n). Columns 3 and 4 identify a typical analog and bit rates, but often there is a wide variation and definitions vary with the user. Both real- and near-real-time services are included (column 5). The holding (or connect) time is given in column 7.

Instead of handling each circuit by itself, it is more economical to combine them and form groups, supergroups, and so on. These combinations may be merged with others to form larger combinations. This process is called multiplexing. The inverse process is called demultiplexing. The equipments are called a multiplexer ("mux") and a demultiplexer ("demux"). Figure 2.1 and Table 2.2 show how these combinations are used in analog telephony.

Frequency division multiplex involves stacking the signals in the frequency domain. The assigned frequency ranges are provided in Table 2.2.

With the passage of time there is an ever-increasing variety of telecommunications services (Figure 2.2). Closer examination reveals that the newer services tend to require broader bandwidths and be digital.

For analog facsimile the channel bandwidth B_{fax} may be estimated from Equation (2.1). See also Tables 2.3 and 2.4. Currently, facsimile uses white-paper compression to send a text page in less than 1 min:

$$B_{fax} = \frac{HWd^2}{120t} \quad (Hz) \tag{2.1}$$

where H, W = height and width (in inches or units) of material to be transmitted
  d = scanning density in lines per inch or unit
  t = transmission time in minutes.

Figure 2.3 is a plot of this equation.

Figure 2.4 shows an alternate technique based on the image quality desired.

Analog systems are rapidly being replaced with digital telecommunications throughout the telephone system. Whenever fiber optics are used, digital electronics are involved. Analog voice and other signals must be converted to a proper digital format. Approximately 8,000–64,000 b/s are required per voice signal (including signaling and other overhead bits). Work is progressing on systems that use fewer bits per voice frequency half-circuit (Green, 1986).

In analog service each switching office and repeater adds noise to the analog signal. The complete circuit is a sum of all the noises inserted along the line. In the case of digital systems it is possible to regenerate the digital signals in the repeaters, thereby eliminating a lot of the noise. The digital systems have two operational states: either a signal (1) or no signal (0). Unsophisticated equipment can tell the difference between these two states. If noise is encountered in the circuit, it produces erroneous bits. In typical services the number of erroneous bits is very small per repeater (one error in many billions of bits). There are ways of detecting the errors and correcting them by use of special codes. The total link performance is determined by the sum of all the errors. In typical services the bit error rate may be between one error per million and one error per billion bits.

For this and other reasons digital telecommunications are rapidly replacing analog facilities.

The international digital building blocks are shown in Table 2.5. Table 2.6 lists typical terrestrial digital transmission elements. Table 2.7 lists digital source rates. Figure 2.5 shows the North American digital hierarchy. Table 2.8 shows transmission standards for fiber optics.

Television may be transmitted in either analog or digital forms. Various digital television methods are under development with bit rates (including sound, sync, etc.) ranging between 3 and 92.5 Mbits/sec. The higher rates may be needed for broadcast "network quality" service.

Figure 2.6 shows the television spectra as broadcast. Multiplexed analog

TABLE 2.1 Telecommunications Classes

Services (1)	Directions (2)	Typical Rates, Hertz (Analog) (3)	Typical Bits per Second (Digital) (4)	Timeliness (5)	Local (L) or Nonlocal (N) (5A)	Switched ? (6)	Duration (Time) (7)	Input Format (8)
Narrowband (Low-Speed) Services								
Alarms (fire, burglar, equipment failure, etc.)[a]	n to 1	Direct current	Negligible	Real	L	No	Full	Direct current[a]
Vehicular traffic counts and control[b]	n to 1	Negligible	Negligible	Real	L	No	Full	Pulses
Hydrographic data collection[b,c]	n to 1	Negligible	100	Near real	N	No	Polled	Levels
Utility meter readings[b,c]	n to 1	Negligible	100	None	L	—	Polled	Levels
Telescratchpad	1 to 1	Negligible	—	Real	—	—	5 min	x, y positions
Meteorological[b,c]	n to 1	Low	Low	Varies	N	—	Polled	Levels
Manual input of data to computer (retail store sales, credit verification, teaching devices from student, etc.)	n to 1	Low	200	Near real	—	—	1 min to several hours	Keystrokes
Teletype, private reservations networks (terminal to computer)[d]	n to 1	Low	24, 48, 96	Real	—	No	1–5 min	Keystrokes
Information retrieval (terminal to data bank)	n to 1	Low	24, 48, 96	Real	N	Yes	Many minutes	Keystrokes
Telegraphy (TTY/Telex/TWX)[d]	n to n	Low	24, 48, 96	None	N	Yes	Tens of seconds	Paper tapes
Teaching devices (computer to terminal)	1 to n	—	900	Real	L	—	Tens of minutes	Data
Mailgram[d]	n to n	—	900	None	N	Yes	—	Teletype (TTY)
Wire services[d]	1 to n	—	96	Near	N	No	Full	TTY
Voice Band Services								
Domestic (U.S.) voice[e,f]	n to n	4×10^3	32×10^3	Real	—	Yes	3 min	Voice
International voice[f,g]	n to n	3×10^3–4×10^3	32×10^3	Real	N	Yes	3 min	Voice

Service	Connection	Bandwidth 1	Bandwidth 2	Timeliness		Memory	Holding time	Signal
Dial-up public telephone[e,f,h]	n to n	$3 \times 10^3 - 4 \times 10^3$	32×10^3	Real	—	Yes	3 min	Voice
Private or leased telephone[f,h]	1 to 1	$3 \times 10^3 - 4 \times 10^3$	$16 \times 10^3 - 32 \times 10^3$	Real	N	No	5 min	Voice
Computer-to-man with voice answerback	1 to n	3×10^3	—	Real	—	—	1 min	Segmented voice tapes
Voice data dissemination (time, weather, etc.)	1 to n	3×10^3	—	Real or near real	L	—	2 min	Voice tapes
Radio program sound material, AM broadcast material[i,j]	1 to n	$3 \times 10^3 - 8 \times 10^3$	—	Real	—	No	Full	Voice
Television broadcast sound[j]	1 to n	8×10^3	—	Real	N	No	Full	Voice
FM broadcast material[j]	1 to n	15×10^3	—	Real	—	No	Full	Voice
Background (subcarrier) material[k]	1 to n	$4 \times 10^3 - 5 \times 10^3$	$0 - 9.6 \times 10^3$	None	—	—	Hours	Voice
Remote sound pickup[l]	1 to 1	$3 \times 10^3 - 15 \times 10^3$	—	Real	N	—	—	Voice
Low-speed data[m]	1 to 1	$3 \times 10^3 - 4 \times 10^3$	$0 - 9.6 \times 10^3$	Real	—	—	—	Digital
Computer to local terminal (e.g., stock and commodity data trading devices)[m,n]	1 to n	$3 \times 10^3 - 4 \times 10^3$	$0 - 9.6 \times 10^3$	Real	—	—	—	Digital
Electronic funds transfer (EFT) from small terminals to node[n]	n to 1	$3 \times 10^3 - 4 \times 10^3$	$0 - 9.6 \times 10^3$	Near real	—	—	—	Keystrokes
Retail store inventory management stocks, new orders[o]	n to 1	$3 \times 10^3 - 4 \times 10^3$	$0 - 56 \times 10^3$	None	—	—	Long	Digital
Credit card verification	n to 1	—	1.2×10^3	Near real	N	No	Burst	Card reader
Facsimile, interbusiness[e,n]	n to n	$3 \times 10^3 - 4 \times 10^3$	$1.2 \times 10^3 - 9.6 \times 10^3$	Near	N	Yes	3–4 min per page	Video
Facsimile Group III[e,p]	n to n	$3 \times 10^3 - 4 \times 10^3$	$4.8 \times 10^3 - 9.6 \times 10^3$	Near	N	Yes	0.2–12 min per page	Digitized video
Facsimile Group IV	n to n	—	56×10^3	Near	N	Yes	1 sec per page	Digitized video
Facsimile, weather[q]	1 to 1	$3 \times 10^3 - 4 \times 10^3$	$1.2 \times 10^3 - 9.6 \times 10^3$	Near	N	No	Full	Video
Telemetry, scientific and engineering[c]	1 to 1	$3 \times 10^3 - 4 \times 10^3$	$1.2 \times 10^3 - 4.8 \times 10^3$	Near	N	—	Full	Levels
Telemetry, medical	1 to 1	3×10^3	$2.4 \times 10^3 - 9.6 \times 10^3$	Real	—	—	—	Heart, etc.
Slow-scan TV ("freeze-frame")	1 to 1	$3 \times 10^3 - 4 \times 10^3$	$4.8 \times 10^3 - 9.6 \times 10^3$	Real	—	—	—	Video

TABLE 2.1 (*Continued*)

Services (1)	Directions (2)	Typical Rates, Hertz (Analog) (3)	Typical Bits per Second (Digital) (4)	Timeliness (5)	Local (L) or Nonlocal (N) (5A)	Switched ? (6)	Duration (Time) (7)	Input Format (8)
Medium-Speed Services								
Data dissemination	1 to n	3×10^3	2.4×10^3	Near real	—	Broadcast	—	TTY or digits
Seismic data collection	n to 1	2×10^3	Negligible	Real	N	No	Full	Analog
Telephone	n to n	4×10^3	8×10^3–64×10^3	Real	N	Yes	Full	VF
Telephone (private tie lines)[r]	n to n	2 at 4×10^3	2 at 64×10^3	Real	N	No	Full	2 at VF
Banking (EFT between banks)[b,c,m,s]	1 to 1	—	56×10^3	Near real	—	—	—	Digital
Data[m,s]	1 to 1	—	19.2×10^3–56×10^3	Real	—	—	—	Digital
Information retrieval (response path from data bank)[m,s]	1 to 1	—	To 56×10^3	Near real	N	—	—	Digital
High-speed facsimile[s] Group IV	n to n	—	56×10^3	Near real	N	Yes	—	Video
ISDN 2B+D[t]	n to n	2×10^3–64×10^3	144×10^3	Real	—	Yes	—	Pulse code modulation (PCM)
Computer graphics and computer aided manufacturing	1 to 1	—	56×10^3–10×10^6	Near real	—	No	—	Digital
Packet switching (intermodal)[u]	n to n	—	56×10^3	Real	N	Yes	—	Digital
Corporate teleconferencing[u]	1 to 1	—	56×10^3–515×10^3	Real	—	—	—	Video
High-Speed Services								
Computer to computer[v]	1 to 1	—	1.544×10^6	Real	—	No	Long	Digital
Telephony (bulk)[v]	n to n	—	1.544×10^6	Real	N	Yes	Long	PCM
ISDN (23 B + D)[t,v,w]	1 to 1	—	1.544×10^6	Real	L	Yes	—	PCM
Commercial TV[x]	1 to n	4.2×10^6–6×10^6	—	Real	N	No	Long	Analog
Closed-circuit TV[x]	1 to 1	1×10^6–10×10^6	—	—	N	—	—	Analog
High-definition TV[y]	1 to n	10×10^6–100×10^6	—	—	N	—	—	Varies
Data[u,z]	1 to 1	—	60×10^6–120×10^6	Real	N	—	Long	Digital

Telephony in bulk[u,z]	n to n	—	60×10^6–274×10^6	Real	N	Full	Digital
Analog commercial TV[x]	1 to n	4.2×10^6	16×10^6–36×10^6	Real	N	—	Analog
Digital TV[u]	1 to n	—	44×10^6–92.5×10^6	Real	N	—	Digital

[a] Some equipment of this class communicates an alarm in case of an event. Often automatic dial-up equipment is used to send this alarm.

[b] Data of this type may be accumulated and transmitted occasionally rather than continuously.

[c] This form of data is often polled by a central station. The bit rate is for the response burst and includes the overhead bits.

[d] Time column assumes a prepunched tape is used for transmission from the terminal.

[e] Public switched telephone service.

[f] The data rate assumes digital voice at 32 kbits/sec per half-circuit. Other rates are also used (e.g., 64 and 16 kbits/sec).

[g] Submarine cable channel bandwidths tend to be 4 kHz when first installed. As the cable gets busy, the bandwidth may be reduced to 3 kHz (by changing the filters at the cableheads) to obtain additional channels. Intelsat's satellite channels are 4 kHz wide. These may be virtual bandwidths.

[h] Domestic systems using in-band signaling are typically 3 kHz. North American service uses out-of-band signaling and may provide 3–4 kHz service using copper, microwave fiber, or satellite.

[i] 8–15 kHz for studio transmitter links and music pickup.

[j] Double for stereophonic service.

[k] May be recorded by broadcaster during off hours for later playback.

[l] Bandwidth depends on station type (AM or FM) and use (low or high fidelity). This category includes studio transmitter links.

[m] Some data types may be near-real time and some may be packet switched.

[n] Local terminal is assumed to be simple and thus speed limited.

[o] Off hours (middle of night, weekends, etc.) burst rate of 56 kbits/sec may be encountered in VSAT networks.

[p] Facsimile terminal is assumed to be intelligent and CCITT group III. If poor line conditions are encountered, the data speed may be lowered automatically.

[q] Primarily charts and computer-corrected satellite photographs designed to be compatible with facsimile equipment.

[r] Two voice frequency (VF) bands multiplexed at 32 kbits/sec.

[s] May be either full-time lease or public-switched (dial-up) lines. VSATs may also be used.

[t] ISDN uses voice (B or bearer) and delta (D or data) channels via a common transmission media. In North America there are two total bit rates (144 kbits/sec and 1.544 Mbits/sec).

[u] Bit rate is typical and includes overhead bits.

[v] Outside North America this rate is generally 2.048 Mbits/sec.

[w] Outside North America this is 30B + D.

[x] Raw color video rate before modulation.

[y] May be analog or digital source.

[z] Time division multiple access (TDMA) rates.

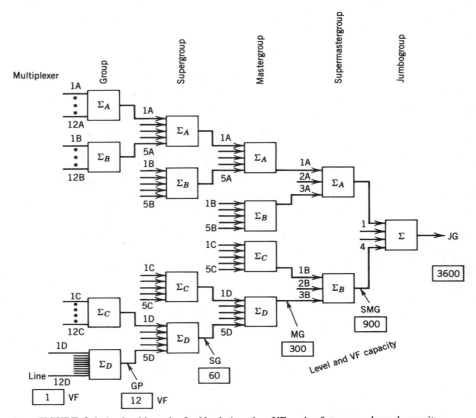

FIGURE 2.1 Analog hierarchy for North America: VF, voice frequency channel capacity.

component (MAC) systems have been designed with separated chromanance (color) luminance (black and white) and audio/data segments.

The majority of the pre-1985 terrestrial circuits were established to handle voice; the application of data to these circuits has had mixed results. All digital networks and modifications of existing equipment (such as digital under voice, or DUV) have provided digital data services at speeds from a few kilobits per second to T1 line rates (1.544 Mbits/sec) and in some instances higher. Analog services may be converted to utilize digital facilities.

Typically the analog input signal is sampled at least twice per cycle.

A voice signal in the range of 300–3400 Hz is generally sampled 8000 times per second. Each sample is encoded using n bits. In a typical example, as shown in Figure 2.7, each sample's numerical value is encoded using 7 bits, and one additional bit per sample is used for signaling and supervision: In this case the 8 bits per sample multiplied by 8000 samples per second results in a bit rate of 8000 $\times$ 8 = 64 kbits/sec.

TABLE 2.2 Analog Building Blocks[a]

Quantity Name	Abbreviation	Equivalent Voice Channels	Bandwidth (kHz)	Bands Occupied (kHz) Intelsat	Terrestrial
Voice frequency pregroup	VF	1	3–4	—	0.300–3.400
Group (12 VF)	GP	12	48 or 56	—	12–60 or 60–108
Supergroup					
5GP	SG	60	240	12–252	12–252, 60–300, or 312–552
10GP	SG	120	480	—	12–552 or 60–552
Basic	SG	300	1200	12–1300[b]	60–1364 or 812–2044
Mastergroup	MG	600	2400	12–2540[b]	60–2792 or 60–2540
Basic supermastergroup	SMG	900	3600	12–4028[b]	8516–12,388
		900	3716		312–4028
		960	4227	12–4028[b]	60–4287 or 8516–12,388
Jumbogroup	JG	3600	—	—	564–17,548
Video North American, Japan (NTSC)	—	—	4200	—	60–4287
Most of rest of world	—	—	6000		6 MHz
Data under voice (DUV), SG or MG/1.544 Mbits/sec (adds a SG to an existing MG)	—	—	—		312–2044

[a]Underline denotes preferred quantities.
[b]Bands occupied for 312, 612, 972, and 972 channels, respectively.

19

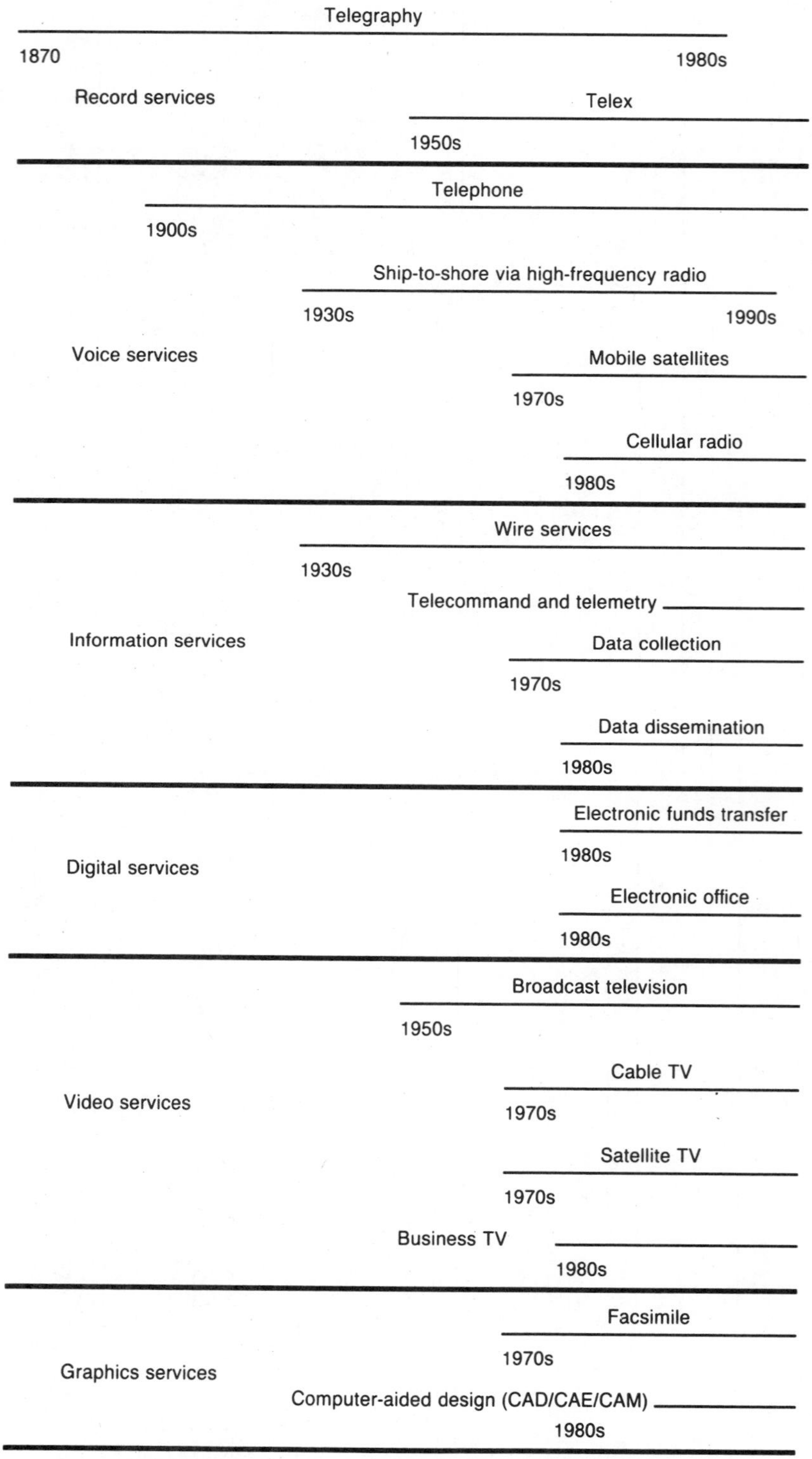

Telegraphy

1870   1980s

Record services   Telex

1950s

Telephone

1900s

Ship-to-shore via high-frequency radio

1930s   1990s

Voice services   Mobile satellites

1970s

Cellular radio

1980s

Wire services

1930s

Telecommand and telemetry

Information services   Data collection

1970s

Data dissemination

1980s

Electronic funds transfer

Digital services   1980s

Electronic office

1980s

Broadcast television

1950s

Cable TV

Video services   1970s

Satellite TV

1970s

Business TV

1980s

Facsimile

Graphics services   1970s

Computer-aided design (CAD/CAE/CAM)
1980s

FIGURE 2.2 Introduction and duration of telecommunications services.

TABLE 2.3 Communications Capacity for Various Analog Facsimile Services (Precompression)

Material	Size (in.) Height	Width	Resolution Lines (in.$^{-1}$)	Time (min)	Typical Bandwidth	Signal-to-noise Ratio dB
First-class mail Pica type[a,b] (3×10^5 bits/page)	11	$8\frac{1}{2}$	67–130	3	3 kHz	10
Weather maps	18	22	96–190	15–30	3 kHz	16
Newspaper plates[c]	15	23	800	4	1.544 MHz	20
No-gray scale Facsimile[d]	11	$8\frac{1}{2}$	700×500	dot matrix	—	—
Half-tone photographs[e]	10	8	3×10^6 per page	100–1000	—	
Envelope data[f]			4×10^5 per front	—	—	
Fingerprints			200	—	—	
Group III Facsimile[g]	10	$8\frac{1}{2}$	200	Under 1	3 kHz	

[a] By 1990, 3×10^{16} bits/yr

[b] Alpha-coded text may be as low as 3×10^4 bits per page at 8 bits per character.

[c] For example, the *Wall Street Journal.*

[d] 700×500 dot matrix yielding quality comparable to that of conventional copying machine.

[e] High-quality facsimile.

[f] 1056 bits per line.

[g] Bit rate depends on the telecommunications line quality (up to 9600 bits/sec).

TABLE 2.4 Advanced Facsimile Methods

Transmission methods	Bits per $8\frac{1}{2} \times 11$-in. Page
Raw facsimile transmission	500,000 (typical)
White-paper compression (skips nonblack elements)	100,000
Pattern recognition compression	50,000
Hybrid optical character recognition and matrix	25,000
Optical character recognition (no graphic capability)	15,000

The 64 kbits/sec is a common digital rate (also referred to as T0 or D0). Through special encoding techniques that use the statistical properties of speech, the number of transmitted bits per sample (n) may be reduced. Through the use of these techniques, 16 and 32 kbits/sec per voice channel are often used. Lower bit rates (to 2400 bits/sec and perhaps less) are being considered for certain applications. Figure 2.8 converts analog to digital services.

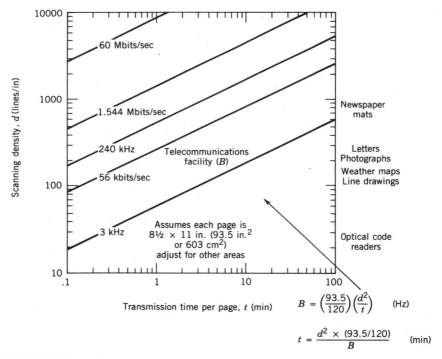

$$B = \left(\frac{93.5}{120}\right)\left(\frac{d^2}{t}\right) \quad \text{(Hz)}$$

$$t = \frac{d^2 \times (93.5/120)}{B} \quad \text{(min)}$$

FIGURE 2.3 Facsimile transmission time versus scanning density for various telecommunications facilities.

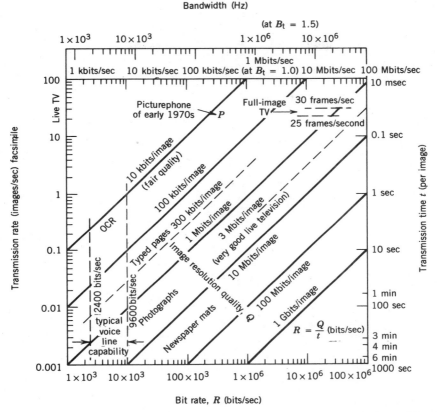

FIGURE 2.4 Bit rate and bandwidth for image services. (*Note*: Does not include provision for overhead bits and coding or take advantage of any inherent redundancy in the source matter (e.g., television).)

2.2 INTERFACES BETWEEN TERRESTRIAL AND SATELLITE SYSTEMS

Figure 2.9 shows the intertwining of computers, typewriters, and communications in the modern business world. In the United States the connections between two locations of the same business have become quite complex, as outlined in Figure 2.10 (Table 2.9 gives examples for Figure 2.10.). The originating location (1) may use a local area network (LAN) within the building or industrial park. The LAN connects to a local loop (2), which may serve other LANs. The local loop provides the connection to the local exchange carrier (LEC), which does the switching at the telephone company level. If the long-distance carrier (5) has an office in the same local area transport arrangement (LATA) as the LEC elements

TABLE 2.5 Digital Building Blocks

Order or Level	Name(s)[a]	Bit Rate (Mbits/sec)	Voice Channels	Preferred Unit or Use
		United States, Canada and Japan		
0	T0, DS0	0.64	1[c]	
ISDN[b]	2B + D	0.144	2	ISDN
ISDN[b]	23B + D	1.488	23	ISDN
1	T1, D1, DS1	1.544	24	Yes
1C	DS1C	3.152	48	
2	T2, D2, DS2	6.312	96	Yes
3	—	32.064	480	Japan
	D3	44.736	672	United States
4	—	97.728	1440	Japan
	D4, DS4	274.176	4032	United States and Canada
5	—	397.200 or 400.352	5760	Japan
	—	562.0	8064	United States
		Conférence Europeene des Postes et Télécommunications *(European Conference on Posts and Telecommunications or CEPT)*		
1	E1 (30B + D)	2.048	30[d]	Yes
2	E2	8.448	120	Yes
3	E3	34.368	480	Yes
4	—	120.00 or 139.264	1920	Yes
5	—	565.148	7680	Yes

[a] The D, DS (digital service), T (transmission), and FT (fiber transmission) prefixes are used interchangeably.

[b] ISDN designates integrated services digital network, B is for the bearer (generally voice), 64 kbit/s channels, and D is the 16 kbit/s delta (digital data) service.

[c] At 64 kbits/sec per channel

[d] Plus a delta channel for signaling and synchronization.

(3), the LEC elements may be directly connected to 5: Otherwise, an inter-LATA carrier (e.g., AT&T) must be used as shown in element 4.

At the distant end of the long-distance service the inverse path (6, 7, 8, and 9) is necessary.

Small microterminal earth stations or VSATs (A and D) and a satellite service (B) with an intermediate hub earth station (C) provides a direct connection between the customer premises 1 and 9. Equipment may be owned by the users with a

TABLE 2.6 Terrestrial Digital Transmission Elements

Order	Name(s)	Bit Rate (Mbits/sec)	Facilities	
0	DS0	0.064	Open lines	United States, Canada
1	D1, DS1	1.544	T1 digital lines, data under voice (DUV), symmetrical wire pair cables, fiber optics	United States, Canada
	D1C	3.152	T1C digital line, symmetrical wire pair cables	United States, Canada
2	D2, DS2	6.312	T2 digital lines	United States, Canada
2		8.448	Sector-screened wire pairs	Europe
		15	2-GHz digital radio, TAT-6 submarine cable (SG)	Japan North Atlantic
		20	4-GHz digital radio (TD3)	United States, Canada
		24	4 GHz-digital radio	Japan, United States
3		34.368	13-GHz digital radio MS43 cables	Italy
3	D3, DS3	44.736	11-GHz digital radio, 6-GHz digital radio (TH), L3 cables, Fiber optics	United States, Canada
		89.472	11-GHz digital radio	United States, Canada
		97.728	Cables	Japan
		120.000	4B2T cables	Europe
		139.264	Class IV, MS43 cables	Europe
		162	L4 cables	United States
		200	11- and 15-GHz digital radio	Japan
4	D4, DS4	274.176	18-GHz digital radio (DR18), T4M, and B3ZS cables, Fiber optics	United States, Canada
		400.352	20-GHz digital radio cables	Japan
5		540	L5 cables, fiber optics	United States

TABLE 2.7 Digital Source Rates

Data Rate	Characteristics	CCITT Recommended
	Low-Speed Terminals	
50–200 bits/sec (50–200 baud)	7.5–12 units per character, start/stop, asynchronous	X
75 baud	Transparent	
200 bits/sec	11 units per character start/stop	X
300 bits/sec	Various formats	
600 bits/sec	Synchronous	X
1.2 kbits/sec	—	
	Medium-Speed Terminals	
2.4 kbits/sec	Synchronous	X
4.8 kbits/sec	—	
9.6 kbits/sec	Synchronous	X
	High-Speed Terminals	
19.6 kbits/sec	—	
48 kbits/sec	Synchronous	X
50/56 kbits/sec[a]	D0, DS0 or T0	
64/56 kbits/sec[b]	D0	
244/256 kbits/sec[a]	—	
1.544 Mbits/sec	DS1 or T1	

[a]With forward error control (FEC).

[b]56 kbits/sec with rate 7/8 coding.

TABLE 2.8 Transmission Standards for Fiber Optics

Name	Mbits/sec	Voice Channels[a]	
		64 kbits/sec	32 kbits/sec
FT1	1.544	24	48
FT1C	3.152	48	96
FT2	6.312	96	192
FT3	44.736 (45)[b]	672	1344
FT3C	89.472 (90)[b]	1440	2880
FT4	274.176 (270)[b]	4032	8064
	432	—	—

[a]Half-circuits (one way). Channel capacity is net (after signaling, supervision, etc.).

[b]Sometimes the exact rate (e.g., 44.736) is rounded off (e.g., 45).

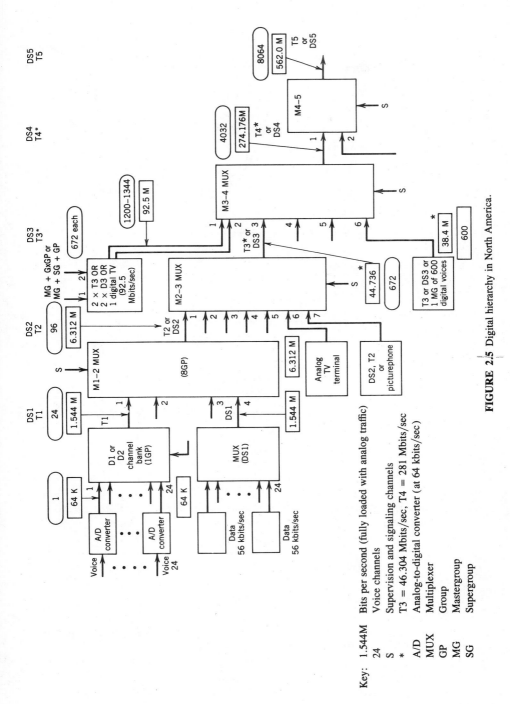

FIGURE 2.5 Digital hierarchy in North America.

Key:
1.544M	Bits per second (fully loaded with analog traffic)	
24	Voice channels	
S	Supervision and signaling channels	
*	T3 = 46.304 Mbits/sec, T4 = 281 Mbits/sec	
A/D	Analog-to-digital converter (at 64 kbits/sec)	
MUX	Multiplexer	
GP	Group	
MG	Mastergroup	
SG	Supergroup	

27

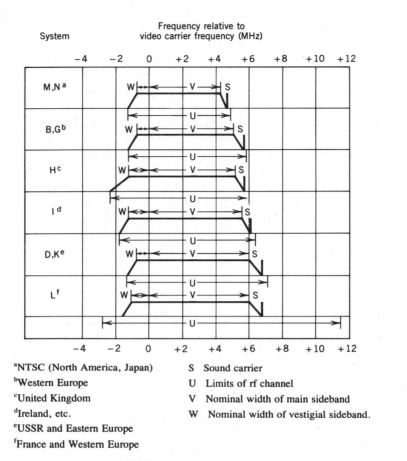

aNTSC (North America, Japan)   S   Sound carrier

bWestern Europe   U   Limits of rf channel

cUnited Kingdom   V   Nominal width of main sideband

dIreland, etc.   W   Nominal width of vestigial sideband.

eUSSR and Eastern Europe

fFrance and Western Europe

FIGURE 2.6 Television spectra.

satellite transponder capacity, and hub stations facilities may be either owned or leased by the user.

2.3 SOURCES OF SATELLITE TRAFFIC

Table 2.10 shows the different types of satellite services and some of the principal influence factors.

A geostationary satellite is in view of about 40% of the earth's surface principally between latitudes 70° N and 70° S. See Figure 2.11.

Satellites in the various geostationary positions around the earth see, respectively, various differing amounts of potential traffic. Figure 2.12 shows the unequal distribution of telephones among the world population (as seen from the geosta-

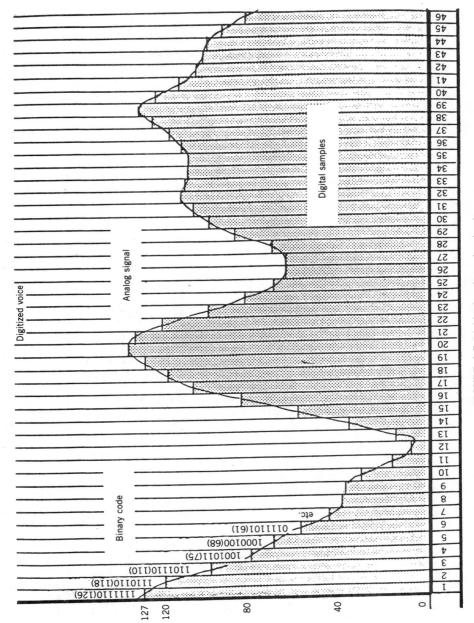

FIGURE 2.7 Digitized analog voice signal.

29

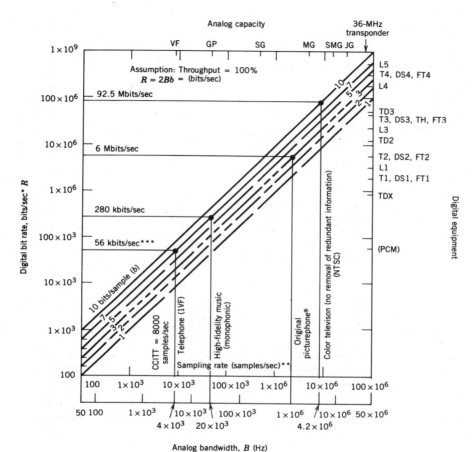

FIGURE 2.8 Analog-to-digital formal conversion chart.

*Without source encoding (rate = 1 : 1)

**At two samples per hertz

***After 1 bit per sample is added for signaling and supervision, this becomes 64 kbits/sec

tionary orbit). A plot of the gross domestic product (GDP) versus telephone quantities, is shown in Figure 2.13.

The number of Intelsat circuits varies, as shown in Figure 2.14. The number of earth stations follows a similar pattern. This is due to the concentration of traffic sources in a few areas (Figure 2.15). Figure 2.16 shows the circuit capacity of an earth station in terms of its relative rank. In the case of the Atlantic Ocean satellites the traffic is as shown in Table 2.11. Television traffic followed a different pattern (Figure 2.17).

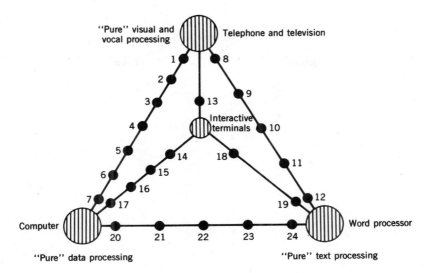

"Pure" visual and vocal processing

Telephone and television

Interactive terminals

Computer

Word processor

"Pure" data processing

"Pure" text processing

1. Digital telephone
2. Message switching
3. Paging
4. Process control
5. Message services
6. Remote job entry
7. Computer graphics and CAD
8. Television via satellite
9. Remote data bases
10. Dial-up dictation
11. Telemail
12. Facsimile
13. Building security
14. Building energy and environmental controls
15. Shared data bases
16. Portable computers
17. Data collection devices
18. Shared text services
19. Desktop publishing
20. Personal computers
21. Calculators
22. Spread sheets and data bases
23. Integrated software packages
24. Laser printers

FIGURE 2.9 Interacting technologies.

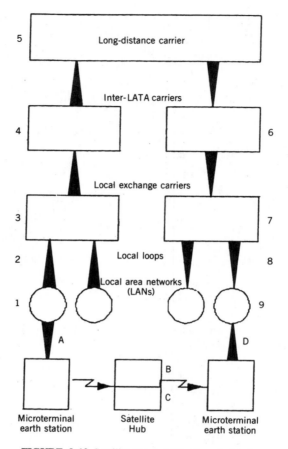

FIGURE 2.10 Satellite-connected business network.

In many cases the traffic is proportional to the population assuming the GNP per capita is evenly spread (as in most developed nations). Figure 2.18 and Table 2.12 show the U.S. population as a function of rank in the top urbanized areas. This approach criterion is more useful than the mere traditional political population because satellite spot beams tend to cover areas rather than arbitrary political boundaries (see also *shaped beams* in the index).

Figure 2.18 illustrates that a few areas contain sizable populations and that the product of rank R and size S is fairly constant (about 40 million people beyond R = 17) as predicted by Ziph's law. The population of the Rth community will be approximately $1/R$ times the size of the largest community. Subsequent census information or application to other countries will change the RS product, but the shape of the figure should remain the same.

$$M = 0.288R^{-1.79} \ (\% \text{ of 1990 U.S. population}) \qquad (2.2)$$

where R and S are defined in the preceding, and M is the market share. Figure

TABLE 2.9 Examples for Figure 2.10

Element	Type	Example
1	Local area network	Wangnet, DECNet, etc.
1–2	Convert digital to analog modulation	—
2	Local loop	Provided by local exchange carrier (e.g., C&P Telephone of Maryland)
3	Local exchange carrier	Provided by local exchange carrier (e.g., C&P Telephone of Maryland)
3–4	Convert to digital for DS1 carrier transmission	AT&T Communications
4	Inter-LATA service	AT&T Communications
5	Long-distance carrier	AT&T Communications, US Sprint, MCI, etc.
6	Inter-LATA carrier	AT&T Communications
6–7	DS1 carrier	AT&T Communications
7	Local exchange carrier	Bell of Pennsylvania
8	Local loop	Bell of Pennsylvania
8–9	Convert analog to digital	Bell of Pennsylvania
9	Local area network	Wangnet, DECNet, etc.
A	Microterminal earth station	Procured from vendor
B	Satellite segment	Procured from vendor
C	Hub earth station	Procured from vendor
D	Microterminal earth station	Procured from vendor

TABLE 2.10 Principal Influences on Satellite Services[a]

Service Type	Sensitive to Number of			
	People	GNP/Capita	Telephones	Television Sets
Telephony				
International			X	
Domestic/Regional		X		
Television				
Direct Broadcast				X
Education TV		X		
CATV			X	
Business Services				
Domestic				
Data		X		
Communications		X		
Private	X			
International		X		
Electronic Funds Transfer		X		
Facsimile		X		
Video Telephone		X		

[a]Only one X has been entered per row. Selection is based on principal factor. In most cases additional factors are also important. For non-U.S. domestic services, there may be other overriding issues.

FIGURE 2.11 Earth as seen by satellite in geostationary orbit at 53° W (307° E) longitude.

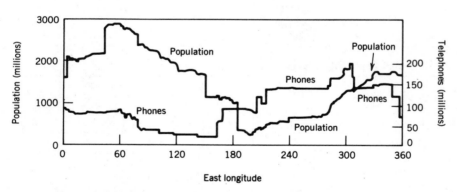

FIGURE 2.12 Number of people and telephones seen by geostationary satellite. © 1975, *TELECOM-MUNICATIONS Magazine*.

2.19 expresses S in terms of the percentage of the total national population. To cover one-third of the national population, at least 15 cities must be included. As the percentage to be covered grows, the number of cities expands very rapidly. Whereas the first 20% can be covered by including five cities, the last 20% requires an enormous number of cities.

Figure 2.20 and Table 2.13 give a specific example (the United States) for a multibeam satellite with separate beams for individual metropolitan areas.

Most telephony traffic is proportional to the number of telephones, which in turn is closely approximated by the population (see Figure 2.21).

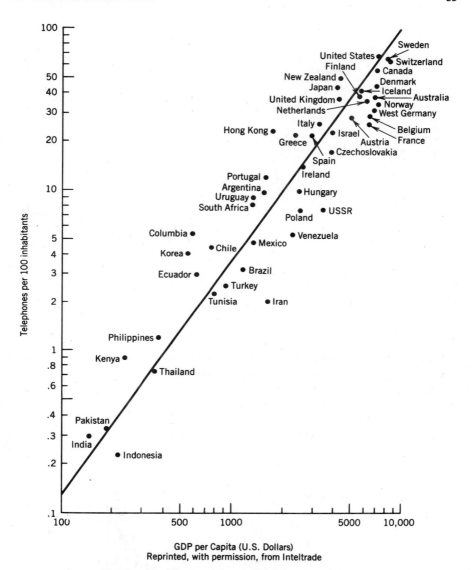

FIGURE 2.13 Telephones versus GDP. Reprinted with permission from Inteltrade.

The traffic (T_{12}) between two locations may be approximated by

$$T_{12} = \left[s(P_1 \times P_2) \right] / d_{12} \quad \text{(voice channels)} \tag{2.3}$$

where P_1, P_2 = populations of places 1 and 2
  d_{12} = distance between places
  s = scaling constant (in voice channels for telephony)

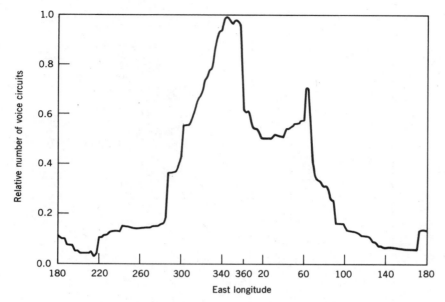

FIGURE 2.14 Relative number of Intelsat circuits versus longitude.

This is the "population potential" equation and applies to parcels, motor truck traffic, and other commerce as well as telephony (see Goode and Machol, 1957).

Equation (2.2) is useful for multipoint services, whereas Equations (2.2) and (2.3) may be used to construct matrices of traffic between earth stations, satellite beams, and the like.

Although it is widely stated that satellites are distance insensitive, the economics

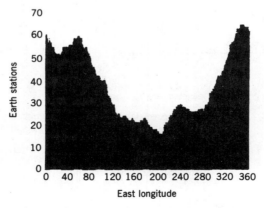

FIGURE 2.15 Intelsat's international telephone traffic versus satellite location. (From Podraczky, 1975.) © 1975, *TELECOMMUNICATIONS Magazine.*

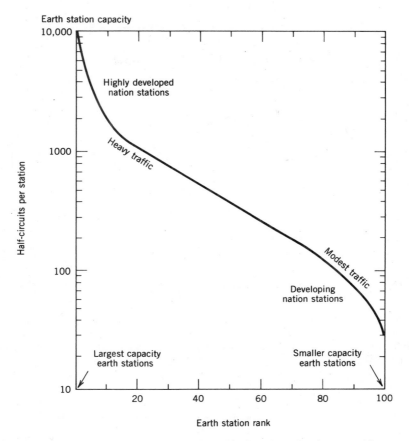

FIGURE 2.16 Amounts of traffic handled by earth stations of various capacities.

TABLE 2.11 Atlantic Ocean Traffic: Intelsat Traffic Distribution

Between big-five nations[a]	30%
From one of big-five to others	55%
Between others	15%

[a]Canada, France, Germany, United Kingdom, and United States.

are such that terrestrial circuits are less expensive for short distances, especially along the more efficient high-traffic-density routes between major centers. Thus, some areas may have city pairs that are served more economically by nonsatellite means such as fiber optics.

The population within a specific beam pattern can be determined from a variety of means. Census information and postal data are often available. With the advent

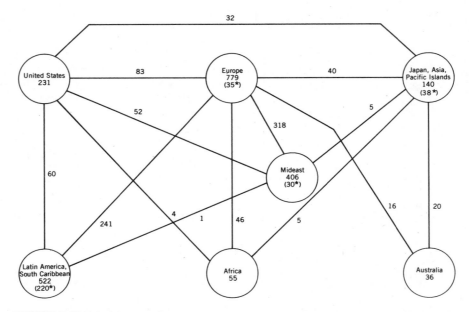

FIGURE 2.17 Television service in Intelsat network (February 1978). Numbers in circles are television hours within each region; numbers along lines are interregional television. *Note*: Additional traffic is carried on domestic and regional satellites and is not shown. *Hours of traffic within the region. © 1978 IEEE.

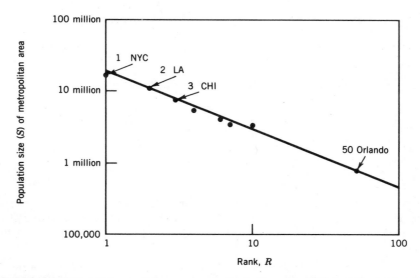

FIGURE 2.18 United States population versus urbanized area rank in 1990 (based on Table 2.12).

TABLE 2.12 Twenty Five Largest Metropolitan Areas,[a] 1985

Metropolitan Area		Population	
		1985, Total (millions)	Annual Percentage Change, 1980–85
New York–Northern New Jersey–Long Island, NY–NJ–CT	CMSA	17.931	0.4
Los Angeles–Anaheim–Riverside, CA	CMSA	12.738	2.0
Chicago–Gary–Lake County (IL), IL–IN–WI	CMSA	8.085	0.4
San Francisco–Oakland–San Jose, CA	CMSA	5.809	1.5
Philadelphia–Wilmington–Trenton, PA–NJ– DE–MD	CMSA	5.776	0.3
Detroit–Ann Arbor, MI	CMSA	4.581	−0.7
Boston–Lawrence–Salem–Lowell–Brockton, MA	NECMA	3.712	0.3
Houston–Galveston–Brazoria, TX	CMSA	3.623	3.0
Dallas–Fort Worth, TX	CMSA	3.512	3.4
Washington, DC–MD–VA	MSA	3.490	1.3
Miami–Fort Lauderdale, FL	CMSA	2.878	1.6
Cleveland–Akron–Lorain, OH	CMSA	2.776	−0.4
Atlanta, GA	MSA	2.472	2.8
St. Louis, MO–IL	MSA	2.412	0.3
Pittsburgh–Beaver Valley, PA	CMSA	2.337	−0.7
Minneapolis–St. Paul, MN–WI	MSA	2.262	1.1
Baltimore, MD	MSA	2.253	0.5
Seattle–Tacoma, WA	CMSA	2.247	1.4
San Diego, CA	MSA	2.133	2.6
Tampa–St. Petersburg–Clearwater, FL	MSA	1.869	2.8
Phoenix, AZ	MSA	1.847	3.8
Denver–Boulder, CO	CMSA	1.827	2.3
Cincinnati–Hamilton, OH–KY–IN	CMSA	1.680	0.2
Milwaukee–Racine, WI	CMSA	1.550	−0.2
Kansas City, MO–KS	MSA	1.494	0.8

[a]As defined June 30, 1986. CMSA = Consolidated metropolitan statistical area; NECMA = New England country metropolitan area; MSA = Metropolitan statistical area.

From: U.S. Department of Commerce, (1986).

of postal codes the task of finding the population in a beam is easier. In the United States, for instance, the first three digits of the zip code identify the regional sorting center of the U.S. Postal Service. Another method involves using telephone area codes. The beam of interest may cover all of some codes and portions of others. The all-inclusive cases are handled by adding their populations. The partial cases have to be handled on a community-by-community basis.

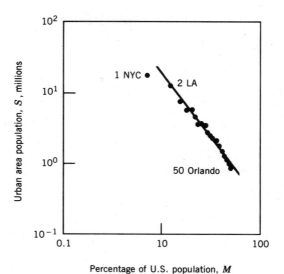

FIGURE 2.19 Percentage of U.S. population in the top R areas.

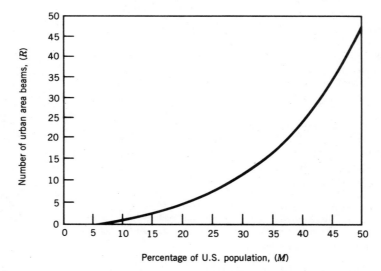

FIGURE 2.20 United States population coverage. *Note*: Each beam covers only *one* urban area with no spillover to adjacent areas.

To find the location of a customer, the postal code locator may be overlain with telephone routing information (city or area codes) and Telex/TWX codes. Since many of these locators are often independent, the location of a town or city can often be found through overlap.

Terrestrial distances between two points (often the fiber or telephone switching

TABLE 2.13 U.S. Population Coverage (1990)

Percentage of U.S. Population Included In beams (M)	Number of City Beams (R)	Total Covered, M (%)
First 20%	5	20
First quarter	8	25
First third	15	33
Second 20%	24	40
First half	46	50

Estimated 1990 U.S. Population: 250 million

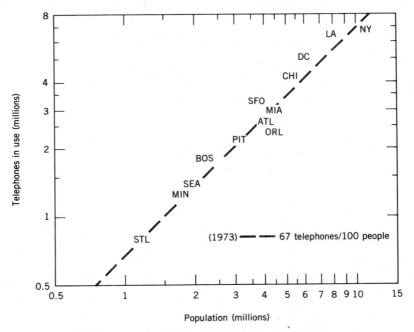

FIGURE 2.21 Population versus telephones for U.S. cities.

offices) is generally needed in an economic analysis of any distance-sensitive telecommunications service.

Two methods are provided for determining the distance from either the longitude and latitude or AT&T's H and V coordinates [see Equations (2.5) and (2.6)].

The great circle distance is the shortest path between any two points (which will be denoted by the subscripts 1 and 2) on earth. This distance, d_{12}, is

$$d_{12} = \cos^{-1}\left[\sin(\text{LAT}_1)\sin(\text{LAT}_2) + \cos(\text{LAT}_1)\cos(\text{LAT}_2)\cos(\text{LNG}_2 - \text{LNG}_1)\right]k_d \qquad (2.4)$$

TABLE 2.14 Distance Constant (k_d)

Unit of Distance	Magnitude of k_d
Kilometers	111.195
Nautical miles	60.000
Statute miles	69.096
Great circle degrees	1.000

where LAT_1, LNG_1 = latitude and longitude of location 1
$\quad\quad$ LAT_2, LNG_2 = latitude and longitude of location 2
$\quad\quad\quad\quad$ k_d = constant to convert to desired units of distance (see Table 2.14)

For example, New York City (40°43' N, 74°1' W) and London (51°30' N, 0°10' W) are 5599.24 km (or 3021.29 nautical miles) apart.

Within the United States the Bell System has established a grid for their rate center offices. The grid coordinates are known as the V and H coordinates. Figure 2.22 shows the contiguous (48 states) grid. The distance between rate centers is given by

$$d_{1,2} = \sqrt{\frac{(V_1 - V_2)^2 + (H_1 - H_2)^2}{10}} \text{ (statute miles)} \tag{2.5}$$

$$= \frac{1}{3.16} \sqrt{(V_1 - V_2)^2 + (H_1 - H_2)^2} \text{ (statute miles)} \tag{2.6}$$

Table 2.15 lists the V and H coordinates for selected cities.

Some populations follow entirely different distributions. Figure 2.23 shows the distribution of merchant ships (including fishing vessels). This is a combination of established trade routes (e.g., from the Gulf states to Europe around the tip of Africa) and the fishing grounds (such as off the west coast of South America).

The distribution of telephone calls by distance has been undergoing a gradual change as long-distance calls becomes more commonplace. Table 2.16 and Figure 2.24 show the distribution of these calls as a function of distance and time. These data pertain to voice toll calls (and thus exclude the many local calls covered by the flat rate for service).

2.4 TELETRAFFIC

The amount of telephone traffic between centers is often expressed in the form of a traffic matrix. Figure 2.25 shows the 10 regional switching centers (RSCs) in the American Telephone and Telegraph network.

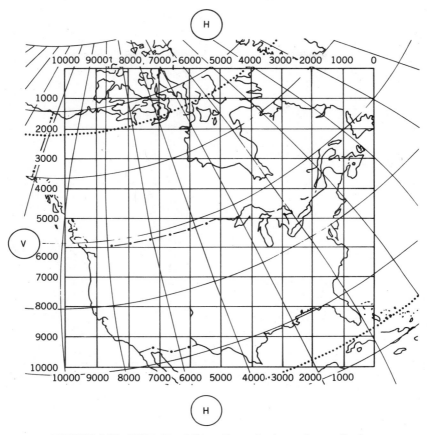

FIGURE 2.22 AT&Ts *V* and *H* coordinates for the contiguous 48 states.

There are $(n(n-1))/2$ possible paths; thus, there are 45 paths between the RSCs. If a satellite switch is used to connect all 10 centers using individual interbeam beams, the 10×10 switch has 45 active connections for each direction of traffic.

Using Figure 2.25 and Table 2.16, the mileage and amount of interregional traffic may be estimated. Table 2.17 is the result. Only half of the matrix is shown since the circuits are two-way and thus the other half is a mirror image.

Table 2.18 is an example of the flow of traffic among the top 10 U.S. cities. Information of this type might be used to size the individual capacities for a multibeam satellite. From this material and the population distribution (see the previous section) it is apparent that the traffic flow in a multibeam satellite is very uneven. As soon as one beam is saturated, the satellite *may* have reached its practical capacity limit even if all of the other beams have substantial unused capacity. Beam shaping, intersatellite links, adaptive coding, and other techniques can be employed to ameliorate this situation.

TABLE 2.15 V and H for Selected Cities

City	V	H
Amarillo, TX	8266	5076
Atlanta, GA	7160	2083
Baltimore, MD	5510	1575
Boston, MA	4422	1249
Chicago, IL	5986	3426
Dallas, TX	8436	4034
Denver, CO	7501	5899
El Paso, TX	9231	5665
Houston, TX	8938	3536
Kansas City, MO	7027	4203
Little Rock, AR	7721	3451
Los Angeles, CA	9213	7878
Miami, FL	8351	0527
New York, NY	4997	1406
Omaha, NE	6687	4595
Phoenix, AZ	9135	6748
Philadelphia, PA	5251	1458
Pittsburgh, PA	5621	2185
Salt Lake City, UT	7576	7065
San Diego, CA	9468	7629
San Francisco, CA	8492	8719
Seattle, WA	6336	8896
Tulsa, OK	7707	4173
Washington, DC	5622-- --	1583

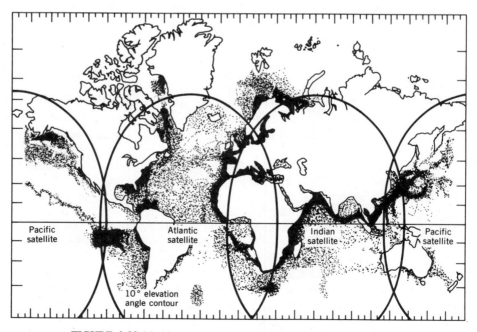

FIGURE 2.23 Maritime population distribution (including fishing vessels).

TABLE 2.16 Distribution of U.S. Calls by Mileage (%)

Mileage Band	1950	1960	1968	1975	1980	1985[a]
0–200	70.3	61.9	49.3	43.0	38.2	34.4
201–500	17.5	19.1	21.5	22.0	23.0	24.0
501–1000	7.9	11.2	15.7	18.0	19.2	20.6
1001–2000	3.2	6.0	9.9	12.0	12.5	13.0
Over 2000	1.1	1.8	3.6	5.0	7.1	8.0
	100	100	100	100	100	100

[a]Projected.

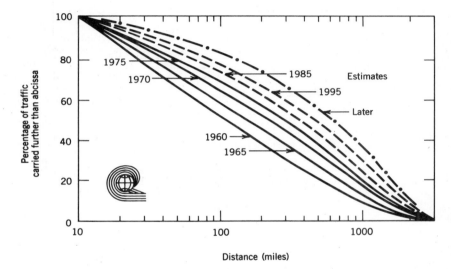

FIGURE 2.24 Traffic versus distance versus date.

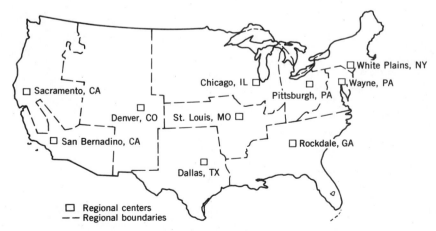

FIGURE 2.25 Ten AT&T regions and their switching centers.

TABLE 2.17 AT&T Interregional Trunk Traffic[a]

Regional Center	PIT 1	PHL 2	NY 3	ATL 4	STL 5	CHI 6	DAL 7	DEN 8	SAC 9	LA 10
1		(12.72) 200	(12.72) 123	(1.20) 530	(1.20) 565	(4.6) 406	(0.78) 1033	(0.78) 1291	(1.42) 2116	(0.78) 1980
2			(12.72) 123	(1.20) 644	(1.20) 787	(1.20) 615	(0.78) 1230	(0.78) 1525	(1.42) 2337	(1.42) 2214
3				(1.20) 775	(1.20) 886	(1.20) 701	(0.78) 1340	(0.78) 1599	(1.42) 2410	(1.42) 2300
4					(4.60) 492	(1.20) 554	(1.20) 664	(0.78) 1156	(0.78) 1980	(0.78) 1759
5						(4.60) 246	(4.60) 492	(1.20) 763	(0.78) 1599	(0.78) 1427
6							(1.20) 775	(1.20) 910	(0.78) 1722	(0.78) 1599
7								(1.20) 640	(0.78) 1389	(0.78) 1132
8									(1.20) 836	(1.20) 713
9										(4.6) 418
10										

[a]Numbers in parentheses represent percentage of total traffic miles. The second number represents mileage.

TABLE 2.18 Traffic in Top 10 U.S. Cities

Metropolitan Area[a]		Percentage of Total Traffic in Top 10 Cities	Percentage of Traffic to New York City
Rank	City		
1	New York City	22.4	—
2	Chicago	17.4	5.4
3	Los Angeles	17.3	5.4
4	San Francisco	8.4	2.3
5	Washington, DC	8.0	2.2
6	Dallas	6.3	1.7
7	Houston	6.3	1.7
8	Minneapolis/St. Paul	5.3	1.4
9	Atlanta	4.8	1.3
10	Denver	3.8	1.0
		100.0	22.4

[a]Includes suburbs and nearby cities.

TABLE 2.19 U.S. Traffic Base

Switched telephone lines in United States[b] 1986	86 million
Residence telephone line[b]	62 million (or 72% of the total)
Business telephone lines[b]	
Switched single lines[b]	16 million (or 19%)
PBX trunk lines[b]	3 million (or 3%)
Centrex lines[b]	5 million (or 6%)
Access lines[b]	111.7 million
Private lines in 1986[b]	2.8 million
Calls	
Local[a]	189×10^9
Toll[a]	10×10^9
Local exchange[b]	84% of all call minutes
Intrastate intralata toll[b]	5% of all call minutes
Intrastate interlata toll[b]	3% of all call minutes
Interstate[b]	8% of all call minutes
Calls per day[a]	
Residence	
Flat rate	4.5 per business day
Usage (measured service)	2 per business day
Business (measured service)	
Single line	9.5 per business day
With PBX trunks	11 per business day
Call duration[a]	
Business	5 min and increasing at 2% per year
Residence	16.5 min and increasing at 4% per year
Number of calls[a]	
Local[a]	Increasing at 2% per year
Long distance[b]	Increasing at 8% per year
Erlang load[a]	
Business	Increasing at 4% per year
Residence	Increasing at 6% per year

[a]Collins (1975).
[b]Huber (1987).

Traffic studies may be carried out on selected portions of the telecommunications marketplace (e.g., video and data). Table 2.19 provides basic data on the AT&T network.

The traffic is not uniform throughout the day, week, or year. Figure 2.26 shows the voice traffic in a 3600-line exchange during a typical day. Business traffic is even more concentrated. Assuming a 10-hr communications day, the peak traffic is typically three times the average level. Extensive use of the network to link microcomputers to a host computer is warping these, especially when the duration

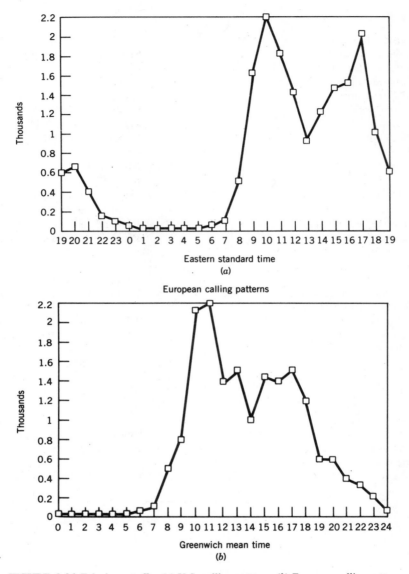

FIGURE 2.26 Telephone traffic: (*a*) U.S. calling patterns; (*b*) European calling patterns.

of the computer "call" is much longer than that of a voice conversation. Figures 2.27 and 2.28 show weekly and annual trends for voice.

Much of the satellite traffic may be from one time zone to another time zone. The traffic between these zones is influenced by each of the diurnal (daily) patterns. Figure 2.29 shows, for example, the composite busy hours for a Saudi Arabia-to-

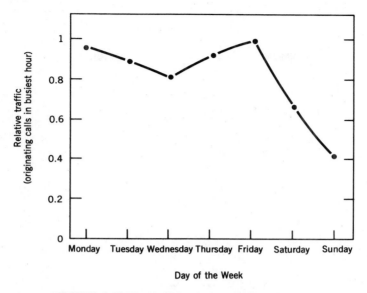

FIGURE 2.27 Weekly U.S. telephone traffic variations.

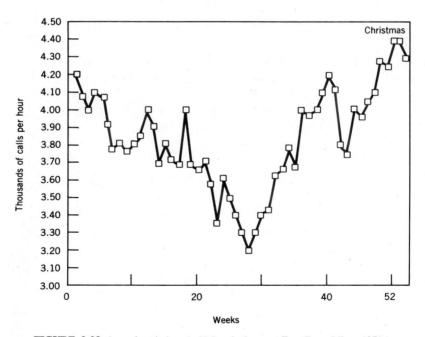

FIGURE 2.28 Annual variations in U.S. telephone traffic. (From Mina, 1971.)

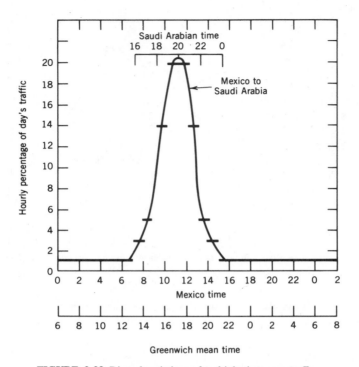

FIGURE 2.29 Diurnal variations of multiple time zone traffic.

Mexico link. These peaks affect the demand-assigned services. If full-time service is elected, these peaks will not show at the satellite but users may get a busy signal at the busy hours. This has the effect of spreading the peaks over a longer time as the subscribers attempt to replace calls blocked during the busy hours.

The international unit of telephone traffic is the erlang:

$$\text{Erlangs} = CH \tag{2.7}$$

where C = average calling rate in calls per hour
H = average call duration or holding time in hours

In North America traffic is often measured in hundreds of call-seconds (CCS) per hour during a busy hour

$$\text{CCS}/\text{hr} = C_s H_s \quad (\text{hundreds of call-seconds per hour}) \tag{2.8}$$

where C_s = average calling rate in calls per hour
H_s = average call duration in hundreds of seconds

For example, if C_s is 5 calls per hour and H_s is 6 min (360 sec, or 3.6 hundreds-of-seconds), then CCS = 5 × 3.6 = 18.

Figure 2.30 is a traffic nomogram for determining the number of satellite transponders needed for telephone service.

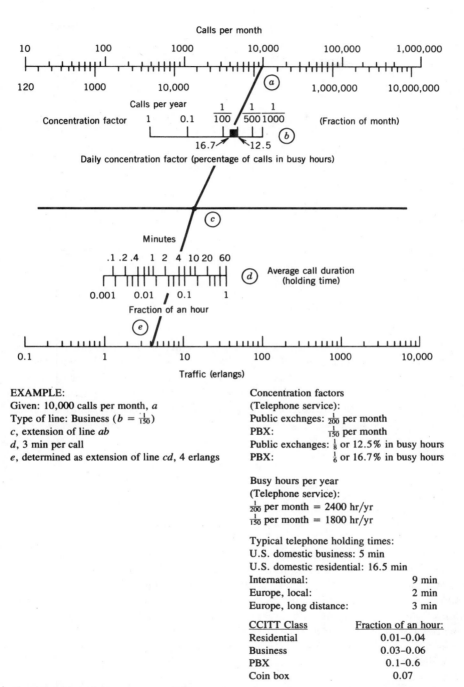

Calls per month

| 10 | 100 | 1000 | 10,000 | 100,000 | 1,000,000 |

(a)

| 120 | 1000 | 10,000 | | 1,000,000 | 10,000,000 |

Calls per year

Concentration factor   1   0.1 $\frac{1}{100}$ $\frac{1}{500}\frac{1}{1000}$ (Fraction of month)

(b)

16.7   12.5

Daily concentration factor (percentage of calls in busy hours)

(c)

Minutes

.1 .2 .4 1 2 4 10 20 60

(d)   Average call duration (holding time)

0.001   0.01   0.1   1

Fraction of an hour

(e)

Traffic (erlangs)

| 0.1 | 1 | 10 | 100 | 1000 | 10,000 |

EXAMPLE:
Given: 10,000 calls per month, *a*
Type of line: Business ($b = \frac{1}{150}$)
c, extension of line *ab*
d, 3 min per call
e, determined as extension of line *cd*, 4 erlangs

Concentration factors
(Telephone service):
Public exchnges: $\frac{1}{200}$ per month
PBX: $\frac{1}{150}$ per month
Public exchanges: $\frac{1}{8}$ or 12.5% in busy hours
PBX: $\frac{1}{6}$ or 16.7% in busy hours

Busy hours per year
(Telephone service):
$\frac{1}{200}$ per month = 2400 hr/yr
$\frac{1}{150}$ per month = 1800 hr/yr

Typical telephone holding times:
U.S. domestic business: 5 min
U.S. domestic residential: 16.5 min

International:	9 min
Europe, local:	2 min
Europe, long distance:	3 min

CCITT Class	Fraction of an hour:
Residential	0.01–0.04
Business	0.03–0.06
PBX	0.1–0.6
Coin box	0.07

FIGURE 2.30 Telephone traffic nomogram. Caution: May not apply to data calls (they may have much longer holding times and different busy hours).

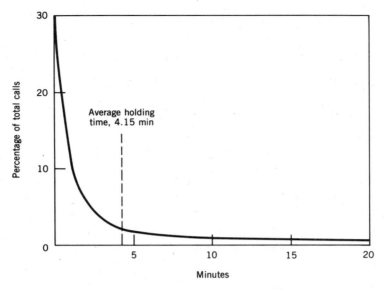

FIGURE 2.31 Typical U.S. telephone call holding times.

Figure 2.31 shows one network's experience with holding time. Some calls are very brief (e.g., a single burst of data), whereas others may tie up the network for extended periods. Still others may require full-time (usually leased-line) service.

Other units of teletraffic intensity are the TU (traffic unit), the equated busy-hour call (EBHC), the unit call (UC), hundreds of call-minutes per hour (C_{min}/hr) and call-seconds per hour (C_{sec}/hr). These are related as follows:

$$1 \text{ erlang } = 1 \text{ TU} \qquad\qquad\qquad 1 \text{ CCS}/hr = 0.0278 \text{ erlang } = 0.0278 \text{ TU}$$
$$= 30 \text{ EBHC} \qquad\qquad\qquad\qquad\quad = 0.833 \text{ EBHC}$$
$$= 36 \text{ CCS}/hr = 36 \text{ UC} \qquad\qquad\quad = \text{UC}$$
$$= 60 \text{ } C_{min}/hr = 3600 \text{ } C_{sec/hr} \qquad = 1.67 \text{ } C_{min}/hr = 100 \text{ } C_{sec/hr}$$

As indicated in the diurnal traffic patterns and the number of calls per day (Figure 2.26), the number of end users seeking access to the telecommunications facilities at any one time is very small. Thus, it is realistic to have a telephony capacity significantly lower than that which would allocate one two-way link per pair of subscribers. The trade-off is between the number of circuits made available and the number of subscriber pairs. This equates to the provision of enough service to satisfy the customers and provide operating revenue at a reasonable and fair cost. The limit or standard of provision is called the grade of service ($P.$). This may be expressed as

$$P. = C_b/C_m \qquad (\text{ratio}) \qquad\qquad\qquad (2.9)$$

where C_b = number of calls blocked (incompleted or "busied-out") by the telecommunications system because of congestion

C_m = number of call attempts made

Usually $P.$ is expressed as the decimal ratio. Thus, $P.10$ indicates that 10% of the calls made failed to be completed because of equipment congestion. Conversely, $P.10$ also means that 90% of the calls were successful.

Figures 2.32 and 2.33 show how the grade of service is related to the calling volume and the number of lines needed. The calling volume is usually measured at the busiest hour. The Poisson distribution is used in Figure 2.32.

Two equations are used to describe the number of trunk lines (n) needed to provide a given grade of service ($P.$) to a specified traffic level (E) as expressed in erlangs.

In North America the Poisson equation is preferred:

$$P. = e^{-E} \sum_{n+1}^{\infty} \frac{E^n}{n!} \qquad (2.10)$$

In Europe (and the CCITT) the erlang B equation is used:

$$P. = \frac{E^n / n!}{\sum_{k=0}^{n} (E^k / k!)} \qquad (2.11)$$

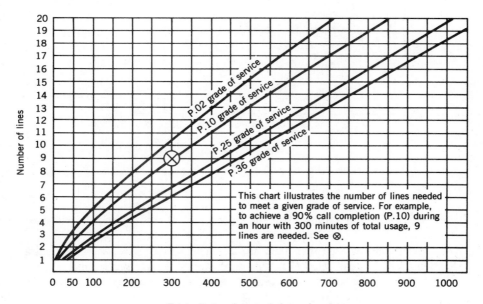

Total minutes of usage during a given hour

FIGURE 2.32 Number of lines needed to meet various grades of service (P.02 grade of service: approximately two busy signals per 100 calls).

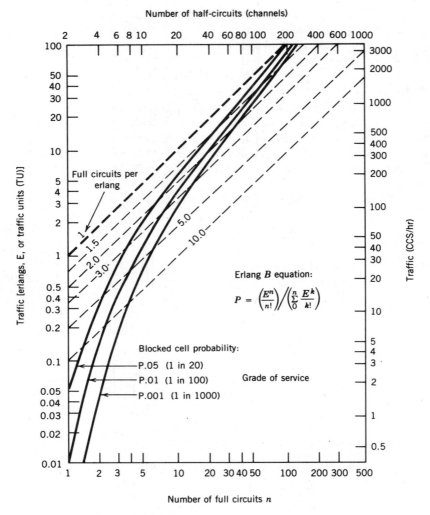

FIGURE 2.33 Traffic level needed to accommodate grade of service.

The erlang *B* equation assumes that the calls arrive at random and that all calls receiving a busy signal are lost; that is, they are not redialed.

As an example, assume that there are 20 circuits in a trunk ($n = 20$) and a traffic density of 12 calls ($E = 12$). This means that if the average number of calls per hour (both completed and blocked) is multiplied by the average call length, the traffic intensity is approximately 12. Therefore, the probability of a call being blocked is determined from Equation (2.11) to be 1 out of 100 ($P. = 0.01$).

Equation 2.11 may be expressed as

$$P. = \frac{E^n}{n!} \left(1 + \frac{E}{1} + \frac{E^2}{2!} + \cdots + \frac{E^{n-1}}{(n-1)!} + \frac{E^n}{n!} \right)^{-1} \qquad (2.12)$$

A method for the efficient computation of the erlang B function is given in Gordon and Dill (1978). The following is an excerpt.

For small trunk groups there are only a few terms, and these are of the same order of magnitude. For large trunks (e.g., $n = 300$ and $E = 277.1255$) there are many terms (301), and the terms near the end of the series are very large. (The final term is 10^{118}.) For efficient calculation, the order of the series must be reversed, so that the more significant terms are computed first. This is accomplished by setting $k = n - j$ as follows:

$$P. = \frac{E^n}{n!} \left(\sum_{j=0}^{n} \frac{E^{n-j}}{(n-j)!} \right)^{-1}$$

$$= \left(\frac{E^n}{n!} \right) \left(\frac{E^n}{n!} + \frac{E^{n-1}}{(n-1)!} + \cdots + \frac{E}{1} + 1 \right)^{-1} \qquad (2.13)$$

If the terms are calculated in this order, the calculations can be terminated when the terms become negligible. The next step is to take the reciprocal and multiply through by $n!/E^n!$:

$$\frac{1}{P.} = \sum_{j=0}^{n} \frac{n!}{E^j (n-j)!}$$

$$= 1 + \frac{n}{E} + \frac{n}{E} \frac{n-1}{E} + \frac{n}{E} \frac{n-1}{E} \frac{n-2}{E} + \cdots + \frac{n!}{E^n} \qquad (2.14)$$

The individual terms now start at unity, increase in magnitude slowly, and then decrease to a very small value. For $n = 300$ and $E = 277.1255$ the largest term is the 24th, whose magnitude is only 2.6; the 150th term is only 10^{-15}; and the remaining terms are even smaller in magnitude.

The series can now be expressed so that each term is a product of the previous term and a simple factor, thereby eliminating the need to calculate powers and factorials of large numbers:

$$\boxed{\begin{array}{l} \dfrac{1}{P.} = \displaystyle\sum_{j=0}^{n} T_j \\[2ex] T_j = \dfrac{n - j + 1}{E} T_{j-1} \\[2ex] T_0 = 1 \end{array}} \qquad (2.15)$$

During the computations the individual terms eventually become very small and have no significant effect on the series sum. A test determines when the calculations can be terminated.

Given a trunk group of n circuits and a traffic intensity of E erlangs, Equation (2.15) may be employed to efficiently calculate the blocking probability or grade of service, P, for any size trunk. Frequently, the grade of service is specified, and the unknown is the maximum traffic intensity that may be served by a given trunk group of circuits. An erlang table for a single grade of service P. and a range of trunks from $n = 1$ to a certain limit, can be calculated as follows:

The first two values ($n = 1$ and $n = 2$) by solving Eq. 2.14 algebraically

An initial estimate of E for the next entry in the table determined by linear extrapolation, and

calculating the grade of service P. with Equation (2.15) and iterating with Newton's Method.

The FORTRAN subroutine and program shown below will generate tables of the erlang B function. The following terms are used:

$A = E =$ traffic intensity in erlangs.

$B = P. =$ probability of call being blocked.

$C = n =$ total number of circuits in trunk group.

The results are shown in Table 2.20.

```
 31            SUBROUTINE ERLF(A, B, MIN, MAX, INC)
       C GDGORDON - APRIL 77 - FINDS A AS A FUNCTION OF B AND C
       C     C IS NUMBER OF CIRCUITS IN A TRUNK
       C     B IS THE GRADE OF SERVICE (FRACTION OF BUSY SIGNALS)
       C     A IS THE TRAFFIC THE TRUNK CAN HANDLE
       C     MIN, MAX, INC ARE THE RANGE AND INCREMENT OF C
       C     IF MIN IS NOT ONE, VALUES OF A FOR C = MIN AND MIN + INC
       C        MUST BE FURNISHED BY THE CALLING PROGRAM
       C     PROCEDURE FINDS TRIAL A (AN) BY EXTRAPOLATION FROM LAST
       C        TWO VALUES, AND THEN FINDS A BY NEWTON'S METHOD.
       C     IF CONVERGENCE FAILS, A IS SET TO -666.
       C
 32            DOUBLE PRECISION A(1), AN, TERM, F, FP, B
 33            INTEGER C
 34            IF (MIN .GT. 1) GO TO 5
 35            A(1) = B/(1.D0 - B)
 36            F = 1.D0 - 1.D0/B
 37            A(2) = (1.D0+DSQRT(1.D0+2.D0*(-F)))/(-F)
 38          5 MIN2 = MIN + 2*INC
       C              OUTER LOOP FINDS INCREMENTS C FROM MIN2 TO MAX
 39            DO 10 C = MIN2, MAX, INC
 40            NC = 1 + (C - MIN)/INC
 41            AN = 2.D0*A(NC-1) - A(NC - 2)
       C              MIDDLE LOOP FINDS VALUE OF AN BY NEWTON'S METHOD
 42            DO 25 J = 1,20
 43            F = 1.D0 - 1.D0/B
 44            FP = 0.
 45            TERM = 1.D0
       C              INNER LOOP SUMS THE TERMS TO FIND F AND FP
 46            DO 20 K = 1,C
 47            TERM = TERM*(C-K+1)/AN
 48            F = F + TERM
```

```
49          FP = FP + K * TERM
      C          WHEN TERM LESS THAN 1.E-12 SERIES SUMMATION ENDS
50          IF (TERM .LT. 1.D-12) GO TO 30
51       20 CONTINUE
52       30 AN = AN*(1 + F/FP)
      C          WHEN CHANGE LESS THAN 1.E-10 NEWTON'S ITERATION ENDS
53          IF ( DABS(F/FP) .LT. 1.D-10) GO TO 15
54       25 CONTINUE
55          AN = -666.
56       15 CONTINUE
57       10 A(NC) = AN
58          RETURN
59          END

      C GDGORDON - APRIL 78 - CALCULATION OF ERLANG TABLE
      C   KMIN & KMAX ARE LIMITS OF C FOR EACH PAGE
      C   KFACT IS THE INCREMENT FOR EACH PAGE
      C   PROGRAM GENERATES 8 PAGES, 51 LINES, 5 COLUMNS
 1          DOUBLE PRECISION A(52,5), B(5)
 2          INTEGER KMIN(8), KMAX(8), KFACT(8), C
 3          DATA INC, C /1,1/, B /1.D-3, 3.D-3, 1.D-2, 3.D-2, 1.D-1/
 4          DATA KMIN /  2,  50, 100, 150, 250, 500, 1500, 4000 /
 5          DATA KMAX / 52, 100, 150, 250, 500,1500, 4000, 9000 /
 6          DATA KFACT/  1,   1,   1,   2,   5,  20,   50,  100 /
 7          DO 80 IPAGE = 1,8
 8          INC = KFACT(IPAGE)
 9          MIN = KMIN(IPAGE) - INC
10          DO 20 KOL = 1, 5
11       20 CALL ERLF (A(1,KOL), B(KOL), MIN, KMAX(IPAGE), INC)
      C          OUTPUT OF RESULTS
12          WRITE(6,40) B
13       40 FORMAT(1H1 //'    C ', 5('    B =',F5.3)/)
14          IF(IPAGE .EQ. 1) WRITE(6,50)C,(A(1,KOL),KOL=1, 5)
15          DO 30 LINE = 1, 51
16          C = LINE*INC + MIN
17       30 WRITE(6,50) C,  (A(LINE + 1,KOL), KOL = 1, 5)
18       50 FORMAT(1X, I5, 5F12.4)
      C          SET FIRST TWO VALUES OF NEXT PAGE
19          IF(IPAGE .GE. 8) GO TO 90
20          KI = -KFACT(IPAGE + 1) / KFACT(IPAGE)
21          FC = KI + FLOAT(KFACT(IPAGE + 1)) / KFACT(IPAGE)
22          DO 80 KOL = 1, 5
23          A(1,KOL) = FC*A(KI + 51,KOL) + (1 - FC)*A(KI + 52,KOL)
24          A(2, KOL) = A(52, KOL)
25          IF (IPAGE .EQ. 1) A(1, KOL) = A(49, KOL)
26          IF (IPAGE .EQ. 1) A(2, KOL) = A(50, KOL)
27       80 CONTINUE
28       90 WRITE(6,40)
29          STOP
30          END
```

Program execution is faster for small numbers of circuits (small C) and for poorer grades of service (large B). There are five arguments for the subroutine call: A (output), B (grade of service), minimum C, maximum C, and increment of C. All variables are integer or double precision.

To model a communications network, the following information is needed for each potential subscriber:

1. Type of service: voice, telegraph, data, image, etc.
2. Service quality: grade of service, error rate, distortion, bandwidth, data rate, etc.
3. Switched or nonswitched.
4. Real time?

TABLE 2.20 Erlang Table for $C = 1$ to $C = 9000$

C	B =0.001	B =0.003	B =0.010	B =0.030	B =0.100
1	0.0010	0.0030	0.0101	0.0309	0.1111
2	0.0458	0.0806	0.1526	0.2816	0.5954
3	0.1938	0.2885	0.4555	0.7151	1.2708
4	0.4393	0.6021	0.8694	1.2589	2.0454
5	0.7621	0.9945	1.3608	1.8752	2.8811
6	1.1459	1.4468	1.9090	2.5431	3.7584
7	1.5786	1.9463	2.5009	3.2497	4.6662
8	2.0513	2.4837	3.1276	3.9865	5.5971
9	2.5575	3.0526	3.7825	4.7479	6.5464
10	3.0920	3.6480	4.4612	5.5294	7.5106
11	3.6511	4.2661	5.1599	6.3280	8.4871
12	4.2314	4.9038	5.8760	7.1410	9.4740
13	4.8305	5.5588	6.6072	7.9667	10.4699
14	5.4464	6.2290	7.3517	8.8035	11.4735
15	6.0772	6.9129	8.1080	9.6500	12.4838
16	6.7215	7.6091	8.8750	10.5052	13.5001
17	7.3781	8.3164	9.6516	11.3683	14.5217
18	8.0459	9.0339	10.4369	12.2384	15.5480
19	8.7239	9.7606	11.2301	13.1150	16.5787
20	9.4115	10.4958	12.0306	13.9974	17.6132
21	10.1077	11.2389	12.8378	14.8853	18.6512
22	10.8121	11.9893	13.6513	15.7781	19.6925
23	11.5241	12.7465	14.4705	16.6755	20.7367
24	12.2432	13.5100	15.2950	17.5772	21.7836
25	12.9689	14.2795	16.1246	18.4828	22.8331
26	13.7008	15.0545	16.9588	19.3922	23.8850
27	14.4385	15.8347	17.7974	20.3050	24.9390
28	15.1818	16.6199	18.6402	21.2211	25.9950
29	15.9304	17.4097	19.4869	22.1402	27.0529
30	16.6839	18.2039	20.3373	23.0623	28.1126
31	17.4420	19.0023	21.1912	23.9870	29.1740
32	18.2047	19.8047	22.0483	24.9144	30.2369
33	18.9716	20.6108	22.9087	25.8442	31.3013
34	19.7426	21.4205	23.7720	26.7763	32.3672
35	20.5174	22.2337	24.6381	27.7106	33.4343
36	21.2960	23.0501	25.5070	28.6470	34.5027
37	22.0781	23.8697	26.3785	29.5854	35.5722
38	22.8636	24.6922	27.2525	30.5258	36.6429
39	23.6523	25.5177	28.1288	31.4679	37.7147
40	24.4442	26.3459	29.0074	32.4118	38.7874
41	25.2391	27.1767	29.8882	33.3574	39.8612
42	26.0369	28.0101	30.7712	34.3046	40.9359
43	26.8374	28.8460	31.6561	35.2533	42.0114
44	27.6407	29.6842	32.5430	36.2035	43.0878
45	28.4466	30.5247	33.4317	37.1551	44.1650
46	29.2549	31.3674	34.3223	38.1081	45.2430
47	30.0657	32.2122	35.2146	39.0624	46.3218
48	30.8789	33.0591	36.1086	40.0180	47.4012
49	31.6943	33.9080	37.0042	40.9748	48.4813
50	32.5119	34.7588	37.9014	41.9327	49.5621
51	33.3316	35.6114	38.8001	42.8919	50.6435
52	34.1533	36.4659	39.7003	43.8521	51.7256
53	34.9771	37.3221	40.6019	44.8134	52.8082
54	35.8028	38.1800	41.5049	45.7758	53.8914
55	36.6305	39.0396	42.4092	46.7391	54.9751
56	37.4599	39.9007	43.3149	47.7034	56.0594
57	38.2911	40.7634	44.2218	48.6687	57.1441
58	39.1241	41.6276	45.1299	49.6348	58.2294
59	39.9587	42.4933	46.0392	50.6019	59.3151
60	40.7950	43.3604	46.9497	51.5698	60.4013
61	41.6328	44.2290	47.8613	52.5385	61.4880
62	42.4723	45.0988	48.7740	53.5081	62.5750
63	43.3132	45.9700	49.6878	54.4784	63.6625
64	44.1557	46.8425	50.6026	55.4496	64.7504
65	44.9995	47.7163	51.5185	56.4214	65.8387
66	45.8448	48.5912	52.4353	57.3940	66.9274
67	46.6915	49.4674	53.3531	58.3673	68.0164

58

TABLE 2.20 (*Continued*)

C	B =0.001	B =0.003	B =0.010	B =0.030	B =0.100
68	47.5395	50.3447	54.2718	59.3413	69.1058
69	48.3888	51.2232	55.1915	60.3160	70.1956
70	49.2394	52.1028	56.1120	61.2913	71.2857
71	50.0913	52.9835	57.0335	62.2673	72.3761
72	50.9444	53.8653	57.9558	63.2439	73.4668
73	51.7987	54.7480	58.8789	64.2211	74.5579
74	52.6542	55.6319	59.8028	65.1989	75.6492
75	53.5108	56.5167	60.7276	66.1773	76.7409
76	54.3685	57.4025	61.6531	67.1562	77.8328
77	55.2274	58.2892	62.5794	68.1358	78.9250
78	56.0873	59.1769	63.5065	69.1158	80.0175
79	56.9483	60.0655	64.4343	70.0964	81.1103
80	57.8104	60.9550	65.3628	71.0775	82.2033
81	58.6734	61.8454	66.2920	72.0591	83.2966
82	59.5375	62.7366	67.2219	73.0412	84.3901
83	60.4025	63.6287	68.1524	74.0238	85.4839
84	61.2685	64.5216	69.0837	75.0069	86.5778
85	62.1354	65.4154	70.0156	75.9904	87.6721
86	63.0033	66.3099	70.9481	76.9744	88.7665
87	63.8721	67.2052	71.8812	77.9589	89.8612
88	64.7417	68.1013	72.8150	78.9438	90.9561
89	65.6123	68.9982	73.7494	79.9291	92.0512
90	66.4837	69.8958	74.6843	80.9149	93.1465
91	67.3559	70.7941	75.6198	81.9010	94.2420
92	68.2290	71.6931	76.5560	82.8876	95.3376
93	69.1029	72.5929	77.4926	83.8746	96.4335
94	69.9776	73.4933	78.4298	84.8619	97.5296
95	70.8531	74.3944	79.3676	85.8497	98.6259
96	71.7294	75.2962	80.3059	86.8378	99.7223
97	72.6064	76.1987	81.2447	87.8263	100.8189
98	73.4842	77.1018	82.1840	88.8151	101.9157
99	74.3627	78.0055	83.1238	89.8043	103.0126
100	75.2420	78.9099	84.0642	90.7939	104.1098
101	76.1220	79.8149	85.0050	91.7838	105.2070
102	77.0026	80.7205	85.9463	92.7741	106.3045
103	77.8840	81.6266	86.8880	93.7646	107.4021
104	78.7661	82.5334	87.8303	94.7555	108.4998
105	79.6488	83.4408	88.7729	95.7468	109.5977
106	80.5322	84.3487	89.7161	96.7383	110.6958
107	81.4163	85.2572	90.6597	97.7302	111.7940
108	82.3010	86.1662	91.6037	98.7224	112.8923
109	83.1863	87.0758	92.5481	99.7148	113.9908
110	84.0723	87.9859	93.4930	100.7076	115.0894
111	84.9588	88.8966	94.4383	101.7006	116.1881
112	85.8460	89.8078	95.3840	102.6940	117.2870
113	86.7338	90.7195	96.3301	103.6876	118.3860
114	87.6222	91.6317	97.2766	104.6815	119.4851
115	88.5112	92.5444	98.2235	105.6757	120.5843
116	89.4007	93.4576	99.1707	106.6702	121.6837
117	90.2908	94.3712	100.1184	107.6649	122.7832
118	91.1815	95.2854	101.0664	108.6599	123.8828
119	92.0727	96.2000	102.0148	109.6551	124.9825
120	92.9645	97.1151	102.9636	110.6506	126.0824
121	93.8568	98.0307	103.9128	111.6464	127.1823
122	94.7496	98.9467	104.8622	112.6423	128.2824
123	95.6430	99.8632	105.8121	113.6386	129.3826
124	96.5369	100.7801	106.7623	114.6351	130.4828
125	97.4312	101.6974	107.7128	115.6318	131.5832
126	98.3261	102.6152	108.6637	116.6287	132.6837
127	99.2215	103.5334	109.6149	117.6259	133.7843
128	100.1174	104.4521	110.5664	118.6233	134.8850
129	101.0138	105.3711	111.5183	119.6209	135.9858
130	101.9106	106.2906	112.4705	120.6188	137.0866
131	102.8080	107.2104	113.4230	121.6169	138.1876
132	103.7058	108.1307	114.3758	122.6151	139.2887
133	104.6040	109.0513	115.3289	123.6136	140.3898
134	105.5028	109.9724	116.2823	124.6123	141.4911

TABLE 2.20 (*Continued*)

C	B =0.001	B =0.003	B =0.010	B =0.030	B =0.100
135	106.4019	110.8938	117.2360	125.6113	142.5924
136	107.3015	111.8156	118.1900	126.6104	143.6939
137	108.2016	112.7378	119.1443	127.6097	144.7954
138	109.1021	113.6604	120.0989	128.6092	145.8970
139	110.0030	114.5833	121.0538	129.6089	146.9987
140	110.9044	115.5066	122.0090	130.6088	148.1004
141	111.8062	116.4303	122.9645	131.6089	149.2023
142	112.7084	117.3543	123.9202	132.6092	150.3042
143	113.6110	118.2787	124.8762	133.6097	151.4062
144	114.5140	119.2034	125.8325	134.6103	152.5083
145	115.4174	120.1284	126.7890	135.6112	153.6105
146	116.3212	121.0538	127.7458	136.6122	154.7127
147	117.2255	121.9796	128.7029	137.6134	155.8150
148	118.1301	122.9056	129.6602	138.6148	156.9174
149	119.0351	123.8320	130.6178	139.6163	158.0199
150	119.9404	124.7588	131.5756	140.6180	159.1224
152	121.7523	126.6132	133.4920	142.6220	161.3277
154	123.5657	128.4688	135.4094	144.6266	163.5332
156	125.3805	130.3258	137.3277	146.6318	165.7390
158	127.1968	132.1839	139.2470	148.6377	167.9451
160	129.0144	134.0432	141.1672	150.6442	170.1514
162	130.8335	135.9037	143.0883	152.6513	172.3580
164	132.6538	137.7653	145.0102	154.6590	174.5648
166	134.4755	139.6280	146.9330	156.6673	176.7719
168	136.2985	141.4918	148.8567	158.6761	178.9792
170	138.1227	143.3568	150.7812	160.6855	181.1867
172	139.9482	145.2227	152.7065	162.6954	183.3944
174	141.7749	147.0897	154.6326	164.7059	185.6023
176	143.6028	148.9577	156.5595	166.7169	187.8105
178	145.4319	150.8267	158.4872	168.7284	190.0188
180	147.2621	152.6967	160.4156	170.7404	192.2273
182	149.0935	154.5677	162.3447	172.7529	194.4361
184	150.9260	156.4396	164.2746	174.7659	196.6450
186	152.7595	158.3124	166.2051	176.7793	198.8541
188	154.5942	160.1862	168.1364	178.7932	201.0634
190	156.4299	162.0608	170.0684	180.8076	203.2728
192	158.2667	163.9363	172.0010	182.8224	205.4825
194	160.1045	165.8127	173.9343	184.8376	207.6923
196	161.9433	167.6900	175.8682	186.8533	209.9022
198	163.7831	169.5680	177.8028	188.8694	212.1124
200	165.6239	171.4470	179.7380	190.8859	214.3226
202	167.4657	173.3267	181.6739	192.9028	216.5331
204	169.3084	175.2072	183.6103	194.9201	218.7437
206	171.1520	177.0885	185.5473	196.9378	220.9544
208	172.9965	178.9706	187.4850	198.9559	223.1653
210	174.8420	180.8534	189.4232	200.9744	225.3763
212	176.6883	182.7370	191.3620	202.9932	227.5874
214	178.5355	184.6214	193.3013	205.0124	229.7987
216	180.3836	186.5065	195.2412	207.0320	232.0102
218	182.2326	188.3923	197.1816	209.0519	234.2217
220	184.0823	190.2787	199.1226	211.0722	236.4334
222	185.9329	192.1659	201.0641	213.0928	238.6452
224	187.7844	194.0538	203.0061	215.1137	240.8571
226	189.6366	195.9424	204.9487	217.1350	243.0692
228	191.4896	197.8316	206.8917	219.1566	245.2814
230	193.3434	199.7215	208.8353	221.1785	247.4937
232	195.1980	201.6120	210.7793	223.2008	249.7061
234	197.0533	203.5032	212.7238	225.2233	251.9186
236	198.9094	205.3950	214.6688	227.2462	254.1312
238	200.7662	207.2874	216.6143	229.2694	256.3439
240	202.6238	209.1805	218.5602	231.2928	258.5568
242	204.4821	211.0741	220.5066	233.3166	260.7697
244	206.3411	212.9684	222.4534	235.3406	262.9827
246	208.2008	214.8632	224.4007	237.3649	265.1959
248	210.0612	216.7587	226.3484	239.3895	267.4091
250	211.9222	218.6547	228.2965	241.4144	269.6225
255	216.5779	223.3971	233.1687	246.4778	275.1562

TABLE 2.20 (*Continued*)

C	B =0.001	B =0.003	B =0.010	B =0.030	B =0.100
260	221.2376	228.1430	238.0435	251.5428	280.6905
265	225.9013	232.8922	242.9208	256.6095	286.2254
270	230.5689	237.6447	247.8005	261.6776	291.7608
275	235.2402	242.4003	252.6825	266.7472	297.2967
280	239.9152	247.1589	257.5668	271.8183	302.8331
285	244.5936	251.9205	262.4534	276.8907	308.3699
290	249.2756	256.6850	267.3420	281.9645	313.9072
295	253.9608	261.4523	272.2328	287.0396	319.4449
300	258.6494	266.2223	277.1255	292.1159	324.9830
305	263.3411	270.9949	282.0203	297.1935	330.5216
310	268.0359	275.7701	286.9169	302.2722	336.0605
315	272.7337	280.5478	291.8154	307.3521	341.5997
320	277.4344	285.3280	296.7157	312.4331	347.1394
325	282.1380	290.1105	301.6178	317.5153	352.6793
330	286.8444	294.8954	306.5215	322.5984	358.2196
335	291.5535	299.6825	311.4270	327.6826	363.7603
340	296.2653	304.4719	316.3341	332.7678	369.3012
345	300.9797	309.2634	321.2428	337.8539	374.8424
350	305.6966	314.0570	326.1530	342.9411	380.3839
355	310.4160	318.8527	331.0647	348.0291	385.9258
360	315.1378	323.6504	335.9780	353.1180	391.4678
365	319.8620	328.4501	340.8926	358.2078	397.0102
370	324.5886	333.2517	345.8087	363.2985	402.5528
375	329.3174	338.0552	350.7262	368.3900	408.0956
380	334.0484	342.8605	355.6450	373.4823	413.6387
385	338.7816	347.6677	360.5651	378.5753	419.1820
390	343.5170	352.4766	365.4865	383.6692	424.7256
395	348.2544	357.2872	370.4092	388.7638	430.2693
400	352.9939	362.0996	375.3331	393.8592	435.8133
405	357.7354	366.9136	380.2583	398.9553	441.3575
410	362.4789	371.7292	385.1846	404.0521	446.9019
415	367.2243	376.5464	390.1120	409.1495	452.4465
420	371.9717	381.3652	395.0407	414.2477	457.9912
425	376.7208	386.1856	399.9704	419.3465	463.5362
430	381.4718	391.0074	404.9012	424.4459	469.0813
435	386.2246	395.8307	409.8331	429.5460	474.6266
440	390.9792	400.6555	414.7660	434.6467	480.1721
445	395.7355	405.4817	419.7000	439.7480	485.7178
450	400.4934	410.3093	424.6350	444.8499	491.2636
455	405.2531	415.1382	429.5709	449.9524	496.8096
460	410.0144	419.9686	434.5079	455.0554	502.3557
465	414.7773	424.8002	439.4458	460.1590	507.9020
470	419.5418	429.6331	444.3846	465.2632	513.4484
475	424.3078	434.4674	449.3243	470.3679	518.9949
480	429.0754	439.3029	454.2650	475.4731	524.5416
485	433.8445	444.1396	459.2065	480.5788	530.0885
490	438.6150	448.9775	464.1490	485.6850	535.6355
495	443.3870	453.8167	469.0922	490.7917	541.1826
500	448.1605	458.6570	474.0364	495.8989	546.7298
520	467.2682	478.0297	493.8210	516.3324	568.9199
540	486.3971	497.4197	513.6180	536.7729	591.1118
560	505.5461	516.8262	533.4267	557.2201	613.3054
580	524.7139	536.2481	553.2463	577.6734	635.5005
600	543.8996	555.6845	573.0763	598.1325	657.6970
620	563.1022	575.1348	592.9159	618.5970	679.8947
640	582.3209	594.5981	612.7648	639.0665	702.0937
660	601.5548	614.0739	632.6224	659.5409	724.2937
680	620.8033	633.5615	652.4882	680.0197	746.4948
700	640.0656	653.0603	672.3619	700.5029	768.6968
720	659.3412	672.5699	692.2431	720.9900	790.8997
740	678.6295	692.0897	712.1313	741.4810	813.1034
760	697.9299	711.6194	732.0264	761.9756	835.3079
780	717.2419	731.1584	751.9280	782.4736	857.5132
800	736.5651	750.7064	771.8358	802.9750	879.7191
820	755.8989	770.2631	791.7495	823.4794	901.9256
840	775.2431	789.8281	811.6689	843.9868	924.1328
860	794.5972	809.4011	831.5938	864.4971	946.3405

TABLE 2.20 (*Continued*)

C	B =0.001	B =0.003	B =0.010	B =0.030	B =0.100
880	813.9608	828.9817	851.5239	885.0100	968.5487
900	833.3336	848.5698	871.4590	905.5256	990.7575
920	852.7153	868.1649	891.3990	926.0436	1012.9667
940	872.1056	887.7670	911.3437	946.5640	1035.1764
960	891.5042	907.3757	931.2929	967.0867	1057.3865
980	910.9108	926.9908	951.2464	987.6116	1079.5970
1000	930.3251	946.6121	971.2041	1008.1386	1101.8079
1020	949.7470	966.2394	991.1658	1028.6676	1124.0191
1040	969.1761	985.8726	1011.1314	1049.1985	1146.2308
1060	988.6123	1005.5113	1031.1008	1069.7314	1168.4427
1080	1008.0553	1025.1556	1051.0739	1090.2660	1190.6549
1100	1027.5050	1044.8051	1071.0505	1110.8023	1212.8675
1120	1046.9611	1064.4597	1091.0305	1131.3404	1235.0804
1140	1066.4235	1084.1194	1111.0139	1151.8800	1257.2935
1160	1085.8920	1103.7839	1131.0004	1172.4212	1279.5069
1180	1105.3664	1123.4530	1150.9901	1192.9639	1301.7205
1200	1124.8466	1143.1268	1170.9828	1213.5081	1323.9344
1220	1144.3324	1162.8051	1190.9785	1234.0537	1346.1485
1240	1163.8236	1182.4876	1210.9770	1254.6006	1368.3628
1260	1183.3202	1202.1744	1230.9783	1275.1489	1390.5773
1280	1202.8220	1221.8653	1250.9823	1295.6984	1412.7921
1300	1222.3289	1241.5602	1270.9889	1316.2492	1435.0070
1320	1241.8407	1261.2590	1290.9981	1336.8012	1457.2221
1340	1261.3573	1280.9617	1311.0097	1357.3543	1479.4374
1360	1280.8786	1300.6680	1331.0238	1377.9086	1501.6529
1380	1300.4046	1320.3780	1351.0403	1398.4640	1523.8686
1400	1319.9350	1340.0915	1371.0591	1419.0204	1546.0844
1420	1339.4698	1359.8085	1391.0801	1439.5779	1568.3003
1440	1359.0089	1379.5288	1411.1033	1460.1364	1590.5164
1460	1378.5522	1399.2525	1431.1286	1480.6958	1612.7327
1480	1398.0996	1418.9794	1451.1561	1501.2562	1634.9491
1500	1417.6511	1438.7094	1471.1855	1521.8175	1657.1656
1550	1466.5467	1488.0479	1521.2678	1573.2247	1712.7075
1600	1515.4656	1537.4047	1571.3619	1624.6372	1768.2501
1650	1564.4068	1586.7789	1621.4670	1676.0546	1823.7934
1700	1613.3690	1636.1696	1671.5827	1727.4766	1879.3373
1750	1662.3514	1685.5759	1721.7083	1778.9030	1934.8818
1800	1711.3530	1734.9973	1771.8434	1830.3335	1990.4268
1850	1760.3730	1784.4330	1821.9876	1881.7678	2045.9724
1900	1809.4105	1833.8823	1872.1403	1933.2058	2101.5184
1950	1858.4649	1883.3448	1922.3012	1984.6473	2157.0648
2000	1907.5355	1932.8197	1972.4700	2036.0921	2212.6116
2050	1956.6216	1982.3067	2022.6462	2087.5400	2268.1588
2100	2005.7226	2031.8052	2072.8296	2138.9908	2323.7064
2150	2054.8379	2081.3149	2123.0198	2190.4444	2379.2543
2200	2103.9671	2130.8351	2173.2166	2241.9008	2434.8025
2250	2153.1096	2180.3657	2223.4197	2293.3596	2490.3510
2300	2202.2650	2229.9061	2273.6289	2344.8209	2545.8998
2350	2251.4328	2279.4561	2323.8439	2396.2846	2601.4488
2400	2300.6125	2329.0153	2374.0645	2447.7504	2656.9981
2450	2349.8039	2378.5834	2424.2905	2499.2184	2712.5476
2500	2399.0064	2428.1600	2474.5217	2550.6884	2768.0974
2550	2448.2199	2477.7450	2524.7580	2602.1603	2823.6473
2600	2497.4438	2527.3380	2574.9990	2653.6341	2879.1974
2650	2546.6779	2576.9388	2625.2447	2705.1097	2934.7477
2700	2595.9219	2626.5472	2675.4950	2756.5870	2990.2982
2750	2645.1755	2676.1628	2725.7496	2808.0660	3045.8489
2800	2694.4384	2725.7856	2776.0084	2859.5466	3101.3997
2850	2743.7104	2775.4152	2826.2713	2911.0287	3156.9507
2900	2792.9911	2825.0515	2876.5382	2962.5122	3212.5018
2950	2842.2804	2874.6943	2926.8089	3013.9972	3268.0531
3000	2891.5781	2924.3434	2977.0834	3065.4835	3323.6044
3050	2940.8838	2973.9987	3027.3614	3116.9712	3379.1560
3100	2990.1974	3023.6599	3077.6429	3168.4601	3434.7076
3150	3039.5187	3073.3269	3127.9279	3219.9502	3490.2593
3200	3088.8475	3122.9996	3178.2161	3271.4415	3545.8112
3250	3138.1836	3172.6778	3228.5076	3322.9340	3601.3631

TABLE 2.20 (*Continued*)

C	B = 0.001	B = 0.003	B = 0.010	B = 0.030	B = 0.100
3300	3187.5268	3222.3614	3278.8021	3374.4276	3656.9152
3350	3236.8770	3272.0501	3329.0997	3425.9222	3712.4673
3400	3286.2339	3321.7441	3379.4002	3477.4179	3768.0196
3450	3335.5975	3371.4429	3429.7037	3528.9146	3823.5719
3500	3384.9675	3421.1467	3480.0099	3580.4122	3879.1243
3550	3434.3439	3470.8552	3530.3188	3631.9108	3934.6768
3600	3483.7264	3520.5683	3580.6304	3683.4103	3990.2294
3650	3533.1150	3570.2859	3630.9445	3734.9107	4045.7821
3700	3582.5095	3620.0080	3681.2612	3786.4120	4101.3348
3750	3631.9098	3669.7344	3731.5804	3837.9140	4156.8876
3800	3681.3157	3719.4650	3781.9019	3889.4169	4212.4405
3850	3730.7271	3769.1998	3832.2258	3940.9206	4267.9934
3900	3780.1440	3818.9386	3882.5520	3992.4250	4323.5464
3950	3829.5662	3868.6814	3932.8804	4043.9301	4379.0995
4000	3878.9936	3918.4280	3983.2109	4095.4360	4434.6526
4100	3977.8636	4017.9326	4083.8784	4198.4499	4545.7590
4200	4076.7531	4117.4518	4184.5541	4301.4663	4656.8656
4300	4175.6616	4216.9849	4285.2375	4404.4853	4767.9724
4400	4274.5881	4316.5315	4385.9284	4507.5067	4879.0793
4500	4373.5322	4416.0910	4486.6264	4610.5303	4990.1865
4600	4472.4931	4515.6630	4587.3313	4713.5561	5101.2938
4700	4571.4703	4615.2470	4688.0427	4816.5839	5212.4013
4800	4670.4632	4714.8426	4788.7605	4919.6136	5323.5089
4900	4769.4714	4814.4494	4889.4843	5022.6452	5434.6166
5000	4868.4942	4914.0669	4990.2140	5125.6786	5545.7245
5100	4967.5312	5013.6949	5090.9493	5228.7137	5656.8325
5200	5066.5820	5113.3330	5191.6900	5331.7503	5767.9406
5300	5165.6462	5212.9808	5292.4360	5434.7886	5879.0488
5400	5264.7233	5312.6381	5393.1870	5537.8283	5990.1571
5500	5363.8130	5412.3046	5493.9428	5640.8694	6101.2655
5600	5462.9149	5511.9799	5594.7034	5743.9118	6212.3740
5700	5562.0286	5611.6639	5695.4685	5846.9556	6323.4826
5800	5661.1538	5711.3562	5796.2380	5950.0006	6434.5913
5900	5760.2902	5811.0567	5897.0118	6053.0468	6545.7000
6000	5859.4375	5910.7650	5997.7897	6156.0942	6656.8089
6100	5958.5954	6010.4810	6098.5716	6259.1427	6767.9177
6200	6057.7636	6110.2045	6199.3574	6362.1922	6879.0267
6300	6156.9418	6209.9352	6300.1469	6465.2428	6990.1357
6400	6256.1299	6309.6730	6400.9400	6568.2943	7101.2448
6500	6355.3274	6409.4176	6501.7367	6671.3469	7212.3540
6600	6454.5343	6509.1690	6602.5368	6774.4003	7323.4632
6700	6553.7503	6608.9269	6703.3402	6877.4546	7434.5725
6800	6652.9751	6708.6911	6804.1469	6980.5098	7545.6818
6900	6752.2085	6808.4615	6904.9567	7083.5658	7656.7911
7000	6851.4504	6908.2380	7005.7696	7186.6226	7767.9005
7100	6950.7005	7008.0204	7106.5855	7289.6801	7879.0100
7200	7049.9587	7107.8086	7207.4042	7392.7384	7990.1195
7300	7149.2248	7207.6023	7308.2258	7495.7975	8101.2290
7400	7248.4986	7307.4016	7409.0501	7598.8572	8212.3386
7500	7347.7799	7407.2062	7509.8771	7701.9176	8323.4482
7600	7447.0686	7507.0161	7610.7067	7804.9787	8434.5579
7700	7546.3646	7606.8312	7711.5389	7908.0404	8545.6676
7800	7645.6675	7706.6512	7812.3735	8011.1027	8656.7773
7900	7744.9774	7806.4762	7913.2105	8114.1656	8767.8871
8000	7844.2941	7906.3059	8014.0499	8217.2291	8878.9969
8100	7943.6175	8006.1404	8114.8916	8320.2931	8990.1067
8200	8042.9473	8105.9794	8215.7356	8423.3577	9101.2166
8300	8142.2835	8205.8230	8316.5817	8526.4229	9212.3265
8400	8241.6259	8305.6710	8417.4300	8629.4885	9323.4364
8500	8340.9745	8405.5233	8518.2804	8732.5546	9434.5463
8600	8440.3292	8505.3798	8619.1328	8835.6212	9545.6563
8700	8539.6897	8605.2406	8719.9872	8938.6883	9656.7663
8800	8639.0560	8705.1053	8820.8436	9041.7559	9767.8763
8900	8738.4280	8804.9741	8921.7019	9144.8239	9878.9864
9000	8837.8056	8904.8468	9022.5621	9247.8923	9990.0964

Reprinted with permission of the *COMSAT Technical* Review, The COMSAT Corporation. The eight original pages have been reformatted.

5. Number of call-minutes per destination per day.

6. Erlangs or CCS/hr during busy hours.

7. Destinations and sources.

8. Priorities and hierarchy (if any).

9. Demand or full-time (24 hr/day) service requirements.

2.5 TELEGROWTH

The growth of telecommunications services, which has paralleled that of many other emerging services and products, is shown in Figure 2.34. The data sources and scaling for this figure are also shown. The overall growth rate (for the first 20

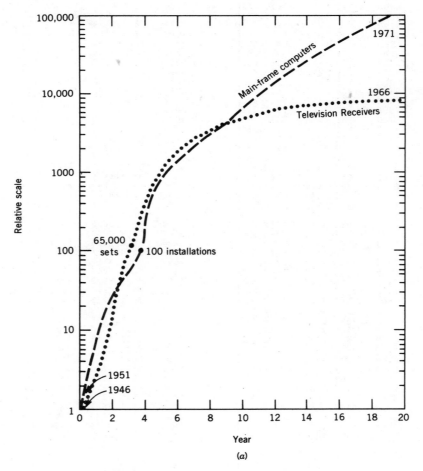

FIGURE 2.34 Growth data for first 20 years.

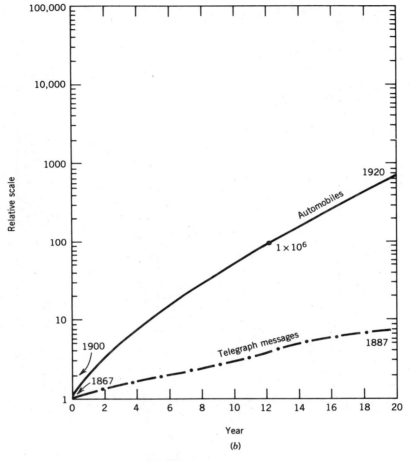

FIGURE 2.34 (*Continued*)

years a product or service is available) is 28% yr^{-1}. For television receivers and computers the growth rates were at least twice as great. Pocket calculators follow a similar growth curve.

Figure 2.35 shows three generic market growth patterns. The explosive class is exemplified by the early years of both voice and data communications, the saturating pattern by the number of television receivers (Figure 2.34), and the collapsing market by the decrease in commercial telegraph service in the United States.

The number of circuits (n) required y years into the future may be extrapolated from the current traffic n_0 if the growth rate g is known:

$$n = n_0(1 + g)^y \quad \text{(circuits)} \tag{2.16}$$

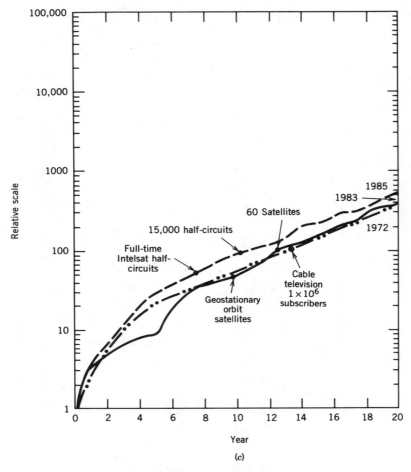

FIGURE 2.34 (*Continued*)

$$n/n_0 = (1 + g)^y \quad \text{(ratio)} \tag{2.17}$$

$$g = (n/n_0)^{1/y} - 1 \quad \text{(decimal)} \tag{2.18}$$

where a g of 0.285 corresponds to a 28.5% per year growth rate.

Table 2.21 shows the number of years for the traffic to increase by one order of magnitude (a factor of 10) as a function of the annual growth rate (here expressed as a percentage).

Figures 2.36 and 2.37 are growth rate charts. Semilog scales are used in this and most other growth-related curves. By making year 0 the base year and plotting

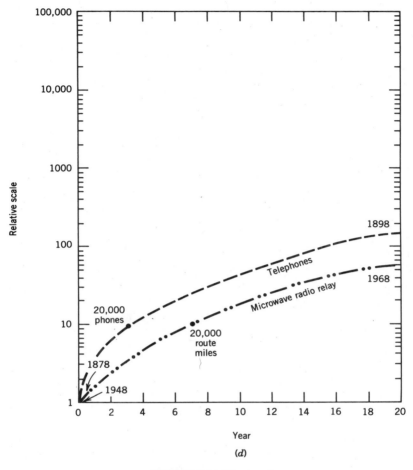

FIGURE 2.34 (*Continued*)

the traffic versus time, the growth rate in percent per year can be read directly. Extrapolation to future years is also possible.

Extensive data exists on the growth of the telephone services. Figure 2.38 shows the world growth rate in the number of instruments. At one time it was estimated that the ceiling to the number of telephone instruments was the number of households plus an allowance for business uses. Later it was popular to estimate the ceiling at one per man, woman, or child. In the United States there were 71.8 telephone instruments per hundred people on January 1, 1977. In Washington, DC, the telephones (principal instruments plus extensions in the home plus office phones) now exceeds 150 per 100 people; thus, both ceilings have been broken.

Several observations may be made about Figure 2.38. The largest number of

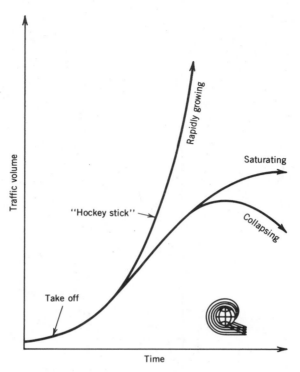

FIGURE 2.35 Three general growth patterns.

TABLE 2.21 Time for Traffic to Increase by Factor of 10

Year to Grow by Factor of 10	Growth Rate, % yr^{-1}	Year to Grow by Factor of 10	Growth Rate, % yr^{-1}
1	900	11	23.3
2	216	12	21.1
3	114	13	19.4
4	77.8	14	17.9
5	58.5	15	16.6
6	46.8	20	12.2
7	38.9	25	9.64
8	33.3	30	7.98
9	29.2	40	5.93
10	25.9	50	4.71

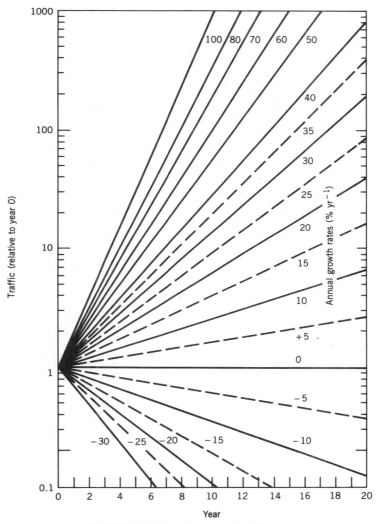

FIGURE 2.36 Growth rate overlay (20 years).

telephones has traditionally been in the developed countries of North America and Europe. The increase in the number of telephones has been most dramatic in Asia and other developing areas. In the late 1970s the average age in developed countries was 29, whereas in the lesser developed areas it was only 15. This indicates that the demand for telephones (and eventually satellite circuits) will continue to grow in these lesser developed areas. Figure 2.39 shows a prediction on the shift in the world population distribution over a 100-yr period (1978–2078).

Figure 2.40 correlates the pattern of international telephone calls from the United

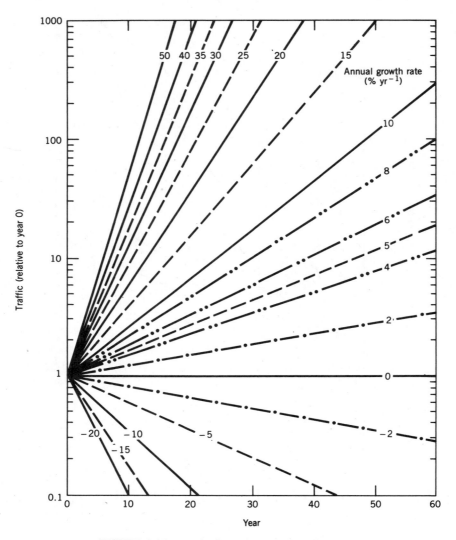

FIGURE 2.37 Growth rate overlay (60 years from year 0).

States compared to the GNP. During the Great Depression of the 1930s the traffic remained essentially constant for a few years and then rose with the resumption of world trade and business. International telephone traffic was handled exclusively by high-frequency radio until the first trans-Atlantic telephone (TAT-1) cable was laid in 1957. The radiotelephone circuits were expensive, were unreliable, had unpredictable waiting times, and were subject to the uncertainties of the ionosphere (which in turn is affected by the time of day and the phase of the sunspot cycle) (see Section 3.1.5). By 1987, 10% of AT&T communications revenue was from international circuits.

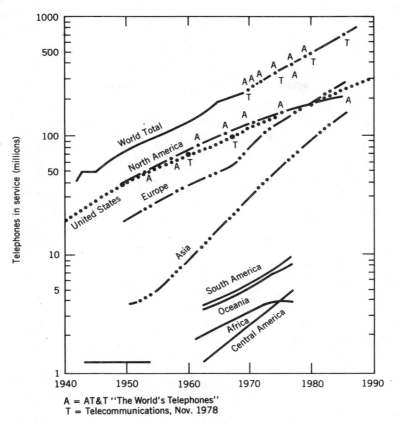

FIGURE 2.38 World's telephones: A, AT&T, *the World's Telephones*: T, *Telecommunications*, November 1978.

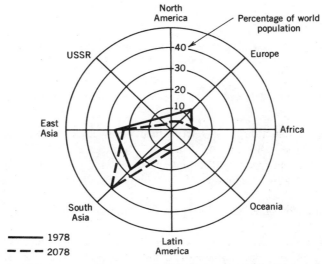

FIGURE 2.39 Predicted change in world population distribution.

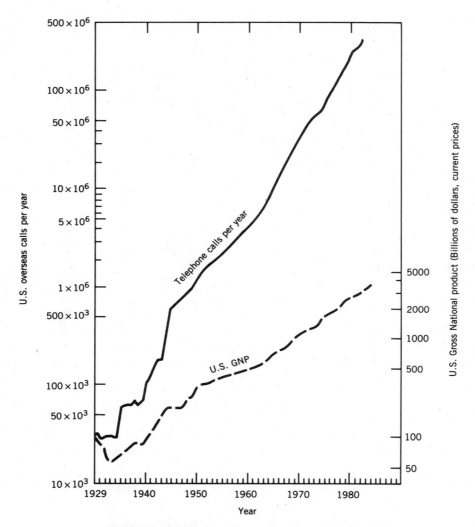

FIGURE 2.40 United States overseas traffic and U.S. GNP.

With the introduction of new services and conveniences (such as international direct dialing) the traffic continues to grow. During the 1980s United States–international traffic was growing at the rate of about 21% yr^{-1}.

Figure 2.41 shows the predictions of telephone traffic requirements in the North Atlantic.

The trends for other paths or types of communications are shown in Figure 2.42.

A competitive means of growing transoceanic communications is the submarine cable (using either copper conductors or fiber optics). Table 2.22 lists some of the submarine cable characteristics.

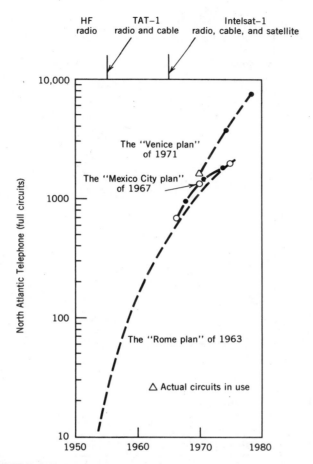

FIGURE 2.41 Historical forecasts of north atlantic telephone traffic.

The subject of U.S. domestic telecommunication was studied in depth at the Stanford Research Institute. Figure 2.43 is based on the work performed under a NASA contract. It shows the projected total and long-distance traffic in the United States. The unit of measure is the bit rate. The conversion basis is given in Table 2.23.

Figure 2.44 shows the growth in telephone conversations in the United States. Further details of the sources involved in U.S. domestic voice traffic are given in Table 2.24.

The average call duration varies from 5 min (domestic business) to 9 min (international) to 16.5 min (domestic residential). Because of the concentration of traffic into the busy hours of the work week, it is possible (for the purposes of facility capacity planning) to assume the traffic is concentrated over the equivalent of 2400 busy hours per year. For every 1×10^9 call-minutes (16.67×10^6 call-hours)

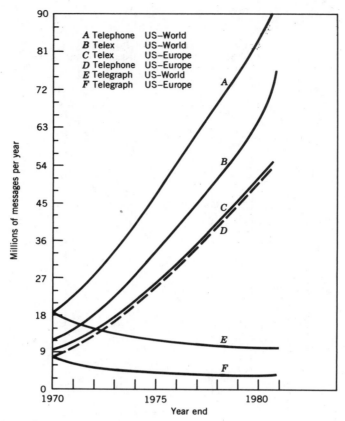

FIGURE 2.42 Worldwide communications services. (From White, 1976.) © 1976 *TELECOMMU-NICATIONS Magazine*.

TABLE 2.22 Submarine Cable Parameters by AT&T Designation

	SB (1956)	SD (1963)	SF (1968)	SG (1976)	SH (1983)	TAT-8 (1988)
Top Frequency (MHz)	0.164	1.05	5.88	30	125	not applicable (280 Mb/s)
TAT number[a]	1, 2	3, 4	5	6	7	8
3 kHz channels	48	138	845	4000	16,000	75,600
3 kHz with TASI[b]	84	—	—	—	16,000	—
Repeaters	48	210	420	815	—	—
Amplifier type	Tube	Tube	Germanium transistors	Silicon transistors	—	Optical
Length of cable, km	3700	7413	7413	7413	—	6657
Armored or not[c]	Yes	No	No	No	No	Semi
Diameter, cm	1.59	2.54	3.81	4.32	5.72	—
Investment, × 10⁶	$44.9	$46.4	$87.8	$145	—	$335

[a]Trans-Atlantic Telephone Cable.

[b]TASI, time-assigned speech interpolation (a circuit multiplier).

[c]Deep ocean portion

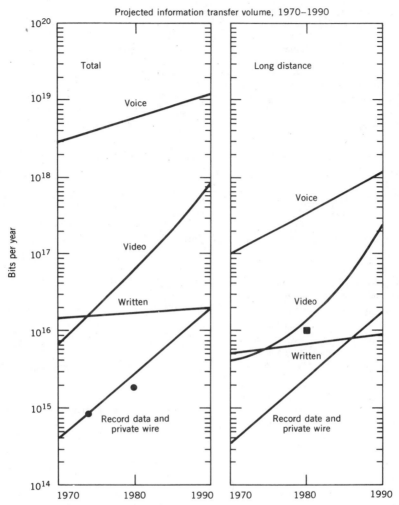

FIGURE 2.43 Projected information transfer in United States. (From Anonymous, 1976.) © 1976 *TELECOMMUNICATIONS Magazine*.

there are 6944 erlangs (equal to 16.67×10^6 call-hours/2400 hr) or 111×10^6 calls (1×10^9 call-minutes/9 min per international call).

The number of circuits needed to provide a given grade of service varies (see Figure 2.33). If the ratio of full circuits to erlangs is 1.25, then 8680 full circuits will be needed (6944 erlangs $\times$ 1.25 = 8680).

The market for data communications is expanding rapidly and is particularly suited to satellites and fiber optics. Unlike the previous telephone services based on analog (voice) processing and containing old equipment and links, both satellites and fiber optics provide a distance-independent, wideband facility and the capability of an all-digital facility. Forecasting the needs for a digital facility is more complex than for the older well-established voice services because many of

TABLE 2.23 Bits per Year for Analog Services

Service	Units for Next Column	Magnitude in 1990	Rate	Bits per Year in 1990
Voice, telephone	Calls/yr at 6 min/call	482×10^9	16 kbits/sec	2.8×10^{18}
Video				
Videotelephone	Calls/yr at 6 min/call	1.0×19^9	515 kbits/sec	1.9×10^{17}
TV transmission channels (in use 12 hr/day)		220	22 Mbits/sec	2.1×10^{14}
Record				
Telegraph and Telex	Messages/yr, 30 words/message	35×10^6	50 bits/words	5.2×10^{10}
Facsimile first-class mail	Letters/yr	1×10^{11}	300 kbits/letter	3.0×10^{16}
"Mug shots," etc.	Cases/yr	25×10^6	300 kbits/page	7.5×10^{12}
Remote title and abstract searches	Searches/yr, 10 pages/search	20×10^6	32 kbits/page	6.4×10^{12}
Remote medical literature searches	Searches/yr, 30 pages/search	2×10^8	32 kbits/page	1.9×10^{14}
Patent searches	Searches/yr, 6 pages/search	7×10^6	32 kbits/page	1.3×10^{12}
National Crime Information Center	Searches/yr	7×10^7	300 kbits/transmission	2.1×10^{13}
National Legal Information Center	Searches/yr, 20 pages/search	3×10^7	32 kbits/page	1.9×10^{13}
Data				
Checks and credit transactions	Transactions/yr	3.4×10^{11}	50 char, 8 bits/char	1.4×10^{14}
Stock exchange quotations	Transactions/yr	3×10^9	100 bits/trans	3×10^{11}
Stock transfers	Transactions/yr	4.9×10^9	3 kbits/trans	1.5×10^{13}
Airline reservations	Transactions/yr	4.2×10^9	200 char, 8 bits/char	6.7×10^{12}
Auto rental reservations	Reservations/yr	4×10^7	4 kbits/reservation	1.6×10^{11}
Hotel/motel reservations	Reservations/yr	1×10^8	4 kbits/reservation	4×10^{11}
Remote medical diagnosis tests	tests/yr	4×10^8	30 kbits/test	1.2×10^{13}

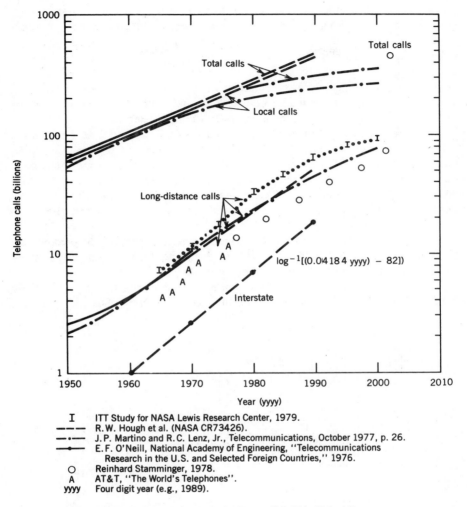

I   ITT Study for NASA Lewis Research Center, 1979.
––– R. W. Hough et al. (NASA CR73426).
–·– J. P. Martino and R. C. Lenz, Jr., Telecommunications, October 1977, p. 26.
–•– E. F. O'Neill, National Academy of Engineering, "Telecommunications
     Research in the U.S. and Selected Foreign Countries," 1976.
O   Reinhard Stamminger, 1978.
A   AT&T, "The World's Telephones".
yyyy Four digit year (e.g., 1989).

FIGURE 2.44 Growth of telephone calls within United States.

TABLE 2.24 1980 Voice Communications with Predicted Values for 1990 and 2000

Service[a]	Thousands of Half-Circuits		
	1980	1990	2000
Public MTS	1,033	2,442	5,244
Business MTS	776	2,280	7,308
Business private line	1,246	2,895	6,368
All other	8	44	88
Total	3,063	7,661	19,008

[a]MTS, message telephone service.

the services that originate in the analog domain (voice and radio programs) get converted to digital signals during their transmission. At the far end they are reconverted to analog signals. Conversely, personal computers use modems to impress a digital signal on local analog facilities. Under ISDN all services will be digitized at the source.

As shown in Figure 2.45, the amount of digital traffic varies depending on where the determination is made. Figures 2.46 and 2.47 show projections of the digital (data) communications requirements for nonlocal services.

The projected number of data terminals is shown in Figure 2.48 for both the United States and Europe. An ambiguity in data sometimes occurs in counting terminals. In some cases (such as factory shipments) all terminals are counted. In other cases only those connected to a host via public telecommunications facilities are counted. With the advent of local area networks (LANs) and remote telecomputing in industry, some of these commercial terminals are being tied directly to the host computer. A contrary trend is evident in personal computers. Most of these were not originally connected to communications networks. Minicomputers and microcomputers are being converted to ''smart terminals'' with modems and software to make use of packet-switched networks.

The communications speed of data terminals is increasing with time. The 300 bits/sec of the early 1980s became 1200 bits/sec in the mid-1980s and is moving

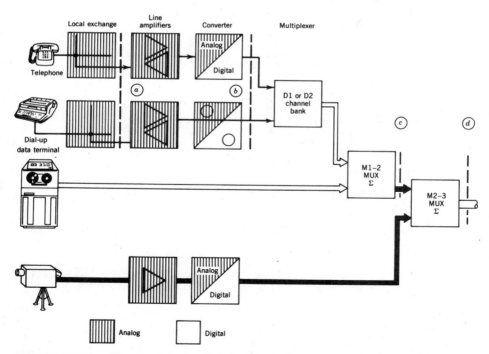

FIGURE 2.45 Locations for measuring digital traffic. (a–d represent places where traffic can be measured; digital portion increases at higher levels of multiplexing).

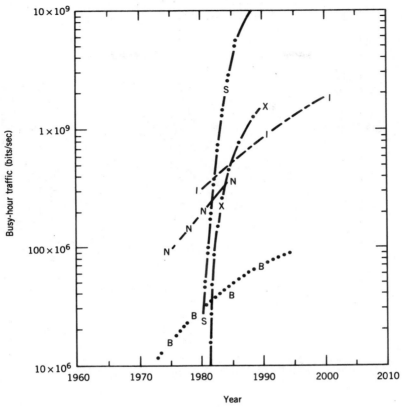

Notes:

I	ITT Study for NASA Lewis Research Center, 1979
N	NASA Study
B	Baran and Lapinski, Institute of the Future, 1971
X	Xerox Corporation, 1978 FCC filing data
S	Xerox data modified by Staelin and Harvey (MIT, 1979)

FIGURE 2.46 Data projections for United States.

toward 9600 bits/sec. With ISDN, 16 kbits/sec will be available. This tends to drive the bits transmitted and received per year per terminal up at a faster rate. Cathode-ray tubes (CRTs) and other soft display devices with good resolution (as opposed to the hard copy provided by teletypewriters) consume much more communications capability than the older types of terminals. One of the key elements in data transmission may be computer-aided manufacturing (CAM). CAM and computer-aided design and computer-aided engineering (CAD, CAE) require high bit rates for transmission between a design location and the manufacturing plant which may be on opposite ends of a nation or the world. As industry continues

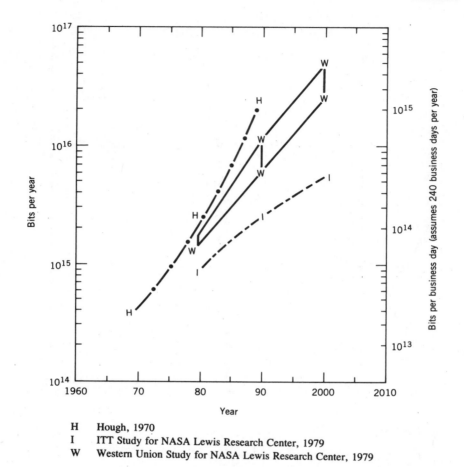

H   Hough, 1970
I   ITT Study for NASA Lewis Research Center, 1979
W   Western Union Study for NASA Lewis Research Center, 1979

FIGURE 2.47 Digital data traffic (United States).

to automate, this could be an important application of fiber and satellite communications.

Table 2.25 is based on studies of European data traffic (EURODATA) conducted by PA International, Quantum Science Corporation, Generale de Service Informatique and Italsiel. Over 100 forms of data traffic were studied. As is so often the case, a small group of traffic forms (in this case nine forms) accounted for almost two-thirds of the total traffic volume. The table shows the variation in the mix of the traffic in 1972 and 1985. The annual growth rates are shown.

The number of computers has been increasing rapidly (see Figure 2.34) as new applications are being found. Increasingly these computers are being tied to remote terminals and now to one another for load sharing and intercomputer communications such as data base sharing and updating. New applications such as the "office

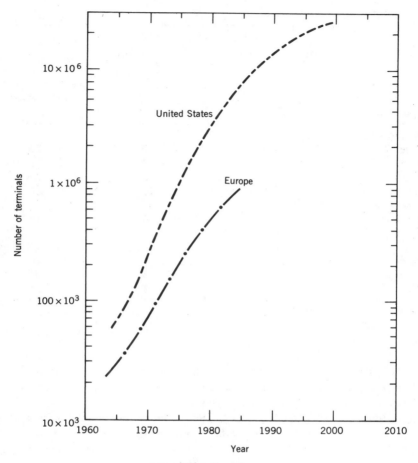

FIGURE 2.48 Data terminals.

of the future'' and electronic funds transfer (EFT) have accelerated the need for data transmission. The speed of these computers has also been increasing dramatically. With most local access loops configured for voice, the computers have used time-slowdown input–output buffers to match their rates to the telephone lines; thus, traffic for the higher speed services may have been deferred. Because the high-data-rate telecommunications facilities were not available, these applications were not developed.

The amount of satellite video traffic grew at 20% per year during the 1980s. In 1988 there were three well-established (ABC, CBS, and NBC) and one educational (PBS) television networks in operation in the United States. A fourth commercial network (Fox) is attempting to get established. Programs from independent television ''superstations'' are being distributed by satellite, and several full-time religious channels are in operation. Pay-TV is being disseminated by satellite to

TABLE 2.25 Eurodata Predictions

Application	Data Traffic Volume (Percentage of Total Data)		Annual Growth Rate (1972–1985) (%)
	1972	1985[a]	
Data-processing services[a]			
Scientific and engineering services	12	18	26
General business services	7	18	31
Other	2	4	28
Engineering and research scientific and engineering[a]	10	7	18
Manufacturing[a]			
Processing manufacturing, administrative and financial	6	4	17
Discrete manufacturing, other	8	6	19
Sales, planning and control	4	2	19
Banking and finance	8	4	14
All other sectors, news agencies	3	3	21

[a]Extrapolated.
From White (1976).

cable television headends. Sports, news, and other teleconferencing is available in a variety of frame rates. Some management levels may use freeze-frame video (almost facsimilelike as far as the service is concerned), whereas higher levels may wish full-motion (*NTSC* quality) video. The advent of very large television screens and projectors (for meetings) shows the limitations of the present 525-line NTSC and 625-line PAL and SECAM systems. Higher resolution systems are emerging for special applications and eventually for the home.

One forecast (prepared for the NASA Lewis Research Center by the International Telephone and Telegraph Corporation) estimated there would be around 5000 video channels in use in the year 2000. The bulk of this is in video teleconferencing. With modern bit rate reduction techniques the actual number of bits or transponders is far less than if a single transponder was devoted to each video channel. We therefore feel the estimate is far too high.

The video telephone is another form of teleconferencing.

With the advent of low-cost satellite and fiber digital services capable of distance-insensitive high-data-rate performance, the reduced bit rate video telephone may eventually prove to be a substantial source of telecommunications traffic.

As a guideline the video service requirements are given in Table 2.26. As can be observed, one full definition video channel may consume the capacity of many voice channels.

TABLE 2.26 Video Requirements

Video Form	Video Rate	Equivalent Voice Channels[a]	Approximate Data Rate (Mbits/s)
NTSC (525 lines)	4.2 MHz	340	22–45
625 lines	5 MHz	400	26–45
T1 quality	—	24	1.544
Videotelephone	—	8	0.515
Limited motion	—	6	0.386
Videovoice	36 kHz	1	0.064

[a]At 64 kbits/s.

The basic ingredients of future telecommunications growth are

1. increased user population;
2. increasing sophistication of communications devices available at reduced costs;
3. more innovative offerings by carriers;
4. degradation of older competing services with time (e.g., telegraphy, mail service, and mobility);
5. increased automation at industrial, retail, and home levels;
6. interactive effects of integrated circuits, computers/processors, and communications;
7. reductions in the cost of telecommunications; and
8. deregulation of what services can be offered.

The truly distance-insensitive pricing of satellite services causes profound and long-lasting changes in the traditional distribution of traffic and the performance of services.

Figure 2.49 shows how cost, traffic, and prices are related.

Equipment costs (A) are under decreasing pressures due to improved technology, whereas inflation tends to drive the price of an identical product upward with time. Technological advances have been more dramatic than inflation, so equipment costs have traditionally been dropping. This permits a reduction in the per-channel charges B, which makes the service more attractive, thus increasing the traffic base C and amount of price-sensitive traffic D, which causes a diversion E from alternatives. Lower prices (B) stimulate both the traditional (G) and new (F) traffic types. The total traffic H increases. By virtue of the economy of scale I, more equipment is ordered and the cost drops (A and B), and so on.

If a link in this chain changes, either due to costs or more attractive alternatives (e.g., fiber optics), the process can change and costs increase rather than decrease.

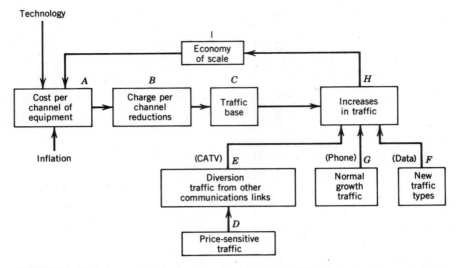

FIGURE 2.49 Economy-of-scale machine. (From Morgan, 1980. Reprinted with permission.)

The terrestrial alternatives (especially copper or microwave based) are conventionally priced on a per-mile basis (Figure 2.50), whereas satellites tend to be distance insensitive. As the cost of satellite systems continues to drop, the crossover distance (where the costs of satellite and terrestrial communications are equal) keeps getting shorter. As was shown earlier in this section, most of the communication is at the lower ends of the mileage range. This indicates that as the cross-

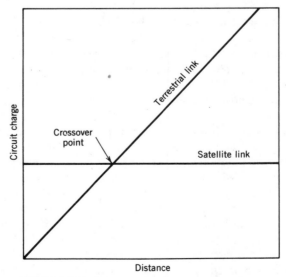

FIGURE 2.50 Crossover distance for communications.

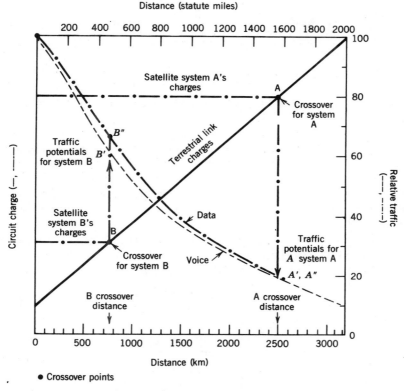

FIGURE 2.51 Traffic potential versus relative charges.

over distance gets shorter, the share of the total communications requirements increases rapidly. Figure 2.51 compares the traffic potential for two hypothetical satellite systems A and B. An increase in the traffic makes satellite communications more efficient and less costly, which moves the crossover distance to still shorter mileage bands. In many corporate/governmental applications the savings that may be realized on long-distance connections by satellite could be used to offset the higher costs on short-haul communications. The trade-off may be away from the least cost network toward one that is homogeneous and easy to use even if the cost is slightly more.

2.6 TELECOMMUNICATIONS NETWORKS

2.6.1 Introduction to Networking

Telecommunication networks enable one user to communicate with one or more distant users out of the thousands or millions of other users. The network permits equipment to be shared among the many nonsimultaneous users.

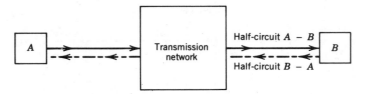

FIGURE 2.52 Point-to-point transmission.

The most elementary network consists of a single point-to-point link. This is shown in Figure 2.52 and is the connection from point A to point B. This type of circuit is typified by the "unswitched," or tie, lines. In most networks the common carrier usually reserves the right to alter the routing (via switches) at any time for maintenance, traffic, engineering, or financial reasons. Submarine cables, coaxial telephone cables, and long-haul microwave radio all are inherently point-to-point devices. (See Figure 2.53.)

Satellites are most efficient if sufficient traffic can be collected so that the repeaters ("transponders") can be operated on a single-carrier basis in the point-to-point service. Sometimes this mode is called the "cable in the sky." Figure 2.53 illustrates this mode and compares its simplicity to a terrestrial network equivalent.

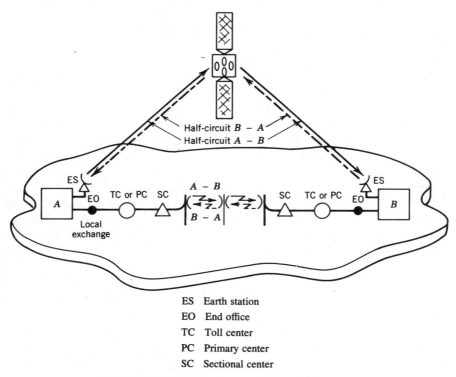

ES Earth station
EO End office
TC Toll center
PC Primary center
SC Sectional center

FIGURE 2.53 Practical point-to-point transmission routes.

Most satellite transponders are operated on a multipoint basis. An earth station may transmit traffic destined for not one but multiple earth stations. The satellite may simply repeat this mixed traffic back to the earth where each earth station sorts out the traffic destined for itself. This sorting is done on a frequency (FDMA), time slot (TDMA), or unique-code (CDMA) basis. Chapter 4 discusses these forms in further detail (see also *FDMA*, *TDMA*, and *CDMA* in the index).

Some traffic is "broadcast" to multiple points. Such traffic includes the radio and television broadcast signals for AM/FM/TV stations and cable systems on earth. Other applications may be new price lists for grocery stores, interoffice memoranda for dissemination to all the field offices, and so on. Figure 2.54 shows one such point-to-multipoint application. In this case a single satellite provides service to all authorized receivers in its coverage area (state, nation, or region), whereas the terrestrial transmitters (which are usually at VHF or UHF frequencies) are limited to the line of sight (about 80–110 km). The network may be fed from any point, but generally a commercial television network will use only one or two locations (for example, New York and California) for uplinks.

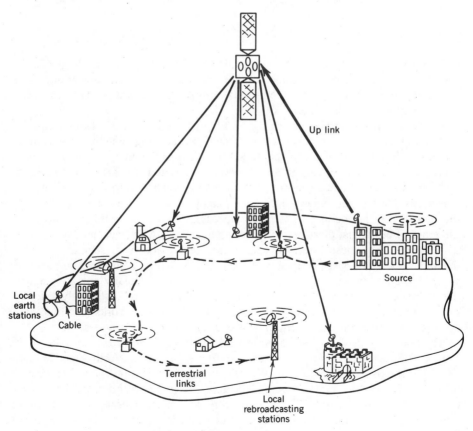

FIGURE 2.54 Point-to-multipoint services.

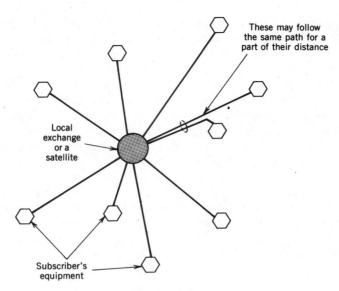

These may follow
the same path for a
part of their distance

Local
exchange
or a
satellite

Subscriber's
equipment

FIGURE 2.55 Star network.

The star network (Figure 2.55) is typified by an end-office local loop. Traffic sources (the subscribers equipments) are connected by communications lines (wires, radio, or fiber optics) to the nearest central office exchange. The subscriber lines radiate out from this hub.

Other forms of networks are shown in Figures 2.56 and 2.57. Devices can be placed along a network to concentrate traffic by performing switching and routing functions. Equipment may also be added to permit different types of equipment users (whose needs may vary in speed, service type, and format) to communicate with one another, thus making the network seem transparent to any user's particular needs. This is particularly desirable if many standards are present (such as in the case of data protocols and facsimile speeds). When used within a building or a campus of buildings, these networks are often called LANs (local area networks).

In some cases these interface equipments select one from several available transmission means based on cost, network traffic patterns, delay, and engineering considerations. Since networks may be very dynamic (especially at a busy hour), this alternate routing may cause one data burst to travel a path substantially different from the next. This is often the case in packet-switching networks (see following discussion). Microprocessing equipment is usually used at the nodes.

The hierarchies of the North American and CCITT networks are shown in Figures 2.58 and 2.59. Note that there are differences in the North American and CCITT symbols (see Table 2.27). With the advent of fiber optics between central telephone offices, the importance of the hierarchy is diminishing, particularly at the class 3, 4, and 5 levels. The CTI (Centre du Transit International) switching areas are shown in Figure 2.60. The figure also shows the first digit of the national dial code.

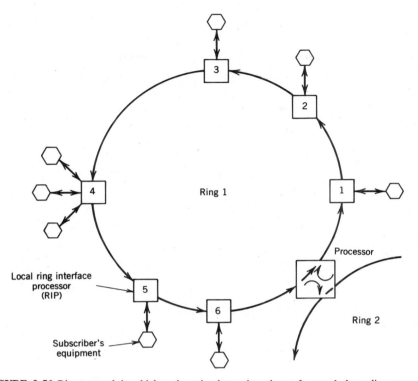

FIGURE 2.56 Ring network in which each station has a time share of network depending on need.

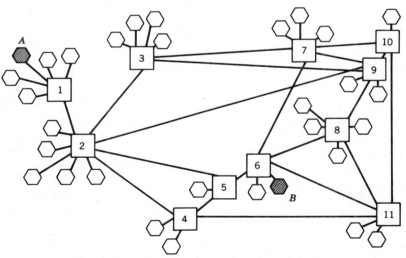

The path from A to B may be 1–6, 1–2–5–6, 1–3–7–9–8–6, etc.,
and may vary from time to time.

FIGURE 2.57 Partially interconnected network.

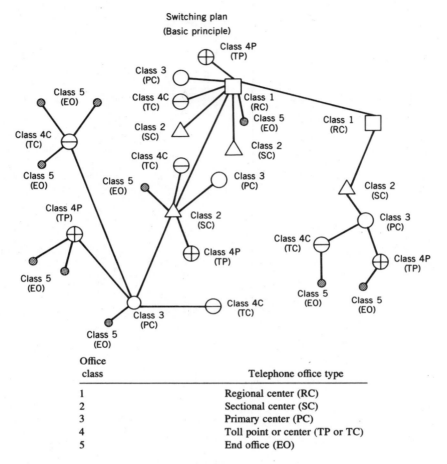

Office class	Telephone office type
1	Regional center (RC)
2	Sectional center (SC)
3	Primary center (PC)
4	Toll point or center (TP or TC)
5	End office (EO)

FIGURE 2.58 North American hierarchy.

Some elements of the network may use two wires (especially the portions from the subscribers equipment to the end office where an individual, unshared path is usually needed). Higher levels of the network may share facilities and typically use four-wire circuits. Four-wire circuits (Figure 2.61) have separate forward and return paths. Radio, cable, and satellite facilities are inherently equivalents of four-wire facilities.

Local loops tend to use two wires (thus reducing copper and cable costs), as shown in Figure 2.62. Traditional hybrid telephone elements permit the local two-wire loop to carry both sides of a conversation between the end office and the telephone instrument. Half of the voice power is sent down the transmit wire pair. The other half of the power is dissipated in a resistor–capacitor balancing network. Switching is often done at the four-wire point. The location of the actual transition between two and four wires may be dictated by facilities rather than the hierarchy.

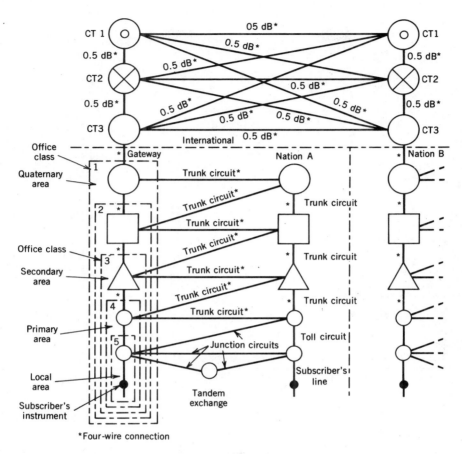

FIGURE 2.59 CCITT hierarchy.

Two-wire circuits (one-way, simplex, or half-duplex, as shown in Figures 2.63 and 2.64) may also be used for communications. In some cases the direction may be reversed by a user. Full-duplex circuits (Figure 2.65) permit traffic to flow both ways simultaneously. Either four-wire or two-wire/four-wire circuits (Figure 2.62) may be used.

A circuit is a loop. A full circuit may be viewed as a full loop, whereas a half-circuit is only a part of a loop. Figure 2.66 shows the connections between earth stations A and B. A satellite S is included. A full circuit is the path from A to B (via S) plus B to A (again via S). A half-circuit or channel is only half the full-circuit path. Which half may depend on the perspective of the provider or user (see Figure 2.66). In the case of the space segment operator he may use up-links as his measurement of traffic (often there may be more down-link destinations than up-link sources). The half-circuit may be SA with AS. The station A operator is

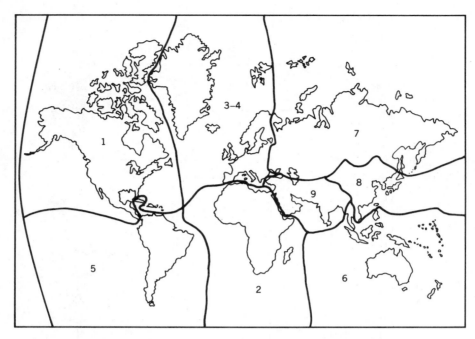

FIGURE 2.60 Locations of CT-1 international switching centers.

TABLE 2.27 Telecommunication Office Classes

	CCITT		North America
	Office Class Symbol		
		Name	Name
1	○	Quaternary center[a]	Regional center (RC)[a]
2	□	Tertiary center[a]	Sectional center (SC)[a]
3	△	Secondary center[a]	Primary center (PC)[a]
4	○	Primary center	Toll center (TC)
			Toll point (TP)
5	○	Local exchange	End office (EO)
Final	●		Trunk group
Subscribers instrument		Subscribers instrument	

Centre du Transit (CT) Symbol	CCITT Name
CT-1	International to international[a]
CT-2	International to international[a]
CT-3	International gateway exchange (international to national)[a]

[a]Four-wire input–output.

92

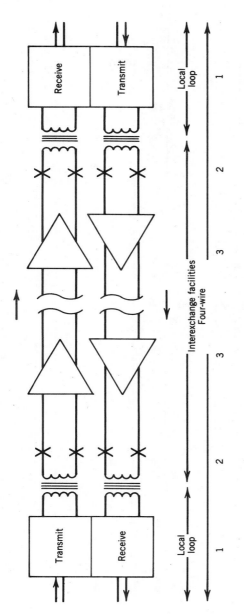

Key	Element	
	Voice	Data
1	Four-wire telephone	Four-wire modem
2	Switching	
3	Amplifiers	

FIGURE 2.61 Four-wire telephone network.

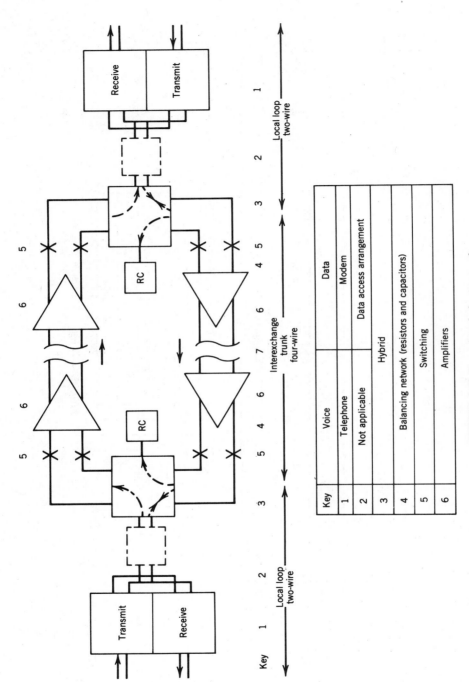

FIGURE 2.62 Two-wire telephone network.

Key	Voice		Data	
1	Telephone		Modem	
2	Not applicable		Data access arrangement	
3		Hybrid		
4		Balancing network (resistors and capacitors)		
5		Switching		
6		Amplifiers		

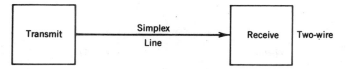

FIGURE 2.63 Simplex.

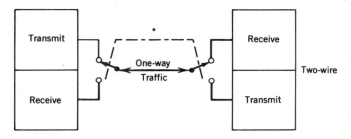

FIGURE 2.64 Half-duplex.

only interested in its links (AS or SA). The user generally does not care (or know) what the transmission network looks like. He is concerned with the AB (or BA) half-circuit.

A full circuit is two half-circuits to the transmission facility provider. To the user it is AB paired with BA. Many facilities (including satellites) have capacities rated in terms of voice channels, simultaneous conversations, or half-circuits. Most voice facilities require a forward and a reverse half-circuit. Thus, a facility rated at 1000 channels can handle only 500 normal connections (500 forward channels plus 500 return channels makes the 1000-channel-capacity total).

2.6.2 Networking Forms

Figure 2.67 shows various forms of networking in use for connecting a factory with its headquarters. In Figure 2.67(*a*) an AT&T tariff is used to provide private-line service entirely along the lines of a traditional terrestrial common carrier. The

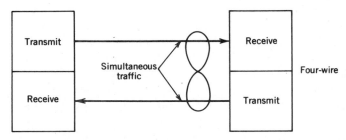

FIGURE 2.65 Full duplex. Only one direction may be used at a time, but direction may be reversed.

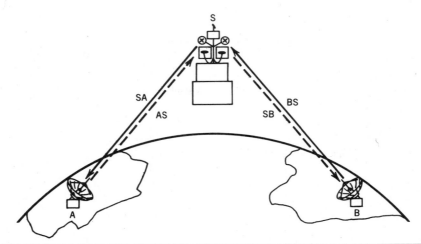

| Name | Satellite operator | Perspective | | Telecommunications user |
		Station A operator	Station B operator	
Channel or half-circuit	SA & AS or SB & BS	AS or SA	BS or SB	AB or BA
Full Circuit		Two Half-circuits		AB and BA

FIGURE 2.66 Full and half-circuits.

total bill is computed on the basis of mileage (Section 2.3) and facilities utilized. There is a different tariff section for each element and a separate mileage rate for the different routes. In the United States multiple carriers may be involved in various portions of a network. Long-haul transmission (5) may be by cable, microwave radio, or satellite.

In the United States the FCC has permitted alternative forms of communications. The specialized common carriers (SCCs) lease local lines from a local exchange carrier to bring traffic to the SCCs central offices (14 and 16), where it is packaged for transmission over their own (15) or leased facilities.

In general, traffic for Satellite Common Carriers is collected in the same way as for the specialized common carriers. It is routed to an earth station. These stations may be as much as 65 km (40 miles) outside a metropolitan area for certain satellites. This link may be owned by the satellite carrier, by a specialized carrier, or even by a conventional common carrier using copper, fiber, or microwave. The earth station may be owned or leased by the satellite carrier, a specialized carrier, or a private owner. The satellite may be owned by the satellite common carrier. In some instances one satellite common carrier may lease part of its satellites' capacity to other telecommunications carriers, thus making satellite transponders widely available. These carriers in turn may sublease all or part of the capability to others.

A message common carrier may use a local two-way microwave or fiber-optics link to connect users into a center city message office (30 or 32) for redistribution locally or to a distant location via terrestrial (15) or satellite (22) facilities. In this and subsequent cases the traditional local loops and exchanges are circumvented by alternate links. Since the local facilities tend to be heavily biased toward voice-grade service (vs. high-speed digital), and since they are major cost elements and subject to circuit impairments, they have been impediments to the growth of high-speed data services. The advent of local fiber is changing this situation.

Customer premises earth stations (34), including VSATs, may connect users directly, with no intervening terrestrial plant (Morgan and Rouffet, 1988).

A middle path is to place the earth stations at the head ends of cable television networks (CATV) and to use the large amounts of unutilized capacity in the cables (which are inherently broadband two-way communication media) for two-way business communications. CATV networks already have receive-only earth stations for satellite-provided television programs. Transmit and baseband equipment must be added to permit two-way operation.

The transmission channel may be allocated on a fixed (preassigned) basis. This is particularly useful for applications where there is a continuous flow of communications, such as in the case of radio or television program material. In other communications the amount and type of traffic may change throughout the day (see Section 2.4), and the destinations may change as the sun moves across a nation. Instead of static assignment of trunks and facilities (which often leads to underutilization during most of the day and busy signals during the peak busy hours), a dynamic allocation method may be used. In this case the users contend among themselves for the facility. This is called demand assignment.

User satisfaction may be measured in terms of availability, cost, connection delay, engineering performance, priority, or other parameters. Routes may be influenced by the time of day and the instantaneous loads on the network.

A communications network manager (now generally a microcomputer) may decide to route the circuit outside a leased network if the leased equipment is overloaded. It may "busy-out" some lower priority attempts during congested periods in an attempt to spread the load over a broader time base. It may decide how to dynamically allocate the resource to fit the type of communications being transmitted (voice, data, video facsimile, etc.). It may monitor the available circuits to detect links that are out of specification. If one is found, an alarm is sounded to alert the user and the carrier. A detected out-of-specification or out-of-service link may result in a rebate from the carrier.

Advanced channel allocation methods include the use of various demand assignment methods (see Chapter 4), digital speech interpolation (to permit a greater number of voice channels to share a given path), and packet switching (which allows substantial flexibility in transporting bursts of data).

Packet switching may incorporate teleprocessors at nodes along the network with store-and-forward capabilities to keep the fill factor of the network high. The packet-switching concept differs from the traditional circuit switching of the voice telephone network operations in that it breaks digital bit streams into short bursts,

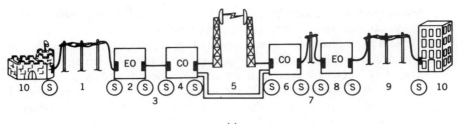

(a)

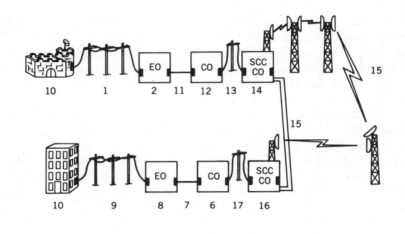

(b)

(c)

FIGURE 2.67 Forms of private-line networks: (a) AT&T; (b) specialized common carrier; (c) satellite common carrier using large earth stations; (d) connected-message common carrier; (e) customer premises earth stations; (f) cable television distribution.

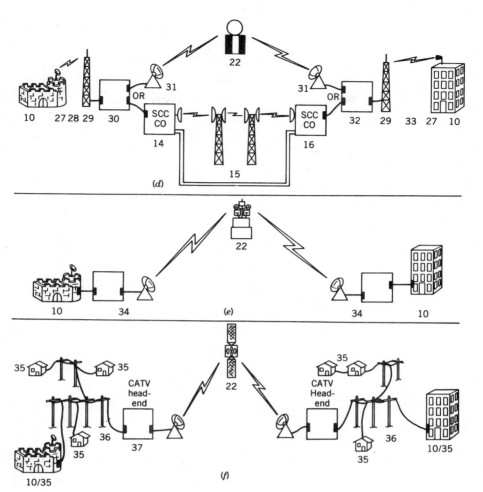

(d)

(e)

(f)

Key	Description	Owner	Key	Description	Owner	Key	Description	Owner
1	Short-haul route	a	13	Short-haul route	a	25	Interfacility link	c
2	Class 5 end office	a	14	SCC central office	b	26	Satellite CC office	c
3	Low-density route	a	15	SCC network	b	27	2-way station	d/u
4	Higher level toll office	a	16	SCC central office	b	28	Microwave link	d
5	High-density route	a	17	Short-haul route	a	29	Omnidirectional antenna	d
6	Toll office	a	18	Satellite CC office	c	30	Message CC office	d
7	Low-density route	a	19	Interfacility link	e	31	Earth station and antenna	d/c
8	Class 5 end office	a	20	Earth station	c–f	32	Message CC office	d
9	Short-haul route	a	21	Antenna	e–f	33	Microwave link	d
10	Users	u	22	Satellite	c–f	34	Customer premises earth station	c/u
11	Short-haul or foreign exchange	a	23	Antenna	c–f	35	CATV subscribers	u
12	Foreign exchange (FX)	a	24	Earth station	c–f	36	Broadband coax cable	e
						37	CATV headend and earth station	e

NOTE: Letters a–f match legend on page 98, u denotes user.

FIGURE 2.67 (*Continued*)

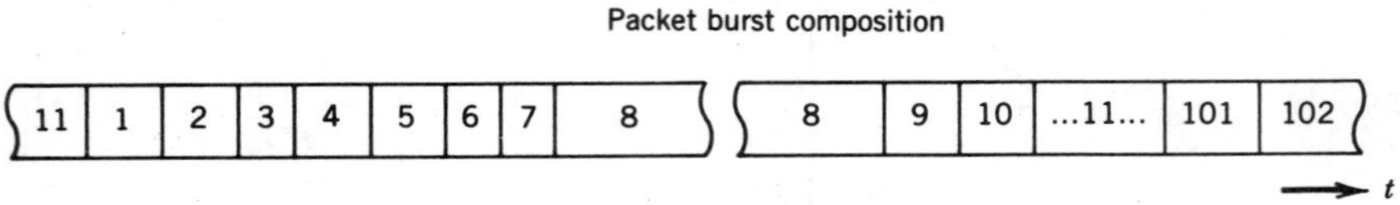

Packet burst composition

Key	Function	Bits	Key	Function	Bits	Key	Function	Bits
1	Start of packet framing sequence*	16	6	Link number	8	11	Gap between bursts (if any)	—
2	Priority, diagnostics, acknowledgments	16	7	Packet number	8	101	Start of next packet framing sequence*	16
3	Message number and last packet indicator	8	8	Text	—	102	Priority, diagnostics, acknowledgments	16
4	Destination address	16	9	End-of-packet framing sequence*	16			
5	Source address	16	10	Error-detecting coding redundancy check*	24		*Key: Inserted (or modified) by node	

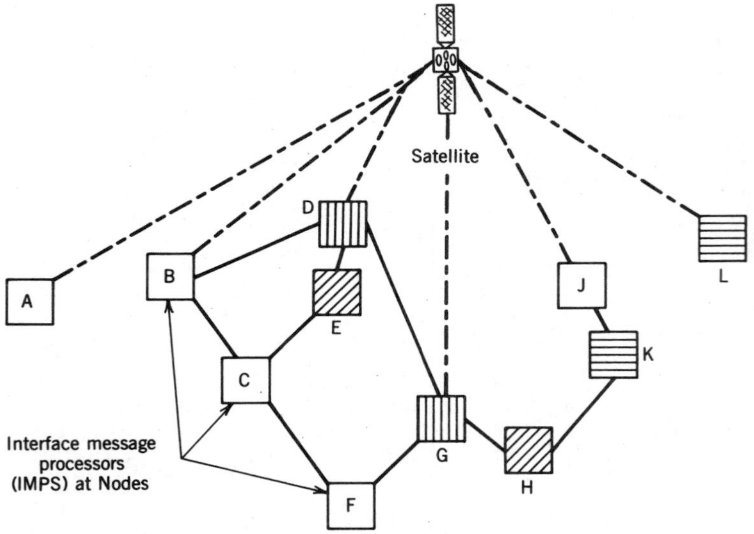

Route	Path Options
E-H	E-D-G-H, E-D-satellite-G-H, E-C-F-G-H, E-C-B-satellite-G-H, E-D-satellite-J-K-H, etc.
K-L	K-J-satellite-L, K-H-G-satellite-L, K-H-G-D-satellite-L, etc,
D-G	D-G, D-satellite-G, D-E-C-F-G, etc.

FIGURE 2.68 Packet-switched network.

100

adds header information about the destination, priority, type, and so on, and then forwards this packet of data forward along a fixed full-time network of leased lines (including satellite connections). Unlike the analog voice service, which has a network that is variable with time (but a fixed route for the duration of a call), the packet-switched network is a fixed one, but it may have a time-varying routing during a call; that is, sequential bursts between two users may take different paths (see Figure 2.68). Instead of circuit switching, the packet network uses microprocessors to route the traffic at the nodes. Since the bursts tend to be very brief, a mechanical switch would have great difficulty making and breaking the connections. Figure 2.69 shows one packet network.

Another method of network construction utilizes polling. In this case a computer at a node may sequentially ask each terminal if it has traffic for the network. When it finds a terminal with traffic, the computer assigns a circuit or time slot. Polling also permits the node to continuously verify the status of terminals and the links for fault detection purposes.

2.6.3 Equipment

As indicated in Section 2.6.2, typical satellite communications networks may contain substantial parts of terrestrial facilities. The tails or local loops tend to constrain the types and amount of traffic that can be fed into the satellite portion of the network. Experience has also shown that in some networks these local loops contribute a disproportionate amount of trouble (e.g., bit errors or noise) and cost to the overall performance.

Figure 2.70 lists the elements that comprise a typical terrestrial circuit.

Figures 2.71 and 2.72 show the elements that may comprise a local loop (including the customer premises, local outside plant, end office, and interexchange plant). If ISDN is successful the digital interface will move out of the central office and into the place of business or residence.

A local CATV plant may appear as in Figure 2.73. The cable may use a form of frequency domain multiple access (FDMA), as shown in Figure 2.74. (See Chapter 4 for FDMA.)

Figure 2.75 shows the growing capability of the telecommunication highways. The rate of growth is such that the capacity increases by a factor of 10 about every 16.7 yr.

2.6.4 Performance

The availability a of a communications facility is measured in terms of the ratio of the time it is operable (t_a) (available) to the total time t:

$$a = t_a/t \quad (\text{ratio}) \qquad (2.19)$$

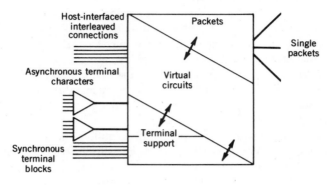

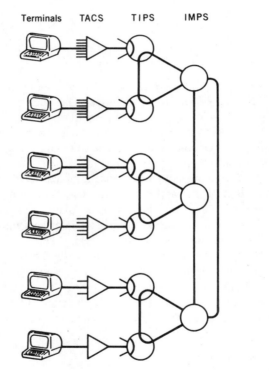

IMP   Interface message processor
TIP   Terminal interface processor
TAC   Telenet access controller

FIGURE 2.69 GTE Telenet hierarchy for packet switching network. (Modified from Roberts, 1976.)

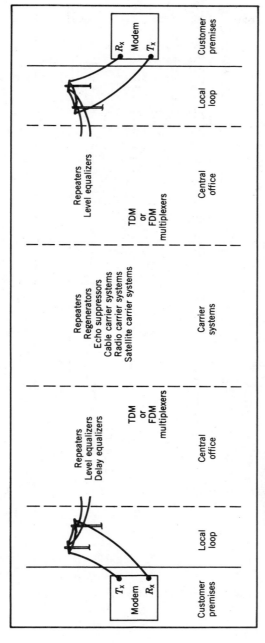

FIGURE 2.70 Terrestrial network elements. (*Source:* P. Archambeault, "Maintaining Data Communications Networks, *Telecommunications,* October 1978.) © 1978 *TELECOMMUNICATIONS* Magazine.

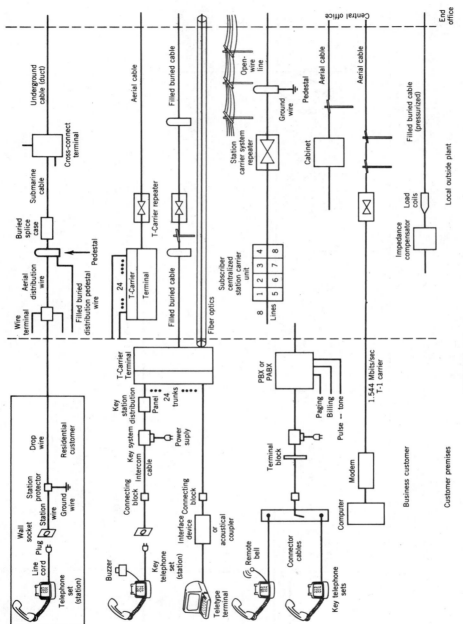

FIGURE 2.71 Local loop elements.

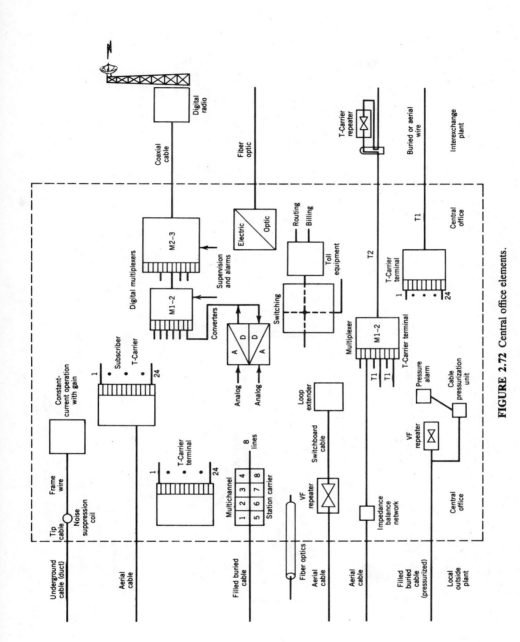

FIGURE 2.72 Central office elements.

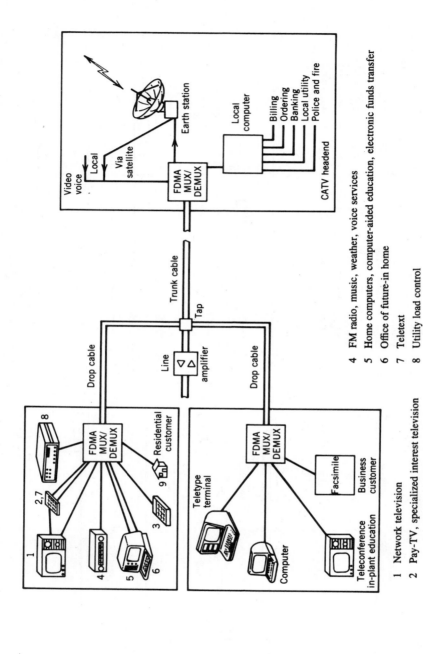

FIGURE 2.73 CATV local loop elements. *Note*: In some systems second cable is added exclusively for return (user-to-headend) traffic.

1 Network television
2 Pay-TV, specialized interest television
3 Shop-at-home and interactive television
4 FM radio, music, weather, voice services
5 Home computers, computer-aided education, electronic funds transfer
6 Office of future-in home
7 Teletext
8 Utility load control
9 Smoke and intrusion alarms

106

mHz

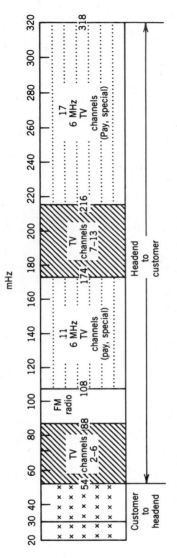

Forty-television channels, frequency plan

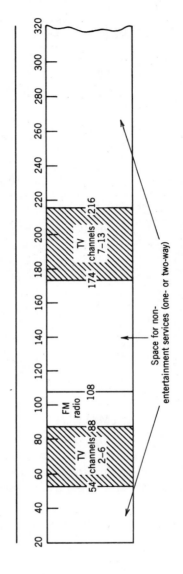

Multiservice frequency plan

FIGURE 2.74 Typical CATV frequency plans for North America.

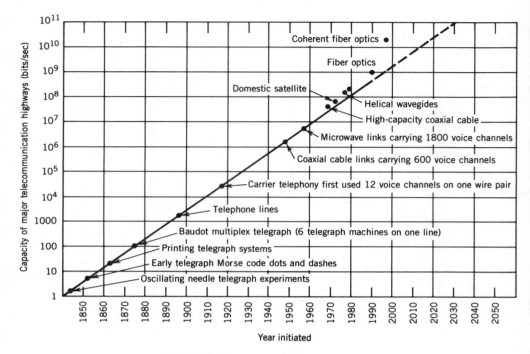

FIGURE 2.75 Telecommunications capabilities.

or

$$a = (\text{MTBF})/(\text{MTBF} + \text{MTTR}) \quad (\text{ratio}) \qquad (2.20)$$

where MTBF = mean time to failure in hours
  MTTR = mean time to restore in hours.
  A system may consist of several subsystems in series:

$$a_s = a_1 \times a_2 \times a_3 \times \cdots \times a_n \quad (\text{ratio}) \qquad (2.21)$$

where a_s = system availability
  $a_1, \cdots, a_n$ = individual availabilities of n individual subsystems
In some cases the ratio is converted to a percentage. Typical goals range from
99.00 to 99.99%.

  Alternative terms are reliability and $1 - a$, or outage. In-depth treatment of
component and system reliability is contained in Chapter 5.

  The definition of what constitutes a failure may vary with application and
between the telecommunications supplier and the user. Because of the diurnal
variations in traffic (see Section 2.4), an outage at a busy hour may be more

troublesome than one at 2 a.m. local time. Brief outages are sometimes excluded from the computation of availability in some services by the carriers in their tariffs.

Terrestrial lines have been specified for voice frequency (VF) and conditioned (C) for special services. Table 2.28 lists the basic specifications for conditioned lines.

Signaling for routing and control of the network may be done on a single-frequency (SF) or a multifrequency (MF) basis. Signaling by multifrequency tones is used to control the operation of networks. One example of these tones can be found in the hypothetical situations indicated in Table 2.29.

Signaling may occur at frequencies below (low-frequency out-band), within (in-band), or above (high-frequency out-band) the nominal 300–3400-Hz voice band (Figure 2.76).

In-band signaling may take place accidently if the voice energy is concentrated at a particular frequency. This is the "talk-down" situation that can result in an unintentional disconnection. Figure 2.77 shows the particular regions of the in-band frequency spectra that may cause this disconnection. The telephone office inspects the line for the condition that represents an on-hook condition. Typically this frequency is 2600 Hz in North America (2280 Hz in the United Kingdom). This is the tone-on-when-idle condition. The equipment also inspects another portion of the in-band spectrum. If energy is present in both portions, it assumes the 2600 Hz is part of the speech spectrum. If there is insufficient energy in the other portion, it assumes the 2600-Hz component is a valid signal that one of the instruments has gone on-hook and thus that the circuit should be torn down. In the out-band case there is no conflict between signaling and voice energy components.

Table 2.30 lists the more common U.S. and international tone frequencies.

The low frequencies lie below the voice frequency band and are primarily ring signals with 20 Hz being the most frequently used in the United States. 48 V direct current is used for local dial and on-hook/off-hook indications (higher voltage for longer local circuits). These dc signals and some low-frequency signals (below 200 Hz) cannot be accommodated in carrier or other ac-coupled systems; therefore, conversion to other signaling frequencies is necessary. For these purposes either in-band or out-band high-frequencies must be used.

Unlike the dc pulses created by the rotary telephone dial, the push-button telephone creates tone pairs. Most residential and commercial touch tone (TT) telephones use a keyboard with 12 keys (0–9, *, and #). Some military and business applications have four more keys (A–D). Figure 2.78 depicts the keyboard and tone pairs. In all cases a tone matrix is used (Table 2.31). Another set of tone pairs (also known as multifrequency) is used for operator and intrasystem signaling. In the CCITT plans, still another set of tones is used for signaling in the forward direction (from calling party to called party), and still another set for the reverse direction.

The use of one path for voice and another separate path for supervisory, routing, and toll information has become common. The second path is used only briefly at the start and end of each call and is shared with other calls.

TABLE 2.28 Summary of Bell System Private-Line Specification Options

Parameter	Basic Leased Line	Linear Conditioning Specifications			Nonlinear Conditioning Specification
		C1	C2	C4	D1
Envelope delay distortion (EDD)	800–2600 Hz, 1750 μsec	1000–2400 Hz, 100 μsec; 800–2600 Hz, 1750 μsec	1000–2600 Hz, 500 μsec; 600–2600 Hz, 1500 μsec; 500–2800 Hz, 3000 μsec	1000–2600 Hz, 300 μsec; 800–2800 Hz, 500 μsec; 600–3000 Hz, 1500 μsec; 500–3000 Hz, 3000 μsec	Does not apply
Frequency response relative to 1004 Hz	500–2500 Hz, −2 to +8 dB; 300–3000 Hz, −3 to +12 dB	1000–2400 Hz, −1 to +3 dB; 300–2700 Hz, −2 to +6 dB; 300–3000 Hz, −3 to +12 dB	500–2800 Hz, −1 to +3 dB; 300–3000 Hz, −2 to +6 dB	500–3000 Hz, −2 to +3 dB; 300–3200 Hz, −2 to +6 dB	Does not apply
Harmonic distortion	Not applicable	Not applicable	Not applicable	Not applicable	Fundamental to second harmonic, 35 dB minimum, to third harmonic, 40 dB minimum
C-notched noise with 1004 Hz tone	Not applicable	Not applicable	Not applicable	Not applicable	Noise at least 28 dB below received 1004-Hz test tone

TABLE 2.29 Call Signaling Functions

Location and Event	Significance or Action	Signal (Typical)	Notes[a]
Subscriber's Instrument			
Calling party, telephone on-hook	Calling party on-hook	0 V	—
Subscriber lifts receiver, telephone off-hook	Calling party off-hook	48 V dc	—
End Office	Recognizes off-hook condition		
Dial tone placed on subscribers line	End office can accommodate call request	350 and 440 Hz	If exchange cannot accommodate request, a busy signal (480 and 620 Hz at 60 ipm)
Subscriber			
Rotary dial: dials desired number	Dial pulses	48 V, dc pulses	—
Tone: keys in desired number	Two-frequency tone	See Table 2.31	—
End Office			
Examine number dialed	Determines if call is intraexchange or interexchange (first three digits)	—	Stores number in temporary memory
Assume interexchange call	Converts number to multifrequency (MF) tone pair sequence (may go via separate common-channel signaling path); sequence starts with KP tone pair and ends with ST tone pair (some digits may be dropped if unnecessary, i.e., area code)	See Table 2.31	If intraexchange, go to called line switch; for short interexchange links dc dial pulses may be used; low-frequency signaling may also be used

TABLE 2.29 (*Continued*)

Location and Event	Significance or Action	Signal (Typical)	Notes[a]
	Places 2600-Hz tone on voice line to indicate off-hook condition		Frequency varies with country and facility
Intermediate offices	Examines MF tones to determine routing; may generate new stream of MF tones to send to next exchange; if all trunks are busy, it returns busy signal	480 and 620 Hz	Alternative: voice announcement (recorded), 120 ipm
Called end office	Switches call to called subscriber's line	—	
Called subscriber's line			
On-hook	Called party on-hook	0 V to local end office	Indicates instrument is not in use
	Local end office places ring signal on called line, returns ringback signal to caller	Varies, 440 and 480 Hz	Local ring frequency varies (locally generated)
Called party lifts receiver	Called party answers	48 V dc, 2600 Hz	To local end office; to calling end office (starts toll charges)
If called party was in use (busy)	Conventional, returns busy signal to caller	480 and 620 Hz	Caller terminates call
	Call waiting, local end office places call-waiting signal on called party's line	440 Hz	Not audible to calling party
	Automatic transfer		Call transferred to another number (secretary or answering service)

[a]Abbreviation: ipm, interruptions per minute.

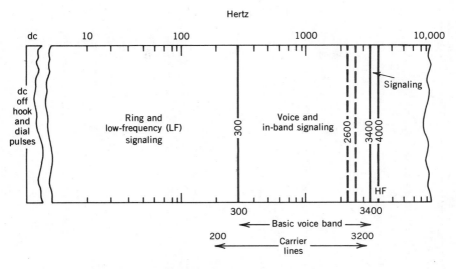

FIGURE 2.76 Signaling frequency bands for North America.

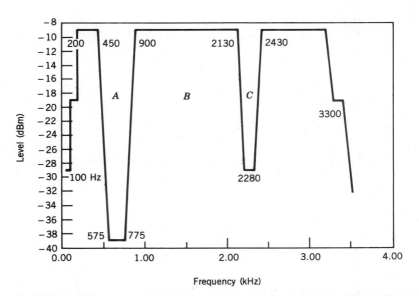

1. Signals permitted in area *A* provided there is no false operation of trunk signaling equipment.
2. Signals permitted in area *C* only if accompanied by signal in area *B* at power level not lower than 12 dB below area *C* power level.

FIGURE 2.77 United Kingdom telephone spectrum.

113

TABLE 2.30 Signaling Frequencies[a]

Frequency (Hz)	Modulation (ipm)	Class	Use	Typical Areas in Use
Low-Frequency Out-Band Signaling				
dc	dc or pulse	Dial	Local dial and signaling	All
16–25	10	Ring	Local party lines	RBOC
16.6	10	Ring	Local party lines	RBOC
20	10	Ring	Local lines	RBOC
25	Pulses	Dial	LF interexchange dial pulses	Germany
33.33	10	Ring	Local/party lines	RBOC
50	Pulses	Dial	Interexchange dial pulses	Italy, France Germany, Spain
50	10	Ring	Local party lines	RBOC
66.6	10	Ring	Local party lines	RBOC
80	Pulses	Dial	Interexchange dial	Spain
120 + 600	—	CP	Dial tone	North America
In-Band Signaling				
350 + 440[b]	—	CP	Dial tone	North America
420	40	CP	Audible ring-back	North America
440[b]	0.3 sec, every 10 sec	CP	Call waiting	North America
440 + 350[b]	—	CP	Dial tone	North America
440 + 480[b]	40	CP	Audible ring-back	North America
440–450	500-msec bursts	CP	Busy signal	Other countries
480	Spurts	CP	Order tones (manual service)	North America
480	—	CP	Line temporarily out of service	North America

Frequency (Hz)		Type	Description	Standard
480	—	—	Services in permanent off-hook condition	North America
480 + 440[b]		CP	Audible ring-back follows ring	North America
480 + 620[b]	60	CP	Busy signal	North America
480 + 620[b]	120	CP	Overflow (all paths busy)	North America
480 + 620	1 pulse	CP	Preemption tones	Private
500 (±2%)	2 sec long, 20 ipm	CP	Signaling current and manual working	CCITT No. 1
540+	—	MF	See Table 2.31	CCITT R2
600[c]	CW and pulsed	[c]	Mobile telephone dialing (not cellular)	R1
600	120	CP	Busy and overflow	North America
600 + 750	—	CP	Operator dialing	CCITT No. 2
620 + 440[b]	—	CP	See 440 + 620	—
620 + 480[b]	—	CP	See 480 + 620	—
660+	—	MF	See Table 2.31	CCITT R2
697(A) + (B)	—	TT	Subscriber dialing	CCITT R1
700+	—	MF	See Table 2.31	CCITT R2
750	—	—	Control	Europe
770(A) + (B)	—	TT	Subscriber dialing	CCITT R1
780+	—	MF	See Table 2.31	CCITT R2
825(A) + (B)	—	TT	Subscriber dialing	CCITT R1
900+	—	MF	See Table 2.31	CCITT R2
941(A) + (B)	—	TT	Subscriber dialing	CCITT R1
950(±50)	—	CP	Sequence of 3 tones: 950, 1400 and 1800 Hz: Special information tone	CCITT
1000	20 Hz	Ring	In-band signaling	RBOC

TABLE 2.30 (Continued)

Frequency (Hz)	Modulation (ipm)	Class	Use	Typical Areas in Use
1004	—	Test	Level test	—
1020+	—	MF	See Table 2.31	CCITT R2
1100+	—	MF	See Table 2.31	R1 and No. 5
1140+	—	MF	See Table 2.31	R2
1209(B) + (A)	—	TT	Subscriber dialing	CCITT R1
1300+	—	MF	See Table 2.31	R1 and No. 5
1336(B) + (A)	—	TT	Subscriber dialing	R1
1380+	—	MF	See Table 2.31	R2
1400 ± 1.5%	0.5 sec, every 15 sec	MF	Recording warning	CCITT
1400 ± 50	—	CP	See 950 Hz	R1
1477(B) + (A)	—	TT	Subscriber dialing	R1 and No. 5
1500+	—	MF	See Table 2.31	R2
1500^c	CW and pulsed	^{c}	Mobile telephone dialing	RBOC
1620+	—	MF	See Table 2.31	R2
1633(B) + (A)	—	TT	Subscriber dialing	R1
1700+	—	MF	See Table 2.31	R1 and No. 5
1740+	—	MF	See Table 2.31	R2
1800'(±50)	—	CP	See 950 Hz	CCITT
1850	—	TASI	TASI Locking	No. 5 bis
2000–2250 (2250 preferred)	400 msec/sec	Echo	Echo suppression disable	—
2040 ± 6	—	Rcv	With 2400 designates receiver is on-hook	No. 4

Frequency				System
2100 ± 15	CW or pulsed	Echo	Echo suppression disable	CCITT
2250	400 msec	Echo	Echo suppression disable	RBOC
2280 ± 6	CW or pulsed	—	Receiver status and line and register signals	No. 3 and Intelsat
2400 ± 6	—	Rcv	With 2040, receiver on-hook	No. 4
2400	CW or pulsed	Rcv	Receiver status	RBOC
2400 ± 6	—	—	Line signals	No. 5
2450–2750	—	—	Signaling	MCI
2600 ± 5	CW or pulsed	Rcv	Receiver status	R1
2600 ± 6	—	—	Line signals	No. 5
2810 ± 14	—	TASI	TASI locking	CCITT
High-Frequency Out-Band Signals (3400–4000 Hz)				
3400	CW or pulsed		Receiver on-hook	Lenkurt
3550	CW or pulsed		Receiver off-hook	Lenkurt
3700	CW or pulsed		Receiver on-hook	RBOC
3825 6	CW or pulsed		Link-by-link continuous low-level line signaling	R2
3850	CW or pulsed		Signaling	Germany
3995–4005	—		Signaling	MCI

[a] Abbreviations: ipm, interruptions per minute; CP, call progress signals; CW, continuous (wave) tone; MCI, Microwave Communications, Inc.; MF, multifrequency tones (see Table 2.31); RBOC, regional Bell operating companies and AT&T; Rcv, receiver status; SF, single-frequency tone; TT, touch-tones; TASI, time-assignment speech interpolation. Plus sign means tones at the same time.

[b] Precision tones (±0.5%).

[c] For dialing from noncellular mobile radio telephone subscribers 600 and 1500 Hz are used sequentially.

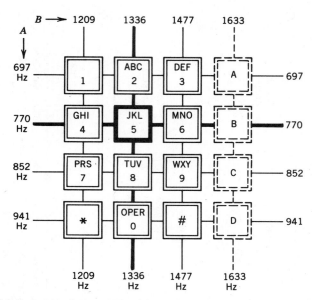

Example: Depressing 5 causes 770 and 1336 Hz tones to be mixed and transmitted.

FIGURE 2.78 Subscriber tone matrix.

In-band signaling may use either single-frequency tones or multifrequency signaling. This signaling operates only during the time the call is being set up or torn down. Use at other times would create interference with the speech (or data) signals. The only exception is the 2040–2600-Hz range (with 2400 and 2600 Hz being the most popular in North America). During a telephone call a sustained tone in this range without other voice frequency energy results in the talk-off condition mentioned.

Out-band signaling may occur at any time during the call without interference to the user. In some cases an out-band tone (on while in use) is used to maintain the circuit.

The push-button (e.g., touch-tone) telephone uses the two-frequency tones listed in Table 2.31. Once the call is established, these tones may be used for subscriber-to-subscriber signaling (completely independent of the telephone network). Typical examples are the remote control of computers (data entry and inquiry), dictation machines, and so on.

Call progress tones (dial tone, busy signal, etc.) are generally a blend of two frequencies or tones. These may be interrupted at various rates. For example, if the called subscriber's line is busy, a signal composed of 480- and 620-Hz tones is returned to the calling party with 60 interruptions per minute (ipm). If all paths to the distant phone are busy, the same tone pair is used but with 120 ipm. This is also called the overflow, or reorder, signal. A prerecorded or synthesized voice response may be used to inform the calling party of the progress of the call.

TABLE 2.31 Multifrequency and Touch-Tones

Character Region	Meaning	Simultaneous Frequency (Hz) Pairs	
		Forward	Reverse
North American (R1) Region: Subscriber Dialing Touch Tones (See Figure 2.78)			
1	1	697, 1209	—
2	2 (ABC)	697, 1336	—
3	3 (DEF)	697, 1477	—
4	4 (GHI)	770, 1209	—
5	5 (JKL)	770, 1336	—
6	6 (MNO)	770, 1477	—
7	7 (PRS)	852, 1209	—
8	8 (TUV)	852, 1336	—
9	9 (WXY)	852, 1477	—
10 or 0	0 (OPER)	941, 1336	—
11	#	941, 1477	—
12	*	941, 1209	—
13	A	697, 1633	—
14	B	770, 1633	—
15	C	852, 1633	—
16	D	941, 1633	—
(all frequencies are ±1.8%), 2 out of 6 codes			
Multifrequency (See Figure 2.78)			
1	1	700, 900	
2	2	700, 1100	
3	3	900, 1100	
4	4	700, 1300	
5	5	900, 1300	
6	6	1100, 1300	
7	7	700, 1500	
8	8	900, 1500	
9	9	1100, 1500	
0	0 or 10	1300, 1500	
11	(ST) End of pulsing	1500, 1700	
12	(KP) Start pulsing	1100, 1700	
13	Spare	700, 1700	
14	Spare	900, 1700	
15	Spare	1300, 1700	
(all are ±1.5%) 2 out of 6 codes; rate, 7 digits/sec; burst, duration 68 msec.			
Multifrequency CCITT No. 5			
1	1	700, 900	
2	2	700, 1100	
3	3	900, 1100	
4	4	700, 1300	

TABLE 2.31 (*Continued*)

Region	Character Meaning	Simultaneous Frequency (Hz) Pairs	
		Forward	Reverse

Multifrequency CCITT No. 5

Region	Meaning	Forward	Reverse
5	5	900, 1300	
6	6	1100, 1300	
7	7	700, 1500	
8	8	900, 1500	
9	9	1100, 1500	
0	0 or 10	1300, 1500	
11	(ST) End of pulsing	1500, 1700	
12	(KP1) Start of pulsing (terminal traffic)	1100, 1700	
13	11	700, 1700	
14	12	900, 1700	
15	(KP2) Start of pulsing (transit traffic)	1300, 1700	

2 out of 6 codes

MFC Bern Region: Multifrequency R2

Region	Meaning	Forward	Reverse
1	1 (send next digit)	1380, 1500	1020, 1140
2	2	1380, 1620	900, 1140
3	3	1500, 1620	900, 1020
4	4	1380, 1740	780, 1140
5	5	1500, 1740	780, 1020
6	6	1620, 1740	780, 900
7	7	1380, 1860	660, 1140
8	8	1500, 1860	660, 1020
9	9	1620, 1860	660, 900
0	0 or 10	1740, 1860	660, 780
11	—	1380, 1980	540, 1140
12	—	1500, 1980	540, 1020
13	Test	1620, 1980	540, 900
14	Echo	1740, 1980	540, 780
15	End	1860, 1980	540, 660
	Congestion at international exchange or at its output	—	
	Accuracy:	±4 Hz, 2 out of 6 codes	±4 Hz, 2 out of 6 codes

TABLE 2.32 Typical Pilot Tones

Frequency (kHz)	Modulation	Use	Type of Link	Used by
20	CW	AGC	Line and microwave	AT&T
40	CW	AGC	Carriers	Lenkurt 45A
64	CW	AGC	Carriers	WE L-1
80	CW	AGC	Carriers	Lenkurt 45A
84.080	−20 dBm0 CW	AGC	Basic group B	CCITT
84.140	−25 dBm0 CW	AGC	Basic group B	CCITT
104.080	−20 dBm0 CW	AGC	Basic group B	CCITT
150	CW	AGC	Carriers	Lenkurt 45A
308	CW	AGC	Carriers	L-3, TH, TD-2, TJ, TL
411.860	−25 dBm0 CW	AGC	Basic supergroup	CCITT
411.920	−20 dBm0 CW	AGC	Basic supergroup	CCITT
547.920	−20 dBm0 CW	AGC	Basic supergroup	CCITT
1552	−20 dBm0 CW	AGC	Basic mastergroup	CCITT
			Basic 15-supergroup	
		AGC	Assembly (No. 1)	CCITT
11096	−20 dBm0 CW	AGC	Basic supermastergroup	CCITT

Key: WE (Western Electric) L-1 and L-3 are cables; TH, TD-2, TJ & TL are microwave radio systems; dBm0 is the mean noise power in decibels (referred to one milliwatt) measured at the zero relative transmission level; CW is continuous wave; and CCITT is The International Consultive Committee for Telegraphy and Telephony. GTE Lenkurt makes the 45A microwave equipment.

TABLE 2.33 Intelsat FDM Spectrum

No. of Channels	Frequency Band (kHz)
12	12–60
24	12–108
36	12–156
48	12–204
60	12–252
72	12–300
96	12–408
132	12–552
192	12–804
252	12–1052
312	12–1300
372	12–1548
432	12–1796
492	12–2044
552	12–2292
612	12–2540
792	12–3284
972	12–4028
1092	12–4892
1332	12–5884

Various pilot tones can be applied to carrier and microwave systems for automatic level (or gain) control (ALC or AGC) purposes (Table 2.32). These are filtered out at the circuit ends.

Freeman's *Reference Manual for Telecommunications Engineering* (1985) contains excellent additional material on signaling and FDMA. There are many frequency domain multiplex channel stacking schemes. Some of the schemes used in Intelsat are depicted in Table 2.33.

In Intelsat, frequencies below 12 kHz are reserved for engineering service channels (e.g., between earth stations) and for energy dispersal. Different energy dispersal waveforms have been devised. When a variety of dispersal frequencies or waveforms are used, it is possible to place a unique signature on each originating earth station carrier.

Intelsat also specifies that as far as possible all basebands shall be assembled by means of standard CCITT 12-channel groups and 60-channel supergroups, with the carriers nominally being spaced on 4-kHz centers.

REFERENCES

Anonymous (1976). Satellite-computers-communications. *Telecommunications*, Oct.

Collins, F. R. (1975). Some facts, thoughts, and considerations related to usage sensitive pricing. *Bus. Commun. Rev.*, Sept./Oct.

Freeman, R. L. (1985). *Reference Manual for Telecommunications Engineering*, Wiley (Interscience), New York.

Gabriszeki, T., Reiner, P., Rogers, J., and Terbo, W. (1979). 18/30 GHz Fixed Communications System Service Demand Assessment, by Western Union Telegraph Company, for NASA, Washington, D.C.

Goode, H. H., and Machol, R. E. (1957). *System Engineering*, an Introduction to the Design of Large-Scale Systems, McGraw-Hill, New York, p. 135.

Gordon, G. D., and Dill, G. D. (1978). Efficient Computation of Erlang Loss Functions *COMSAT Tech. Rev.*, **8,** no. 2, pp. 353-370.

Green, J. H. (1986). *The Dow Jones-Irwin Handbook of Telecommunications*, Dow Jones-Irwin, Homewood, IL, p. 561-577.

Huber, P. W. (1987). *The Geodesic Network*, U.S. Govt. Printing Office, Washington, D.C.

Mina, R. R. (1971). The theory and reality of teletraffic engineering. *Telephony*, April 12.

Morgan, W. L. (1980). The economics of satellite telecommunications in the 1980's. *Satellite Communications*, January.

Morgan, W. L., and Rouffet, D. D. (1988). *Business Earth Stations for Telecommunications*, Wiley, New York.

Podraczky, E. (1975). Utilization of the geostationary satellite orbit. *Telecommunications*, Jan.

Roberts, L. G. (1976). Development of pocket switching networks worldwide. *Telecommunications*, Oct.

U.S. Department of Commerce (1986). *USA Statistics in Brief*, USDC, Bureau of the Census, Washington, D.C.

White, C. E. (1976). International carriers looking up and looking forward. *Telecommunications*, Aug.

COMMUNICATIONS SATELLITE SYSTEMS

Introduction

This chapter examines the fundamental communications satellite parameters. The initial sections deal with the basic resources: spectrum, link medium, satellite, and earth stations.

Having established this basic framework, the effects of these and associated parameters on the performance and utility of the link are analyzed. Link impairments are discussed, particularly precipitation and equipment-caused outages.

Link budgets are constructed for both analog and digital services using nonregenerative repeaters. Several methods of determining the performance capabilities of a link are provided using equations, graphs, a pocket calculator, and actual examples, so that cross-checks may be performed.

Another section deals with modeling techniques for the simulation of actual satellite circuits and estimation tools for the design of transponders.

3.1 BASIC RESOURCES

The basic resources available for communications satellite transponders are the radio frequency (rf) spectrum, power, mass, and cost. Frequency, and to some extent rf power, are covered by international treaty agreements between the members of the International Telecommunications Union (the ITU is an agency of the United Nations) and its Consultive International Committee on Radio Communications (the CCIR). These matters are considered at international meetings called World Administrative Radio Conferences (WARCs). Usually these WARCs consider a specific topic (e.g., in 1987 the WARC-87, for instance, considered mobile space telecommunications). Approximately every 20 years a general WARC is held to consider the entire frequency spectrum and rf power levels. The most recent general WARCs were held in 1959 and 1979.

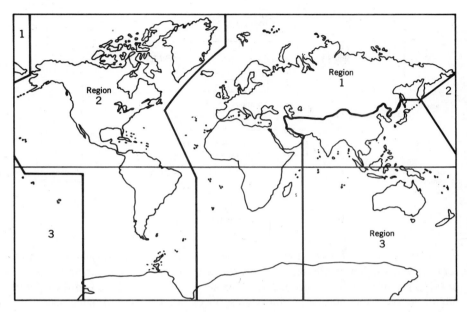

FIGURE 3.1 ITU regions.

The ITU has divided the world into three regions (see Figure 3.1). The Americas, for example, are in region 2. Occasionally, there are Regional Administrative Radio Conferences (RARCs) to consider matters that are localized to one or two regions.

At these world and regional meetings the allocation, use, and sharing of the radio spectrum is considered, and revisions are made in the form of *Final Acts* of the WARC or RARC; the revisions alter the *ITU Radio Regulations*, and the *Table of Frequency Allocations*.

The demand for frequencies and bandwidth for any service (particularly new services with rapid growth) may exceed the supply, such as has often been the case with space communications. Principles have evolved for assigning various levels of priorities (see Figure 3.2) where frequency assignments are shared among different types of users.

There may be multiple primary, permitted, or secondary allocations. The highest classification is an exclusive allocation. In this case the user category is allowed sole use of a specified band of frequencies (e.g., the fixed–satellite service; see Table 3.1). The exclusive frequency band will not be subject to interference to or from other classes of users. Geography is also a factor in allocations (see Fig. 3.3).

Frequency bands may be reassigned: During the transition there exists the possibility of interference between new user requirements and the previously authorized users.

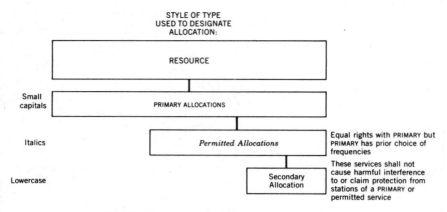

FIGURE 3.2 ITU allocations by priority classification.

The other types of allocations involve some form of frequency sharing. Fixed services using nonmovable transmitters and receivers on earth are generally coordinated to minimize interference between the sharing services. This is difficult with mobile services and, aeronautical mobile services cannot be coordinated with other satellite services. Additionally, most national administrations decide what uses may be made of the ITU allocated spectrum within their boundaries. In the United

TABLE 3.1 ITU User Definitions

Service Name	Functions
Space system	Any group of cooperating earth and/or space stations employing space radio communication for specific purposes
Satellite network	Satellite system or part of satellite system consisting of only one satellite and cooperating earth stations
Satellite link	Radio link between transmitting earth station and receiving earth station through one satellite; satellite link comprises one up link and one down link.
Multisatellite link	Radio link between transmitting earth station and receiving earth station through two or more satellites without any intermediate earth station; multisatellite link comprises one up link, one or more satellite-to-satellite links, and one down link
Fixed-satellite service	Radio communication service: between earth stations at specified fixed points when one or more satellites are used; in some cases includes satellite-to-satellite links, which may also be affected in intersatellite service; for connection

TABLE 3.1 ITU (*Continued*)

Service Name	Functions
	between one or more earth stations at specified fixed points and satellites used for service other than fixed-satellite service (e.g., mobile satellite service, broadcasting-satellite service, etc.)
Mobile-satellite service	Radio communication service: between mobile earth stations and one or more space stations or between space stations used by this service; between mobile earth stations by means of one or more space stations; if system so requires, for connection between these space stations and one or more earth stations at specified fixed points
Aeronautical mobile-satellite service	Mobile-satellite service in which mobile earth stations located on board aircraft: survival craft stations and emergency position indicating radiobeacon stations may also participate in this service
Maritime mobile-satellite service	Mobile-satellite service in which mobile earth stations located on board ships: survival craft stations and emergency position indicating radiobeacon stations may also participate
Land mobile-satellite service	Mobile-satellite service in which mobile earth stations located on land
Broadcasting-satellite service	Radio communication service in which signals transmitted or retransmitted by space stations intended for direct reception[a] by general public
Space research service	Radio communication service in which spacecraft or other objects in space used for scientific or technological research purposes
Intersatellite service	Radio communication service providing links between artificial earth satellites
Geosynchronous satellite	Earth satellite whose period of revolution equal to period of rotation of earth about its axis
Geostationary satellite	Satellite whose circular orbit lies in plane of earth's equator and turns about polar axis of earth in same direction and with same period as those of earth's rotation; orbit on which satellite should be placed to be geostationary satellite called *geostationary satellite orbit*

[a]In broadcasting-satellite service the term *direct reception* shall encompass both individual reception and community reception. The definition of *individual reception* is the reception of emissions from a space station in the broadcasting-satellite service by means of simple domestic installations and in particular those possessing small antennae. The definition of *community reception* is the receipt of emissions from a space station in the broadcasting-satellite service by receiving equipment, which in some cases may be complex and have antennas larger than those used for individual reception and intended for use by a group of the general public at one location or through a distribution system covering a limited area.

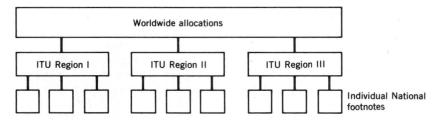

FIGURE 3.3 ITU allocations by geographical region © 1978 IEEE.

States the national agency is the Federal Communications Commission (FCC). The FCC further subdivides the ITU allotments among various civil and governmental users.

Signals from satellites often fail to respect national boundaries, especially when antenna sidelobes are considered. The International Frequency Registration Board (or IFRB) is an agency of the ITU. New satellites are to be described in filings with the IFRB long before launch. These data are circulated among the ITU members. If there is a potential interference problem, the parties involved enter into a formal coordination procedure. This procedure has become more complex as more and more satellites seek access to the geostationary orbit and as earth stations become less discriminating against other signals (due to the smaller antennas).

International satellite assignments sometimes are made on the basis of one or more ITU regions (shown in Figure 3.1). Individual nations are sometimes identified only by a letter code (see Table 3.2). These are called country symbols.

At the other extreme of priority is the "noninterference basis" classification allowed by some administrations. Users with this classification must cease operation if they cause interference to other, higher priority users.

The most common class of satellite allocations is the multiple primary user (or shared) category. These allocations permit two or more services (e.g., fixed-satellite services and point-to-point fixed terrestrial services) to use the same frequency bands. Coordination between these users is usually done at the national administration level. For example, to establish a U.S. 6-GHz transmit earth station, the FCC requires the applicant to show that it will not interfere with existing (or planned) 6-GHz common-carrier services in the terrestrial service. To license a receiving station, it must be shown that any existing (or planned) 4-GHz point-to-point terrestrial beams will not cause interference to the earth station. For receive-only 4-GHz operation, unlicensed operation is permitted in the United States, but no protection is afforded against other current or future 4-GHz users. There is a minimum-size, receive-only, 4-GHz antenna diameter that can be licensed. No licenses are required for reception in some exclusive bands in the United States, and blanket licenses are being issued by the FCC for certain small two-way earth stations.

In some cases footnotes to the ITU *Radio Regulations* and *Table of Frequency Allocations* further restrict the uses of frequencies in certain countries.

TABLE 3.2 Country Symbols

Symbol[a]	Country or Geographical Area[b]	Region
AFG	Afghanistan, Republic of	3
AFS	South Africa, Republic of	1
AGL	Angola, People's Republic of	1
ALB	Albania, Socialist People's Republic of	1
ALG	Algeria, Algerian Democratic and Popular Republic	1
ALS	State of Alaska, United States	2
AMS	St. Paul and Amsterdam Islands	3
AND	Andorra	1
AOE	Spanish Saharian Territory	1
ARG	Argentine Republic	2
ARS	Kingdom of Saudi Arabia	1
ASC	Ascension	1
ATG	Antigua and Barbuda	2
ATN	Netherlands Antilles	2
AUS	Australia	3
AUT	Austria	1
AZR	Azores	1
B	Brazil, Federative Republic of	2
BAH	Bahamas, Commonwealth of the	2
BDI	Republic of Burundi	1
BEL	Belgium	1
BEN	People's Republic of Benin	1
BER	Bermuda	2
BGD	Bangladesh	3
BHR	Bahrain, State of	1
BLR	Byelorussian Soviet Socialist Republic	1
BLZ	Belize	2
BOL or BLO	Bolivia, Republic of	2
BOT	Botswana, Republic of	1
BRB	Barbados	2
BRM	Burma, Socialist Republic of the Union of	3
BRU	Brunei	3
BUL	People's Republic of Bulgaria	1
CAF	Central African Empire	1
CAN	Canada	2
CAR	Caroline Islands	3
CBG	Khmer Republic (Cambodia)	3
CHN	China, People's Republic of	3
CHR	Christmas Island, Indian Ocean	3
CKH	Cook Island	3
CKH or CKN	Cook Islands	3
CHL or CLH	Chile	2
CLM	Republic of Colombia	2
CLN	Sri Lanka (Ceylon), Republic of	3
CME	Republic of Cameroon	1

TABLE 3.2 (*Continued*)

Symbol[a]	Country or Geographical Area[b]	Region
CNR	Canary Islands	1
COG	People's Republic of the Congo	1
COM	Comoros Islands	1
CPV	Cape Verde, Republic of	1
CRO	Crozet Archipelago	1
CTI	Republic of the Ivory Coast	1
CTR	Costa Rica	2
CUB	Cuba	2
CVA	Vatican City State	1
CYP	Republic of Cyprus	1
D	Germany, Federal Republic of	1
DDR	German Democratic Republic	1
DJI	Djibouti, Republic of	1
DMA	Dominica	2
DNK	Denmark	1
DOM	Dominican Republic	2
E	Spain	1
EGY	Egypt, Arab Republic of	1
EQA	Ecuador	2
ETH	Ethiopia	1
F	France	1
FJI	Fiji Islands	3
FLK	Falkland Islands and Dependencies	2
FNL	Finland	1
G	United Kingdom of Great Britain and Northern Ireland	1
GAB	Gabon Republic	1
GCA	Territories and Colonies of the United Kingdom	1
GCC	Territories and Colonies of the United Kingdom	3
GDL	French Department of Guadeloupe	2
GHA	Ghana	1
GIB	Gibraltar	1
GIL	Gilbert Islands	3
GMB	Gambia, Republic of	1
GNB	Guinea-Bissau, Republic of	1
GNE	Republic of Equatorial Guinea	1
GRC	Greece	1
GRD	Grenada	2
GRL	Greenland	2
GTM	Guatemala, Republic of	2
GUB or GUY	Guyana	2
GUF	Guyana, French Department of	2
GUI	Republic of Guinea	1
GUM	Guam	3

TABLE 3.2 (*Continued*).

Symbol[a]	Country or Geographical Area[b]	Region
HKG	Hong Kong	3
HND	Republic of Honduras	2
HNG	Hungarian People's Republic	1
HOL	Netherlands, Kingdom of the	1
HTI	Republic of Haiti	2
HVO	Republic of Upper Volta	1
HWA	State of Hawaii, United States	2
HWL	Howland Island	3
I	Italy	1
ICO	Cosos Keeling Islands	3
IFB	Reserved by IFRB (ITU)	N/A
IND	India, Republic of	3
INS	Indonesia, Republic of	3
IOB or BIO	British West Indies	2
IRL	Ireland	1
IRN	Iran	3
IRQ	Republic of Iraq	1
ISL	Iceland	1
ISR	State of Israel	1
J	Japan	3
JAR	Jarvis Island	3
JMC	Jamaica	2
JON	Johnston Island	2
JOR	Haschemite Kingdom of Jordan	1
KEN	Kenya, Republic of	1
KER	Kerguelen Islands	3
KIR	Kiribati	3
KOR	Korea, Republic of	3
KRE	Democratic People's Republic of Korea	3
KWT	Kuwait, State of	1
LAO	Lao People's Democratic Republic	3
LBN	Lebanon	1
LBR	Republic of Liberia	1
LBY	Libya, Socialist People's Libyan Arab Jamahiriya	1
LCA	St. Lucia	2
LIE	Liechtenstein	1
LSO	Kingdom of Lesotho	1
LUX	Luxembourg	1
MAC	Macao	3
MAU	Mauritius	1
MCO	Monaco	1
MDG	Madagascar, Democratic Republic of	1
MDR	Madeira	1

TABLE 3.2 (*Continued*)

Symbol[a]	Country or Geographical Area[b]	Region
MDW	Midway Islands	2
MEX	Mexico	2
MLA	Malaysia	3
MLD	Republic of Maldives	3
MLI	Republic of Mali	1
MLT	Malta, Republic of	1
MNG	Mongolian People's Republic	1
MOZ	Mozambique, People's Republic of	1
MRA	Mariana Islands	3
MRC	Kingdom of Morocco	1
MRL	Marshall Islands	3
MRN	Marion Island	3
MRT	French Department of Martinique	2
MTN	Islamic Republic of Mauritania	1
MWI	Malawi	1
MYT	Mayotte Island	1
NCG	Nicaragua	2
NCL	New Caledonia and Dependencies	3
NGR	Republic of the Niger	1
NHB	New Hebrides, British-French Condominium	3
NIG	Nigeria, Federal Republic of	1
NIU	Niue Island	3
NMB	Namibia	1
NOR	Norway	1
NPL	Nepal	3
NRU	Nauru Island	3
NZL	New Zealand	3
OCE	French Polynesia	3
OMA	Sultanate of Oman	1
ORB	Orbit	N/A
PAK	Pakistan, Islamic Republic of	3
PAQ	Easter Island, Chile	2
PHL	Philippines, Republic of the	3
PHX	Phoenix Islands	3
PLM	Palmyra Island	3
PNG	Papua New Guinea	3
PNR	Panama, Republic of	2
POL	Poland, People's Republic of	1
POR	Portugal	1
PRG	Paraguay, Republic of	2
PRU	Peru	2
PTC	Pitcairn Island	2
PTR	Puerto Rico	2

TABLE 3.2 (*Continued*)

Symbol[a]	Country or Geographical Area[b]	Region
QAT	Qatar, State of	1
REU	French Department of Reunion	1
RHS	Rhodesia	1
ROD	Rodriguez	3
ROU	Roumania, Socialist Republic of	1
RRW	Republic of Rwanda	1
S	Sweden	1
SDN	Democratic Republic of the Sudan	1
SEN	Senegal, Republic of the	1
SEY	Seychelles	1
SHN	St. Helena	1
SLM	Solomon Islands	3
SLV	Republic of El Salvador	2
SMA	American Samoa	3
SMO	Samoa (Western)	3
MR	San Marino	1
SNG	Singapore	3
SOM	Somali Democratic Republic	1
SPM	S. Pierre and Miquelon	2
SRL	Sierra Leone	1
STP	Sao Tome and Principe (Democratic Republic of)	1
SUI	Switzerland	1
SUR	Surinam (Republic of)	2
SWZ	Kingdom of Swaziland	1
SYR	Syrian Arab Republic	1
TCD	Republic of the Chad	1
TCH	Czechoslovak Socialist Republic	1
TGK or TZA	United Republic of Tanzania, Tanganyika, Zanzibar	1
TGO	Togolese Republic	1
THA	Thailand	3
TKL	Tokelau Islands	3
TMP	Portuguese Timor	3
TON	Tonga, Kingdom of	3
TRC	Tristan de Cunha, Station of the Republic of South Africa	2
TRD	Trinidad and Tobago	1
TUN	Tunisia	1
TUR	Turkey	1
TUV	Tuvalu	3
UAE	United Arab Emirates	1
UGA	Uganda, Republic of	1
UKR	Ukrainian Soviet Socialist Republic	1
URG	Oriental Republic of Uruguay	2
URS	Union of Soviet Socialist Republics	1

TABLE 3.2 (*Continued*)

Symbol[a]	Country or Geographical Area[b]	Region
USA	The 48 contiguous states of United States (excluding Alaska and Hawaii and its territories)	2
VCT	St. Vincent and Grenadines	2
VEN	Republic of Venezuela	2
VIR	Virgin Islands of the United States	2
VRG	British Virgin Islands	2
VTN	Viet Nam, Socialist Republic of	3
VUT	Vanuatu	3
WAK	Wake Island	2
WAL	Wallis and Futuna Islands	3
YEM	Yemen Arab Republic	1
YMS	Yemen, People's Democratic Republic of	1
YUG	Yugoslavia, Socialist Federal Republic of	1
ZAI	Zaire, Republic of	1
ZMB	Republic of Zambia	1
ZWE	Republic of Zimbabwe	1

[a]Alternate symbols are shown.
[b]Country symbols have geographical significance only and do not signify territorial status.

3.1.1 Frequencies

The frequencies available have varying degrees of attractiveness to potential space users. Some are presently beyond the range of commercially available equipment. Others have propagation (transmission) and/or precipitation (rain) problems. Still others are exposed to a high level of natural or man-made interference. These effects are considered in later sections of this chapter.

The remaining frequencies (generally between 1 and 15 GHz for communications satellite services) have become increasingly crowded and valuable. Section 3.1.4 discusses methods for reusing these frequency allocations several times while remaining within the allocated limits. Figure 3.4 lists general frequency assignments of up and down links for the Western Hemisphere.

The use of capitals letters, italics, and lowercase names follows the ITU method (see Figure 3.2 and Tables 3.3–3.10). Footnotes are used to further define the use of bands. These are shown as FOOTNOTE STATUS in Tables 3.3–3.10. The ITU and national (e.g., FCC) footnotes should be very carefully checked, as variations in allocations occur even within a region. Some services are allowed in the footnotes to the ITU radio regulations, which often impose limitations on their use. Where indicated, this is the case for Tables 3.3–3.10. In some cases national

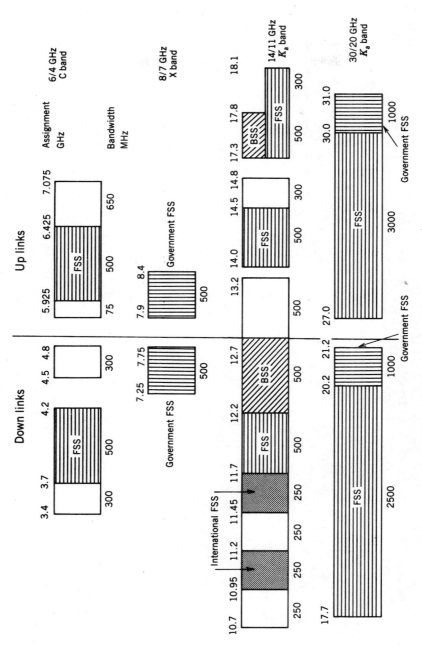

FIGURE 3.4 Western Hemisphere frequency assignments.

135

TABLE 3.3 Earth-to-Space Frequency Allocations in Fixed Satellite Service

Band Name(s)	Allocation Frequency and Wavelength	Total Bandwidth Available (MHz)	Typical Satellite Uses	ITU Region 1	2	3	United States	Status[a]	Sharing Services
S	2.655–2.690 GHz, 12 cm	35	Broadcasting		X	X		PRI	BROADCASTING SATELLITE, FIXED, MOBILE (Except Aeronautical) Earth Exploration Satellite, Radio Astronomy, Space Research (Passive)
Total		35							
C	4.5–4.8 GHz, 6.4 cm	300		X	X	X	X	PRI	FIXED, MOBILE
Total		300							
C	5.725–5.85 GHz, 5.2 cm	125	USSR	X				PRI	RADIOLOCATION Amateur
	5.85–5.925 GHz, 5.1 cm	75	USSR	X				PRI	FIXED, MOBILE,
	5.85–5.925 GHz, 5.1 cm	(75)			X			PRI	FIXED, MOBILE, Amateur, Radiolocation
	5.85–5.925 GHz, 5.1 cm	(75)				X		PRI	FIXED, MOBILE, Radiolocation
							X	PRI	RADIOLOCATION, Amateur
C	5.925–7.075 GHz, 5.1 cm	1150	(Many)	X	X	X		PRI	FIXED, MOBILE
	5.925–6.425 GHz, 4.8 cm	(500)	(Most)				X	PRI	FIXED (Domestic public point-to-point microwave, local television)
	6.425–6.525 GHz, 4.7 cm	(100)					X	PRI	MOBILE
	6.525–6.875 GHz, 4.5 cm	(350)					X	PRI	FIXED
	6.875–7.075 GHz, 4.3 cm	(200)					X	PRI	FIXED, MOBILE
					X				EARTH EXPLORATION SATELLITE, Space to Earth, FIXED, MOBILE,

Band	Frequency, wavelength	Total (MHz)	Use					Status	Allocation
Total		1350					X	PRI	METEOROLOGICAL SATELLITE, Earth to Space / GOVERNMENT
	8.215–8.4 GHz, 3.6 cm	185	Government Satellites	X		X		PRI	FIXED, MOBILE, Earth Exploration Satellite (Space to Earth)
	8.215–8.4 GHz, 3.6 cm	(185)	Government Satellites			X		PRI	FIXED, MOBILE, EARTH EXPLORATION SATELLITE (Earth to Space)
	8.215–8.4 GHz, 3.6 cm	(185)	Government Satellites				X	PRI	FIXED, EARTH EXPLORATION SATELLITE (Earth to Space), Mobile Satellite (Earth to Space, no Aeronautical transmissions)
Total		185					X	PRI	GOVERNMENT, DEEP-SPACE RESEARCH
K_u, 11	10.7–11.7 GHz, 2.7 cm	1000	Feeder links to broadcast satellites only	X				PRI	FIXED, FIXED SATELLITE, Space to Earth, MOBILE, (Except Aeronautical)
Total		1000						PRI	FIXED SATELLITE, Space to Earth
K_u, 13	12.5–12.75 GHz, 2.4 cm	250		X		X	X	PRI	FIXED, MOBILE, (Except Aeronautical)
Total		250						PRI	FIXED, MOBILE, Space Research (Deep Space & Space to Earth)
	12.75–13.25 GHz, 2.3 cm	500	Broadcast satellite feeders	X	X	X	X	PRI	FIXED, MOBILE
Total		500					X	PRI	RADIONAVIGATION, Space Research
K_u, 14	14.0–14.25 GHz, 2.1 cm	250	International and Domestic Satellites	X	X	X	X	PRI	RADIONAVIGATION, Space Research
	14.0–14.2 GHz, 2.1 cm	200	International and Domestic Satellites		X	X	X		RADIONAVIGATION, Space Research
	14.2–14.5 GHz, 2.1 cm	300	International and Domestic Satellites				X	Exclusive	
	14.25–14.3 GHz, 2.1 cm	50	International and Domestic Satellites	X	X	X	X	PRI	RADIONAVIGATION, Space Research

TABLE 3.3 (*Continued*)

Band Name(s)	Allocation Frequency and Wavelength	Total Bandwidth Available (MHz)	Typical Satellite Uses	ITU Region 1	2	3	United States	Status[a]	Sharing Services
	14.3–14.4 GHz, 2.1 cm	100	International and Domestic Satellites	X		X		PRI	FIXED, MOBILE, (Except Aeronautical), Radionavigation Satellite
			International and Domestic Satellites		X			PRI	Radionavigation Satellite
	14.4–14.47 GHz, 2.1 cm	70	International and Domestic Satellites	X	X	X		PRI	FIXED, MOBILE (Except Aeronautical), Space Research
	14.47–14.5 GHz, 2.1 cm	30	International and Domestic Satellites	X	X	X		PRI	FIXED, MOBILE (Except Aeronautical,) Radio Astronomy
Total		$\overline{500}$							
K_u, 14	14.5–14.8 GHz, 2.0 cm	300	Broadcast satellite feeder links	X	X	X		PRI	FIXED, MOBILE, Space Research (outside of Europe and Malta)
Total		$\overline{300}$							
K_a, 17	17.3–17.7 GHz, 1.7 cm	400	Broadcast satellite feeder links	X	X	X		none	Radiolocation
							X	PRI	
Total		$\overline{400}$							
K_a, 30	27–27.5 GHz, 1.1 cm	500	Japan	X	X	X		PRI	FIXED, MOBILE, Earth Exploration Satellite, (Space to Space)
	27.5–29.5 GHz, 1.05 cm	2000	Japan, Europe	X	X	X	X	PRI	FIXED, MOBILE FIXED, Domestic Public MOBILE, Domestic Public Land Mobile, (Except Aeronautical)
	29.5–30 GHz, 1.0 cm	500	Japan, Europe	X	X	X	X	PRI	Mobile Satellite, (Earth to Space)
	30.0–31.0 GHz, 1.0 cm	1000	Japan, Europe	X	X	X	X	PRI	MOBILE SATELLITE (Earth to Space), Standard Frequency and Time Signal Satellite (Space to Earth)
Total		$\overline{4000}$							

		Government				Service
K_a, 144	42.5–43.5 GHz, 7 mm	1000	X X X	X	PRI	FIXED, MOBILE (Except Aeronautical), RADIO ASTRONOMY
Total		1000				
K_a, 50	47.2–50.2 GHz, 6 mm	3000	X X	X	PRI	FIXED, MOBILE
	50.4–51.4 GHz, 6 mm	1000	X X		PRI	FIXED, MOBILE, Mobile Satellite (Earth to Space)
				X	PRI	FIXED, MOBILE, MOBILE SATELLITE (Earth to Space)
Total		4000				
75	71.0–74.0 GHz, 4.1 mm	3000	X	X	PRI	FIXED, MOBILE, MOBILE SATELLITE (Earth to Space)
	74.0–75.5 GHz, 4.1 mm	1500	X	X	PRI	FIXED, MOBILE
Total		4500				
93	92.0–95.0 GHz, 3.2 mm	3000	X	X	PRI	FIXED, MOBILE, RADIOLOCATION
Total		3000				
	202–217 GHz, 1.4 mm	15,000	X	X	PRI	FIXED, MOBILE
Total		15,000				
	265–275 GHz, 1.1 mm	10,000	X	X	PRI	FIXED, MOBILE, RADIO ASTRONOMY
Total		10,000				

[a]Allocation: SMALL CAPITALS = primary; *italics* = permitted; Lower Case = secondary.

Caution: These allocations are subject to change. Consult the Current *Table of Frequency Allocations* (ITU, 1988).

TABLE 3.4 Space-to-Earth Frequency Allocations in Fixed Satellite Service

Band Name(s)	Allocation Frequency and Wavelength	Total Bandwidth Available (MHz)	Typical Satellite Uses	ITU Regions 1	ITU Regions 2	ITU Regions 3	United States	Status[a]	Sharing Services
S, 2.5	2.5–2.535 GHz, 12 cm	35	Broadcasting			X		PRI	FIXED, MOBILE (Except Aeronautical) BROADCASTING SATELLITE
S, 2.5	2.5–2.655 GHz, 12 cm	155			X			PRI	BROADCASTING SATELLITE, FIXED MOBILE (Except Aeronautical)
	2.655–2.690 GHz, 11.2 cm	35			X			PRI	FIXED, FIXED SATELLITE (Earth to Space), MOBILE (Except Aeronautical) BROADCASTING SATELLITE Earth Exploration Satellite (Passive), Radio Astronomy, Space Research (Passive)
Total		$\overline{190}$							
C, 4	3.4–3.5 GHz, 8.8–8.6 cm	100	USSR		X	X		PRI	FIXED, Mobile, Radiolocation, Amateur
	3.4–3.6 GHz, 8.8–8.3 cm	200	USSR	X		X		PRI	FIXED, Mobile, Radiolocation
	3.5–3.7 GHz, 8.1–8.6 cm	200	USSR		X	X		PRI	FIXED, MOBILE (Except Aeronautical), Radiolocation
	3.6–4.2 GHz, 7.1–8.3 cm	600	Intelsat	X				PRI	FIXED, Mobile
	3.7–4.2 GHz, 8.1–7.1 cm	500	Intelsat and many domestic satellites		X	X	X	PRI	FIXED, MOBILE (Except Aeronautical) FIXED, Domestic Public Local Television, point-to-point microwave
Total	4.5–4.8 GHz, 6.5 cm	300 $\overline{1100}$		X	X	X	X	PRI	FIXED, MOBILE

Band	Frequency	Allocation					PRI	Service	No.
X, 7	7.25–7.3 GHz, 4.1 cm	Government Satellites	X	X	X		PRI	FIXED, MOBILE	50
	7.25–7.3 GHz, 4.1 cm					X	PRI	MOBILE SATELLITE (Space to Earth), Fixed	50
	7.3–7.45 GHz, 4.0–4.1 cm	Government Satellites	X	X	X		PRI	FIXED, MOBILE (Except Aeronautical)	150
	7.3–7.45 GHz, 4.0–4.1 cm					X	PRI	FIXED, Mobile Satellite (Space to Earth)	150
	7.45–7.55 GHz, 4.0 cm	Government Satellites	X	X	X		PRI	FIXED, METEOROLOGICAL SATELLITE, Space to Earth, MOBILE, (Except Aeronautical)	100
	7.45–7.55 GHz, 4.0 cm					X	PRI	FIXED, FIXED SATELLITE (Space to Earth), METEOROLOGICAL SATELLITE (Space to Earth), Mobile Satellite (Space to Earth)	100
	7.55–7.75 GHz, 3.9–4.0 cm	Government Satellites	X	X	X		PRI	FIXED, MOBILE, (Except Aeronautical)	200
	7.55–7.75 GHz, 3.9–4.0 cm					X	PRI	FIXED, Mobile Satellite (Space to Earth)	200
								Total	500
Ku, 11	10.7–11.7 GHz, 2.7 cm	International	X				PRI	FIXED, FIXED SATELLITE, (Earth to Space), MOBILE, (Except Aeronautical)	1,000
				X	X		PRI	FIXED, MOBILE (Except Aeronautical)	
						X	PRI	FIXED, (Domestic public point-to-point microwave, local television)	
								Total	1000
Ku, 12	11.7–12.2 GHz, 2.5 cm	Domestic Satellites		X			PRI	FIXED, Mobile (Except Aeronautical)	500
						X	PRI	Mobile, (Except Aeronautical), TV pickup	
								Total	500

TABLE 3.4 (Continued)

Band Name(s)	Allocation Frequency and Wavelength	Total Bandwidth Available (MHz)	Typical Satellite Uses	ITU Regions 1	2	3	United States	Status[a]	Sharing Services
K_u, 13	12.5–12.75 GHz, 2.4 cm	250	Business Satellites	X		X		PRI	FIXED SATELLITE, (Earth to Space)
Total		250							
K_a, 20	17.7–18.6 GHz, 1.7 cm	900	Japan, Europe	X	X	X	X	PRI	FIXED, MOBILE
	18.6–18.8 GHz, 1.6 cm	200	Japan, Europe	X		X		PRI	FIXED, MOBILE, Except Aeronautical, Earth Exploration, Satellite (Passive), Space Research (Passive)
					X		X	PRI	EARTH EXPLORATION SATELLITE (Passive), FIXED, MOBILE (Except Aeronautical), SPACE RESEARCH (Passive).
	18.8–19.7 GHz, 1.56 cm	900	Japan, Europe	X	X	X	X	PRI	FIXED, MOBILE
	19.7–20.2 GHz, 1.5 cm	500	Japan, Europe	X	X	X	X	PRI	Mobile Satellite (Space to Earth)
	20.2–21.2 GHz 1.4 cm	1,000	Japan	X	X	X	X	PRI	MOBILE SATELLITE (Space to Earth), Standard Frequency and Time Satellite (Space to Earth)
Total		3,500							
K_a, 40	37.5–39.5 GHz, 7.8 mm	2,000		X	X	X	X	PRI	FIXED, MOBILE
	39.5–40.5 GHz, 7.5 mm	1,000		X	X	X	X	PRI	FIXED, MOBILE, MOBILE SATELLITE (Space to Earth)
Total		3,000							

4 mm, 82	81–84 GHz, 3.6 mm	3,000	X	X	X	X	PRI	FIXED, MOBILE, MOBILE SATELLITE (Space to Earth)
Total		3,000						
3 mm	102–105 GHz, 3 mm	3,000	X	X	X	X	PRI	FIXED, MOBILE
Total		3,000						
2 mm	149–150 GHz, 2 mm	1,000	X	X	X	X	PRI	FIXED, MOBILE
	150–151 GHz, 2 mm	1,000	X	X	X	X	PRI	EARTH EXPLORATION SATELLITE (Passive), FIXED, MOBILE, SPACE RESEARCH (Passive)
	151–164 GHz, 1.9 mm	13,000	X	X	X	X	PRI	FIXED, MOBILE
Total		15,000						
1.3 mm	231–235 GHz, 1.3 mm	4,000	X	X	X	X	PRI	FIXED, MOBILE, Radiolocation
	235–238 GHz, 1.3 mm	3,000	X	X	X	X	PRI	EARTH EXPLORATION SATELLITE (Passive), FIXED, MOBILE, SPACE RESEARCH (Passive)
	238–241 GHz, 1.2 mm	3,000	X	X	X	X	PRI	FIXED, MOBILE, Radiolocation
Total		10,000						

[a]Allocation: SMALL CAPITALS = primary; italics = permitted; Lower Case = secondary.

TABLE 3.5 Intersatellite Frequency Allocations

Allocation Frequency and Wavelength	Total Bandwidth Available (MHz)	Typical Satellite Uses	ITU Regions			United States	Status[a]	Terrestrial Services[b]
			1	2	3			
22.55–23.0 GHz, 13 mm	450		X				PRI	FIXED, MOBILE
22.55–23.0 GHz, 13 mm	(450)			X	X	X	PRI	FIXED, MOBILE, BROADCASTING SATELLITE
23.0–23.55 GHz, 13 mm	550		X	X	X	X	PRI	FIXED, MOBILE
Total	1000							
32.0–32.3 GHz, 9 mm	300		X	X	X		PRI	RADIONAVIGATION, Space Research
32.0–32.3 GHz, 9 mm	(300)					X	PRI	RADIONAVIGATION
32.3–33.0 GHz, 9 mm	700		X	X	X	X	PRI	RADIONAVIGATION
Total	1000							
54.25–58.20 GHz, 5.1–5.5 mm	3,950		X	X	X	X	PRI	EARTH EXPLORATION SATELLITE, FIXED, MOBILE, SPACE RESEARCH (Passive)
59–64 GHz, 4.7–5.1 mm	5,000		X	X	X	X	PRI	FIXED, MOBILE, RADIOLOCATION
116–126 GHz, 2.5 mm	10,000		X	X	X	X	PRI	EARTH EXPLORATION SATELLITE (Passive), FIXED, MOBILE, SPACE RESEARCH (Passive)
126–134 GHz, 2.3 mm	8,000		X	X	X	X	PRI	FIXED, MOBILE, RADIOLOCATION
170–174.5 GHz, 1.7 mm	4,500		X	X	X	X	PRI	FIXED, MOBILE
174.5–176.5 GHz, 1.7 mm	2,000		X	X	X	X	PRI	FIXED, EARTH-EXPLORATION SATELLITE (Passive), SPACE RESEARCH (Passive), MOBILE
176.5–182 GHz, 1.7 mm	5,500		X	X	X	X	PRI	FIXED, MOBILE
185–190 GHz, 1.6 mm	5,000		X	X	X	X	PRI	FIXED, MOBILE

[a]Allocations: SMALL CAPITALS = primary; *italics* = permitted; Lower Case = secondary.

[b]In general, the earth's atmosphere blocks terrestrial signals from interfering with the intersatellite links.

Caution: These allocations are subject to change. Consult the current TABLE OF FREQUENCY ALLOCATIONS.

restrictions also apply. In the United States the FCC and NTIA frequency tables have further sets of footnotes. Do not ignore the footnotes! Always use current allocation tables as the assignments are subject to change. The allocations shown herein are as of mid-1987.

Tables 3.3 and 3.4 define the basic up and down links for the fixed–satellite service. For the Americas Figure 3.4 shows the major allocations.

Intersatellite links (ISL, or cross-links) permit two or more satellites to pass traffic between themselves without the need for an intermediate earth station. These are listed in Table 3.5. Optical links are also possible for the intersatellite service.

Other allocations of interest are listed in Tables 3.6–3.8.

Certain frequencies have been set aside for radio astronomy because of their unique properties (often an emission line, or window in the earth's atmosphere). Table 3.9 lists frequencies and bands that should be avoided by active radiators wherever possible to protect the fragile nature of the radio astronomy service.

The channels assigned to the broadcasting-satellite service are shown in Table 3.10 (for regions 1 and 3) and Table 3.11 (for region 2). Table 3.12 provides data on the orbit locations in the frequency range 11.7–12.5 GHz in regions 1 and 3 and 12.2–12.7 GHz in region 2. Figures 3.5–3.13 illustrate the geographical locations. Table 3.13 indicates the satellite longitudes and channels assigned to countries for all three regions; country symbols are keyed to Table 3.10. Figure 3.14 indicates the number of television channel assignments for satellites at various longitudes. Not all of these assignments will be used in the near future.

In regions 1 and 3 the channels are assigned numbers from 1 to 40 (only the lower 25 of which are usable in region 3).

The center frequency (f_{13}) of any rf channel for regions 1 and 3 can be found from

$$f_{13} = 11.7083 + (N_{c13} - 1) \times 0.01918 \quad \text{(GHz)} \quad (3.1)$$

where N_{c13} is the channel number (from 1 to 40) in region 1 or 3.

The rf channel spacing is 19.18 MHz. The total FM television-plus-sound bandwidth is 27 MHz. Thus, each signal occupies two rf channels.

In region 2 (the Americas) the bandwidth is 24 MHz, and the spacing is 14.58 MHz; thus, the channel numberings are different.

For region 2 the center frequency (f_2) of any rf channel can be found from

$$f_2 = 12.22401 + (N_{c2} - 1) \times 0.01458 \quad \text{(GHz)} \quad (3.2)$$

where f_2 is the region 2 center frequency for channel N_{c2}.

3.1.1.1 Harmonics

Since a satellite may operate in several frequency bands simultaneously, care must be taken to suppress harmonics so that the transmitter of one service does not interfere with the receiver of another service.

TABLE 3.6 Frequency Allocations for Mobile and Navigation Satellite Services

Band	Allocation Frequency and Wavelength	Bandwidth (MHz)	Satellite Service	Up Link, Earth to Space
VHF	149.9–150.05 MHz, 2 m	0.15	RADIONAVIGATION SATELLITE	
UHF	235–322 MHz, 1 m	87.0	*Mobile Satellite*	
	335.4–399.9 MHz, 82 cm	64.5	*Mobile Satellite*	
	399.9–400.05 MHz, 75 cm	0.15	RADIONAVIGATION SATELLITE	
	406–406.1 MHz, 75 cm	0.1	MOBILE SATELLITE	X
	608–614 MHz, 49 cm	6.0	Mobile Satellite (Except Aeronautical Mobile Satellite)	X
	890–896 MHz, 34 cm	6	Mobile Satellite	
L	1.215–1.24 GHz, 25 cm	25	RADIONAVIGATION SATELLITE	
	1.24–1.26 GHz, 25 cm	20	RADIONAVIGATION SATELLITE	
L	1.53–1.533 GHz, 19 cm	3.0	MARITIME MOBILE SATELLITE	
			LAND MOBILE SATELLITE	
			MARITIME MOBILE SATELLITE, LAND MOBILE SATELLITE	
			MARITIME MOBILE SATELLITE, LAND MOBILE SATELLITE	
	1.533–1.535 GHz, 19 cm	2.0	MARITIME MOBILE SATELLITE	
			Land Mobile Satellite MARITIME MOBILE SATELLITE	

Down Link, Space to Earth	Regions 1	2	3	United States	Status[a]	Sharing Services
X	X	X	X		Exclusive	
	X	X	X	X	Permitted	FIXED, MOBILE
	X	X	X	X	Permitted	FIXED, MOBILE
X	X	X	X	X	Exclusive	
X	X	X	X	X	Exclusive	Emergency Position Indicating Radio Beacon (EPIRB)
		X			SEC	RADIO ASTRONOMY
				X	Footnote U.S., Canada, Brazil only	Cellular reserve
X	X	X	X	X	PRI	RADIOLOCATION
X	X	X	X		PRI	RADIOLOCATION, Amateur
				X	Not Allocated to Satellites	
X	X				PRI	SPACE OPERATION (Space to Earth), Earth Exploration Satellite, Fixed, Mobile (Except Aeronautical)
X	X				PRI	SPACE OPERATION (Space to Earth), Earth Exploration Satellite, Fixed, Mobile (Except Aeronautical)
X		X	X		PRI	SPACE OPERATION (Space to Earth), Earth Exploration Satellite, Fixed, Mobile, Aeronautical Telemetering
X				X	PRI	SPACE OPERATION (Space to Earth), Earth Exploration Satellite, Fixed, Mobile, Aeronautical Telemetering
X	X				PRI	SPACE OPERATION (Space to Earth), Earth Exploration Satellite, Fixed, Mobile (Except Aeronautical)
					SEC	
X		X	X	X	PRI	

TABLE 3.6 (*Continued*)

Band	Allocation Frequency and Wavelength	Bandwidth (MHz)	Satellite Service	Up Link, Earth to Space
	1.535–1.544 GHz, 19 cm	9.0	Land Mobile Satellite MARITIME MOBILE SATELLITE	
	1.544–1.545 GHz, 19 cm	1.0	Land Mobile Satellite MARITIME MOBILE SATELLITE	
	1.545–1.555 GHz, 19 cm	10	AERONAUTICAL MOBILE SATELLITE (R)	
	1.555–1.559 GHz, 19 cm	4.0	LAND MOBILE SATELLITE	
	1.559–1.61 GHz, 19 cm	51	RADIONAVIGATION SATELLITE	
	1.593–1.594 GHz, 19 cm	1.0	RADIONAVIGATION SATELLITE	
L	1.610–1.6255 GHz, 19 cm	15.5	Land Mobile Satellite AERONAUTICAL MOBILE SATELLITE	X
	1.6255–1.6265 GHz, 18 cm	1	MARITIME MOBILE SATELLITE	X
	1.6265–1.6315 GHz, 18 cm	5	Land Mobile Satellite MARITIME MOBILE SATELLITE	X X
	1.6315–1.6345 GHz, 18 cm	3	Land Mobile Satellite MARITIME MOBILE SATELLITE LAND MOBILE SATELLITE	X X X
	1.6345–1.6455 GHz, 18 cm	11	MARITIME MOBILE SATELLITE	X
	1.6455–1.6465 GHz, 18 cm	1	MOBILE SATELLITE	X
	1.6465–1.6565 GHz, 18 cm	10	AERONAUTICAL MOBILE SATELLITE (R)	X
	1.6565–1.66 GHz, 18 cm	3.5	LAND MOBILE	X
	1.66–1.6605 GHz, 18 cm	0.5	LAND MOBILE	X
C, 5	5–5.25 GHz, 6 cm	250	AERONAUTICAL MOBILE SATELLITE	X
X	7.3–7.75 GHz, 4 cm	450	Mobile Satellite, Government	
K_u	14.3–14.4 GHz, 2.1 cm	100	Radionavigation Satellite	

Down Link, Space to Earth	Regions			United States	Status[a]	Sharing Services
	1	2	3			
X		X	X	X	SEC	See page 147
X	X	X	X	X	PRI	
X	X	X	X	X	SEC	
X	X	X	X	X	Exclusive	
X	X	X	X	X	Exclusive (R) = reserve	
X	X	X	X	X	Exclusive	
X	X	X	X		PRI	AERONAUTICAL RADIONAVIGATION
X	X	X	X	X	PRI	AERONAUTICAL RADIO NAVIGATION
X				X	Footnote	
	X	X	X		Footnote status only	AERONAUTICAL RADIONAVIGATION
	X	X	X		Exclusive	
				X	SEC	
	X	X	X		PRI	
X	X	X	X	X	SEC	
X	X	X	X	X	PRI	
X	X	X	X	X	PRI	
X	X	X	X	X	PRI	LAND MOBILE SATELLITE
	X	X	X	X	Exclusive	
	X	X	X	X	Exclusive (R) = reserve	
	X	X	X	X	Exclusive	
	X	X	X	X	PRI	RADIO ASTRONOMY
	X	X	X		Footnote status	AERONAUTICAL RADIONAVIGATION
X				X		See Table 3.4
	X		X		SEC	FIXED, FIXED SATELLITE, MOBILE (Except Aeronautical)
		X			SEC	FIXED SATELLITE
				X	Not Allocated to Radionavigation Satellites	FIXED SATELLITE

TABLE 3.6 (*Continued*)

Band	Allocation Frequency and Wavelength	Bandwidth (MHz)	Satellite Service	Up Link, Earth to Space
	15.4–15.7 GHz, 2.0 cm	300	AERONAUTICAL MOBILE SATELLITE	
K_a	19.7–20.2 GHz, 1.5 cm	500	Mobile Satellite	
	20.2–21.2 GHz, 1.5 cm	1,000	MOBILE SATELLITE	
	29.5–30 GHz, 1.0 cm	500	Mobile Satellite	X
	30–31 GHz, 1.0 cm	1,000	MOBILE SATELLITE	X
	39.5–40.5 GHz, 7.5 mm	1,000	MOBILE SATELLITE	
	43.5–47 GHz, 6.7 mm	3,500	MOBILE SATELLITE, RADIONAVIGATION SATELLITE	
	43.5–45.5, 6.7 mm	2,000	MOBILE SATELLITE (Gov't)	X
	45.5–47 GHz, 6.7 mm	1,500	MOBILE SATELLITE, RADIONAVIGATION SATELLITE	X
	50.4–51.4 GHz, 6 mm	1,000	Mobile Satellite	X
	66–71 GHz, 4 mm	5,000	MOBILE SATELLITE, RADIONAVIGATION SATELLITE	
	71–74 GHz, 4 mm	3,000	MOBILE SATELLITE	X
	81–84 GHz, 3.7 mm	3,000	MOBILE SATELLITE	
	95–100 GHz, 3 mm	5,000	MOBILE SATELLITE, RADIONAVIGATIONAL SATELLITE	
	134–142 GHz, 2 mm	8,000	MOBILE SATELLITE, RADIONAVIGATION SATELLITE	
	190–200 GHz, 1.5 mm	10,000	MOBILE SATELLITE, RADIONAVIGATION SATELLITE	
	252–265 GHz, 1.2 mm	13,000	MOBILE SATELLITE, RADIONAVIGATION SATELLITE	

*Allocations: SMALL CAPITALS = primary; *italics* = permitted; Lower Case = secondary.
Caution: These allocations are subject to change. Consult the current *Table of Frequency Allocations*

Down Link, Space to Earth	Regions			United States	Status[a]	Sharing Services
	1	2	3			
	X	X	X		Footnote Status	AERONAUTICAL RADIONAVIGATION
X	X	X	X	X	SEC	FIXED SATELLITE
X	X	X	X	X	PRI	FIXED SATELLITE, Standard Frequency and Time Signal Satellite
	X	X	X	X	SEC	FIXED SATELLITE
	X	X	X	X	PRI	FIXED SATELLITE, Standard Frequency and Time Signal Satellite
X	X	X	X	X	PRI	FIXED, FIXED SATELLITE, MOBILE
	X	X	X			MOBILE, RADIONAVIGATION
				X	PRI	MOBILE, RADIONAVIGATION
				X	PRI	FIXED, FIXED SATELLITE (Earth to Space), MOBILE (Except Aeronautical Mobile), RADIO ASTRONOMY
	X	X	X	X	SEC	FIXED, FIXED SATELLITE, MOBILE
	X	X	X	X	PRI	MOBILE, RADIONAVIGATION
	X	X	X	X	PRI	FIXED, FIXED SATELLITE, MOBILE
X	X	X	X	X	PRI	FIXED, FIXED SATELLITE, MOBILE
	X	X	X	X	PRI	MOBILE, RADIONAVIGATION, Radiolocation
	X	X	X	X	PRI	MOBILE, RADIONAVIGATION, Radiolocation
	X	X	X	X	PRI	MOBILE, RADIONAVIGATION
	X	X	X	X	PRI	MOBILE, RADIONAVIGATION

TABLE 3.7 Frequency Allocations for Other Satellite Services

Allocation Frequency and Wavelength	Satellite Service	Earth to Space	Space to Earth	ITU Region 1	ITU Region 2	ITU Region 3	United States	Status[a]	Sharing Services
7.0–7.100 MHz, 40 m	AMATEUR SATELLITE	X		X	X	X	X	PRI	AMATEUR
14.0–14.250 MHz, 20 m	AMATEUR SATELLITE	X		X	X	X	X	PRI	AMATEUR
18.068–18.168 MHz, 17 m	AMATEUR SATELLITE	X		X	X	X	X	PRI	AMATEUR
21.0–21.450 MHz, 14 m	AMATEUR SATELLITE	X		X	X	X	X	PRI	AMATEUR
24.890–24.990 MHz, 13 m	AMATEUR SATELLITE	X		X	X	X	X	PRI	AMATEUR
28.0–29.700 MHz, 10 m	AMATEUR SATELLITE	X		X	X	X	X	PRI	AMATEUR
30.005–30.01 MHz, 10 m	SPACE OPERATION (ID)		X	X	X	X		PRI	FIXED, MOBILE
136–137 MHz, 2.2 m	SPACE RESEARCH		X	X	X	X		PRI	FIXED, MOBILE
	SPACE RESEARCH		X	X	X	X	X	SEC	Until January 1, 1990
137–138 MHz, 2.2 m	SPACE OPERATIONS		X	X	X	X	X	PRI	After January 1, 1990
	SPACE RESEARCH		X	X	X	X	X	PRI	Fixed, Mobile (Except Aeronautical)
	SPACE OPERATION		X	X	X	X	X	PRI	FIXED, MOBILE, RADIOLOCATION
	METEOROLOGICAL SATELLITE		X		X			SEC	FIXED, MOBILE
138–143.6 MHz, 2.1–2.2 m	Space Research		X	X				SEC	AERONAUTICAL MOBILE
143.6–143.65 MHz, 2.1 m	SPACE RESEARCH		X	X				PRI	FIXED, MOBILE, RADIOLOCATION
143.65–144 MHz, 2 m	Space Research		X		X			PRI	FIXED, MOBILE
	Space Research		X			X		SEC	FIXED, MOBILE, RADIOLOCATION
	Space Research		X			X		SEC	FIXED, MOBILE
144–146 MHz, 2 m	AMATEUR SATELLITE	X		X	X	X	X	PRI	AMATEUR
146–149.9 MHz, 2 m	Space Telecommand	X		X	X			Footnote	FIXED, MOBILE (Except Aeronautical)
148–149.9 MHz, 2 m	Space Telecommand	X			X	X		Footnote	FIXED, MOBILE
149.9–150.05 MHz, 2 m	RADIO NAVIGATION SATELLITE		X	X	X	X	X	Exclusive	

Frequency Band	1	2	3	4	5	Status	Service / Notes
399.9–400.05 MHz, 75 cm			X	X	X	Exclusive	RADIONAVIGATION SATELLITE
400.05–400.15 MHz, 75 cm		X	X	X	X	Exclusive	STANDARD FREQUENCY & TIME SIGNAL SATELLITE
400.15–401 MHz, 75 cm			X	X	X	PRI	METEOROLOGICAL SATELLITE — METEOROLOGICAL AIDS, Radiosonde
		X	X	X	X	PRI	SPACE RESEARCH — METEOROLOGICAL AIDS, Fixed, Mobile, Except Aeronautical Mobile
401–402 MHz, 75 cm		X	X	X	X	PRI	SPACE OPERATION — METEOROLOGICAL AIDS, Fixed, Mobile, Except Aeronautical Mobile
	X	X	X	X	X	SEC	Meteorological Satellite — METEOROLOGICAL AIDS, Fixed, Mobile, Except Aeronautical Mobile
	X	X	X	X	X	SEC	Earth Exploration Satellite — METEOROLOGICAL AIDS, Fixed, Mobile, Except Aeronautical Mobile
402–403 MHz, 75 cm	X	X	X	X	X	SEC	METEOROLOGICAL SATELLITE — METEOROLOGICAL AIDS, Fixed, Mobile, Except Aeronautical Mobile
	X	X	X	X	X	SEC	Earth Exploration Satellite — METEOROLOGICAL AIDS, Fixed, Mobile, Except Aeronautical Mobile
435–438 MHz, 69 cm	X	X	X	X	X	Footnote	Amateur Satellite — AMATEUR, RADIONAVIGATION
460–470 MHz, 64 cm			X	X	X	SEC	Meteorological Satellite — FIXED, MOBILE
1.26–1.27 Ghz, 24 cm	X		X	X	X	Footnote	Amateur Satellite — RADIOLOCATION, Amateur
1.400–1.427 GHz, 20 cm			X	X	X	PRI	EARTH EXPLORATION (Passive) / SPACE RESEARCH (Passive) — RADIO ASTRONOMY
1.427–1.429 GHz, 20 cm	X		X	X	X	PRI	SPACE OPERATION (command) — FIXED, MOBILE, (Except Aeronautical)

TABLE 3.7 (*Continued*)

Allocation Frequency and Wavelength	Satellite Service	Earth to Space	Space to Earth	ITU Regions 1	2	3	United States	Status[a]	Sharing Services
1.525–1.535 GHz, 20 cm	SPACE OPERATIONS (Telemetry)		X	X				PRI	FIXED, MARITIME MOBILE SATELLITE, Mobile, (Except Aeronautical) Mobile
			X		X			PRI	Earth Exploration Satellite, Fixed, Mobile
			X			X		PRI	FIXED, Earth Exploration Satellite, Mobile
1.525–1.535 GHz, 20 cm	Earth Exploration Satellite		X	X	X	X		SEC	FIXED, MARITIME MOBILE SATELLITE
1.6605–1.6684 GHz, 18 cm	SPACE RESEARCH (Passive)			X	X	X	X	PRI	RADIO ASTRONOMY, Fixed, Mobile (Except Aeronautical)
1.670–1.690 GHz, 18 cm	Meteorological Satellite		X	X	X	X	X	PRI	METEOROLOGICAL AIDS, FIXED (Radiosones), MOBILE (Except Aeronautical)
1.690–1.700 GHz, 18 cm	METEOROLOGICAL SATELLITE		X	X				PRI	METEOROLOGICAL AIDS, Fixed, Mobile, (Except Aeronautical)
			X		X	X	X	PRI	METEOROLOGICAL AIDS, Radiosones
1.700–1.710 GHz, 18 cm	METEOROLOGICAL SATELLITE		X	X				PRI	FIXED, Mobile (Except Aeronautical)
			X		X	X	X	PRI	FIXED, MOBILE (Except Aeronautical)
2.290–2.300 GHz, 13 cm	SPACE RESEARCH		X	X				PRI	FIXED, Mobile (Except Aero)
			X		X	X		PRI	FIXED, MOBILE

Frequency allocation table (entries read left-to-right: Allocation — Footnote/Status — Region marks — Service — Band/Wavelength):

Band / Wavelength	Service	Region marks	Status	Allocation
4.202 GHz, (±2 MHz), 7.1347 cm	Time Signal Satellite	X X X		AERONAUTICAL RADIONAVIGATION
5.250–5.255 GHz, 5.7 cm	Space Research	X X X	SEC	RADIOLOCATION
5.670–5.725 GHz, 5.3 cm	Deep-Space Research	X X X	SEC	RADIOLOCATION, Amateur
6.427 GHz, (±2 MHz), 4.6647 cm	Time Signal Satellite	X (X X X)	Footnote	FIXED, MOBILE
7.450–7.550 GHz, 4.0 cm	METEOROLOGICAL SATELLITE	X X X	PRI	FIXED, FIXED SATELLITE (Space to Earth), MOBILE (Except Aeronautical)
8.025–8.175 GHz, 3.7 cm	Earth Exploration Satellite	X X	SEC	FIXED, FIXED SATELLITE (Earth to Space), MOBILE
8.175–8.215 GHz, 3.7 cm	EARTH EXPLORATION	X X	PRI	FIXED, FIXED SATELLITE (Earth to Space), MOBILE
	METEOROLOGICAL SATELLITE	X X	PRI	FIXED, FIXED SATELLITE, (Earth to Space)
8.215–8.46 GHz, 3.6 cm	Earth Exploration Satellite	X X	SEC	FIXED, FIXED SATELLITE, (Earth to Space)
	Earth Exploration Satellite	X X	SEC	FIXED, FIXED SATELLITE, (Earth to Space) MOBILE
		X	PRI	FIXED, FIXED SATELLITE, government Mobile satellite (no airborne transmissions)
8.4–8.5 GHz, 3.6 cm	SPACE RESEARCH	X X X	PRI	FIXED, MOBILE (except aeronautical)
		X	PRI	FIXED
10.45–10.5 GHz, 2.9 cm	Amateur Satellite	X X X	SEC	RADIOLOCATION, Amateur
10.6–10.68 GHz, 2.8 cm	SPACE RESEARCH (Passive)	X X X	PRI	FIXED, MOBILE (except aeronautical), RADIO ASTRONOMY
	EARTH EXPLORATION SATELLITE (Passive)	X	PRI	FIXED

TABLE 3.7 (*Continued*)

Allocation Frequency and Wavelength	Satellite Service	Earth to Space	Space to Earth	ITU Regions 1	2	3	United States	Status[a]	Sharing Services
10.68–10.7 GHz, 2.8 cm	EARTH EXPLORATION SATELLITE (Passive), SPACE RESEARCH (Passive)			X	X	X	X	PRI	RADIO ASTRONOMY
12.75–13.25 GHz, 2.3 cm	Deep Space Research		X	X	X	X	X	SEC	FIXED, MOBILE, FIXED SATELLITE
14.0–14.3 GHz, 2.1 cm	Space Research		X		X	X	X	SEC	FIXED SATELLITE (Earth to Satellite), RADIONAVIGATION
14.4–14.47 GHz, 2.1 cm	Space Research		X	X	X	X		SEC	FIXED, FIXED SATELLITE (Earth to Space), MOBILE
			X				X	SEC	FIXED SATELLITE, Earth to Space
14.5–14.8 GHz, 2.0 cm	Space Research		X	X	X	X		SEC	FIXED, FIXED SATELLITE, MOBILE
14.8–15.35 GHz, 2.0 cm	Space Research		X	X	X	X		SEC	FIXED, MOBILE
16.6–17.1 GHz, 1.8 cm	Deep Space Research	X		X	X	X		SEC	Radiolocation
17.2–17.3 GHz, 1.7 cm	Earth Exploration Satellite (Active), Space Research (Active)		X	X	X	X	X X	SEC SEC	RADIOLOCATION
18.6–18.8 GHz, 1.6 cm	Earth Exploration Satellite (Passive), Space Research (Passive)			X				SEC	FIXED, FIXED SATELLITE, MOBILE (Except Aeronautical)
	EARTH EXPLORATION SATELLITE (Passive), SPACE RESEARCH (Passive)				X		X	PRI	FIXED, FIXED SATELLITE, MOBILE (Except Aeronautical)

Frequency	Service							PRI/SEC	Service
21.2–21.4 GHz, 1.4 cm	EARTH EXPLORATION SATELLITE (Passive)			X	X	X	X	PRI	FIXED, MOBILE, SPACE RESEARCH (Passive)
22.21–22.5 GHz, 1.3 cm	EARTH EXPLORATION SATELLITE (Passive), SPACE RESEARCH (Passive)			X	X	X	X	PRI	FIXED, MOBILE (Except Aeronautical), RADIO ASTRONOMY
24.0–24.05 GHz, 1.2 cm	AMATEUR SATELLITE	X			X	X	X	PRI	AMATEUR
24.05–24.25 GHz, 1.2 cm	Earth Exploration Satellite (Active)					X	X	SEC	RADIOLOCATION, Amateur
					X			SEC	Radiolocation, Amateur Industrial, Scientific and Medical (24.125 GHz ± 125 MHz)
25.25–27 GHz, 1.2 cm	Earth Exploration Satellite			X	X	X	X	SEC	FIXED, MOBILE, Standard Frequency Time Signal Satellite (Earth to Space)
27–27.5 GHz, 1.1 cm	Earth Exploration Satellite	X [b]			X	X		SEC	FIXED, MOBILE
27–27.5 GHz, 1.1 cm	Earth Exploration Satellite				X	X	X	SEC	FIXED, MOBILE, FIXED SATELLITE (Earth to Space)
27–27.5 GHz, 1.1 cm	Earth Exploration Satellite	[b]			X	X	X	SEC	FIXED, MOBILE
31–31.3 GHz, 9.6 mm	Space Research			X	X	X	X	SEC	FIXED, MOBILE, Standard Frequency and Time Signal Satellite (Space to Earth)
31.3–31.5 GHz, 9.6 mm	EARTH EXPLORATION SATELLITE (Passive), SPACE RESEARCH (Passive)			X	X	X	X	PRI	RADIO ASTRONOMY
31.5–31.8 GHz, 9.6 mm	EARTH EXPLORATION SATELLITE (Passive), SPACE RESEARCH (Passive)			X	X	X	X	PRI	RADIO ASTRONOMY, Fixed, Mobile (except Aeronautical)
31.8–32 GHz, 9.4 mm	Space Research			X	X	X	X	PRI	RADIO ASTRONOMY
32–32.3 GHz, 9.3 mm	Space Research			X	X	X	X	SEC	RADIONAVIGATION
					X	X	X	SEC	RADIONAVIGATION, INTER-SATELLITE

TABLE 3.7 (*Continued*)

Allocation Frequency and Wavelength	Satellite Service	Earth to Space	Space to Earth	ITU Region 1	ITU Region 2	ITU Region 3	United States	Status[a]	Sharing Services
34.2–35.2 GHz, 8.6 mm	Space Research			X	X	X		SEC	RADIOLOCATION
36–37 GHz, 8.2 mm	SPACE RESEARCH (Passive)						X	SEC	Radiolocation
	EARTH EXPLORATION SATELLITE (Passive)			X	X	X	X	PRI	FIXED, MOBILE,
47–47.2 GHz, 6.38 mm	AMATEUR SATELLITE			X	X	X	X	PRI	AMATEUR
50.2–50.4 GHz, 6 mm	EARTH EXPLORATION SATELLITE (Passive)			X	X	X	X	PRI	FIXED, MOBILE
	SPACE RESEARCH (Passive)								
51.4–54.25 GHz, 5.7 mm	EARTH EXPLORATION SATELLITE (Passive)			X	X	X	X	PRI	FIXED, INTER-SATELLITE MOBILE
	SPACE RESEARCH (Passive)								
54.25–58.2 GHz, 5.3 mm	EARTH EXPLORATION SATELLITE (Passive)			X	X	X	X	PRI	FIXED, INTER-SATELLITE MOBILE
	SPACE RESEARCH (Passive)								
58.2–59 GHz, 5.2 mm	SPACE RESEARCH (Passive)			X	X	X	X	PRI	EARTH EXPLORATION SATELLITE (Passive)
	EARTH EXPLORATION SATELLITE (Passive)								
64–65 GHz, 4.6 mm	SPACE RESEARCH (Passive)			X	X	X	X	EXCLUSIVE	
	EARTH EXPLORATION SATELLITE (Passive)								
65–66 GHz, 4.6 mm	EARTH EXPLORATION SATELLITE			X	X	X	X	PRI	Fixed, Mobile
	SPACE RESEARCH								
75.5–76 GHz, 3.95 mm	AMATEUR SATELLITE			X	X	X	X	PRI	Fixed, Mobile
				X	X	X	X	PRI	AMATEUR
76–81 GHz, 3.82 mm	Amateur Satellite			X	X	X	X	SEC	RADIOLOCATION, Amateur

Frequency band	Allocation					Allocation
86–92 GHz, 3.3–3.5 mm	SPACE RESEARCH (Passive) EARTH EXPLORATION SATELLITE (Passive)	X	X	X	X PRI	RADIO ASTRONOMY
100–102 GHz, 3.0 mm	SPACE RESEARCH (Passive)	X	X	X	X PRI	FIXED, MOBILE
105–116 GHz, 2.7 mm	SPACE RESEARCH (Passive) EARTH EXPLORATION SATELLITE (Passive)	X	X	X	X PRI	RADIO ASTRONOMY
116–126 GHz, 2.5 mm	SPACE RESEARCH (Passive) EARTH EXPLORATION SATELLITE (Passive),	X	X	X	X PRI	FIXED, INTER-SATELLITE, MOBILE
				X	X PRI	Industrial, Scientific & Medical: 122.5 GHz ± 500 MHz
142–144 GHz, 2.1 mm	AMATEUR SATELLITE Amateur Satellite	X	X	X	X PRI	AMATEUR
144–149 GHz, 2.04 mm	SPACE RESEARCH (Passive)	X	X	X	X SEC	RADIOLOCATION, Amateur
150–151 GHz, 2.0 mm	EARTH EXPLORATION SATELLITE (Passive)	X	X	X	X PRI	FIXED, FIXED SATELLITE (Space to Earth), MOBILE
164–168 GHz, 1.8 mm	SPACE RESEARCH (Passive) EARTH EXPLORATION SATELLITE (Passive)	X	X	X	X PRI	RADIO ASTRONOMY
174.5–176.5 GHz, 1.7 mm	SPACE RESEARCH (Passive) EARTH EXPLORATION SATELLITE (Passive) INTER-SATELLITE	X	X	X	X PRI	FIXED, MOBILE
182–185 GHz, 1.6 mm	SPACE RESEARCH (Passive) EARTH EXPLORATION SATELLITE (Passive)	X	X	X	X PRI	RADIO ASTRONOMY
200–202 GHz, 1.5 mm	SPACE RESEARCH (Passive) EARTH EXPLORATION SATELLITE (Passive)	X	X	X	X PRI	FIXED, MOBILE, SPACE RESEARCH (Passive)

TABLE 3.7 (Continued)

Allocation Frequency and Wavelength	Satellite Service	Earth to Space	Space to Earth	ITU Regions			United States	Status[a]	Sharing Services
				1	2	3			
217–231 GHz, 1.3 mm	SPACE RESEARCH (Passive) EARTH EXPLORATION SATELLITE (Passive)			X	X	X	X	PRI	RADIO ASTRONOMY
235–238 GHz, 1.3 mm	SPACE RESEARCH (Passive) EARTH EXPLORATION SATELLITE (Passive)			X	X	X	X	PRI	FIXED, FIXED-SATELLITE (space to Earth), MOBILE
241–248 GHz, 1.22 mm	Amateur Satellite			X	X	X	X	PRI	AMATEUR
250–252 GHz, 1.2 mm	SPACE RESEARCH (Passive) EARTH EXPLORATION SATELLITE (Passive)			X	X	X	X	EXCLUSIVE	

[a] Allocations: SMALL CAPITALS = primary; *italics* = permitted; Lower Case = secondary.
[b] Space to space links.

160

TABLE 3.8 Broadcasting Satellite Service Frequency Allocations

Allocation Frequency and Wavelength	Bandwidth MHz	Earth to space	Space to Earth	ITU Regions 1	2	3	United States	Status	Sharing Service[a]
620–790 MHz, 38–48 cm	170		X	X	X	X		Footnote status	BROADCASTING (UHF-TV), FIXED, MOBILE
2.5–2.655 GHz, 12 cm	155		X	X				PRI	FIXED, MOBILE, (Except Aeronautical Mobile)
2.5–2.655 GHz, 12 cm	155		X		X		X	PRI	FIXED, FIXED SATELLITE (Space to Earth),
2.5–2.535 GHz, 12 cm	35					X		PRI	FIXED, FIXED SATELLITE (Space to Earth), MOBILE (Except Aeronautical)
2.535–2.655 GHz, 12 cm	120		X			X		PRI	FIXED, MOBILE, (Except Aeronautical Mobile)
2.655–2.69 GHz, 11 cm	35	X		X				PRI	FIXED, MOBILE, (Except Aeronautical) Earth Exploration Satellite (Passive), Radio Astronomy Space Research (Passive)
	35	X			X	X		PRI	FIXED, FIXED SATELLITE, MOBILE (Except Aeronautical) Earth Exploration Satellite (Passive), Radio Astronomy, Space Research (Passive)

TABLE 3.8 (Continued)

Allocation Frequency and Wavelength	Bandwidth MHz	Earth to space	Space to Earth	ITU Regions 1	ITU Regions 2	ITU Regions 3	United States	Status	Sharing Service[a]
	35	X					X	PRI	FIXED, Instructional Television Fixed, Operational Fixed, Earth Exploration Satellite (Passive), Radio Astronomy, Space Research (Passive)
11.7–12.5 GHz, 2.5 cm	800			X				PRI	FIXED, Mobile (Except Aeronautical) BROADCASTING
11.7–12.1 GHz, 2.5 cm	400		X		X			Footnote status	FIXED, FIXED SATELLITE (Space to Earth), Mobile (Except Aeronautical),
11.7–12.2 GHz, 2.5 cm	500		X			X		PRI	FIXED, MOBILE (Except Aeronautical), BROADCASTING
11.7–12.2 GHz, 2.5 cm	500		X				X	Footnote status	FIXED SATELLITE, Space to Earth, Mobile (Except Aeronautical)
12.1–12.3 GHz, 2.5 cm	200		X		X			PRI	FIXED, FIXED SATELLITE, MOBILE (Except Aeronautical) BROADCASTING

Band	BW (MHz)				Pri	Service allocations
12.2–12.7, 2.5 cm	500	X				FIXED
12.3–12.7 GHz, 2.4 cm	400	X	X		PRI	FIXED, MOBILE (Except Aeronautical), BROADCASTING
12.5–12.75 GHz, 2.4 cm	250		X	X	PRI	FIXED, FIXED SATELLITE (Space to Earth), MOBILE (Except Aeronautical)
22.5–22.55 GHz, 1.3 cm	50	X	X	X	PRI	FIXED, MOBILE
22.55–23 GHz, 1.3 cm[b]	450	X	X	X	PRI	FIXED, MOBILE, INTER SATELLITE
40.5–42.5 GHz, 7.1 mm	2000	X	X	X	PRI	BROADCASTING, *Fixed, Mobile*
84–86, 3.5 mm[b]	2000	X	X	X	PRI	FIXED, MOBILE, BROADCASTING

[a] Allocations: SMALL CAPITALS = primary; *italics* = permitted; Lower Case = secondary.

[b] Satellite service direction not specified explicitly but may be used for space to earth (earth to space link is often in the fixed-satellite service allocations).

Caution: These allocations are subject to change. Consult the current *Table of Frequency Allocations* (ITU, 1988).

TABLE 3.9 Radio Astronomy Frequencies

Band Frequency and Typical Wavelength	Line Name[a] and Rest Frequency	Status[b] (ITU)
25.55–25.67 MHz, 11.7 m		Exclusive
37.5–38.25 MHz, 7.89 m		SECONDARY
73.00–74.60 MHz, 4 m		EXCLUSIVE in ITU region 2 only
150.05–153.00 MHz, 1.98 m		PRIMARY in region 1
322–328.6 MHz, 92 cm	Deuterium (H), 327 MHz	PRIMARY
406.1–410.0 MHz, 73 cm		PRIMARY
608–0.614.0 MHz, 49 cm		PRIMARY in Region 2
608.0–614.0 MHz, 49 cm		FOOTNOTE STATUS, region 1
610.0–614.0 MHz, 49 cm		FOOTNOTE region 3
1.350–1.400 GHz, 22 cm		FOOTNOTE STATUS
1.400–1.427 GHz, 21 cm	Neutral Hydrogen (H^1), 1.4204 GHz	PRIMARY with passive EARTH EXPLORATION SATELLITE and SPACE RESEARCH
1.5585–1.6365 GHz, 19 cm	Hydroxyl $(O^{18}H^1)$, 1.584 GHz	FOOTNOTE (Extraterrestrial listening frequency)
1.6106–1.6138 GHz, 18.6 cm	Hydroxyl $(O^{16}H^1)$, 1.612231 GHz	SECONDARY (FOOTNOTE STATUS)
1.660–1.670 GHz, 18 cm	Hydroxyl (OH) Hydroxyl $(O^{16}H^1)$, 1.665 GHz Hydroxyl $(O^{18}H^1)$, 1.667 GHz	PRIMARY
1.7188–1.7222 GHz, 17 cm	Hydroxyl (OH)	SECONDARY (FOOTNOTE STATUS)
	Hydroxyl $(O^{16}H^1)$, 1.720530 GHz	
~2.4 GHz, ~12.5 cm	SiH, ~2.4 GHz	None
2.655–2.690 GHz, 11 cm		SECONDARY
2.690–2.700 GHz, 11 cm		PRIMARY
3.263794 GHz, 9.19 cm	CH	None
3.335481 GHz, 8.99 cm	CH	None
3.349193 GHz, 8.96 cm	CH	None
4.8–4.99 GHz, 6 cm	Formaldehyde (HCHO), 4.829649 GHz	SECONDARY
4.990–5.000 GHz, 6 cm		PRIMARY
5.009 GHz, ~6 cm	Ionized hydrogen (HII)	None
5.736 GHz, ~5.2 cm	Ionized hydrogen (HII)	None
8.872 GHz, ~3.4 cm	Ionized hydrogen (HII)	None

TABLE 3.9 (*Continued*)

Band Frequency and Typical Wavelength	Line Name[a] and Rest Frequency	Status[b] (ITU)
10.68–10.70 GHz, 2.8 cm		Shared with Passive Services
~2.6 cm	Carbon monoxide (CO)	None
14.47–14.5 GHz, 2.1 cm	Formaldehyde (HCHO), 14.489 GHz	SECONDARY
15.35–15.40 GHz, 1.9 cm		Shared with Passive Services
22.21–22.5 GHz, 1.3 cm	Water vapor (H_2O), 22.235 GHz	PRIMARY
23.6–24.0 GHz, 1.25 cm		Shared with Passive Services
31.2–31.3 GHz, 1 cm	Window in the earth's atmosphere	FOOTNOTE STATUS
31.3–31.8 GHz, 1 cm		Shared with Passive Services
36.023 GHz, ~8.3 mm	Ozone (O_3)	None
36.43–36.5 GHz, 8.2 mm	Hydroxyl (OH)	FOOTNOTE STATUS
37.832 GHz, ~7.9 mm	Ozone (O_3)	None
43.5 GHz, ~6.9 mm	Silicon monoxide (SiO)	None
51.4–54.25 GHz, 5.8 mm		FOOTNOTE STATUS
58.2–59 GHz, 5.1 mm		FOOTNOTE STATUS
~60 GHz, 5 mm	Oxygen (O_2)	
64–65 GHz, 4.7 mm		FOOTNOTE STATUS
72.77–72.91 GHz, 4.1 mm		FOOTNOTE STATUS
86–92 GHz, 3.26–3.49 mm	Window in the earth's atmosphere	Shared with Passive Services
	Methyl alcohol (CH_3OH), 86 GHz, 3.49 mm	
	Hydrogen cyanide (HCN), 88.2 GHz, 3.40 mm	
	Silicon monoxide (SiO), 88.2 GHz, 3.40 mm	
~3 mm	Dinitrogen oxide (N_2O), 100.492 GHz, 3 mm	
105–116 GHz, 2.8 mm		Shared with Passive Services
	Cyanogen (CN), ~113 GHz, 2.6 mm	None
	Carbon monoxide (CO), 115.271 GHz, 2.6 mm	
118.364 GHz, ~2.5 mm	Ozone (O_3)	None
118,746 GHz, ~2.5 mm	Oxygen (O_2)	None

TABLE 3.9 *(Continued)*

Band Frequency and Typical Wavelength	Line Name[a] and Rest Frequency	Status[b] (ITU)
~2 mm	Window in the earth's atmosphere	
130 GHz	Silicon monoxide (SiO)	
150.176–150.644 GHz, ~2 mm	Nitric oxide (NO)	None
164–168 GHz, 1.8 mm	Hydrogen sulfide (H_2S), 167 GHz	Shared with Passive Services
182–185 GHz, 1.6 mm	Water vapor (H_2O), 183.3 GHz, 1.6 mm	Shared with Passive Services
217–231 GHz, 1.3 mm	Window in the earth's atmosphere	PRIMARY
265–275 GHz, 1.1 mm		PRIMARY

[a]Other lines of interest to radio astronomers (all are in the millimeter or decimeter regions): HC_3N (cyanoacetylene), HCOOH (formic acid), and NH_2CHO (formamide).
[b]Allocations: SMALL CAPITALS = primary; *italics* = permitted; Lower Case = Secondary.
Caution: These allocations are subject to change. Consult the current *Table of Frequency Allocations* (ITU, 1988).

TABLE 3.10 Broadcasting Satellite Service Channel Locations (Regions 1 and 3)

		Assignments	
Country			East Longitude[a]
Symbol	Name	RF Channels	(degrees)
AFI	Afars and Issacs (Djibouti)	21, 25, 29, 33, 37	23
AFG	Afghanistan	1, 3, 5, 7, 9, 11, 13, 15	50
ALB	Albania	22, 26, 30, 34, 38	−7
ALG	Algeria	2, 4, 6, 8, 10, 12, 14, 16, 18, 20	−25
AND	Andorra	4, 8, 12, 16, 20	−37
AGL	Angola	23, 27, 31, 35, 39	−13
AUS	Australia	1, 2, 3, 5, 6, 7, 9, 10, 11, 13, 14, 15, 17, 18, 19, 21, 22, 23	98
		2, 3, 4, 6, 7, 8, 10, 11, 12, 14, 15, 16, 18, 19, 20, 22, 23, 24	128
AUT	Austria	4, 8, 12, 16, 20	−19
AZR	Azores	3, 7, 11, 15, 19	−31
BHR	Bahrain	27, 31, 35, 39	17
BGD	Bangladesh	15, 18, 20, 22, 24	74
BEL	Belgium	21, 25, 29, 33, 37	−19

TABLE 3.10 (*Continued*)

		Assignments	
Country			East Longitude[a] (degrees)
Symbol	Name	RF Channels	
BLR	Byelorussia	21, 25	23
BEN	Benin	3, 7, 11, 15, 19	−19
BOT	Botswana	2, 6, 10, 14, 18	−1
BRU	Brunei	12, 14	74
BUL	Bulgaria	4, 8, 12, 16, 20	−1
BRM	Burma	17, 19, 21, 23	74
BDI	Burundi	22, 26, 30, 34, 38	11
CME	Cameroon	1, 5, 9, 13, 17	−13
CNR	Canary Islands	23, 27, 31, 35, 39	−31
CPV	Cape Verde Islands	4, 8, 12, 16, 20	−31
CAR	Caroline Islands	1, 5, 9, 13, 17	122
CAF	Central African Empire	24, 28, 32, 36, 40	−13
TCD	Chad	2, 6, 10, 14, 18	−13
TCH	Czechoslovakia	3, 7, 11, 15, 19	−1
CHN	China	1, 2, 3, 4, 5, 6, 7, 8, 9, 10, 11, 12, 13, 14, 15, 16, 18, 20, 22, 24	62
		1, 5, 9, 10, 12, 14, 15, 17, 18, 19, 20, 21, 22, 23, 24	80
		1, 2, 3, 4, 5, 6, 7, 8, 9, 10, 11, 12, 13, 14, 15, 16, 17, 19, 21, 22, 24	92
COM	Comoros	3, 7, 11, 15	29
COG	Congo Republic	22, 26, 30, 34, 38	−13
CKH	Cook Islands, New Zealand	2, 6, 10, 14	158
CKN	Cook Islands, Northern Group	4, 8, 12, 16	158
CYP	Cyprus	21, 25, 29, 33, 37	5
DNK	Denmark[b]	12, 16, 20, 24, 27, 35, 36	5
DJI	Djibouti	21, 25, 29, 33, 37	23
EGY	Egypt	4, 8, 12, 16, 20	−7
ETH	Ethiopia	22, 26, 30, 34, 38	23
GNE	Equatorial Guinea	23, 27, 31, 35, 39	−19
FJI	Fiji Islands	1, 5, 9	152
FNL	Finland[b]	2, 6, 10, 22, 26	5
F	France	1, 5, 9, 13, 17	−19
OCE	French Polynesia	*See* Tahiti	
GAB	Gabon	3, 7, 11, 15, 19	−13
GMB	Gambia	3, 7, 11, 15, 19	−37
GHA	Ghana	23, 27, 31, 35, 39	−25
D	West Germany	2, 6, 10, 14, 18	−19
DDR	German Democratic Republic	21, 25, 29, 33, 37	−1

TABLE 3.10 (*Continued*)

Country		Assignments	
Symbol	Name	RF Channels	East Longitude[a] (degrees)
G	Great Britain	4, 8, 12, 16, 20	−31
GRC	Greece	3, 7, 11, 15, 19	5
GUM	Guam	2, 6, 10, 14, 18	122
GUI	Guinea	1, 5, 9, 13, 17	−37
GNB	Guinea-Bissau	2, 6, 10, 14, 18	−31
HNG	Hungary	22, 26, 30, 34, 38	−1
ISL	Iceland[b]	21, 25, 29, 33, 37	−31
	Iceland and Faroe Islands	23, 31, 39	5
IFB	IFRB; This assignment has been included in plan by the conference	22, 26, 30, 34, 38 21, 25, 29, 33, 37	−1 5
IND	India	1, 2, 3, 4, 5, 6, 7, 8, 9, 10, 11, 12, 13, 14, 15, 16, 17, 18, 19, 20, 21, 22, 23, 24	56
		1, 2, 3, 4, 5, 6, 7, 8, 9, 10, 11, 12, 13, 14, 15, 16, 17, 18, 19, 20, 21, 22, 23, 24	68
INS	Indonesia	2, 4, 6, 8, 17, 18, 19, 20, 21, 22, 23, 24	80
		1, 3, 5, 7, 9, 11, 13, 15, 17, 19	104
IRN	Iran	3, 7, 11, 15, 19	34
IRQ	Iraq	24, 28, 32, 36, 40	11
IRL	Ireland	2, 6, 10, 14, 18	−31
ISR	Israel	25, 29, 33, 37	−13
I	Italy	24, 28, 32, 36, 40	−19
CTI	Ivory Coast	22, 26, 30, 34, 38	−31
J	Japan	1, 3, 5, 7, 9, 11, 13, 15	110
JOR	Jordan	23, 27, 31, 35, 39	11
KEN	Kenya	21, 25, 29, 33, 37	11
CBG	Khmer Republic (Cambodia)	18, 20, 22, 24	68
KOR	Republic of Korea	2, 4, 6, 8, 10, 12	110
KRE	Democratic People's Republic of Korea	14, 16, 18, 20, 22	110
KWT	Kuwait	22, 26, 30, 34, 38	17
LAO	Laos	2, 4, 6, 8, 10	74
LBN	Lebanon	3, 7, 11, 15, 19	11
LBR	Liberia	3, 7, 11, 15	−31
LBY	Libya	1, 3, 5, 7, 9, 11, 13, 15, 17, 19	−25

TABLE 3.10 (*Continued*)

Country		Assignments	
Symbol	Name	RF Channels	East Longitude[a] (degrees)
LIE	Liechtenstein	3, 7, 11, 15, 19	−37
LSO	Lesotho	24, 28, 32, 36, 40	5
LUX	Luxembourg	3, 7, 11, 15, 19	−19
MDG	Madagascar	1, 5, 9, 13, 17	29
MLA	Malaysia	2, 4, 6, 8, 10, 16, 18, 20, 22, 24	86
MLD	Maldives	12, 16	44
MLI	Mali	2, 4, 6, 8, 10, 12, 14, 16, 18, 20	−37
MWI	Malawi	24, 28, 32, 36, 40	−1
MLT	Malta	4, 8, 12, 16	−13
MRA	Mariana Islands	3, 7, 11, 15, 19	122
MRL	Marshall Islands	2, 6, 10, 14, 18	146
MTN	Mauritania	22, 24, 26, 28, 30, 32, 34, 36, 38, 40	−37
MAU	Mauritius	2, 4, 6, 8, 10, 12, 14, 16, 18	29
MYT	Mayotte Island	24, 28, 32, 36, 40	29
MCO	Monaco	21, 25, 29, 33, 37	−37
MNG	Mongolia	25, 29, 33, 37, 39	74
MRC	Morocco	21, 25, 29, 33, 37	−25
MOZ	Mozambique	4, 8, 12, 16, 20	−1
NMB	Namibia (Southwest Africa)	25, 29, 33, 37	−19
NRU	Nauru	3, 7, 11, 15	134
NPL	Nepal	17, 19, 21	50
HOL	Netherlands	23, 27, 31, 35, 39	−19
NCL	New Caledonia	2, 6, 10, 14	140
NHB	New Hebrides Islands	3, 7, 11, 15	140
NZL	New Zealand	13, 17, 21	128
		1, 5, 9, 13	158
NGR	Niger	24, 28, 32, 36, 40	−25
NIG	Nigeria	22, 26, 30, 34, 38	−19
NIU	Niue Island	19, 23	158
NOR	Norway[b]	14, 18, 28, 32, 38	5
OMA	Oman	24, 28, 32, 36, 40	17
PAK	Pakistan	2, 4, 6, 8, 10, 12, 14, 18, 20, 22, 24	38
PLM	Palmyra Island	1, 5, 9, 13, 17	170
PNG	Papua New Guinea	2, 6, 10, 14	110
		4, 8, 12	128
PHL	Philippines	16, 18, 20, 22, 24	98
POL	Poland	1, 5, 9, 13, 17	−1

TABLE 3.10 (*Continued*)

Country		Assignments	
Symbol	Name	RF Channels	East Longitude[a] (degrees)
POR	Portugal	3, 7, 11, 15, 19	−31
QAT	Qatar	1, 5, 9, 13, 17	17
REU	Reunion	22, 26, 30, 34, 38	29
ROU	Romania	2, 6, 10, 14, 18	−1
RRW	Rwanda	4, 8, 12, 16, 20	11
STP	Sao Tome and Principe	4, 8, 12, 16, 20	−13
ARS	Saudia Arabia	2, 4, 6, 8, 10, 12, 14, 16, 18, 20, 23	17
SMA	Samoa Islands (American)	1, 5, 9, 13, 17	170
SMO	Samoa (Western)	3, 7, 11, 15	158
SMR	San Marino	1, 5, 9, 13, 17	−37
SEN	Senegal	21, 25, 29, 33, 37	−37
SRL	Sierra Leone	23, 27, 31, 35, 39	−31
SNG	Singapore	3, 7, 11, 15	74
SOM	Somalia	3, 7, 11, 15, 19	23
E	Spain	23, 27, 31, 35, 39	−31
CLN	Sri Lanka (Ceylon)	2, 6, 10, 14	50
SDN	Sudan	22, 23, 24, 26, 27, 28, 30, 31, 32, 34, 35, 36, 38, 39, 40	−7
S	Sweden[b]	4, 8, 30, 34, 40	5
SUI	Switzerland	22, 26, 30, 34, 38	−19
SWZ	Swaziland	1, 5, 9, 13, 17	−1
SYR	Syria	22, 26, 30, 34, 38	11
OCE	Tahiti, French Polynesia	4, 8, 12, 16	−160
TGK	Tanzania	23, 27, 31, 35, 39	11
THA	Thailand	1, 5, 9, 13	74
TGO	Togolese Republic	2, 6, 10, 14, 18	−25
TKL	Tokelau Islands	20, 24	158
TON	Tonga Islands	4, 8, 12, 16	170
TUN	Tunisia	22, 26, 30, 34, 38	−25
TUR	Turkey	1, 5, 9, 13, 17	5
UGA	Uganda	3, 7, 11, 15, 19	11
UKR	Ukraine	29, 33, 37	23
UAE	United Arab Emirates	21, 25, 29, 33, 37	17
HVO	Upper Volta	21, 25, 29, 33, 37	−31
URS	USSR	1, 3, 4, 5, 7, 8, 9, 11, 12, 13, 15, 16, 17, 19, 20, 23,[c] 23,[c] 27, 31, 35, 39	23
		1, 3, 5, 7, 9, 12, 13, 16, 18, 20, 22, 24, 26, 28, 32, 34, 36, 38, 40	44

TABLE 3.10 (*Continued*)

Country		Assignments	
Symbol	Name	RF Channels	East Longitude[a] (degrees)
		26, 28, 32, 34, 38	74
		19, 23, 25, 27, 31, 35, 39	110
		20, 22, 24, 26, 28, 32, 34, 36, 38, 40	140
CVA	Vatican State	23, 27, 31, 35, 39	−37
VTN	Vietnam	3, 7, 11, 15	86
WAK	Wake Island	1, 5, 9, 13, 17	140
WAL	Wallis Islands	2, 6, 10, 14	140
YEM	Yemen Arab Republic	2, 6, 10, 14, 18	11
YMS	People's Democratic Republic of South Yemen	1, 5, 9, 13, 17	11
YUG	Yugoslavia	21, 23, 25, 27, 29, 31, 33, 35, 37, 39	−7
ZAI	Zaire	2, 4, 6, 8, 10, 12, 14, 16, 18, 20	−19
ZMB	Zambia	3, 7, 11, 15, 19	−1

[a]Location in geostationary orbit; negative number denotes west longitude; positive number denotes east longitude.

[b]Members of the Nordic Satellite Group: Denmark, Iceland, Finland, Norway and Sweden.

[c]These beams cover different areas of the Soviet Union from a single orbit location.

Figure 3.15 on page 195 shows these relationships for the various harmonics. For instance, the second harmonic of a 4-GHz transmitter falls atop the passband of an 8-GHz receiver; also, a 7-GHz transmitter's fourth harmonic can produce interference in the 30-GHz band.

In most situations odd harmonics are generated more easily than the even harmonics. Transmitters use band-limiting filters, or a waveguide that is beyond cutoff, to attenuate the harmonics.

3.1.1.2 Optical Links

In 1986 there were no official allocations for optical links between satellites, to or from a satellite from or to earth, or to or from a satellite from or to below the seas.

Some of the natural limitations to these links are as follows:

☐ Light scattering (due to clouds, haze, dust, rain, snow, etc.).

☐ Blockage (water and waves, trees, buildings, etc.).

☐ Interference from other optical sources (primarily the sun).

☐ Light-bending mechanisms.

TABLE 3.11 Broadcasting Satellite Service Orbital Position, Channel, and Polarization Assignments for Each Area in Region 2

Example: 138/32A = 138° W, all 32 transponders, polarization A
53/8 (3,7) A = 53° W, 8 transponders using channel families 3 and 7,[b] polarization A[c]
129/32A, 91/32A = 2 locations (129 and 91° W), each with 32 transponders using polarization A

Service Area	Nominal Orbital Position,[a] Number of Channels (Channel Family[b]), and Polarization Plan[c]	Quantity of Orbital Positions
	North America	
Canada		
British Columbia	138/32A	1
Alberta/Saskatchewan	138/32A, 129/32A, 72.5/32A	3
Manitoba	129/32A, 91/32A	2
Ontario	129/32A, 91/32A, 82/32A	3
Quebec	91/32A, 82/32A	2
Eastern Provinces	82/32A, 70.5/32A	2
Greenland (Denmark)	53/8(3,7)A	1
Mexico		
North	136/32A, 78/32A	2
South	127/32A, 69/16(1,3,5,7)A	2
St. Pierre and	53/8(1,5)A	1
Miquelon (France)		
United States		
Alaska	175/32A, 166/32A	2
Hawaii	175/32A, 166/32A	2
Pacific service area	175/32A, 166/32A	2
Western half	157/32A, 148/32A	2
Eastern half	119/32A, 110/32A, 101/32A, 61.5/32A	4

172

Central America

Belize	116/4(6)A, 92.5/16(2,4,6,8)B	2
Costa Rica	131/8(2,6)A	1
El Salvador	107.5/4(4)B	1
Guatemala	107.5/4(8)B	1
Honduras	107.5/4(2)B	1
Nicaragua	107.5/4(6)B	1
Panama	121/4(8)B	1

Western Caribbean

Bahamas	92.5/16(2,4,6,8)B, 87/4(1)A	2
Bermuda (United Kingdom)	96/16(1,3,5,7)B, 92.5/16(2,4,6,8)B, 31/5(1 + channel 17)	3
British Virgin Islands (United Kingdom)	79.5/4(1)A	1
Cayman Islands (United Kingdom)	116/4(4)A	1
Cuba	89/8(3,7)A	1
Dominican Republic	83.5/4(4)A	1
Haiti	83.5/4(2)A	1
Jamaica	92.5/20(2,4,5,6,8)B, 34/4(4)A	2
Montserrat (United Kingdom)	79.5/4(7)A	1
Puerto Rico/Virgin Islands (United States)	110/32A, 101/32A	2
St. Christopher, Nevis, (United Kingdom)	79.5/4(5)A	1
Turks and Caicos Islands (United Kingdom)	116/4(2)A	1

TABLE 3.11 (Continued)

Example: 138/32A = 138° W, all 32 transponders, polarization A
53/8 (3,7) A = 53° W, 8 transponders using channel families 3 and 7,[b] polarization A[c]
129/32A, 91/32A = 2 locations (129 and 91° W), each with 32 transponders using polarization A

Service Area	Nominal Orbital Position,[a] Number of Channels (Channel Family[b]), and Polarization Plan[c]	Quantity of Orbital Positions
Eastern Caribbean		
Antigua and Barbuda	79.5/4(3)A	1
Barbados	92.5/8(3,7)B	1
Dominica	79.5/4(6)A	1
Eastern Caribbean	92.5/16(2,4,6,8)B	1
Grenada	79.5/4(8)A, 57/8(3,7)A, 42/8(1,3)A	3
Netherlands Antilles	53/8(2,6)A	1
St. Lucia	79.5/4(4)A	1
St. Vincent and The Grenadines	79.5/4(2)A	1
Trinidad and Tobago	84.5/4(1)A	1
South America		
Argentina		
North	94/16(2,4,6,8)A, 55/16(2,4,6,8)A	2
South	94/16(1,3,5,7)A, 55/12(3,5,7)A	2
Islands	94/16(1,3,5,7)A, 55/12(3,5,7)A	2
Bolivia	115/16(1,3,5,7)A, 87/8(3,7)A	2
Brazil		
South	81/32A, 45/32A	2
East central	81/32A, 45/32A	2

174

Northeast	64/32A, 45/32A	2
South central	64/32A, 45/32A	2
North central	64/32A	1
Southwest	74/32B	1
Northwest	74/32B	1
West	74/32B	1
Southeast	102/16(2,4,6,8)A	1
Chile		
North	106/16(2,4,6,8)A	1
Central	106/16(1,3,5,7)A	1
South	106/16(2,4,6,8)A	1
Pacific Islands	106/16(1,3,5,7)A	1
Colombia	115/16(1,3,5,7)A, 103/16(1,3,5,7)A	2
Easter Island (Chile)	106/16(1,3,5,7)A	1
Ecuador[a]		
Continental	115/16(1,3,5,7)A, 95/16(2,4,6,8)B	2
Galapagos Islands	115/16(1,3,5,7)A, 95/16(2,4,6,8)B	2
Falkland Islands (United Kingdom)	31/5(1 + Channel 17)A	1
Falklands, Antarctica (United Kingdom)	57/4(1)A	1
French Guiana, Guadeloupe, Martinique (France)	53/8(4,8)A	1
Guyana	84.5/4(7)A, 34/4(2)A	2
Paraguay	99/16(1,3,5,7)A	1
Peru	115/16(1,3,5,7)A, 86/16(2,4,6,8)A	2
Suriname	84.5/8(3,5)A	1
Uruguay	71.5/16(1,3,5,7)A	1

TABLE 3.11 (Continued)

Example: 138/32A = 138° W, all 32 transponders, polarization A

53/8 (3,7) A = 53° W, 8 transponders using channel families 3 and 7,[b] polarization A[c]

129/32A, 91/32A = 2 locations (129 and 91° W), each with 32 transponders using polarization A

Service Area	Nominal Orbital Position,[a] Number of Channels (Channel Family[b]), and Polarization Plan[c]	Quantity of Orbital Positions
	South America	
Venezuela		
Continental	115/16(1,3,5,7)A, 104/16(2,4,6,8)A	2
Islands	104/4(8)A	1

Total service areas: 74   Total coverage areas: 112   Total channel assignments: 2050

[a]Nominal orbital location in degrees west longitude. Actual orbital position assignment is usually 0.2° W of nominal odd-numbered channels and 0.2° E for nominal even-numbered channels.

[b]When number of assigned channels is less than 32:

Channel Family	Channels
1	1, 5, 9, 13
2	2, 6, 10, 14
3	3, 7, 11, 15
4	4, 8, 12, 16
5	17, 21, 25, 29
6	18, 22, 26, 30
7	19, 23, 27, 31
8	20, 24, 28, 32

[c]A = odd channels (clockwise) or even channels (counterclockwise).

B = odd channels (counterclockwise) or even channels (clockwise).

[d]The geographical areas are so far apart that individual (nonoverlapping) beams and programs could be used

From Reinhart (1985).

TABLE 3.12 Orbit Locations for 12-GHz Broadcasting Satellites

Nominal Orbit Location	ITU Country Symbol and Number of Television Channels per Country	Total Television Channels per Location
175° W	USA (96)	96
166° W	USA (96)	96
160° W	OCE (4)	4
157° W	USA (32)	32
148° W	USA (32)	32
138° W	CAN (64)	64
136° W	MEX (32)	32
131° W	CTR (8)	8
129° W	CAN (96)	96
127° W	MEX (32)	32
121° W	PNR (4)	4
119° W	USA (32)	32
116° W	Turks and Caicos (4), Cayman Islands (4), Belize (4)	12
115° W	BLO (16), CLM (16), EQA (32), PRU (16), VEN (16)	96
110° W	USA (64)	64
107.5° W	HND (4), SLV (4), NCG (4), GTM (4)	16
106° W	CLH (80)	80
104° W	VEN (20)	20
103° W	CLM (16)	16
102° W	B (16)	16
101° W	USA (64)	64
99° W	PRG (16)	16
96° W	Bermuda (16)	16
95° W	EQA (32)	32
94° W	ARG (48)	48
92.5° W	Bermuda (16), Belize (16), JMC (20), BAH (16), Eastern Caribbean (16), Barbados (8)	92
91° W	CAN (96)	96
89° W	CUB (8)	8
87° W	BAH (4), BLO (8)	12
86° W	PRU (16)	16
84.5° W	TRD (4), SUR (8), GUB (4)	16
83.5° W	HTI (4), DOM (4)	8
82° W	CAN (96)	96
81° W	B (64)	64
79.5° W	British Virgin Islands (4), St. Vincent and The Grenadines (4), Antigua and Barbuda (4), St. Lucia (4), St. Christopher Nevis (4), Dominica (4), UK, Montserrat (4), Grenada (4)	32

TABLE 3.12 (*Continued*)

Nominal Orbit Location	ITU Country Symbol and Number of Television Channels per Country	Total Television Channels per Location
78° W	MEX (32)	32
74° W	B (96)	96
72.5° W	CAN (32)	32
71.5° W	Uruguay (16)	16
70.5° W	CAN (32)	32
69° W	MEX (16)	16
64° W	B (96)	96
61.5° W	USA (32)	32
57° W	UK, Falklands/Antartica (4), Grenada (8)	12
55° W	ARG (40)	40
53° W	S. Pierre and Miquelon (8), Netherlands Antilles (8), GRL (8), French Guiana/Guadeloupe/Martinique (8)	32
45° W	B (128)	128
42° W	GRD (8)	8
37° W	AND (5), CVA (5), GMB (5), GUI (5), LIE (5), MCO (5), MLI (10), MTN (10), SEN (5), SMR (5)	60
34° W	GUB (4), JMC (4)	8
31° W	AZR (5), CNR (5), CPV (5), CTI (5), E (5), G(5), GNB (5), HVO (5), IRL (5), ISL (5), LBR (4), POR (5), SRL (5), UK, Falklands (5), Bermuda (5)	74
25° W	ALG (10), GHA (5), LBY (10), MRC (5), NGR (5), TGO (5), TUN (5)	45
19° W	AUT (5), BEL (5), BEN (5), D (5), F (5), GNE (5), HOL (5), I (5), LUX (5), NIG (5), NMB (4), SUI (5), ZAI (10)	69
13° W	AGL (10), CAF (5), CME (5), COG (5), GAB (5), MLT (4), STP (5), TCD (5)	44
7° W	ALB (10), EGY (5), SDN (16), YUG (10)	41
1° W	BOT (5), BUL (5), DDR (5), HNG (5), IFB (5), MOZ (5), MWI (5), POL (5), ROU (5), SWZ (5), TCH (5), ZMB (5)	60
5° E	CYP (5), DNK (7), FNL (5), GRC (5), IFB (5), ISL (3), LSO (5), NOR (5), S (5), TUR (5)	50
11° E	BDI (5), IRQ (5), JOR (5), KEN (5), LBN (5), RRW (5), SYR (5), TGK (5), UGA (5), YEM (5), YMS (5)	55

TABLE 3.12 (*Continued*)

Nominal Orbit Location	ITU Country Symbol and Number of Television Channels per Country	Total Television Channels per Location
17° E	ARS (11), BHR (4), KWT (5), OMA (5), QAT (5), UAE (5)	35
23° E	AFI (5), BLR (2), ETH (5), SOM (5), UKR (3), URS (21)	41
29° E	COM (4), MAU (9), MDG (5), REU (5), MYT (5)	28
34° E	IRN (5)	5
38° E	PAK (11)	11
44° E	MLD (2), URS (19)	21
50° E	AFG (8), CLN (4), NPL (3)	15
56° E	IND (24)	24
62° E	CHN (20)	20
68° E	CBG (4), IND (24)	28
74° E	BGD (5), BRM (4), BRU (2), LAO (5), MNG (5), SNG (4), THA (4), URS (5)	34
80° E	CHN (15), INS (12)	27
86° E	MLA (10), VTN (4)	14
92° E	CHN (21)	21
98° E	AUS (18), PHL (5)	23
104° E	INS (10)	10
110° E	J (8), KOR (6), KRE (5), PNG (4), URS (7)	30
116° E	(none assigned)	0
122° E	CAR (5), GUM (5), MRA (5)	15
128° E	AUS (18), NZL (3), PNG (3)	24
134° E	NRU (4)	4
140° E	NCL (4), NHB (4), URS (11), WAL (4), WAK (5)	28
146° E	MRL (5)	5
152° E	FJI (3)	3
158° E	CKH (4), CKN (4), NIU (2), NZL (4), SMO (4), TKL (2)	20
164° E	(none assigned)	0
170° E	PLM (5), SMA (5), TON (4)	14

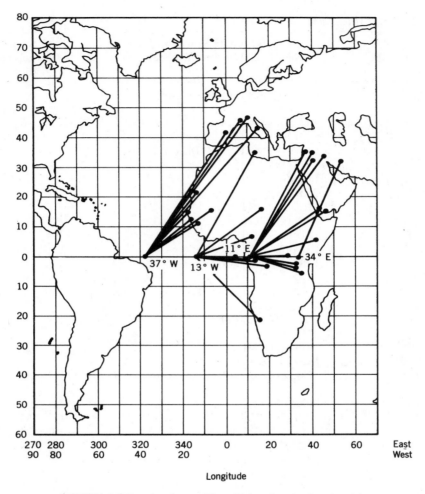

FIGURE 3.5 Broadcasting satellite orbit locations (regions 1 and 3).

3.1.2 Band and Emission Designations

Figure 3.16 on page 210 is a nomogram for converting frequency to wavelength. The relationship between frequency f and wavelength λ is

$$c = f\lambda \qquad (3.3)$$

where the units are meters per second for c (c is the velocity of light and equals 2.99793×10^8 m/sec), hertz for f, and meters for λ. For simplicity, 2.99793 is often rounded to 3.

The spiral chart in Figure 3.17 provides convenient conversions over the range 100–300 GHz (or 3 m to 0.1 cm).

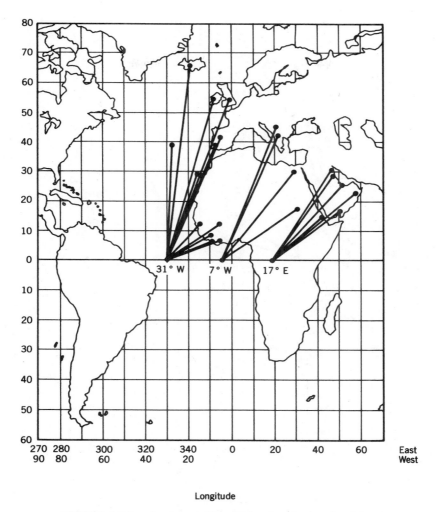

Longitude

FIGURE 3.6 Broadcasting satellite orbit locations (regions 1 and 3).

Frequency bands are sometimes designated by letters (e.g., C band and K_u band). These stem from World War II attempts to keep secret exact radar frequencies. The intent was to confuse the enemy. Subsequent years have seen these designators declassified, redefined several times, and abused. The meanings of the letter designations vary with the user and listener, with confusion resulting if their band uses are different. Table 3.14 gives some of the designations; the last column provides the popular usage. Since there are variations, the use of a band's frequency limits (e.g., 11.7–12.2 GHz) is suggested in place of its letter designation (e.g., K band), which can lead to confusion.

Emissions and modulations are identified in accord with ITU and FCC desig-

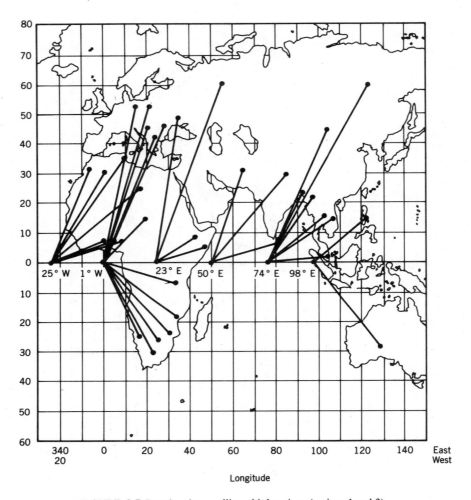

FIGURE 3.7 Broadcasting satellite orbit locations (regions 1 and 3).

nations (Page 214). The first four characters designate the bandwidth. Additional symbols, up to five, specify the type of modulation, and the nature of the information being transmitted. Thus, the designation 15M0F3F can be used for television, and decyphered through the use of Table 3.15 as follows:

15M0	F	3	F
Bandwidth	Modulation	Signal	Content
15.0 MHz	Frequency (FM)	Analog, single	Video

Additional symbols may specify details of signals and use of multiplex. An alter-

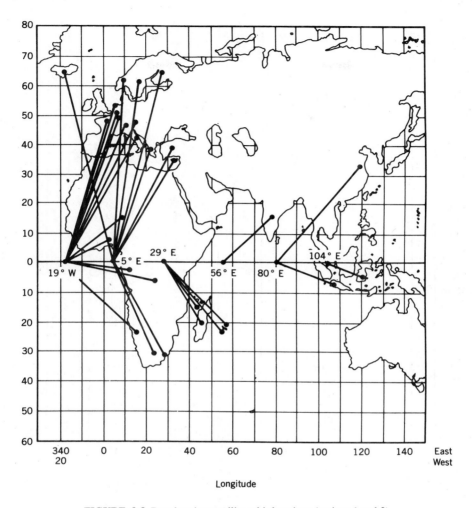

FIGURE 3.8 Broadcasting satellite orbit locations (regions 1 and 3).

nate (older) designation omitted the letter in the bandwidth, and kilohertz was implied.

The designation A0 refers to continuous wave (CW) (such as might be used for tracking or research), in which there is no modulation.

Table 3.16 shows the ITU station class codes. These codes are used for preparing or understanding filings to the IFRB (see page 217).

3.1.3 Illumination Level and Power Flux Density

As shown in the Tables 3.3–3.13, many of the satellite frequency allocations are based on the principle of sharing between terrestrial and space applications. To

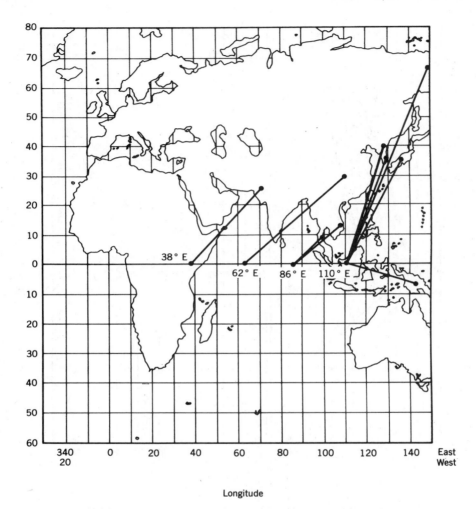

FIGURE 3.9 Broadcasting satellite orbit locations (regions 1 and 3).

protect one sharing service from another, upper limits have been established by the ITU on the amount of rf power (from a transmitting station) that can be present at and interfere with a receiving station. Figure 3.18 shows examples that produce interference. Figure 3.19 shows the angles where terrestrial service can interfere with (or receive interference from) satellites.

This power is expressed as a power flux density (PFD) at the receiving station in terms of the ratio (in decibels) of rf power with respect to 1 W/m^2 (in the worst case) in some specified rf bandwidth. The PFD is often expressed in units of $-N$ dBW/m^2 per 4 kHz; this is read as a power level of N decibels below one watt per square meters across, for example, a 4-KHz-wide frequency band.

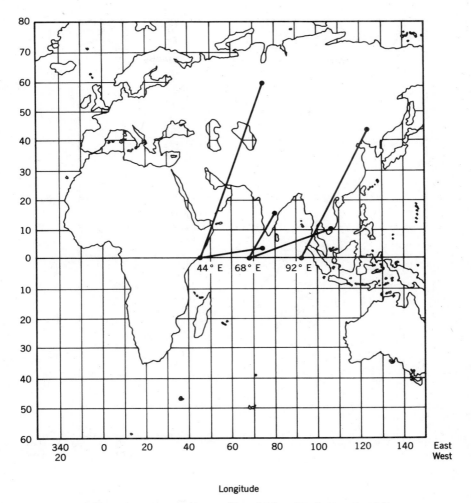

FIGURE 3.10 Broadcasting satellite orbit locations (regions 1 and 3).

The decibel (dB) notation is used extensively in this book:

$$\text{Decibels} = 10 \log_{10} (P_1/P_2) \quad (\text{dBW}) \qquad (3.4)$$

where $\log_{10}$ is the logarithm to the base 10, P_1 is the power being compared to the reference P_2, and $P_2 = 1$ W (as indicated by the W in dBW.)

The illumination level W is expressed in terms of dB W/m^2 at the receiving station and is the total power in the *full* transmission bandwidth (note the absence

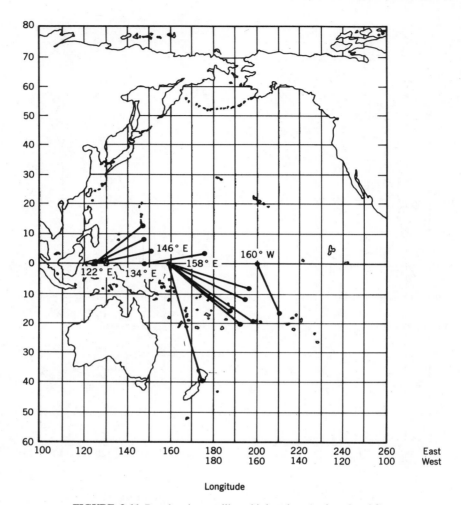

FIGURE 3.11 Broadcasting satellite orbit locations (regions 1 and 3).

of the per-specific-unit-of-bandwidth term used in power flux density measurements). See page 221.

Figure 3.20 illustrates this difference. The illumination level is the total power to (or from) the satellite. It includes spectral sidelobes such as might come from a nonfiltered time division multiple access (TDMA) signal. The power flux density is the integrated power in the worst (cross-hatched) segment with a bandwidth of B hertz.

The illumination level W is the transmitted power expressed in terms of the equivalent isotropically radiated power (EIRP) in dBW minus path losses L (dB) plus the gain G of a theoretical antenna with an area of 1 m^2 (which is equivalent

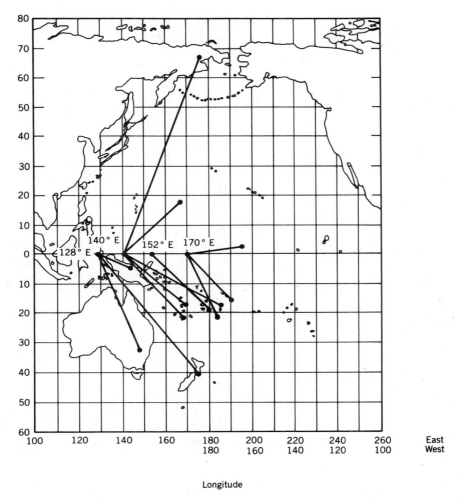

FIGURE 3.12 Broadcasting satellite orbit locations (regions 1 and 3).

to an ideal circular antenna with a diameter of 1.13 m):

$$W = \text{EIRP} - L + G \quad (\text{dBW}/\text{m}^2) \tag{3.5}$$

Substituting in the decibel equations for L (losses) from equation 3.56 and G (antenna gain) from equation 3.34 yields

$$W = \text{EIRP} - (92.45 + 20 \log_{10} S + 20 \log_{10} f) + 20.4$$

$$+ 20 \log_{10} f + 20 \log_{10}(1.13) \quad (\text{dBW}/\text{m}^2) \tag{3.6}$$

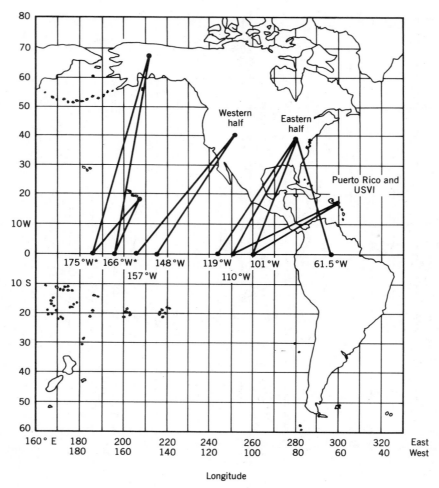

*Also U.S. Pacific Service Area

(a)

FIGURE 3.13 (a) BSS orbit locations (Region 2);

where f = frequency in gigahertz
             S = link distance in kilometers
92.45, 20.4 = constant and unit conversions
      1.13 = diameter of idealized, circular, antenna of area 1.0 m².

The frequency terms ($20 \log_{10} f$) cancel out, showing that W is frequency independent. Simplifying Equation (3.6), we obtain

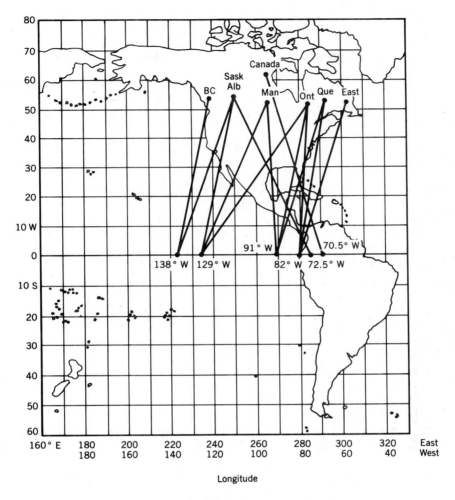

FIGURE 3.13 (*Continued*) (*b*) BSS orbit locations (region 2):

$$W = \text{EIRP} - 20 \log_{10} S - 92.45 + 20.4$$
$$+ 20 \log_{10}(1.13) \quad (\text{dBW/m}^2) \tag{3.7}$$

$$W = \text{EIRP} - 20 \log_{10} S - 71 \quad (\text{dBW/m}^2) \tag{3.8}$$

For geostationary satellites Equations (3.9) and (3.10) and Figure 3.21 represent the limit conditions on the illumination level. The equation for the edge-of-

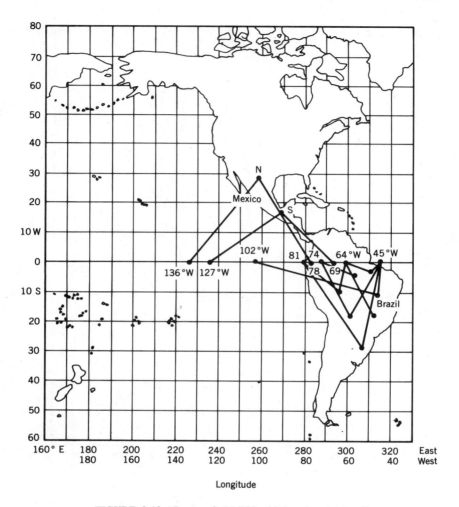

FIGURE 3.13 (*Continued*) (*c*) BSS orbit locations (region 2);

earth coverage illumination level is

$$W = \text{EIRP} - 163.3 \quad (\text{dBW/m}^2) \tag{3.9}$$

The EIRP is the power (dBW) in the direction of the earth station (or satellite). The 163.3 includes both the slant range attenuation factor and the gain of a 1-m^2 ideal antenna.

Since most communications satellites are geostationary we will use the 163.3 term (and derivatives) throughout the rest of this chapter. Tables 3.59 and 3.64 provide means to use equations in nongeostationary situations.

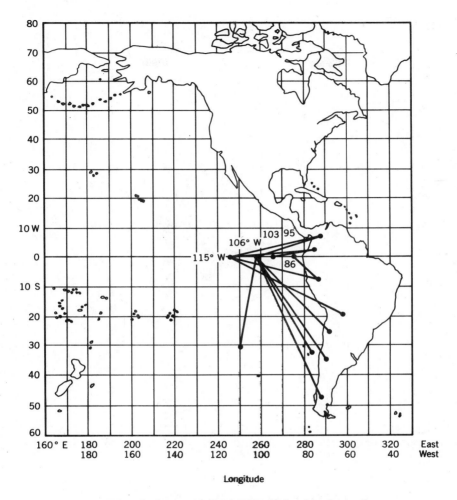

FIGURE 3.13 (*Continued*) (*d*) BSS orbit locations (region 2);

The subsatellite point is located directly below the satellite, on the equator. The range is shorter by 1.3 dB [i.e., 20 $\log_{10}$ (41,680 km/35,786 km) due to the distance-squared term in Equation (3.7)]. Thus, the equation for a 90° elevation angle is

$$W = \text{EIRP} - 162 \quad (\text{dBW/m}^2) \tag{3.10}$$

A more general equation is

$$W = \text{EIRP} - k_\text{w} \quad (\text{dBW/m}^2) \tag{3.11}$$

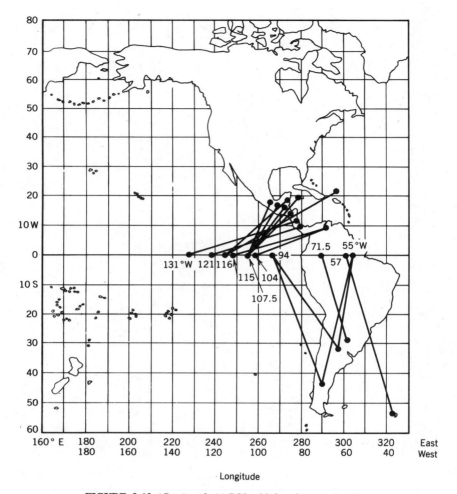

FIGURE 3.13 (*Continued*) (*e*) BSS orbit locations (region 2);

where k_w is defined in Figure 3.22 for both intersatellite (i.e., between-satellite) links and deep-space missions (see page 222).

The elevation angle of the satellite (from the earth station) is measured between the line from the earth station to the satellite and the local horizontal. At the point directly beneath the satellite the elevation is 90° (the earth station antenna points straight up). At the extreme limit of coverage the elevation angle is 0° (the earth station antenna beam is horizontal).

Figure 3.23 shows the basic geometry of a geostationary satellite and the earth.

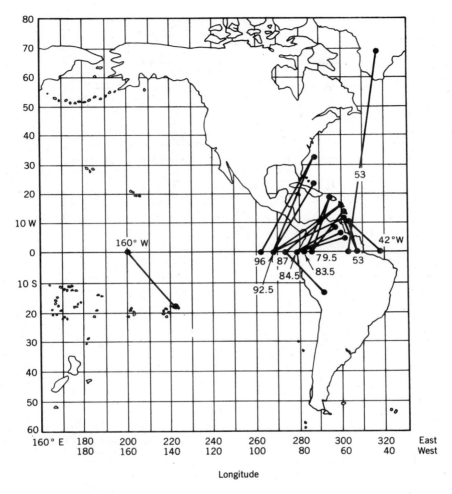

FIGURE 3.13 (*Continued*) (*f*) BSS orbit locations (region 2).

Figures 3.23 to 3.25 show the basic equations that relate the vital angles and distances (see pages 220 to 230).

To find the elevation angle from an earth station (located at latitude LA_{es} and longitude LO_{es}) to a satellite located over the equator at LO_{sat} use the form in Table 3.17.

Equation 2 in Figure 3.25 will always yield the correct azimuth if it can be programmed as ATAN2[-TAN(LOES-LOSAT), -SIN(LAES)] using east longitudes. If calculations are done by hand (or as a check), the correct azimuth can be found with the tables on pages 210 and 211.

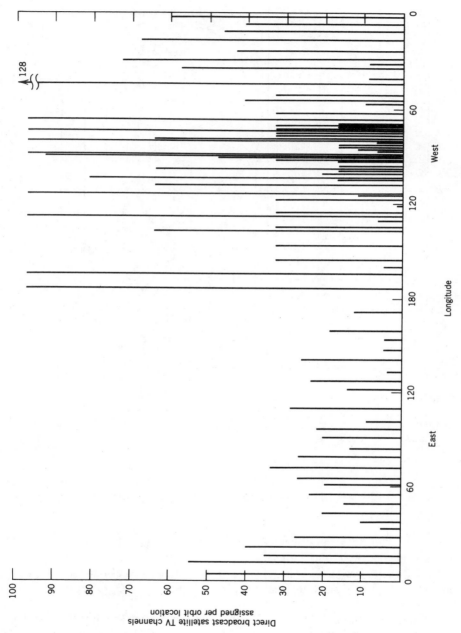

FIGURE 3.14 Broadcasting satellite assignments versus satellite longitude.

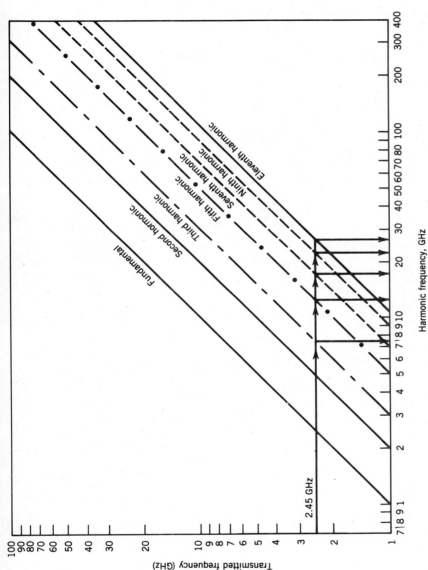

Example of use:

1. Assume 2.45-GHz transmitter.
2. Its third harmonic (3×2.45) falls at 7.35 GHz, which is in space-to-earth transmit band.
3. Fifth harmonic (5×2.45) falls at 12.25 GHz, which is in another space-to-earth band.
4. etc.

FIGURE 3.15 Harmonics that may cause interference.

TABLE 3.13 Country Channel Assignments, Channels 1–23[a]

Assigned Satellite Longitude	Channel Numbers[b]										
	1	2	3	4	5	6	7	8	9	10	11
175° W	←										
175° W	←										
175° W	←										
166° W	←										
166° W	←										
160° W	←			OCE				OCE			
157° W	←										
148° W	←										
138° W	←										
138° W	←										
136° W											
131° W		CTR				CTR				CTR	
129° W	←										
129° W	←										
129° W	←										
127° W	←										
121° W	←										
119° W	←										
										GCC	
116° W		T		C		T		C		T	
115° W	BLO		BLO		BLO		BLO		BLO		BLO
115° W	CLM		CLM		CLM		CLM		CLM		CLM
115° W	EQA		EQA		EQA		EQA		EQA		EQA
115° W	PRU		PRU		PRU		PRU		PRU		PRU
115° W	VEN		VEN		VEN		VEN		VEN		VEN
115° W	EQA		EQA		EQA		EQA		EQA		EQA
110° W	←										
110° W											
107.5° W		HND		SLV		HND		SLV		HND	
106° W	C	N	C	N	C	N	C	N	C	N	C
	P	S	P	S	P	S	P	S	P	S	P
	E		E		E		E		E		E
104° W		V		V		V		V		V	
103° W	CLM		CLM		CLM		CLM		CLM		CLM
102° W		B		B		B		B		B	
101° W	←										
101° W	←										
99° W	PRG		PRG		PRG		PRG		PRG		PRG
96° W		GCC		GCC		GCC		GCC		GCC	
95° W		C		C		C		C		C	
		G		G		G		G		G	

[a]Many locations have additional channels (see pages 202–209 for data and additional information.)

12	13	14	15	16	17	18	19	20	21	22	23	a
				USA								
				USA								
				USA								
				USA								
				USA								
				OCE								
				USA								
				USA								
				CAN								
				CAN								
				MEX								
		CTR				CTR				CTR		
				CAN								
				CAN								
				CAN								
				MEX								
								PNR				
				USA								
C		T		C		BLZ				BLZ		
	BLO		BLO		BLO		BLO		BLO		BLO	
	CLM		CLM		CLM		CLM		CLM		CLM	
	EQA		EQA		EQA		EQA		EQA		EQA	
	PRU		PRU		PRU		PRU		PRU		PRU	
	VEN		VEN		VEN		VEN		VEN		VEN	
	EQA		EQA		EQA		EQA		EQA		EQA	
				USA (PTR & VIR)								
				USA								
SLV		HND		SLV		NCG		GTM		NCG		
				(CLH)								
N	C	N	C	N	C	N	C	N	C	N	C	
				(CLH)								
S	P	S	P	S	P	S	P	S	P	S	P	
	E		E		E		E		E		E	
				(VEN)								
V		V		V		V		V		V		
									I			
	CLM		CLM		CLM		CLM		CLM		CLM	
B		B		B		B		B		B		
				USA								
				USA								
	PRG		PRG		PRG		PRG		PRG		PRG	
GCC		GCC		GCC		GCC		GCC		GCC		
				(EQA)								
C		C		C		C		C		C		
G		G		G		G		G		G		
				(ARG)								

[b]Channel numbering determined by the ITU region. The bandwidth and center frequency plan for region 2 is different from regions 1 and 3.

TABLE 3.13 (*Continued*)

Assigned Satellite Longitude	Channel Number[b]										
	1	2	3	4	5	6	7	8	9	10	11
94° W	S	N	S	N	S	N	S	N	S	N	S
	I		I		I		I		I		I
92.5° W		A	D	A		A	D	A		A	D
		B		B		B		B		B	
		J		J		J		J		J	
		H		H		H		H		H	
		E		E		E		E		E	
91° W	←————————————————										
91° W	←————————————————										
91° W	←————————————————										
89° W			CUB				CUB				CUB
87° W	BAH		BOL		BAH		BOL		BAH		BOL
86° W		PRU		PRU		PRU		PRU		PRU	
84.5° W	TRD		SUR		TRD		SUR		TRD		SUR
83.5° W		HTI		DOM		HTI		DOM		HTI	
82° W	←————————————————										
82° W	←————————————————										
82° W	←————————————————										
81° W	←————————————————										
81° W	←————————————————										
79.5° W	V	VCT	ATG	LCA	V	VCT	ATG	LCA	V	VCT	ATG
78° W	←————————————————										
74° W	←————————————————										
74° W	←————————————————										
74° W	←————————————————										
72.5° W	←————————————————										
71.5° W	URG		URG		URG		URG		URG		URG
70.5° W	←————————————————										
69° W	S		S		S		S		S		S
64° W	←————————————————										
64° W	←————————————————										
64° W	←————————————————										
61.5° W	←————————————————										
57° W	FLK		GRD		FLK		GRD		FLK		GRD
55° W		N	I	N		N	I	N		N	I
			S				S				S
53° W	SPM	N	GRL	F	SPM	N	GRL	F	SPM	N	GRL
45° W	←————————————————										
45° W	←————————————————										
45° W	←————————————————										
45° W	←————————————————										
42° W	GRD		GRD		GRD		GRD		GRD		GRD
37° W	GUI	MLI	GMB	AND	GUI	MLI	GMB	AND	GUI	MLI	GMB
	SMR		LIE	MLI	SMR		LIE	MLI	SMR		LIE

12	13	14	15	16	17	18	19	20	21	22	23 [a]
N	S	N	S	N	S	N	S	N	S	N	S
	I		I		I		I		I		I
A		A	D	A		A	D	A		A	D
B		B		B		B		B		B	
J		J		J	J	J		J	J	J	
H		H		H		H		H		H	
E		E		E		E		E		E	
				CAN							
				CAN							
				CAN							
			CUB				CUB			CUB	
	BAH		BOL				BOL			BOL	
PRU		PRU		PRU		PRU		PRU		PRU	
	TRD		SUR		SUR		GUB		SUR		GUB
DOM		HTI		DOM							
				CAN							
				CAN							
				CAN							
				B							
				B							
LCA	V	VCT	ATG	LCA	C	DMA	M	GRD	C	DMA	M
				MEX							
				B							
				B							
				B							
				CAN							
	URG		URG		URG		URG		URG		URG
				CAN							
				(MEX)							
	S		S		S		S		S		S
				B							
				B							
				B							
				USA							
	FLK		GRD				GRD				GRD
				(ARG)							
N		N	I	N	I	N	I	N	I	N	I
			S		S		S		S		S
F	SPM	N	GRL	F	SPM	N	GRL	F	SPM	N	GRL
				B							
				B							
				B							
				B							
	GRD		GRD								
AND	GUI	MLI	GMB	AND	GUI	MLI	GMB	AND	MCO	MTN	CVA
MLI	SMR		LIE	MLI	SMR		LIE	MLI	SEN		

TABLE 3.13 (Continued)

Assigned Satellite Longitude	1	2	3	4	5	6	7	8	9	10	11
						Channel Number[b]					
34° W		GUB		JMC		GUB		JMC		GUB	
31° W	FLK BER				FLK BER				FLK BER		
31° W		GNB IRL	AZR LBR POR	CPV G		GNB IRL	AZR LBR POR	CPV G		GNB IRL	AZR LBR POR
25° W	LBY	ALG TGO	LBY	ALG	LBY	ALG TGO	LBY	ALG	LBY	ALG TGO	LBY
19° W	F	D ZAI	BEN LUX	AUT ZAI	F	D ZAI	BEN LUX	AUT ZAI	F	D ZAI	BEN LUX
13° W	CME	TCD	GAB	MLT STP	CME	TCD	GAB	MLT STP	CME	TCD	GAB
7° W				EGY				EGY			
1° W	POL SWZ	BOT ROU	TCH ZMB	BUL MOZ	POL SWZ	BOT ROU	TCH ZMB	BUL MOZ	POL SWZ	BOT ROU	TCH ZMB
5° E	TUR	FNL	GRC	S	TUR	FNL	GRC	S	TUR	FNL	GRC
11° E	YMS	YEM	LBN UGA	RRW	YMS	YEM	LBN UGA	RRW	YMS	YEM	LBN UGA
17° E	QAT	ARS		ARS	QAT	ARS		ARS	QAT	ARS	
23° E	URS		SOM URS	URS	URS		SOM URS	URS	URS		SOM URS
29° E	MDG	MAU	COM	MAU	MDG	MAU	COM	MAU	MDG	MAU	COM
34° E			IRN				IRN				IRN
38° E		PAK		PAK		PAK		PAK		PAK	
44° E	URS		URS		URS		URS		URS		
50° E	AFG	CLN	AFG		AFG	CLN	AFG		AFG	CLN	AFG
56° E	IND	IND	IND	IND	IND	IND	IND	IND	IND	IND	IND
62° E	CHN	CHN	CHN	CHN	CHN	CHN	CHN	CHN	CHN	CHN	CHN
68° E	IND	IND	IND	IND	IND	IND	IND	IND	IND	IND	IND
74° E	THA	LAO	SNG	LAO	THA	LAO	SNG	LAO	THA	LAO	SNG
80° E	CHN	INS		INS	CHN	INS		INS	CHN	CHN	
86° E		MLA	VTN	MLA		MLA	VTN	MLA		MLA	VTN
92° E	CHN	CHN	CHN	CHN	CHN	CHN	CHN	CHN	CHN	CHN	CHN

[a]Many locations have additional channels (see pages 202–209 for data and additional information.)

	Channel Number[b]											[a]
12	**13**	**14**	**15**	**16**	**17**	**18**	**19**	**20**	**21**	**22**	**23**	
JMC		GUB		JMC								
	FLK				FLK							
	BER				BER							
CPV		GNB	AZR	CPV		GNB	AZR	CPV	HVO	CTI	CNR	
G		IRL	LBR	G		IRL	POR	G	ISL		E	
			POR								SRL	
ALG	LBY	ALG	LBY	ALG	LBY	ALG	LBY	ALG	MRC	TUN	GHA	
		TGO				TGO						
AUT	F	D	BEN	AUT	F	D	BEN	AUT	BEL	NIG	GNE	
ZAI		ZAI	LUX	ZAI		ZAI	LUX	ZAI		SUI	HOL	
MLT	CME	TCD	GAB	MLT	CME	TCD	GAB	STP		COG	AGL	
STP				STP								
EGY				EGY				EGY	YUG	ALB	SDN	
											YUG	
BUL	POL	BOT	TCH	BUL	POL	BOT	TCH	BUL	DDR	HNG		
MOZ	SWZ	ROU	ZMB	MOZ	SWZ	ROU	ZMB	MOZ		IFB		
DNK	TUR	NOR	GRC	DNK	TUR	NOR	GRC	DNK	CYP	FNL	ISL	
									IFB			
RRW	YMS	YEM	LBN	RRW	YMS	YEM	LBN	RRW	KEN	BDI	JOR	
			UGA				UGA			SYR	TGK	
ARS	QAT	ARS		ARS	QAT	ARS		ARS	UAE	KWT	ARS	
URS	URS		SOM	URS	URS		SOM	URS	DJI	ETH	URS	
			URS				URS		BLR			
MAU	MDG	MAU	COM	MAU	MDG	MAU				REU		
			IRN				IRN					
PAK		PAK				PAK		PAK		PAK		
MLD	URS			MLD		URS		URS		URS		
URS				URS								
	AFG	CLN	AFG		NPL		NPL		NPL			
IND	IND	IND	IND	IND	IND	IND	IND	IND	IND	IND	IND	
CHN	CHN	CHN	CHN	CHN		CHN		CHN		CHN		
IND	IND	IND	IND	IND	IND	CBG	IND	CBG	IND	CBG	IND	
						IND		IND		IND		
BRU	THA	BRU	BGD		BRM	BGD	BRM	BGD	BRM	BGD	BRM	
			SNG									
CHN		CHN	CHN		CHN	CHN	CHN	CHN	CHN	CHN	CHN	
					INS	INS	INS	INS	INS	INS	INS	
			VTN	MLA		MLA		MLA		MLA		
CHN	CHN	CHN	CHN	CHN	CHN		CHN		CHN	CHN		

[b]Channel numbering determined by the ITU region. The bandwidth and center frequency plan for region 2 is different from regions 1 and 3.

TABLE 3.13 (Continued)

Assigned Satellite Longitude	Channel Number[b]										
	1	2	3	4	5	6	7	8	9	10	11
98° E	AUS	AUS	AUS		AUS	AUS	AUS		AUS	AUS	AUS
104° E	INS		INS		INS		INS		INS		INS
110° E	J	KOR PNG	J	KOR	J	KOR PNG	J	KOR	J	KOR PNG	J
122° E	CAR	GUM	MRA		CAR	GUM	MRA		CAR	GUM	MRA
128° E		AUS	AUS	AUS PNG		AUS	AUS PNG	AUS		AUS	AUS
134° E			NRU				NRU				NRU
140° E	WAK	NCL WAL	NHB		WAK	NCL WAL	NHB		WAK	NCL WAL	NHB
146° E		MRL				MRL				MRL	
152° E	FJI				FJI				FJI		
158° E	NZL	CKH	SMO	CKN	NZL	CKH	SMO	CKN	NZL	CKH	SMO
170° E	PLM SMA			TON	PLM SMA			TON	PLM SMA		

TABLE 3.13 Country Channel Assignments (Continued), Channels 24–40

Assigned Satellite Longitude	Channel Number[b]										
	24	25	26	27	28	29	30	31	32	33	34
175° W	─────────────────────────────────────→										
175° W	─────────────────────────────────────→										
175° W	─────────────────────────────────────→										
166° W	─────────────────────────────────────→										
166° W.	─────────────────────────────────────→										
166° W	─────────────────────────────────────→										
160° W	─────────────────────────────────────→										
157° W	─────────────────────────────────────→										
148° W	─────────────────────────────────────→										
138° W	─────────────────────────────────────→										
138° W	─────────────────────────────────────→										
136° W	─────────────────────────────────────→										
131° W			CTR				CTR				
129° W	─────────────────────────────────────→										
129° W	─────────────────────────────────────→										
129° W	─────────────────────────────────────→										
127° W	─────────────────────────────────────→										
121° W	PNR				PNR				PNR		
119° W	─────────────────────────────────────→										
116° W			BLZ				BLZ				
115° W		BLO		BLO		BLO		BLO			

[b]Channel numbering determined by the ITU region. The bandwidth and center frequency plan for region 2 is different from regions 1 and 3.

					Channel Number[b]							[a]
12	13	14	15	16	17	18	19	20	21	22	23	
	AUS	AUS	AUS	PHL	AUS	AUS PHL	AUS	PHL	AUS	AUS PHL	AUS	
KOR	INS J	KRE PNG	INS J	KRE	INS	KRE	INS URS	KRE		KRE	URS	
AUS PNG	CAR NZL	GUM AUS	MRA AUS	AUS	CAR NZL	GUM AUS	MRA AUS	AUS	NZL	AUS	AUS	
	WAK	NCL WAL	NRU NHB		WAK			URS		URS		
		MRL				MRL						
CKN	NZL	CKH	SMO	CKN			NIU	TKL			NIU	
TON	PLM SMA		TON	PLM SMA								

		Channel Number[b]				Country
35	36	37	38	39	40	
						United States, Hawaii (region 2 plan)
						United States, Alaska
						United States, Pacific service area
						United States, Hawaii (region 2 plan)
						United States, Alaska
						United States, Pacific service area (region 2 plan)
						French Polynesia (region 3 plan)
						United States, western half (region 2 plan)
						United States, western half
						Canada, British Columbia
						Canada, Alberta and Saskatchewan
						Mexico, north
						Costa Rica
						Canada, Alberta and Saskatchewan
						Canada, Manitoba
						Canada, Ontario
						Mexico, south
						Panama
						United States, eastern half
						BLZ, Belize; C, Cayman Islands; T, Turks and Caicos Islands (GCC)
						Bolivia

TABLE 3.13 (*Continued*)

Assigned Satellite Longitude	Channel Number[b]										
	24	25	26	27	28	29	30	31	32	33	34
115° W		CLM		CLM		CLM		CLM			
115° W		EQA		EQA		EQA		EQA			
115° W		PRU		PRU		PRU		PRU			
115° W		VEN		VEN		VEN		VEN			
115° W		EQA		EQA		EQA		EQA			
110° W	──→										
110° W	──→										
107.5° W	GTM		NCG		GTM		NCG		GTM		
106° W	N	C	N	C	N	C	N	C	N		
	S	P	S	P	S	P	S	P	S		
		E		E		E		E			
104° W	V		V		V		V		V		
	I				I				I		
103° W		CLM		CLM		CLM		CLM			
102° W	B		B		B		B		B		
101° W	──→										
101° W	──→										
99° W		PRG		PRG		PRG		PRG			
96° W		GCC		GCC		GCC		GCC			
95° W	C		C		C		C		C		
	G		G		G		G		G		
94° W	N	S	N	S	N	S	N	S	N		
		I		I		I		I			
92.5° W	A		A	D	A		A	D	A		
	B		B		B		B		B		
	J	J	J		J	J	J		J		
	H		H		H		H		H		
	E		E		E		E		E		
91° W	──→										
91° W	──→										
91° W	──→										
89° W				CUB				CUB			
87° W				BOL				BOL			
86° W	PRU		PUR		PRU		PRU		PRU		
84.5° W		SUR		GUB		SUR		GUB			
83.5° W											
82° W	──→										
82° W	──→										
82° W	──→										
81° W	──→										
81° W	──→										
79.5° W	GRD	C	DMA	M	GRD	C	DMA	M	GRD		
78° W	──→										
74° W	──→										

		Channel Numbers[b]				
35	36	37	38	39	40	Country
						Colombia
						Equador
						Peru
						Venezuela
						Equador, Galapagos
						United States, Puerto Rico and Virgin Islands
						United States, eastern half
						Central America: GTM, Guatamela; HND, Honduras; NCG, Nicaragua; SLV, El Salvador
						Chile: C, central; N, north; P, Pacific Islands; S, south; E, Easter Island
						V, Venezuela; I, Venezuela islands
						Colombia
						Brazil
						United States, Puerto Rico and Virgin Islands
						United States, eastern half
						Paraguay
						Bermuda (GCC)
						Ecuador: C, continental; G, Galapagos
						Argentina: N, north; S, south; I, islands
						A, Bermuda; B, Belize; D, Barbados; E, eastern Caribbean; H, Bahamas; J, Jamaica (JMC)
						Canada, Manitoba
						Canada, Ontario
						Canada, Quebec
						Cuba
						Bahamas and Bolivia
						Peru
						GUB, Guyana; SUR, Surinam; TRD, Trinidad and Tobago
						Haiti and Dominican Republic
						Canada, Ontario
						Canada, Quebec
						Canada, eastern provinces
						Brazil, South
						Brazil, East Central
						V, Virgin Islands (United Kingdom); VCT, St. Vincent and Grenadines; ATG, Antigua and Barbuda; LCA, St. Lucia; C, St. Christopher Nevis; DMA, Dominica; M, Montserrat (United Kingdom); GRD, Grenada
						Mexico, north
						Brazil, southwest

TABLE 3.13 (*Continued*)

Assigned Satellite Longitude	Channel Number[b]										
	24	25	26	27	28	29	30	31	32	33	34
74° W	──────────────────────────────────────→										
74° W	──────────────────────────────────────→										
72.5° W	──────────────────────────────────────→										
71.5° W		URG		URG		URG		URG			
70.5° W	──────────────────────────────────────→										
69° W		S		S		S		S			
64° W	──────────────────────────────────────→										
64° W	──────────────────────────────────────→										
64° W	──────────────────────────────────────→										
61.5° W	──────────────────────────────────────→										
57° W				GRD				GRD			
55° W	N	I S	N	I S	N	I S	N	I S	N		
53° W	F	SPM	N	GRL	F	SPM	N	GRL	F		
45° W	──────────────────────────────────────→										
45° W	──────────────────────────────────────→										
45° W	──────────────────────────────────────→										
45° W	──────────────────────────────────────→										
42° W											
37° W	MTN	MCO SEN	MTN	CVA	MTN	MCO SEN	MTN	CVA	MTN	MCO SEN	MTN
34° W											
31° W											
31° W		HVO ISL	CTI	CNR E SRL		HVO ISL	CTI	CNR E SRL		HVO ISL	CTI
25° W	NGR	MRC	TUN	GHA	NGR	MRC	TUN	GHA	NGR	MRC	TUN
19° W	I	BEL NMB	NIG SUI	GNE HOL	I	BEL NMB	NIG SUI	GNE HOL	I	BEL NMB	NIG SUI
13° W	CAF	ISR	COG	AGL	CAF	ISR	COG	AGL	CAF	ISR	COG
7° W	SDN	YUG	SDN ALB	SDN YUG	SDN	YUG	SDN ALB	SDN YUG	SDN	YUG	SDN ALB
1° W	MWI	DDR	HNG IFB		MWI	DDR	HNG IFB		MWI	DDR	HNG IFB

[a]Many locations have additional channels (see pages 202–209 for data and additional information.)

35	36	37	38	39	40	Country
						Brazil, northwest
						Brazil, west
						Canada, Alberta and Saskatchewan
						Uruguay
						Canada
						Mexico: S, South
						Brazil
						Brazil, south central
						Brazil, north central
						United States, eastern half
						FLK, Falklands/Antartica; GRD, Grenada
						Argentina: N, north; S, south; I, islands
						SPM, St. Pierre and Miquelon; N, Netherlands Antilles; GRL, Greenland; F, French Guiana (GUF)/ Guadeloupe (GDL)/Martinique (MRT)
						Brazil, south
						Brazil, east central
						Brazil, northeast
						Brazil, south central
						Grenada (Region 2 plan)
CVA	MTN	MCO SEN	MTN	CVA	MTN	GUI, Republic of Guinea; SMR, San Marino; CVA, Vatican; MLI, Mali; GMB, Gambia; LIE, Liechtenstein; AND, Andorra; MCO, Monaco; SEN, Senegal; MTN, Mauritania (Region 1 plan)
						GUB, Guyana; JMC, Jamaica (Region 2 plan)
						FLK, Falkland Islands; BER, Bermuda (Region 2 plan)
CNR E SRL		HVO ISL	CTI	CNR E SRL		GNB, Guinea-Bissau; IRL, Ireland; AZR, Azores; LBR, Liberia; G, United Kingdom; POR, Portugal; CPV, Cape Verde; HVO, Upper Volta; ISL, Iceland; CTI, Ivory Coast; CNR, Canaries; E, Spain; SRL, Sierra Leone (Region 1 plan), Falklands (5); Bermuda (5) (Region 2 plan)
GHA	NGR	MRC	TUN	GHA	NGR	ALG, Algeria; LBY, Lybia; TGO, Togolese Republic; MRC, Morocco; TUN, Tunisia; GHA, Ghana; NGR, Niger
GNE HOL	I	BEL NMB	NIG SUI	GNE HOL	I	F, France; D, Germany (Federal Republic); ZAI, Zaire; BEN, Benin; LUX, Luxembourg; AUT, Austria; SUI, Switzerland; HOL, Netherlands; NMB, Namibia; GNE, Equatorial Guinea; I, Italy; BEL, Belgium; NIG, Nigeria
AGL	CAF	ISR	COG	AGL	CAF	CME, Cameroon; AGL, Angola; TCD, Czechoslovakia; GAB, Gabon; MLT, Malta; STP, Sao Tome and Principe; COG, Congo; ISR, Israel; CAF, Central African Republic
SDN YUG	SDN	YUG	SDN ALB	SDN YUG	SDN	ALB, Albania; EGY, Egypt; YUG, Yugoslavia; SDN, Sudan
	MWI	DDR	HNG IFB		MWI	POL, Poland; SWZ, Swaziland; BOT, Botswana; ROU, Romania; TCH, Czechoslovakia; ZMB, Zambia; BUL, Bulgaria; MOZ, Mozambique; DDR, Germany (Democratic Republic); HNG, Hungary; IFB, IFRB Reserve; MWI, Malawi

[b]Channel numbering determined by the ITU region. The bandwidth and center frequency plan for region 2 is different from regions 1 and 3.

TABLE 3.13 (*Continued*)

Assigned Satellite Longitude	Channel Number[b]										
	24	25	26	27	28	29	30	31	32	33	34
5° E	DNK LSO	CYP IFB	FNL	DNK	LSO NOR	CYP IFB	S	ISL	LSO NOR	CYP IFB	S
11° E	IRQ	KEN	BDI SYR	JOR TGK	IRQ	KEN	BDI SYR	JOR TGK	IRQ	KEN	BDI SYR
17° E	OMA	UAE	KWT	BHR	OMA	UAE	KWT	BHR	OMA	UAE	KWT
23° E		DJI BLR	ETH	URS		DJI UKR	ETH	URS		DJI UKR	ETH
29° E	MYT		REU		MYT		REU		MYT		REU
34° E											
38° E	PAK										
44° E	URS		URS		URS				URS		URS
50° E											
56° E	IND										
62° E	CHN										
68° E	CBG IND										
74° E	BGD	MNG	URS		URS	MNG			URS	MNG	URS
80° E	CHN INS										
86° E	MLA										
92° E	CHN										
98° E	PHL										
104° E											
110° E		URS		URS				URS			
122° E											
128° E	AUS										
134° E											
140° E	URS		URS		URS		URS		URS		URS
146° E											
152° E											
158° E	TKL										
170° E											

		Channel Number[b]				
35	36	37	38	39	40	Country
DNK	DNK LSO	CYP IFB	NOR	ISL	LSO S	TUR, Turkey; FNL, Finland; GRC, Greece; S, Sweden; DNK, Denmark; NOR, Norway; CYP, Cyprus; IFB, IFRB Reserve; LSO, Lesotho; ISL, Iceland
JOR TGK	IRQ	KEN	BDI SYR	JOR TGK	IRQ	YMS, Yemen (P.D.R.); YEM, Yemen (Arab Republic); LBN, Lebanon; UGA, Uganda; RRW, Rwanda; KEN, Kenya; BDI, Burundi; SYR, Syria; JOR, Jordan; TGK, Tanzania; IRQ, Iraq
BHR	OMA	UAE	KWT	BHR	OMA	QAT, Qatar; ARS, Saudi Arabia; UAE, United Arab Emirates; KWT, Kuwait; BHR, Bahrain; OMA, Oman
URS		DJI UKR	ETH	URS		URS, USSR; SOM, Somali; DJI, Djibouti; BLR, USSR (Byelorussia); ETH, Ethiopia; UKR, USSR (Ukrania)
	MYT		REU		MYT	MDG, Madagascar; MAU, Mauritius; COM, Comoros; REU, Reunion; MYT, Mayotte Island
						Iran
						Pakistan
	URS		URS		URS	URS, USSR; MLD, Maldives
						AFG, Afghanistan; CLN, Sri Lanka; NPL, Nepal
						India
						People's Republic of China
						IND, India; CBG, Khmer Republic
		MNG	URS	MNG		LAO, Laos; SNG, Singapore; THA, Thailand; BRU, Brunei; BGD, Bangladesh; BRM, Burma; MNG, Mongolia; URS, USSR
						CHN, People's Republic of China; INS, Indonesia
						MLA, Malasia; VTN, Viet Nam
						People's Republic of China
						AUS, Australia; PHL, Philippines
						Indonesia
URS				URS		J, Japan; KOR, Republic of Korea; PNG, Papua New Guinea; KRE, Democratic People's Republic of Korea; URS, USSR
						CAR, Caroline Islands; GUM, Guam; MRA, Mariana
						AUS, Australia; PNG, Papua New Guinea; NZL, New Zealand
						Nauru
	URS		URS		URS	NCL, New Caledonia; NHB, New Hebrides; URS, USSR, WAL, Wallis; WAK, Wake
						Marshall Islands
						Fiji
						NZL, New Zealand; CKH, Cook Islands Northern Group; SMO, Western Samoa; NIU, Niue Island; TKL, Tokelau Islands
						PLM, Palmyra Island; SMA, American Samoa; TON, Tonga (region 3 plan)

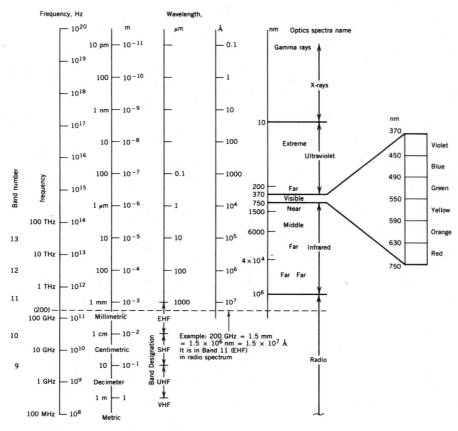

FIGURE 3.16 Electromagnetic Spectrum, including radiowaves and light. Use either scale for frequency; other scale is the wavelength.

Northern Hemisphere Earth Stations

Location of Earth Station with Respect to Satellite	Azimuth Range (degrees)	Formula for Finding True Azimuth (degrees)	Figure (as Seen From Satellite)
West and north, L negative	90–180	Az = 180 − Azm	
East and north, L positive	180–270	Az = 180 + Azm	

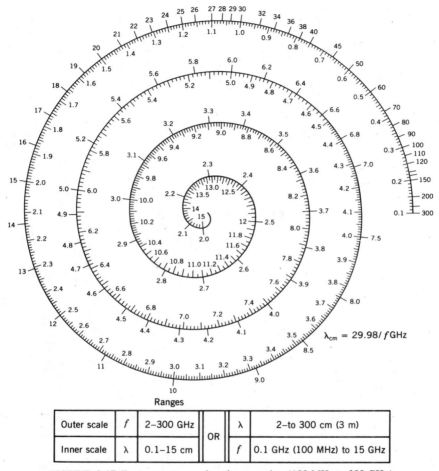

$$\lambda_{cm} = 29.98/f\,\text{GHz}$$

Ranges

Outer scale	f	2–300 GHz	OR	λ	2-to 300 cm (3 m)	
Inner scale	λ	0.1–15 cm		f	0.1 GHz (100 MHz) to 15 GHz	

FIGURE 3.17 Frequency-to-wavelength conversion (100 MHz to 300 GHz).

Southern Hemisphere Earth Stations

Location of Earth Station with Respect to Satellite	Azimuth Range (degrees)	Formula for Finding True Azimuth (degrees)	Figure (as Seen From Satellite)
East and south, L positive	270–360	Az = 360 − Azm	
West and south, L negative	0–90	Az = Azm	

TABLE 3.14 Band Designation Frequency Range

Band Designation	IEEE RDE General[a]	Standard 521 Radar[b]	Old English[a]	NASA Space[c]	U.S. Congress[c]	CCIR World	Commercial Communications Frequencies
				Source			
A	—	—	—	—	30–250 MHz	—	—
B	—	—	—	—	250–500 MHz	—	—
C	3.9–6.2 GHz	4–8 GHz	3.95–5.85 GHz	3.9–6.2 GHz	500–1000 MHz	—	3.7–6.425 GHz
D	—	—	—	—	1–2 GHz	—	—
E	—	—	—	—	2–3 GHz	—	—
F	—	—	—	—	3–4 GHz	—	—
G	150–225 MHz	—	—	—	4–6 GHz	—	—
H	—	—	—	—	6–8 GHz	—	—
I	100–150 MHz	—	—	—	8–10 GHz	—	—
J	—	—	—	—	10–20 GHz	—	—
K	10.9–36.0 GHz	18–27 GHz	12.4–40 GHz	10.9–36 GHz	20–40 GHz	—	10.7–18 GHz
K_a	33–36 GHz	27–40 GHz	26.5–40 GHz	—	—	—	18–304 GHz
K_u	15.35–17.25 GHz	12–18 GHz	12.4–18 GHz	—	—	—	10.7–18 GHz
L	390–1550 MHz	1–2 GHz	1.1–1.4 GHz	390–1050 MHz	40 GHz	—	1.5–1.6 GHz
P	225–390 MHz	—	—	225–390 MHz	—	—	—
Q	36–46 GHz	—	—	—	—	—	44 GHz
S	1.55–5.2 GHz	2–4 GHz	2.6–3.95 GHz	1.05–3.9 GHz	—	—	2.5–2.7 GHz
V	46–56 GHz	—	—	—	—	—	—
W	56–100 GHz	—	—	—	—	—	—
X	5.2–10.9 GHz	8–12 GHz	8.2–12.4 GHz	6.2–10.9 GHz	—	—	7.25–8.4 GHz

212

Band							
7[d]	—	—	—	—	3–30 MHz (HF)	—	3–30 MHz (decameter waves = HF)
8[d]	—	—	—	—	30–300 MHz (VHF)	—	30–300 MHz (meter waves = VHF)
9[d]	—	—	—	—	300–1000 MHz (UHF)	—	300–3000 MHz (decameter waves = UHF)
10[d]	—	—	—	—	—	—	3–30 GHz (centimeter waves = SHF)
11[d]	—	—	—	—	40–300 GHz (millimeter waves)	—	30–300 GHz (millimeter waves = EHF)
12[d]	—	—	—	—	—	—	300–3000 GHz (decimillimeter waves)

[a]From Jordan (1985).

[b]From Anonymous (1979).

[c]From Anonymous (1977).

[d]Band n extends from 0.3×10^n to 3×10^n Hz.

TABLE 3.15 ITU Emission Designations

Example:	3	6	M	0	F	3	C	C	N
	Number	Number	Letter	Number	Letter	Number	Letter	Letter	Letter
					Modulation	Signal Number	Information Letter		
	←——— BANDWIDTH ———→				Letter Mod.	Sig.	Info.	Details	Mux.

First Alphanumeric Set (Necessary Bandwidth)

Consists of three numbers and one letter. The letter is the decimal place.

Letter Code	Range	Example
H	0.001–999 Hz	$\overline{400H = 400\ Hz,\ H100 = 0.1\ Hz,\ 1H23 = 1.23\ Hz}$
K	1–999 kHz	9K60 = 9.6 kHz, 64K0 = 64.0 kHz, 144K = 144 kHz
M	1–999 MHz	1M54 = 1.54 MHz, 36M0 = 36.0 MHz, 500M = 500 MHz
G	1–999 GHz	1G30 = 1.30 GHz, 2G49 = 2.49 GHz

Three-Symbol Classification Designation

First Symbol, Main Carrier Modulation

Letter Code	Modulation Type
N	Unmodulated
A	AM/DSB (Amplitude modulation with double sidebands)
H	AM/SSB (Amplitude modulation with a single sideband) full carrier
R	AM/SSB (Amplitude modulation with a single sideband) reduced or variable carrier
J	AM/SSB suppressed carrier
B	AM independent sideband
C	AM vestigial sideband
F	FM
G	Phase modulation
D	Main-carrier amplitude and angle modulated either simultaneously or in preestablished sequence
P	Unmodulated sequence of pulses
K	PAM (Pulse amplitude modulation)

214

L PWM (PDM) (Pulse width or duration modulation)

M Pulses modulated in position/phase

Q Pulse sequence in which carrier is angle modulated during period of pulse

V Sequence of pulses that is combination of K, L, M, and Q produced by other means

W Combinations of amplitude, angle, and phase modulation

X Cases not otherwise covered

Note: For quantized signals (e.g., PCM) use A, H, R, J, B, C, F, G, or D.

Second Symbol, Nature of Signal(s) Modulating Main Carrier

Alphanumeric	Modulating Signal
0	No modulating signal
1	Single channel containing quantized or digital information without use of modulating subcarrier (excludes TDM)
2	Single channel containing quantized or digital information with use of modulating subcarrier (excludes TDM)
3	Single channel containing analog information
7	Two or more channels containing quantized or digital information
8	Two or more channels containing analog information
9	Composite system with one or more channels containing quantized or digital information together with one or more channels containing analog information
X	Cases not otherwise covered

Third Symbol, Information Being Transmitted

Letter Code	Information Content
N	No information transmitted
A	Telegraphy, for aural reception
B	Telegraphy, for automatic reception
C	Facsimile
D	Data transmission, telemetry, and command
E	Telephony (including sound broadcasting)
F	Television (video)
W	Combination of above
X	Cases not covered

TABLE 3.15 (*Continued*)

Fourth Symbol, Details of Signal(s)

Letter Code	Signals
A	Two-condition code with elements of differing numbers and/or durations
B	Two-condition code with elements of same number and duration without error correction
C	Two-condition code with elements of same number and duration with error correction
D	Four-condition code in which each condition represents single element (of one or more bits)
E	Multicondition code in which each condition represents single element (of one or more bits)
F	Multicondition code in which each condition or combination of conditions represents a character
G	Sound of broadcast quality (monophonic)
H	Sound of broadcast quality (stereophonic or quadraphonic)
J	Sound of commercial quality (but not K or L)
K	Sound of commercial quality with use of frequency inversion or band splitting
L	Sound of commercial quality with separate FM signals to control level of demodulated signal
M	Monochrome
N	Color
W	Combination of preceding
X	Cases not otherwise covered

Fifth Symbol, Nature of Multiplexing

Letter Code	Multiplex
N	None
C	CDM (Code division multiplex, including bandwidth expansion techniques)
F	FDM (Frequency division multiplex)
T	TDM (Time division multiplex)
W	Combinations of FDM and TDM
X	Other types of multiplexing

TABLE 3.16 ITU Station Class Codes

Code	Station Class
Space Stations, Satellites	
EA	Space station in amateur-satellite service
EB	Space station in broadcasting-satellite service (sound broadcasting)
EC	Space station in fixed-satellite service
ED	Space telecommand space station
EE	Space station in standard frequency-satellite service
EF	Space station in radiodetermination-satellite service
EG	Space station in maritime mobile-satellite service
EH	Space research space station
EJ	Space station in aeronautical mobile-satellite service
EK	Space-tracking space station
EM	Meteorological-satellite space station
EN	Radionavigation-satellite space station
EO	Space station in aeronautical radionavigation-satellite service
EQ	Space station in maritime radionavigation-satellite service
ER	Space-telemetering space station
ES	Station in intersatellite service
ET	Space station in the space operation service
EU	Space station in land mobile-satellite service
EV	Space station in broadcasting-satellite service (television)
EW	Space station in earth exploration satellite service
EY	Space station in time signal satellite service
Terrestrial Earth Stations	
TA	Space operation earth station in amateur-satellite service
TB	Fixed earth station in aeronautical mobile-satellite service
TC	Earth station in fixed-satellite service
TD	Space telecommand earth station
TE	Typical transmitting earth station for emergency position-indicating radio beacon in mobile-satellite service
TF	Fixed earth station in radiodetermination-satellite service
TG	Mobile earth station in maritime mobile-satellite service
TH	Earth station in space research service
TI	Fixed earth station in maritime mobile-satellite service
TJ	Mobile earth station in aeronautical mobile-satellite service
TK	Space-tracking earth station
TL	Mobile earth station in radiodetermination-satellite service
TM	Earth station in meteorological-satellite service
TN	Earth station in radionavigation-satellite service
TO	Mobile earth station in aeronautical radionavigation-satellite service
TQ	Mobile earth station in maritime radionavigation-satellite service
TR	Space-telemetering earth station

TABLE 3.16 (*Continued*)

Code	Station Class
	Terrestrial Stations
TT	Earth station in space operation service
TU	Mobile earth station in land mobile-satellite service
TW	Earth station in earth exploration-satellite service
TX	Fixed earth station in maritime radionavigation-satellite service
TY	Fixed earth station in land mobile-satellite service
TZ	Fixed earth station in aeronautical radionavigation-satellite service
	Nature of Service Symbols
CO	Station open to official correspondence exclusively
CP	Station open to public correspondence
CR	Station open to limited public correspondence
CV	Station open exclusively to correspondence of private agency
	Other Designators
ME	Space Station
EX	Experimental station
OT	Station open exclusively to operational traffic of service concerned
RA	Radio astronomy station

Note that the view point for this check table on page 211 ''as seen from the satellite'') is the opposite of that in Figure 3.25.

The relationships of the more common elements are given in Figures 3.26 and 3.27 and in Section 6.6.2.

Figure 3.26 shows how to relate the earth station location (relative to a geostationary satellite) to the antenna pointing in azimuth and elevation.

Figures 3.27 and 3.28 relate the relative satellite location to the spacecraft-pointing angles.

Figure 3.29 shows how azimuth and elevation are determined.

Figure 3.30 converts the elevation angle to range and the variation in the slant range path loss.

Figure 3.31 shows the separation between two satellites (1 and 2). The orbital separation angle θ_g is measured from the center of the earth in the equatorial plane. A satellite located over 100° W and another at 102° W have a separation θ_g of 2°.

Since the earth station is *not* at the center of the earth, the angular separation of the two satellites (measured at the earth station) is different: This topocentric separation θ_t is slightly greater. As shown in Table 3.18, the elevations for satellite 1 have been preselected (0°, 5°, 10°, etc.) and the elevation angle for satellite 2 has been calculated (1.5°, 6.5°, 11.5°, etc.). The slant range S distance for both

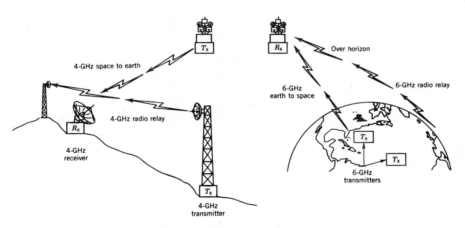

FIGURE 3.18 Examples of interference between services.

satellites has been calculated. The topocentric angle between the two satellites is shown. At low elevation angles the topocentric angle and the satellite separation are nearly the same. As the elevation increases, the separation angle also increases. The equation for the topocentric angle θ_t is

$$\theta_t = \cos^{-1}\left\{\left(\frac{S_1^2 + S_2^2 - [84332 \sin (\theta_g/2)^2]}{2S_1 S_2}\right)\right\} \quad \text{(degrees)} \quad (3.12)$$

where S_1, S_2 = distance (in km) to satellites 1 and 2 from the earth station
θ_g = orbital separation of satellites 1 and 2 (degrees of longitude)

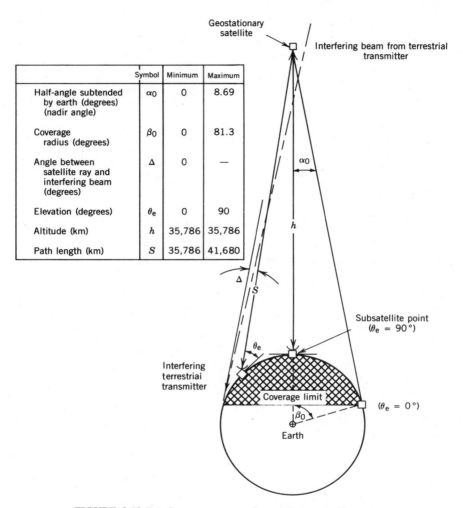

	Symbol	Minimum	Maximum
Half-angle subtended by earth (degrees) (nadir angle)	α_0	0	8.69
Coverage radius (degrees)	β_0	0	81.3
Angle between satellite ray and interfering beam (degrees)	Δ	0	—
Elevation (degrees)	θ_e	0	90
Altitude (km)	h	35,786	35,786
Path length (km)	S	35,786	41,680

FIGURE 3.19 Interference geometry of geostationary satellite paths.

Figure 3.32 shows ratio θ_g/θ_t vs the longitudinal separation between the satellites and an earth station at 40 degrees latitude.

Table 3.18 also provides the actual distances between the satellites (one degree of orbital separation is 736 km or 457 statute miles) when they are both in equatorial plane, for an earth station at 40° latitude.

Table 3.19 relates several other parameters to the elevation angle.

Figure 3.33 shows how to convert the illumination level to field strength (in microvolts per meter) at the subsatellite point and the edge of coverage. The term L_{add} is used for any additional losses (rain, polarization, rotation, blockages, etc.). As in the case of the illumination level, these values are for the field strength of the *entire* transmitted bandwidth (e.g., there is no per-4-kHz term).

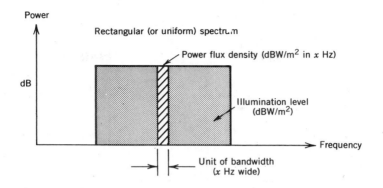

Power

Rectangular (or uniform) spectrum

Power flux density (dBW/m² in x Hz)

dB

Illumination level
(dBW/m²)

Frequency

Unit of bandwidth
(x Hz wide)

Power spectrum density for 16-phase fast FSK

Power flux density (dBW/m² in x Hz)

Illumination level (dbW/m²)

dB

-2.0   -1.0   0   1.0   2.0

Frequency

Total (integrated) power from transmitter
(illumination level)

Integrated power in worst x Hz segment (power flux density)

FIGURE 3.20 Determination of illumination level: f = frequency (Hz); f_c = nominal carrier frequency (Hz); T = time per bit (sec).

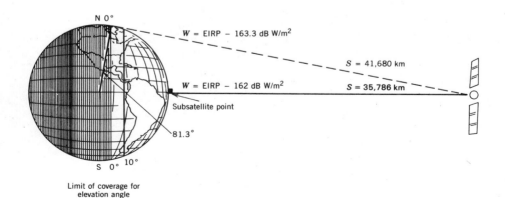

N 0°

W = EIRP − 163.3 dB W/m²

S = 41,680 km

W = EIRP − 162 dB W/m² S = 35,786 km

Subsatellite point

81.3°

S 0° 10°

Limit of coverage for
elevation angle

FIGURE 3.21 Illumination level for geostationary satellite.

221

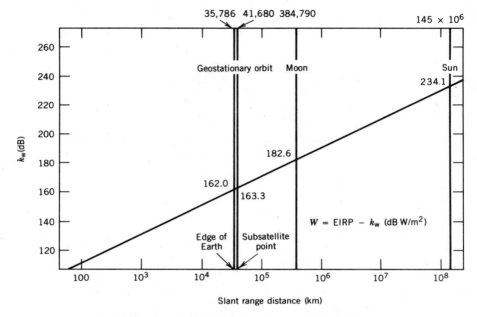

FIGURE 3.22 Illumination level factor as function of distance.

TABLE 3.17 Azimuth and Elevation Determination

1. Earth station latitude: LA_{es} = _____ degrees north or south.
2. Earth station longitude: LO_{es} = _____ degrees east.
3. Geostationary sat. longitude: LO_{sat} = _____ degrees east.
4. (a) Longitudinal difference (2 − 3): L = _____ degrees
4. (b) If either 1 or 4(a) is greater than 81.3 degrees, or if cos (LA_{es}) cos (L) is less than 0.151, the satellite is below the horizon, and this procedure must be stopped.

5. (a) Elevation = $\tan^{-1}\left(\dfrac{\cos LA_{es} \cos L - 0.151}{SQRT(1 - \cos^2 LA_{es} \cos^2 L)}\right)$

    Note: SQRT denotes square root
5. (b) Elevation magnitude: _____ degrees.
6. Azimuth magnitude: Azm = $\tan^{-1}(\tan L/\sin LA_{es})$
    Azm = _____ degrees (magnitude only).
7. (a) Determine azimuth quadrant using tables on pages 210 and 211.
7. (b) True Az = _____ degrees (from true north).

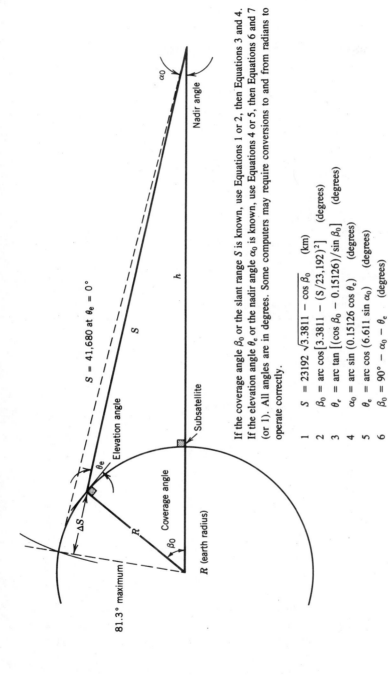

If the coverage angle β_0 or the slant range S is known, use Equations 1 or 2, then Equations 3 and 4. If the elevation angle θ_e or the nadir angle α_0 is known, use Equations 4 or 5, then Equations 6 and 7 (or 1). All angles are in degrees. Some computers may require conversions to and from radians to operate correctly.

1    $S = 23192 \sqrt{3.3811 - \cos \beta_0}$ (km)

2    $\beta_0 = \text{arc} \cos [3.3811 - (S/23,192)^2]$ (degrees)

3    $\theta_e = \text{arc} \tan \left[(\cos \beta_0 - 0.15126)/\sin \beta_0 \right]$ (degrees)

4    $\alpha_0 = \text{arc} \sin (0.15126 \cos \theta_e)$ (degrees)

5    $\theta_e = \text{arc} \cos (6.611 \sin \alpha_0)$ (degrees)

6    $\beta_0 = 90° - \alpha_0 - \theta_e$ (degrees)

7    $S = 6378 \sin \beta_0 / \sin \alpha_0$ (km)

Constants calculated from: $h = 35,787$ km, $R = 6378$ km, $h + R = 42,165$ km
Path loss difference from 0° elevation angle loss

$$\Delta S = -20 \log (S/41,680) \quad \text{(dB)}$$
$$\Delta S = -20 \log S + 92.4 \quad \text{(dB)}$$

FIGURE 3.23 Geometry of geostationary satellites and the earth.

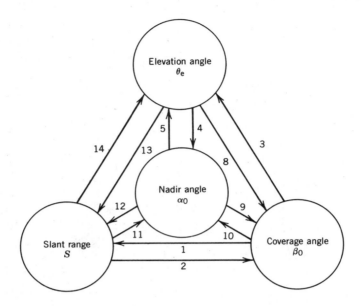

Equations listed below, plus those in Figure 3.23, can be used to calculate any of the four variables, starting with any of the remaining three. See also Section 5.7.1 for more discussion on these equations, and a numerical example (Table 5.15).

8      $\beta_Q = \text{arc cos} \ (0.15126 \cos \theta_e) - \theta_e$ (degrees)

9      $\beta_0 = \text{arc sin} \ (6.611 \sin \alpha_0) - \alpha_0$ (degrees)

10      $\alpha_0 = \text{arc tan} \ [\sin \beta_0/(6.611 - \cos \beta_0)]$ (degrees)

11      $\alpha_0 = \text{arc cos} \ (20{,}600/S + S/84{,}330)$ (degrees)

12      $S = 42{,}164 \ (\cos \alpha_0 - \sqrt{0.02288 - \sin^2 \alpha_0})$ (km)

13      $S = 6378 \ (\sqrt{43.705 - \cos^2\theta_e} - \sin \theta_e)$ (km)

14      $\theta_e = \text{arc sin} \ (136{,}188/S - S/12{,}756)$ (degrees)

Constants calculated from: $h = 35{,}787$ km, $R = 6378$ km, $h + R = 42{,}165$ km

FIGURE 3.24 Geometric equations.

For the case of a direct broadcasting or mobile satellite operating at VHF or UHF, Figure 3.34 provides a nomogram to determine the field strength at a conventional lower frequency antenna (e.g., a yagi or helix). For comparison with normal (terrestrial) FM and television signals, Table 3.20 is provided.

The power at the output of the receiving antenna (or the receiver input if the

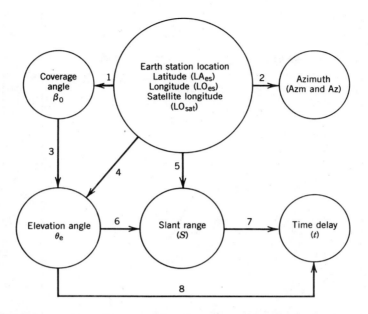

The following equations are useful in calculating the coverage angle β_0, the azimuth Azm, the elevation angle θ_e, slant range S, and one-way time delay t. For azimuth calculations, east longitudes are assumed. If the earth station latitude is negative (southern hemisphere), add $180°$ to the azimuth; after this, add $360°$ if azimuth is negative.

1     $\beta_0 \;= \text{arc cos} \left[\cos \text{LA}_{es} \cos(\text{LO}_{es} - \text{LO}_{sat}) \right]$ (degrees)

2     $\text{Azm} = \text{arc tan} \left[\tan(\text{LO}_{es} - \text{LO}_{sat})/\sin \text{LA}_{es} \right]$ (degrees)

3     $\theta_e \;= \text{arc tan} \left[(\cos \beta_0 - 0.15126)/\sin \beta_0 \right]$ (degrees)

4     $\theta_e \;= \text{arc tan} \left(\dfrac{\cos(\text{LO}_{es} - \text{LO}_{sat}) \cos \text{LA}_{es} - 0.15126}{\sqrt{1 - \cos^2(\text{LO}_{es} - \text{LO}_{sat}) \cos^2(\text{LA}_{es})}} \right)$ (degrees)

5     $S \;= 23192 \sqrt{3.3811 - \cos \text{LA}_{es} \cos(\text{LO}_{es} - \text{LO}_{sat})}$ (km)

6     $S \;= 6378 \left(\sqrt{43.705 - \cos^2\theta_e} - \sin \theta_e \right)$ (km)

7     $t \;= S/299793$ (sec)

8     $t \;= \left(\sqrt{43.705 - \cos^2\theta_e} - \sin \theta_e \right)/47.007$ (sec)

Constants calculated from: $h = 35{,}787$ km, $R = 6378$ km, $h + R' = 42{,}165$ km

FIGURE 3.25 Transmission relationships.

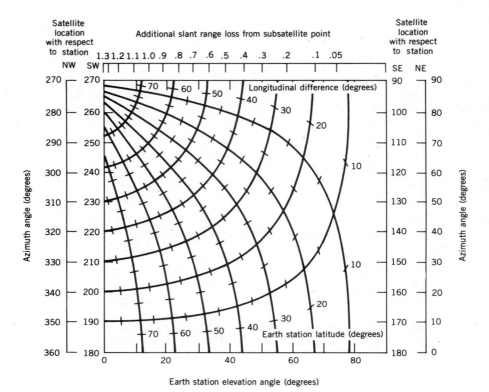

FIGURE 3.26 Azimuth and elevation angles to geostationary satellite. Reprinted with permission of the COMSAT Technical Review, the COMSAT Corporation.

line losses are zero) is

$$P = \text{EIRP} - L + G \quad (\text{dBW}) \tag{3.13}$$

where L = path loss in decibels (see equation 3.56)

G = antenna gain in dBi (see equation 3.34)

For a parabolic-type antenna

$$P = \text{EIRP} - (92.45 + 20 \log_{10} S + 20 \log_{10} f)$$
$$+ (20.4 + 20 \log_{10} f + 20 \log_{10} D + \log_{10} \eta) \quad (\text{dBW}) \tag{3.14}$$

where S = distance in kilometers

D = antenna diameter in meters

η = reflector efficiency, typically 0.50–0.80

20.4 = see equations 3.29–3.34 and Table 3.30.

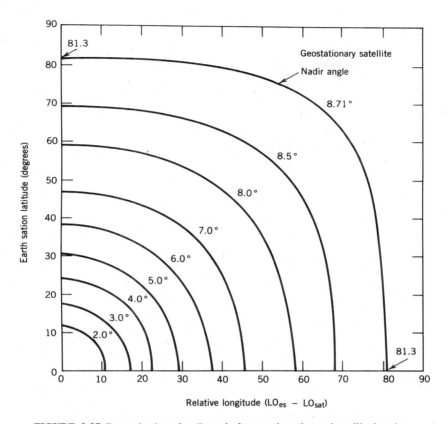

FIGURE 3.27 Determination of nadir angle from earth station and satellite locations.

If there are additional losses (rainfall in path, cable losses, etc.), these should be added to L in Equation (3.13).

The frequency terms cancel; thus,

$$P = \text{EIRP} - 20 \log_{10} S + 20 \log_{10} D + 10 \log_{10} \eta - 72.05 \quad (\text{dB}) \quad (3.15)$$

Let $S = 41,679$ km (edge of earth for a geostationary satellite) and $\eta = 0.60$ then,

$$P = \text{EIRP} + 20 \log_{10} D - 166.67 \quad (\text{dBW}) \quad (3.16)$$

This equation is plotted in Figure 3.35 on page 234.

For a *rectangular* spectrum signal the power flux density Equation (3.17) for the PFD$_\text{B}$ is the illumination level W (or EIRP $- k_\text{w}$) divided by the ratio of the

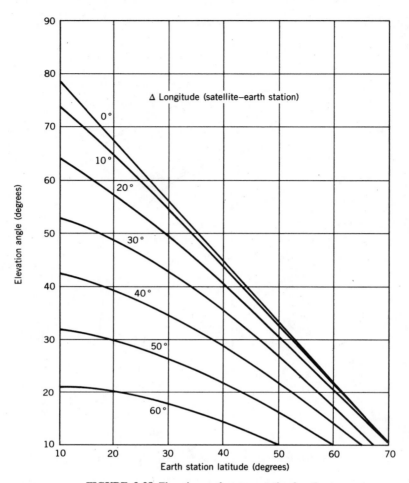

FIGURE 3.28 Elevation angle versus station location.

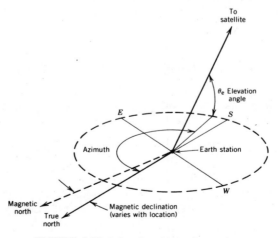

FIGURE 3.29 Azimuth and elevation angles.

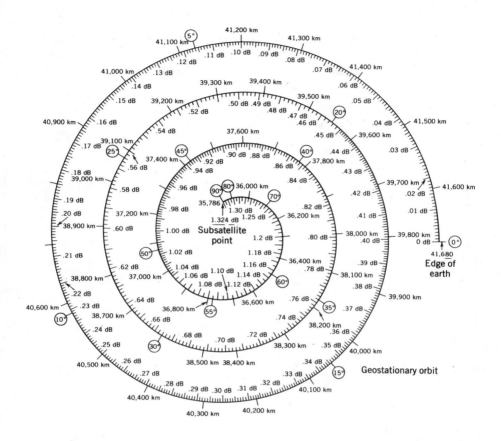

Scale		Range and Units	
Inner	ΔS	0–1.3424 dB	Slant range loss variation
Outer	S	35,786–41,680 km	Range (geostationary)
	θ	0°–90°	Elevation angle

FIGURE 3.30 Nomogram for satellite elevation angle, range, and range loss.

TABLE 3.18 Satellite Separation Seen from Earth Station at 40°N Latitude

θ_g (degrees)	Elevation 1 (degrees)	Distance 1 (km)	Elevation 2 (degrees)	Distance 2 (km)	θ_t (degrees)	Satellite Separation	
						km	Statute Miles
2.0	0	41687.9	1.5	41510.5	2.02	1472	914
2.0	5	41135.5	6.5	40961.5	2.05	1472	914
2.0	10	40594.6	11.5	40426.4	2.07	1472	914
2.0	15	40069.0	16.5	39909.1	2.10	1472	914
2.0	20	39562.4	21.5	39413.5	2.13	1472	914
2.0	25	39078.1	26.5	38943.4	2.16	1472	914
2.5	0	41687.9	1.9	41468.4	2.53	1840	1142
2.5	5	41135.5	6.9	40920.3	2.56	1840	1142
2.5	10	40594.6	11.9	40386.7	2.60	1840	1142
2.5	15	40069.0	16.9	39871.6	2.63	1840	1142
2.5	20	39562.4	21.9	39378.7	2.66	1840	1142
2.5	25	39078.1	26.9	38911.9	2.70	1840	1142
3.0	0	41687.9	2.3	41426.4	3.03	2207	1371
3.0	5	41135.5	7.3	40879.2	3.07	2207	1371
3.0	10	40594.6	12.3	40347.1	3.11	2207	1371
3.0	15	40069.0	17.3	39834.1	3.15	2207	1371
3.0	20	39562.4	22.3	39344.2	3.19	2207	1371
3.0	25	39078.1	27.3	38880.9	3.23	2207	1371
3.5	10	40594.6	12.6	40307.7	3.63	2575	1599
4.0	10	40594.6	13.0	40268.4	4.15	2943	1828
4.5	10	40594.6	13.4	40229.3	4.67	3311	2056
5.0	10	40594.6	13.8	40190.3	5.19	3678	2284
5.5	10	40594.6	14.1	40151.4	5.71	4046	2512
6.0	10	40594.6	14.5	40112.8	6.23	4413	2741
7.0	10	40594.6	15.3	40035.9	7.28	5148	3197
8.0	10	40594.6	16.0	39959.7	8.33	5882	3653

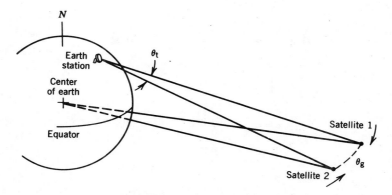

FIGURE 3.31 Satellite separations.

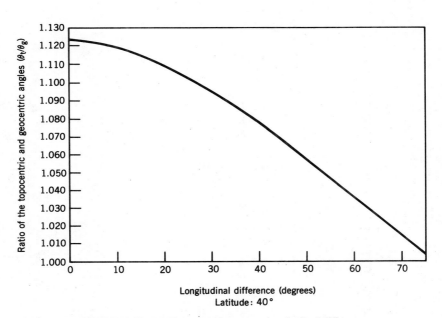

FIGURE 3.32 Satellite separation versus longitudinal difference.

TABLE 3.19 Elevation Angle Values

Elevation (degrees)	Slant Range (km)	Slant Range Attenuation[a] (dB from 0°)	Nadir, α_0 (degrees)	Central Angle,[b] β_0 (degree)
0	41676	0	8.694	81.31
10	40585	−0.2311	8.561	71.44
20	39554	−0.4545	8.166	61.83
30	38613	−0.6637	7.522	52.48
40	37781	−0.8529	6.649	43.35
50	37079	−1.0158	5.576	34.42
60	36521	−1.1475	4.335	25.66
70	36116	−1.2444	2.963	17.04
80	35870	−1.3040	1.504	8.50
90	35788	−1.3240	0	0

[a]Due to decreasing slant range, slant range attenuation is less at higher elevation angles.

[b]This angle also corresponds to maximum latitude (or maximum difference in longitude from subsatellite point) for this elevation angle.

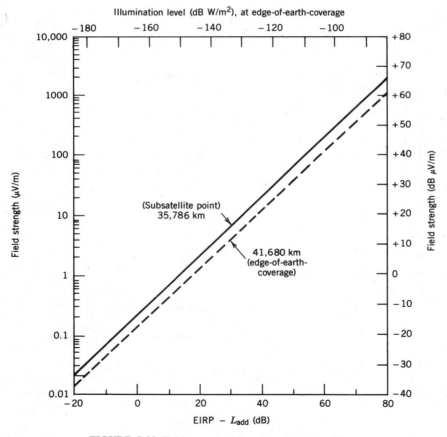

FIGURE 3.33 Field strengths from geostationary satellites.

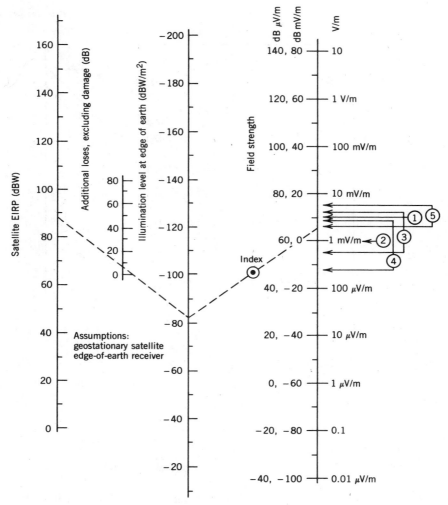

FIGURE 3.34 Field strength nomogram.

TABLE 3.20 FCC Signal Levels

	FM Radio Broadcasting[a]			
88–108 MHz	3.16 mV/m (principal community),			①
	1 mV/m (outer service contour)			②
	Television Broadcasting			
54–88 MHz	Channels 2–6	68 dBµV/m	Grade A	③
		47 dBµV/m	Grade B	③
174–216 MHz	Channels 7–13	71 dBµV/m	Grade A	④
		56 dBµV/m	Grade B	④
470–806 MHz	Channels 14–69[b]	74 dBµV/m	Grade A	⑤
		64 dBµV/m	Grade B	⑤

[a]For use with Figure 3.34

[b]Some UHF television channels are no longer available for new television broadcast stations in United States due to sharing with land mobile services.

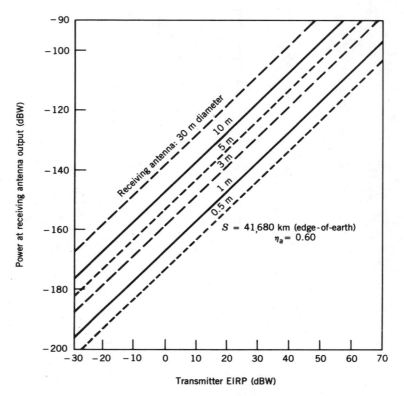

FIGURE 3.35 Receiving antenna power output.

actual transponder bandwidth B_t to the CCIR-specified bandwidth (see Table 3.21):

$$\text{PFD}_B = (\text{EIRP} - k_w) - 10 \log_{10} (B_t/B_{CCIR}) \qquad (\text{dBW/m}^2 \text{ per } B_{CCIR})$$

$$(3.17)$$

where PFD_B = power flux density at edge of earth coverage (dBW/m² per B_{CCIR})

k_w = 163.3 dB/m² for edge of earth; otherwise, see Figure 3.22

B_t = transponder bandwidth actually occupied in hertz

B_{CCIR} = CCIR-specified bandwidth (Table 3.22)

The following should be considered:

1. In place of EIRP $- k_w$ the illumination level W may be substituted. For locations other than the edge-of-earth coverage, add the slant range attenuation (decibels from 0° elevation) as indicated in Table 3.19.

2. In radio astronomy terms the unit for flux density is the jansky: 1 jansky = 1×10^{-26} W/(m²Hz). This term has not been used with artificial satellites because many services would be in the range 10^6–10^8 janskies. It is included for use in determining the interference to radio astronomy services.

TABLE 3.21 4-GHz Power Flux Density Limits

Elevation Angle (degrees)	Delta Longitude at 40° (degrees)[b]	Power Flux Density (4 kHz) (dBW/m^2)	Uniform EIRP[a] By Bandwidth			
			40 kHz (56 kbits/sec)	200 kHz Teleconference	2 MHz (DS-1)	36 MHz (TV)
0	84	−152	21	28	38	51
5	77	−152	21	28	38	51
10	70	−149.5	24	31	41	53
15	63	−147	26	33	43	56
20	57	−144.5	29	36	46	58
25	49	−142	31	38	48	61
30	43	−142	31	38	48	61
40	28	−142	31	38	48	61
46–90	0	−142	31	38	48	61

[a]Assumes absolute uniform spectrum spreading, an ideal situation that is sometimes approached in real world.

[b]For an earth station at a latitude of 40 degrees.

235

TABLE 3.22 Power Flux Density Limitations, Space to Earth[a]

Frequency (GHz)	Power Flux Density[b] (dBW/m^2)			θ_e Elevation		B_{CCIR} in any x Hz	Article 26 ITU Reference
	$0° \leq \theta_e \leq a°$	$a° \leq \theta_e \leq b°$	$b° \leq \theta_e \leq 90°$	a	b		
0.62–0.79	−129	$-129 + 0.4(\theta_e - 20)$	−113	20	60	Not specified	Rec. Spa 2-10
1.67–1.7	−133	−133	−133	0	90	1.5 MHz	6050 (Meteorological)
1.525–2.500	−154	$-154 + \left(\dfrac{\theta_c - 5}{2}\right)$	−144	5	25	4 kHz	6054
2.5–2.69	−152	$-152 + 1.5\left(\dfrac{\theta_c - 5}{2}\right)$	−137	5	25	4 kHz	6059
3.4–7.75	−152	$-152 + \left(\dfrac{\theta_c - 5}{2}\right)$	−142	5	25	4 kHz	6063
8.025–11.7[c]	−150	$-150 + \left(\dfrac{\theta_c - 5}{2}\right)$	−140	5	25	4 kHz	6067
12.5–12.75	−148	$-148 + \left(\dfrac{\theta_c - 5}{2}\right)$	−138	5	25	4 kHz	6072
17.7–19.7	−115	$-115 + \left(\dfrac{\theta_c - 5}{2}\right)$	−105	5	25	1 MHz	6075

[a]Consult *ITU Radio Regulations* and footnotes for national and other limitations. These limits can be exceeded under special conditions with permission of administrations affected.

[b]In B_{CCIR}

[c]No limit between 11.7 and 12.5 GHz for fixed-satellite services.

3. If the spectrum is not uniform, find the portion with the highest power portion over the bandwidth of B_{CCIR} hertz.

Table 3.22 gives the power flux density limitations for the space-to-earth path. These limitations are given as a function of elevation angle (see also Figure 3.36).

Figures 3.37–3.41 give the maximum EIRP as a function of the occupied transponder bandwidth for the principal bands with PFD limits. In certain types of audio, data, or teletype distribution systems, only a small fraction of a 36-MHz transponder might be occupied so as to concentrate the available EIRP into a narrow band, thus raising the EIRP density (dBW/Hz), thereby raising the PFD (dBW/m² Hz). The PFD limits establish how little bandwidth can actually be used.

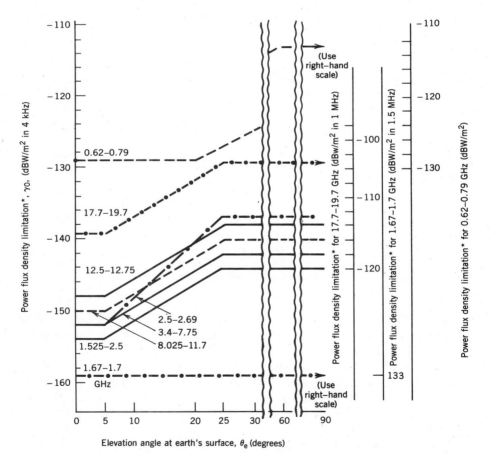

*"Limits may exceed on the territory of another country, the administration of which has so agreed."
Note: all measurements performed on earth's surface. See the ITU's, *Radio Regulations*, Sections 2552 to 2585 for further details.

FIGURE 3.36 Power flux density limits vs elevation angle for space-to-earth links.

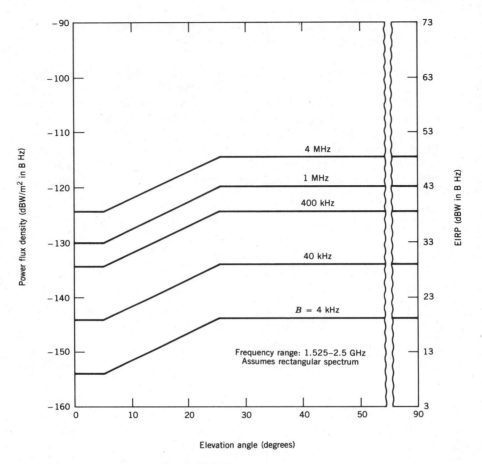

Elevation angle (degrees)

FIGURE 3.37 EIRP limits for 1.525–2.5 GHz.

Since 4 kHz is the B_{CCIR} for the bands shown, the EIRP limitations may be determined as a function of the elevation angle. The figures take into account the variation in slant range.

Figure 3.38 includes the case of 3.7–4.2 GHz. A 40-MHz transponder in the 3.7–4.2-GHz band could be as powerful as 51.3 dBW (EIRP) at the zero degree elevation angle if the signal spectrum is rectangular and fully occupies the transponder. If only 4 MHz is occupied, the maximum EIRP toward the horizon is 41.3 dBW. The figure shows other elevation angles and bandwidths. See Table 3.23 for other bandwidth combinations. The satellite's transmit antenna can be designed to provide this EIRP variation.

Except in the case of remodulation aboard a satellite, the down-link power flux density must be controlled by the up links. Thus, earth stations must insert an rf

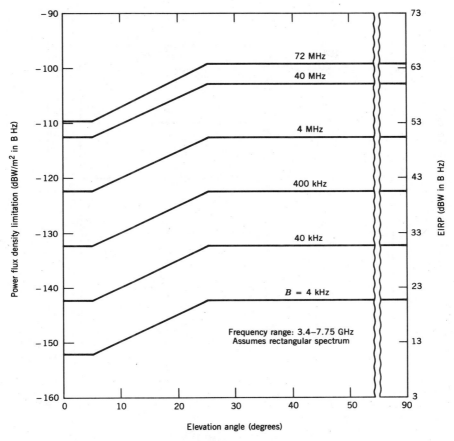

FIGURE 3.38 EIRP limits for 3.4–7.75 GHz.

energy dispersion waveform on their carriers to produce the desired down-link frequency deviations.

Assuming a constant-power satellite transmitter, a single-narrowband (CW) carrier has the highest power flux density because of the concentration of energy. As the transmitter is modulated or additional carriers are added, the power is spread over a wider bandwidth. Thus, the power flux density is decreased while the illumination level remains constant (assuming no output backoff). In many cases a deliberate spectrum-spreading waveform is used to produce energy dispersion, thus reducing the PFD to legal limits. Obviously, an unmodulated (CW) carrier should not be permitted (even under failure modes) because the PFD limits could be exceeded.

A low-frequency symmetrical triangular waveform is generally added to the baseband signal prior to the earth station's FM modulator.

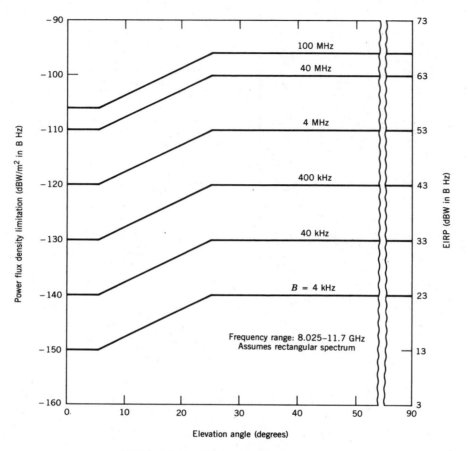

FIGURE 3.39 EIRP limits for 8.025–11.7 GHz.

For frequency division multiplex using frequency modulation (FDM FM) telephony carriers Intelsat requires a triangular frequency dispersal waveform with an amplitude such that the maximum EIRP per 4 kHz of the transmitted carriers does not exceed the maximum EIRP per 4 kHz of the fully loaded carrier by more than 2 dB. This dispersal waveform is in the range 20–150 Hz, an exact frequency being assigned to each station and maintained by ±1 Hz. This unique frequency assignment permits identification of individual earth stations.

For FM television transmissions a fixed-amplitude symmetrical triangular waveform is generally added to produce a 1-MHz peak-to-peak deviation (resulting in a 24-dB reduction in interference compared to an unmodulated carrier in a 4-kHz band). In the case of two television carriers per transponder it is important that both be adequately modulated to reduce the intermodulation of the two signals. Whenever a video signal is not present, the peak-to-peak frequency deviation

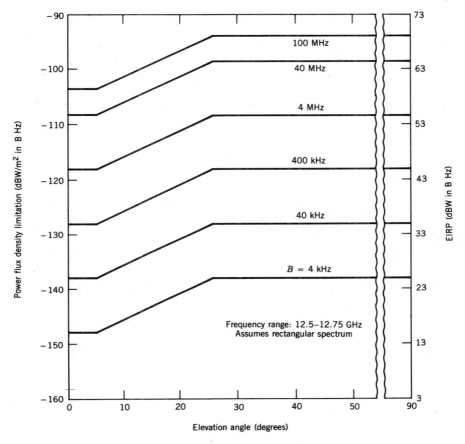

FIGURE 3.40 EIRP limits for 12.5–12.75 GHz.

should be automatically increased from, for instance, 1 to 2 MHz by increasing the amplitude of the triangular waveform. The frequency and phase of the triangular waveform should be in accordance with the standards of television being transmitted by the station; that is, the frequency of 25 Hz (for most 625-line television systems) or 30 Hz (for NTSC and other 525-line television systems) and phase should be chosen so as to have the points of inflection occurring during the field-blanking intervals. When these TV carriers are seen on a spectrum analyzer they will seem to sweep back and forth in frequency.

Frequency spreading (and hopping) allow a single, otherwise narrow band signal, to use the full power of a transponder for communications to a mobile receiver. In the reverse direction the same technique (often with spread spectrum multiple access, or SSMA, see Chapter 4) is used to transmit with broad beamed antennas and not interfere with the adjacent satellite uplinks.

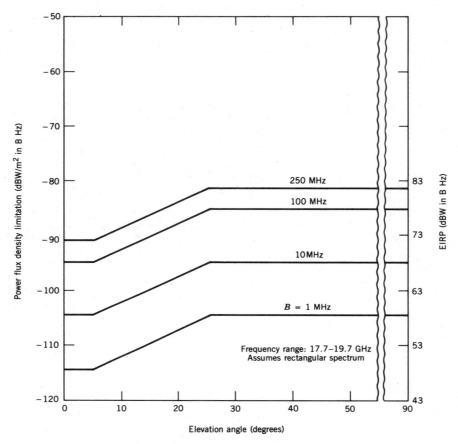

FIGURE 3.41 EIRP limits for 17.7–19.7 GHz.

3.1.4 Frequency Reuse

Since the amount of spectrum available for space communications is limited, it is desirable to employ techniques that allow the same frequencies to be utilized multiple times. This is called frequency reuse. This generally takes the form of a spatial separation of the antenna beam "footprints" (Figure 3.42) or by employing two orthogonal polarizations (Figure 3.43), or both.

Polarization reuse involves the use of either combination shown in Table 3.24. If the isolation between the two polarizations (referred to as the cross-polarization isolation) is sufficient (often 30 dB or more), independent signals may be simultaneously transmitted or received. The cross-polarization isolation between a perfect circularly polarized (CP) signal and a linearly polarized (LP) signal is 3 dB (and vice versa). The use of dual polarizations alone can result in a doubling of the useful bandwidth. Since there is finite signal coupling (because the cross-

TABLE 3.23 Maximum EIRP Limits[a]

Power Flux Density, dBW/m^2	B (dB with reference to 4 kHz)	Maximum EIRP,[b] dBW
-152	4 kHz, 0 dB	11.3
-150	4 kHz, 0 dB	13.3
-148	4 kHz, 0 dB	15.3
-152	40 kHz, 10 dB	21.3
-150	40 kHz, 10 dB	23.3
-148	40 kHz, 10 dB	25.3
-152	2 MHz, 27.0 dB	38.3
-150	2 MHz, 27.0 dB	40.3
-148	2 MHz, 27.0 dB	42.3
-152	10 MHz, 34 dB	45.3
-150	10 MHz, 34 dB	47.3
-148	10 MHz, 34 dB	49.3
-152	25 MHz, 38.0 dB	49.3
-150	25 MHz, 38.0 dB	51.3
-148	25 MHz, 38.0 dB	53.3
-152	36 MHz, 39.5 dB	50.8
-150	36 MHz, 39.5 dB	52.8
-148	36 MHz, 39.5 dB	54.8
-152	45 MHz, 40.5 dB	51.8
-150	45 MHz, 40.5 dB	53.8
-148	45 MHz, 40.5 dB	55.8
-152	72 MHz, 42.6 dB	53.9
-150	72 MHz, 42.6 dB	55.9
-148	72 MHz, 42.6 dB	57.9

[a]EIRP = PFD + 163.3 + 10 $\log_{10}(B/4\text{ kHz})$.

[b]In direction of specified PFD measurement point. Higher levels may be permitted at higher elevation angles. Assumes $B_{\text{CCIR}} = 4$ kHz.

polarization isolation is never perfect), some cochannel noise is coupled. This represents a loss in the channel capacity. The cochannel noise contribution is present in both the up and down links; thus, if each had a 30-dB isolation, the composite is 27 dB. Whereas the bandwidth may double, the capacity gain is actually slightly less (the amount depending on factors such as the system cross-polarization isolations, how the cochannel frequencies are assigned, etc.). In addition to the satellite antennas, the two earth station antenna polarizations must be considered along with the precipitation depolarization and antenna misalignments. See section 4.4 for applications of dual polarization and multiple beams.

It would be nice if all linearly polarized satellites were arranged in an orderly manner with adjacent satellites having carrier frequencies aligned (see the top portion of Figure 3.44). In reality, not all satellites even have their "horizontally polarized" signals parallel to the earth's equator and the "vertical" signals aligned

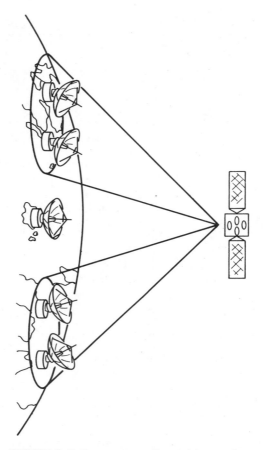

FIGURE 3.42 Frequency reuse by spatial separation.

with the poles. Variations of 5°–30° have been used. This deviation from ideal has several effects: (1) the earth station antenna's feed horn should be aligned to each satellite; (2) the isolation between adjacent satellites will be less than if the transponders were *cross-polarized* (see Figure 3.44); and (3) even in the case of perfect polar/equatorial alignment there is a slight polarization shift due to the *"rainbow"* seen by a typical earth station.

Circularly polarized systems avoid problems 1, 2, and 3, but the antenna feeds are slightly more expensive.

Practical systems have employed either linearly or circularly polarized signals with similar overall results. The choice is generally dictated by the selection of the particular frequency band, its regulations, and the nearby satellites.

The cross-polarization isolation may suffer because of rotation of the polarization vectors as the signal passes through precipitation (principally rain; rarely snow, hail, or fog). See Figure 3.45 for a pictorial of how the wave's polarization is

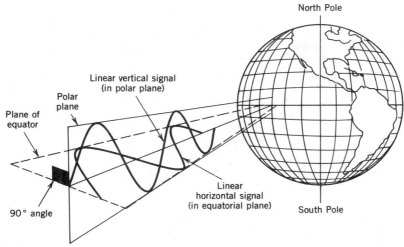

Notes:

1. In some satellites horizontal and vertical planes plane may be rotated with respect to earth's equator.
2. Beam may be pointed at some other spot on earth (e.g., United States).
3. Circular polarizations may be used in place of linear polarizations.

FIGURE 3.43 Frequency reuse by orthogonal polarizations.

TABLE 3.24 Isolation (dB) Combinations Available for Frequency Reuse by Orthogonal Polarization

Second Polarization	First Polarization			
	LHCP	RHCP	VLP	HLP
LHCP	0 dB	[a]	3 dB	3 dB
RHCP	[a]	0 dB	3 dB	3 dB
VLP	3 dB	3 dB	0 dB	[b]
HLP	3 dB	3 dB	[b]	0 dB

[a]Isolation depends on design (usually 30 dB or more in region of interest).
[b]Estimated as $-20 \log_{10} \cos(\Delta\phi)$, where $\Delta\phi$ is the angle between the two linear polarizations.
[c]Assumption: LHCP (left-hand circular polarization) and RHCP (right-hand circular polarization) signals are perfect (no ellipticity). VLP, vertical linear polarization; HLP horizontal linear polarization.

rotated as it passes through a raindrop. Figure 3.46 shows how often and how great this effect is in a loopback link at 6 GHz (earth to space) and at 4 GHz (space to earth). The signal was looped back to the transmit station via the satellite; thus, each curve represents a double pass through the rain.

At higher frequencies the isolation degradation due to precipitation is greater (see Figure 3.47).

If both ends of a link have fixed linearly polarized antennas, this rotation can be detrimental. If one end (usually the earth end) can be rotated, it can be made

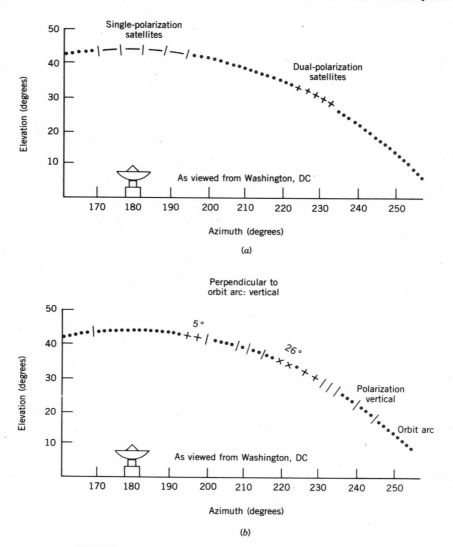

FIGURE 3.44 Adjacent satellite polarizations: (*a*) ideal; (*b*) actual.

to cancel out the precipitation rotation error and thus restore most of the lost isolation.

The isolation between the polarizations is a function of the two signals radiated (the axial ratio). Figure 3.48 shows a one-way link (one each earth and space terminal) for circular polarization. Figure 3.49 presents the linear polarization case. The satellite antenna discrimination is shown in parentheses. If either the spacecraft antenna or earth station antenna is unintentionally rotated, the linear discrimination degrades, as shown in Figure 3.49. For a satellite this usually occurs

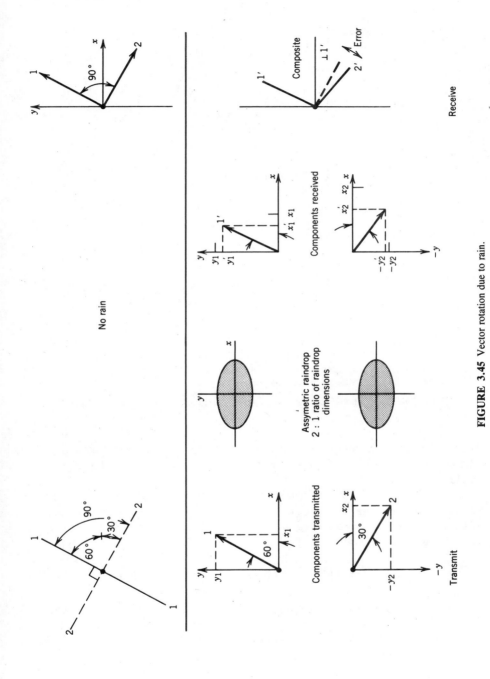

FIGURE 3.45 Vector rotation due to rain.

247

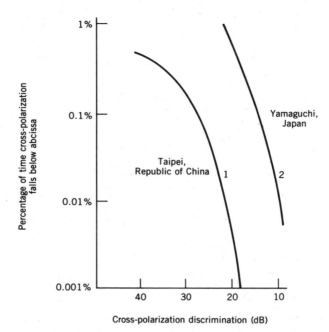

Curve 1 shows data recorded at Taipei, Republic of China, earth station for 1 year starting July 1975.

Curve 2 shows data recorded at Yamaguchi, Japan, earth station for 1 year (November 1975 to December 1976); several typhoons were noted during this period.

FIGURE 3.46 Rain depolarization at 6 and 4 GHz. (Dicks and Brown, 1978 © 1978 IEEE.)

because of an attitude control error; in an earth station it may be caused by improper alignment. The signal amplitude also varies with the cosine of the misalignment angle.

Figures 3.48 and 3.49 are for a single link. Practical satellite systems employ at least two links (earth to space and space to earth). The polarization discriminations of each must be combined to find the system performance. Equation (3.18) is useful for this purpose:

$$(PI_t)^{-1} = (PI_u)^{-1} + (PI_d)^{-1} + (PI_x)^{-1} \quad (\text{ratio}) \qquad (3.18)$$

where (PI) is polarization isolation (expressed as a ratio, not in decibels). The u, d, and x subscripts represent the up-link, down-link, and cross-link isolations, respectively. The t subscript is the total isolation. The (PI) ratio may be converted from decibels by

$$(PI) = 10^{(dB/10)} \quad (\text{ratio}) \qquad (3.19)$$

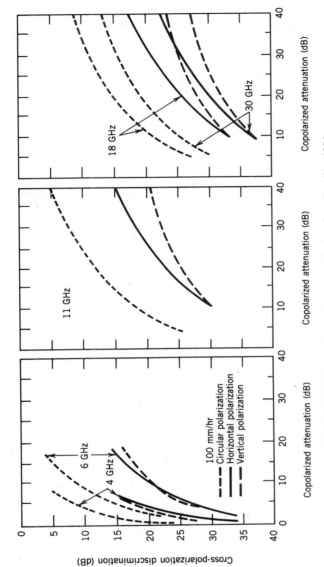

FIGURE 3.47 Depolarization due to precipitation. From Ippolito (1986).

249

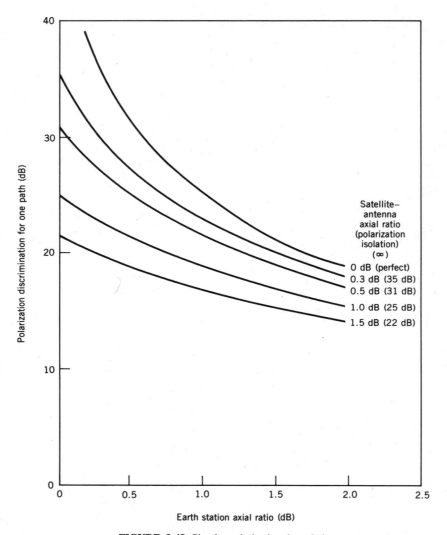

FIGURE 3.48 Circular polarization degradation.

Thus,

$$(\text{PI}_t) = 10^{-(\text{dB}_u/10)} + 10^{-(\text{dB}_d/10)} + 10^{-(\text{dB}_x/10)} \quad (\text{ratio}) \quad (3.20)$$

The imperfect isolation causes interference (expressed as the ratio of the carrier power to the interference power or C/I) that must be combined with the more conventional C/N or C/N_0 to determine overall system performance.

Frequencies may also be reused by having a satellite with multiple spot beams.

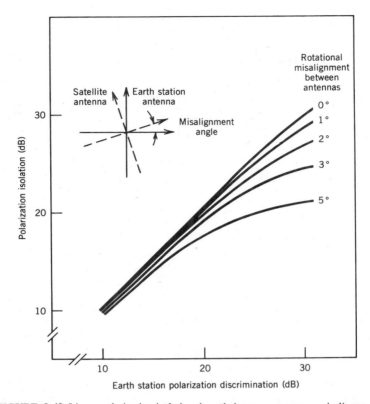

FIGURE 3.49 Linear polarization isolation degradation versus antenna misalignment.

In the simplest case each of the n small beams is mutually exclusive; that is, there is no overlap between the beams, and therefore, reuses are possible. If each beam carries two orthogonal signals (see preceding discussion), $2n$ reuses are possible. See Figure 3.50.

This simplistic case is usually unrealistic. Practical antenna beams have sidelobes that overlap other beams and have imperfect polarization isolations for off-axis angles. Since major population centers often tend to be contiguous (e.g., Boston to Washington), deliberate adjacent beam overlap is often required. This also avoids the missed coverage between the beams. Each beam usually will have its own unique traffic requirements (a beam covering a desert area will have smaller need for spectrum than the same-sized beam covering a major city such as New York. These influences combine to reduce the theoretical number of reuses and thus the practical channel capacity.

Figures 3.51 and 3.52 show two examples of frequency reuse by beam separation. In the first case both polarization and spatial reuse are involved.

Figure 3.52 illustrates a simplistic multiple-spot-beam coverage. The frequency band is split into four subbands (to avoid having adjacent cofrequency beams).

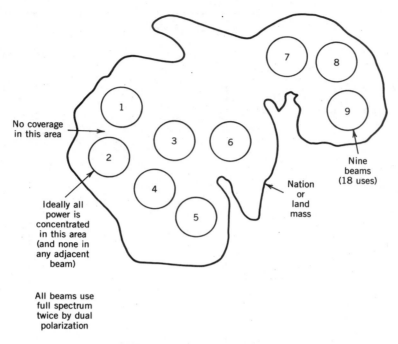

FIGURE 3.50 Ideal spot beam coverage.

Isolation is increased between the widely separated cofrequency beams by using orthogonal polarizations. If there are m beams and the frequency is divided into four equal parts (based on the four-color map analogy), the total theoretical reuse of the spectrum is $m/4$. In the case shown, $m = 20$; thus, five reuses are available.

3.1.5 Utilization of Various Frequency Assignments

This section discusses the basic effects of the propagation anomalies as they influence the communications satellite system performance. The greatest difference between the bands above 10 GHz and those between 1 and 10 GHz is the influence of the atmosphere and its precipitation (particularly rain) on the passage of the signals and the noise levels.

At frequencies below about 1 GHz the galactic and man-made noise levels (see pages 307–309) increase substantially, and the ionospheric absorption and scintillation reduce the signal level. These frequencies have been so extensively utilized and allocated that little spectrum can be found for a new service such as space communications.

The 1–10-GHz range is already extensively used by both terrestrial microwave and satellite services. Although the noise level and attenuation are lower than at the higher frequencies, the potential for interference from terrestrial point-to-point services has limited earth station locations. As fiber optics replaces the long-haul

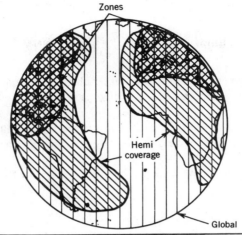

Beam	Direction	Polarizations	
		Up Link	Down Link
Hemi	East	·LHC	RHC
Hemi	West	LHC	RHC
Zone	East	RHC	LHC
Zone	West	RHC	LHC

FIGURE 3.51 Frequency reuse in Intelsat (6/4 GHz).

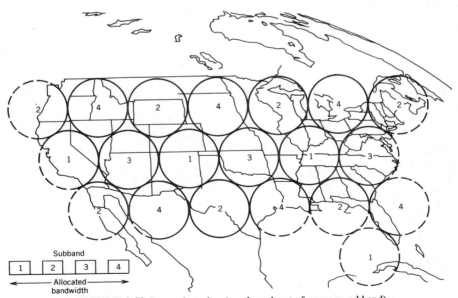

FIGURE 3.52 Beam clustering (numbers denote frequency subband).

terrestrial microwave links, these frequencies are becoming more useful for satellites.

Above about 10 GHz the rain attenuation increases, but the likelihood of finding terrestrial sources of interference in a satellite band becomes less due to better regulatory planning.

At certain wavelengths signals encounter absorption bands due to atmospheric components (like water vapor and oxygen). The so-called window for space communications is in the 1–10-GHz region.

Frequencies above 30 GHz have been underutilized; thus, there is spectrum available, especially for services that do not pass through the atmosphere (such as intersatellite links).

The fundamental equation for the free-space portion of the slant range losses (L_{range}) is

$$L_{range} = \left(\frac{4\pi S}{\lambda}\right)^2 = \left(\frac{4\pi f S}{2.998 \times 10^8}\right)^2 \quad \text{(ratio)} \qquad (3.21)$$

where S = slant range in meters
$\quad \lambda$ = wavelength in meters
$\quad f$ = frequency in Hertz
The 2.998×10^8 m/sec is the speed of light.

A more practical expression is given in Equation (3.22). The constant values in Equation (3.21) (i.e., 4π and the conversion constant from wavelength to frequency) have been combined into a single constant k_r, and the ratio value has

TABLE 3.25 Values for k_r in Range Loss Equation

Units of Distance S	Units of Frequency f	Value for Constant k_r
Kilometers	Gigahertz	92.45
	Megahertz	32.45
Statute miles	Gigahertz	96.58
	Megahertz	36.58
Nautical miles	Gigahertz	97.80
	Megahertz	37.80
Earth radii	Gigahertz	168.53
Astronomical unit	Gigahertz	255.94
Light-second[a]	Gigahertz	201.99
Geostationary orbit to subsatellite point (35,788 km)[b]	Gigahertz	183.52
Geostationary orbit to edge of earth (41,679 km)[b]	Gigahertz	184.85

[a] 1 light-second = 299,800 km.
[b] In this case the $20 \log_{10} S$ term is dropped, and the equation becomes $L_r = k_r + 20 \log_{10} f$.

been converted to decibels of range loss:

$$L_{\text{range}} = k_{\text{r}} + 20 \log_{10} S + 20 \log_{10} f \quad (\text{dB}) \qquad (3.22)$$

where k_{r} is the conversion constant (see Table 3.25) including 4π and the 2.998 term from Equation (3.21).

Figure 3.53 provides a graphical solution. For finer detail see Figure 3.30.

A convenient rule of thumb is that at 6 GHz the slant range attenuation is about 200 dB. Each time the frequency is doubled (or halved) 6 dB is added (or subtracted). Thus, 1.5 GHz (one-fourth of 6 GHz) is 200 − 6 − 6, or about 188, dB. At 12 GHz the loss is 200 + 6, or about 206 dB.

In addition, there are atmospheric effects. Figure 3.54 shows the magnitude of these losses. Rain (shown as dotted lines) is the predominant loss element below 60 GHz. Fog is shown as the dashed–dotted line (at 0.1 g/m^3). The solid line is the clear-sky gaseous absorption. The total-link attenuation is the sum of the losses due to the slant range [Equation (3.22)], the atmosphere, precipitation, and any additional losses (such as Faraday rotation, scintillation, etc.).

As may be noted from Figure 3.54, there are peaks (or absorption lines) in the atmospheric attenuation curve.

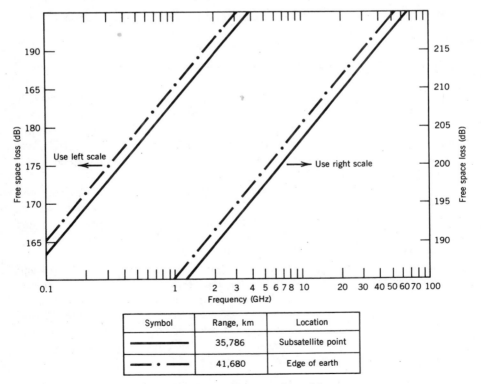

Symbol	Range, km	Location
————	35,786	Subsatellite point
— · —	41,680	Edge of earth

FIGURE 3.53 Geostationary orbit path losses.

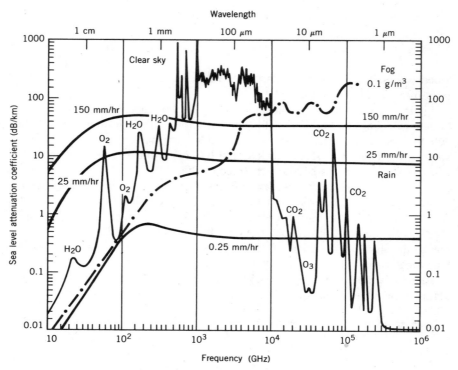

FIGURE 3.54 Atmospheric path losses at sea level.

At lower frequencies (e.g., 1.5 and 2.5 GHz) ionospheric effects may be encountered, particularly scintillation. The magnitude of these losses varies considerably with the time of day and the sunspot activity level (that affects the ionosphere). Figure 3.55 shows the monthly sunspot number level during two cycles (each cycle is about 11.5 years long). Peaks occur around 1969, 1980, 1991, 2003, and so on.

Figure 3.56 shows the Crane rain regions of the world. Figures 3.57 and 3.58 provide closeup views for the United States (Crane model) and Europe (CCIR model).* The rainfall rate (in millimeters per hour) is shown as a function of time in Figure 3.59. Figure 3.60 converts rainfall rate to attenuation and the percentage of time that attenuation exceeds a given level at 12 GHz.

Tables 3.26 and 3.27 give typical year values for several cities at 12, 14 and 20 GHz. It should be noted that, in general, the attenuation experienced in Europe is less than in the East Coast of the United States. As shown in Figure 3.60, rainfall regions in the United States have a variety of attenuation situations. For this reason some of the satellites providing coverage at frequencies greater than 10 GHz use specially shaped beams that place more of their radiated power in those areas (e.g., rain region E) where there is greater attenuation.

*These models are not interchangeable.

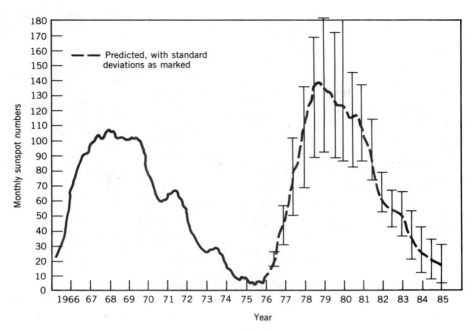

FIGURE 3.55 Sunspot numbers.

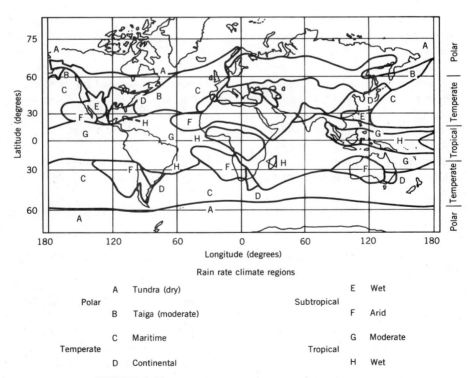

Rain rate climate regions

Polar	A	Tundra (dry)	Subtropical	E	Wet
	B	Taiga (moderate)		F	Arid
Temperate	C	Maritime	Tropical	G	Moderate
	D	Continental		H	Wet

FIGURE 3.56 Worldwide rain regions using the Crane model.

257

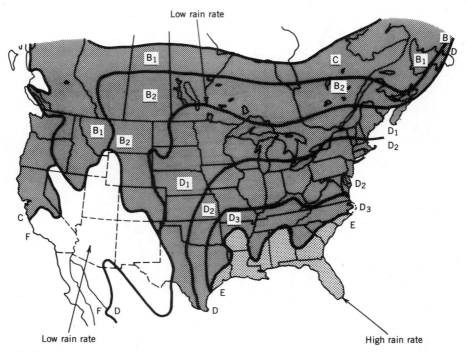

FIGURE 3.57 Crane model of rain regions in United States. Reprinted with permission from Morgan and Rouffet, *Business Earth Stations for Telecommunications*, Wiley, 1988.

There are a number of different methods that have been devised for calculating the anticipated attenuation at different frequencies. See, for instance, *Radiowave Propagation in Satellite Communications*, Ippolito (1985).

Figure 3.61 converts availability and outage to the cumulative outage time over either a monthly or annual period.

Rain also affects the performance of a receiving earth station antenna. The rain falling in the antenna beam between the satellite and the station has a temperature that approaches that of the earth (290 K) if the rain is heavy.

3.1.6 Space Resource

The frequency domain is a basic communications resource. In space there are other forms of resources, the satellite and the orbit.

Most current satellites for communications are placed in geosynchronous or geostationary orbits. Although these terms are similar and often used interchangeably, there is a difference.

Geosynchronous (from *geo*, meaning earth, and *synchronous*) satellites have an orbit whose period (the time to make one full revolution around the earth as viewed from some distant star) is the same as one sidereal day (the time required for the

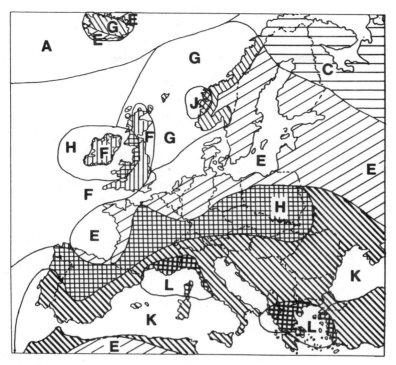

FIGURE 3.58 CCIR model of rain regions for Europe. Reprinted with permission from Morgan and Rouffet, *Business Earth Stations for Telecommunications*, Wiley, 1988.

earth to make one full rotation as seen from the same distant star). This period is 23 hr, 56 min, and 4.1 sec (usually referred to simply as "24 hours"), or about 1436 min per revolution. The reason for the difference is due to the earth's own revolution around the sun. See Chapter 6.

The satellite must be in a prograde orbit (i.e., the satellite moves in the same direction as the earth revolves). These are the only requirements for a geosynchronous orbit; the orbit may be inclined at any angle with respect to the earth's equatorial plane. *Syncom 1* was placed in an orbit inclined by 33°. During one orbital day the point directly below the satellite (the subsatellite point) oscillated between 33° N and 33° S. See Figure 3.62. Twice a day it passed over the equator. Tracking earth station antennas were needed to follow the satellite. Geosynchronous satellites are still used where tracking earth station antennas are required for other reasons (e.g., on mobile platforms like ships and planes), where the high expense of a tracking antenna can be afforded, where the satellite has global or broad area coverage beams, or where the satellite has run out of its north–south station keeping fuel. This orbit may also be used to store satellites prior to being put into service.

A *geostationary* satellite has the earth synchronous period, and the direction of

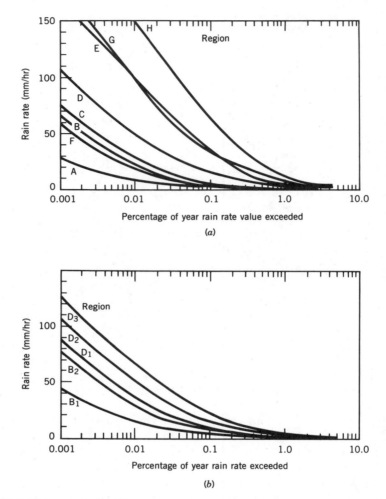

FIGURE 3.59 Point rain rate distributions as function of year rain rate exceeded: (*a*) all regions; (*b*) subregions of United States (Crane Model).

rotation of a geosynchronous satellite *and* the satellite's orbit is confined to the immediate vicinity of the equatorial plane of the earth and the orbit inclination is essentially zero. In commercial applications the satellite's orbit is virtually circular (the high point, apogee, and the low point, perigee are nearly the same height), so the orbit eccentricity is zero. In this condition the satellite remains motionless over a single spot on the earth's equator. The satellite continuously views the same portion of the earth. Both the satellite beams and earth station beams are motionless. This simplifies the design and operation requirements for both the satellite and the earth station.

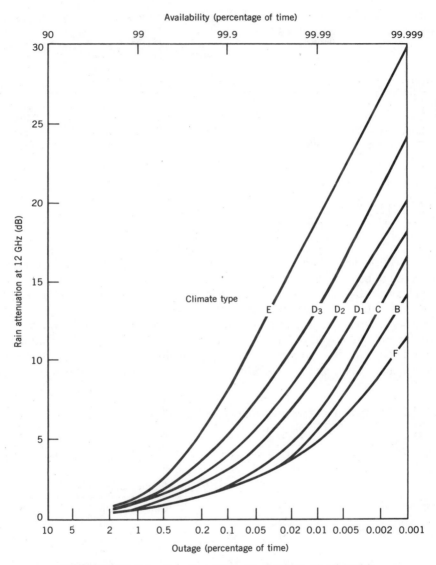

FIGURE 3.60 Rain attenuation allowances at 12 GHz (Crane model).

Because the satellite is essentially motionless with respect to an earth station, TDMA timing errors (which are caused by the Doppler effect due to satellite motion) are small.

A geostationary satellite is always geosynchronous, but the reverse is not necessarily true. See also Table 3.28.

TABLE 3.26 Comparison of U.S. and European Rain Attenuation Margins at 14/11 GHz to Obtain 99.99% Annual Signal Availability

Country	Elevation Angle (degrees)	14.5 GHz Attenuation (dB)	11.7 GHz Attenuation (dB)
1. Paris, France	23	7.9	5.1
2. Usingen, Germany	19	11.7	7.6
3. Lario, Italy	21	10.6	6.8
4. Madley, United Kingdom	23	8.4	5.5
Approximate average of 1–4	22	10.0	6.0
5. Stroudsburg (U.S.)	15	20.3	12.8

Source: Dicks and Brown (1978). © 1978 IEEE

In practice, few, if any, satellites really meet all of the theoretical conditions necessary to be truly geostationary. If a satellite were to achieve this condition, it would not remain geostationary for any long period because of the disturbing effects (e.g., solar, lunar, and nonsymmetric gravitational forces) that perturb the orbit parameters. On-board propulsion (the reaction control system, or RCS) is needed to counteract these natural forces (see the index). Although no rigid definition of the boundary between geosynchronous and geostationary orbits exists, any satellite that is maintained at an assigned orbit longitude to within a few tenths of a degree

TABLE 3.27 Precipitation Margins Required for Various Signal Availabilities[a]

Frequency, GHz	20	20	20	30	30	30
Elevation angle, degrees	20	20	20	20	20	20
Time availability, %	99.5	99.9	99.99	99.5	99.9	99.99
Location			Link Margin required (dB)			
Washington, DC	(5)	17	64	(10)	32	132
Norfolk, VA	(5)	22	78	(12)	42	>140
Omaha, NB	(5)	15	70	(9)	27	133
San Francisco, CA	(4)	9	26	(6)	16	50
London, United Kingdom	(2)	6	17	(6)	12	40
Heidelberg, Germany	(3)	8	22	(4)	14	42
Naples, Italy	(5)	12	34	(10)	26	95
Madrid, Spain	(2)	5	17	(5)	11	32
Manila, Philippines	(12)	53	>140	(23)	98	>140
Okinawa	(10)	32	>140	(20)	60	>140
Guam	(13)	41	>140	(26)	78	>140

[a]Numbers in parentheses are extrapolated values.

Source: Brandinger, (1978). © Copyright American Institute of Aeronautics and Astronautics; reprinted with permission from the 7th AIAA Communications Satellite Systems Conference.

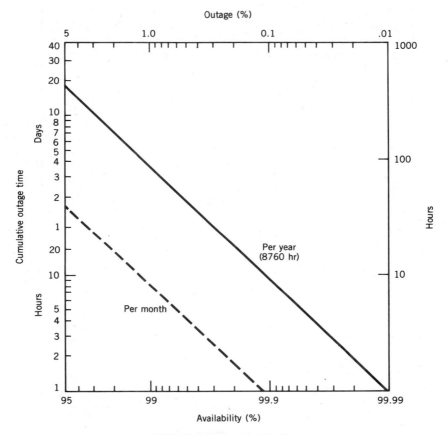

FIGURE 3.61 Service reliability.

with an inclination of less than ~1° is generally considered as being geostationary.

It should be noted that for a satellite's earth-pointing antennas to keep "facing" the earth during geostationary orbiting, either the satellite itself must rotate about its own axis (the axis parallel to the earth's axis) at the sidereal rate or it must have antenna systems that rotate about that axis at the sidereal axis. See Chapter 5 for further details (including consideration of solar power array pointing) and Chapter 6 for discussion of orbital mechanics.

Figure 3.63 shows the conditions for these orbits and illustrates the vast amount of space in which a satellite spends its life.

The Soviet Union established the Molniya satellite series using a 65°-orbit (later changed to 62.9°). This orbit is very elliptical, typically with a perigee of 1000 km and an apogee of 39,400 km, and has a period of about 12 hr. Through the use of on-board propulsion, one of the two daily high points was made to occur over the Soviet Union in the Northern Hemisphere. This method requires at least

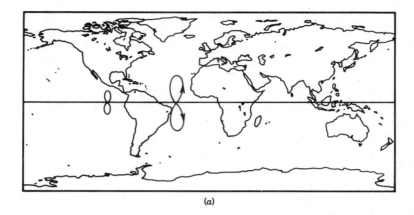

(a)

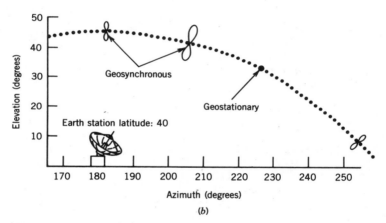

(b)

FIGURE 3.62 Geosynchronous satellites: (*a*) Inclined geosynchronous subsatellite point traces; (*b*) earth station views of geostationary and geosynchronous satellites.

three satellites to be maintained in the orbit and phased so that one (or more) is in view of the earth stations at all times. Tracking antennas are required at earth stations. Unless each station has two tracking antennas, there is an outage while one antenna points over from the setting satellite to the next Molniya.

The Molniya orbit has the advantage of high-elevation-angle coverage of the Northern Hemisphere (even in polar areas where geostationary coverage is poor to nonexistent), long dwell times, and launch economies.

Its disadvantages include the need for multiple satellites, the very poor Southern Hemisphere capabilities, and the need for two tracking antennas at each earth station. Because the slant range keeps changing, there is a Doppler effect, and there may be a need for an automatic up-link power control. Since all earth stations being connected must view the same satellite simultaneously, scheduling is needed.

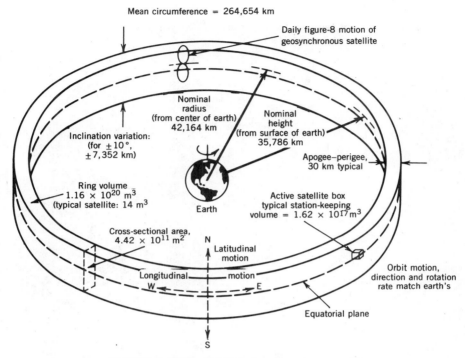

FIGURE 3.63 Nominal limits for geosynchronous orbit.

Since the orbit altitude varies, the beam coverage also varies. The Molniya satellites carry a tracking antenna that must be moved in orbit to point the beam at the area containing the earth stations.

As the number of satellites launched or planned to be launched to the geostationary orbit continues to increase, contention between nations and users has emerged (see Figures 3.64 and 3.65).

TABLE 3.28 Primary Telecommunications Orbits

			Typical Parameters			
Type of Orbit	Apogee[a] (km)	Perigee[a] (km)	Eccentricity[a]	Period[a]	Inclination[a] (degrees)	Doppler Shift
Geostationary	35,786	35,786	Essentially 0	1 sidereal day[b]	0.0[c]	None or low
Geosynchronous	35,786	35,786	Usually near 0	1 sidereal day[b]	0–90	Low
Molniya	39,400	1,000	High	$\frac{1}{2}$ sidereal day[b]	62.9	High
Low	Various	Various	0 to high	100 min or more	0–90	Highest

[a]Consult Chapter 6 for use and definition of terms.

[b]1 sidereal day = 23 hr, 56 min, 4.1 sec.

[c]Typical practical value, 0.1°.

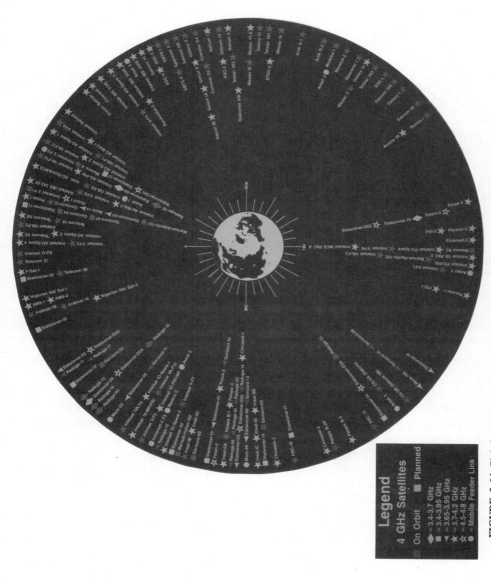

FIGURE 3.64 Global satellite stations (6/4 GHz). Reprinted with permission of *Satellite Communications*, 6300 S. Syracuse Way, Englewood CO.

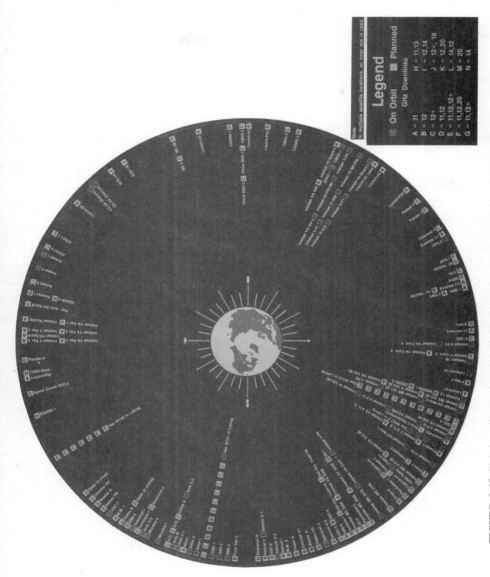

FIGURE 3.65 Global satellite stations (11–18 GHz). Reprinted with permission of *Satellite Communications*, 6300 S. Syracuse Way, Englewood CO.

Geostationary satellites ''see'' the earth, as shown in Figure 3.66. This view, as seen from the satellite, is called a hodocentric projection. Its advantage is that the cross section of the antenna beam from the satellite is reproduced, without distortion, on the hodocentric map. The beam coverage is then easily identified. The outlines of nations look different from the traditional illustrations in geography books because of the spherical shape of the earth as viewed from 35,786 km above the equator.

Figure 3.67 shows a generalized view of the earth from a geostationary satellite. This is called a hodocentric projection. Lines on the sphere represent (vertically) latitude and (horizontally) the longitude relative to the satellite's longitude. Each grid line is 5°.

Hodocentric projections are useful for determining pointing angles for antenna beams and for plotting their coverages. The area covered on the earth's surface by a circular beam is shown as a circle on a hodocentric map regardless of where on the earth the beam is pointed. Furthermore, any cross-sectional shape of a beam pattern and the shape of that beam's coverage area on a hodocentric map are identical. If conventional two-dimensional maps (such as the familiar Mercator projection) were used, a circular beam projection would be rounded (but not circular) if the beam center coincides with the subsatellite point. For any other location on the Mercator map, the beam will appear egg-shaped. The hodocentric projection may also be used to estimate the beamwidths needed to provide service

FIGURE 3.66 View of earth from geostationary satellite.

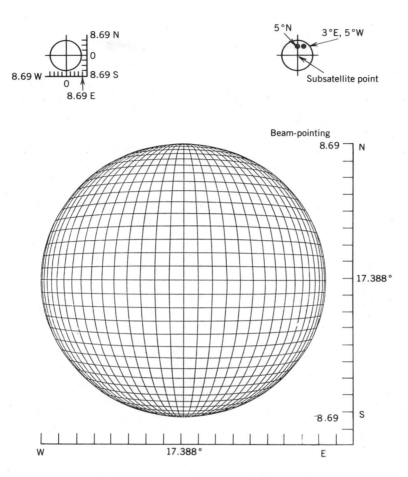

FIGURE 3.67 Hodocentric projection.

to a geographical area. In the case of spacecraft with antenna-beam-pointing insta-bilities, the coverage error may be plotted as for the extremes. These errors come from the attitude control, thermal control, and stationkeeping subsystems. An alternate method is to place the target in an error box (e.g., $0.1° \times 0.1°$).

These errors may be represented graphically in either of two manners. The first employs transparent map overlays that are moved with respect to the map to repre-sent the spacecraft errors (the earth being considered motionless). Such overlays would indicate the actual coverage area of the mispointed or displaced beam. For small angles this closely represents the true displacement or motion of the beam as seen by the satellite. Another approach is to effect the same motion except that the satellite is assumed to be the stable reference point and the locations of the earth stations are allowed to move with respect to the satellite.

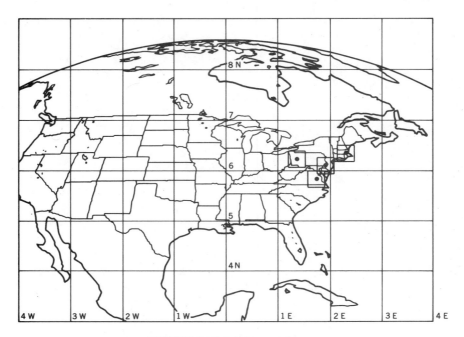

FIGURE 3.68 Pointing error boxes.

The amount of motion is the same in both cases. In the second case little error figures (typically boxes, rectangles, or circles) are added around each nominal earth station (see Figure 3.68). The illustration is for a satellite at 90° W. The antenna-pointing grid (referred to the satellite) provides the reference for the error figures (here represented as 0.3° × 0.3° boxes). The center of a narrow beam launched from the satellite and aimed at the nominal center of a box may actually intercept the earth anywhere in the box.

In the construction of the error figures the following must be considered: (1) inclination variation (north–south or latitude) produces errors in the vertical direction (perpendicular to the equatorial line); (2) east–west pointing (or longitude drift) variations produce errors in a horizontal direction (parallel to the equator); (3) if the inclination is more than a few degrees, a very thin figure eight is produced; (4) the orbit may not be circular (see the index); (5) a pitch error produces an up–down error; and (6) a roll error is a left–right motion and a yaw error is a rotation of the error figure about its center (marked with a dot). In addition, the antenna may have thermal distributions (usually related to the time at the subsatellite location) and other pointing error sources.

3.1.7 Satellite Link Delays

A radio signal travels at the speed of light ($c = 2.997925 \times 10^8$ m/s), and this determines the time delay encountered between two points. The maximum time delay from an earth station to a satellite occurs when the satellite is on the earth station's horizon (thus having the maximum slant range). For other ranges the time delay for the path from the earth station to a satellite is

$$t = \frac{1}{c} \sqrt{R^2 + r^2 - 2Rr \cos \beta_0} \quad (\text{sec}) \qquad (3.23)$$

where R = earth radius, = 6378.14 km
$\quad r = R + h$ (km), = 42,164 km
$\quad h$ = satellite altitude, typically 35,786.43 km
$\quad \beta_0$ = coverage angle centered at center of earth, see Figure 3.19
$\quad c$ = speed of light, see Table 3.29

The minimum and maximum delays from an earth station to a geostationary satellite are 119.3 msec (from the subsatellite point) and 138.9 msec (at the horizon), respectively. Figures 3.69 and 3.70 give time delays for multiple links.

Terrestrial extensions add additional delays. A CCIR–CCITT formula for these delays is

$$t_{\text{terr}} = 12 + (0.004 \times \text{distance in kilometers}) \quad (\text{msec}) \qquad (3.24)$$

The typical range for t_{terr} is 10–50 msec, with 30 being a suggested average.

Time delay is a problem in voice circuits, especially if echo is present. Echo control devices (preferably echo cancelers) are used to minimize these effects. A "half-hop" circuit is one wherein a two-way call is routed via satellite in one direction and by a terrestrial connection in the other direction. While this reduces the satellite echo, it complicates the circuit connections due to the need to use several different types of facilities. A multihop circuit is one where the signal goes from an originating earth station to a first satellite, then to another ("via") earth station, then to a second satellite, and finally to the destination earth station. The multihop cases are limited to traffic being passed from one satellite system (e.g., domestic or maritime) to another (e.g., long-haul international satellite circuits such as Buenos Aires to Tokyo via Italy), to some thin-route traffic (ship to ship in Marisat and bush services in Telesat-Canada), and very small aperture satellite terminals (VSAT) services (see Morgan and Rouffet, 1988).

TABLE 3.29 Units for Determining Time Delay

Units of Distance (S_1, S_2)	Units of Time	Units of Speed of Light (c)
Kilometers	Second	2.997925×10^5 km/sec[a]
Meter	Second	2.997925×10^8 m/sec[a]
Kilometers	Millisecond	2.997925×10^2 km/msec[a]
Foot	Millisecond	9.836×10^5 ft/msec
Nautical mile	Millisecond	1.6187×10^2 nautical miles/msec
Statute mile	Millisecond	1.8626×10^2 statute miles/msec

[a]2.997925 is usually approximated by 3.

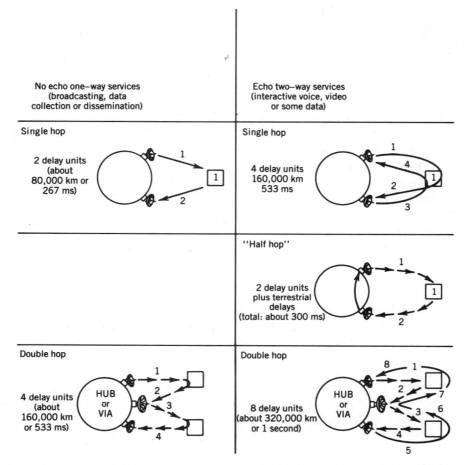

FIGURE 3.69 Delays encountered in satellite services.

Double hops often involve a single satellite. Many of these cases enable one small earth station to contact another via a large station (often called the hub). The inadequacies for small-to-small terminal service are made up by the hub, but the signals must pass through the satellite at least twice (four times for the two-way services).

A final condition is a circuit involving an intersatellite (or cross-) link. In this case the time delay(s) for the cross-link(s) must be added to the up and down links. Figure 3.71 may be used to find the cross-link delays.

Figure 3.71 shows the delay time (exclusive of t_{terr}) for links including an intersatellite link. In general, the intersatellite link delays are lower than a double hop. Beyond 162.5° the earth blocks the path between the two satellites, and the cross-link cannot be accomplished.

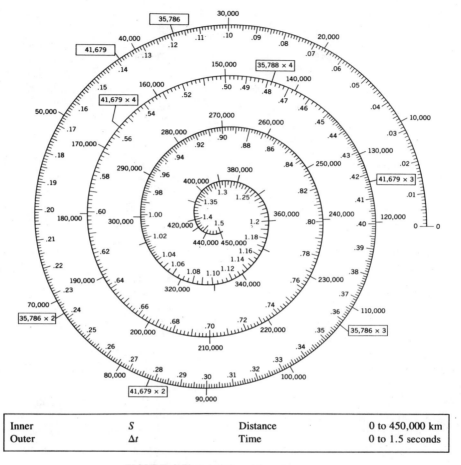

| Inner | S | Distance | 0 to 450,000 km |
| Outer | Δt | Time | 0 to 1.5 seconds |

FIGURE 3.70 Spiral time delay nomogram.

The time delay is also important in digital communications. Some early data systems required the sending terminal to retain data until the distant terminal acknowledged an error-free receipt (the ARQ method). A preferred method is to use forward error correction (FEC) coding and delay-tolerant protocols.

3.1.8 Earth Segment Resource

The experimental communications satellite systems of the 1950s and early 1960s employed a variety of frequencies, receiver and antenna types, and transmitter powers. By the late 1960s Intelsat had established its requirement for the original A standard trunk telephone earth stations to have a figure of merit (G/T_s) of at least 40.7 dBi/K at 4.000 GHz. In practical terms this meant a 30-m antenna and

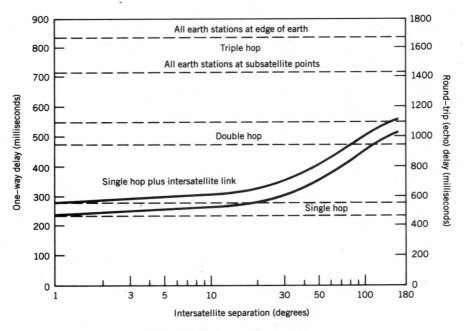

FIGURE 3.71 Intersatellite time delays.

an 80-K system noise temperature (typically a cryogenically cooled parametric amplifier with a noise temperature of about 30 K).

By the mid-1970s more powerful satellites could be built, and requirements emerged for lighter traffic stations, digital communications, and leased domestic services. In response Intelsat established performance specifications for the smaller, less expensive B class of earth stations. The G/T_s was reduced to 31.7 dBi/K at 4.000 GHz. Typical stations use 11- or 12-m antennas and a system noise temperature of 85–100 K. In 1986 Intelsat published a revised Standard A, which reduced the G/T_s to 35.0 dBi/K and permitted the use of smaller (13–18-m) antennas.

Domestic systems with the still higher satellite EIRPs are possible through the use of less-than-global beams and improvements in low-cost receivers permitted still simpler earth stations.

A quality factor Q can be defined as

$$Q = \text{EIRP} + G/T_s \quad \text{(dB)} \tag{3.25}$$

Figure 3.72 shows the value of Q for various services. For a given bandwidth, modulation, and signal-to-noise (S/N) ratio objective, trade-offs can be made along an iso-Q line. The trade most frequently made is transmitter power (EIRP) for receiver figure of merit (G/T_s). By going to a higher Q line, S/N can be improved.

As earth station antennas became smaller, the overall cost of the stations

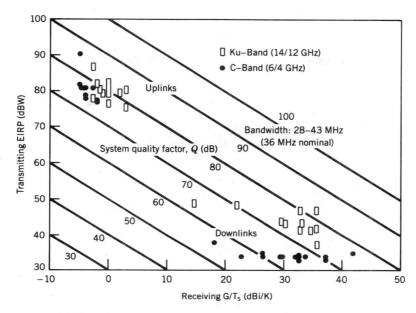

FIGURE 3.72 Relationship between earth station G/T_s and satellite EIRP.

decreased. In general, each of these smaller stations handled less traffic than their larger counterparts. In some countries it was found economical to use many smaller and decentralized stations in place of one or a few large stations.

For broadband stations in the 4- and 6-GHz bands, locations are limited by the already extensively developed terrestrial point-to-point microwave services. Many of these stations are located in natural valleys and canyons, others in man-made pits, behind buildings, and so on, where they are shielded from the terrestrial services. See Figure 3.73.

Narrow-band stations (i.e., with one or a few voice channels and often with one-way traffic) have been coordinated in city centers by using the guard band gaps in the terrestrial frequency allocations.

The trend toward smaller earth station antennas has limitations. As the antennas grow smaller, the beamwidth broadens, thus increasing the chances for interference from adjacent satellites or terrestrial services. At the same time that the earth stations were becoming more susceptible to interference, more satellites were being crowded into the orbit arc, thus adding more potential sources of interference.

3.2 SYSTEM PERFORMANCE

The total system quality is the result of a long chain of individual processes. This section examines each step.

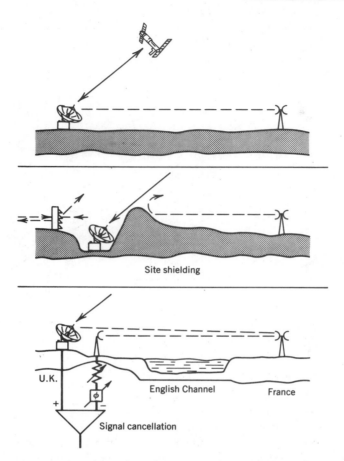

FIGURE 3.73 Terrestrial interference, site shielding and interference cancellation.

3.2.1 Antennas

Figure 3.74 shows the typical performance of an earth station antenna having a figure of merit (G/T_s) of 31.7 dBi/K at 4 GHz. The main lobe has an on-axis gain of 55.4 dBi. The dBi means decibels (dB) referenced to the gain of a theoretical isotropic (i) antenna. An isotropic antenna radiates equal power in all directions. The sun is an optical isotropic radiator because it sends the same amount of light into all parts of the universe. The curved dashed lines are two sidelobe limits $(29 - 25 \log_{10} \theta$ and $32 - 25 \log_{10} \theta)$, where θ is the off-axis angle in degrees measured at the antenna. The figure shows the result of one sweep (or "cut") across the antenna beam. Note that the two sides are more or less symmetrical and not necessarily identical. Note also the two near-in sidelobes at approximately $\pm 0.5°$ that are at about $+42$ dBi (or ~ 13.5 dB down from the peak gain). The off-axis angle in degrees is measured at the antenna. The lower set of numbers

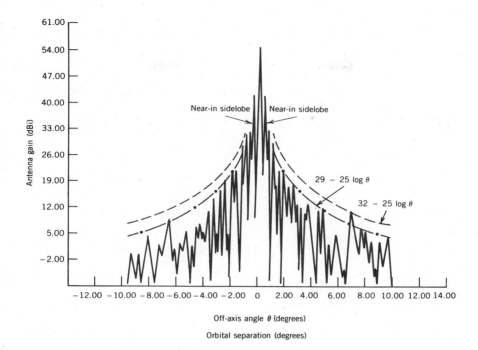

FIGURE 3.74 Typical earth station antenna pattern.

shows the geostationary orbit satellite separation angles for a typical earth station location.

Antenna dimensions are sometimes expressed in feet instead of meters. The conversion is

$$\text{Meters} = \text{feet} \times 0.30480 \qquad (3.26)$$

The conversion from frequency to wavelength is

$$\lambda = c/f \quad (\text{cm}) \qquad (3.27)$$

$$f = c/\lambda \quad (\text{GHz}) \qquad (3.28)$$

where λ = wavelength in centimeters
$\quad f$ = frequency in gigahertz
$\quad c$ = 29.979250 cmGHz (speed of light in these units)

The underlying equation for the gain ratio of an antenna referred to an isotropic radiator is

$$G = \frac{4\pi A \eta_a}{\lambda^2} \quad \text{(ratio w.r.t. isotropic)} \quad (3.29)$$

where A = antenna area in square centimeters
  η_a = antenna efficiency
  λ = wavelength in centimeters
This equation is useful for flat plate antennas.
For an alternate discussion see section 5.3.5.
For a circular antenna

$$A = \pi r^2 = \frac{\pi D^2}{4} \quad (\text{cm}^2) \quad (3.30)$$

After substituting equations 3.27 and 3.30 into 3.29:

$$G = \left(\frac{\pi D f}{c}\right)^2 \eta_a \quad \text{(ratio with respect to isotropic)} \quad (3.31)$$

where D is the antenna diameter in centimeters.
  Expressed in the more practical terms of dBi, this becomes

$$G = (20 \log_{10} \pi - 20 \log_{10} c) + 20 \log_{10} D + 20 \log_{10} f$$
$$+ 10 \log_{10} \eta_a \quad (\text{dBi}) \quad (3.32)$$

  For convenience, $20 \log_{10} \pi - 20 \log_{10} c$ and units for D and f may be combined into a constant k_a with a magnitude selected from Table 3.30:

$$G = k_a + 20 \log_{10} D + 20 \log_{10} f + 10 \log_{10} \eta_a \quad (\text{dBi}) \quad (3.33)$$

Using the preferred units, the equation becomes

$$G = 20.4 + 20 \log_{10} D + 20 \log_{10} f + 10 \log_{10} \eta_a \quad (\text{dBi}) \quad (3.34)$$

where D is in meters, f is in gigahertz, and 20.4 is k_a. This form of the equation is more convenient for users of computers and pocket calculators.

TABLE 3.30 Values for k_a

Units of Diameter (D)	Units of Frequency (f)		
	Hz	MHz	GHz[a]
Meter[a]	−159.6[b]	−39.6	20.4
Centimeter	−199.6	−79.6	−19.6
Foot	−169.9	−49.9	10.1

[a]Preferred units.
[b]$20 \log_{10}[\pi/(2.998 \times 10^8)] = -159.6$.

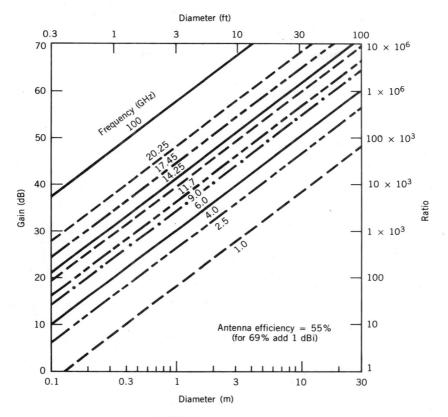

FIGURE 3.75 Antenna gain.

Figure 3.75 provides gain estimates for earth station antennas.

If a square flat plate antenna is used (where the width and length are set equal to the diameter D) approximately 2 dB more gain will be achieved due to the increased area (D^2 vs $\dfrac{\pi D^2}{4}$) and efficiency.

The antenna efficiency η_a depends on the applications. For conventionally illuminated parabolic reflectors such as used in large earth stations, a value of 0.65–0.75 (which is 65–75%) is typical. Flat plate antennas are in the 75% range. Superconductive surfaces may raise these values further. Spacecraft antennas tend to have a lower efficiency (0.4–0.55), and even lower values (0.2–0.3) may be encountered when they are used in a multibeam service. To achieve a desired gain, the multibeam antenna diameter must be slightly oversized. This also causes slight changes in the beamwidth and sidelobe structure.

Compared to the 0.55 value often used for modeling, this range represents +1 dBi (for $\eta_a = 0.7$) to −4.4 dBi (for $\eta_a = 0.2$), or −2.6 dBi (for $\eta_a = 0.3$).

For elliptical antennas the gain is still governed by Equation (3.31). The area

A of an ellipse is

$$A = \pi \left(\frac{D_1}{2}\right)\left(\frac{D_2}{2}\right) \quad (\text{cm}^2) \tag{3.35}$$

where D_1 and D_2 are the lengths of the minor and major axes.

Using the preferred units of meters and gigahertz, the equation becomes

$$G = 20.4 + 10 \log_{10} D_1 + 10 \log_{10} D_2 + 20 \log_{10} f$$

$$+ 10 \log_{10} \eta_a \quad (\text{dBi}) \tag{3.36}$$

The half-power (or -3-dB) beamwidth (θ_3) is approximately

$$\theta_3 \approx k_b/fD \tag{3.37}$$

where k_b depends on the units selected. See Table 3.31.

For elliptical beams, D may be D_1 or D_2, in which case θ_3 becomes θ_{31} or θ_{32}, the angles measured along that particular direction.

Here k_b is in reality related to the sidelobe level control and antenna efficiency and thus is not a true constant. Depending on the sidelobe levels, k_b may vary from 16.5 to 22.5 for sidelobe levels of 20–35 dB. The choice of 21 is based on current technology.

Using the preferred units of gigahertz, meters, and degrees,

$$\theta_3 = 21/fD \quad (\text{degrees}) \tag{3.38}$$

Figure 3.76 shows the relationship between the antenna diameter and the beamwidth as a function of frequency. Figure 3.77 shows the on-axis gain (for conventional antennas) versus beamwidth. It also converts degrees to milliradians (1 mrad = $0.057°$). This figure shows the trade between coverage (beamwidth) and gain. The optimum trade-off is reached when the edge of coverage is set at

TABLE 3.31 Antenna Beamwidth, Value for $k_b{}^a$

Units of Half-Power Beamwidth (θ_3)	Units of Diameter (D)	Units of Frequency (f)		
		Hz	MHz	GHz[b]
	Meter[b]	2.1×10^{10}	2.1×10^4	21
Degree	Centimeter	2.1×10^{12}	2.1×10^6	2.1×10^3
	Foot	6.9×10^{10}	6.9×10^4	69
	Meter	3.7×10^{11}	3.7×10^5	367
Milliradian	Centimeter	3.7×10^{13}	3.7×10^7	3.7×10^4
	Foot	1.2×10^{12}	1.2×10^6	1203

^{a}Approximate values for antenna efficiencies of 0.55 to 0.65

^{b}Preferred values

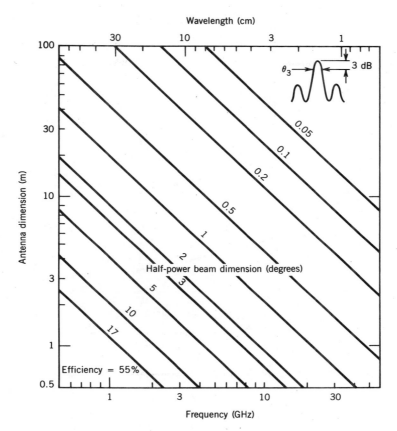

FIGURE 3.76 Beam dimensions.

the −4.35-dB (actually 10 $\log_{10} e$) point with respect to the peak gain. Figure 3.78 shows the relationship between antenna size, beamwidth, and gain.

The important parameter in many systems is the coverage requirement. This is dictated by the size of a nation, a time zone, or the region to be covered. Figure 3.79 shows the gain as a function of the transmitting (or receiving) beam size. For convenience the beam is shown in terms of square degrees, thus permitting the figure to be used for circular and elliptical beams. This figure may be used to estimate the gain of a shaped beam if the square degrees of coverage are known within the 3-dB contours. For a circular beam the cross-sectional area A_b is that of a circle:

$$A_b = \tfrac{1}{4}\pi(\theta_3)^2 \quad \text{(square degrees)} \tag{3.39}$$

where θ_3, the half-power beamwidth, is measured in degrees.

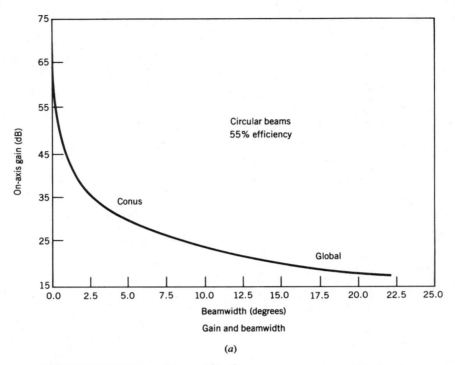

FIGURE 3.77 (a) Gain and beamwidth; (b) antenna coverage versus on-axis gain.

For an elliptical beam the beam cross-sectional area is

$$A_b = \frac{1}{4}\pi(\theta_{31} \times \theta_{32}) \quad \text{(square degrees)} \quad (3.40)$$

where θ_{31} and θ_{32} are the minor and major beam axes (in degrees), respectively. (For conversion to steradians, see Figure 3.77b.)

Note that these angles are measured from the satellite (or earth station) antenna and are *not* degrees of longitude or latitude at the earth's surface or orbital separation angles. (see also Table 3.18).

Antennas transmit and receive energy on the main beam and the many sidelobes. A typical main beam shape is shown in Figure 3.80 (which may also be used to estimate the effect of antenna-pointing losses to off-axis earth stations).

Three cases are given: The top curve is the on-axis gain reduction (this is used where there is only one satellite or earth station per beam); the lower two curves are for multiple earth stations. The half-power (−3-dB) contour is provided.

For narrow satellite antenna beams the coverage (in square kilometers) increases

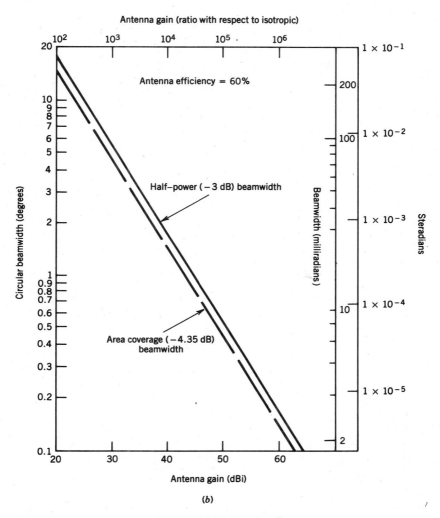

(b)

FIGURE 3.77 (*Continued*)

substantially as the beam center is moved away from the subsatellite point. Figure 3.81 shows that the area covered by a circular beam of a satellite antenna varies from a circle (when aimed at the subsatellite point) to an egg-shaped oval. The orientation of the oval is such that its long axis points to the subsatellite point. At low elevation angles a portion of the beam may miss the earth as it passes over the horizon.

Figure 3.82 shows two presentations of beam clustering. The choice of the map projection is important. In the top figure the satellite view is provided.

The gain produced by an antenna depends on its root-mean-square (rms) surface

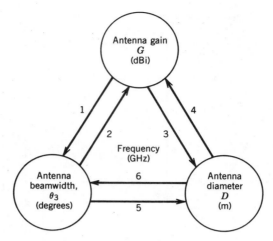

1     $\theta_3 = 10^{[(46.8 - G + 10\log_{10}\eta_a)/20]}$ (degrees)

2     $G = 46.8 - 20 \log_{10} \theta_3 + 10 \log_{10} \eta_a$ (dBi)

3     $D = 10^{[(G - 20.4 - 20\log_{10}f - 10\log_{10}\eta_a)/20]}$ (m)

4     $G = 20.4 + 20 \log_{10} D + 20 \log_{10} f + 10 \log_{10} \eta_a$ (dBi)

5     $D = 20.94/(f\theta_3)$ (m)

6     $\theta_3 = 20.94/(fD)$ (degrees)

Note: the "constants" used here are representative of typical antennas, small variations exist in practice.

FIGURE 3.78 Interrelationship of basic antenna equations.

tolerance e:

$$G = \eta_a \left(\frac{\pi D}{\lambda}\right)^2 \exp\left[-\left(\frac{-4\pi e}{\lambda}\right)^2\right] \quad (\text{ratio}) \qquad (3.41)$$

Both e and λ must be in the same units (generally centimeters). (see Figure 3.83). These tolerances pose limits to the achievable gain that can physically be realized with economic structures (Figure 3.84). The maximum gain is

$$G = \frac{\eta_a}{43}\left(\frac{D}{e}\right)^2 \quad (\text{ratio}) \qquad (3.42)$$

The CCIR requirements for antennas where the ratio of the diameter to the operating wavelength is greater than 100 (see Figure 3.85) specify that the sidelobe gain be confined with an envelope defined by the function $32 - 25 \log_{10} \theta$ (dBi). Where θ is the off axis angle in degrees. For ratios less than 100, the allowable gain is $52 - 10 \log_{10} (D/\lambda) - 25 \log_{10}\theta$. In each case $G = -10$ dBi for 48°–

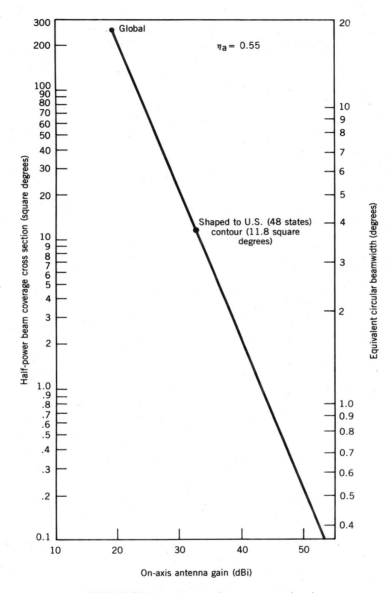

FIGURE 3.79 Beam cross section versus on-axis gain.

285

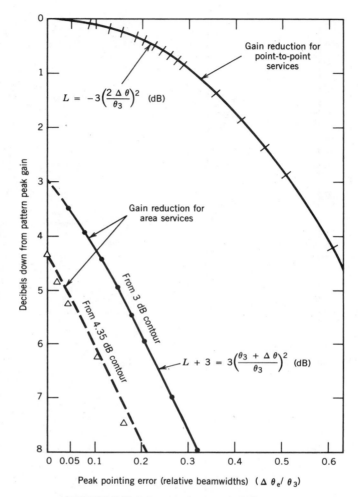

FIGURE 3.80 Influence of antenna pointing errors.

$180°$. The FCC requires that some antennas meet the $29 - 25 \log_{10} \theta$ contour. These contours are shown in Figure 3.74. For further information consult Maral and Bousquet (1986).

Figure 3.86 portrays the off-axis gain loss for four antennas meeting the specifications.

For a geostationary satellite $1.1°$ of off-axis angle from an earth station corresponds to approximately $1°$ of orbital separation. For example, assume a 4.5-m antenna is being used at 4 GHz (gain = 43 dBi on axis). If an adjacent satellite is two orbit degrees away from the satellite being worked (or $2.2°$, as seen by the earth station), the FCC pattern's sidelobe gain may be as high as $29 - 25 \log_{10}$

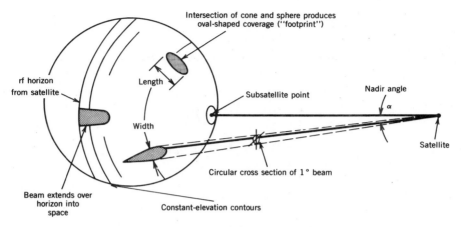

FIGURE 3.81 Antenna footprint geometry.

2.2, or 20.4 dBi. The undesired satellite signal will be only 22.6 dB (i.e., 43 − 20.4 = 22.6) below the desired signal, assuming that the adjacent satellite has an identical EIRP and modulation, and there are no other satellites. Figure 3.87 shows a case where there are three satellites.

In another example, assume adjacent and identical satellites at −6°, −4°, −2°, +2°, +4°, and +6° along the orbit arc. Table 3.32 shows the antenna gains. The combined entry carrier-to-interference ratio (C/I) assumes all satellites' signals overlap. This is the power sum (not the arithmetic sum) of the interference into the desired carrier. The power sum is designated by the plus sign in a circle (see Section 3.2.6). The two immediately adjacent satellites produce a combined C/I of 19.6 dB. The combination of the closest four interfering satellites reduce the combined C/I to 18.9 dB. The combination of the six satellites drops the C/I only another 0.3 dB (to 18.6 dB). Satellites still farther away have even less effect.

In an actual situation the interfering satellites may have different power (EIRP) levels on the station and use various modulation methods. It is also unlikely that *exactly* the same frequency plans will be used.

Sometimes it is desirable to use elliptical earth station beams. The gain equation [(3.35)] and Figures 3.76 and 3.88 are usable for both earth station and satellite elliptical beams. Observe that the major axis of the physical antenna dish corresponds to the beam's minor cross-sectional axis, and vice versa. By orienting the earth station antenna properly, the minor beam axis can be aligned with the orbit arc. Since all the interference comes from satellites along this arc, this orientation causes the antenna gain to drop fastest in those directions. The orientation (with respect to the horizon) varies along the arc. A polar antenna mount will keep the minor beam axis properly aligned as the antenna is moved from one satellite to another.

Shaped and weighted pattern beams may also be generated using a more complex

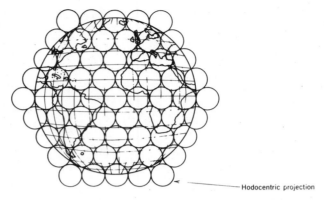

Hodocentric projection

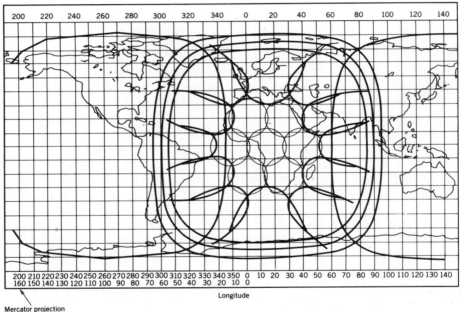

Mercator projection

FIGURE 3.82 Two different examples of global beam clustering. © Copyright American Institute of Aeronautics and Astronautics; reprinted with permission from the 7th AIAA Communications Satellite Systems Conference.

feed system. In these cases the pattern deviates sufficiently from the conventional patterns as to make use of the general parabolic gain and beamwidth equations unwise.

In weighted patterns the gain is usually specified at a series of cities (often at the edge of the coverage area) and not by the peak gain.

Table 3.33 illustrates several complex antenna feed systems and shows the important characteristics of each type. (See also Figure 5.33)

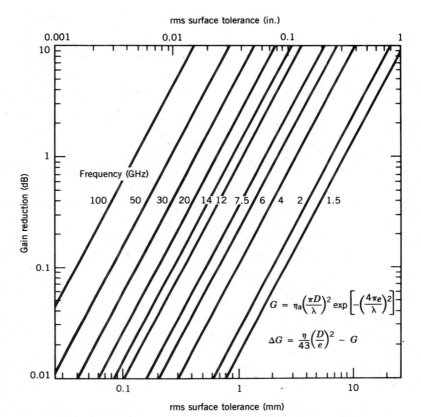

FIGURE 3.83 Gain reduction due to antenna surface tolerance.

The formulas shown in the figure are:

$$G = \eta_a \left(\frac{\pi D}{\lambda}\right)^2 \exp\left[-\left(\frac{4\pi e}{\lambda}\right)^2\right]$$

$$\Delta G = \frac{\eta}{43}\left(\frac{D}{e}\right)^2 - G$$

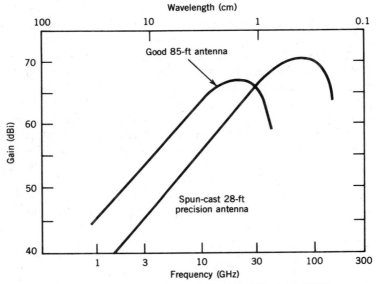

FIGURE 3.84 Antenna limitations. © 1966 IEEE, Ruze (1966).

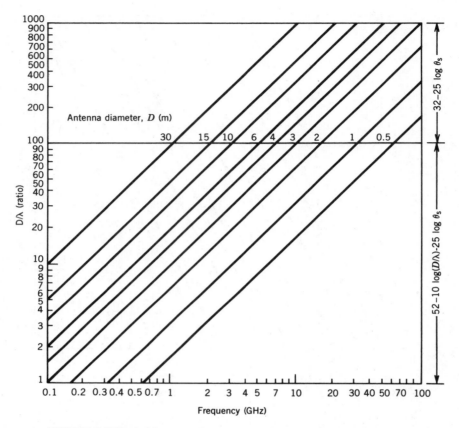

FIGURE 3.85 Ratio of antenna diameter to wavelength (D/λ) versus frequency.

Figures 3.89 and 3.90 show various forms of coverage. Figures 3.91 and 3.92 show that the population distribution is not uniform and therefore suggest weighted antenna patterns or many small spot beams.

Figures 3.93–3.99 show how the coverage becomes increasingly localized as a circular beam is made smaller. Figure 3.93 depicts an overall view of the geographic relation between a satellite and the coverage areas shown in Figures 3.94–3.99. Table 3.34 lists the cities covered by each beam size. The beam center has been optimized in each case for maximum population coverage. This is the most densely populated region of the United States.

The theoretical gain of an *ideal* antenna with the area A of 1 m^2 is used in illumination level and power flux density measurements. The gain of such an antenna is

$$G = 4\pi A\eta_a/\lambda^2 \qquad (\text{ratio w.r.t. isotropic}) \qquad (3.43)$$

where η_a = efficiency, =1.0 (100%) for this *ideal* case.
   λ = wavelength, =c/f

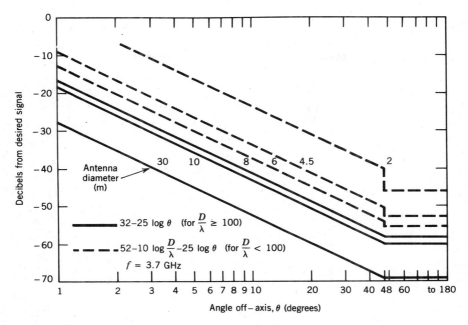

FIGURE 3.86 Off-axis gain for antennas meeting specifications.

Thus,

$$G_{1m^2} = 4\pi/\lambda^2 \quad \text{(ratio)} \tag{3.44}$$

Figure 3.100 shows this theoretical gain over a range of frequencies.

A 1-m² circle has the diameter of 1.128 m. This may be used with the previous gain equations requiring D, but remember to set η_a at 1.0 and $10 \log_{10} 1.0$ equals zero.

3.2.2 Receiving Equipment Parameters

The equipment aspects of receivers is discussed in section 5.3.2. The figure of merit (G/T_s) of an earth station's receiving equipment is stated as the ratio of the antenna gain G to the system noise temperature T_s. The unit of the figure of merit is dBi/K.

$$G/T_s = 10 \log_{10}(\text{antenna gain ratio})$$

$$- 10 \log_{10}(\text{system noise temperature}) \quad \text{(dBi/K)}$$

$$= G - 10 \log_{10} T_s \quad \text{(dBi/K)} \tag{3.45}$$

where G is in dBi and T_s is in kelvin. Absolute zero is 0 K. Room temperature is ~290 K, or +16.8°C or +62°F.

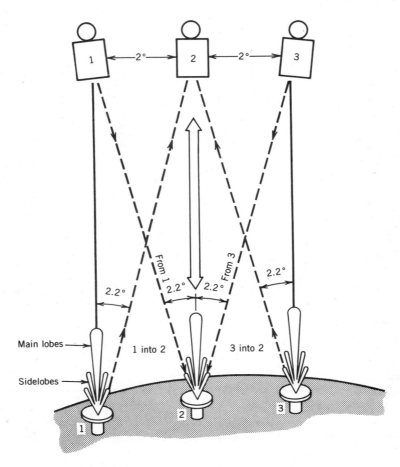

FIGURE 3.87 Sources of interference in satellite systems.

The antenna gain was derived in section 3.2.1 and 5.3.5. The system noise temperature is a weighted composite of the antenna (sky) noise consisting of the following:

1. cosmic background noise at rf is approximately 2.78 K;
2. galactic noise;
3. noise temperature due to precipitation in the path;
4. solar noise (either in the main lobe or a sidelobe);
5. presence of the earth (typically at 290 K) in a sidelobe;
6. contribution of nearby objects, buildings, radomes, and so on, and
7. temperature of blockage items in the antenna subsystems such as booms, feeds, sidelobe absorbers, and so on.

TABLE 3.32 Interference from Adjacent Satellites[a]

Satellite Number	Orbital Separation between Satellites (degrees)	Earth Station Off-Axis Angle (degrees)	Antenna Gain (dBi)	Gain with Respect to On-Axis (dBi)	Carrier-to-Interference Ratio			
					Single Entry dB	Multiple-Entry Combinations		
						dB	dB	dB
1	−6	−6.6	8.5	−34.5	34.5			
2	−4	−4.4	12.9	−30.1	30.1			
3	−2	−2.2	20.4	−22.6	22.6			
4 (desired)	0	0	43	0	n.a.	19.6	18.9	18.6
5	+2	+2.2	20.4	−22.6	22.6			
6	+4	+4.4	12.9	−30.1	30.1			
7	+6	+6.6	8.5	−34.5	34.5			

[a]Assumptions: Antenna patterns meet $29 - 25 \log_{10} \theta$. Identical power and modulation in all satellites.

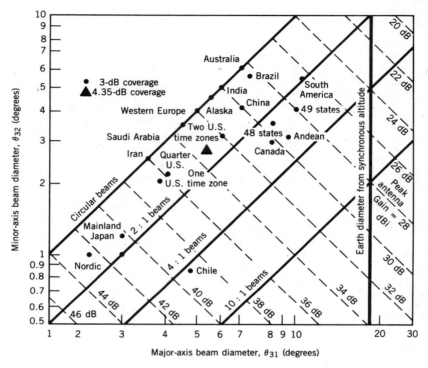

FIGURE 3.88 Antenna coverage and peak gain for area.

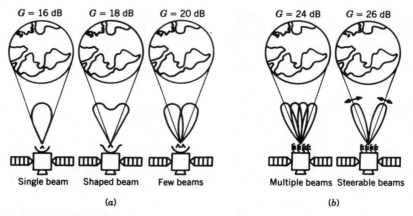

FIGURE 3.89 Global coverage beams and gains: (*a*) reflector systems; (*b*) array systems. Aasted and Roederer (1978). © Copyright American Institute of Aeronautics and Astronautics; reprinted with permission from the 7th AIAA Communications Satellite Systems Conference.

TABLE 3.33 Various Complex Antenna Feed Systems

Antenna Type	Phased Array	Lens Antennas		Single-Reflector Antennas		Double-Reflector Antennas	
		Dielectric Lens	Bootlace	On-Axis Feed	Offset Feed	Cassegrain	Bifocal Aplanatic
Sidelobe level	Up to −35 dB	Up to −20 dB	Up to −35 dB	Up to −20 dB	Up to −25 dB	Up to −20 dB	Up to −30 dB
Losses	High	Low	High	Low	Low	Low	Low
Number of beams	Any	Limited	Any	Limited	Limited	Limited	Any
Weight	Large	Large	Problematic	Small	—	Small	Small
Size	Acceptable	Large focal distance	Acceptable	Large focal distance	—	Large focal distance	Acceptable
Cost	High	Low	High	Low	—	Low	Acceptable

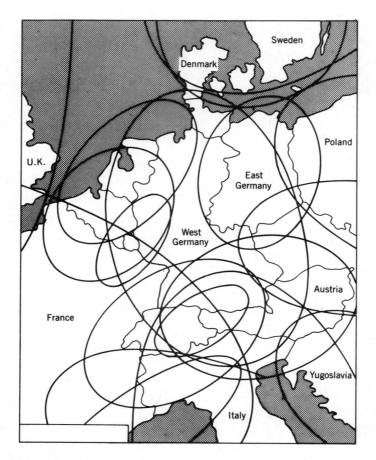

FIGURE 3.90 WARC-77 national coverage patterns for Central Europe. (1978) © Copyright American Institute of Aeronatuics and Astronautics; reprinted with permission from the 7th AIAA Communications Satellite Systems Conference.

The usual temperature seen by a perfect earth station antenna is that of the sky. The clear sky temperature is frequency dependent, as shown in Figures 3.101 and 3.102. The first figure provides information on the cosmic and galactic noise temperature (the maximum is in the direction of the Milky Way). The tropospheric noise temperature is shown in Figure 3.102. The total noise level varies as the antenna pointing is changed (see Figure 3.103).

A practical antenna includes noise contributions from the surroundings, as exaggerated in Figure 3.104.

The antenna pattern (see Figure 3.105 for typical performance) consists of the main lobe (1), the sidelobes (2), and the backlobes (3).

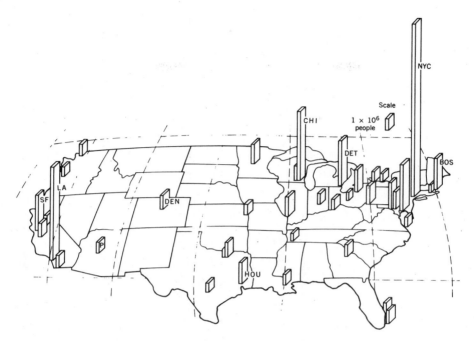

FIGURE 3.91 1984 populations for top 25 urbanized areas. Hodocentric projection from 90° W.

Any loss mechanism produces a noise temperature according to

$$T_{\text{atten}} = (1 - a_{\text{atten}})T_m \quad (\text{K}) \quad (3.46)$$

where a_{atten} = attenuation of loss mechanism (decimal ratio between 0 and 1)
T_m = ambient temperature of loss mechanism (290 K for terrestrial objects)

Then,

$$T_{\text{atten}} = T_m[1 - 10^{-(L_{\text{atten}}/10)}] \quad (\text{K}) \quad (3.47)$$

where L_{atten} is the loss in decibels. Figures 3.106 and 3.107 show this relationship.

In the case of antenna temperatures it has become common to talk in terms of a clear-sky temperature (no precipitation in the path from the satellite). At frequencies above 10 GHz a system noise temperature under specified precipitation (rainfall) conditions should also be estimated. In these cases the precipitation attenuation L_{rain} causes an increase in the sky temperature, as shown in Figure 3.107. The limit condition (for *very* high rain rates, as in a monsoon) is between 270 and 290 K. It is important to recognize that the precipitation has *two* effects on the signals passing through: (1) the attenuation (denoted as L_{rain}) that results in

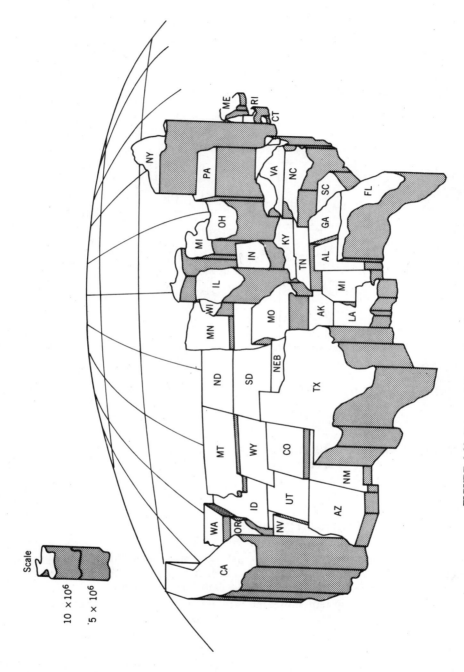

FIGURE 3.92 State populations are seen from geostationary satellite at 90° W.

Scale

10×10^6

5×10^6

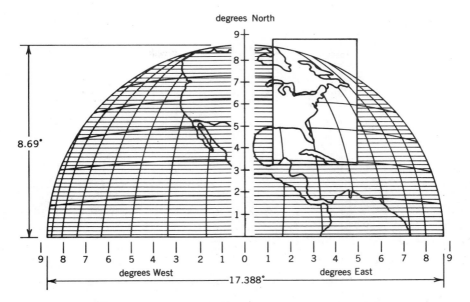

FIGURE 3.93 Satellite-based coordinate systems for antenna pointing.

lowering the signal level and (2) the increase in noise temperature (T_{atten}) due to the increased sky noise which lowers the $(G/T_s)_d$.

Precipitation may occur in the form of fog, rain, snow, or hail. Of these only the first two (fog and especially heavy rain) are of concern. For frequencies below 20 GHz only rain is usually considered. Sleet sometimes starts as rain at high cloud altitudes. If melting takes place aloft, attenuation (and thus noise) may result. Most snows do not produce appreciable attenuation (unless it accumulates on the antenna surface, thus creating a new reflective surface that distorts the antenna pattern and gain). Hail usually does not present a problem. Icing of the antenna can produce gain loss (which, though not an increase in noise temperature, reduces the gain G in the figure of merit G/T_s).

The antenna noise (as measured at the antenna output flange) is the temperature noise distribution over all antenna angles taking gain into account. This relationship is shown empirically in Figure 3.104 and Tables 3.35 and 3.36.

The antenna noise temperatures of two antenna sizes are shown in Figures 3.108 and 3.109.

The antenna noise temperature may rise substantially if the antenna main lobe is intersected by a hot object such as the moon or especially the sun.

Figure 3.110 shows the conditions when an earth station antenna is pointed directly at the sun. This is the worst possible condition. A small antenna has a wider beamwidth, and thus, the 0.5° sun diameter can occupy only a small fraction of the beamwidth. The sun is surrounded by cold (2.78 K) space which dilutes the effect of the sun. As the antenna diameter increases, the beamwidth decreases.

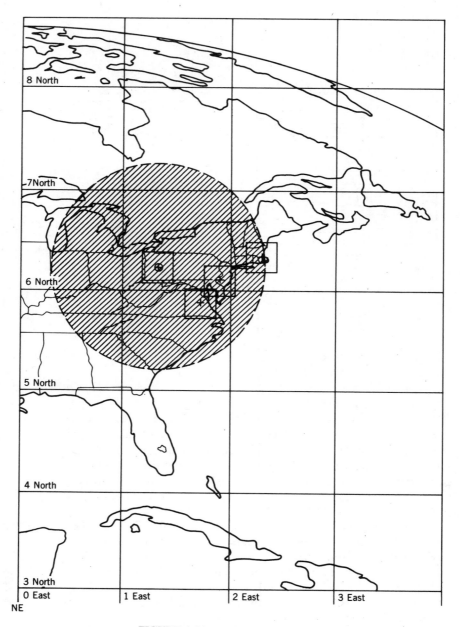

FIGURE 3.94 2° beam on East Coast.

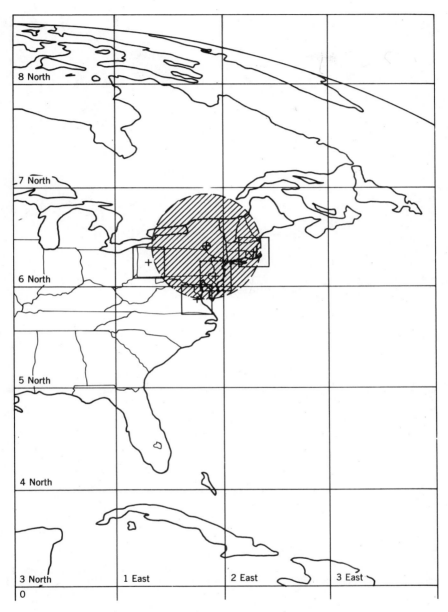

FIGURE 3.95 1° beam on East Coast.

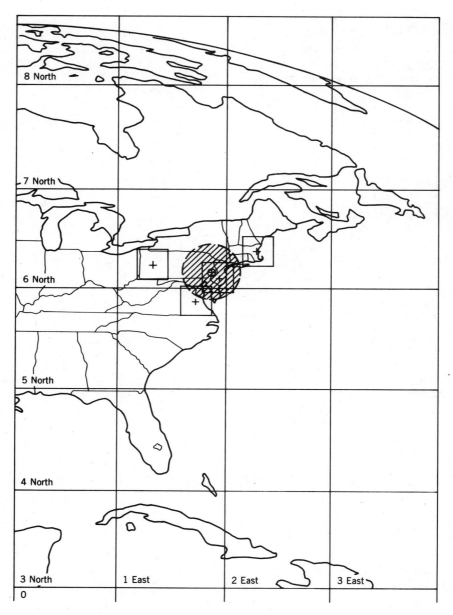

FIGURE 3.96 0.5° beam on East Coast.

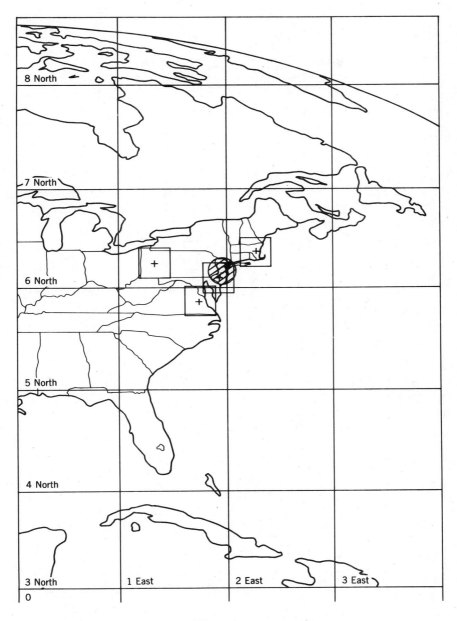

FIGURE 3.97 0.2° beam on East Coast.

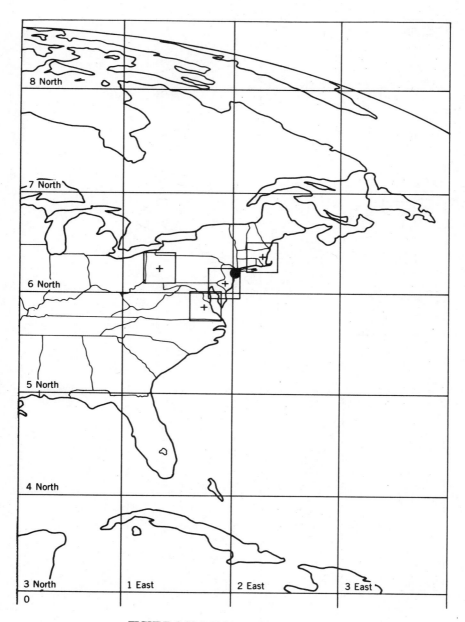

FIGURE 3.98 0.1° beam on East Coast.

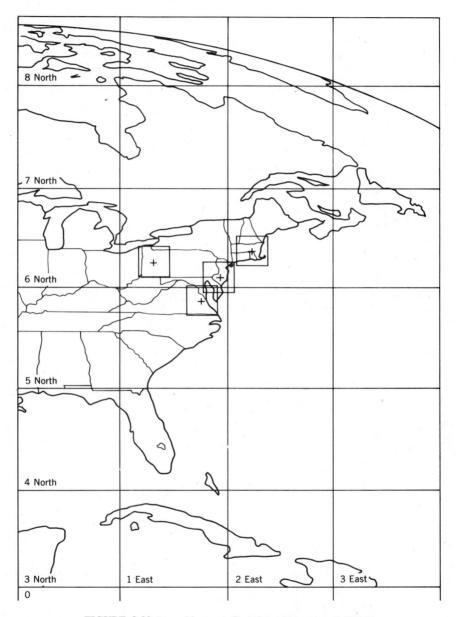

FIGURE 3.99 1-mrad beam on East Coast (1 mrad = 0.0573°).

TABLE 3.34 Eastern U.S. City Coverage

	Beam Diameter			
	2°	1°	0.5°	0.2°
		City near the Beam Center		
City	Pittsburgh, PA	Binghamton, NY	Philadelphia, PA	Newark, NJ
New York, NY	X	X	X	X
Newark, NJ	X	X	X	X
White Plains, NY	X	X	X	X
Trenton, NJ	X	X	X	X
Philadelphia, PA	X	X	X	X
Allentown, PA	X	X	X	X
Baltimore, MD	X	X	X	
Washington, DC	X	X	X	
Hartford, CT	X	X		
Pittsburgh, PA	X			
Buffalo, NY	X	X		
Albany, NY	X	X		
Providence, RI	X	X		
Boston, MA	X	X		
Cleveland, OH	X	X		
Cincinnati, OH	X			
Chicago, IL	X			
Richmond, VA	X			
Detroit, MI	X			
Indianapolis, IN	X			
Nashville, TN	X			
Columbia, SC	X			
Atlanta, GA	X			

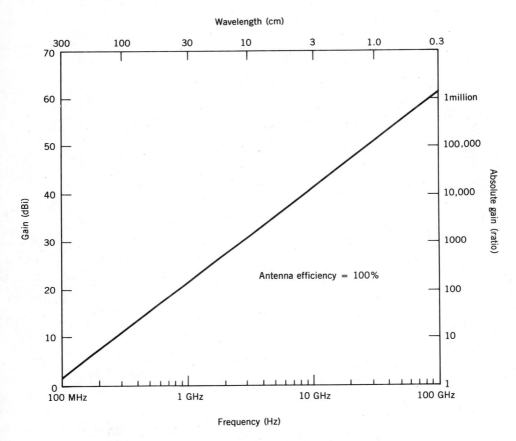

FIGURE 3.100 Gain of 1-m² antenna.

A 4-m, 12-GHz antenna "sees" the full sun disk. The solar noise temperature varies considerably with the sunspot number, and peaks of several orders of magnitude higher may be experienced during solar flares. Figure 3.111 indicates the solar noise temperatures at sunspot maxima and minima as a function of frequency. It also shows those conditions where there can be a transitory problem.

The sky noise temperature may rise from a normal 50 K to 10,000–30,000 K (a change of 23–28 dB). As the sun moves off axis, the sidelobe attenuation quickly reduces the amount of solar noise received by the receiver. When the sun passes through the main lobe of a large antenna, the G/T_s degradation is so severe that it generally and briefly renders the station useless. Spread spectrum earth stations often have enough margin and the capability to spread the solar noise (Morgan and Rouffet, 1988). This event is seasonal, very predictable as to timing, and afflicts only one city at a time (different earth stations have different view angles); thus,

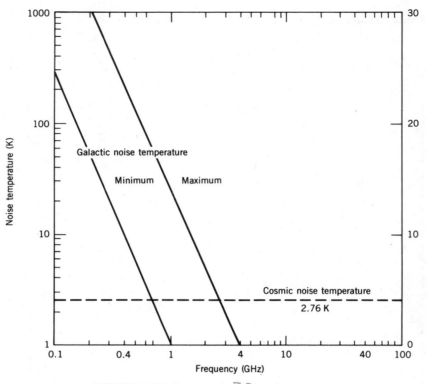

FIGURE 3.101 Cosmic and galactic noise temperatures.

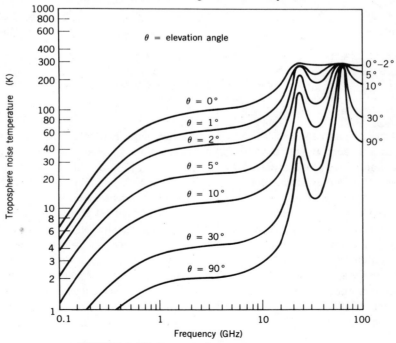

FIGURE 3.102 Tropospheric noise temperatures.

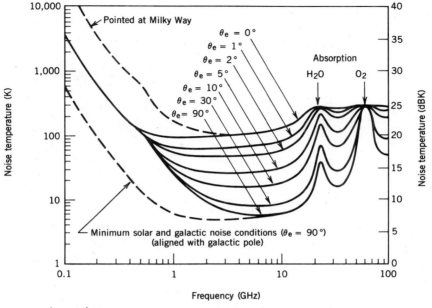

FIGURE 3.103 Earth station antenna noise temperatures.

the outage will occur at different times in different cities. The sun moves at an apparent rate of $15°$ hr^{-1} ($0.25°$ min^{-1} of time), and thus the event is brief.

The steradian is a convenient unit to use in determining noise contributions. Table 3.37 gives the cone angles subtended from a geostationary satellite (or, in the cases of the sun and moon, and the earth station). There are 4π (or 12.566) steradians in a sphere. Thus, from an earth station the sun occupies (6.791×10^{-5})/12.566 or 5.4 millionth's of the total spherical area.

3.2.2.1 Transmission Line Noise Temperature

The transmission line that connects the antenna to the receiver (or preamplifier) also contributes to the system noise temperature due to the losses, as indicated in Figure 3.112.

The sum of the losses in the waveguide, coaxial cable, joints, connectors, and directional couplers may be 0.25 dB, which contributes about 15 K.

The combination of the antenna and the transmission line produces a noise

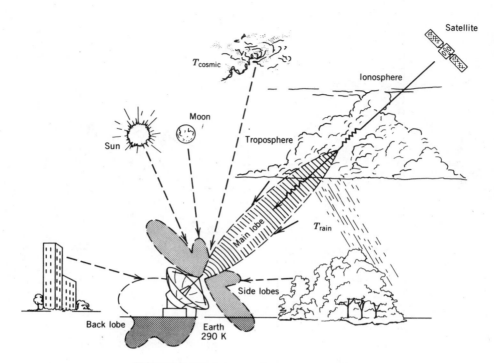

FIGURE 3.104 Antenna noise temperature pickup.

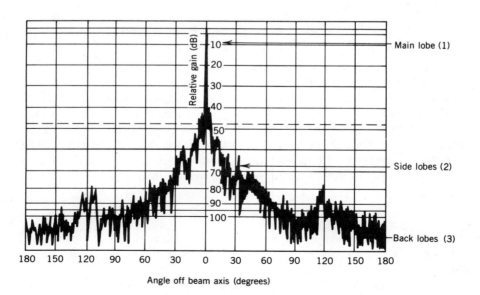

FIGURE 3.105 Typical antenna pattern in one plane.

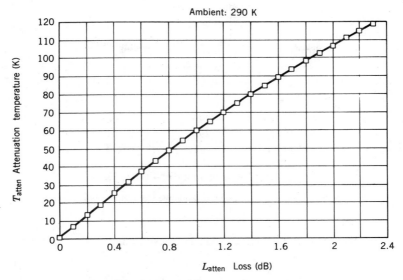

FIGURE 3.106 Loss element noise temperature.

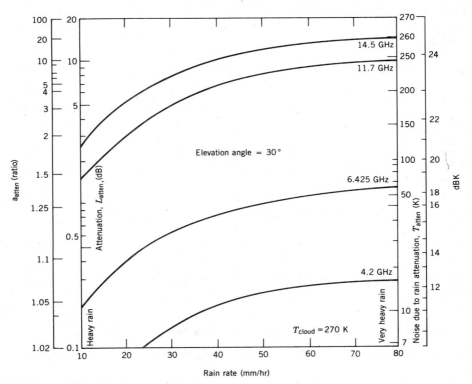

FIGURE 3.107 Sky noise increase due to attenuation in heavy and very heavy rainfall.

TABLE 3.35 Effective Clear-Sky Antenna Temperature

Element	Lobe	Temperature (K)	Weighting[a] Factor	Weighted Temperature (K)
Atmospheric absorption (tropospheric) noise	Main	5	0.8	4
Earth and objects	Side	290	0.05	14.5
	Back	290	0.1	29
Galactic noise	Main	3	0.8	2.4
	Side	3	0.05	0.15
	Effective temperature at 4 GHz, T_{ant}:			50 K

[a]Weighting is based on the integrated antenna gain towards the element at a low elevation angle.

TABLE 3.36 Effective Antenna Temperature in Heavy Rain

Element	Lobe	Temperature (K)	Weighting Factor	Weighted Temperature (K)
Atmospheric precipitation noise due to heavy rain (40 mm/hr), at 12 GHz[a]	Main	215	1	215
Earth and objects[b]	Side	290	0.05	14.5
	Back	290	0.1	29
Galactic noise[b]	Main	3	0.8	2.4
	Side	3	0.05	0.15
Effective temperature at 12 GHz, T_{ant}				261 K

[a]From Figure 3.107
[b]From Table 3.35

temperature at the receiver input flange:

$$T = (T_{ant}/a_r) + T_R[(a_r - 1)/a_r] \quad (K) \quad (3.48)$$

The value for a_r can be found in Table 3.38, and T_R is the reference temperature (usually 290 K).

3.2.2.2 Amplifier Noise Temperature

The active equipment (defined as having a gain greater than unity) is the final portion of the total system noise temperature.

The noise temperature of the receiving equipment shown in Figure 3.113

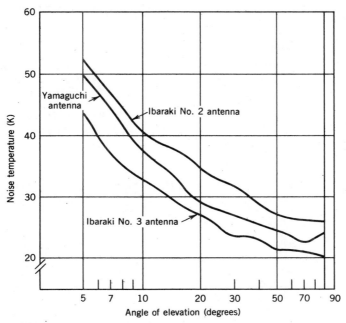

FIGURE 3.108 Antenna sky noise temperature versus elevation angle for large earth station antennas ($D = 30$ m). (Miya, 1975)

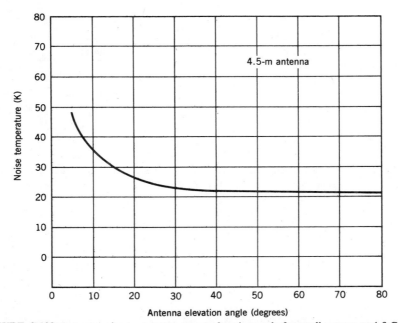

FIGURE 3.109 Antenna noise temperature versus elevation angle for small antenna at 4.0 GHz.

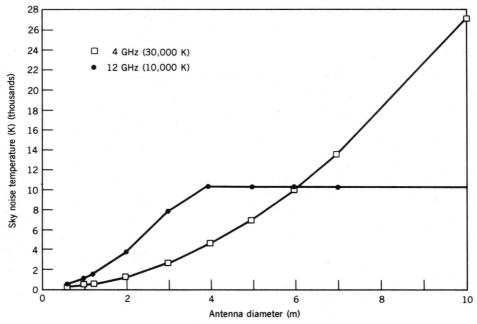

FIGURE 3.110 Noise temperature during sun transit at time antenna is pointed at sun.

TABLE 3.37 Angular Sizes of Earth, Moon, and Sun

Object	Symbol	Typical Angle Subtended from Geostationary Orbit			
		Degrees	Arc	mrad	sr
Earth		17.4	17°23'14"	303	7.21×10^{-2}
Moon[a]		0.52	0°31'5"	9.04	6.42×10^{-5}
Sun[a]		0.53	0°31'58"	9.30	6.79×10^{-5}

[a]These values may also be used for earth station antennas.

including the low noise section is represented by

$$T_{rx} = T_{A1} + \frac{T_{A2}}{G_1} + \frac{T_{A3}}{G_1 G_2} + \frac{T_{final}}{G_1 G_2 G_3} \quad (\text{K}) \qquad (3.49)$$

If G_1, G_2, and G_3 are 11 dB each (12.6:1), and T_{A1} is 30 K, T_{A2} and T_{A3} are both 100 K, and T_{final} is a 2000 K driver transistor, then the value for the entire

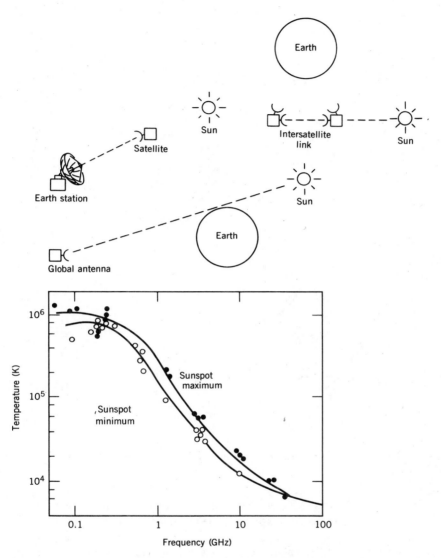

FIGURE 3.111 Solar outage and noise temperature.

receiver (T_{rx}) is:

$$T_{\text{rx}} = 30 + \left(\frac{100}{12.6}\right) + \left(\frac{100}{12.6 \times 12.6}\right) + \left(\frac{2000}{12.6 \times 12.6 \times 12.6}\right)$$

$$= 30 + 7.9 + 0.6 + 1.0$$

$$= 40 \text{ Kelvin} \tag{3.50}$$

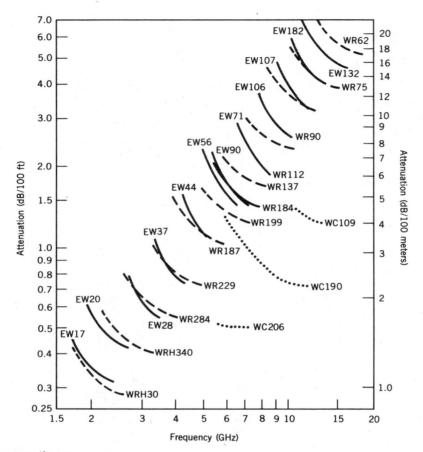

FIGURE 3.112 Waveguide attenuation. Attenuation curves based on: VSWR, 1.0; ambient temperature, 290K.

As shown by the example, the input stage predominates in T_{rx}. Figure 3.114 shows typical noise temperatures for various rf amplifiers and mixers.

Instead of noise temperatures, many amplifiers and converters are rated in terms of their noise figure (NF). The noise figure is usually expressed in dBNF and may be converted into Kelvin as follows:

$$F = 10^{(dBNF/10)} \quad (\text{ratio}) \tag{3.51}$$

$$T_{rx} = T_R(F - 1) \quad (\text{Kelvin}) \tag{3.52}$$

where T_R is the 290 K reference temperature, F is the absolute value of the receiver's noise factor, and T_R is as defined for Equation (3.48); for convenience use Figure 3.115 to convert the F to K (or dB referenced to 1 K, which is referred to as dBK).

**TABLE 3.38 Line Noise Temperature versus Line
Attenuation**

Line Attenuation[a] L_{line} (dB)	a_r	T_{line} (K)[b]
0.05	1.01	3.3
0.10	1.02	6.6
0.20	1.05	13.1
0.22	1.05	14.3
0.25	1.06	16.2
0.30	1.07	19.4
0.50	1.12	31.5
0.70	1.17	43.2
1.00	1.26	59.6
2.00	1.58	107.0
3.00	2.00	144.7

[a]Including joints, directional couplers, and connectors.
[b]$T_R = 290$ K

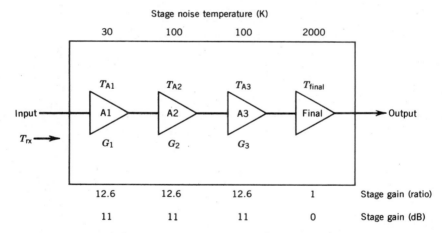

FIGURE 3.113 Receiver noise temperature components.

$$T_{rx} = 10 \log_{10}\left[T_R(F - 1) \right]$$
$$= 24.624 + 10 \log_{10}(F - 1) \quad \text{(dBK)} \qquad (3.53)$$

3.2.2.3 Overall Figure of Merit G/T_s

The figure of merit of an earth station is its G/T_s. The antenna gain (G in dBi) is available from Section 3.2.1, and T_s is the sum of the noise temperatures calculated in the preceding. Figure 3.116 shows the G/T_s versus T_s for several antennas.

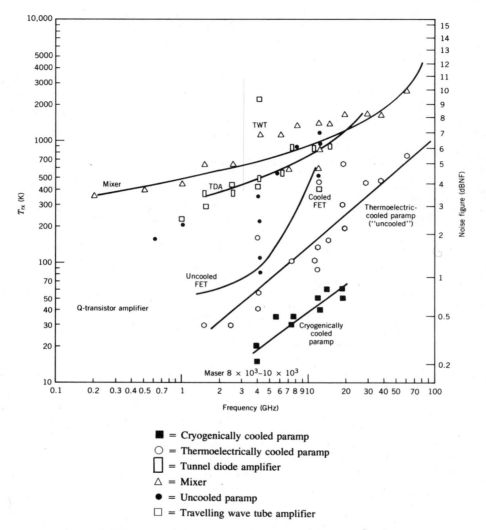

FIGURE 3.114 Noise temperature for rf amplifiers and mixers.

The system noise temperature T_s may also be expressed in terms of dBK and subtracted directly from G to obtain G/T_s.

Figure 3.117 depicts the steps to determine the system noise temperature T_s. Also indicated, in tabular form, are the system elements that contribute to the overall noise temperature, along with the sources for calculating their respective noise figures. Table 3.39 provides typical values of system and element noise temperatures.

Figures 3.118–3.121 show the combination of T_s and antenna diameter required

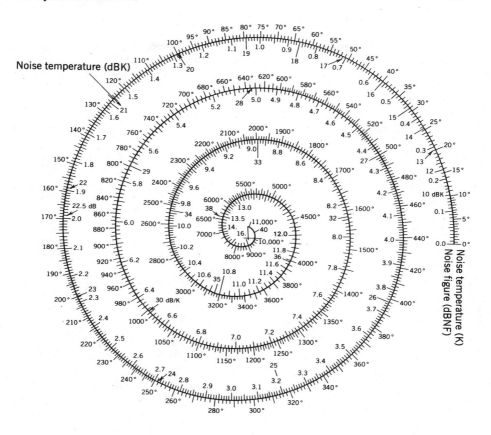

Noise temperature (dBK)

Scale	Element	Range	Units
Outer	Noise temperature	0–11,000	K
Inner	Noise figure	0–16	dBNF
Inner	Noise temperature	0–40.4	dBK

FIGURE 3.115 Spiral noise figure to noise temperature chart.

at four selected frequencies to achieve a desired G/T_s. Large antennas tend to be an expensive choice in obtaining the desired G/T_s. Small antennas, while least expensive, may require a good low-noise amplifier. The smaller antennas have the broader beams and thus may pick up interference from adjacent satellites or terrestrial transmitters.

For many applications a satellite transponder is served by the antenna and receiving amplifier combination shown in Figure 3.122.

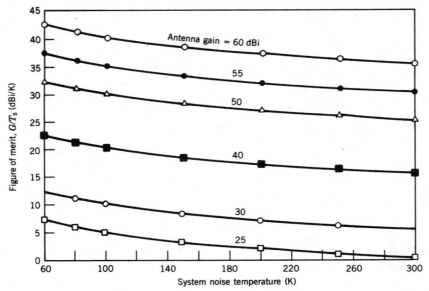

FIGURE 3.116 G/T_s versus T_s for several antennas.

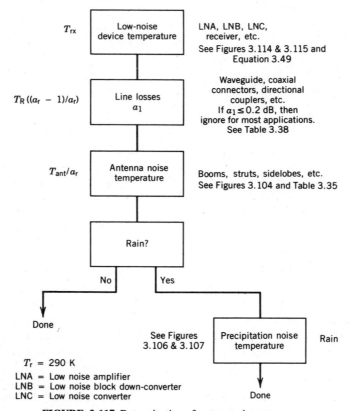

T_{rx} — Low-noise device temperature — LNA, LNB, LNC, receiver, etc. See Figures 3.114 & 3.115 and Equation 3.49

$T_R((a_r - 1)/a_r)$ — Line losses a_1 — Waveguide, coaxial connectors, directional couplers, etc. If $a_1 \leq 0.2$ dB, then ignore for most applications. See Table 3.38

T_{ant}/a_r — Antenna noise temperature — Booms, struts, sidelobes, etc. See Figures 3.104 and Table 3.35

Rain?

No — Done

Yes

Precipitation noise temperature — Rain

See Figures 3.106 & 3.107

Done

$T_r = 290$ K
LNA = Low noise amplifier
LNB = Low noise block down-converter
LNC = Low noise converter

FIGURE 3.117 Determination of system noise temperature.

TABLE 3.39 Typical Values of System and Element Temperature

Frequency (GHz)	Sky Condition	T_{ant} (K)	T_{atten} (K)	Line Loss (dB)	T_{line} (K)	Amplifier Type	T_{rx} (K)	T_s (K)	dBK
0.7	n.a.	30	Negligible	0.5	35	Transistor	200	265	24.2
1.5	n.a.	30	Negligible	0.2	14	Transistor	300	344	25.4
2.5	n.a.	45	Negligible	0.1	7	Transistor	400	452	26.6
4	n.a.	25	Negligible	0.05	3	Maser	7	35	15.5
		40	Negligible	0.05	3	Cooled paramp	18	61	17.9
		40	Negligible	0.1	7	FET	60	107	20.3
7.5	n.a.	40	Negligible	0.1	7	Cooled paramp	35	82	19.1
						FET	100	147	21.7
						Mixer	1200	1247	31.0
11.7	Heavy rain, 30 mm/hr	40	170	0.1	7	Paramp	100	317	25.0
	Clear sky	40	0			FET	225	442	26.5
				0.1	7	Paramp	100	147	21.7
						FET	225	272	24.3
19	Clear sky	55	0	0.2	14	Paramp	300	369	25.7

[a]Elevation angle, 30°. Antenna temperature depends on sidelobe and backlobe; typical values used. Abbreviations: n.a., not applicable; FET, field-effect transistor; Maser = microwave amplifier; Paramp = parametric amplifier.

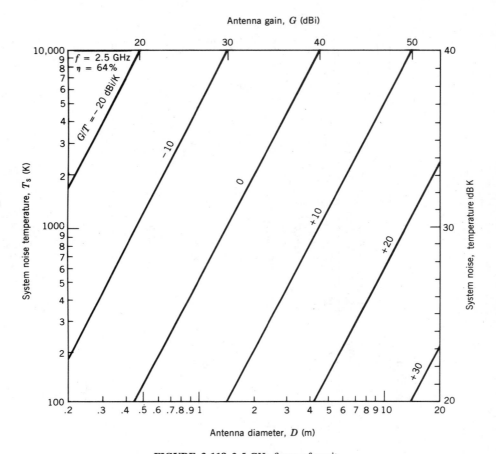

FIGURE 3.118 2.5-GHz figure of merit.

In phased-array and shaped-beam applications individual feed horns or antennas may be connected to individual amplifiers, as shown in Figure 3.123, and their outputs summed.

In a satellite the G/T_s is determined in a similar manner. For antenna beams that "see" only the earth, it is common to assume that the antenna noise temperature T_{ant} is 290 K. At 6 GHz the receivers are often deliberately less sensitive than state-of-the-art would permit in order to reduce the satellite's sensitivity to stray beams from the terrestrial point-to-point services that share this band.

An intersatellite link antenna will normally see the 2.87-K cosmic temperature behind the other satellite. In rare cases the sun or moon will be in the main lobe sidelobe or backlobe. (see Figure 3.111)

Precipitation occurs in the up-link path, but only in the case of a tiny pencil

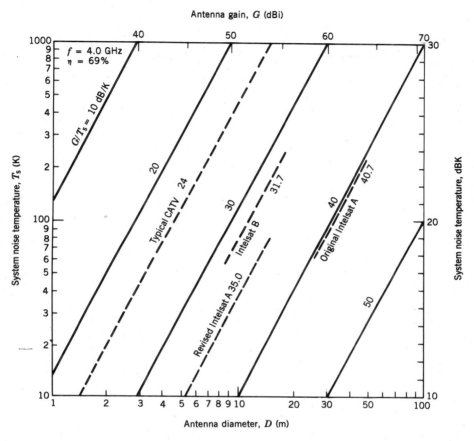

FIGURE 3.119 4-GHz figure of merit.

beam when there is rain throughout most of the beam is this significant. In this case 290 K may be too high because of the cooler cloud tops.

Ignoring the transmission line noise temperature and assuming (for simplicity) a 290-K earth and a 310-K amplifier, the approximate combined T_s of Figure 3.123 would be 600 K (or 27.8 dBK). Assuming 10 identical antennas (each with a bandwidth of 1° wide and having a gain of 44 dBi), the composite gain is 34 dBi (the shaped beam covers 10 times the area of a single beam; thus, the gain is lower than for a single beam). The noise temperature power seen by each antenna is 290 K and must be assumed to be incoherent because the beams are pointed in slightly different directions and thus see different parts of the earth. The amplifiers each contribute an additional 310 K of noise power, which again, as each is an independent source, is incoherent. The combiner power sums the noise of the ten 600-K noise sources into an equivalent noise power of 1897 K (or 32.8 dBK). This is

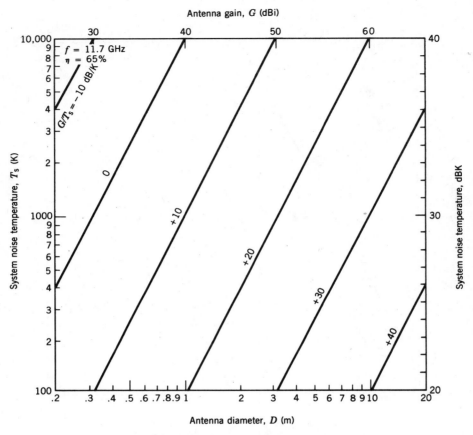

FIGURE 3.120 11.7-GHz figure of merit.

derived by taking the square root of the sum of the squares, that is, $\text{SQRT}[(600)^2 \times 10]$. The net G/T_s is then 34 dBi − 32.8 dBK or +1.2 dBi/K. Additional phase shifters and attenuators may be needed to obtain the desired beam shape. These will lower the composite G/T_s due to their noise contributions.

3.2.3 Transmitting Equipment Parameters

The equivalent isotropically radiated power (EIRP) is a measure of the power transmitted by an antenna by referring it to the power input requirement of a theoretical isotropic antenna to produce the same level of rf illumination.

The EIRP is the product of the power fed to an antenna and its gain:

$$\text{EIRP} = G^* \times P \quad \text{(W)} \tag{3.54}$$

where G^* = antenna gain (ratio to isotropic)
P = input power in watts

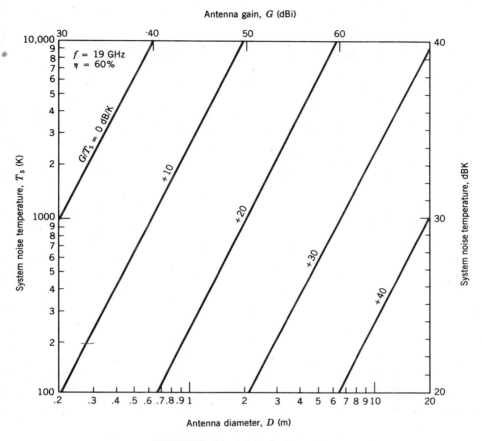

FIGURE 3.121 19-GHz figure of merit.

A more useful relationship is

$$\text{EIRP} = G + 10 \log_{10} P \quad (\text{dBW}) \tag{3.55}$$

where G = antenna gain in dBi
P = antenna input power in watts

The input power is the net value (after losses due to feed lines, transmitter backoff, equipment maintenance, etc.).

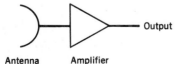

Antenna   Amplifier   **FIGURE 3.122** Single-element satellite receiver.

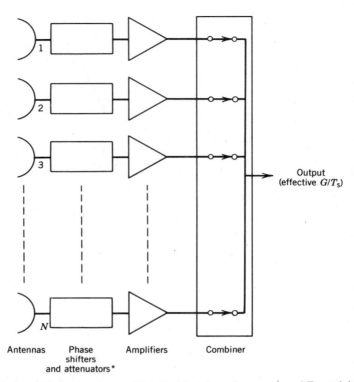

Antennas   Phase   Amplifiers   Combiner
              shifters
         and attenuators*

*These can be located before or after amplifier. Location shown increases a_r and T_{line} and thus system noise temperature. If located after amplifier, effect of noise is diluted by gain of amplifier.

FIGURE 3.123 Multielement satellite receiver.

Figures 3.124–3.131 provide an easy method of determining the EIRP for several frequency bands.

Figures 3.132 provides the range of EIRP allocated for broadcasting satellites at the World Administrative Radio Conferences for ITU regions 1 and 3.

Tables 3.40 and 3.41 provide typical EIRP budgets for satellites and earth stations.

The backoff is a reduction in input (and thus output) power. This is necessitated in multicarrier operation because of the intermodulation distortion (IM) produced in the amplifier. Figure 3.133 shows the saturation characteristics of a typical TWTA. See also Section 3.4.2.

In general the EIRP of the earth station or satellite is fixed. When rain attenuation is a problem the EIRP may be varied (see Seta & Ayukawa, 1988).

3.2.4 Links and Link Losses

The path loss is proportional to the squares of the distance and frequency (see Figure 3.134 on page 338):

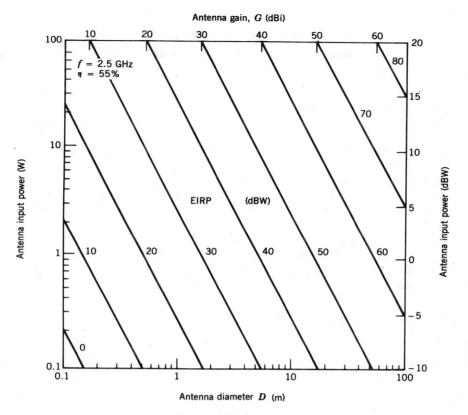

FIGURE 3.124 EIRP at 2.5 GHz.

$$L = 92.45 + 20 \log_{10} S + 20 \log_{10} f \quad \text{(dB)} \quad (3.56)$$

where S is in kilometers and the 92.45 is a constant for $(20 \log_{10} 4\pi/c)$ in kilometers and gigahertz. See page 254 for the derivation.

Attenuation in satellite-to-satellite (intersatellite) links may also be estimated. Both distance (kilometers) and longitudinal separation (along the geostationary orbit) are given. Beyond 162.6°, the earth intersects and thus blocks the path:

$$L_x = 191.0 + 20 \log_{10}\left[\sin\left(\tfrac{1}{2}\alpha\right)\right] + 20 \log_{10} f_x \quad \text{(dB)} \quad (3.57)$$

where α is the longitudinal separation (in degrees) of the two geostationary satellites. See Figures 3.135 and 3.158 on pages 339 and 377.

There are additional losses present in every system, as indicated in Table 3.42.

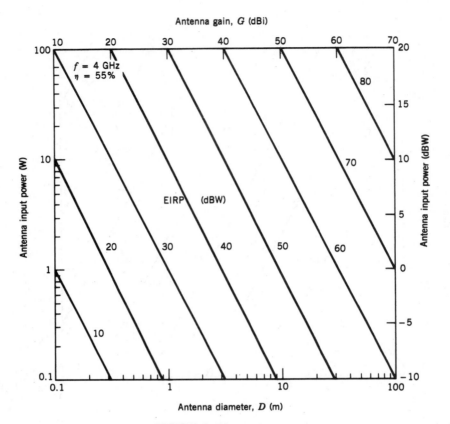

FIGURE 3.125 EIRP at 4 GHz.

3.2.5 Carrier and Noise Levels

The ratio of the received carrier level (C) and the noise (N) due to the single link and the receiver is expressed as C/N. For the purpose of a single-link calculation it is assumed that the signal radiated from the transmitting antenna is "pure" and that all noise is added en route or in the receiver. As will be seen later, each of the several links involved in a satellite service adds noise, and the overall, end-to-end or system carrier-to-noise ratio is the composite of each of these links.

Noise may be expressed in various but interrelated ways, depending on where the ratio is measured. The block diagram of Figure 3.136 shows where noise measurements are made in a typical earth station.

The carrier-to-thermal-noise ratio (C/T) is defined as

$$C/T = W + (G/T_s) - 21.5 - 20 \log_{10} f \quad (\text{dBW}/\text{K}) \quad (3.58)$$

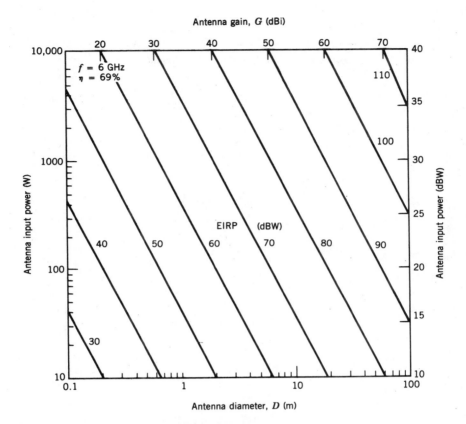

FIGURE 3.126 EIRP at 6 GHz.

where W = illumination level (see Section 3.1.2) in dBW/m^2

G/T_s = receiving system figure of merit in dBi/K

f = frequency in gigahertz

The C/kT ratio is the C/T with Boltzmann's constant (k). Boltzmann's constant is 1.3803×10^{-23} W sec/K. This is the same as 1.3803×10^{-23} W/K Hz, or 1.3803×10^{-23} J/K. In decibel terms 10 log 1.3803×10^{-23} is -228.6 dB W/K Hz. Then

$$C/kT = C/T - (-228.6) = C/T + 228.6 \quad \text{(dBHz)} \quad (3.59)$$

If bandwidth B is set at 1 Hz, the resulting carrier-to-noise-density ratio (known as C/N_0) is the same as C/kT; that is, $C/N_0 = C/kT$ (dBHz) for $B = 1$ Hz.

When a practical noise bandwidth is included, the carrier-to-noise ratio (C/N

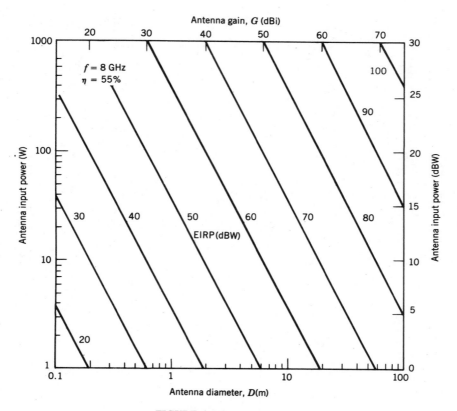

FIGURE 3.127 EIRP at 8 GHz.

or C/kTB) may be found:

$$C/N = C/kTB = C/kT - 10 \log_{10} B \quad \text{(dB)}$$

$$C/N = C/N_0 - 10 \log_{10} B \quad \text{(dB)} \tag{3.60}$$

The bandwidth is usually the predetection (IF) filter bandwidth B_{if}, which can be estimated using Carson's rule for FM:

$$B_{if} = 2(B_{high} + d_{peak}) \quad \text{(Hz)} \tag{3.61}$$

where B_{high} = highest baseband modulation frequency in hertz
$\quad d_{peak}$ = peak FM deviation in hertz

The signal-to-noise ratio (S/N) appears at the detector output. Traditional modulation systems require positive C/N levels (with 10–12 dB being a minimum

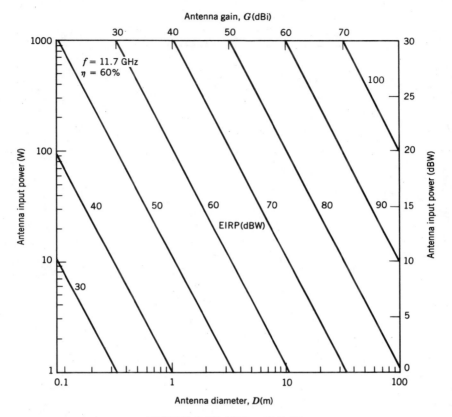

FIGURE 3.128 EIRP at 11.7 GHz.

threshold for good reception). Threshold extension reduces this to about 7 dB. Spread spectrum can operate with negative C/N levels because of the large processing gain (see Chapter 4).

Digital systems need a further conversion to find the ratio of the energy per bit (E_b) to the noise density (N_0) or E_b/N_0.

3.2.6 System Performance

The overall link is composed of up, down, and intersatellite (or cross-) links. Each contributes noise to the overall (system) carrier-to-noise ratio.

The individual link noises are power summed on the basis of

$$\frac{1}{(C_t/N_t)^*} = \frac{1}{(C_u/N_u)^*} + \frac{1}{(C_d/N_d)^*} \quad \text{(ratio)} \quad (3.62)$$

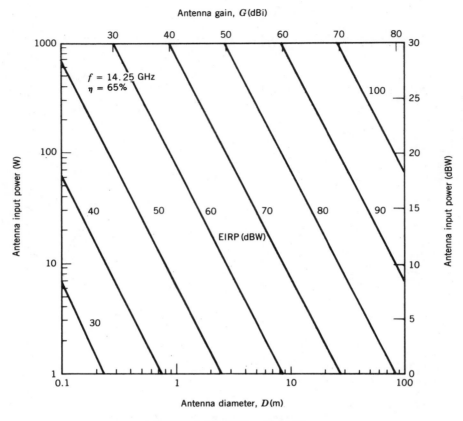

FIGURE 3.129 EIRP at 14.25 GHz.

The $(C/N)^*$ are the numeric ratio values (e.g., $3.16:1$ in place of 5 dB). Here and in what follows the subscripts t, u, x, and d stand for total, up link, cross-link, and down link, respectively.

If C_u/N_u, etc. are in terms of decibels then use this form:

$$\frac{C_t}{N_t} = -10 \log_{10}\left[\left(10^{-(C_u/N_u)/10}\right) + \left(10^{-(C_d/N_d)/10}\right)\right] \quad \text{(dB)} \quad (3.63)$$

Figures 3.137 and 3.138 provide charts for accomplishing these functions using the difference between the various link carrier-to-noise ratios. Thus, if one link is 20 dB and the other is 15 dB, the difference $(20 - 15)$ is 5 dB. The reduction (from Figure 3.138) is 1.2 dB. Thus, the combined C/N for these two links is $15 - 1.2$ dB, or 13.8 dB. The result is always poorer than the poorest single item; thus, the subtraction is from the lowest value.

The total C_t/kT_t, C_t/N_t, C_t/N_{0t}, and S_t/N_t may be found by using consistent

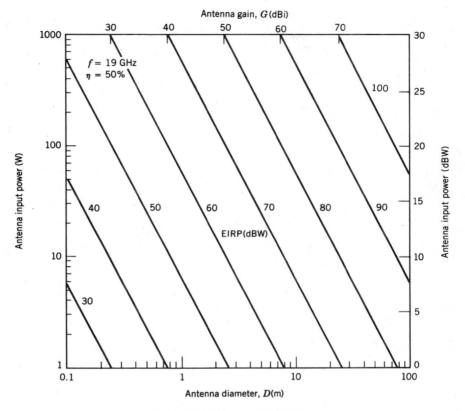

FIGURE 3.130 EIRP at 19 GHz.

values (ratio or decibels but not mixed) for the C/kT's, C/N's, C/N_0's, and S/N's, respectively, for the individual links.

In some calculations it is desirable to consider additional sources of noise (such as intermodulation noise in transponders, interference from other systems, or the effect of intersatellite or terrestrial links). These may be accommodated on a piecemeal basis provided all units are consistent. In a system consisting of an uplink (u), an intersatellite link (x), and a down link (d), the first two may be combined using the figures to find C_{ux}/T_{ux}, and then that answer is combined with the down link to obtain C_{uxd}/T_{uxd}, which, in this example, is C_t/T_t.

If the system carrier-to-interference ratio (C/I) is known (see Section 3.2.1), it may be combined with the C_t/N_t using Figures 3.137 and 3.138 to obtain $C_t/(N_t + I)$. It may also be determined on a piecemeal basis and combined the same way thermal noise is combined (see the previous paragraph).

In digital communications systems the transmission capacity R is defined as

$$10 \log_{10} R = (C_t/N_t) - (E_b/N_{0\text{Th}}) \quad \text{(bits/sec)} \quad (3.64)$$

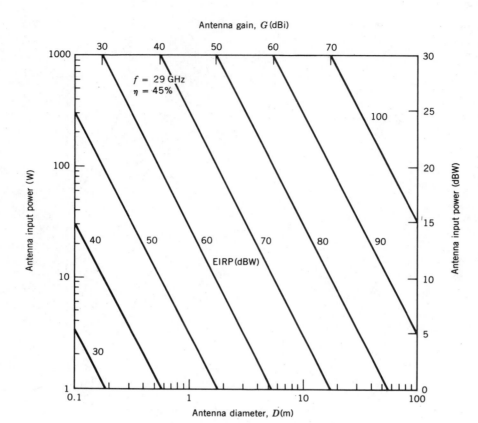

FIGURE 3.131 EIRP at 29 GHz.

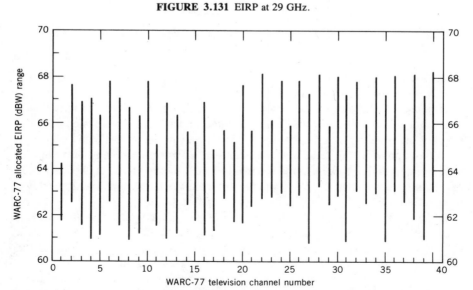

FIGURE 3.132 WARC-77 12-GHz EIRP levels (varies with each country beam; maximum and minimum shown.)

TABLE 3.40 Earth Station EIRP Budgets

Frequency (GHz)	Service Mode[b]	Instantaneous Number of Carriers Present	Transmitter Rating (W)	Transmitter Rating dBW	Output Backoff (dB)	Line Loss (dB)	Antenna Input (dBW)	Antenna Gain[a] (dBi)	EIRP (dBW)
6	SCPC	1	0.13	−8.9	0	0.2	−9.1	54.0	44.9
	FM/TV	1	1350	31.3	0	0.3	31.0	54.0	85.0
	FDM/FM	1	3000	34.8	0	0.5	34.3	54.0	88.3
	TDMA	1	500	27.0	0	0.5	26.5	54.0	80.5
14	SCPC	1	1	0.0	0	0.5	−0.5	45.8	45.3
14	TDMA	1	500	27.0	0	0.9	26.1	61.0	87.1
29	TDMA	1	5	7.0	0	2.0	5	56.6	61.6

[a]All are 10-m antennas except 14-GHz SCPC, which is 1.8 m and the 3m 29-GHz TDMA.

[b]SCPC is single per channel carrier; FDM is frequency division multiplex; and TDMA is time division multiple access.

TABLE 3.41 Satellite ERP Budgets

Frequency (GHz)	Service Mode	Carriers Present	Transmitter Rating (W)	Transmitter Rating (dBW)	Output Backoff (dB)	Line & Multiplexer Losses (dB)	Antenna Input (dBW)	Antenna Gain (dBi)	Satellite EIRP (dBW)
6	SCPC	600	10	10.0	7.0	1.0	2.0	30.0	32.0
	FM/TV	1	10	10.0	0.0	1.0	9.0	30.0	39.0
	FDM/TV	3	10	10.0	5.0	1.0	4.0	30.0	34.0
	TDMA	1	10	10.0	0.0	1.0	9.0	30.0	39.0
12	SCPC	600	40	16.0	7.0	1.0	8.0	30.0	38.0
	TDMA	1	40	16.0	0.0	1.0	15.0	30.0	45.0
	Single TV	1	40	16.0	0.0	1.0	15.0	30.0	45.0
	Dual TV	2	60	17.8	4.0	1.0	12.8	30.0	42.8
30	TDMA	1	20	13.0	0.0	2.0	11.0	40.0	51.0

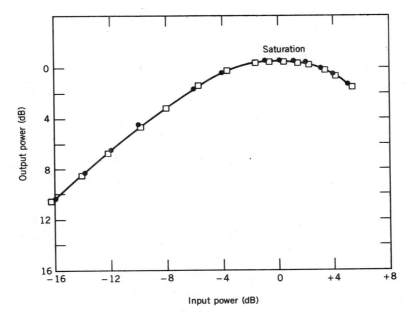

FIGURE 3.133 Saturation characteristics.

where E_b/N_{0Th} is the threshold value of the ratio of the power per bit to the noise power density for a given bit error rate. The required C_t/N_t may be taken as

$$C_t/N_t = (E_b/N_{0Th}) + M_i + 10 \log_{10} R \quad (dB) \quad (3.65)$$

where M_i is the implementation margin for equipment imperfections and assorted interferences (i.e., interchannel noise, intersymbol, terrestrial user, adjacent satellite., etc.)

3.3 SYSTEM MODELING

System modeling is done most often on either an analog or a digital basis and generally with a computer. In some instances a hybrid combination is used.

The system under study can be modeled by connecting physical hardware elements in a laboratory or by digital simulation. In the physical hardware method a signal generator, modulator, transmitter, link, receiver, demodulator, and so on, are connected. The input–output, noise, interference, and other indicators of performance are measured. Some elements (e.g., the link) may be simulated with test equipment such as attenuators, noise generators, and so on. This may be referred to as an analog approach and usually utilizes racks filled with test equipment and (if possible) pieces of actual equipment from the system being tested.

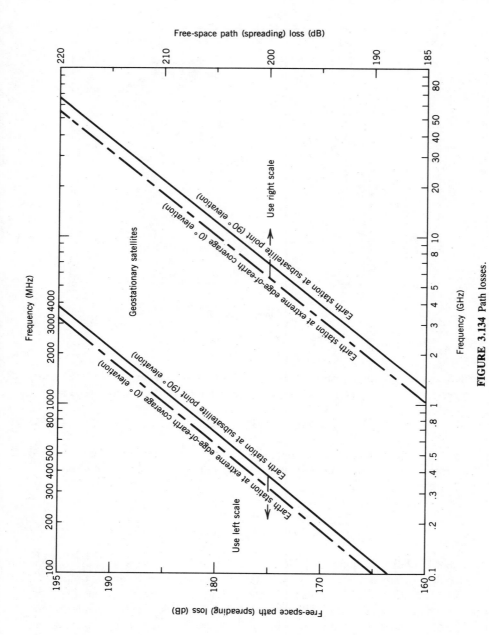

FIGURE 3.134 Path losses.

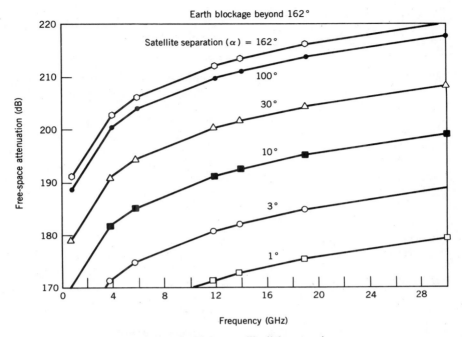

FIGURE 3.135 Intersatellite link attenuation.

The digital simulation approach uses a digital computer that has been programmed with the performance characteristics of the system elements. It is a requirement that the transfer characteristics, noise, spectrum, and so on, of each element be understood so it may be characterized. The system can then be stimucated with input signals. The system outputs resulting from those signals (in the presence of other simulated signals such as encountered in actual practice) are then measured.

A hybrid simulation requires analog-to-digital (A-to-D) and digital-to-analog (D-to-A) converters at appropriate places so that some portions of the equipment may be digitally simulated and others represented by actual equipment that has not been adequately described in a digital subroutine. With the sophistication of digital computer programming, the hybrid approach is infrequently used.

System simulation is useful at the following times in the evolution of a system:

1. When the system is first being conceived, it may be too early for hardware to be available, but it is desirous to find an optimum system configuration. Trade-off analysis can be done between the earth and space segments, power and mass, various modulation methods, transponder frequency plans, and so on, using digital simulations. The output of these studies is a clear picture of what is desired of the satellite earth station or system. A request for proposal (RFP) can then be prepared based on a knowledge of how the system should perform.

TABLE 3.42 Additional Losses

Type	Notes
General	
Transmitter backoff	Depends on number of carriers and type of transmitter; affects intermodulation noise
End-of-life transmitter power	Degradation assumption
Filters	Bandpass, harmonic, multiplexers
Test equipment	Directional couplers and loadings
Transmission line	Coaxial or waveguide losses
Receiver degradation margin	Loss of sensitivity
Maintenance margin	Increase in T_s
Implementation margins	Digital modems
Path Losses	
Atmospheric	Refraction
Precipitation	Rain
Faraday rotation	Scintillation
Path length variation	Nonstationary satellite
Earth Station, Unique	
Antenna-pointing errors	Wind, snow load, human errors, and foundation settling
Antenna aging	Misalignment of feed
Increase in T_s due to rain attenuation	See section 3.2.2
Satellite motion	Gain loss due to mispointing or lack of adequate satellite station keeping
Satellite, Unique	
Antenna-pointing errors	Attitude control
Off-axis location of earth station	Antenna pattern shape

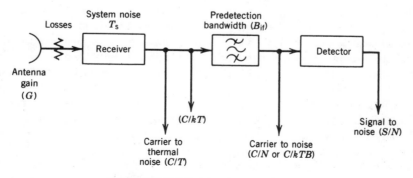

FIGURE 3.136 Noise ratios at receiver.

340

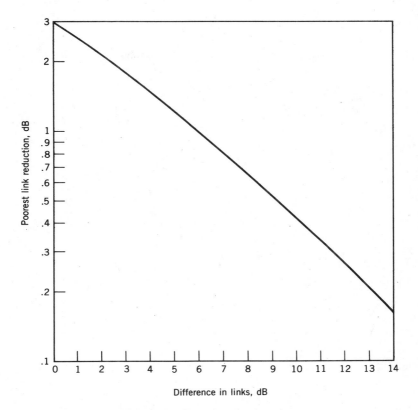

FIGURE 3.137 Total link performance.

2. During the contractor selection process it is important to evaluate the options that are proposed to see what, if any, improvements result in the system performance. An existing model may be used (or modified) for these evaluations.

3. After the contract is let, there may be design alternatives available, changes may be made in the specifications, actual prototype hardware measurements may be better or worse than proposed, alterations may occur in the customer's requirements, and so on. A well-characterized system model can be used to predict the effects of these changes in advance of a launching. If the impact is sufficient, prelaunch changes can be analyzed and, if desirable, made.

4. Before launch, transponder frequency plans can be constructed for the various bandwidth requirements (voice, video, data), modulation methods (FDM/FM, TDMA, SCPC, SCPT, SSMA, etc.), C/N requirements, earth station values, and so on. The simulator allows prelaunch trials without tying up a spacecraft or an earth station.

5. If a problem arises in the operating system after launch, the source of the

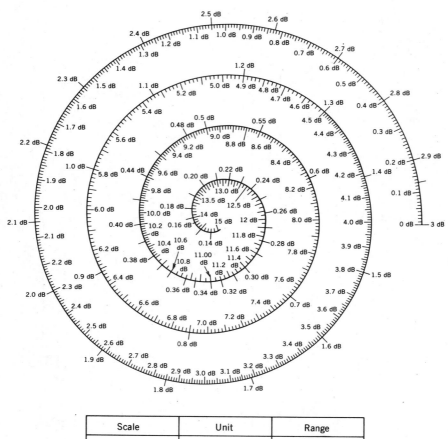

Scale	Unit	Range
Inner	Difference	0–15 dB
Outer	Reduction	0.14–3 dB

FIGURE 3.138 Spiral nomogram for total link.

trouble may be found using the simulator, so that system traffic is not interrupted by potentially interfering experiments.

6. Once the system is well established and is carrying traffic, there is a natural reluctance to disturb the smooth operation with experiments. If it is desired that a new service (e.g., digital television) be introduced, the simulator may be the only feasible method of proving the initial feasibility and discovering the limitations. Once the compatibility with existing equipments and services has been shown on the simulator, the service may be introduced into the operating environment.

7. The simulation approach is also useful for studying the effects of various

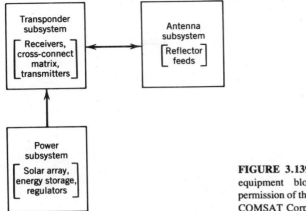

FIGURE 3.139 Communications payload equipment block diagram. (Reprinted with permission of the *COMSAT Technical Review*, the COMSAT Corporation.)

impairments on the system operation. These include precipitation losses, interference, and jamming. Equipment degradations and malperformance modes may also be studied. In-orbit failures may be re-created with the simulator to aid in finding the cause.

3.3.1 Space Segment Modeling

Figure 3.139 is a block diagram showing the interrelationships of the transponder, antennas, and power supply. The effects on the following items can be determined:

1. traffic capacity and distribution,
2. multiple-access method,
3. frequency plan,
4. modulation method,
5. multiplexing method, and
6. earth station G/T_s.

Figure 3.140 shows a block diagram of a typical power subsystem.

For a full-time satellite service (i.e., one that operates through the sunlight *and* eclipse times), the satellite must have an energy storage device and be provided (generally) with solar-powered supplies.

The satellite's dc electrical system generates power and stores energy. Geostationary satellites spend most of their time in the sun, but during the two annual

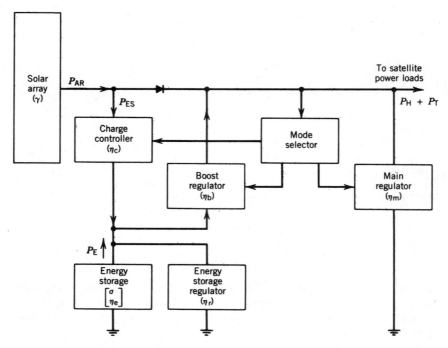

FIGURE 3.140 Power subsystem block diagram. *Note*: Not all elements may be required in specific design. (Reprinted with permission of the *COMSAT Technical Review*, the COMSAT Corporation.)

eclipse seasons (see Section 6.2.2) the satellite's housekeeping (P_H) and transponder (P_T) power requirements must be supplied from stored energy for as long as 1.2 hr/day. The power requirement during an eclipse is

$$P_E = \frac{KP_T + P_H}{\eta_b \eta_r} \quad (\text{W}) \qquad (3.66)$$

where K = eclipse factor; (generally between 0.95 and 1) for partial or full loads
        during eclipse but in some direct broadcast satellites it may be as low
        as 0 if all telecommunications loads are off during the eclipse
    P_T = noneclipse transponder subsystem power load in watts
    P_H = housekeeping power load in watts
    η_b = boost regulator efficiency
    η_r = storage regulator efficiency

These and subsequent terms are summarized in Table 3.43 and given typical ranges. An energy storage device is used during an eclipse, which can last for as much as 1.2 hr. The remaining time is available for recharge. Assuming only 20 of the 22.8 hr are used for recharging, the load presented to the charging array

TABLE 3.43 Parameter Values

Symbol	Meaning	Typical Units	Typical Range
A	Reflector area	m^2	—
a_t	Transmission factor	a	<1
BO_0	Output backoff of transmitter	dB	0–10
	Single carrier per transponder (e.g., TDMA or TV)	dB	0
	Multicarrier per transponder	dB	1–10
	Dual video per TWTA transponder	dB	4
D	Effective reflector diameter	m	—
$EIRP_c$	EIRP per transmitter at saturation in beam center	W	—
f	Transmit frequency	GHz	—
G_t	Satellite transmit antenna gain	a	—
K	Eclipse factor	a	0–1.2
L	Reflector physical diameter	m	—
M_A	Antenna subsystem mass (e.g., includes supports)	kg	—
M_{AR}	Solar array mass (including deployment and orientation)	kg	—
M_{dc}	Mass of each decoder	kg	—
M_{dist}	Mass of the power distribution equipment	kg	—
M_{ES}	Energy storage mass	kg	—
M_f	Antenna feed mass (per beam)	kg	—
M_m	Matrix element mass (per connection)	kg	—
	TDMA	kg	0.05
	FDMA	kg	Filter mass
M_p	Communications payload mass	kg	—
M_R	Reflector mass	kg	—
M_{rc}	Mass of each recoder	kg	—
M_{reg}	Mass of the spacecraft power regulators external to the transponder subsystem	kg	—
M_{rm}	Mass of each remodulator	kg	—
M_{rx}	Receiver mass	kg	—
M_{sup}	Antenna support and hardware mass	kg	—
M_T	Transponder subsystem mass	kg	—
M_{tx}	Transmitter mass	kg	—
M_{t0}	Fixed portion of transmitter mass (includes dc/dc converter, coaxial cables, leads, mounting hardware, etc.)	kg	(TWTA[b], typical 2.3)
M_{dc}	Mass of each decoder	kg	—
M_{dm}	Mass of demodulator (per beam)	kg	—
M_{rm}	Mass of remodulator (per beam)	kg	—
P_a	EIRP per carrier	W (r.f.)	—
P_{AR}	Solar array dc output	W (dc)	—
P_{dm}	Power consumption of each active demodulator	W (dc)	—
P_E	Eclipse power load	W (dc)	—
P_{ES}	Energy storage power loss	W (dc)	—
P_H	Housekeeping power (noncommunications)	W (dc)	—

TABLE 3.43 (*Continued*)

Symbol	Meaning	Typical Units	Typical Range
P_m	Matrix switch element power consumption	W (dc)	—
P_0	Saturated amplifier power output	W (rf)	—
P_{Ob}	Power rating of an amplifier used in an unsaturated mode	W (rf)	—
P_{rm}	Power consumption of each active remodulator	W (dc)	—
P_{rx}	Receiver power consumption (each, includes regulators, etc.)	W (dc)	—
P_T	Transponder subsystem power	W (dc)	—
P_{t0}	Transmitter dc input power with no output rf power (each)	W (dc)	—
P_{tx}	Power per transmitter	W (dc)	—
P_{dc}	Decoder power consumption	W (dc) per beam	—
P_{rc}	Power consumption of each active recoder	W (dc)	—
S_r	Spare reception elements/active reception elements	a	0–1
S_t	Spare transmission elements/active transmission elements	a	0–1
T	Transmitters per beam	—	1–2
X	Number of receiving beams (earth to space)	—	1 or more
Y	Number of transmitting beams (space to earth)	—	1 or more
β_p	dc-to-RF conversion efficiency	a	1 (0.2–0.55)
β_w	Specific mass of transmitter	kg/W (rf)	TWTA typical: 0.10
γ	Solar array specific power lightweight panels (flat)	W (dc)/kg	30
η_a	Antenna efficiency	a	0.2–0.6 0.32^c
η_b	Boost regulator efficiency	a	0.90
η_c	Charge controller efficiency	a	0.90
η_e	Charging efficiency	a	0.69
η_m	Main regulator efficiency	a	0.90
η_r	Storage regulator efficiency	a	0.90
θ_3	Half-power beamwidth	degrees	—
λ	Transmission wavelength	m	—
ρ	Reflector mass density	kg/m^2	—
	Aluminum	—	1.61
σ	Energy storage specific energy		
	NiCd	W hr (dc)/kg	10
	NiH$_2$	W hr (dc)/kg	30

^{a}Ratio, no units.

^{b}TWTA is traveling wave tube amplifier.

^{c}Includes underillumination and blockage.

(P_{ES}) is

$$P_{ES} = \frac{P_E}{20\eta_e\eta_c} \quad (\text{W}) \tag{3.67}$$

where η_e = charging efficiency, a decimal value
$\quad\eta_c$ = charge controller efficiency, a decimal value
The total array power P_{AR} may be obtained by combining Equations (3.66) and (3.67):

$$P_{AR} = \frac{KP_T + P_H}{20\ \eta_b\eta_r\eta_e\eta_c} + \frac{P_H + P_T}{\eta_m} \quad (\text{W}) \tag{3.67a}$$

If $K = 1$, then

$$P_{AR} = (P_T + P_H)\left(\frac{1}{20\ \eta_b\eta_r\eta_e\eta_c} + \frac{1}{\eta_m}\right) \quad (\text{W}) \tag{3.67b}$$

where η_m is the main regulator efficiency.
The mass of the solar array is

$$M_{AR} = \frac{P_{AR}}{\gamma} \quad (\text{kg}) \tag{3.69}$$

and the mass of the energy storage element is

$$M_{ES} = \frac{1.2P_E}{\sigma} \quad (\text{kg}) \tag{3.70}$$

where γ = solar array specific power in watts per kilograms
$\quad\sigma$ = storage element specific energy in watts per kilogram (or watt hours of storage capacity required per kilogram divided by 1.2)
$\quad 1.2$ = maximum eclipse duration in hours
The block diagram of a multibeam transponder is shown in Figure 3.141. Frequency reuse is provided through independent spot beams and/or orthogonal polarizations. This satellite has X up and Y down beams. A satellite-switched interconnection matrix connects traffic arriving on one beam to the transmitter (or transmitters) of another beam. The number of transmitters may outnumber the receivers.

Figures 3.142 and 3.143 show the block diagrams of typical receivers and transmitters. The receivers tend to be broadband devices (e.g., 500 MHz), whereas the transmitters may be of various bandwidths (such as 27, 36, 43, 54, 72 or 241 MHz) depending on the access method, service provided, and most important, the mass available. The output amplifier may be a traveling-wave tube amplifier (TWTA) or a solid-state power amplifier (SSPA).

Traveling-wave tube amplifiers are wideband amplifiers with nonlinear characteristics. Requirements for low intermodulation (IM) distortion make it necessary

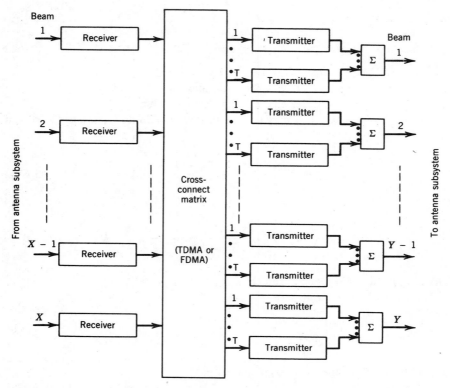

FIGURE 3.141 Transponder block diagram for multibeam satellite. (Reprinted with permission of the *COMSAT Technical Review*, the COMSAT Corporation.)

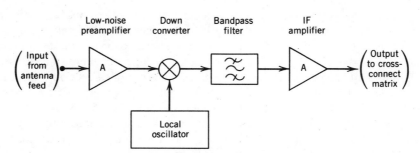

FIGURE 3.142 Receiver block diagram. (Reprinted with permission of the *COMSAT Technical Review*, the COMSAT Corporation.)

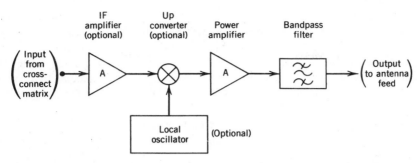

FIGURE 3.143 Transmitter block diagram. (Reprinted with permission of the *COMSAT Technical Review*, the COMSAT Corporation.)

to reduce the output rf power. (See Sections 3.2, 3.4, and Chapter 5 for further details.) This process, called *backoff*, significantly reduces the dc-to-useful rf conversion efficiency. Backoff is used whenever an amplifier amplifies two or more carriers and thus applies in FDMA, SCPC, and SSMA and a partial transponder TDMA or video situations. No backoff is needed if a single carrier occupies a full transponder. This occurs in video and high-bit-rate TDMA or TDM applications.

The transmitter's dc requirements (P_{tx}) may be stated as

$$P_{tx} = P_{t0} + \beta_p P_0 \quad (\text{W}) \tag{3.71}$$

where P_{t0} = transmitter input power when there is zero rf output power (cutoff condition) (dc watts)

β_p = dc-to-rf conversion efficiency (decimal) of the power amplifier

P_0 = saturated amplifier rf power output (rf watts) see equation 3.80

The per-transmitter mass (M_{tx}) is

$$M_{tx} = M_{t0} + \beta_w P_0 \quad (\text{kg}) \tag{3.72}$$

where M_{t0} = fixed portion of the amplifier mass (for $P_0 = 0$ W) in kilograms

β_w = specific mass of amplifier in kilograms per rf watts

A matrix switch may be used with FDMA, TDMA, or remodulating repeaters. For full interconnectivity (any up-link beam to any down-link beam) the matrix switch contains XY elements. In reality not all beams always need to be interconnected, and therefore, a smaller switch might be used. A variable-power divider is a solid-state alternative that can be commanded to split the input power into two (not necessarily equal) parts. Each part goes to a different antenna beam.

For frequency domain multiple-access (FDMA) systems each earth station transmits an individual carrier for each destination beam (except in cases where the satellite contains a demodulator, address recognition, switching, and remodulation equipment). The cross-connect matrix consists of up to XY filters for separating the individual carriers and another array of XY filters (or variable-power dividers) for feeding the combined carriers to the appropriate beam. Additional frequency converters are needed if the switching is done at baseband or a common intermediate frequency. Optical switching and processing is forecast.

Multiple TDMA time slots may be assigned to each up link for addressing each down link. The matrix contains up to XY high-speed switch elements that connect the appropriate receivers to the appropriate transmitters during the assigned time interval.

The assignment of the FDMA filters or the TDMA switch intervals may be controlled from the ground via an on-board microcomputer to match the varying traffic patterns.

If on-board demodulation and remodulation is provided, the up-link noise may be decoupled to a great extent from the down link. The coding and modulation characteristics may also be altered in the regeneration process.

The total transponder electronics subsystem mass M_T is

$$M_T = (X + XS_r)(M_{rx} + M_{dm} + M_{dc}) + XYM_m$$
$$+ (Y + YS_t)(M_{rc} + M_{rm} + TM_{t0} + T\beta_w P_0)(\text{kg}) \qquad (3.73)$$

where X = number of receive beams

$\quad S_r$ = reception equipment spares ratio (e.g., two spares for two active elements is $S_r = 1$)

$\quad M_{rx}$ = mass of each receiver in kilograms

$\quad M_{dm}$ = mass of each demodulator (if present) in kilograms

$\quad M_{dc}$ = mass of each decoder (if present) in kilograms

$\quad Y$ = number of transmit beams

$\quad M_m$ = mass of each cross-connect element in the satellite switch in kilograms

$\quad S_t$ = transmission equipment spares ratio (e.g., one spare for four active elements is $S_t = 0.25$)

$\quad M_{rc}$ = mass of each recoder (if present) in kilograms

$\quad M_{rm}$ = mass of each remodulator (if present) in kilograms

$\quad T$ = transmitters per beam

The mass of ancillary equipment such as connecting cables, switches, and so on, is included in the appropriate transponder element. As indicated, XY connections may not be required. In this case substitute the appropriate value for XY. In FDMA systems M_m is predominately the mass of the filters, whereas in TDMA, it is the individual switch elements (typically diodes, transistors, or optical devices).

The total communications power for all of the transponders is derived in a manner similar to the mass (M_T) except that the spare equipment is not powered.

The total communications power (P_T) is

$$P_T = X(P_{rx} + P_{dm} + P_{dc}) + XYP_m + Y(P_{rc} + P_{rm} + TP_{tx}) \qquad (\text{W}) \qquad (3.74)$$

where P_{rx} = dc input power per active receiver in watts

$\quad P_{dm}$ = power consumption of each demodulator (if present) in watts

P_{dc} = power consumption of each decoder (if present) in watts
P_m = power consumed by each cross-connect element in the switch matrix (if present) in watts
P_{rc} = power consumption of each recoder (if present) in watts
P_{rm} = power consumption of each remodulator (if present) in watts

The antenna is composed of a reflector underilluminated by a series of beams (Figure 3.144). The feed array generates the X independent beams. For geostationary satellites an antenna with the capability of placing beams anywhere on the earth's surface may be characterized as

$$L \simeq 1.4D \quad (m) \tag{3.75}$$

where L = physical diameter of reflector in meters
$\quad\ D$ = effective diameter of reflector in meters
$\quad 1.4$ = oversize allowance for multiple beams
The mass of the reflector (M_R) is

$$M_R = A\rho = \tfrac{1}{4}\pi L^2 \rho \quad (kg) \tag{3.76}$$

where ρ is the density of the reflector in kilograms per meters square.

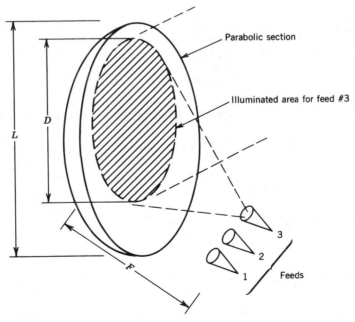

FIGURE 3.144 Multibeam paraboloid reflector antenna. *Note*: L does not equal D. (Reprinted with permission of the *COMSAT Technical Review*, the COMSAT Corporation.)

When each beam is formed by one feed, the total antenna receive/transmit subsystem mass M_A is

$$M_A = \tfrac{1}{4}\pi L^2 \rho + (X + Y)M_f + M_{sup} \quad (kg) \tag{3.77}$$

where M_f = feed mass per beam in kilograms
M_{sup} = supports and mounting hardware
for the case of $L \approx 1.4D$ this may be restated as

$$M_A = \frac{1.4\pi D^2 \rho}{4} + (X + Y) M_f + M_{sup} \quad (kg) \tag{3.78}$$

The transmit antenna gain G_t may be expressed as

$$G_t = \eta_a \pi^2 D^2 / \lambda^2 \simeq 110\eta_a f^2 D^2 \quad (\text{ratio to isotropic}) \tag{3.79}$$

where η_a = antenna efficiency (a decimal)
λ = wavelength in meters
f = frequency in gigahertz
The half-power (-3-dB) beamwidth is

$$\theta_3 \simeq 21/(fD) \quad (\text{degrees}) \tag{3.80}$$

Assuming the EIRP has already been selected (based on a particular modulation method, frequency plan, carrier bandwidth, and earth station G/T_s), the saturated power amplifier requirement (P_0) may be calculated as

$$P_0 = EIRP_c/(G_t a_t) \quad (W) \tag{3.81}$$

where $EIRP_c$ = equivalent isotropically radiated power required at beam center in watts
a_t = the transmission factor calculated from the line loss between amplifier output and antenna feed (a factor less than 1)

For multicarrier operations Equation 3.81 must be modified for the effect of output backoff to operate the amplifier in an unsaturated mode. Section 4.2.1 (see also the Index) discusses output backoff. Once the $EIRP_c$ is determined the P_0 must be increased to account for the backoff.

$$P_{OB} = P_0 \times 10^{(BO_o/10)} \quad (W) \tag{3.82}$$

where P_{OB} = the saturated power amplifier requirement in watts
BO_o = output backoff in decibels

Currently most satellites have a uniform power amplifier level (e.g., 60 watts) per frequency band. Future designs may more closely match the power amplifier to the modulation/multiple access selection and the beam sizing. The use of SSPAs reduces the BO_o by several decibels from the TWTA requirements.

By combining Equations (3.69), (3.70), (3.73), (3.74) and (3.77), the total communications payload mass (i.e., antennas, feeds, transponders, and a pro-rata share of the power supply to support the transponders) for a service may be modeled:

$$M_p = M_T + M_A + \left(\frac{P_T}{P_T + P_H}\right)(M_{AR} + M_{ES} + M_{reg} + M_{dist}) \quad \text{(kg)} \quad (3.83)$$

and where M_{reg} = the total mass of the spacecraft power regulators that are external to the transponder subsystem in kilograms

M_{dist} = the mass of the power distribution facilities (cables, connectors, fuses/circuit breakers, etc. in kilograms

M_P = the total communications (payload) mass requirement in kilograms

3.3.2 Examples of Modeling

By substituting the value for P_{0b} for P_0 and combining Equations (3.73), (3.78), and (3.83) we may find the total mass requirement as a function of the power amplifier rating.

$$M_P = (X + XS_r)(M_{rx} + M_{dm} + M_{dc}) + XYM_m$$

$$+ (Y + YS_t)(M_{rc} + M_{rm} + TM_{t0} + T\beta_w P_{0B})$$

$$+ \frac{1.4\pi D^2 \rho}{4} + (X + Y)M_f + M_{sup}$$

$$+ \left(\frac{P_T}{P_T + P_H}\right)(M_{AR} + M_{ES} + M_{reg} + M_{dist}) \quad \text{(kg)} \quad (3.84)$$

Figure 3.145 shows the optimization of a large multibeam satellite. The system is optimized when the partial mass of the reflector equals the partial mass of the power subsystem that supplies power to the final rf amplifier. The EIRP$_d$ is fixed at 36.6 dBW. Since the rf output power (P_0) must rise as the beamwidth increases (and the gain decreases). Narrow beams (the left side of Figure 3.145) have a high gain and large, heavy, reflectors. On the right side the reflectors are small but the mass of the power amplifiers and their power supply is growing. The optimum is shown.

These equations are easily handled on a computer. They are useful for determining the optimum location of the operating conditions. This information is useful in initial system sizing and later in knowing how far from optimum any compromises may have forced the design.

These operating conditions are coverage, power flux density limitations, G/T_s of earth stations, and so on.

For additional examples, refer to Kiesling et al. (1972) and Koelle & Dodel (1988).

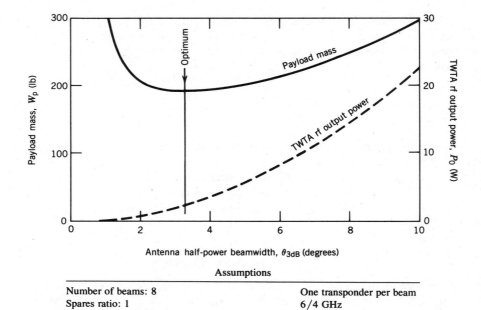

Assumptions

Number of beams: 8	One transponder per beam
Spares ratio: 1	6/4 GHz
Saturated EIRP: 36.6 dBW	Solar array: 10 W/lb
Energy storage: 16 W h/pound	TDMA: four-phase CPSK
G/T_s of earth stations: 40 dBi/K	

FIGURE 3.145 Payload mass optimization. (Reprinted with permission of the *COMSAT Technical Review*, the COMSAT Corporation.)

3.4 OVERALL SYSTEM CALCULATIONS

The up link of a satellite may be characterized as composed of the following elements:

1. A transmitter and antenna;
2. The path between the transmitter and the receiver, consisting of the atmosphere and space; and
3. A receiving antenna and receiver.

Tables 3.44–3.48 provide the items in an up-link budget. Table 3.44 is a condensed, or short, form of Tables 3.45–3.47. Table 3.48 summarizes the up-link budgets.

The down link uses the same elements as the up link. Tables 3.49 and 3.50 indicate the allocation of satellite power. Tables 3.51 and 3.52–3.55 are the short, long, and summary forms of the down-link budgets. Table 3.56 is provided for intersatellite (space-to-space) links budgets. Finally, the net system performance is derived for nonremodulating systems. Section 3.4.4 discusses the advantages of remodulation (regenerative) repeaters. The long forms are designed to provide pessimistic answers by combining the worst combinations (the limit conditions) of each parameter. Most commercial systems operate very well with small margins since the worst case rarely occurs.

3.4.1 Up-Link Budgets

Up-link budgets are prepared to determine the equipment performance trade-offs that may be made, the link capability with a given set of equipments, and its contribution to the overall system performance. Figure 3.146 provides a generalized depiction of the various power levels in a typical up link.

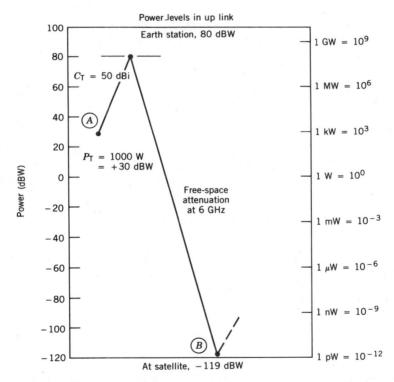

This figures traces power levels in typical up link (earth to space). Earth station (A) power amplifier shown generating 1000 W (+30 dBW, 1 kW, or 10^3 W). Transmit antenna gain (G_T) is 50 dBi (100,000 times better than theoretical isotropic antenna). 30 dBW + 50 dBi = 80 dBW (isotropical equivalent to 100 MW).

Free-space attenuation (due to signal spreading) at 6 GHz reduces power to mere −119 dBW at satellite. This is approximately 1 pW at B. Satellite antenna gain is typically 20 (global) to 30 dBi (U.S. national beam). This concentrates signal into nanowatt range.

FIGURE 3.146 Earth-to-space link.

TABLE 3.44 Up-Link Budget (Short Form)

Earth station location _____

Up-link frequency f = _____ GHz   Earth station antenna diameter D = _____ m

Satellite name _____   Up-link beam _____

Earth Station

(1)	Power at antenna	(P = _____ W), dBW		_____ (a)
(2)	Antenna gain, dBi (Section 3.2.1)		(+)	_____ (b)
(3)	EIRP, dBW (Figures 3.124–3.131)		(=)	_____ (c)

Earth to space

(4)	Free space path loss, dB	(_____ km)	(−)	_____ (d)
	Figure 3.53			

Satellite

(5)	Satellite G/T_s, dB/K		(+)	_____
	(Figures 3.116–3.121)			
			(=)	_____
(6)	$(C/T)_u$, dBW/K (Equation 3.58)			_____
(7)	1/Boltzmann constant, dBW/K/Hz		(−)	−228.6
(8)	$(C/kT)_u$, dBHz			_____ (e)

(a) 10 log P
(b) 18.2 + 20 log D + 20 log f (includes antenna efficiency of 60%)
(c) (1) + (2)
(d) For an elevation angle of 15° (range = 40,000 km) and

f	1.6	6	14	30	GHz
Loss	181	200	206	214	dB

(e) = (6) + (7)

For up-link to geostationary satellites, using frequencies below 10 GHz and tracking earth stations, the computation is quite easy, and Table 3.44 suffices for most applications. When the earth station antennas are fixed (nontracking) and the frequency is high enough that precipitation becomes a significant loss, the longer form is suggested.

Most of the items in Table 3.45 are self-explanatory but some may require further definition.

The transmitter-saturated power output rating (item 1) at the output flange is

TABLE 3.45 Up-Link Budget (Long Form)

Earth station location ———————————————————————

Up-link frequency ——————GHz   Earth station antenna diameter ———— m

Satellite ———————————   Up-link beam ——————————

	Nominal	Worst Case	

Earth Station

	Nominal	Worst Case	
1. Transmitter saturated power rating at output flange (———W), dBW	(+)————	(+)————	
2. Reserve for end-of-life loss, dB	(−)————	(−)————	
3. Output backoff for ———carriers, dB	(−)————	(−)————	BO_o
4. Net power into transmission line, dBW	(=)————	(=)————	P_0
5. Transmission line loss, dB	(−)————	(−)————	L
6. Other feeder losses (directional couplers, switches, filters, antenna feeds, and radomes), dB	(−)————	(−)————	
7. Antenna gain on axis, dBi[a]	(+)————	(+)————	G_{tu}
8. Nominal EIRP of earth station, dBW[b]	(=)————		
9. Maintenance margin, dB		(−)————	
10. Antenna-pointing loss (wind, snow, and foundation settling), dB[c]		(−)————	
11. Antenna-pointing loss due to satellite motion, dB[c]		(−)————	
12. Worst case EIRP, dBW		(=)————	$EIRP_u$

Earth-to-Space Path

	Nominal	Worst Case	
13. Clear sky free-space path loss for ———km, dB[d]	(+)————	(+)————	L_{pu}
14. Atmospheric loss, dB		(+)————	L_{au}
15. Precipitation losses (for ——— % of a ———) propagation effect losses (scintillation, polarization coupling, etc.),[f] dB		(+)————	M_{pu} M_{0u}
16. Nominal up-link losses, dB	————		L_{up}
17. Worst case, dB		(=)————	L'_{up}

Satellite

	Nominal	Worst Case	
18. Satellite antenna peak gain (——— beam), dBi[g]	(+)————		G_{ru}

357

TABLE 3.45 (*Continued*)

Earth station location _____

Up-link frequency _____GHz   Earth station antenna diameter _____ m

Satellite _____   Up-link beam _____

	Nominal	Worst Case

Satellite (continued)

	Nominal	Worst Case	
19. Receiving system noise temperature (_____ K antenna, _____ K feeds, _____ K receiver = _____ K system,[h] dBK	$(-)$_____		T_{ru}
20. Satellite figure of merit (G/T_s), dBi/K[i]	$(=)$_____	$(+)$_____	$(G/T_s)_{sat}$
21. Off-beam center loss, dB	$(-)$_____	$(-)$_____	
22. Pointing error (attitude control, thermal misalignments, etc.), dB[c]		$(-)$_____	
23. Nominal G/T_s, dBi/K	$(=)$_____		$(G/T_s)_u$
24. Worst case G/T_s, dBi/K		$(=)$_____	$(G/T_s)'_u$

[a]See Figures 3.75 & 3.76   [f]See Figures 3.56–3.61
[b]Check using Figures 3.124–3.131   [g]See Figures 3.79 and 3.88
[c]See Figure 3.80   [h]Convert to dBK using 10 log K = dBK.
[d]See Figure 3.53   [i]See Figures 3.116–3.121
[e]See Figure 3.54

sometimes called the nameplate rating. This is the beginning of life rating. In practice, the manufacturer usually ships the power amplifier with a small margin of excess power. This power is determined by coupling the transmitter into a matched resistive load. As the power output device (typically a TWT) and its power supply age, some decline in power may take place, and thus a reserve is needed (item 2): A typical value is 1 dB.

In a multicarrier situation the power amplifier must be operated at less than saturation power to avoid excessive intermodulation noise. The value chosen is a compromise between excessive intermodulation noise (C/IM) and thermal noise (C/N). The selection of the amount of backoff (3) is complex and depends on the frequency plans in the earth station and satellite; the size, number, and spacing of the carriers; the interference susceptibility of the modulation method; the input–output transfer characteristics of the nonlinear devices (transmitters in particular); and so on. See Section 4.2.1 and Figure 4.20. In multicarrier FDMA/FM cases the earth station output backoff is typically in the range of 4–7 dB. TDMA stations usually operate with a single carrier per transmitter, as do certain heavy-traffic

FDMA and single-channel FDMA stations. In these cases the power output may range between 1 dB below saturation (backoff) and 1.5 dB beyond saturation. Typically, 0 dB of output backoff is assumed for TDMA and other single-carrier-per-transponder (SCPT) services such as television.

Other feeder losses (6) include those of the filter and diplexer (which may be significant if the transmit and receive frequencies are nearby or harmonically related), switches (to alternate standby transmitters), combiners (if multiple active transmitters are used), directional couplers (for isolation of power measurements, spectrum analyzers, etc.), and antenna feed element losses. Typical values for line losses (5) and other feeder losses (6) total about 2 dB for large stations and 0.5 dB for compact VSATs.

Radomes were once used with earth station antennas, but the present use is limited to the harsher environments such as encountered in the maritime or aeronautical mobile services. The loss is dependent on the construction and frequency. A radome used in the maritime mobile service is rigid fiberglass and has losses of 0.2 and 0.5 dB (item 16) at 1.6 GHz when dry and wet, respectively.

Antenna-pointing losses (item 10) are divided into two categories: those that happen at the earth station and those due to the satellite. If the station uses some form of tracking to maintain pointing, these items are zero (unless there is an offset in the pointing error loop). For fixed antennas a margin is advisable. The satellite stationkeeping is contained within an error rectangle of A degrees latitude and B degrees longitude. The hypotenuse of the right triangle ABC is given as

$$C = \sqrt{A^2 + B^2} \quad \text{(degrees)} \tag{3.85}$$

where C is the worst case condition. Typical values for A and B are $0.2°$ (which makes C equal to $0.2828°$). Figure 3.147 shows this rectangle as seen from the earth station. If the angles are small a plane triangle may be assumed, otherwise spherical trigonometry should be used. See Section 6.1.4.

The antenna-pointing error loss L_e may be expressed in terms of relative beamwidths. Typical values are a few tenths of a decibel to 1 dB (depending on the satellite stability and the beamwidth of the transmit antenna).

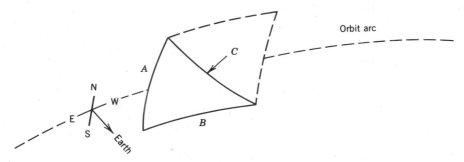

FIGURE 3.147 Typical stationkeeping of satellite.

The free-space path loss (item 13) varies as the square of the distance between the transmitter and the antenna. (See Section 3.1.5.)

At frequencies greater than about 10 GHz there are significant additional losses. The oxygen and water vapor absorption bands cause atmospheric (14) and precipitation (15) losses such as shown in Figure 3.54. The precipitation loss varies with location (with rainfall) and time of year. Often the value selected for item 15 is relatable to a given outage (see Figure 3.60) at the earth station. It is often referred to as rain margin. In the eastern United States a value of 2 dB at 6 GHz and 4–7 dB at 14 GHz is not uncommon.

Depending on the frequency and the climatic conditions, the other link losses often may be neglected.

The derivation of the receiving system noise temperature (19) is given in Section 3.2.2.

The G/T_s of the satellite (20) is generally available from the operator. In case of a shaped-beam satellite receiving antenna it is necessary to consult the actual pattern to determine the off-beam-center losses (21).

Table 3.46 is a summary of the prior links. The numbers under the entry lines designate the line in Table 3.45. Both nominal and worst case columns are provided.

The ratio of the up-link carrier power (C) to thermal noise power (T) is given. Adding the decibel value of the inverse of the Boltzmann constant (-228.6

TABLE 3.46 Up-Link Budget, Summary

	Nominal[a]	Worst Case[a]	
25. Earth station EIRP, dBW	_____	_____	EIRP$_u$
	8	12	
26. Up-link losses, dB	$(-)$_____	$(-)$_____	L_{up}
	16	17	
27. Satellite G/T_s, dBi/K	$(+)$_____	$(+)$_____	$(G/T_s)_u$
	23	24	
28. C_u/T_u at satellite receiver output, dBW/K	$(=)$_____	$(=)$_____	$(C/T)_u$
29. 1/Boltzmann constant dBW/K/Hz	$+228.6$	$+228.6$	
30. Satellite transmitter input backoff, dB	$(-)$_____	$(-)$_____	BO$_i$
31. C_u/kT_u at transmitter input, dBHz	$(=)$_____	$(=)$_____	(C_u/kT_u)
32. Earth station EIRP, dBW	_____	_____	EIRP$_u$
	8	12	
33. From Equation 3.9, dB	-163.3	-163.3	
34. Range correction factor[b], dB	$(+)$_____	$(+)$_____	
35. Illumination level at the satellite, dBW/m^2	$(=)$_____	$(=)$_____	

[a] The small numbers under the lines indicate the origin line in Table 3.45.

[b] See Figure 3.30.

dBW/K/Hz) and subtracting the transmitter input backoff (often 4–12 dB) produces the transmitter input C_u/kT_u (item 31).

Steps 32 to 35 provide an alternate summary that provides the illumination level at the satellite. The elevation angle of the earth station (or the slant range) is used to obtain a precise clear-weather value for the illumination level (35). The maximum error due to elevation angle is 1.3 dB. This occurs at the subsatellite point.

3.4.2 Down-Link Budgets

The down link is basically the inverse of the up link. The satellite transmitter generally has a much lower power device than that of the earth station. Whereas an earth station may handle a single signal, the satellite amplifier must handle one to many signals. A satellite could use one transmitter to handle the entire bandwidth (typically 500 MHz), but modern satellites split the received spectrum into many individual transponders. The channelized repeater has the advantage of amplifying fewer signals per power amplifier. Individual transmitters may also be coupled to separate down-link antennas, thus getting the advantage of the increased antenna gain of the narrow beams.

Multicarrier operation of a power output device requires that the output power be reduced (backed off) from its saturated power (for tubes) or intercept point (for SSPAs) to reduce the intermodulation distortion (that eventually is seen in the carrier-to-intermodulation noise ratio, C/IM). Figure 3.148 illustrates this effect in terms of the input backoff.

In contrast, a single-carrier TDMA tube may be operated at saturation and even slightly overdriven (by 1.5 dB in some cases). Overdrive of a TWTA produces a slight reduction in the output power, but the tube is being operated as a limiter. Therefore, a 5.5-dB variation in the input drive (due to precipitation losses in the up-link path) results in only 0.8 dB change in the output level. See Figures 3.149 & 4.33 and Section 3.2.3.

Figure 3.150 shows the flow of power in an rf power amplifier. Whatever power is not radiated as rf must be changed to heat in the satellite. Table 3.47 shows several ways of dividing 500 MHz among many transponders. Table 3.48 shows how the power in an rf power amplifier may be allocated. ·

As a first approximation, the total power is inelastic and must be consumed by all the users of the transponder. Thus, a change in the demands of one consumer is immediately reflected into the other user's power requirements. This is especially true when the tube is backed off and thus is in the linear region.

How is the power capacity of a transponder actually utilized? The answer depends very much on the application of each transponder and earth station types. Thus, in a satellite with identical transponders, the power output of each transponder may be different.

Although Tables 3.49 and 3.50 are self-explanatory, some additional definition may be helpful. In Table 3.50, if the EIRP is known, skip to item 5. The backoff of a multicarrier satellite transmitter (item 6) is usually greater than that of an earth

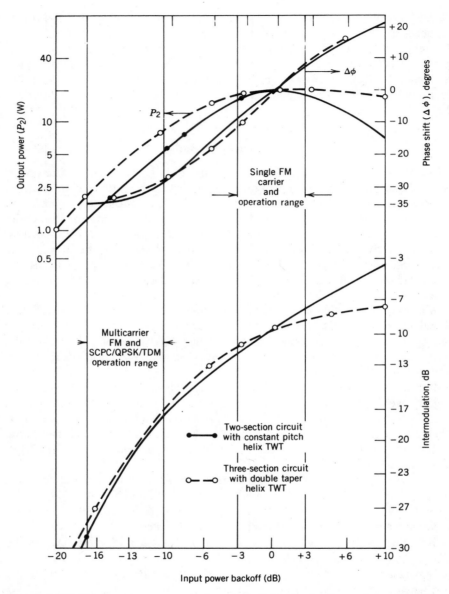

FIGURE 3.148 TWTA performance versus input backoff. Reprinted with permission from *Microwaves & RF Magazine.*

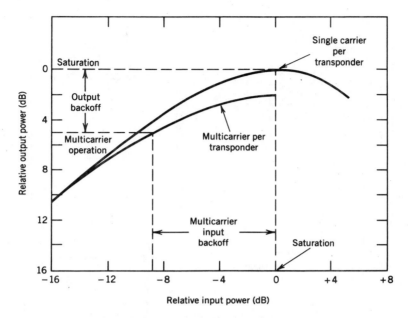

FIGURE 3.149 TWTA characteristics.

station because of the greater number of carriers present: Typical values for output backoff are 4–8 dB.

Figure 3.151 shows data from several in-service Intelsat satellites based on Table 4.10. These transponders carried from one to nine carriers each. The global, hemispheric, and zone beams, are capable of producing EIRP levels of 23–30 dBW under saturated conditions. Transponders with band occupancies of 30–36 MHz are included in the sample. The earth stations were primarily Intelsat's *original* A (G/T_s = 40.7 dBi/K) and standard B (31.7 dBi/K) stations. The down-link frequency is in the 4-GHz band. Intelsat specifies the same power levels for zone and hemi beams even though the zone beam has more gain than the hemi beam.

Figure 3.151 shows output backoff versus number of carriers and antenna. Typical output backoffs are 7 dB for the hemispheric and zone beams.

Since

$$Q_d = \text{EIRP} + G/T_s \quad \text{(dB per carrier)} \tag{3.86}$$

where Q_d is the figure of quality for a down link with a given modulation, bandwidth, and S/N in decibels.

If all of the stations had identical traffic but some had G/T_s of 40.7 dBi/K and others had 31.7 dBi/K, a 9-dB variation in EIRP would be expected. Figure 3.152 shows a fairly simple hypothetical four-carrier situation.

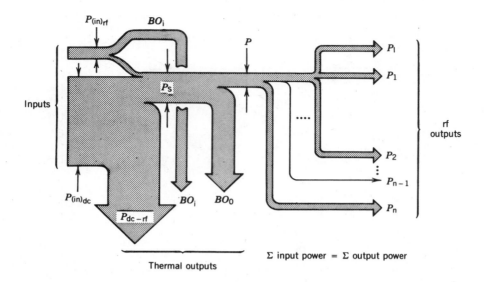

$P_{(in)dc}$	dc input power from solar array or power storage system; for single-collector TWTA this is nearly constant; for multicollector TWTAs, it varies with actual rf output
$P_{(in)rf}$	rf input power from driver stage
P_{dc-rf}	Inefficiencies in converting dc to rf in power amplifier
P_s	Saturated rf power (nameplate) rating of power amplifier; this is measured at output flange of amplifier waveguide
BO_i	Input backoff of amplifier
BO_0	Output backoff
P	Actual rf power output
$P_1, P_2, \cdots, P_n$	rf power in each of n carriers
P_I	rf power lost in intermodulation products, tube noise, and spurious frequencies due to multiple-carrier operation

FIGURE 3.150 Power flow in power amplifier. *Note*: There are additional rf losses between output flange and antenna feed(s) that are not shown.

The off-axis loss (item 8 of Table 3.50) can be estimated from the transponder's transmit antenna pattern.

The pointing error loss (item 9) is obtained from Figures 3.154–3.157 for conventional beams (for shaped beams the error box method is explained in Section 3.1.6). Note that for spacecraft with large antennas, it may be preferable to steer the antenna rather than the spacecraft. In this way thermal and mechanical misalignments between the antenna and spacecraft are more tolerable. Item 10, the nominal EIRP, may often be obtained directly from the EIRP pattern (especially for shaped beams; see, e.g., Figure 3.153). Multicarrier backoff corrections may still be needed.

TABLE 3.47 Spectrum Division among Transponders

Transponders in 500 MHz after 20 MHz offset for Polarization and TT&C[a]	Total allocation per transponder MHz	Guard Band[b] MHz	Useful MHz
1	480	0	480
2	240	24	216
3	160	16	144
4	120	12	108
5	96	10	86
6	80	8	72
8	60	6	54
10	48	5	43
12	40	4	36
14	34	3	31
16	30	3	27

[a]For dual polarization double these values. TT&C = telemetry, tracking, and control.
[b]at 10%

TABLE 3.48 Power Allocation to Each Carrier

Long-Term Effects

1. The C_d/kTB requirement of each down link in transponder
    (a) Each carrier's earth station G/T (T in C/kTB)
    (b) Each carrier's bandwidth (B in C/kTB)
    (c) Bit rate R of each carrier
    (d) Modulation and margin requirements of each carrier
2. Individual slant ranges (maximum variation ± 0.65 dB)
3. Rain margin requirements at various stations
4. Transmit antenna selected and its gain to each station
5. Demands of other carriers
6. Total bandwidth utilized in transponder
    (a) EIRP density
    (b) PFD limitations (especially for narrow-band services)
7. Intermodulation noise, AM-to-PM conversion noise, etc.
8. Diurnal traffic variations
9. Interference (including intersystem noise, electromagnetic interference, rf interference, and jamming).
10. Satellite motion with respect to earth station's antenna beam.

Short-Term Effects (Seconds to Few Minutes)

1. Up-link fading due to rain
2. Any decrease or increase in power demands of other carriers in transponder
3. Any change in bandwidth
4. Voice activation of channels
5. Any intermittent power shifting in transmit station's carrier level
6. Slant range variation for a non-geostationary satellite

TABLE 3.49 Down-Link Budget Approximation, Short Form

Earth Station location _____

Down-link frequency _____ GHz   Earth Station Antenna _____ m

Satellite _____   Down-link beam _____

Satellite

(1) Satellite EIRP, dBW   (+) _____ (a)

Satellite to Earth

(2) Free space path loss in clear sky
     (_____ km), dB   (−) _____ (b)

Earth Station

(3) Earth station G/T_s, dBi/K   (+) _____ (c)

(4) C_d/T_d, dBW/K   (=) _____ (d)

(5) 1/Boltzmann constant, $1/k$, dBW/K/Hz   (+) 228.6

(6) C_d/kT_d, dBHz   (=) _____ (e)

(a) Obtain from the satellite carrier. Use a backed off value per carrier in multicarrier
    cases.
(b) For an elevation angle of 15° (range = 40,000 km) and

f	1.5	4	11	12	19	GHz
Loss	187	197	205	206	210	dB

(d) = (1) − (2) + (3)
(e) = (4) + 228.6

Faraday rotation will cause an error in nonpolarization tracking earth stations, which appears in item 16 (see Section 3.1.4).

Item 17 is the effect of the increased sky temperature (and hence an increased T_s and a lowered G/T_s of the earth station) during precipitation. It is associated with the precipitation (rain) margin (14). See Figure 3.107.

The earth station maintenance margin (23) is a trade-off between the operating conditions, the maintenance schedule, and the ability to measure the actual in-field performance of an operating station. Typical values for unattended stations are in the 0.7–1-dB range assuming the antenna was initially aimed properly.

TABLE 3.50 Down-Link Budget, Long Form

Satellite _____ Application _____ Frequency _____ GHz

Earth station location _____

	Nominal	Worst Case	

Satellite

	Nominal	Worst Case	
1. Transmitter saturated power rating (____ W), dBW	(+) _____		P
2. Reserve for end-of-life loss, dB	(−) _____		
3. Losses due to multiplexer, filters, isolators, directional couplers, switches, VSWRs, transmission line hybrids, power dividers, and antenna feeds, dB	(−) _____		
4. Antenna gain, dBi	(+) _____		G_{td}
5. Saturated EIRP, dBW	(=) _____		
6. Output backoff for ____ carriers, dB	(−) _____		
7. Effective EIRP, dBW	(=) _____	(+) _____	
8. Off-beam center loss, dB	(−) _____	(−) _____	
9. Pointing error (attitude control, thermal misalignments, etc.), dB		(−) _____	
10. Nominal EIRP, dBW	(=) _____		$(EIRP)_d$
11. Worst case EIRP, dBW		(=) _____	

Space to Earth

	Nominal	Worst Case	
12. Clear-sky free-space path loss for ____ km, dB	(−) _____	(−) _____	
13. Atmospheric loss,[a] dB	(−) _____		
14. Precipitation loss margin,[a] dB for ____ % of ____	(−) _____		
15. Subtotal, dB	(=) _____	(=) _____	
16. Other propagation effect losses (scintillation, polarization coupling, etc.), dB		(−) _____	
17. Increase in sky noise temperature due to precipitation,[b] dB		_____	
18. Total, nominal case, dB	(=) _____		
19. Total, worst case, dB		(=) _____	

367

TABLE 3.50 (*Continued*)

Satellite _____ Application _____ Frequency _____ GHz

Earth station location _____

	Nominal	Worst Case

Earth Station

	Nominal	Worst Case
20. Earth station clear-sky figure of merit (G/T_s),[c] dBi/K	(+)_____	
21. Pointing error (satellite motion), dB		(−)_____
22. Subtotal, dBi/K	(+)_____	(=)_____
23. Earth Station maintenance margin, dB		(−)_____
24. Earth station performance, dBi/K	(=)_____	
25. Worst case performance, dBi/K		(=)_____

Summary

	Nominal	Worst Case
26. Satellite EIRP toward earth station, dBW	(+)_____ (10)	(+)_____ (11)
27. Total down-link losses, dB	(−)_____ (18)	(−)_____ (19)
28. Earth station G/T_s, dBi/K	(+)_____ (24)	(+)_____ (25)
29. C_d/T_d at earth station receiver output, dBW/K	(=)_____	(=)_____
30. 1/Boltzmann constant, dBW/K/Hz	+228.6	+228.6
31. C_1/kT_d at receiver output, dBHz	(=)_____	(=)_____
32. Satellite EIRP toward earth station, dBW	(+)_____ (10)	(+)_____ (11)
33. From Equation 3.9	(−)163.3	(−)163.3
34. Illumination levels, dBW/m²	(=)_____	(=)_____
35. Constant, dB	+85.7	+85.7
36. Electric field strength, dBμV	(=)_____	(=)_____
37. Earth Station receiver antenna gain, dBi	(+)_____	(+)_____
38. Receiver input signal, dBμV/m	(=)_____	(=)_____

[a]If significant or applicable.

[b]$10 \log_{10}(T_{sc}/T_{sp})$, where T_{sc} and T_{sp} are earth station system noise temperatures under clear sky and the above-specified precipitation conditions.

[c]Degradation of G/T_s due to precipitation handled in step 17.

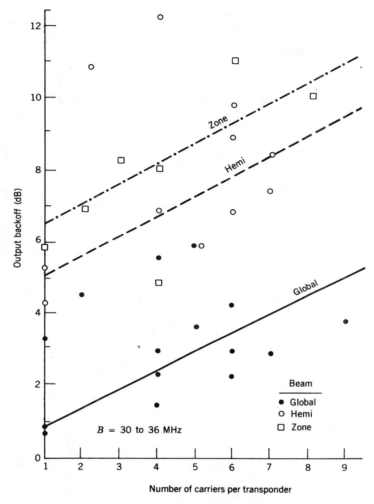

FIGURE 3.151 Multicarrier backoffs.

3.4.3 Intersatellite Link Budgets

Satellite-to-satellite links may be handled in a similar manner to the up and down links. Table 3.51 provides the basic elements. Assuming the path between the satellite is unobstructed by the earth or its atmosphere, the path loss is that of free space. Figure 3.158(a) shows how two widely separated satellites can provide extended coverage to the earth stations. Using Equation (3.87), the loss may be determined using Figure 3.135 or

$$L_{\text{range}} = 191.0 + 20 \log_{10}\left[\sin\left(\tfrac{1}{2}\alpha\right)\right] + 20 \log_{10} f \quad \text{(dB)} \quad (3.87)$$

	Carrier Number				Total	Units
	1	2	3	4		
Voice channels	24	24	432	60	540	
Occupied bandwidth	2	2	20.7	4	28.7	MHz
	63.0	63.0	73.2	66.0		dB Hz
Earth station class[a]	A	B	A	B		
Elevation angle	90	27	32	10	—	degrees
Range loss (with respect to sub-satellite point)	0.0	0.7	0.6	1.3	—	dB
G/T_s deficit (with respect to 40.7 dBi/K)	0.0	9.0	0.0	9.0	—	dBi/K
Satellite Antenna pattern pointing loss (with respect to on-axis gain)	0.0	0.9	0.8	2.8	—	dB
Total down link	63.0	73.6	74.6	79.1	81.3	dB
Power (dB down from total rf power available after backoff)	18.3	7.7	6.7	2.7	—	dB

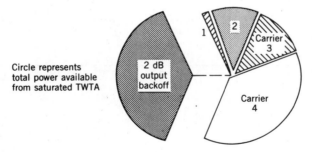

Circle represents total power available from saturated TWTA

2 dB output backoff

[a]Intelsat Standard A (original) required 40.7 dBi/K, Standard B requires 37.1 dBi/K

FIGURE 3.152 Hypothetical four-carrier power split.

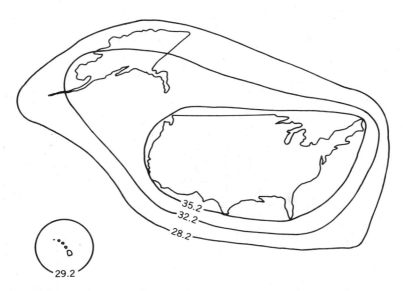

FIGURE 3.153 Down-link Coverage (radiated power level in dBW at 4 GHz).

where α is the longitudinal angular separation (in degrees) between two geosynchronous satellites as shown in Figure 3.158(b). Likewise, the illumination level W may be expressed as

$$W = \text{EIRP} - 169.5 - 20 \log_{10}\left[\sin\left(\tfrac{1}{2}\alpha\right)\right] - L_{\text{add}} \quad (\text{dBW/m}^2) \quad (3.88)$$

and the electric field strength E is

$$E = \text{EIRP} - 83.8 - 20 \log_{10}\left[\sin\left(\tfrac{1}{2}\alpha\right)\right] - L_{\text{add}} \quad (\text{dB}\mu\text{V/m}) \quad (3.89)$$

where L_{add} are the additional losses due to items 8 and 12 of Table 3.51. Figure 3.155 may be used to estimate these antenna-pointing errors.

The path loss (item 10) and the illumination level are obtained using Equations (3.87) and (3.88), respectively.

Figure 3.158(b) illustrates the angular separations between two geosynchronous satellites.

For further information consult Lo & Mizuno (1988), Lee (1988) and Millies-Lacroix & Peters (1988).

3.4.4 Overall Link Performance

The preceding three sections have defined the up-, down-, and cross-link performances. This section looks at the overall system performance.

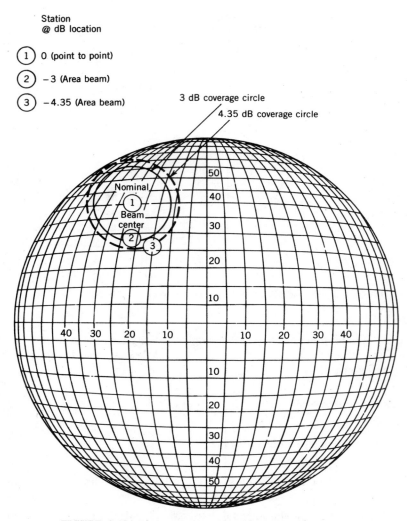

FIGURE 3.154 Earth station location with respect to beam center.

Assume a link between earth stations A and B, as shown in Figure 3.159. The up-link carrier to thermal noise ratio (C_u/T_u) is measured at the output of satellite 1's receiver, the cross-link ratio (C_x/T_x) was determined at the output of receiver 2, and the down-link ratio was determined at the third receiver's output in station B. Each calculation was done in isolation to the others. In reality, this is not the case, as each step adds more thermal noise, so that the earth station B output noise power is the power sum of all three noise sources (unless regeneration is used at some point to remove some of the noise power).

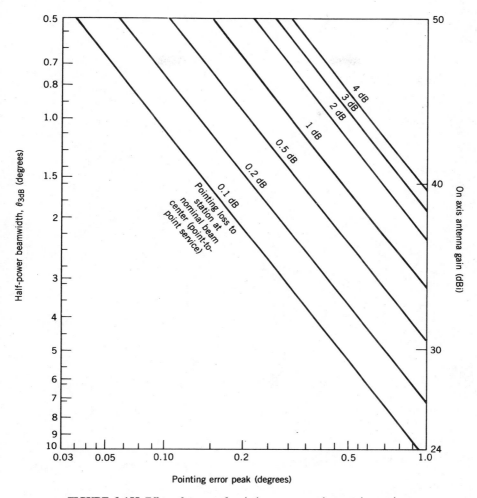

FIGURE 3.155 Effect of spacecraft pointing error on point-to-point service.

To combine noise powers, it is necessary to convert the normal decibel qualities back to absolute values:

$$(C/T)^* = 10^{(C/T)/10} \quad (\text{W/K}) \tag{3.90}$$

or

$$C/T = 10 \log_{10}(C/T)^* \quad (\text{dBW/K}) \tag{3.91}$$

where $(C/T)^*$ is the absolute ratio and C/T is the decibel (dBW/K) value.

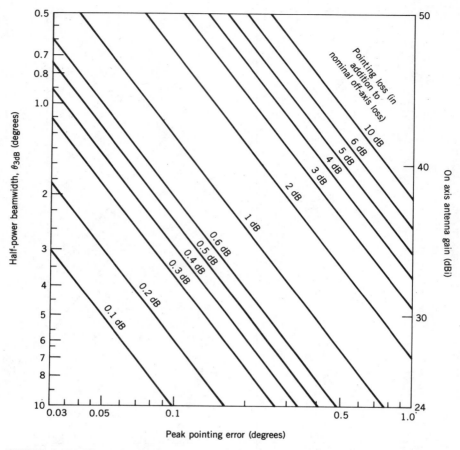

FIGURE 3.156 Effect of spacecraft pointing error on station on edge of nominal 3-dB coverage circle.

The subscripts of Figure 3.159 designate what links are included in each C/T. Single-letter subscripts assume that there is no noise contribution from any prior link. Multiple-letter subscripts illustrate the cumulative effect of the noise. The actual total system ratio for nonregenerative repeaters is

$$\left[\frac{1}{(C_t/T_t)*}\right] = \left[\frac{1}{(C_u/T_u)*}\right] + \left[\frac{1}{(C_x/T_x)*}\right] + \left[\frac{1}{(C_d/T_d)*}\right] \quad (K/W)$$

$$(3.92)$$

Using the more familiar dBW/K units for C/T yields

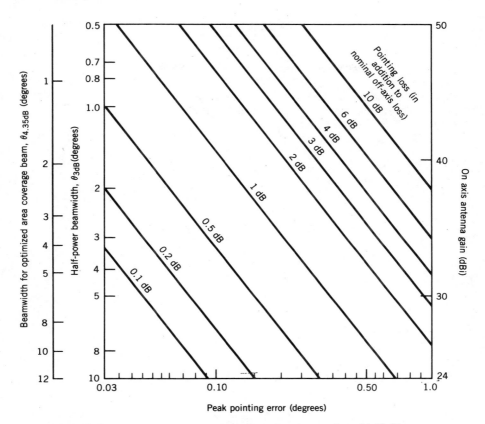

FIGURE 3.157 Effect of spacecraft pointing error on station on edge of 4.35-dB coverage.

$$\left[\frac{1}{10^{[(C_t/T_t)/10]}}\right] = \left[\frac{1}{10^{[(C_u/T_u)/10]}}\right] + \left[\frac{1}{10^{[(C_x/T_x)/10]}}\right]$$

$$+ \left[\frac{1}{10^{[(C_d/T_d)/10]}}\right] \quad (K/W) \tag{3.93}$$

Section 3.2.6 contains instructions on how to combine an up link and a down link using Figure 3.137 or 3.138. To combine three links requires two steps. See Figure 3.159. To illustrate, assume

$$C_u/T_u = -122 \text{ dBW}/\text{K} \quad (\text{or } 6.3 \times 10^{-13}\text{W}/\text{K}:1)$$

$$C_x/T_x = -117 \text{ dBW}/\text{K} \quad (\text{or } 2 \times 10^{-12}\text{W}/\text{K}:1)$$

$$C_d/T_d = -134 \text{ dBW}/\text{K} \quad (\text{or } 4 \times 10^{-14}\text{W}/\text{K}:1)$$

TABLE 3.51 Cross-Link Budget

Transmitter

1. Transmitter saturated power rating
   (_____ W), dBW   _____
2. Reserve for end-of-life loss, dB $(-)$ _____
3. Losses due to switches, power monitor,
   transmission line, etc., $(-)$ _____
4. Antenna gain, dBi   _____

5. Saturated EIRP, dBW   _____
6. Output backoff for _____ carriers, $(-)$ _____

7. Effective EIRP, dBW   _____
8. Pointing error (attitude control and
   uncertainty as to where other satellite is
   located), dB $(-)$ _____

9. Nominal EIRP, dBW   _____

Path

10. Path loss, dB $(-)$ _____

Receiver

11. Figure of merit (G/T_s), dBi/K $(+)$ _____
12. Pointing error (step 8), dB $(-)$ _____

13. C_x/T_x at receiver output, dBW/K   _____
14. 1/Boltzmann constant, dBW/K/Hz $(+)$ 228.6
15. C_x/kT_x at receiver output, dBHz   _____

Then:

1. Combine C_u/T_u and C_x/T_x to obtain C_{ux}/T_{ux} using Figure 3.138:

$$\pm C_u/T_u \pm C_x/T_x = 5\text{-dB difference} \quad (\text{or reduction of 1.2 dB})$$
$$C_{ux}/T_{ux} = -122 - 1.2 = -123.2 \text{ dBW/K}$$

2. Combine C_{ux}/T_{ux} and C_d/T_d to obtain C_{uxd}/T_{uxd} using Figure 3.137:

$$\pm C_{ux}/T_{ux} \pm C_d/T_d = 10.8\text{-dB difference} \quad (\text{or reduction of 0.4 dB})$$
$$C_{uxd}/t_{uxd} = -134 - 0.4 = -134.4 \text{ dBW/K} = C_t/T_t$$

(*Note*: C_d/T_d is smaller than C_{ux}/T_{ux} and thus was used.)

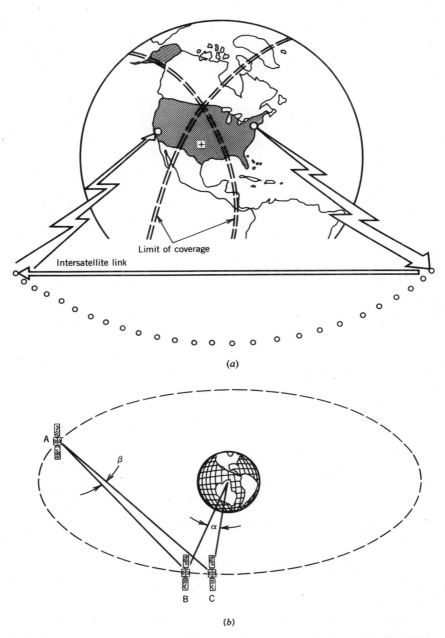

Limit of coverage

Intersatellite link

(a)

β

α

A

B   C

(b)

FIGURE 3.158 Extension of orbit arc using intersatellite link; (a) domestic service (+, station can operate with both satellites); (b) cross-link geometry.

377

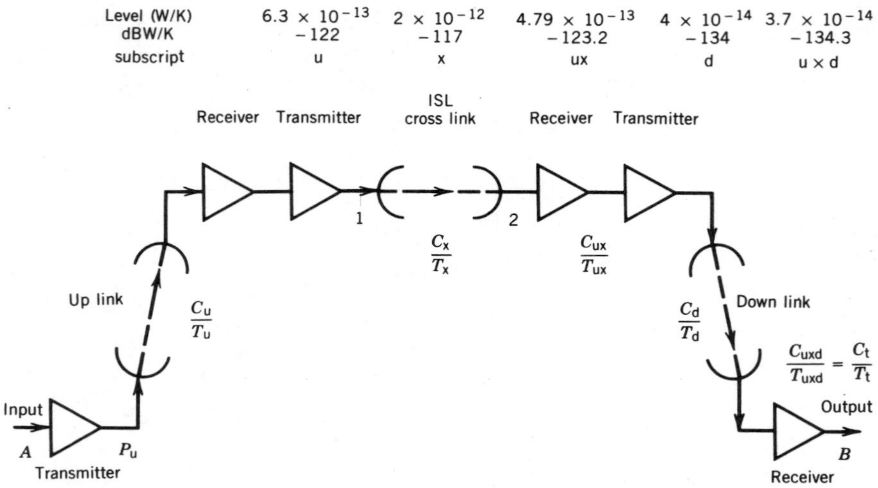

Level (W/K)	6.3×10^{-13}	2×10^{-12}	4.79×10^{-13}	4×10^{-14}	3.7×10^{-14}
dBW/K	-122	-117	-123.2	-134	-134.3
subscript	u	x	ux	d	u × d

FIGURE 3.159 Path with intersatellite link.

The alternate method uses the following:

$$\frac{1}{(C_t/T_t)^*} = \frac{1}{6.3 \times 10^{-13}} + \frac{1}{2 \times 10^{-12}} + \frac{1}{4 \times 10^{-14}} \text{ K/W} \qquad (3.94a)$$

$$\frac{1}{(C_t/T_t)^*} = 2.71 \times 10^{13} \text{ K/W} \qquad (3.94b)$$

$$(C_t/T_t)^* = 3.69 \times 10^{-14} \text{ W/K} \qquad (3.94c)$$

$$C_t/T_t = 10 \log_{10}(C_t/T_t)^* \qquad (3.94d)$$

$$= -134.3 \text{ dBW/K}$$

The same approach may be used whenever noise ratios are being combined. Equations (3.90) and (3.92) or Figure 3.137 may be used by converting each C/T to C/kT or C/N_0 or C/N. *All units must be uniform.*

Tables 3.52–3.54 are illustrative link budgets for international fixed, direct broadcast, and maritime satellite services. See also Tamir & Rappaport (1987).

In Table 3.52, four different beam sizes and three classes of earth stations are employed. Multicarrier backoffs are necessary for the earth station and satellite transmitters.

Figure 3.160 shows the carrier power levels in a television link between 30-m earth stations using an Intelsat satellite. Figure 3.161 shows this same link with the noise power at each step.

TABLE 3.52 *Intelsat* V Link Budget

	Up-Links			
Frequency Band, GHz	6	6	6	14
Earth Station Type[a]	A	B	A	C
Antenna diameter, m	30	11	30	12
Up-link beam (see Figure 3.51)	Global	Hemi	Zone	West Spot
EARTH STATION				
Power into antenna line, dBW	31.6	40.3	31.6	28.9
Antenna gain on axis dBi	63.3	54.6	63.3	62.7
Worst case EIRP, dBW[b]	93.9	93.9	93.9	90.6
EARTH TO SPACE				
Free-space path loss (40,000 km), dB	200.1	200.1	200.1	207.4
Precipitation, dB	—	—	—	5
Margin, dB	3	3	3	3
SATELLITE				
Satellite G/T_s, dBi/K	−18.6	−11.6	−8.6	−3.3
C_u/T_u, dBW/K	−127.8	−120.8	−117.8	−128.1
Illumination level, dBW/m^2	−69.4	−69.4	−69.4	−72.7
	Down-Links			
Frequency band, GHz	4	4	4	11
Down-link beam	Global	Hemi	Zone	East Spot
Earth station type[a]	A	B	A	C
SATELLITE				
Saturated EIRP, dBW	26.5	29.0	⁻29.0	41.1
Output backoff, dB	10	8	7	4
Worst case EIRP, dBW	16.5	21.0	22.0	37.1
SPACE TO EARTH				
Free-space path loss (40,000 km), dB	196.5	196.5	196.5	205.3
Precipitation, dB	—	—	—	4
Margin, dB	3	3	3	3
Sky noise temperature increase, dB	—	—	—	2
EARTH STATION				
Figure of merit, dBi/K	40.7	31.7	40.7	37
C_d/T_d, dBW/K	−142.3	−146.8	−136.8	−140.2
Illumination level, dBW/m^{2c}	−146.8	−142.3	−141.3	−126.2
OVERALL				
C_t/T_t, dBW/K	−142.5	−146.8	−136.9	−140.5
1/Boltzmann's constant, dBW/K/Hz	228.6	228.6	228.6	228.6
C_t/kT_t, dBHz	86.1	81.8	91.7	88.1

[a]Original Intelsat earth station standards: A, 30-m antenna ($G/T_s = 40.7$ dBi/K); B, 11-m antenna ($G/T_s = 31.7$ dBi/K); C, 12-m antenna ($G/T_s = 37$ dBi/K).

[b]Assumes a 1-dB transmission line loss.

[c]Assumes earth station at extreme edge of earth coverage. This value is the EIRP − 163.3 dB

TABLE 3.53 Direct Broadcast Link Budget

Up-Link

Frequency band, GHz	18
Earth station type	Major station (13 m)

EARTH STATION

Transmitter power (100 W), dBW/channel	20.0
Feed losses, dB	−3.5
Antenna gain, dBi	65.2
EIRP, dBW	81.7

EARTH TO SPACE

Clear-sky path attenuation, dB	−209.6

SATELLITE

Satellite antenna gain (on axis), dBi	39.5
Satellite system noise temperature (1000 K), dBK	−30
Satellite G/T on axis, dBi/K	+9.5
Illumination level, dBW/m^2	−81.6
C_u/T_u, dBW/K	−118.4
1/Boltzmann constant, dBHzK/W	228.6
C_u/kT, dBHz	110.2
1/Bandwidth (24 MHz), dBHz	−73.8
C_u/N_u, dB	36.4

Down-Links

Frequency band, GHz	12	12
Satellite beam	Spot	Conus
Earth station antenna diameter, m	0.75	1.8

SATELLITE

Satellite transmitter (200 W/dBW/channel)	23.0	23.0
Feed losses, dB	−1.7	−1.7
Satellite antenna gain, dBi	34.0	30.0
Nominal EIRP, dBW	55.3	51.3

SPACE TO EARTH

Free-space path loss, dB	−205.8	−205.8
Illumination level (no rain), dBW/m^2	−108.0	−112.0
Precipitation loss (for an availability of 99.99% in any month), dB	−7.0	−7.0

EARTH STATIONS

Antenna gain, dBi	37.6	45.2
Earth station noise temperature, K	550	480
System noise temperature, dBK	27.4	26.8
G/T dBi/K	10.2	18.4
C_d/T_d, dBW/K	−147.3	−143.1
1/Boltzmann constant, dBW/K/Hz	228.6	228.6

TABLE 3.53 (*Continued*)

Down-Links (continued)

EARTH STATIONS (continued)

C_d/kT_d, dB/Hz	81.3	85.5
1/bandwidth (24 MHz), dBHz	−73.8	−73.8
C_d/N_d, dB (in rain)	7.5	11.7
Threshold, dB	7.4	7.4
Link margin (in rain), dB	0.1	4.3
Ratio of 4 kHz to 24 MHz,[a] dB	−37.8	−37.8
Power flux density,[a] dBW/m² per 4 kHz	−145.8	−149.8

Summary

C_u/N_u, dB	36.4	36.4
C_d/N_d,[b] dB	17.1, 7.5	21.5, 11.7
C_t/N_t,[b] dB	17.1, 7.5	21.4, 11.7
Threshold, dB	7.4	7.4
Link margin,[b] dB	9.7, 0.1	14.0, 4.3
Application, NTSC	Color TV	Color TV
Weighting, dB	10.2	10.2
FM improvement, dB	23.0	23.0
S_t/N_t (peak-to-peak video/rms noise),[b] dB	50.3, 40.7	54.6, 44.9

[a]Assumes power is uniformly distributed over 24 MHz (*note:* use of 4 kHz in 12-GHz direct broadcast satellite television band is arbitrary).

[b]First number is for clear-sky conditions; the second has 7 dB rain degradation and associated increased sky noise in down link.

TABLE 3.54 Maritime Satellite Service Link Budget[ab]

Parameter	Voice	TDM
Shore to satellite, 6 GHz		
Shore station transmitter power, dBW	11.9	9.5
Combining network and feeder loss, dB	2.1	2.1
Shore station antenna gain, dBi	56.2	56.2
Shore station EIRP, dBW	66.0	63.6
Free space attenuation, dB	200.8	200.8
Absorption with precipitation 10° elevation, dB	0.7	0.7
Polarization coupling loss, dB	0.4	0.4
Satellite G/T, dBi/K	−19.5	−19.5
Saturation flux density, (dBW/m²)	−87.0	−87.0
Up-Link C/N_0, dBHz	73.2	70.8
Satellite to ship, 1.5 GHz		
Intermodulation, noise, and pilot losses, dB	1.4	1.4
Small carrier suppression, dB	0.0	0.6
Satellite EIRP, dBW	15.6	12.6
Free space loss, dB	188.5	188.5
Polarization coupling loss, dB	0.4	0.4
Absorption, 5° elevation, dB	0.4	0.4

TABLE 3.54 (*Continued*)

Parameter	Voice	TDM
Short-term fading, dB	4.0	4.0
Additional wet radome degradation, dB	0.5	0.5
Ship terminal G/T (including losses due to dry radome and pointing error, dB/K	−4.0	−4.0
Down-link C/N_0 with short-term fade, dBHz	46.4	43.4
Down-link C/N_0 without short-term fade, dBHz	50.4	47.4
C/N_0 intermodulation, dBHz	>70.0	>67.0
Total link C/N_0 without fade, dBHz	50.4	47.4

Parameter	Voice or TDMA/ Request
Ship to satellite, 1.6 GHz	
Ship transmitter power, dBW	14.8
Diplexer loss, dB	0.6
Ship antenna gain, dBi	23.0
Dry radome loss, dB	0.2
Ship EIRP, dBW	37.0
Free space attenuation, dB	189.1
Additional wet radome loss, dB	0.3
Absorption for 5° elevation, dB	0.4
Polarization coupling loss, dB	0.4
Short-term fading, dB	4.0
Satellite G/T dB/K	−17.0
Saturation flux density, dBW/m^2	−99.5
Up-Link C/N_0 with short-term fade, dBHz	54.4
Up-Link C/N_0 without Short-Term Fade, dBHz	58.4
Satellite to shore, 4 GHz	
Gain compression and intermodulation loss, worst case, dB	1.8
Satellite EIRP without L-band short-term fade, dBW	−4.6
Free space loss, dB	197.1
Absorption, 10° elevation, dB	0.2
Polarization coupling loss, dB	0.4
Excess attenuation with precipitation, dB	0.5
Excess sky noise degradation with precipitation, dB	1.2
Shore terminal G/T, (clear sky, 10° elevation), dB/K	32.2
Down-link C/N_0 without short-term L-band fade, dB-Hz	56.8
Down-link C/N_0 with short-term L-band fade, dB-Hz	52.8
Worst-case C/N_0 intermodulation, dB-Hz	>62.0
Total link C/N_0 without short-term L-band fade, dB-Hz	53.8
Total link C/N_0 with short-term L-band fade, dB-Hz	49.8
Total link C/N_0 without fade, dBHz	56.0

[a]In the high-power mode intermodulation noise will be greater. Also, the shore-to-ship satellite repeater gain will increase, so that the up-link power per carrier will decrease.

[b]Reprinted with permission of the COMSAT Technical Review, the COMSAT Corporation.

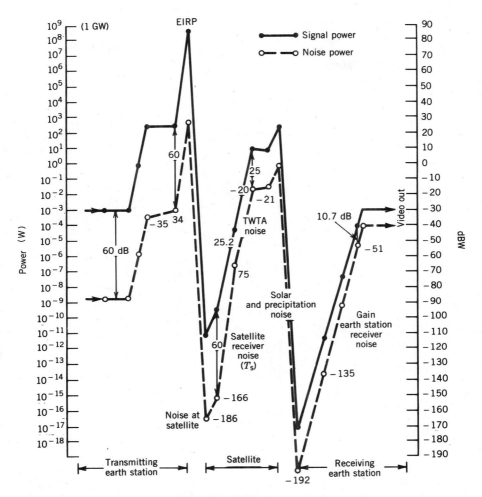

FIGURE 3.160 Power levels in an Intelsat satellite.

3.4.5 Signal-to-Noise Ratios

To derive the signal-to-noise ratio S/N from the carrier-to-noise ratio, more must be known about the signal, its modulation, and how S/N is expressed. The weighting and FM improvements are dependent on the actual television system employed (see Figure 2.6) and the modulation index:

$$S_t/N_t = C_t/T_t - k + 7.8 + 20 \log_{10} d - 10 \log_{10} f_m + Q_p - M \quad \text{(dB)}$$

$$(3.95)$$

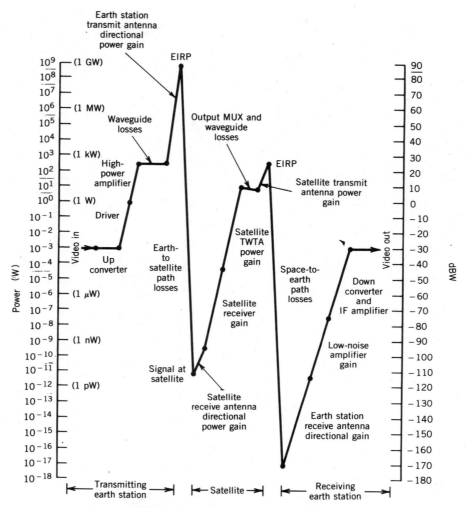

FIGURE 3.161 Carrier-to-noise ratios in link.

where S_t/N_t = ratio of peak-to-peak video signal to rms weighted noise in decibels for television

k = Boltzmann's constant (-228.6 dBW/K/Hz)

d = peak-to-peak modulation index (ratio)

f_m = video signal bandwidth in hertz (4.2×10^6 for standard M)

Q_p = pre-emphasis and weighting factor in decibels (13 dB for standard M)

M = total system operating (including implementation) margin in decibels

7.8 = $10 \log_{10} 6$

6 = conversion factor of rms noise to peak-to-peak luminance signal

Equations (3.96) and (3.97) provide two alternative means of expression. The S_t/N_t value is usually the stated requirement:

$$(C_t/T_t)_{\text{req}} = (S_t/N_t)_{\text{req}} - 236.4 - 20 \log_{10} d$$

$$+ 10 \log_{10} f_m - Q_p + M \quad (\text{dBW/K}) \tag{3.96}$$

$$(C_t/N_t)_{\text{req}} = (S_t/N_t)_{\text{req}} - 7.8 + k - 10_{10} \log B - 20 \log_{10} d$$

$$+ 10 \log_{10} f_m - Q_p + M \quad (\text{dBW/K}) \tag{3.97}$$

where k is Boltzmann's constant (-228.6 dBW/K/Hz)

Figure 3.162 may be used when solving for S_t/N_t for NTSC color television (standard M). Table 3.55 provides information for conversion to other global television systems (see also Chapter 2 for details on these systems).

The CCIR recommended minimum S/N for feeds into a national distribution network is 52 dB for standards B, C, G, H, and I, 57 dB for standards D, K, and L, and 56 dB for standard M. Direct broadcast services (and many indirect broadcast services such as CATV) tend to use lower values.

A more complete set of requirements is contained in EIA and CCIR standards and in Freeman, (1985).

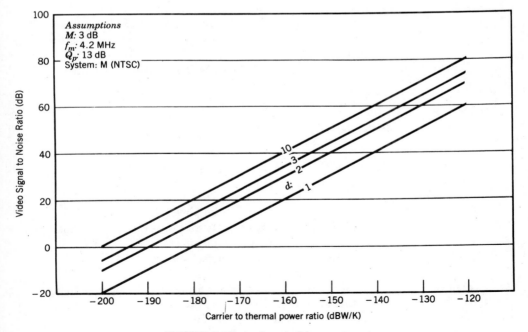

FIGURE 3.162 Analog television nomogram.

TABLE 3.55 Television Transmission Parameters

Bandwidth Used in Receiver B (MHz)	Highest Baseband Frequency, (MHz) f_m	Peak-to-Peak Frequency Deviation, (MHz) d	Modulation Index, (ratio) M	FM Improvement $3M^2(M-1)$, (dB)
36	4.5	13.5	3.0	20.3
	6.0	12	2.0	15.6
27	6.0	7.5	1.25	10.2
25	4.5	8	1.78	9.2
	6.0	6.5	1.08	8.7
19.18	4.5	5.1	1.13	9.1
	6.0	3.6	0.60	2.3
17.5	4.5	4.25	0.94	7.2
	6.0	2.75	0.46	−0.3
17	4.5	4.00	0.89	0.7

For FM telephony, for example, the signal-to-noise ratio may be expressed as

$$S_t/N_t = C_t/T_t + 236.4 + 20 \log d - 10 \log f_m + Q_p - M \text{ (dB)}$$

where S_t/N_t = noise ratio of rms test tone to rms weighted in decibels
$d = f/f_m$, where f is peak deviation
f_m = top baseband frequency in hertz
M = implementation margin (for nonideal demodulation, degradation due to oscillator phase noise, and intermodulation noise for narrowband signals

Television quality is sometimes expressed in TASO units (see Table 3.56).

3.4.6 Digital Link Performance

For digital transmission the required C_t/T_t is

$$\frac{C_t}{T_{t,\text{req}}} = \frac{E_b}{N_0} + M + R - 228.6 \text{ (dBW/K)} \tag{3.98}$$

where E_b/N_0 = ratio of bit energy to channel noise power density
M = implementation margin in decibels for losses due to demodulator performance and nonlinear channel effects
R = information bit rate expressed in decibels above 1 b/s.
228.6 = reciprocal of Boltzmann's constant in dBW/K/Hz

Figures 3.163 and 3.164 are two nomograms for using this equation. In Figure 3.164 the example of a digital television service is used. In this case the implementation and precipitation margins have been combined. Thus, the result is the bit rate that can be supported with rain.

TABLE 3.56 Television Quality Measurements

TASO Grade	Quality Details	S/N (TASO) dB	$S/N_{video,w}$ (CCIR),[a] dB
1 Excellent	Extremely high quality, as good as could be desired	40[b]	49.5
2 Fine	High-quality enjoyable viewing; perceptible interference	38	40.3
3 Passable	Acceptable quality; interference not objectionable	31	32.2
4 Marginal	Poor quality, improvement desired; interference somewhat objectionable	25	25.9
5 Inferior	Very poor quality but could be watched; definitely objectionable interference	19	19.9
6 Unusable	Too bad to be watched	—	—

[a]$S/N_{video,w}$ is the ratio of weighted peak-to-peak video to noise power.
[b]Valid for 65% of viewers. Other figures are based on viewer's percentage of 75%.

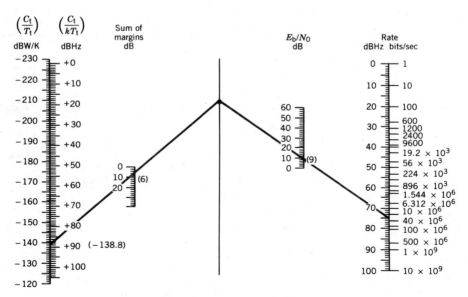

FIGURE 3.163 Digital link performance.

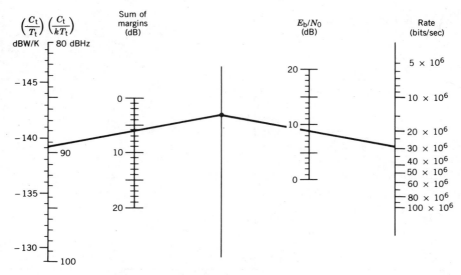

Example: $R = 29$ Mbits/sec $E_b/N_0 = 9$ dB

Precipitation margin = 3 dB

Implementation margin = 3 dB

Solution: $C_t/T_t = -139.0$ dBW/K

FIGURE 3.164 Digital television performance.

3.4.7 Summary of Useful Up-Link Equations

The following geostationary orbit link equations have been collected together for the user's convenience (see also Tables 3.57 and 3.58).

If the satellite is not in the geostationary orbit (and/or the range, S, is not 41,679 km) use Table 3.59 to modify the up link equations.

$$P_u = 10^{[(L_{add} + M_{du} - G/T_s - 20\log_{10}D_u - 10\log_{10}\eta_a + C_u/T_u + 164.4)/10]} \quad \text{(W)} \quad (3.99)$$

$$D_u = 10^{[(C_u/T_u - G/T_s + L_{add} + M_{du} + M_{tu} - 10\log_{10}P_u - 10\log_{10}\eta_a + 164.4)/20]} \quad \text{(m)}$$
$$(3.100)$$

$$\text{EIRP}_u = C_u/T_u - G/T_s + L_{add} + M_{tu} + 20\log_{10}f_u + -184.8 \quad \text{(dBW)}$$
$$(3.101)$$

$$M_{du} + M_{tu} = -\text{EIRP}_u + -G/T_s - C_u/T_u - 20\log_{10}f_u - 184.8 - L_{add} \quad \text{(dB)}$$
$$(3.102)$$

$$G/T_s = L_{add} + M_{tu} - \text{EIRP}_u + C_u/T_u + -20 - \log_{10}f_u + 184.8 \quad \text{(dBi/K)}$$
$$(3.103)$$

TABLE 3.57 Guide to Useful Up-link Equations

Given	Equations
Existing earth station	(3.103)
Existing satellite	(3.99), (3.101) (3.106)
Existing link (earth station and satellite)	
Known $EIRP_u$ and G/T_s	(3.100), (3.105), (3.112), (3.115)
Known P_u, D_u and G/T_s	(3.107), (3.111), (3.113)
Power-limited earth station	(3.108), (3.111), (3.113)
Size-limited earth station antenna	(3.99), (3.107), (3.111), (3.113)
Noise-limited satellite	(3.102)

Requirement	Equations
Maximum bit rate, R_u	(3.115), (3.116)
E_b/N_0	(3.113), (3.114)
C_u/T_u	(3.105), (3.106)
C_u/kT_u or C_u/N_{0u}	(3.107), (3.108), (3.109)
C_u/N_u	(3.110), (3.111), (3.112)
W_u	(3.104)
Q_u	(3.117), (3.118), (3.119), (3.120)

$$W_u = EIRP_u - L_{add} - 163.3 \quad (dBW/m^2) \tag{3.104}$$

$$C_u/T_u = EIRP_u + G/T_s - L_{add} - M_{tu} - 20\log_{10}f_u - 184.8 \quad (dBW/K) \tag{3.105}$$

$$C_u/T_u = W_u + G/T_s - 20\log_{10}f_u - 21.5 - M_{tu} \quad (dBW/K) \tag{3.106}$$

$$C_u/kT_u = C_u/T_u + 228.6 \quad (dBHz) \tag{3.107}$$

$$C_u/kT_u = 20\log_{10}D_u + 10\log_{10}\eta_a + 10\log_{10}P_u$$
$$- L_{add} + G/T_s + 64.2 - M_{tu} - M_{du} \quad (dBHz) \tag{3.108}$$

$$C_u/N_{0u} = C_u/kT_u \quad (dBHz) \tag{3.109}$$

$$C_u/N_u = C_u/kT_u - 10\log_{10}B_u \quad (dB) \tag{3.110}$$

$$C_u/N_u = 20\log_{10}D_u + 10\log_{10}\eta_a + 10\log_{10}P_u - L_{add} - M_{tu} - M_{du}$$
$$+ G/T_s - 10\log_{10}B_u + 64.2 \quad (dB) \tag{3.111}$$

$$C_u/N_u = EIRP_u + G/T_s - L_{add} - 20\log_{10}f_u - 10\log_{10}B_u$$
$$+ 43.8 - M_{tu} - M_{du} \quad (dB) \tag{3.112}$$

$$(E_b/N_0)_u = 20\log_{10}D_u + 10\log_{10}\eta_a + 10\log_{10}P_u - L_{add}$$
$$+ G/T_s - 10\log_{10}R - M_{du} - M_{tu} + 64.2 \quad (dB) \tag{3.113}$$

TABLE 3.58 Symbols for Up-Link Equations

Symbol	Meaning	Units	First used in Equation
M_{tu}	Margins for satellite nonlinearities	dB	
P_u	Earth station transmitter power rating	W	
Q_u	EIRP + G/T_s for up link	dBiW/K	(3.25)
R_u	Up-link bit rate capacity	bits/sec	
S	Path length (assumed to be 41,679 km, distance from edge of earth to geostationary satellite); for subsatellite point adjust answers by 1.3 dB, see Figure 3.30 for other locations	km	
T_s	Satellite receiving system noise temperature	dBK	(3.45)
W_u	Illumination level at satellite (from earth station on edge of earth)	dBW/m²	(3.7–3.11)
η_a	Earth station antenna efficiency	Decimal	(3.33–3.40)
Subscript u	Up link		
B_u	Noise bandwidth of satellite receiver or channel	Hz	(3.60)
C	Carrier power	W	
C_u/kT_u	See C_u/N_{0u}	dBHz	(3.59)
C_u/N_u	Carrier-to-noise ratio of uplink	dB	(3.60)
C_u/N_{0u}	Carrier-to-noise-density ratio	dBHz	(3.60)
C_u/T_u	Up-link carrier-to-thermal-noise power ratio	dBW/K	(3.58)
D_u	Up-link earth station antenna diameter	m	(3.100)
$(E_b/N_0)_u$	Up-link ratio of bit energy to channel noise power density	dB	(3.98)
$EIRP_u$	Earth station equivalent isotropically radiated power	dBW	(3.55)
f_u	Up-link frequency	GHz	(3.101)
G	Satellite receiver antenna gain	dBi	(3.33–3.40)
G/T_s	Satellite figure of merit	dBi/K	(3.45)
L_{add}	Additional up-link losses (rain and atmospheric)	dB	(3.113)
M_{du}	Up-link degradations due to earth station high power amplifier (also includes up-link modem implementation margin for remodulating repeater)	dB	
S	Slant range	km	(3.14)
21.5	Units Conversion	dB[a]	(3.58)
163.3	Units Conversion	dB[a]	(3.9)
164.4	(92.45 − 20.4 + 20 log₁₀ (41679))	dB[a]	(3.99)
184.8	(21.5 + 163.3)	dB[a]	(3.101)
228.6	k	dBW/K/Hz	(3.59)
43.8	(228.6 − 21.5 − 163.3)	dB[a]	(3.112)
64.2	(228.6 − 184.8 + 20.4)	dB[a]	(3.113)

[a]The decibel units are given in the List of Constants.

$$\left(E_b/N_0\right)_u = \text{EIRP}_u + G/T_s - L_{add} - 20\log_{10}f_u$$
$$- 10\log_{10}R_u - M_{du} - M_{tu} + 43.8 \quad (\text{dB}) \tag{3.114}$$

$$R_u = 10^{[(\text{EIRP}_u + G/T_s - L_{add} - 20\log_{10}f_u - (E_b/N_0)_u - M_{tu} + 43.8)/10]} \quad (\text{bits/sec}) \tag{3.115}$$

$$R_u = 10^{([C_u/T_u - (E_b/N_0)_u - M_{du} + 228.6]/10)} \quad (\text{bits/sec}) \tag{3.116}$$

$$Q_u = \text{EIRP}_u + G/T_s \quad (\text{dBiW/k}) \tag{3.117}$$

$$Q_u = C_u/T_u + 20\log_{10}f_u + L_{add} + 184.8 \quad (\text{dBiW/k}) \tag{3.118}$$

$$Q_u = \left(E_b/N_0\right)_u + 10\log_{10}R_u + 20\log_{10}f_u$$
$$+ L_{add} - 43.8 \quad (\text{dBiW/k}) \tag{3.119}$$

$$Q_u = C_u/N_u + 10\log_{10}B_u + 20\log_{10}f_u$$
$$+ L_{add} - 43.8 \quad (\text{dBiW/k}) \tag{3.120}$$

3.4.8 Summary of Useful Cross-Link Equations

These equations may be useful in the design of links between two geostationary satellites separated by α degrees of longitude (see also Equations 3.57, 3.87–3.89, Figure 3.135 and Tables 3.59 and 3.60):

$$L_x = 92.45 + 20\log_{10}f_x + 20\log_{10}S_x + L_{addx} \quad (\text{dB}) \tag{3.121}$$

$$S_x = 2 \times 42161 \sin(\alpha/2) \quad (\text{km}) \tag{3.122}$$

$$L_x = 191.0 + 20\log_{10}\sin(\alpha/2) + 20\log_{10}f_x$$
$$= 191.0 + 20\log_{10}f_x + \beta + L_{addx} \quad (\text{dB}) \tag{3.123}$$

$$t_x = \frac{S_x}{c} \quad (\text{sec}) \tag{3.124}$$

$$c = 2.998 \times 10^5 \; (\text{km/sec}) \tag{3.125}$$

$$\beta = 20\log_{10}\left(\sin(\alpha/2)\right) \tag{3.126}$$

$$\theta_3 = \frac{?1}{f_x D_x} \quad (\text{degrees}) \tag{3.127}$$

$$G_x = 20.4 + 20\log_{10}D_x + 20\log_{10}f_x + 10\log_{10}\eta_x \quad (\text{dBi}) \tag{3.128}$$

$$P_x = 10\log_{10}P_{0x} - L_{linex} \quad (\text{dBW}) \tag{3.129}$$

$$\text{EIRP}_x = P_x + G_x \quad (\text{dBW}) \tag{3.130}$$

$$\text{EIRP}_x = C_x/T_x - G_x/T_{sx} + 20\log_{10}f_x + 191.0 + \beta \quad (\text{dBW}) \tag{3.131}$$

TABLE 3.59 Up-link Modifications for Nongeostationary Satellites

In Equations	Replace	With	Reference Equations
3.99, 3.100	164.4	$72.05 + 20\log_{10}S$	(3.15)
3.104	163.3	$71 + 20\log_{10}S$	(3.9)
3.101–3.103, 3.105, 3.118	184.8	$92.45 + 20\log_{10}S$	(3.56)
3.106	W_u	$EIRP_u - 20\log_{10}S - 71$	[a]
3.108, 3.111, 3.113	64.2	$156.6 - 20\log_{10}S$	(3.8)
3.112, 3.114, 3.115, 3.119, 3.120	43.8	$136.1 - 20\log_{10}S$	[b]

[a] $156.6 = 228.6 + 20.4 - 92.45$

[b] $136.1 = 228.6 - 21.5 - 71$

TABLE 3.60 Guide to Useful Cross-link Equations

Given	Equations
Antenna diameter	(3.127), (3.128)
Satellite separation	(3.121), (3.122), (3.123), (3.124),
	(3.131), (3.132), (3.134), (3.136)
Pointing accuracy	(3.127)
Maximum time delay	(3.124)

Requirement	Equations
Antenna gains	(3.128)
Power amplifier rating	(3.129)
EIRP_x	(3.130), (3.131)
G_x/T_{sx}	(3.133), (3.134)
C_x/N_{0x}	(3.137)
C_x/N_x	(3.138), (3.140)
$(E_b/N_0)_x$	(3.139)

$$W_x = \text{EIRP}_x - 169.5 - 20 \log_{10}\left(\sin\left(\alpha/2\right)\right) - L_{addx}$$

$$= \text{EIRP}_x - 169.5 - L_{addx} - \beta \quad (\text{dBW}/\text{m}^2) \tag{3.132}$$

$$G_x/T_s = G_x - 10 \log_{10} T_{sx} \quad (\text{dBi}/\text{K}) \tag{3.133}$$

$$G_x/T_s = C_x/T_x - \text{EIRP}_x + 20 \log_{10} f_x + 191.0 + \beta + L_{addx} \quad (\text{dBi}/\text{K})$$
$$\tag{3.134}$$

$$Q_x = \text{EIRP}_x + G_x/T_{sx} \quad (\text{dBiW}/\text{K}) \tag{3.135}$$

Link performance:

$$C_x/T_x = G_x + 10 \log_{10} P_{0x} + G_x/T_s - 20 \log_{10} f_x$$

$$- 191.0 - \beta - L_{addx} - L_{line} \quad (\text{dBW}/\text{K})$$

$$= 2G_x + 10 \log_{10} P_{0x} - 10 \log_{10} T_{sx} - 20 \log_{10} f_x$$

$$- 191.0 - \beta - L_{line} - L_{addx} \quad (\text{dBW}/\text{K})$$

$$= W_x + G_x/T_{sx} - 20 \log_{10} f_x - 21.5 \quad (\text{dBW}/\text{K})$$

$$= \text{EIRP}_x + G_x/T_{sx} - 191.0 - 20 \log_{10} f_x - \beta - L_{addx} \quad (\text{dBW}/\text{K})$$
$$\tag{3.136}$$

$$C_x/N_{ox} = C_x/kT_x = C_x/T_{sx} + 228.6 \quad (\text{dBHz}) \tag{3.137}$$

$$C_x/N_x = C_x/T_x + 228.6 - 10 \log_{10} B_x = C_x/kT_x B_x \quad (\text{dB}) \tag{3.138}$$

TABLE 3.61 Symbols for the Cross-Link Equations

Symbol	Meaning	Units	Equation
B_x	Noise bandwidth	Hz	(3.60)
C_x/N_x	Carrier-to-noise ratio	dB	(3.60)
C_x/T_x	Carrier-to-thermal-noise ratio	dBW/K	(3.58)
D_x	Common diameter for cross-link receive and transmit antennas	m	(3.127)
$(E_b/N_0)_x$	Cross-link performance for ratio of bit energy to channel noise power density	dB	(3.98)
$EIRP_x$	Cross-link EIRP	dBW	(3.55)
f_x	Cross-link frequency	GHz	(3.57)
G_x/T_{sx}	Cross-link figure of merit	dBi/K	(3.45)
L_{addx}	Losses due to cross-link antenna-pointing errors plus margin	dB	(3.121)
L_x	Path loss	dB	(3.57)
M_x	Cross-link degradations (principally transmitter power amplifier and receiver nonlinearities and, if applicable, cross-link modem implementation margin)	dB	(any)
P_x	Power input to the transmit antenna	dBW	(3.130)
P_{0x}	Rated transmitter power	W	
R_x	Cross-link bit rate capacity	bits/sec	
T_s	Cross-link receiving system noise temperature	K	(3.45)
W_x	Illumination level	dBW/m^2	(3.7–3.11)
α	Longitudinal separation of satellites (57.3° = 1 radian)	Degrees	(3.122)
β	See Figure 3.158b	Degrees	(3.126)
η_x	Common efficiency for receive and transmit antennas	Decimal	(3.29)
t_x	Cross-link time delay	msec	(3.124)
θ_{3dB}	Half-power beamwidth	Degrees	(3.37–3.38)

Note: See also the List of Symbols.

$$E_b/N_{ox} = C_x/N_x - 10 \log_{10}(B_x/R_x) \quad \text{(dB)} \tag{3.139}$$

$$C_x/N_x = 40 \log_{10}D_x + 20 \log_{10}f_x + 20 \log_{10}\eta_x - 20 \log_{10}S_x - 10 \log_{10}T_{sx}$$

$$- 10 \log_{10}B_x + 10 \log_{10}P_x + 176.95 \quad \text{(dB)} \tag{3.140}$$

Since cross links run in two different directions (eastbound and westbound), each may be further denoted by a numerical subscripts; thus, $(C_x/T_x)_1$ would be for one direction and $(C_x/T_x)_2$ for the return path. The frequencies (f_{x1} and f_{x2} respectively) are slightly different.

Simplifying assumptions: The transmit and receive antennas have the same diameter (D_x), efficiency (η_x), and gain (G_x). This leads to a fourth power influence of D_x (Gagliardi, 1984)

3.4.9 Summary of Useful Down-Link Equations

The down-link equations represent the last of the three sets of geostationary satellite equations (see also Tables 3.62 and 3.63). Table 3.64 permits these equations to be used in nongeostationary situations or if $S \neq 41{,}679$ km.

$$P_d = 10^{[(L_{addd} - (G/T_s)_d + (C_d/T_d) + M_{dd} - 20\ \log_{10}D_{sa} - 10\log_{10}\eta + 164.4)/10]} \quad \text{(W)}$$
(3.141)

$$\text{EIRP}_d = C_d/T_d + L_{addd} - (G/T_s)_{es} + 20\ \log_{10} f_d + 184.8 \quad \text{(dBW)}$$
(3.142)

$$M_{td} = \text{EIRP}_d - C_d/T_d + (G/T_s)_{es} - 20\ \log_{10} f_d$$
$$- L_{addd} + M_{id} + M_{dd} - 184.8 \quad \text{(dB)}$$
(3.143)

$$(G/T_s)_{es} = 184.8 + L_{addd} - \text{EIRP}_d + C_d/T_d + 20\ \log_{10} f_d \quad \text{(dBi/K)}$$
(3.144)

$$W_d = \text{EIRP}_d - L_{addd} - 163.3 \quad \text{(dBW/m}^2)$$
(3.145)

$$C_d/T_d = \text{EIRP}_d + (G/T_s)_{es} - L_{addd} - 20\ \log_{10} f_d - 184.8 \quad \text{(dBW/K)}$$
(3.146)

$$= Q_d - L_{addd} - 20\ \log_{10} f_d - 184.8 \quad \text{(dBW/K)}$$
(3.147)

$$= W_d + (G/T_s)_{es} - 20\ \log_{10} f_d - 21.5 \quad \text{(dBW/K)}$$
(3.148)

$$C/kT)_d = (C/T)_d + 228.6 \quad \text{(dBHz)}$$
(3.149)

$$(C/kT)_d = \text{EIRP}_d - L_{addd} + \left(\frac{G}{T_s}\right)_{es} - 20\ \log_{10} f_d + 43.8 \quad \text{(dBHz)}$$
(3.150)

$$(C/N_0)_d = (C/kT)_d \quad \text{(dBHz)}$$
(3.151)

$$(C/N)_d = (C/kT)_d - 10\ \log_{10} B_d \quad \text{(dB)}$$
(3.152)

$$(C/N)_d = \left(\frac{G}{T_s}\right)_{es} - L_{addd} + 10\ \log_{10}\eta_{ad} + 20\ \log_{10} (D_{sa}) + 10\ \log_{10}(P_d)$$
$$- 10\ \log_{10} (B_d) + 64.2 \quad \text{(dB)}$$
(3.153)

TABLE 3.62 Guide to Useful Down-link Equations

Given	Equations
Existing satellite	(3.144), (3.145)
Existing earth station	(3.141), (3.142)
Existing link (earth station and satellite with known $EIRP_d$ and G/T)	(3.146)–(3.157), (3.160)–(3.163)
Power-limited satellite	(3.146), (3.150), (3.153), (3.155), (3.157)
Size-limited earth station antenna	(3.142), (3.144), (3.124)–(3.148), (3.150), (3.153), (3.155)–(3.158)
Noise-limited receiver	(3.141), (3.142), (3.144), (3.150), (3.154), (3.155), (3.157), (3.158)
Coverage requirement on satellite transmit antenna	(3.143)

Requirement	Equations
Maximum bit rate, R_d	(3.157)
$(E_b/N_0)_d$	(3.155), (3.156)
$(C/T)_d$	(3.146), (3.147)
$(C/kT)_d$ or $(C/N_0)_d$	(3.149)–(3.151)
$(C/N)_d$	(3.152)–(3.154)
W_d	(3.145)
Q_d	(3.160)–(3.163)
B_d	(3.164)

TABLE 3.63 Symbols for Down-Link Equations:

Symbol	Meaning	Units	Equation
Subscript d	Down link		
B_d	Channel noise bandwidth	Hz	(3.60)
C_d/kT_d	See C_d/N_{0d}	dBHz	(3.59)
C_d/N_d	Carrier-to-noise ratio	dB	(3.60)
C_d/N_{0d}	Carrier-to-noise-density ratio	dBHz	(3.60)
C_d/T_d	Carrier-to-down-link-thermal-noise power ratio	dBW/K	(3.58)
D_{es}	Down-link earth station antenna diameter	m	(3.157)
D_{sa}	Satellite down-link transmit antenna diameter	m	(3.141)
$(E_b/N_0)_d$	Down-link ratio of bit energy to channel noise power density	dB	(3.98)
EIRP_d	Satellite down-link equivalent isotropically radiated power	dBW	(3.55)
f_d	Down-link frequency	GHz	(3.141)
$(G/T_s)_{es}$	Earth station figure of merit	dBi/K	(3.144)
L_{add}	Additional down-link losses (excluding range)	dB	(3.142)
M_{dd}	Down-link degradations due to satellite transmitter and earth station receiver non linearities	dB	(3.143)
M_{id}	Down-link modem implementation margin	dB	(3.143)
M_{td}	Total of down-link losses, margins, and M_{dd} (except range of 41,679 km)	dB	(3.143)

397

TABLE 3.63 (Continued)

Symbol	Meaning	Units	Equation
P_d	Satellite down-link transmitter power per carrier	W	(3.141)
Q_d	Down-link EIRP$_d$ + G/T_s	dBiW/K	(3.160)
R_d	Down-link bit rate capacity	bits/sec	(3.155)
W_d	Edge of earth illumination level	dBW/m^2	(3.7–3.11)
η_{ad}	Earth station antenna efficiency	Decimal	(3.29)
η_{sad}	Satellite down-link transient antenna efficiency	Decimal	(3.29)
20.4	Unit conversion	dB^a	(3.6)
21.5	Unit conversion	dB^a	(3.58)
163.3	Unit conversion	dB^a	(3.9)
184.8	(21.5 + 163.3)	dB^a	
228.6	Boltzmann's constant	dBW/K/Hz	(3.59)
227.5	(228.6 − 21.5 + 20.4)	dB^a	
43.8	(228.6 − 21.5 − 163.3)	dB^a	(3.112)
64.2	(228.6 − 184.8 + 20.4)	dB^a	(3.113)

[a]The decibel units are given in the List of Constants.

TABLE 3.64 Down-link Modifications for Nongeosynchronous Satellites

In Equations	Replace	With
3.141–3.144, 3.146, 3.147	184.8	$92.45 + 20 \log_{10} S$
3.145	163.3	$70.9 + 20 \log_{10} S$
3.148	W_d	$\text{EIRP}_d - 20 \log_{10} S - 70.9$
3.150, 3.154	43.8	$136.1 - 20 \log_{10} S$
3.153	64.2	$156.6 - 20 \log_{10} S$

$$(C/N)_d = \text{EIRP}_d - L_{\text{add}_d} - 20 \log_{10} f_d - 10 \log_{10} B_d + (G/T_s)_{es}$$
$$+ \; 43.8 \quad (\text{dB}) \tag{3.154}$$

$$\left(\frac{E_b}{N_o}\right)_d = (C/N)_d + 10 \log_{10}\left(\frac{B_d}{R_d}\right) \quad (\text{dB}) \tag{3.155}$$

$$\left(\frac{E_b}{N_o}\right)_d = (C_d/T_d) - 10 \log_{10} B_d + 10 \log_{10}\left(\frac{B_d}{R_d}\right) \quad (\text{dB}) \tag{3.156}$$

$$D_{es} = 10^{[((C/N)_d - 10 \log_{10} \eta_{ad} + 10 \log_{10} T_{sd} - W_d + 10 \log_{10} B_d - 227.5)/20]} \quad (\text{m}) \tag{3.157}$$

$$D_{es} = 10^{[((E_b/N_o)_d + 10 \log_{10} R_d) - 10 \log_{10} \eta_{ad} + 10 \log_{10} T_{sd} - W_d - 227.5)/20]} \quad (\text{m}) \tag{3.158}$$

$$T_{es} = 10^{[(20.4 + 20 \log_{10} f_d + 20 \log_{10} D_{es} + 10 \log_{10} \eta_{ad} - (G/T_s)_{es})/10]} \quad (\text{K}) \tag{3.159}$$

$$Q_d = \text{EIRP}_d + (G/T_s)_{es} \quad (\text{dBiW/K}) \tag{3.160}$$

REFERENCES

Aasted, J. and Roederer, A. (1978), ''A multibeam array for communication with mobiles,'' *Collected Papers of the 7th AIAA Communications Satellite Systems Conference*, April, San Diego, CA., p. 401.

Anonymous (1977). *World Wide Space Activities*, Library of Congress, Washington, D.C. Sept., p. 518.

Anonymous (1979). *Microwave Systems News (MSN)*, May, p. 40.

Brandinger, P. E. (1978). 20–30 communication satellite systems design. *ICC-78 Conf. Rec.*, p. 10.4.2.

Dicks, J. L., and Brown, M. P., Jr. (1978). Intelsat V satellite transmission design. *ICC-78 Conf. Rec.*, Montreal pp. 2.2.1–2.2.5.

Freeman, R. L. (1985). *Reference Manual for Telecommunications Engineering*, Wiley (Interscience), New York, Chap. 20.

Gagliardi, R. M. (1984) *Satellite Communications*, Lifetime Learning Publications, Belmont, CA, pp. 117–122.

Hilborn, Jr., C. G. et al (1977) ''Implications of Demand Assignment for Future Satellite Communications Systems,'' Defense Communications Agency, NTIA Report ADA043002.

International Telecommunications Union (ITU) (1988). *Radio Regulations*, United Nations, Geneva.

International Telecommunications Union (ITU) (1988). *Table of Frequency Allocations*, United Nations, Geneva.

Ippolito, L. J., Jr. (1985). *Radiowave Propagation in Satellite Communications*, Van Nostrand-Reinhold, New York.

Jordan, E. C. (Ed.) (1985). *Reference Data for Engineers: Radio, Electronics, Computer and Communications*, 7th ed., Howard Sams, Indianapolis, IN.

Kiesling, J. D., Elbert, B. R., Garner, W. B., and Morgan, W. L. (1972). A technique for modeling communications satellites. *COMSAT Tech. Rev.* 2(1), pp. 73–104.

Koelle, D. E. (1978), "Advanced technology for direct broadcasting satellites," *Collected Papers of the 7th AIAA Communications Satellite Systems Conference*, April, San Diego, CA, p. 709.

Koelle, D. E. and Dodel, H. (1988). Cost reduction potential for communication satellite systems and services. *Collected Papers of the AIAA 12th International Communications Systems Conference*, March, Arlington VA, pp. 463–468.

Lee, Y. S. (1988). Cost-effective intersatellite link applications to the fixed satellite services. *Collected Papers of the 12th AIAA International Communications Satellite Systems Conference*, March, Arlington VA, pp. 167–173.

Lo, G. J. P. and Mizuno, T. (1988). The sharing of broadcast satellite services with inter-satellite services in the 23 GHz band. *Collected Papers of the AIAA 12th International Communications Satellite Systems Conference*, March, Arlington VA, pp. 53–62.

Maral, G., and Bousquet, M. (1986). *Satellite Communications Systems*, Wiley, New York, pp. 322–335.

Millies-Lacroix, O. and Peters, R. A. (1988) ESA/Eutelsat/Intelsat intersatellite link program. *Collected Papers of the AIAA 12th International Communications Satellite Systems Conference*, March, Arlington VA, pp. 704–710.

Miya, K. (1975) *Satellite Communications Engineering*, Lattice Co., Tokyo.

Reinhart, E. (1985). Introduction to RARC '83 plan. *IEEE J. Sel. Areas Commun.* SAC-3(1), 17.

Ruze, J. (1966) "Antenna tolerance theory-a review," *Proceedings of the IEEE*, April, 54; pp. 633–640.

Seta, M. and Ayukawa, I. (1988). A study of the transmitting power control for earth stations. *A Collection of Papers of the 12th AIAA International Communications Satellite Systems Conference*, March, Arlington VA, pp. 174–184.

Skolnick, M. E. (1970), *Radar Handbook*, McGraw-Hill, New York.

Tamir, I. and Rappaport, Y. (1987). Generalized satellite link model and its application to the transmission plan. *Int. J. Sat. Comm.*, 5, pp. 49–56.

MULTIPLE-ACCESS TECHNIQUES

4.1 OVERVIEW OF MULTIPLE ACCESS

Satellites usually divide the total available frequency band (e.g., 3.7–4.2 GHz) into several subbands (transponders), each of which is independently amplified. (See also Chapter 3.) Access to a transponder may either be limited to one earth station at a time, or many simultaneous rf carriers may be permitted. Within a satellite both conditions may be found with some transponders carrying only a single carrier (such as television or trunk telephony) and others handling multicarrier traffic (e.g., thin route data such as from a collection of microterminals).

This introductory section discusses the choices available and the methods of obtaining access to/from the satellite. The sections that follow will consider the details of various multiple-access techniques using the frequency, time, space, and code domains.

Variations on these basic methods include the use of access by contention (e.g., ALOHA), the effect of circuit extensions by the use of terrestrial or space circuits, and the influence of on-board signal processing (such as satellite switching, regeneration, and remodulation).

4.1.1 Access

Access is the means by which multiple users may gain entry to share a common facility. In the case of satellite systems the users traditionally have been earth stations of a similar service type. As intersatellite links (ISL) become available, the definition will be broadened to include other satellites and eventually (with the advent of multiservice satellites with intrasatellite switching) other user services.

Various multiple-access (MA) methods are available to enable sharing of the space segment resources by a number of users. Table 4.1 defines the various multiple-access methods.

In the preassigned case channel blocks (ranging from a single voice or data channel to entire transponders) are assigned on a quasi-permanent basis for the exclusive use of an earth station or earth station pair. The terrestrial equivalent is

TABLE 4.1 Multiple-Access Definitions

The following are the basic access methods. Variations, extensions, and combinations are covered in subsequent parts of this chapter. (See also Figure 4.2.) **The words *domain* and *division* may be used interchangably.**

CDMA

Code domain multiple access shares a common transponder spectrum by spreading the signal throughout a bandwidth substantially wider than needed just for the information content. This may be done on either an instantaneous or frequency-hopping basis. Each transmit station uses a unique pseudorandom code to spread its transmitted signal. Each receiving station in the network must have the identical pseudorandom noise (PN) code for despreading and information retrieval. Other networks may operate simultaneously within the same spectrum if a different, noninterfering (orthogonal) code is used for spreading/despreading. Eventually, the interference products limit the information bit error rate. In one case up to six networks share 5 MHz of spectrum. Another CDMA system uses a fixed or varying combination of time and bandwidth for communications (Figure 4.1). CDMA is characterized by its privacy, interference tolerance, poor frequency utilization, and complex, but often low-cost, earth stations.

DAMA

Demand access multiple access systems respond to channel requests from stations (there are no preassigned time, frequency, or space slots). After the request is received, a central station (or satellite) makes a temporary assignment to permit one- or two-way service.

FDMA

Frequency domain multiple access shares a common transponder spectrum (band) by dividing it into subbands and assigning these among the users. Each station pair may then further subdivide their allocation into individual channels. FDMA is characterized by simplicity in earth station equipment, but it is limited by multichannel transmission. These result in needs for up-link power control and complex frequency plans with unique assignments, impairments, and traffic capabilities.

OMA

Orthogonal multiple access permits one channel to be transparent to the others, and there is no contention for channels. CDMA, FDMA, SDMA, and TDMA are examples of OMA.

RMA

Random multiple access operates on the basis of each user contending with others for use of a facility. This is attractive to small and low-duty-cycle users because of the limited (or nonexistent) control equipment used. Demand for the facility may be done on a random (unorganized, uncontrolled, unscheduled) basis, and thus the performance is variable (being sensitive to the amount of traffic). RMA is often done with various levels of control. As the traffic builds, more sophisticated controls are needed to keep the system delays due to retransmissions to find a clear channel at a moderate level.

TABLE 4.1 *(Continued)*

RMA *(Continued)*

Some applications have multiple control methods that may be switched in under various traffic conditions.

Single Access

Single access indicates only one carrier may access a transponder at a time. This is also referred to as single carrier per transponder, (SCPT).

SDMA

Space domain multiple access makes use of the orthogonalities in geometry (either in beam patterns or rf signal polarizations) to permit sharing of the spectrum. The communicating, transmitting, and receiving stations may appear in separate beams (or use separate polarizations (see section 3.1). Filters or switches may be provided on the satellite to permit traffic to flow from one use to another (e.g., from beam N to beam $N + 1$ or from one polarization sense to another).

The two forms of signal polarization used are linear and circular. Each has two orthogonal senses (horizontal and vertical and left and right hand).

SDMA is used in conjunction with FDMA, TDMA, or CDMA. Figures 4.4 and 4.5 show how these signals might be switched among beams.

Another SDMA form is to use satellites in different parts of the geostationary orbit or in diverse units (e.g., the *Molniya* and inclined geosynchronous orbits).

TDMA

Time domain multiple access provides access to the available transponder spectrum on a time-shared orthogonality basis. In a simple example, one carrier fully occupies the transponder at a time. Each station transmits a burst of data in its exclusively assigned time interval. The duration of the burst may be varied to fit the station's instantaneous traffic needs. As soon as the station's burst is completed, another station gets its assigned time slot.

TDMA is characterized by being inherently digital compatible, utilizing the maximum power capability of a transponder and being bandwidth efficient, but it results in more complex (and costly) earth stations than FDMA because of the need to maintain precise network timing and to operate at burst rates of 60–120 Mbits/sec. This extra cost is especially apparent in small-traffic stations.

Hybrid Systems

Small K_u band earth stations (microterminals) often use one multiple access system between the small terminal and the main (hub) station. Access for burst and intermittent services is often on a DAMA or RMA basis. A request channel may be used to indicate that the station has traffic. The actual information may be sent on a DAMA-based FDMA (using SCPC) at 9600 to 64,000 bits/sec.

The link from the main (hub) station to a small terminal is made on the basis of preassigned FDMA channels. Each channel may carry a continuous stream of single-access time-division-multiplexed (TDM, not TDMA) bits. The small station has been alerted to expect a burst of data so many microseconds after the TDM unique word.

a dedicated microwave, fiber, or full-time leased private line. This results in a fairly simple earth station as the communications always take place under the same conditions and slot (be it time, frequency, or code).

Multiple access (MA) is the means for sharing this resource among a number of users on a preassigned or fixed (PAMA or FAMA), demand-assigned (DAMA), or randomly assigned (RMA) basis. Examples of each are given in Table 4.2 and Figure 4.1.

PAMA is best assigned to point-to-point services or to domestic broadcast modes (such as television or radio and data distribution), whose requirements do not

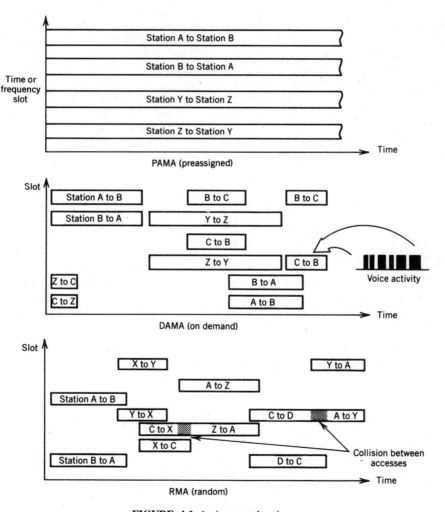

FIGURE 4.1 Assignment durations.

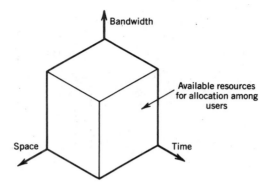

FIGURE 4.2 Basic multiple access resources.

change very often. In the latter cases the assignments may become de facto permanent if a large number of receive-only fixed-channel receiving earth stations have been deployed because of the inconveniences in making a change at each of hundreds or thousands of earth stations.

Demand-assigned multiple access permits a number of users to share a common resource pool. In the terrestrial long-distance dial network a user places a request for a circuit to a specified destination. If all of the trunks are busy, he receives a busy signal from his nearby telephone office. In the case of satellite links the "office" may be remote, but it does the same function by signaling the requesting terminal a "busy" or by assigning a pair of channels or half-circuits—one outbound, the other for the return: The two channels form one complete circuit. These assignments are made from a common resource pool. As soon as the conversation or data stream (circuit or packet) is completed, the assignments are returned to the pool for use by others.

DAMA is useful to both large and small users. Intelsat's SPADE (single channel per carrier pulse code modulation multiple-access demand assignment equipment), for instance, was originally conceived as a way of lowering the costs for what Intelsat considered small users (e.g., a few tens of channels). Instead of being required to pay for 24-hr/day preassigned FDMA channels (which were empty most of the time), the SPADE system permitted these stations to pay for only the time actually used. This is especially attractive on a global basis where there are many time zones to spread the peak traffic over many busy hours. While the preassigned FDMA was restricted to point-to-point service (e.g., from Canada to France), SPADE is a multipoint service (e.g., from Canada to any other SPADE station operating with the satellite). Although the SPADE equipment is more expensive than the simpler preassigned FDMA counterparts, the reduced satellite charges and increased destination possibilities have favored SPADE in many applications.

Large earth stations (e.g., those in Canada and Italy) have discovered that

TABLE 4.2 Transponder Assignment Methods

	Type of Assignment		
	Preassigned Multiple Access (PAMA)	Demand-Assigned Multiple Access (DAMA)	Random Multiple Access (RMA)
Earth station request method	Written document to network operations center	Via request channel to network operations center	None used (contention for channel)
Network operations center (NOC) function	Adds/deletes frequency or time slot assignments; often a written plan	Assignments done by NOC operator and/or computer; satellite channel used to notify each earth station	Monitor network performance to detect lockout conditions
Examples	Intelsat FDMA network television	Microterminals, Inmarsat, SPADE	ALOHA, data collection, and some microterminal systems
Assignment duration	Often months to years	Duration of connection (conversation, data burst, etc.)	Duration of burst or packet.
Domains used	Long-term exclusive assignment in frequency, time, or code	Temporary exclusive assignment in frequency time or code	Shared (contention) basis

Busy-hour effects	Not applicable as capacity is fixed	New accesses will be busied out; makes good use of resources if multiple time zones are involved and traffic has daily peaks	Interference between contending users
Advantages	Low earth station costs; good control over intermodulation and power	Pay by minutes used; dynamic reassignments of resources; makes good use of resources if multiple time zones are involved and traffic has daily peaks	earth station equipment is modest; quality links; no requirement for NOC
Disadvantages	Inflexible to actual traffic patterns; hard to reassign traffic on diurnal basis	More complex earth station and NOC equipment; dependent on the NOC operating	No control over number of simultaneous users; interferences and bistable blocking may occur
Terrestrial telephone equivalent	Diversified full-time circuits	Dial-up lines	Party lines; citizens band and amateur radio
Terrestrial radio equivalent	AM, FM, and TV assignments	Cellular telephones	Party lines; citizens band and amateur radio

NOC = Network Operations Controller

SPADE is an attractive method for handling peak traffic loads that would otherwise require full-time leases of FDMA preassigned channels (see Chapter 2 about erlang loading and busy-hour traffic).

There are several basic options available for multiple access. These are the amplitude (or magnitude), frequency, time (temporal), code, and space domains. These domains may be divided among the users seeking access. The amplitude domain multiple access (ADMA) does not readily lend itself to multiple access (but see Capture ALOHA in Section 4.6). The four remaining domains are discussed further in Tables 4.1 and 4.3, Figures 4.2 and 4.3, and subsequent sections of this chapter. Figures 4.4 and 4.5 show cases where transponders are used for multiple signals.

4.1.2 Up-Link Signals

One method to describe a carrier is by its application. This may be done on a carrier or burst modulation basis. A cryptic description uses a series of W, X, P, Y symbols to identify the earth station transmit signal. The symbols W, X, and Y are defined in Table 4.4, and P means "per." The satellite transponder uses a slightly different set of symbols (W, Z, P, and T, where W and Z are defined in Table 4.5).

Intelsat I (*Early Bird*) carried just two transponders. Neither was capable of handling any form of multiple access and thus was operated on a SCPT basis. One transponder carried east-to-west traffic. The second was used for the other direction. European earth stations in France, Germany, and the United Kingdom took daily turns transmitting the European traffic to the westbound transponder (Italy took the weekends). The station at Andover, Maine, transmitted eastbound traffic on the other transponder. This is an example of single access (SA).

Intelsat II had more linear transponders that could be operated multicarrier, and the operating restrictions on the earth stations were removed. The transponders were operated in a global coverage mode.

SCPT is the most efficient way of operating a transponder provided sufficient traffic can be found at any instant to fill the capacity. Fiber optics has captured much of this bulk traffic, and thus SCPT telephony is less common than in the early 1980s. Television often fills an entire transponder on a SCPT basis.

An interesting variation of SCPT has appeared to permit several high powered audio (or data) channels to use a full transponder power. This involves a single carrier (which may have little or no intelligence content) that is modulated by a series of subcarriers with intelligence. The SCPT permits full saturated power. Earth station receivers detect the carrier and demodulate one or more of the subcarriers. Since the reception process is like normal satellite TV, simple mass produced, equipment may be used. This technique is called FM–FM (FM^2) and is likened to a TV transmission without a picture. Present applications include multiple music/advertising channels for supermarkets.

TABLE 4.3 Domain Multiple-Access Techniques[a]

Domain	Frequency FDMA	Time TDMA	Code CDMA[b]	Space SDMA[c]
Channel separation means	Spectrum	Time	Orthogonal codes or spread spectrum	Spatial (beams, orbits, polarization)
Channels per transponder at any one instant	Many	One[d]	Many	Many
Spectrum use	Nonoverlapping	Sequentially	Overlapping	Reused
Typical application	Low-traffic-level analog	High-traffic-level analog and digital	Military, data collection, and digital	Wherever extra spectrum or capacity is needed
Typical baseband multiplexer	FDM	TDM	Frequency hopping, FSK; pseudo, noise, PSK	[c]
Network discipline at transmission	Frequency and power adherence	Time slot (burst) adherence	None required	[c]
Synchronization	None	Time	Long or changing codes	[c]
Channel capacity limits	Intermodulation (IM) products from multiple carriers; guard bands	Bandwidth-to-band ratio, spectrum and guard times, and TDMA efficiency	Self-interference	Self-interference and number of interbeam connections (switch size)
Limits to number of instantaneous carriers per transponder	Intermodulation (IM) products	One per frequency slot[d]	Self-interference	[c]

TABLE 4.3 (Continued)

Domain	Frequency FDMA	Time TDMA	Code CDMA[b]	Space SDMA[b]
Satellite amplifier operating point	May be substantial backoff in multicarrier situations	Saturation[e]	Similar to FDMA	c
Earth station size	May be varied to fit traffic	Requires a minimum EIRP and G/T_s for all earth stations	Small due to processing gain	c
Network flexibility	Generally low	Very dynamic	Dynamic	c
Blocking	None	When filled	Interference limited	When filled
Down-link security	Least	Inherent	Greatest	Good due to possibility that down link may not be observed
Susceptibility to interference	Greatest	Lower	Least	c
Spectrum conservation	Good	Good	Poor	Best (reuse)
Earth station requirements				
Multiple-access equipment complexity and cost	Least	High	Varies	Slight increase
PA operating mode	Continuous (FM)	Burst (low duty cycle)	Depends on application (often FM-like)	c

410

PA bandwidth	Narrow	Full transponder	Wide	[c]
Antenna	Not critical	Critical	Critical	Smaller
Receiver addressing	By frequency slot	By time slot and address	By preassigned codes	[c]
Up-link power control	Required	Not required	Required	[c]
Satellite requirements				
Antennas	—	—	—	One per beam or polarization[f] [c]
Limiters	No (unless SCPT)	PA saturation or limiters	No	[c]
Trades:	Bandwidth, capacity, and filter weight	Earth station complexity	Spectrum inefficiency	Satellite complexity
For:	Satellite power	Capacity	Security, jam resistance, earth station size, and low cost	Frequency reuse
Spectrum spreading	—	PN noise bits in empty time slots and/or EDW[g]	Hopping and high chip bit rate	varies

[a] Abbreviation: PA, power amplifier.
[b] CDMA may use spread spectrum multiple access (SSMA).
[c] SDMA uses FDMA, TDMA, or CDMA as basic access method via one of spatially separated beams or polarizations; therefore, these techniques influence SDMA performance.
[d] Narrow-band (less than full transponder bandwidth) TDM or TDMA may share a transponder with other services (including other TDM or TDMA users).
[e] For single carrier per transponder
[f] Dual surface antennas may be used.
[g] Energy dispersal waveform

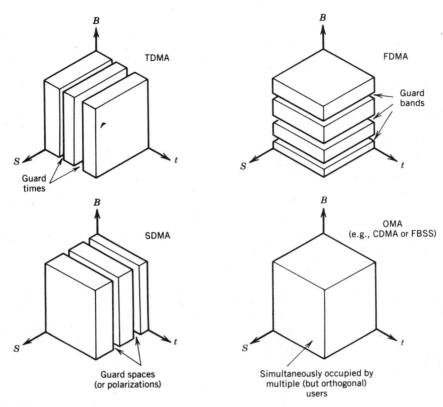

Key: B, bandwidth; S, space; T, time; FBSS, full band spread spectrum

FIGURE 4.3 Multiple-access domains.

4.1.3 Digital Traffic

Digitized communications may be handled on a continuous, burst, or packet basis. Figure 4.6 shows these methods.

Continuous data is provided from a single source and thus is received with a fixed time delay. The delay is the sum of the equipment and distance delays. Once the link is established, there is no need for unique words, addresses, or other supervisory overhead signaling because the channel is not shared. Parity codes may be used. The throughput of useful data is very high. The bit rate is at the input (''natural'') speed.

Burst communications involve time-sharing the satellite with other users on a time division basis. Unique words, synchronization signals, and address (and sometimes routing) information must be added to ''tag'' and route each burst. The bit rate within a burst is much higher than the natural rate because of the time

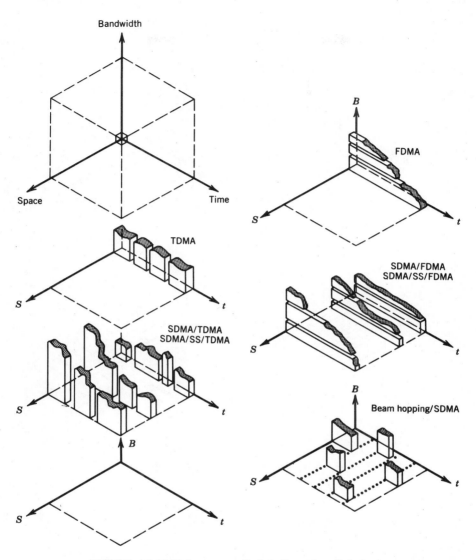

FIGURE 4.4 Multiple-access methods in Space-time (S-t) plane.

sharing and overhead bits and because all earth stations using a transponder must operate at the transponder bit rate (or possibly a subrate) and in synchronism with one another. The burst is made up from a block of bits. There is an additional time delay of up to one frame at the transmitter. At the receive end there is the propagation delay of the up and down links. If low-bit-rate digital voice is being used, there are additional processing delays.

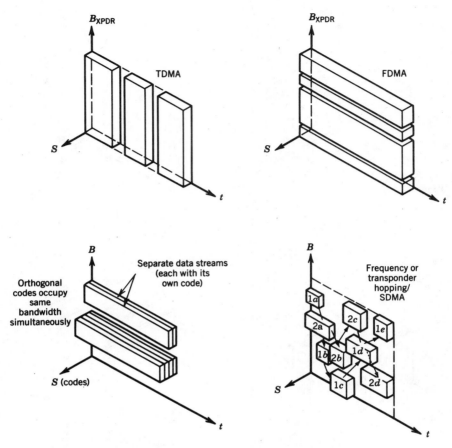

FIGURE 4.5 Multiple-access methods in Bandwidth-time (B-t) plane.

TABLE 4.4 Carrier and Burst Modulation Choices[a]

Inputs, W	S	S	M	M
Modulation, X	C	B	C	B
Per, P	P	P	P	P
Means, Y	C	C	C	C
Result, WXPY	SCPC	SBPC	MCPC	MBPC
Baseband multiplex means	none	TDM	FDM	TDM
Operation per carrier	Saturated	Saturated	Backed off	Backed off
Destinations	Single or multiple	Single or multiple	May be multiple	May be multiple

[a]Key: B, burst (X or Y); C, channel (X) or carrier (Y); M, multiple (W); P, per (P); S, single (W).

TABLE 4.5 Transmitter Operating Modes[a]

Simultaneous transmitter inputs, W	S	S	M	M
Multiple access	FM/FDMA	TDMA	FM/FDMA	TDMA
Modulation, Z	C	B	C	B
Per, P	P	P	P	P
T	T	T	T	T
Result, WZPT	SCPT[b]	SBPT[b]	MCPT	MBPT
Transmitter operating mode	Saturated	Saturated	Backoff	Saturated
Destinations	MM	MM	MM	MM
Contention basis?	No	No	Not usually (preassigned)	Not unless CDMA or RMA

[a]The T stands for transmitter (in the case of earth stations) or transponder (for satellites); MM, may be multiple. See Table 4.4 for other abbreviations.

[b]This may also be called single station per transponder (SSPT).

A packet consists of traffic from one or more sources that has been accumulated, packaged, and readied for transmission to the next common node along the network. Alternate nodes may be employed to bypass busy routes. Routing information is generally added.

Packet communications also use unique words and signaling for access, control, and especially for priority and routing. A computer-to-computer session may

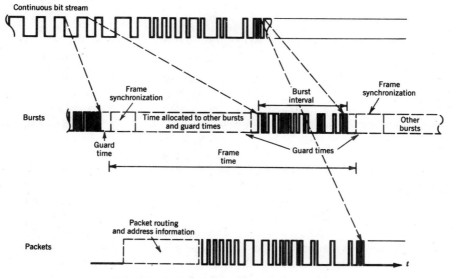

FIGURE 4.6 Digital communications methods.

require many packets of bits. The individual packets may take diverse routes (including one by land and another by space depending on availability of facilities or other considerations), and therefore, the delay time may be variable over a wide range. Packet switching is attractive for many users through the use of store-and-forward techniques, microprocessor selection of the most economical routing paths, and other techniques. Demand access may be done on a packet-by-packet basis with packets being temporarily stored at an earth or space station until an on-going connection can be found and established. Packet switching may be used in contention-type multiple-access systems (e.g., ALOHA) where a transmitting earth station may have to retransmit a blocked packet. Intermediate packet-switching nodes may add routing bits to the packet. Packets may be of variable durations.

Narrow-band pitch-excited vocoders (PEV) for packetized voice tend to use a header with 13 bits, plus f_p bits of error detection or forward error control (FEC), and 80 bits of data. The number of f_p bits is determined by the coding rate. For instance, a coding rate of "one-half" (denoted as $R1/2$), means that half the bits (80) are used for FEC, and 80 bits are used for the digitized voice data. Equation (4.1) enables calculation of f_p for a given number of data bits d_p and for a coding rate R:

$$f_p = \frac{d_p}{R} - d_p \text{ (bits)} \qquad (4.1)$$

The wideband, continuously variable-slope, delta modulation (CVSD) vocoders might use the same number of header bits plus 320 bits of data per packet. In both cases additional signaling bits will be needed.

4.1.4 On-Board Regeneration and Remodulation

Regenerative satellite repeaters may be readily used with packet communications. In a multibeam regenerative satellite store-and-forward techniques can be used to resolve contentions for a down-link beam. In unresolvable cases (e.g., the down link may be already saturated with other traffic), alternative routing may be employed: For instance, if the Paris beam is filled, the packet might be sent down the Marseilles beam with terrestrial routing instructions to direct the packet to Paris. Regeneration is applicable to digitized FDMA data and especially TDMA by making the burst-to-burst times uniform. The bit retiming is used to remove the delay variations of individual up links, intersatellite links, and the inherent time variables in any store-and-forward method. The receiving earth station's synchronization equipment requirements are substantially eased by standardizing the down-link burst and bit locations. This equipment no longer has to search for and lock onto bits that have variable locations and therefore variable delays.

Remodulation is transparent to the multiple-access techniques unless there is a change in modulation form (e.g., from FM to VSB/AM).

4.1.5 Spread Spectrum

Spread spectrum was originally used for military and covert communications. It is available for other users where security and interference immunity are needed, but in satellite communications its ability to resist high levels of interference is employed along with low-cost electronics.

A wide spectrum must be available (generally on a shared basis) and coding, decoding, and synchronization equipment are needed. At one time this equipment was elaborate and expensive. The use of custom large-scale integrated circuits has substantially reduced the size and cost of this equipment. Code domain multiple access with spread spectrum (CDMA/SS) uses this technique, which has gained importance for small civil earth stations. A long pseudorandom noise (PN) code preference may be modulated by the information being transmitted. The information rate may be in the range of 19,700 bits/sec, and the PN rate is around 15 Mbits/sec. The ratio (15 Mbits/sec to 19.7 bits/sec) is $761:1$, or 28.8 dB. This is also referred to as the processing gain, which, when added to C/N, produces the useful E_b/N_0.

In one form of spread spectrum CDMA, carriers are caused to "frequency hop" (CDMA/FH) in an apparently random manner. The users share the transponder spectrum by individually frequency hopping their carriers (which are actually controlled by pseudorandom noise codes). Ideally, the individual carriers are all in different frequency slots at any time, but in reality (since there may be no "slots" per se) some spectra may instantaneously overlap (or occasionally collide in frequency). When this happens, a momentary increase in noise if the codes are not orthogonal (due to proximity in frequency or intermodulation) or even an occasional loss of signal (due to a collision) will occur. The use of the long PN bit streams and extensive error control tolerates this situation. An instant later the user has hopped its frequency to another portion of the transponder spectrum, and a new set of conditions exist.

Provided there are enough simultaneous CDMA/FH users, the down-link transmission looks like white noise. Careful examination with a spectrum analyzer might show a series of modulated carriers whose frequencies are hopping about in a random manner. If the spectra are rectangular and broad (as is the case for some PN modulation forms) and overlapping, it would be difficult to find, let alone track an individual signal on the display. No clue about the sources or destinations of the signals or their data are available to the casual observer.

A CDMA/FH receiving earth station code generator is locked to the transmitting PN code generator. The receiving PN generator drives a frequency synthesizer to dehop (track) the incoming signal. The demodulated signal contains synchronization pulses for further locking. Flywheel synchronization is used to maintain the lock through noisy and collision conditions.

If a PN frequency-spreading code is used, a signal spectrum may be broadened to the full bandwidth available (full band spread spectrum, or CDMA/FBSS). An instantaneous view of one signal on a spectrum analyzer will reveal many peaks and nulls within the overall spectrum. Through the use of unique (orthogonal) codes, each code's spectra will be different at any time. Additional CDMA/FBSS stations may place signals with these different (and thus not totally overlapping) spectra into the same band. Some peaks may collide, and noise will result, as in the CDMA/FH case.

FDMA and TDMA systems have more or less fixed spectrum signatures that might cause interference to adjacent satellite systems. CDMA does not have these consistent spectrum lines and, therefore, may create a different form of intersystem interference (which look like pseudorandom noise). CDMA may also be used on-board large multibeam satellites to ease the beam-to-beam isolation requirements by minimizing the effects of self-interference.

4.1.6 Intersatellite Links

The introduction of ISL and on-board switching between services (e.g., in multi-service satellites) expands the multiple-access possibilities. With the advent of on-board micro- and minicomputers, a satellite can handle different, but convertible, access formats; thus, each service might have a near-optimum access format. An example of this may be an up link from a mobile-service transmitter using a highly spectrum efficient form of amplitude modulation. The down link may use CDMA/SS, which permits modest earth stations at telephone offices. The satellite would need to demodulate the up link, switch the traffic to the right down-link beam, remodulate the signal, and generate the needed rf power for transmission.

4.2 FREQUENCY DOMAIN MULTIPLE ACCESS (FDMA)

4.2.1 Basic FDMA Characteristics

Each transmitting earth station is assigned a portion of the available transponder bandwidth for its use. In the simplest case (Figure 4.7), the earth station collects all of its traffic onto a single carrier by baseband multiplexing (either FDM or TDM) regardless of destination. This single multidestination FM carrier is ampli-fied by an earth station power amplifier (often on the basis of SCPT) and fed to the station's antenna. The satellite's receive antenna collects this and many other carriers (separated in frequency) simultaneously. The satellite amplifiers operate on filter-selected portions of the total receive band in a single- or multicarrier-per-transponder (SCPT or MCPT) mode. Figure 4.8 shows the more general MCPT case. The signal is transmitted to the receiving earth stations, which filter out the desired carriers. The desired carriers and their subcarriers are demodulated and demultiplexed down to individual telephone or data channels.

Several forms of FDMA are possible (see Table 4.6).

The multichannel-per-carrier (MCPC) case just described is designated FDM/FM/FDMA with the order of the sequence of terms being: the baseband multiplex method (FDM, or frequency division multiplex), the modulation (FM, or frequency

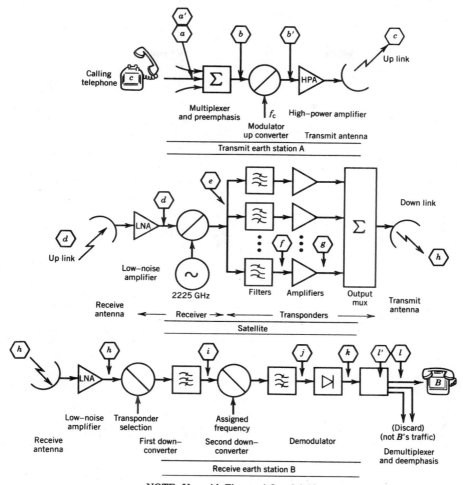

NOTE: Use with Figures 4.8 and 4.10.
FIGURE 4.7 Generalized FDMA operation.

modulation), and finally the multiple-access (FDMA) method. This is the mux–mod–mac nomenclature shown in Figure 4.9.

The other common form is a single channel per carrier (SCPC), wherein one signal (usually data or telephony) frequency or PSK modulates a carrier, which, in turn, accesses a satellite on a FDMA basis. SCPC/FDMA is often done on a FVDA (fully variable demand access) or RMA (random multiple-access) basis.

Another form uses PCM/PSK/FDMA (see Figure 4.10). Telephone channels may arrive at an earth station in a PCM bit stream converted from the analog to the digital format. Typically 16–64 kbits/sec are used per telephone half-circuit. These are multiplexed and are used to phase-shift-key (PSK) a carrier, which then

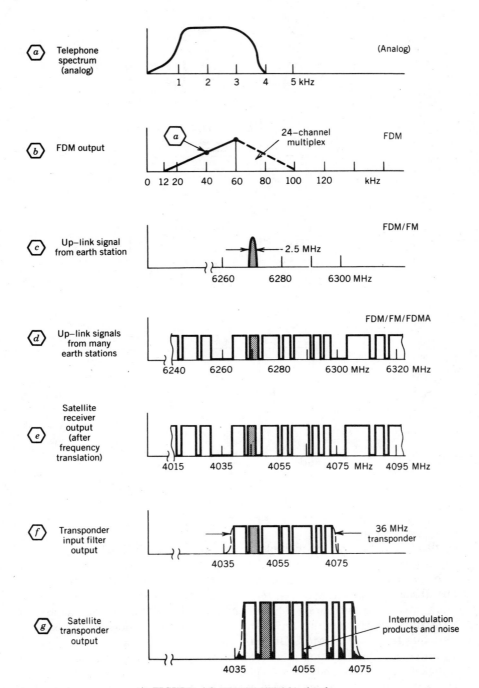

FIGURE 4.8 FDM/FM/FDMA signals.

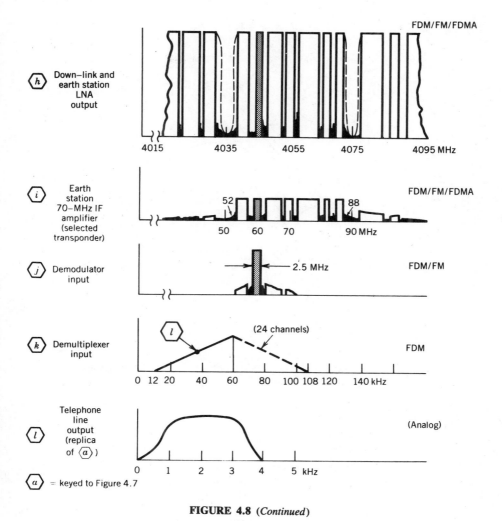

FIGURE 4.8 (*Continued*)

TABLE 4.6 Forms of FDMA

FDMA Type	Advantages
FDM/FM/FDMA	Efficient and inexpensive for very limited access trunk operation
SCPC/FDMA (SPADE)	Efficient for large number of accesses from small users
PCM/PSK/FDMA or TDM/PSK/FDMA	Allows substantial increase in channel capacity using digital speech interpolation (DSI)

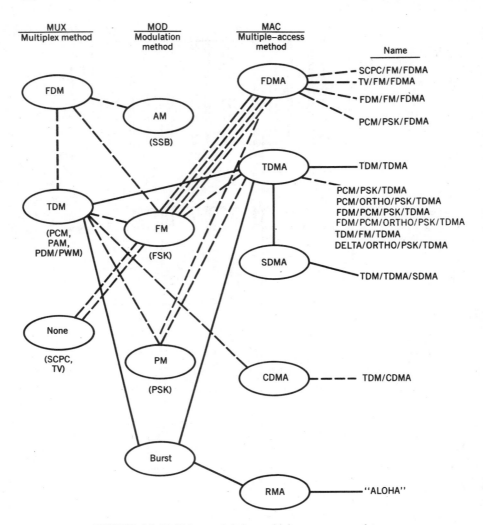

FIGURE 4.9 Multiplex-modulation multiple-access nomenclature.

accesses the satellite via FDMA. If the traffic is all bound for one point, digital speech interpolation (DSI) may be used to augment the channel capacity by a factor of 2. Generally, DSI is thought of as being done at (or before) the transmit earth station. It may also be done in an advanced design satellite after regeneration and remultiplexing and prior to demodulation.

Voice activation, sometimes referred to as "push to talk," can be used with SCPC. On the average each of the two speakers talks 40% of the time (for 60% of the time his outbound channel is silent). Unlike DSI, voice activation does not decrease the average in-use bandwidth in a transponder (since the frequency slots

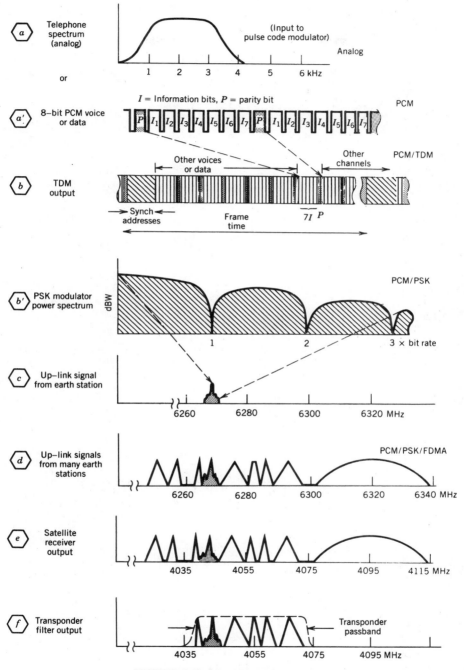

FIGURE 4.10 PCM/PSK/FDMA signals.

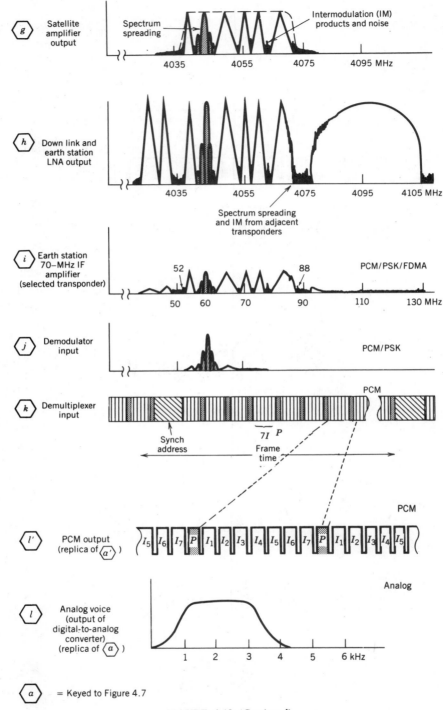

FIGURE 4.10 (*Continued*)

424

are preassigned), but since only 40% of the voice carriers are present at any time, the power requirements are substantially smaller, and intermodulation distortion (IM) is reduced (as less sources of IM are present). The increased use of modems on "voice" lines alters both the DSI and voice activation situations. The modems always have a tone present, even if no information content is present. Since both ends of the connection have modem tones, both half-circuits are effectively in *use* 100% of the time.

Digital data (computer, digital facsimile, etc.) may also be handled by PCM/PSK/FDMA except that the DSI advantage is not possible for that portion.

The earth station is sized for its traffic. Often only one up-link carrier (SCPC or MCPC) is used. The receiving equipment must have filters, demodulators, and demultiplexers for each incoming pathway, but no network timing and the minimum of control are needed. The demodulators have narrow noise bandwidths (matching the carrier) rather than a full transponder width (as in the case of TDMA or CDMA). Both the transmit and receive local oscillators must be stable to stay within the preassigned frequencies.

The satellite transponder is often operated as a linear amplifier on a multicarrier basis, especially if the network has many small-traffic earth stations.

FDMA becomes a complex management problem as the number of carriers grows. Intermodulation (IM) distortion appears as a noise source. As the number of carriers is increased, careful control over the individual up-link carrier powers is required, and frequency assignments must be carefully coordinated to minimize the sum and difference products of carriers causing interference to other users in the transponder. This becomes a very complex situation when the number of carriers is large and requires a computer to manage the IM noise power problem.

Any increase in the total power received (due to an overzealous earth station operator who may think more power is better or to a jammer or interference) will cause an increase in the total noise budget to all users. The excessive signal level takes satellite output power away from the weaker carriers through a process called "power robbing." The FDMA system control center establishes a specific power level for each carrier on a case-by-case basis. The power level is established on the basis of the modulation, bandwidth, and earth station capabilities. Intelsat has established a series of monitoring stations to ensure that its up-link power levels are maintained.

FDMA has the advantage of an inexpensive earth station; but the station must be carefully controlled to use the complex satellite operating environment.

The lack of a timing requirement makes FDMA attractive for fading communication channels (e.g., at frequencies over 10 GHz).

Power amplifiers that handle multiple carriers are summarized in Table 4.7.

The space segment channel capacity per transponder (m) may be determined for both the power, and bandwidth-limited cases, as shown in Tables 4.8 and 4.9.

Figures 4.11 and 4.12 show the transponder channel capacities of various DAMA methods. It is assumed that the DAMA technique involves the use of channel assignments that have been selected to reduce the intermodulation. The SPADE system takes advantage of the gaps in analog telephony. A carrier is only

TABLE 4.7 Power Amplifier Requirements for FDMA Earth Stations with Multiple Carriers

Number of Carriers	Mode[a]	Ratio of Peak to Average Power (dB)	HPA Output Backoff (dB)	PA Power Rating Relative to Average Power per Channel[b] (dB)
1	SCPT	4	0	4
2	2CPT	4	3	10
6	MCPT	3	4.5	15.3
20	MCPT	1	5	19
200	MCPT	0	5	28

[a]Abbreviation: M, multi, PA, Power amplifier.
[b]Column 3 + column 4 + [10 × $\log_{10}$ (column1)].

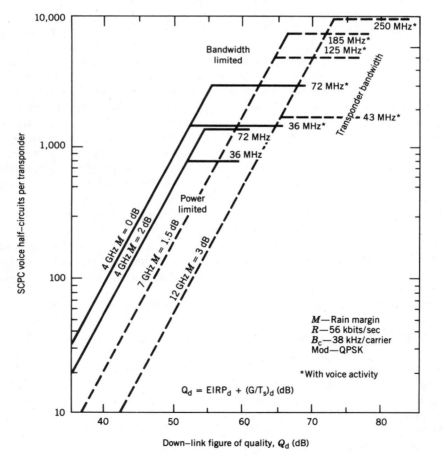

FIGURE 4.11 SCPC transponder capacity with voice activity.

TABLE 4.8 FDMA Capacity

$$m = [\text{EIRP}_d + (G/T_s)_d - \text{BO}_0] - L_d - k + \text{VA}' - (C/N_0)_R - M$$

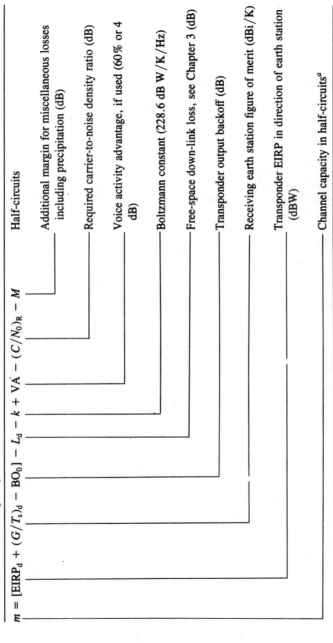

Half-circuits

Additional margin for miscellaneous losses including precipitation (dB)

Required carrier-to-noise density ratio (dB)

Voice activity advantage, if used (60% or 4 dB)

Boltzmann constant (228.6 dB W/K/Hz)

Free-space down-link loss, see Chapter 3 (dB)

Transponder output backoff (dB)

Receiving earth station figure of merit (dBi/K)

Transponder EIRP in direction of earth station (dBW)

Channel capacity in half-circuits[a]

[a]The value of m may be expressed in data channels (at R_v bits/sec each) if $(E_b/N_0)_R$ is used in place of $(C/N_0)_R$.

TABLE 4.9 Typical Values for Parameters in Table 4.8

Access	VA (dB)	$BO_0{}^a$ (dB)	$(C/N_0)_R$ (dBHz)	M(dB) for 4 GHz
Pure				
FDMA	0	3	Varies	3
TDMA	0	0	Varies	3
Demand Access				
FDMA	0	3	Varies	3
TDMA	0	0	Varies	3
SPADE	4	3	Varies	3
ALOHA	4	0	Varies	3
SLOHA (Slotted ALOHA)	4	0	Varies	3

^{a}For TWTAs. Use a lower value for SSPAs and linearized TWTAs (e.g. 1.5 to 2 dB in place of 3 dB).

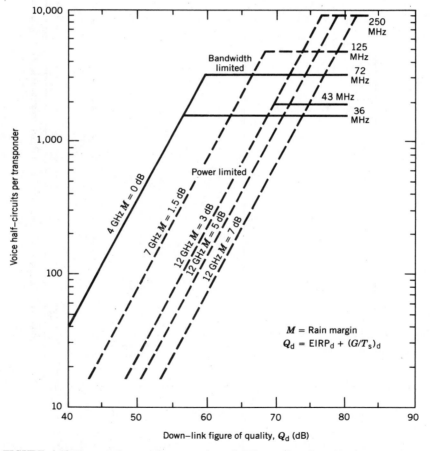

FIGURE 4.12 Transponder capacity versus down-link figure of quality without voice activity.

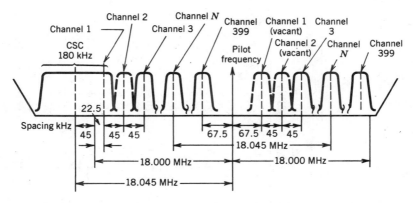

FIGURE 4.13 SPADE transponder plan. (Dicks and Brown, 1974 © 1974 IEEE.)

present when someone is talking. Thus, the 800-channel SPADE system needs power to support only about 320 channels. This is a 4-dB advantage. The SPADE frequency plan is shown in Figure 4.13.

Figure 4.14 shows how the channel capacity drops rapidly as the number of carriers increases in multicarrier FDM/FM/FDMA. This converts a transponder with a theoretical capacity of 1000 half-circuits into one with an effective capacity of only 500 half-circuits if there are six carriers. There are several reasons for this capacity drop. Extra carriers bring more intermodulation products and thus increase the IM-prone frequencies that should not be used for traffic. Another reason is the need for guard bands.

Table 4.10 indicates one particular satellite's circuit loading. See also Figures 3.151 and 3.152.

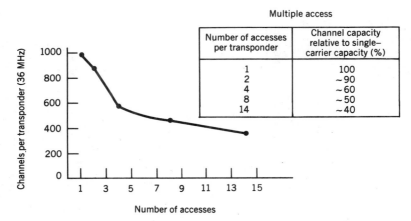

FIGURE 4.14 Transponder capacity for FDM/FM/FDMA. (Dicks and Brown, 1974 © 1974 IEEE.)

TABLE 4.10 Circuit Loadings

Transponder Number	Carriers Per Transponder	Occupied Bandwidth (MHz)	Saturated TWTA Power (dBW)	Output Backoff (dB)	Net Output (dBW)	Antenna Gain (dBi)	EIRP (dBW)	Capacity Half-Circuits[a]
			Satellite 1					
1	4	35.0	6.0	10.1	-4.1	21.0	16.9	612
2	6	35.0	6.2	6.5	-0.3	24.0	23.7	648
3	8	32.5	6.0	9.7	-3.7	24.0	20.3	648
4	5	32.5	6.2	7.9	-1.7	24.0	22.3	792
5	7	35.0	5.4	8.4	-3.0	21.0	18.0	588
6	6	35.0	6.6	6.7	-0.1	21.0	20.9	744
7	4	35.0	6.0	7.5	-1.5	24.0	22.5	912
8	7	35.0	6.2	7.4	-1.2	24.0	22.8	816
9	1	36.0	5.5	3.7	1.8	21.0	22.8	972
10	3	35.0	6.6	5.3	1.3	21.0	22.3	960
11	10	32.5	6.4	5.0	1.4	17.7	19.1	480
12	1	36.0	5.2	4.2	1.0	21.0	22.0	972
13	5	35.0	5.8	5.9	-0.1	21.0	20.9	756
14	6	22.5	6.7	5.5	1.2	17.7	18.9	396
15	7	32.5	5.8	8.8	-3.0	21.0	18.0	672
16	8	35.0	6.6	8.2	-1.6	21.0	19.4	624
17	483 SPADE	36.0	6.5	14.6	-8.1	17.9	9.8	483

18	4	27.5	5.8	7.8	−2.0	21.0	19.0	672
19	7	35.0	5.8	7.4	−1.6	21.0	19.4	672
20	2 video	35.0	6.6	3.6	3.0	17.8	20.8	2 TV
Total								12936
								+ SPADE

Satellite 2

1	1	36.0	6.0	5.8	0.2	24.0	24.2	972
2	1	25.0	6.2	7.9	−1.7	24.0	22.3	792
3	2	35.0	6.1	7.6	−1.5	21.0	19.5	900
4	6	35.0	4.8	2.7	2.1	18.0	20.1	600
5	4	35.0	5.8	10.4	−4.6	24.0	19.4	828
6	1 video	17.5	6.6	9.4	−2.8	21.0	18.2	1 TV
7	5	32.5	7.5	8.4	−0.9	16.6	15.7	432
8	4	25.0	5.4	8.3	−2.9	21.0	18.1	444
9	3	22.5	6.6	10.0	−3.4	21.0	17.6	420
10	2 video	35.0	6.7	3.7	3.0	17.7	20.7	2 TV
11	1	36.0	6.0	5.2	0.8	21.0	21.8	962
12	2	25.0	6.1	–	–	21.0	–	684
13	3 + 2 audio	20.0	7.6	5.3	2.3	16.8	19.1	300
14	4	35.0	5.2	7.5	−2.3	21.0	18.7	492
15	2 video	35.0	6.6	3.6	3.0	17.8	20.8	2 TV
Total								7826

a"Half-circuit audio or video capacity shown "as configured"; actual loading will be smaller. Abbreviation: TV, television video (each at 17–20 MHz).

431

4.2.2 FDMA Impairments

In the early satellites there were only a few transponders, and intermodulation created by traveling-wave tube amplifier (TWTA) amplitude nonlinearity was the dominant factor in limiting system capability. In second-generation satellites more power became available, and the 500 MHz available was split into (typically) 12 channels (or transponders) each with its own TWTA. Operating with only a few carriers near saturation became practical, and the maximum capacity became dependent on a carefully evaluated trade-off analysis such as indicated in the previous section.

As the cost of the space and earth segments declined, there was a dramatic increase in users, stations, and thus carriers. The multicarrier limitations reappeared, and thus the topic of impairment is still vital. Portions of this section follow Dicks & Brown (1974), which should be consulted for further details.

Among the primary impairments are rf power amplifiers. Power amplifiers (of which TWTAs are one form) are inherently nonlinear devices. As shown in Chapter 3, the input–output relationship of a TWTA follows one another for low levels. At intermediate levels the output continues to increase, but not at the same rate as before.

Eventually, the output just does not increase at all (this is called saturation). If the input power continues to increase, the output power actually drops (this is called overdrive).

These amplitude modulation (AM) nonlinearities (and the resulting phase modulation (PM) nonlinearities produce both intermodulation (IM) distortion and intelligible crosstalk when the amplifier is used in multicarrier service.

Linearizing circuits have been designed to overcome the AM and PM nonlinearities over part of the tube's operating range. Figure 4.15 compares a conventional and a modified TWTA performance.

In a single-carrier-per-transponder (SCPT) operation, intermodulation is not a concern (since there is no second carrier). In some SCPT cases up-link fades are of great concern (especially if the up-link frequency is about 10 GHz). The first

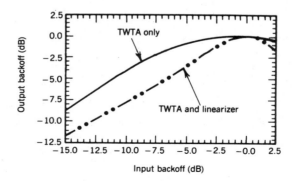

FIGURE 4.15 Conventional and modified TWTAs.

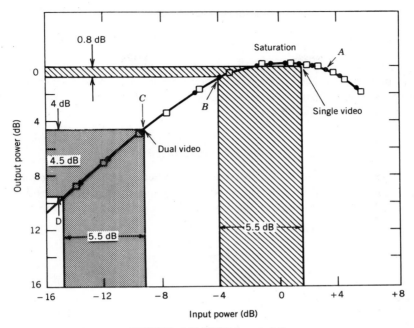

FIGURE 4.16 TWT characteristics.

defense may be to overdrive the tube (to A) in Figure 4.16. When there is an up-link fade the input signal drops to B. A wide range of input signal levels (e.g., 6 dB) produces a modest drop (about 1 dB) in the output. (See also Figure 4.33.) The same fade for dual video (C to D) is more severe.

Through the use of a limiter stage the maximum output signal may be established (often by ground command). Limiters are extremely nonlinear and thus are not used in commercial FDMA service.

Usually ground commands are provided with modified TWTAs. This allows the following:

Linearizer, on/off.

Limiter, on/off clip level selection.

Maximum output level selection.

If both the limiter and the linearizer circuits are off, the TWTA operates as usual.

The newer solid-state power amplifiers (SSPAs) are much more linear. Less (often 2 dB) backoff is needed for the same IM levels, but an overdriven SSPA looks even more like a limiter than a TWTA.

4.2.3 Interference and Noise Sources

4.2.3.1 Polarization Impairments

Cross-polarized interference can occur as a result of imperfect polarization alignments between the earth station and the satellite, polarization rotation due to
precipitation in the path, and less-than-perfect antennas at the earth stations or
satellite (see also Chapter 3).

4.2.3.2 Interference

Cochannel interference may also come from other sources, for example, from
adjacent satellites or from terrestrial users of the frequency band. Some earth station
antennas (particularly the older designs) have poor polarization discrimination off
the main beam.

4.2.3.2.1 SPURIOUS EMISSIONS

Earth stations may degrade with time, and emissions may appear outside of the
assigned SCPC or FDMA channel bandwidth. These factors may be due to the
improper modulator settings (often overmodulation or overdeviation) or spurious
products in the up converter. One of the functions of both the earth station operator
and the network operations controller is to check for this condition and to have
equipment adjustments made at the offending earth station or satellite power amplifier.

4.2.3.2.2 SPECTRUM OVERLAP

If the spectra of two adjacent carriers overlap, there may be interference that will
appear as baseband impulse noise or convoluted signals. A transponder with IM
and a poor output filter cutoff may cause interference to the adjacent copolarized
transponders.

4.2.3.2.3 DUAL-PATH DISTORTIONS

If the transponder input filters do not have good cutoffs, and a carrier is placed
near or at the highest frequency of one transponder, some of the energy may pass
through the input filter of the adjacent frequency copolarized transponder.

The second transponder has an extra carrier and thus an opportunity for IM. It
also has a carrier to amplify, which consumes some rf power.

Eventually the outputs of the first and second transponders are combined
(assuming they use the same down-link beam) in the output multiplexer. When
this happens, there may be trouble. Each signal has been independently amplified
and may have a different power level. Each has passed through its own filters (and
in a different part of the passbands) and through a different power amplifier. The
phase shifts are independent, not controlled, and are unpredictable. Both paths

have the identical frequency but differing amplitudes and phases. In the extreme case the two signals would be 180° out of phase and would try to cancel one another. In more practical cases the amplitude and phase combine in unpredictable ways. This is referred to as adjacent-transponder dual-path distortion.

Dual-satellite-path distortion can also occur if an up-link earth station has too broad a beam or a sidelobe in the wrong place, thereby illuminating not only its assigned satellite with correct polarization and with most of its power, but also an adjacent satellite with some of its power. The adjacent satellite could pick up, frequency convert, amplify, and reradiate this spurious signal back to earth.

The signals may be combined in a broad-beamed earth station receive antenna with unpredictable results. One signal (the intended one) should be on axis; the other one may be in the first sidelobe. Some of the effects are similar to the adjacent-transponder dual-path distortion, but there are additional variables. The magnitude of the interference depends on the main-lobe and sidelobe performance of the small earth stations, pointing accuracy, relative satellite sensitivities and gains, and the difference in the satellite local oscillators.

Typical 6/4-GHz satellites use a nominal 2.225 GHz for frequency conversion.

If the desired satellite's conversion is exactly 2,225,000,000 Hz and the adjacent is 2,225,002,000 Hz, the difference is 2 kHz (about one part in a million).

Assume the up link is 6.0 GHz exactly. After conversions one signal is 6.0–2.225 = 3.775 GHz and the other is 2 kHz lower in frequency (6.0–2.225002 GHz). If both of these signals reach the earth station demodulator, the stronger will be selected, but the weaker will produce a continuous beat interference tone at the frequency difference (2 kHz, which is very audible and annoying). With time the two local oscillators (one per satellite) will drift with respect to one another, thus changing the beat tone. The level of interference can change as a result of subtle variations in the earth station antenna surfaces (due to heating and distortion by the sun, wind, etc.) and by the satellite position.

4.2.3.3 Formulas

This section provides a description of the interdependence of the various factors relating to FDMA system design. The intermodulation power is discussed in a previous section and is dependent on the quantity, power, and frequency plan of the multiple carriers. It is also influenced by the type of power amplifier and its operation.

Up- and down-link interference consists of the satellite's cross-polarized components (often 33 dB down, or 1:2000), and interference from adjacent satellites and terrestrial transmitters.

The interference tends to be unique to each satellite and transponder.

The carrier-to-noise ratios (C/N) are easier to predict, as shown in Tables 4.11 and 4.12.

The major factor that determines the utility of a FDM/FM/FDMA system is the ratio of the received carrier power to the power sum of the noise (N) and interference (I) components. These components are:

TABLE 4.11 Up-link Carrier-To-Noise Ratio

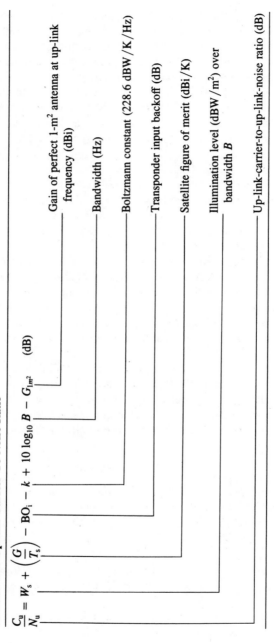

$$\frac{C_u}{N_u} = W_s + \left(\frac{G}{T_s}\right) - BO_i - k + 10 \log_{10} B - G_{1m^2} \quad \text{(dB)}$$

Gain of perfect 1-m² antenna at up-link frequency (dBi)

Bandwidth (Hz)

Boltzmann constant (228.6 dBW/K/Hz)

Transponder input backoff (dB)

Satellite figure of merit (dBi/K)

Illumination level (dBW/m²) over bandwidth B

Up-link-carrier-to-up-link-noise ratio (dB)

TABLE 4.12 Down-Link Carrier-To-Noise Ratio

$$\frac{C_\mathrm{d}}{N_\mathrm{d}} = W_\mathrm{s} + \left(\frac{G}{T_\mathrm{s}}\right) - \mathrm{BO_i} - k + 10\log_{10} B - G_{1\mathrm{m}^2} \quad (\mathrm{dB})$$

Gain of perfect 1-m^2 antenna at down-link frequency (dBi)

Bandwidth (Hz)

Boltzmann constant (228.6 dBW/K/Hz)

Transponder output backoff (dB)

Earth station figure of merit (dBi/K)

Saturated down-link illumination level (dbW/m^2) over bandwidth B

Down-link-carrier-to-down-link-noise ratio (dB)

☐ The up link (designated by the subscript u)
☐ The up-link interference (I_u)
☐ Satellite intermodulation (IM)
☐ The down link (subscript d)
☐ Down-link interference (I_d)

These comprise the total (subscript t) performance, as shown in Table 4.13. Note that each term in parentheses with an asterisk is a ratio while each term in parentheses without the asterisk is in decibels. To convert a quantity in decibels to the ratio use,

$$\left(\frac{C}{N}\right)^* = 10^{(C/N)/10} \quad \text{and} \quad \left(\frac{C}{I}\right)^* = 10^{(C/I)/10} \tag{4.2}$$

After determination of the $(C_t/N_t)^*$ available from the satellite at the receiving earth station, the transmission parameters for each carrier can be calculated from the FM equation and Carson's rule bandwidth (see Table 4.17). Table 4.15 shows the difference between "flat" (unfiltered) and various audio-weighting networks for voice.

Once the number of channels per carrier and the desired S/N are selected, the (C_t/N_t) can be traded for the rf bandwidth to maximize total transponder capacity (see Tables 4.14, 4.16, and 4.17). From these relationships an optimum transponder operating point ("backoff") can be derived by a trade-off between up-path noise, intermodulation noise, and down-path noise contributions, as shown in Figure 4.17. By controlling the amount of attenuation inserted before the input to the transponder (see the satellite gain steps in Figure 4.18), the (C_u/N_u) ratio can be varied while maintaining a constant input level to the traveling-wave tube (TWT). The (C_u/N_u) is thus varied by adjusting the carrier power at the earth station without affecting the intermodulation and down-path thermal noise. Control over most of the remaining impairments is mainly achieved by equipment design factors.

The effect of amplifier intermodulation products is shown in Figure 4.19. Comsat Laboratories, for example, has developed mathematical models that consider both the phase and amplitude situations.

Typical C/IM values for a transponder are shown in Figure 4.20. The carrier-to-intermodulation power ratio is plotted against input backoff for five transponder carrier configurations.

These computer models have produced close correlation between predicted and measured results.

A typical model separates its analysis into two parts, one representing amplitude and group delay frequency response and the other an AM/PM conversion device. The crosstalk carrier is assumed to be modulated with a multichannel baseband, while the wanted carrier is represented by an unmodulated carrier. Intelligible crosstalk therefore originates from the baseband channel of the modulated carrier

TABLE 4.13 Total Link Performance

$$\frac{1}{(C_\text{t}/(N + I)_\text{t})^*} = \frac{1}{(C_\text{u}/N_\text{u})^*} + \frac{1}{(C_\text{u}/I_\text{u})^*} + \frac{1}{(C_\text{d}/IM)^*} + \frac{1}{(C_\text{d}/N_\text{d})^*} + \frac{1}{(C_\text{d}/I_\text{d})^*} \quad \text{(inverse ratio)}$$

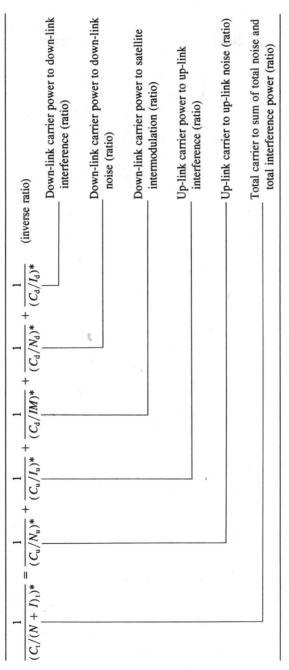

Down-link carrier power to down-link interference (ratio)

Down-link carrier power to down-link noise (ratio)

Down-link carrier power to satellite intermodulation (ratio)

Up-link carrier power to up-link interference (ratio)

Up-link carrier to up-link noise (ratio)

Total carrier to sum of total noise and total interference power (ratio)

TABLE 4.14 Total Signal-to-Noise Ratio[a]

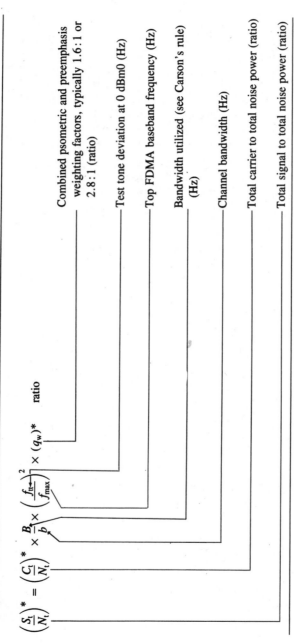

$$\left(\frac{S_t}{N_t}\right)^* = \left(\frac{C_t}{N_t}\right)^* \times \frac{B_r}{b} \times \left(\frac{f_{tt}}{f_{max}}\right)^2 \times (q_w)^* \qquad \text{ratio}$$

Combined psometric and preemphasis weighting factors, typically 1.6:1 or 2.8:1 (ratio)

Test tone deviation at 0 dBm0 (Hz)

Top FDMA baseband frequency (Hz)

Bandwidth utilized (see Carson's rule) (Hz)

Channel bandwidth (Hz)

Total carrier to total noise power (ratio)

Total signal to total noise power (ratio)

[a] Asterisks denote ratios, not decibels. dBm0 is the noise power measured at the zero transmission level reference point.

TABLE 4.15 Noise Weighting Systems

Weighting Network Name	Noise Reduction (dBm minus)	Picowatts[a]
Flat	0	1000
Psophometric (CCITT)	2.5	562
C-Message	1.5	708
F1A	3	500
144	8	158

[a]1 pW $= -90$ dBm $= -120$ dBW.

The zero transmission level point is 1 mW $= 10^9$ pW $= -30$ dBW. If voltages are used, the reference impedance is 600 Ω. The voice band is 300–3400 Hz. The reference frequency is 800 Hz (CCITT) or 1000 Hz (U.S.)

Terms: C-message weighting is used a device to measure line noise where the circuit is terminated in a Type 500 noise measurement set (or equivalent).

and is transferred to the wanted carrier. Since crosstalk increases proportionally with baseband frequency, the analysis is often performed at the top baseband frequency.

A simplified equation for crosstalk ratio (XTR) that can be usefully employed under normal multicarrier operation (-10 to -20 dB input backoff) and small gain slopes is given in Table 4.18.

When the carriers are placed too close, the adjacent channel noise has two forms:

1. When the desired-to-undesired carrier power ratio within the desired channel is less than 1 (under 0 dB), convolution noise is generated.
2. When this ratio exceeds 1 (over 0 dB), the desired detector produces a click sound (in analog service) or an error burst (in data communications). This is also called impulse noise and usually is the dominating effect.

Common practice is to place a guard band (equal to about 10% of the carrier bandwidth) between the carriers. If the adjacent carriers are equal in amplitude and bandwidth, this will produce less than 18 noise counts in 15 min (the CCITT data service requirement).

Filters (including transponders) have group delay. This is caused by the phase variation across the passband. Figure 4.21 shows the typical group delay response versus frequency for a 45-MHz-wide transponder at 12 GHz. As carriers are placed further away from the center frequency, the group delay increases. Some portions of an FM signal thus may be delayed more than others, thereby causing distortion. To compensate, group delay equalization filters—having a delay characteristic inverse to that of the satellite transponder—can be installed in an earth station.

The CCIR states that during 80% of any month the *1-min mean noise* power should not exceed 10,000 pW. For 0.3% of any month it should not exceed 50,000

TABLE 4.16 Total Signal-to-Noise (in Decibels)

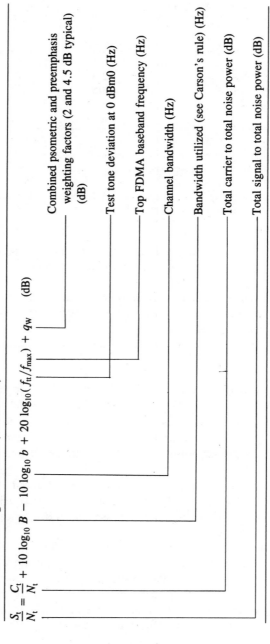

$$\frac{S_t}{N_t} = \frac{C_t}{N_t} + 10 \log_{10} B - 10 \log_{10} b + 20 \log_{10}(f_{tt}/f_{max}) + q_w \qquad (\text{dB})$$

Combined psometric and preemphasis weighting factors (2 and 4.5 dB typical) (dB)

Test tone deviation at 0 dBm0 (Hz)

Top FDMA baseband frequency (Hz)

Channel bandwidth (Hz)

Bandwidth utilized (see Carson's rule) (Hz)

Total carrier to total noise power (dB)

Total signal to total noise power (dB)

TABLE 4.17 Carson's Rule

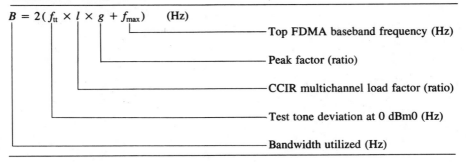

$$B = 2(f_{tt} \times l \times g + f_{max}) \quad \text{(Hz)}$$

── Top FDMA baseband frequency (Hz)

── Peak factor (ratio)

── CCIR multichannel load factor (ratio)

── Test tone deviation at 0 dBm0 (Hz)

── Bandwidth utilized (Hz)

pW and not more than 1×10^6 pW for 0.01% of any year (excluding sun outages). The noise power $(N_w)^*$ (in picowatts) may be converted to N_w (in dBW) using $1 \text{ pW} = 10^{-12}$ W:

$$N_w = -120 + 10\left[\log_{10}(N_w)^*\right] \quad \text{(dBW)} \qquad (4.3)$$

This is the noise at the receiver output. If it is assumed that all of the noise is in the low-noise amplifier and the "sky" (see Section 3.2.2), the amplifier input noise can be determined using its gain (see Figure 4.22). Most link budgets stop

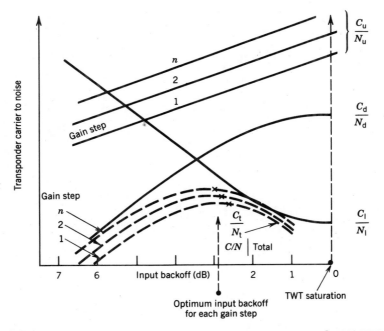

FIGURE 4.17 Optimum TWTA input power. (Dicks and Brown, 1974 © 1974 IEEE).

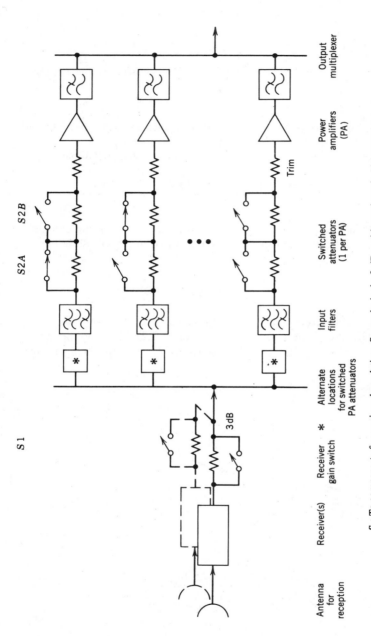

S 1

S2A   S2B

Antenna
for
reception

Receiver(s)   Receiver   *   Alternate   Input   Switched   Power   Output
             gain switch   locations   filters   attenuators   amplifiers   multiplexer
                              for switched   (1 per PA)   (PA)
                              PA attenuators

3 dB

Trim

S_1: To compensate for receiver degradation, S_1 may be in the 3-dB position at launch. If the receiver
gain degrades, S_1 may be placed in the 0-dB position.

$S_{2A,B}$ (etc.): Individual transponder attenuators. Various switch settings yield 0, 3, 6, and 9 dB
attenuation.

Asterisk refers to alternate location for $S_{2A,B}$ (etc.).

Trim: To match power amplifier levels.

FIGURE 4.18 Satellite attenuator locations.

444

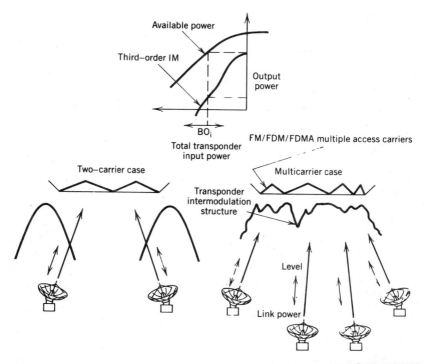

FIGURE 4.19 Transponder intermodulation patterns. (Dicks and Brown, 1974 © 1974 IEEE.)

at the receiver input; thus the conversion shown in Figure 4.22 is needed. To find the noise level at the antenna, the noise contribution of the LNA/LNB/LNC must be removed and the result diminished by the LNA/LNB/LNC gain (usually 45–60 dB).

In the satellite a typical transponder has a gain of 100–125 dB.

A discussion of the Intelsat noise budgets is given in Miya (1975).

Table 4.19 summarizes an Intelsat noise budget.

4.2.4 Single Channel per Carrier

Instead of a multichannel earth station transmission based on frequency domain multiplexing (FDM), each channel could be modulated on its own rf carrier (hence the single channel per carrier, or SCPC). At the transponder the individual SCPC signals are combined into a series of carriers in the transponder bandwidth.

To determine the optimized operating point for a satellite transponder, a number of items must be considered, as shown in Figure 4.23. Many of the steps are similar to those considered in the FDMA section.

For small earth stations the down-link noise may predominate and require a higher power per channel (less input backoff). Figure 4.23 shows the optimum

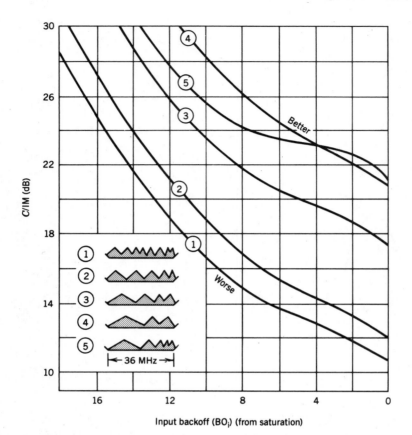

FIGURE 4.20 Typical intermodulation (C_i/IM) values for various transponder carrier configurations (Dicks and Brown, 1974 © 1974 IEEE.)

operating point for TWTAs. Approximately 3–5 dB less backoff is needed for SSPAs because of their lower intermodulation distortion noise. Linearized TWTAs require less output backoff and thus can deliver more power to the antenna at a given C/I ratio. See Figure 4.24.

The down link figure of quality, Q_d, may be expressed as

$$Q_d = \mathrm{EIRP}_d + \left(\frac{G}{T_s}\right)_d \quad (\mathrm{dB}) \tag{4.4}$$

This term is used in Figures 4.11 and 4.12.

A popular international form of SCPC is known as SPADE. Figure 4.13 shows a system where 45 kHz is set aside for each SPADE channel (half-circuit). The PCM/QPSK/FDMA channels are paired with an 18.045-MHz separation between

TABLE 4.18 Crosstalk Ratio[a]

$$XTR = 20 \log \left\{ [57.3/(K_p S f_{max})] \times (f_w/f_i) \times (P_t/P_i) \right\} \quad (dB)$$

- Interfering power (W)
- Total power (W)
- Peak deviation of interfering carrier (Hz)
- Peak deviation of wanted carrier (Hz)
- Maximum frequency (Hz)
- Linear gain slope occupied by first two sidebands (ratio)
- Average AM/PM coefficient (degrees/dB)
- Radians conversion $(180/\pi)$
- Crosstalk ratio for backed-off multicarrier operation (dB)

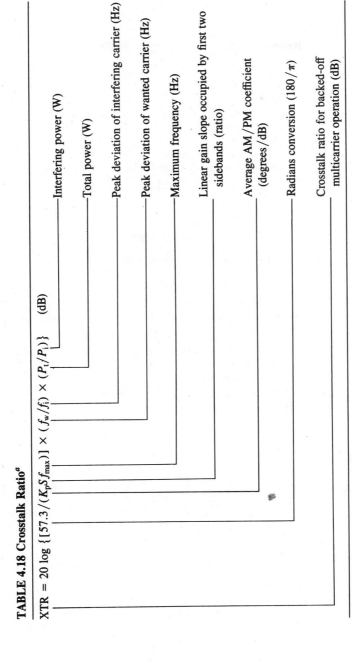

[a]BO_i range: -20 to -10 dB with respect to saturation.

447

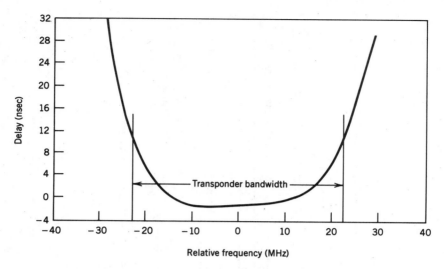

FIGURE 4.21 Typical transponder group delay response.

the to and from channels. A common signaling channel (CSC) is used by stations requesting a connection to another station and for the assignment of the working SPADE frequencies. The frequency pairs are then assigned randomly on demand so the IM products are randomized and appear as though they were noise. When the circuit is in operation, the carrier is present only while speech is actually present and then (usually) only in one direction at a time. When viewed on a spectrum analyzer, the SCPC SPADE carriers appear to pop in and out as would be expected with a voice-activated (sometimes called VOX, or push-to-talk) operation.

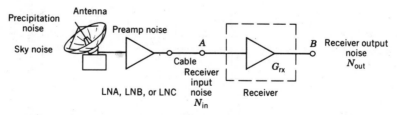

Note: * denotes ratio. If N_{in}^* is in picowatts, N_{out}^* is in picowatts. Assumes a receiver signal output of 0 dBm ($= -30$ dBW, or 1 mW). Asterisk denotes ratio.

$$N_{out} = N_{in} + G_{rx} \quad \text{(dBW)}$$

$$N_{out}^* = N_{in}^* \times G_{rx}^* \quad \text{(W)}$$
Typical value: $G_{rx} = 50$ dB

$$G_{rx}^* = 100{,}000$$

FIGURE 4.22 Receiver noise levels.

TABLE 4.19 Intelsat IV, IV-A, and V Noise Budgets for FDM/FM/FDMA

Impairment	Intelsat IV (pWp 0)[c]	Intelsat IV-A (pWp 0)	Intelsat V (pWp 0)	Intelsat V[b] (pWp 0)
Space Segment				
1. Thermal noise (up and down link)	7,500 }	6,500 }	7,500	7,500
2. Transponder intermodulation	—	—		
3. Cochannel interference	— } 7,500	500 } 7,500[a]	500	2,500
4. Adjacent transponder interference due to intermodulation	—	500		
Earth Segment				
1. Earth station rf out-of-band emission	500	500	500	
2. Earth station transmitter noise excluding multicarrier intermodulation and group delay noise	250 } 2,500	250 } 2,500	2,500	2,500
3. Noise due to total system group delay and dual-path distortion after any necessary equalization	500	500	500	
4. Other earth station receiving noise	250 } 1,000	250 } 1,000	250 } 1,000	———
5. Interference from terrestrial systems sharing same frequency bands	1,000	1,000	1,000	
Total system per channel noise budget (CCIR)	10,000	10,000	10,000	10,000
General Satellite Capability				
1. Transponder bandwidth (MHz)	36	36	36 41 72 77	72[b] 77[b] 241[b]
2. Number of transponders	12	20	4 1 13 4	2 2 2 2
3. Total bandwidth (MHz)	432	720	1429	780
4. Nominal channel capacity (Atlantic primary) (plus one transponder each for SPADE and TV)	7,500	12,500	24,000 voice channels	

[a]Objectives may be exceeded as long as total equals 7,500 pWp 0.

[b]At 11 GHz (all others are at 4 GHz).

[c]pWp 0 is the psophometrically weighted power with reference to the zero transmission level point (in picowatts).

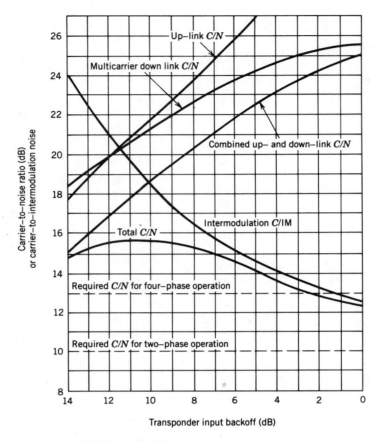

FIGURE 4.23 SCPC optimization characteristics.

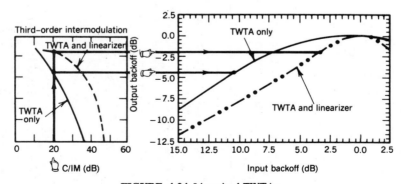

FIGURE 4.24 Linearized TWTAs.

450

Although the transponder capacity is 800 SPADE channels, it is unlikely that even half would be active at any instant even at a 100% fill factor because of the voice activity. Typically, only 40% of the channels (or 320) are active. This assumes no modems are used on the voice lines.

Three impairments must be considered: thermal noise, adjacent channel inter- ference, and intermodulation noise.

The adjacent channel noise is a function of the transmit low-pass filter and the receive bandpass filter in the earth stations. The voice activity and random assign- ment of channels help to reduce this problem. Because of the equally spaced channel assignments, the intermodulation noise is such that the even and odd products (amplitude and phase) fall atop other channels. Again, advantage is taken of both the voice activity and random assignments to reduce the noise magnitude. The CSC is located 18.045 MHz below the center of the SPADE transponder and occupies 160 kHz using two-phase PSK/TDMA.

The typical threshold C_t/N_t of the PCM-based SPADE is about 13 dB. For Intelsat TWTAs the input backoff is about 10 dB, which requires an earth station EIRP of 60.5 dBW per carrier at 6 GHz into a global beam at a 10° elevation angle. SCPC is often used for 56-kbits/sec digital voice or data. Various forms of modulation may be used on each carrier. In addition forward error control (FEC) may be used to improve the signal's resistance to noise and the various forms of interference.

Table 4.20 shows the characteristics and unfiltered bandwidths of these combi- nations.

4.2.5 TV/FM/FDMA and Narrow-Band TV/FM/FDMA

Television program video and audio material are distributed by satellite using one or more carriers per transponder. It has been common to use one television carrier per domestic 36-MHz-wide transponder. As the demand for transponders approaches the supply, alternates may be needed. MCPT television have been implemented where the economics and performance were right. This requires an improved figure of merit (G/T_s) in the earth stations or lowered quality require- ment.

In the case of the Intelsat transponders the use of 2CPT television (2 video) was first instituted in the Atlantic Ocean region, as shown in Figure 4.25, for inter- national television. Similar frequency plans are used for domestic satellite transponders.

Since the transponder power is limited, the use of MCPT television reduces the power 3–5 dB per carrier (by half, or 3 dB, in the case of 2CPT television), *and* a backoff must be employed with satellite TWTAs to reduce the IM products. The television carrier bandwidth may have to be reduced (e.g., to 17.5 to 20 MHz) to fit two carriers in a 36–41-MHz transponder. In Intelsat a global transponder may have a saturated EIRP of 22.5 dBW when used for a single television carrier, the same transponder when used for 2CPT television can be operated at 14.5–16.5

TABLE 4.20 Typical Modulation Bandwidths

Phase Shift Keyed Modulation	Forward Error Correction (FEC) Rate	Phase States	Information Bits per Symbol (b/S)	Hertz per Bit	E_b/N_0 for BER of 10^{-4} (dB)	Bandwidths for an Information Rate of:[a]			
						56 kb/s (kHz)	64 kb/s (kHz)	1.544 Mb/s (MHz)	2.048 Mb/s (MHz)
BPSK (2PSK)	R 1	2	1	1	8.2	56.0	64.0	1.544	2.048
	R 7/8					64.0	73.1	1.765	2.341
	R 3/4					74.7	85.3	2.059	2.731
	R 2/3					84.0	96.0	2.316	3.072
	R 1/2					112.0	128.0	3.088	4.096
QPSK (4PSK)	R 1	4	2	0.5	9.0	28.0	32.0	0.772	1.024
	R 7/8					32.0	36.6	0.882	1.170
	R 3/4					37.3	42.7	1.029	1.365
	R 2/3					42.0	48.0	1.158	1.536
	R 1/2					56.0	64.0	1.544	2.048
8PSK	R 1	8	3	0.33	12.5	18.7	21.3	0.515	0.683
	R 7/8					21.3	24.4	0.588	0.780
	R 3/4					24.9	28.4	0.686	0.910
	R 2/3					28.0	32.0	0.772	1.024
	R 1/2					37.3	42.7	1.029	1.365
16PSK	R 1	16	4	0.25	17.2	14.0	16.0	0.386	0.512
32PSK	R 1	32	5	0.20	22.1	11.2	12.8	0.309	0.410
64PSK	R 1	64	6	0.17	27.0	9.3	10.7	0.257	0.341

[a]The receiver noise bandwidth should be wider to accommodate local oscillator drifting in the two earth stations and the satellite. Through the use of filtering and advanced modulation methods these bandwidths can be further reduced.

452

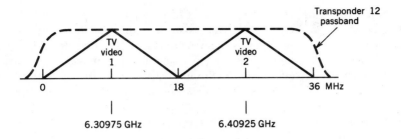

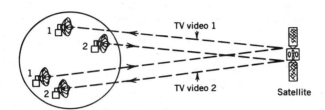

FIGURE 4.25 Two television carriers per transponder in Intelsat satellites.

dBW per carrier. For 2CPT television service via Intelsat between standard B stations (6/4 GHz, $G/T = 31.7$ dBi/K), the up link is 85 dBW (into the satellite's global beam). The receive bandwidth is 15.75–20 MHz.

The audio may be handled as an FM subcarrier with the video or as stand-alone service using a carrier at the upper end of the transponder. Domestic (U.S.) practice allows multiple audio subcarriers in a video transponder. In Intelsat the extra audio carriers may need separate transponders (often FDMA) or even separate satellites.

The two television signals may come from diverse earth stations. Usually access to an Intelsat video transponder (number 12 in their numbering system) is provided on the basis of program duration (including setup, cue, and tear-down times) and as such is a DAMA service (the control center is at Intelsat, Washington, DC).

Domestic television services go to simpler earth stations, and the multichannel audio virtually always accompanies the video. Several methods are available.

1. One or several subcarriers may be used to provide monaural, stereo, or multilingual sound. Cueing may also be provided. These subcarriers are typically located at 5.41, 6.2, or 6.8 MHz. Additional audio or audio-bandwidth program material that is unrelated to the video can be added at these or other frequencies. A relatively simple single-channel, receive-only earth station may provide diverse services such as television (network or pay) plus audio programs.

2. The audio may be digitized and inserted into or adjacent to the horizontal

retrace time (which occurs 15,750 and 15,625 times per second for the 525- and 625-line systems). See Figure 4.26. The two pulse-width-modulated (PWM) audio signals may be used for two narrow-band audio channels or one 10–15-kHz audio channel. The MAC (multiplexed analog component) systems time share the horizontal interval between video, audio and data signals. In this way only one signal is being transmitted at a time, thus avoiding intermodulation distortion, crosstalk and other problems.

3. The U.S. Public Broadcasting Service (PBS) utilizes another method for distributing audio channels in the same transponder as the video. Up to four high-quality channels (15 kHz each at ±1.5 dB) are sampled at 34.42 kHz each. The samples are then converted to a 14-bits-per-sample (481.88-kbits/sec) word. This in turn is companded to 11 bits plus sign plus parity (447.46 kbits/sec). The four channels produce a 1.78984-Mbits/sec continuous data stream that is used to QPSK modulate a 5.5-MHz subcarrier. Figure 4.27 shows the basic block diagram. Figure

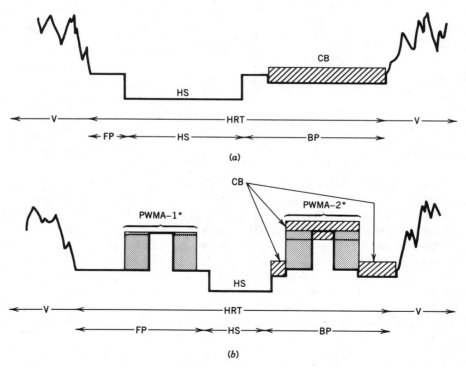

Key: V, video; HS, horizontal sync pulse; CB, color burst; HRT, horizontal return time; FP, front porch of HRT; BP, back porch of HRT; PWMA-1, pulse-width-modulated audio number 1; PWMA-2, pulse-width-modulated audio number 2.

FIGURE 4.26 Television sound in sync: (a) conventional color video signal; (b) sound in sync color video signal.

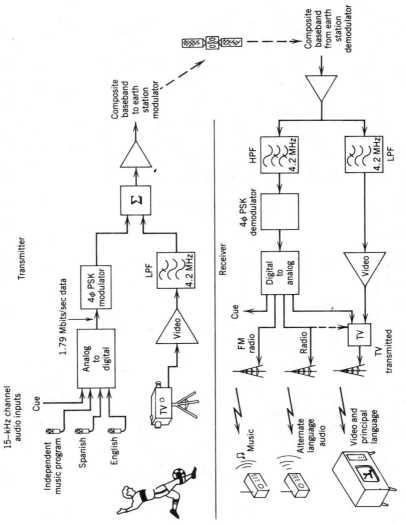

FIGURE 4.27 Analog video with four-channel digital audio.

455

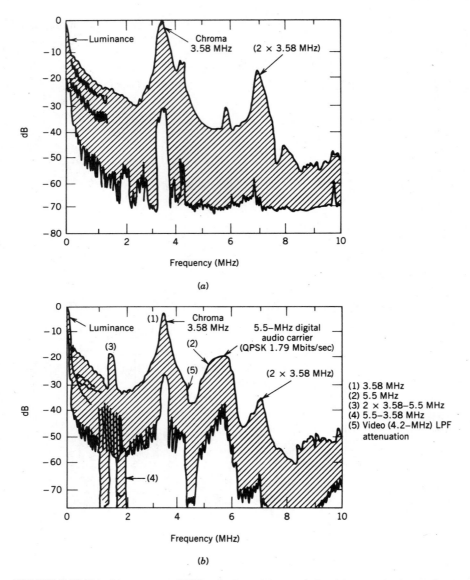

FIGURE 4.28 Television spectra of 75% color bars: (*a*) normal signal (no sound subcarriers); (*b*) with 5.5-MHz subcarrier for four-channel sound.

4.28 illustrates the television spectra with and without the digital audio subcarrier. The video is an analog signal, whereas the subcarrier is digital. The use of QPSK reduces the PSK spectrum to 5.5 MHz ($\pm 1.78984/2$ MHz).

The four audio channels are phase related for quadraphonic (four-channel) or stereo (two-channel) audio applications. Alternatively, four independent signals may be used simultaneously. Among the PBS applications are multilingual program channels, pickaback audio for AM and FM stations (mono, stereo, or quad), cueing channels, order wires, facsimile, and data (to 1.79 Mbits/sec in place of audio).

4. Many modern television systems use scrambling to provide security for the video and audio contents. One method is to digitize the audio (and any data channels) and then to encrypt them, often using the U.S. National Bureau of Standard's DES (Digital Encryption Standard) codes. These are mixed (usually on a TDM basis) with the scrambled video signal for transmission. Companded single-sideband amplitude modulation (CSSB/AM) is still another method of audio transmission. The word *compander* is derived from the combination of two words: *compressor* and *expander*. A voice signal covers a wide dynamic range from a whisper to a shout. See Figure 4.29. In the case of frequency domain FM systems, a shout may cause the frequency spectrum to overdeviate, expand, and potentially cause intereference to a neighboring call. Bandwidth is consumed, and some power must be set aside in allowing for this occasional event, thus limiting the transponder channel capacity. For a whisper the transmission noise may be as high as the signal. In digital systems many bits are needed to express the wide dynamic range.

A compander—which acts like two funnels end to end—cuts this dynamic range by reducing the shout and increasing the loudness of the whisper. The wide input range (65 dB, or over $3 \times 10^6 : 1$) is adjusted. The gain of the originating compressor is nonlinear (in accordance with the level of the incoming signal), while the distant expander has an inverse nonlinearity so it can recreate the full dynamic range (65 dB, or over $3 \times 16^6 : 1$). The transmission channel has a reduced range (1778 : 1) and thus may operate very efficiently and at increased capacity.

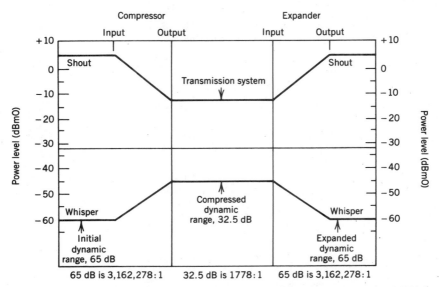

FIGURE 4.29 Companders. Reprinted with permission of *Satellite Communications*, 6300 S. Syracuse Way, Englewood, CO.

4.3 TIME DOMAIN MULTIPLE ACCESS (TDMA)

4.3.1 Basic TDMA Characteristics

By its very nature, TDMA is a digital service readily compatible with terrestrial digital telecommunications networks, fiber optics, digital video, packet communications, and computer-to-computer load-sharing operations.

Digital techniques depend on integrated circuit (IC) chips. These include speech-processing methods such as digital speech interpolation (DSI), digital echo control, bit regeneration at various points along the communications chain (including on-board a satellite), switching (see also SDMA/SS/TDMA), routing, and pulsed laser intersatellite communications.

For simplicity this section deals with single-carrier-per-transponder (SCPT) TDMA. It should be realized that MCPT/TDMA is also possible. In these multi-carrier cases some of the impairments are the same as multicarrier FDMA.

TDMA divides the time resource into small segments with brief gaps (called guard times) between them (see Figure 4.30), just as FDMA divides the transponder's frequency band resource into small subbands with guard bands between them. Whereas FDMA provides simultaneous access to the transponder for several users (typically 1–9), TDMA operates with the one carrier at a time on a sequential access basis with its users (often 5–50 in number).

In TDMA, each earth station (Figure 4.31) is assigned an exclusive nonoverlapping use of a time slot during which it transmits its accumulated traffic as a digital bit stream burst (Figure 4.32). The beginning time of the burst is established with respect to a synchronization burst received from a central control station. The burst bit rate and burst duration must be controlled. The duration is assigned by

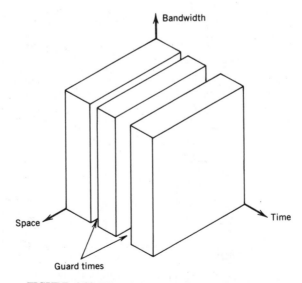

FIGURE 4.30 Time division multiple access (TDMA).

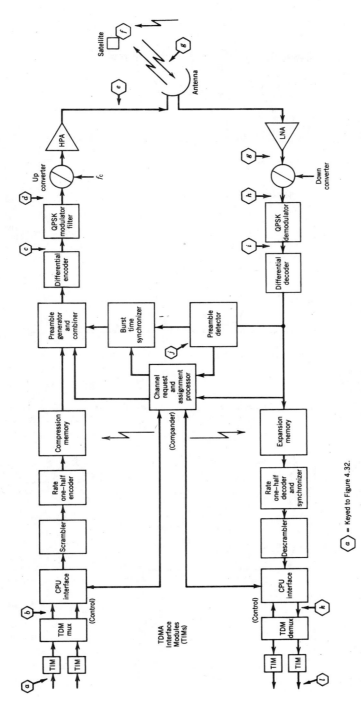

FIGURE 4.31 TDMA earth station.

ⓐ = Keyed to Figure 4.32.

459

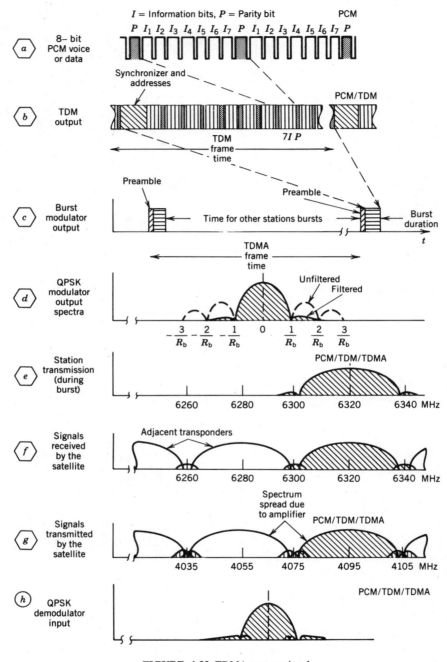

FIGURE 4.32 TDMA system signals.

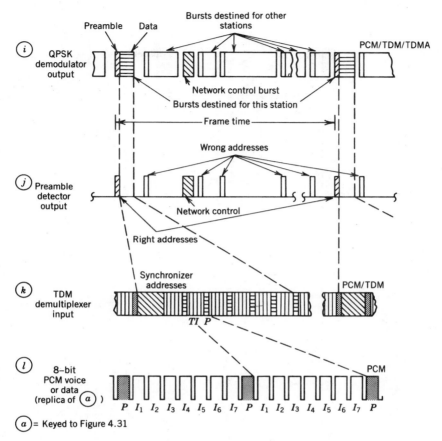

FIGURE 4.32 (*Continued*)

the control station (Table 4.21) to match the traffic demands of each communications stations; thus, there is a station-to-station variation in the burst duration (or length). The duration may be rapidly altered to match the station's instantaneous traffic demands. Any station has access to all other stations for interconnectivity. For FDMA changing the station bandwidth assignment is costly and impractical on a continuous basis (except in the special case of SPADE). TDMA offers a better dynamic operational use of the transponder.

After one station completes its burst, a guard time it provided before the next station's burst commences. The length of this guard time is established on the basis of the anticipated variable delays in equipment, synchronization acquisition, and variations in earth-station-to-satellite slant ranges. The guard time prevents burst collisions due to simultaneous transmissions.

TDMA has the practical advantage of many more accesses per transponder than FDMA. The bit capacity of TDMA is nearly independent of the number of

TABLE 4.21 TDMA Assignment Methods

Assignment Protocol	System Control Function	Frame Efficiency
1. Preassigned, fixed TDMA	None needed	Highest, since there is no need for station addressing of bursts
2. Preassigned	System control monitors traffic and transmits a new program of assignments periodically	High, since there is little or no need for individual station addressing of burst; control burst is needed for networking
3. Demand assigned	System control assigns each frame a time slot for connection duration in response to demand from earth station	Lower, as each burst must contain station address; synchronization may be poorer, and queues may result during busy hours
4. Limited preassignment, demand overflow	Similar to 1 and 2 except that overflow (busy hours) handled by demand assignments from pool (often premium priced)	Combination of 1 and 2 (for normal traffic volume) and 3 (for overflow)

accesses, but overhead signaling and the multiplicity of guard times can reduce the throughput of useful communications when there is a large number of users.

Since only one carrier is present at a time in an earth station transmitter or the satellite transponder (for SCPT), the power amplifiers may be operated at their full saturated power because there is no second carrier present to produce IM distortion. The satellite amplifier may be used as (or may have) a limiter because of the SCPT operation. Figure 4.33 shows the typical characteristics of a satellite TWTA when subjected to a fading up-link signal and burst-to-burst power variations. This illustrates the possible relaxation of the up-link power control. The slight overdrive (typically 1.2 dB) of the TWTA reduces the up-link variations that are translated into the down link.

Figure 4.34 shows the operating modes of the earth station and satellite transmitters. Whereas FDMA requires its transmitters to operate on a continuous basis usually in the multiple-carrier basis (MCPT), TDMA requires pulsed (SCPT) earth station and satellite transmitters. The duty cycle of the earth station transmitter is low, whereas the satellite duty cycle varies with changes in the transponder fill factor, from low (at little or no traffic) to nearly continuous (for a fully loaded

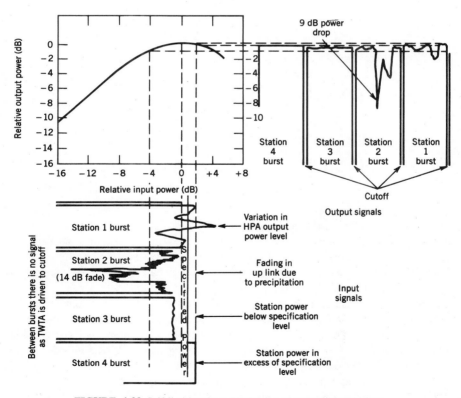

FIGURE 4.33 Self-limiting from TWTA (or other nonlinear device).

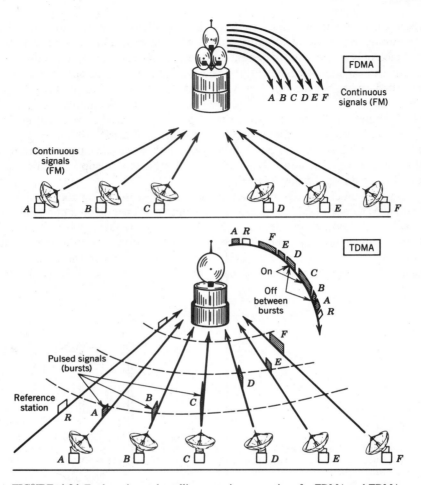

FIGURE 4.34 Earth station and satellite transmitter operations for FDMA and TDMA.

TDMA transponder). Both daily and long-term variations should be considered. In some TDMA systems one station fills any unused time slots with pseudorandom bits to keep the spectrum looking like noise to reduce interference to other systems. For this reason a TDMA carrier as radiated from a satellite may look the same at 2 AM and at 2 PM of a business day.

Since the SCPT satellite transmitter is generally driven to saturation (instead of some backed-off input level as in the case of MCPT), more earth station peak power is required. The satellite amplifier duty cycle is discussed in the preceding. The on–off nature of the power output introduces design problems into the amplifier's electric power conditioners and causes power surges into the main spacecraft power supply. Single-collector TWTAs and some solid state power amplifiers (SSPAs) tend to draw a nearly constant amount of dc power regardless of their rf power output. Multicollector TWTAs and many SSPAs more closely match input dc power with output rf power.

Figure 4.31 shows a TDMA earth station. The block diagram is more complex than its FDMA counterpart because of the analog-to-digital conversions, accumulation of bits between bursts, synchronization recovery equipment, burst bit rate clocks, and the modems, which must operate at the system bit rates (in the range of one to hundreds of megabits per second) regardless of the size of the station.

A basic TDMA challenge is to get a high-speed PSK or FSK channel through a highly nonlinear and bandwidth-limited channel without the spectrum spreading into adjacent transponders and the use of expensive earth station equipment (Figure 4.35).

Although a TDMA signal usually fills a transponder on a SCPT basis, it is possible to share the transponder resources (power and frequency) with another service (e.g., FDMA or CDMA, as shown in Figure 4.36) or with independent TDMA signals. The MCPT operation requires backoff to reduce intersymbol distortion and adjacent transponder interference due to spectrum spreading. The pulse-to-pulse power demand variations of TDMA may be deleterious to analog FDMA signals sharing a transponder. Up-link TDMA power variations (including the slack time gaps) appear directly in the down link because it is operating in the linear amplifier mode.

Table 4.22 defines some of the digital modulation terms in the following section.

4.3.2 Types of TDMA

Figure 4.37 shows four methods of demand access. Satellite-switched TDMA (SDMA/SS/TDMA) uses the space domain (SDMA) and orthogonal beams and/or polarizations with an on-board switch matrix for connections. SDMA/SS/TDMA is discussed in Section 4.4.3. Regenerative repeaters remove up-link noise, reshape, retime, and reassemble the bit streams for the down-link.

Tables 4.23 and 4.24 indicate the relationship between the bit rate (R_b or $1/t_b$), noise bandwidth per channel (B), ratio of the energy per bit to the noise power density (E_b/N_0), total link carrier-to-noise ratio (C_t/N_t), and total carrier-to-noise power density ratio (C_t/N_{ot}).

Figure 4.38 relates E_b/N_0 to the bit error rate for several digital systems. By increasing the forward error control bits to rates under 1 (see Section 4.1.3), the system becomes more tolerant of noise, but the total number of bits increases, and thus the bandwidth is broadened, thereby allowing more noise to get through the filters. Rate $\frac{1}{2}$ QPSK, for instance, has a FEC bit for every information bit. For a given bit error rate (BER) it requires 3 dB less E_b/N_0 than BPSK (but there is 3 dB more noise due to the doubled bandwidth). It is 6 dB better than uncoded QPSK at a BER of 10^{-4}.

The (C_t/N_t) and (C_t/N_{ot}) are the levels required at the point of demodulation. By demodulating the signal in the satellite and regenerating the signal, the up-link and down-link degradations may be separated so that the receiving earth station is

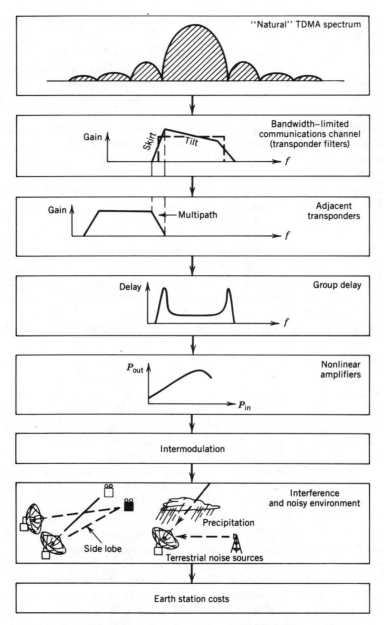

FIGURE 4.35 TDMA challenges to maximize performance and minimize costs.

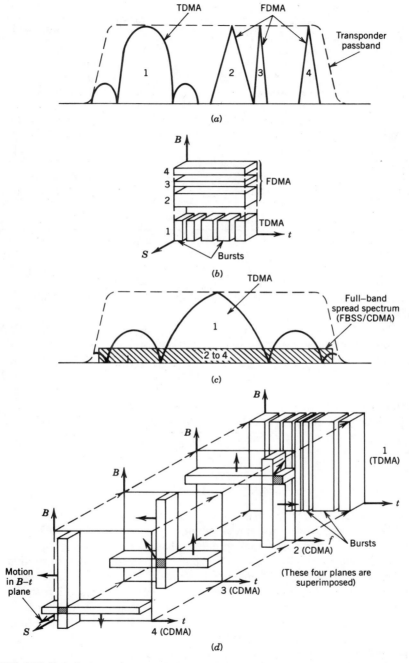

FIGURE 4.36 Shared transponder examples: (*a*) Four CPT, mixed TDMA, and FDMA; (*b*) multiple-access diagram for (*a*), (*c*) MCPT, mixed TDMA, and FBSS/CDMA; (*d*) multiple-access diagram for (*c*).

467

TABLE 4.22 Digital Modulation Terms

Symbol	Meaning	Fundamental Units[a]
B	Bandwidth	Hz, dBHz
R_b	Bit rate	bits/sec
t_b	Time per bit	sec/bit
E_b	Energy per bit	watts/bit, W/Hz, dBW/bit
N_0	Noise power density	W/Hz, dBW/Hz
E_b/N_0	Energy per bit/noise power density	Ratio, dB
R_b/B	Bits per hertz $(=1/Bt_b)$	b/Hz, dBb/Hz
Bt_b	Figure of modulation merit	Hz/b, dBHz/b
S	Symbols	Symbols
R_b/S	Bits per symbol	bits/sec/symbol
t_{fr}	Frame time	sec
t_{bu}	Burst duration	sec

[a]Prefixes may be added: k, kilo; M, mega; G, giga; m, milli; μ, micro; n, nano. The second set of units are the decibel representation of the first set.

concerned with only the down link (instead of the entire link as in the case of nonregenerative satellites).

Table 4.25 shows various FSK and PSK keying methods used for TDMA and other digital services.

The bit rate R_b is the inverse of the time per bit ($R_b = 1/t_b$); thus, the R_b/B ratio of Table 4.20 becomes $1/Bt_b$. In some cases, the Bt_b product is used as a figure of merit. It is the reciprocal of R_b/B. Figure 4.39 is a nomogram for finding these ratios.

Practical PSK modems handle a Bt_b product of about 1.2. The amount of integrated energy after filtering for various keying methods is shown in Figure 4.40.

4.3.3 TDMA Operations

Table 4.26 summarizes the important operational attributes of TDMA. A typical sequence for acquiring a time slot assignment is shown in Figure 4.41. For clarity it has been simplified. An incoming digital channel request is received (a) from a terrestrial source or facility (e.g., a computer or a telephone administration). The originating earth station is assigned an idle time slot by the network control center (b). The earth station transmits a channel request to the destination earth station using the earth station's code. A call identification "tag" is added for later use in pairing the forward and return paths (c). This tag will be used on each subsequent burst. This procedure is recognized by two stations: the TDMA control center as it assigns channels from its pool of idle channels and starts the billing clock (d) and the destination earth station (e). The destination station seizes a matching reply time slot channel that has been assigned and establishes the return path (f).

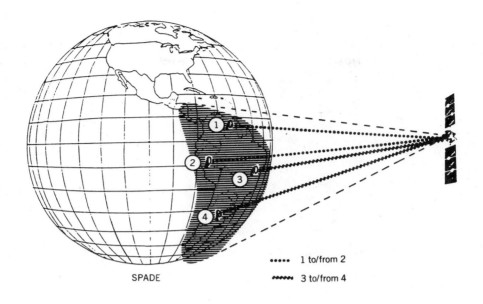

SPADE

••••• 1 to/from 2

~~~~~ 3 to/from 4

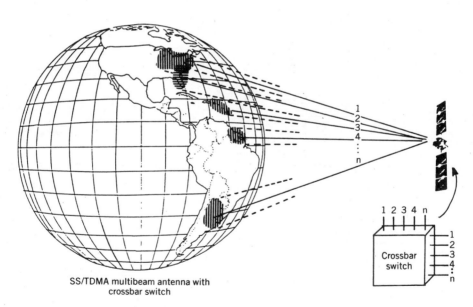

SS/TDMA multibeam antenna with
crossbar switch

FIGURE 4.37 Time domain transmission systems.

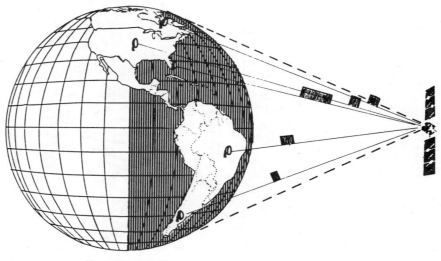

Conventional TDMA

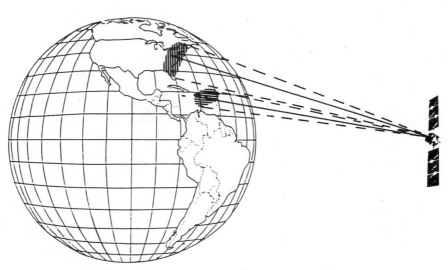

On-board regeneration, retiming and remodulation

FIGURE 4.37 (*Continued*)

TABLE 4.23 Carrier to Noise Ratio

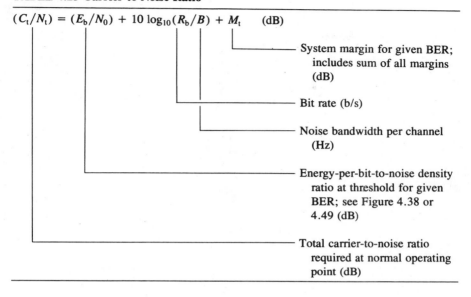

$(C_t/N_t) = (E_b/N_0) + 10 \log_{10}(R_b/B) + M_t$ (dB)

— System margin for given BER; includes sum of all margins (dB)

— Bit rate (b/s)

— Noise bandwidth per channel (Hz)

— Energy-per-bit-to-noise density ratio at threshold for given BER; see Figure 4.38 or 4.49 (dB)

— Total carrier-to-noise ratio required at normal operating point (dB)

TABLE 4.24 Carrier-to-Noise Power Density Ratio

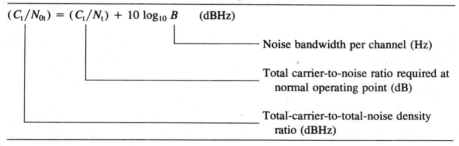

$(C_t/N_{0t}) = (C_t/N_t) + 10 \log_{10} B$ (dBHz)

— Noise bandwidth per channel (Hz)

— Total carrier-to-noise ratio required at normal operating point (dB)

— Total-carrier-to-total-noise density ratio (dBHz)

While this has been going on, the incoming call is stripped of its supervisory information, which consists of international and local area codes, central office, and line numbers, special billing and routing codes, echo control defeat tones, bit rate indicators, voice recognition, passwords, and so on. This information is stored (g) until the local supervisory equipment (h) indicates that a two-way connection has been established and the destination is ready to receive this supervisory and control data (i) and pass it out to the distant terrestrial facility (j). When this is completed, the call proceeds (k). When either party hangs up, the supervisory equipment signals a call completion, and each station returns its time slot to the pool (l). The control center computer recognizes this, stops the billing clock (m),

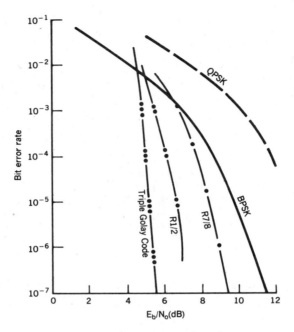

FIGURE 4.38 Error performance (R, coding rate).

and adds the utilized resource charges to the station's bill (n). In some applications the operating conditions for each earth station may be changed by downloading new software (or commands to select among the already present software) from the network operations center via the satellite. A very adaptive network can result.

This simplistic view of channel assignments has deliberately avoided such embellishments as variable burst lengths, analog-to-digital and digital-to-analog interfaces, priorities, details of finding and using idle channels, multiple networks within a transponder, network preassigned time slots, and premium-priced overflow channels (available from a pool), time delays, details of the control channel operation, and so on. For microterminal (VSAT) applications a more complex series of events may occur (1988, Morgan & Rouffet).

The general procedure also applies to many other forms of demand assignment multiple access.

A TDMA earth station has its own set of monitors and alarms. Some of these are located in Figure 4.41 at (b), (e), (f), (i), (l), and (m).

Notice that the destination TDMA earth station (and all others) must "listen" and decode the address for each request burst even if none of the traffic is for the station unless there are fixed time slots. In contrast, a FDMA station listens only on its preassigned filter frequency(ies).

The CCIR requirements for an 8-bit PCM call for a bit error rate of 10^{-6} in

TABLE 4.25 Digital Transmissions[a]

Modulation	Phases, Bits per Symbol (R_b/S)	E_b/N_0, BER = 10^{-5} (dB)	E_b/N_0, BER = 10^{-7} (dB)
FM	2	13.4 dB	15.1
	4		21.2
	8		25.1
DPSK, differentially detected PSK	2 (BPSK)	10.3	12.1
	2 ($R\frac{1}{2}$)	6.5^b	—
	4 (QPSK)	12.3	14.1
	4 ($R\frac{7}{8}$)	9.4	10.6
	4 ($R\frac{3}{4}$)	7.1^b	8.5
	4 ($R\frac{1}{2}$)	6.3	7.4
	8	16.2	18.2
	16	21.0	23.2
	32	26.0	28.2
CPSK, coherent PSK	2	8.3	11.3
	2 ($R\frac{7}{8}$)	7.3	—
	2 ($R\frac{3}{4}$)	4.6	8.6
	2 ($R\frac{1}{2}$)	3.6	5.8
	4	12.5	11.3
	8	17.6	14.9
	16	23.0	19.6
	32	28.2	24.6
	64		29.9
QASK, four-phase ASK	2	1.6–2	—
CPFSK, continuous-phase FSK	2	9.5	12
	3	8.6	10.4
	5	8.4	10.2
MPSK, matched filter PSK	2	9.8	11.2
	4	12.6	14.3
QASK, MAMSK, multitone AM	4	13.4	—
VSB/SC or QAM/SC,	4	—	15.1^c
Vestigial sideband or four-phase AM supressed	16	—	21^c
Carrier	64	—	27.3^c

[a]Coding: Rate 1 (no coding) unless otherwise noted.

[b]Viterbi decoder.

[c]At 10^{-8} BER; two-phase shift keying = BPSK, four-phase = QPSK, M-phase = MPSK, ASK = amplitude shift keying.

Note: These data come from many sources with various margins. This table should be used only as a guide for comparison.

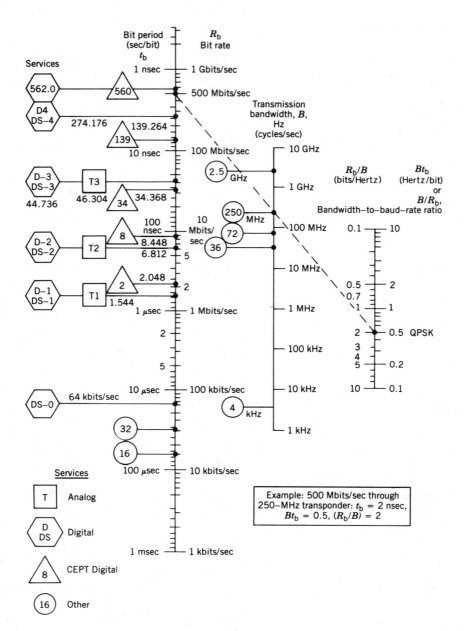

FIGURE 4.39 TDMA spectra nomogram.

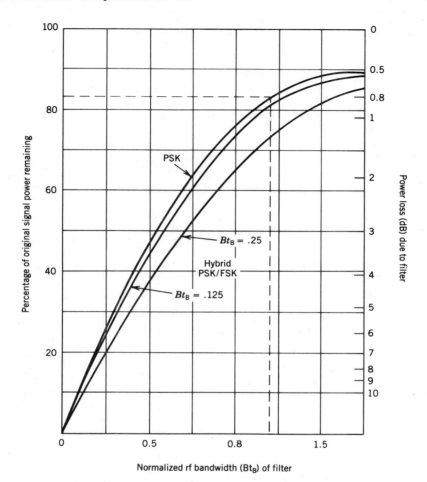

FIGURE 4.40 Signal power remaining after filtering.

clear weather. Under adverse conditions (e.g., rain) the outage (when the bit error rate is 10^{-4}) shall be less than 0.3% in any month. The "any month" practically translates to the "worst month of the typical year." Typically, a 3-dB difference (improvement) in the value of E_b/N_0 results in a bit error rate change of from 10^{-4} to 10^{-6}.

The formula for determining the earth station power is derived as shown in Tables 4.27 and 4.28.

The 207.1 term of $4\pi k/c^2$ in Table 4.27 is a combination of constants, including Boltzmann's constant k and the velocity of light c. The 163.3 term of $\pi(2000\ S)^2$ in Table 4.28 includes the maximum slant range $S = 41,679$ km (for $0°$ elevation angle), a $\pi/4$ conversion term for a square-meter antenna, and a term of 1000 for conversion between kilometers and meters.

TABLE 4.26 Important Characteristics of TDMA

Transponder power	a. May operate at saturation or slightly beyond.
	b. When bandwidth limited allows efficient bandwidth power trade-off to increase information rate using high-order modulation.
	c. Amplifier may be operated as limiter to remove minor up-link fades and eliminate need for rigorous up-link power control.
Transponder bandwidth	When power limited, allows 2–5 dB FEC coding gain.
Frequency plan	All accesses use same frequency band.
Reconfiguration	Allows easy adjustment of capacity for each access.
Traffic loading	a. Efficient for medium to high loading.
	b. Inefficient for light traffic loading.
	c. Satellite amplifier duty cycle and dc power vary with traffic loading.
	d. Diurnal duty cycle variations should be eliminated through use of dummy (PN) bits.

TABLE 4.27 Uplink Illumination Level

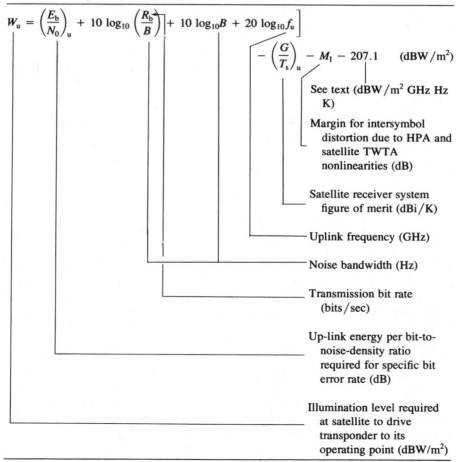

$$W_u = \left(\frac{E_b}{N_0}\right)_u + 10\log_{10}\left(\frac{R_b}{B}\right) + 10\log_{10}B + 20\log_{10}f_u$$
$$- \left(\frac{G}{T_s}\right)_u - M_I - 207.1 \quad (\text{dBW}/\text{m}^2)$$

See text (dBW/m² GHz Hz K)

Margin for intersymbol distortion due to HPA and satellite TWTA nonlinearities (dB)

Satellite receiver system figure of merit (dBi/K)

Uplink frequency (GHz)

Noise bandwidth (Hz)

Transmission bit rate (bits/sec)

Up-link energy per bit-to-noise-density ratio required for specific bit error rate (dB)

Illumination level required at satellite to drive transponder to its operating point (dBW/m²)

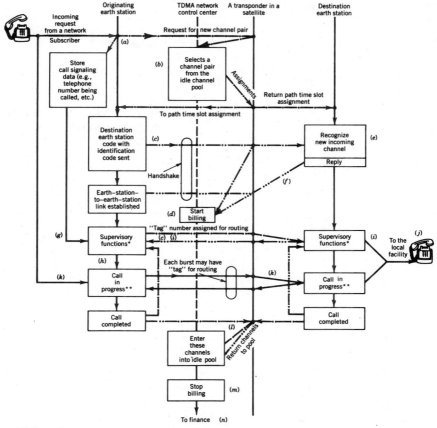

[——] Information

— · · · — · · · —Housekeeping (order wires)

*CCITT MF dial tones, ringing, answer, disconnect, preemption, billing codes, etc.

**Digital speech interpolation (DSI) may be inserted here to time-share the connection among several users.

(a) to (n)—see text

FIGURE 4.41 Channel assignments for TDMA

Typical values are given in Table 4.29. The output backoff BO_0 is much smaller than the typical 5–10 dB for FDMA.

Figure 4.42 shows the composition of a typical TDMA frames and an individual burst. The function of each element is given in Table 4.30.

The preamble and the guard times reduce the capacity of the fully loaded TDMA system just as the guard bands reduce the FDMA capacity. The burst bit rate R_b is shown in Table 4.31. Sample values are given in Table 4.32. Table 4.33 shows the maximum bit rate under two limit conditions.

Figure 4.43 shows the ratio of information bit rate to the burst bit rate as a function of frame period F. Other aspects in the selection of the frame time are

TABLE 4.28 Uplink Power

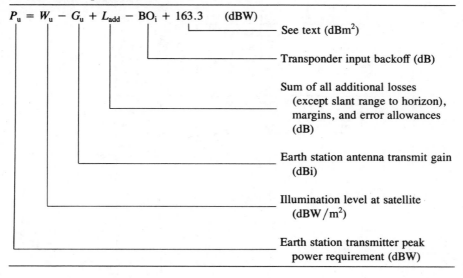

$$P_u = W_u - G_u + L_{add} - BO_i + 163.3 \quad (dBW)$$

See text (dBm^2)

Transponder input backoff (dB)

Sum of all additional losses (except slant range to horizon), margins, and error allowances (dB)

Earth station antenna transmit gain (dBi)

Illumination level at satellite (dBW/m^2)

Earth station transmitter peak power requirement (dBW)

TABLE 4.29 Typical TDMA Values[a]

$(E_b/N_0)_u$	See Figure 4.38	
R_b/B	See Table 4.20	
$f, R_b, (G/T_s)_u, G_u$	System values ("givens")	

BO_i (dB)	BO_0 (dB)	Notes
−4	−0.8	See Figure 4.33
−0	0.0	Saturation
+1.2	0.0	Self-limiting mode

f_u (GHz)	L_{add} (dB)	
6.0	4.0	
7.5	6.5	
14.0	10.0	

[a]Typical value for M_I, 3 dB.

478

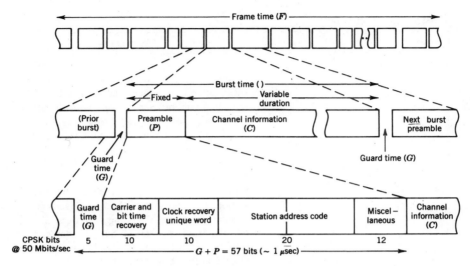

FIGURE 4.42 TDMA frame and burst organization.

TABLE 4.30 TDMA Burst Elements

Element	Purpose
Guard time, t_g or G	Time between last bit of nth burst and first bit of preamble of $(n + 1)$th burst; permits some variation in timing of bit streams of individual bursts (due to differences in slant range and equipment delays); guard time ensures that successive bursts do not collide in time
Preamble or P	Synchronization, timing, address, and miscellaneous bits; preamble and guard times represent TDMA system "overhead"
Synchronization carrier recovery[a]	Pattern of pulses (often 1111 or 1010 · · ·) used by receiving modem to quickly recover each new carrier (each new burst may be new carrier); carrier lock loop uses these pulses for locking; number of pulses set by lock loop acquisition time
Synchronization bit-timing recovery[a]	During or after carrier locked up these bits permit modem to perform bit rate synchronization; individual transmit stations may have very slightly different bit rates; due to variations in distance between transmit station and satellite, relative delays between bursts (and thus bit phasing) will vary

479

TABLE 4.30 (*Continued*)

Element	Purpose
Synchronization clock recovery	Unique word (uw) the receiving modem uses to produce timing reference so subsequent information pulses identified accurately; FEC encoding may be used to improve recognition of word under degraded link conditions; if this word cannot be found, information bits will be lost. In some cases station addresses and clock recovery combined with each station having its own unique word.
Station address	Identification code provides address of receiving station. In some systems each earth station receives all bursts, decoding each burst in search of its address; discards all bursts for other stations and processes its own bursts; code may also identify sending station; may be incorporated into clock recovery unique word. If TDMA system has fixed time slot format or burst that announces all time slot assignments periodically, station address may be eliminated from individual bursts; for SDMA/SS/TDMA bits may be used by satellite switch for routing; for regenerating/remodulating repeater may be used on-board satellite to change incoming bit format to one compatible with chosen down link.
Miscellaneous	Used for various housekeeping functions such as new burst request, station telemetry, order wires, power adjustments, transmission, bit error rate sequences, station-to-station toll data, and identifies down-link antenna beams for SDMA, SS/TDMA, etc.
Information or C	Data bits; block of bits that may contain one or several trains of information destined for one or several circuits; parity bits and forward error correction (FEC) may be included. Digital speech interpolation (DSI) may be used.

[a]On-board regeneration with retiming of the down link may permit easing requirement for these preamble portions. Such retiming produces bursts with uniform bit rates and positions by removing up-link and earth station time delay variations due to range variation.

TABLE 4.31 TDMA Bit Rate

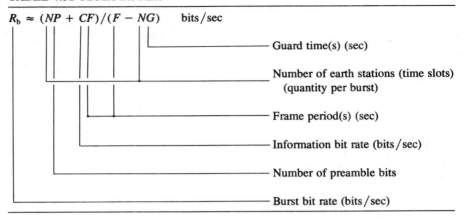

$$R_b \approx (NP + CF)/(F - NG) \quad \text{bits/sec}$$

- Guard time(s) (sec)
- Number of earth stations (time slots) (quantity per burst)
- Frame period(s) (sec)
- Information bit rate (bits/sec)
- Number of preamble bits
- Burst bit rate (bits/sec)

TABLE 4.32 Sample Values of Parameters in Table 4.31

Symbol	Definition	Sample Values	
C	Information rate or symbol rate	45 Mbits/sec	45×10^6 bits/sec
G	Guard time	200 nsec	200×10^{-9} sec
P	Preamble	200 bits	200 bits
N	Time slots or earth stations	10 time slots or earth stations	10
F	Frame length or period	1 msec	1×10^{-3} sec
R_b	Burst bit rate	47.094 Mbits/sec	47.094×10^6 bits/sec

the amount of terminal storage (between bursts), queueing delays, voice delays, and how long the station equipment can maintain frame synchronization.

The channel information may be subdivided into one or more channels or packets of uniform or nonuniform length. In general, each burst length (composed of preamble and information bits) is uniform or it may be a multiple of one standard burst length (sometimes with the guard times included).

Forward error correction (FEC) coding is used to overcome the long time delays and high storage requirements inherent in the old automatic retransmission request (ARQ) techniques.

FEC can reduce the E_b/N_0 requirement by 5–6 dB only if sufficient extra bandwidth (bit rate capacity) is available. The Q_d or [EIRP + $(G/T_s)_d$] may be lowered by FEC. If the downlink is power limited, FEC can maintain the throughput in spite of the redundancy bits added by FEC (assuming the higher system bit rate, R_b can be accommodated in the transponder bandwidth). Conversely, a system unconstrained by either power or bandwidth can use a smaller antenna (and thus less expensive rf equipment) or survive noisier conditions, thus operating longer in rain, interference, and so on.

TABLE 4.33 TDMA Channel Capacity

Bandwidth limited case

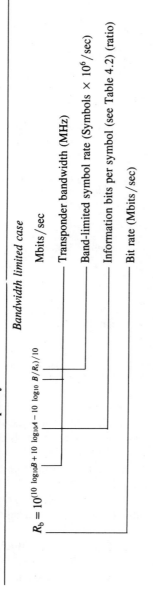

$$R_b = 10^{(10\,\log_{10}B + 10\,\log_{10}A - 10\,\log_{10} B/R_s)/10}$$

- Mbits/sec
- Transponder bandwidth (MHz)
- Band-limited symbol rate (Symbols × 10⁶/sec)
- Information bits per symbol (see Table 4.2) (ratio)
- Bit rate (Mbits/sec)

Power limited case

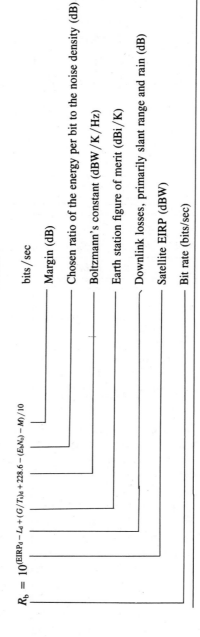

$$R_b = 10^{(\text{EIRP}_d - L_d + (G/T_s)_d + 228.6 - (E_b N_0) - M)/10}$$

- bits/sec
- Margin (dB)
- Chosen ratio of the energy per bit to the noise density (dB)
- Boltzmann's constant (dBW/K/Hz)
- Earth station figure of merit (dBi/K)
- Downlink losses, primarily slant range and rain (dB)
- Satellite EIRP (dBW)
- Bit rate (bits/sec)

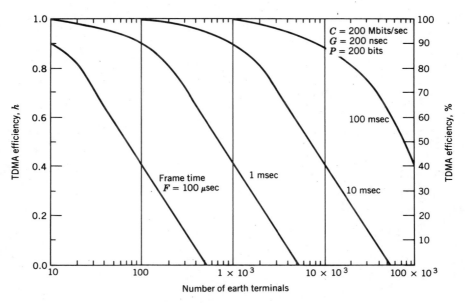

FIGURE 4.43 TDMA efficiency for various frame durations.

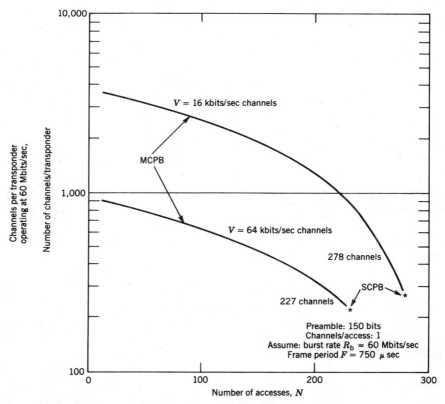

Key: SCPB, single channel per burst; MCPB, multiple channel per burst

FIGURE 4.44 Transponder capacity versus number of accesses.

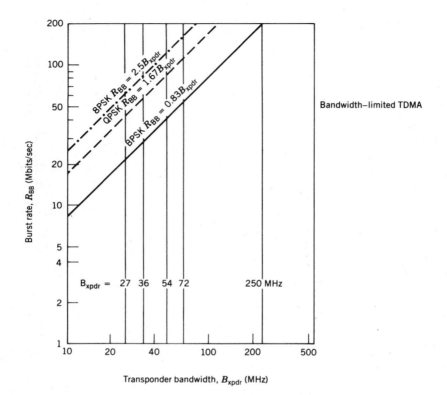

R_{BB} = baseband rate, B_{xpdr} = transponder bandwidth

FIGURE 4.45 TDMA burst rates in bandwidth-limited transponders.

The transponder capacity is shown in Figure 4.44. When the number of channels (N_c) equals the number of time slots (N), the service is single channel per burst (SCPB). All other cases involve at least some multiple channels per burst (MCPB).

Figures 4.45 and 4.46 show the TDMA channel capacities of various transponders.

4.3.4 TDMA Impairments

Table 4.34 shows the principal sources of TDMA degradation.

Satellite transponder filters are a primary limitation. The PSK spectrum is wide (see Figure 4.47). Figure 4.48 shows the effect of two filter types on the bit error rates.

Figure 4.49 shows that the bit error rate is very sensitive to the E_b/N_0 level. A small decrease in carrier energy per bit (E_b) or an increase in the noise density (N_0) can change the bit error rate substantially. A change of 1 or 2 dB can result

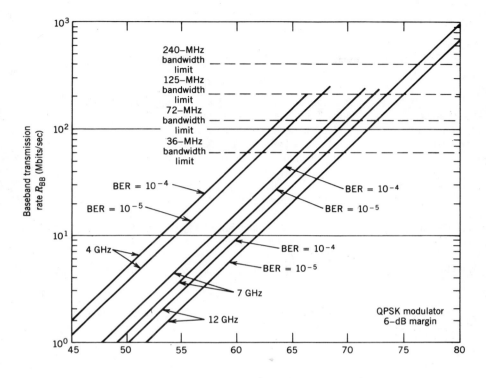

Down–link figure of quality, Q_d (dB/W/K)

FIGURE 4.46 Power-limited TDMA capacity.

in a degradation in the BER of several orders in magnitude (e.g., clear sky at BER of 10^{-7}). Path fading due to precipitation is a predominant source of BER degradation, but any increase in N_0 (due to interference, intersymbol distortion, etc.) is also damaging.

A loss of clock synchronization may destroy the ability to find bits in the unique words, tags, and channel information until the next burst or frame synchronization occurs. In this case the burst is lost.

Since there is only one rf spectrum to share, it is important that only one station transmit at a time. If a station loses frame timing and transmits at random, it will block the transmission(s) from the assigned station(s), and none of these users will be satisfied. If an earth station transmitter does not turn its carrier off between bursts (or if rf power leaks to the antenna), it will interfere with all other users by unintentionally jamming the up link. If each of a hundred earth stations leaks even a little power, the summed power as received by the satellite may cause a noticeable increase in N_0.

TABLE 4.34 Sources of TDMA Degradations

Element	Effect	Location
Earth Station Transmit Chain		
Fast or slow clock	Out-of-tolerance bit rates	Modem
Bandwidth	Restricts amount of sidelobe energy	IF filters
Group delay	Distorts individual pulses, makes them harder to decode, worsens BER	Filters and HPA
AM/AM distortion	Distorts individual pulses, makes them harder to decode, worsens BER	HPA
AM/PM distortion	Distorts individual pulses, makes them harder to decode, worsens BER	HPA
Nonlinear channel	Spectrum spreading into adjacent transponder (out-of-band energy intersymbol distortion)	Upconverter and HPA
rf signal leakage between bursts from earth station	Unintentional up-link jamming interference to cross-polarized services	Burst modulator
Poor antenna polarization	Same as above	Antenna
Up Link		
Interference from cofrequency adjacent satellite system or cross-polarized cosatellite system	Transponder interference and power-sharing noise	Antenna sidelobes, out-of-band ES radiation, poor cross-polarization isolation
Fading	Nonoptimum use of transponder C_u/N_u degrades	Precipitation
Jamming, spoofing	Capture of channel assignment and control means	Other earth stations (including unintentional channel requests due to malfunction in earth station)
Delay variation	Phasing of bits between bursts	Difference in up-link path lengths from various earth stations and satellite motion
Satellite		
Bandwidth	Restricts amount of sidelobe energy	Input filters
Nonlinear channel	Intersymbol distortion	Amplifier and spectrum overlap

TABLE 4.34 (*Continued*)

Element	Effect	Location
	Satellite (Continued)	
Group delay	Distorts individual pulse, worsens BER	Amplifiers
AM/AM distortion	Distorts individual pulse, worsens BER	Amplifiers
AM/PM distortion	Distorts individual pulse, worsens BER	Amplifiers
Improper equalization	Distorts individual pulse, worsens BER	Input filters
Cochannel isolation	Interference (noise) from cross-polarized service	Antenna beam isolation, beam pointing errors due to attitude control
Multipath	Interference from adjacent channels	Antenna beam isolation, filters
	Down Link	
Precipitation	Fading (reduction in carrier level), increased noise (reduction in G/T_s)	In path
Delay variation	Instantaneous bit and burst variations, Doppler shifts	Satellite longitude/ latitude maintenance
	Earth Station Receive Chain	
Sidelobes	Interference from other satellites or services	Antenna
Linearity	Spectrum spreading and intersymbol distortion	LNA, IF
Threshold	Preamble detection (station address)	Modem
	Earth Station Control Equipment	
Synchronization	Improper burst bit rate due to jitter or loss	Control
Clock recovery	Loss of synchronization	Control
Burst delay	Burst transmitted in wrong time slot (blockage)	Control
Carrier recovery	Carrier jitter	Control

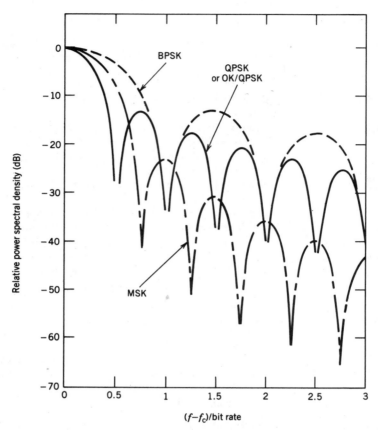

FIGURE 4.47 Power spectral densities for unfiltered BPSK, QPSK, or OK/QPSK and MSK (f_c, carrier frequency). Reprinted with permission from Maral and Bousquet, *Satellite Communications Systems*, Wiley, 1986.

The time delay between a satellite and the earth station is proportional to the distance between them. The geostationary satellite may be anywhere about its nominal location (Figure 4.50). While an orbit that holds a ± 15-km accuracy (out of 42,165 km) may be thought of as being nearly stationary, the satellite actually moves substantially in a day through a volume about 10^8 times as large as the typical satellite's own volume. This will be of particular concern if the satellite does not use North-South stationkeeping where the values will be substantially larger than shown in Figures 4.50 and 4.51.

The delay variation with range is shown in Figure 4.51.

If the ground station can "see" its own burst (repeated by the satellite), it might measure the propagation delay and compensate for its burst delay time.

The most pressing need is to reduce the high costs of TDMA (with respect to

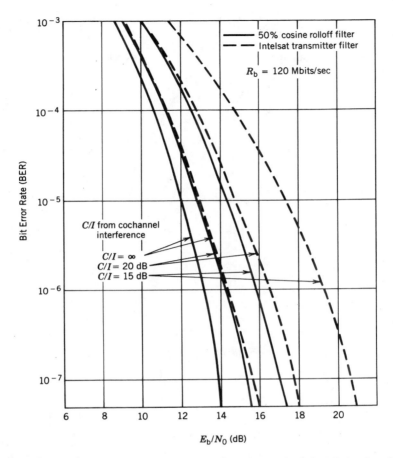

FIGURE 4.48 TDMA error rates for two types of filters. Reprinted with permission from *Microwave Systems News*, March 1978, © EW Communications, Inc.

FDMA) for users with small amounts of traffic. Figure 4.52 shows one approach to lower cost TDMA. See also Morgan & Rouffet (1988)

Spectrum-shaping techniques (to fit within an analog transponder channel and to prevent spectrum components from falling into adjacent transponders), digital speech interpolation, and microprocessor control of earth station and satellite TDMA functions are likely improvements.

Higher bit rates are possible with improved, TDMA-unique transponders and MPSK.

Unmanned earth stations (with status reports to a central control station) have reduced operating costs.

Unmanned earth stations (Figure 4.53) represent a trade-off between cost and control complexity and operating costs.

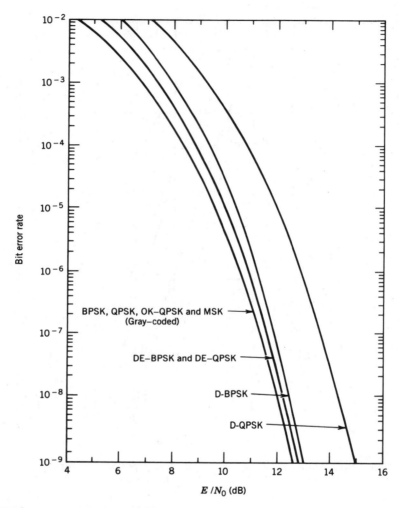

FIGURE 4.49 Ideal bit error rate performance of various digital modulation schemes (E, energy per bit; N_0, one-sided noise spectral density in watts per hertz; $E = E_b$ if no coding, $E = E_c$ if coding). Reprinted with permission from Maral and Bousquet, *Satellite Communications Systems*, Wiley, 1986.

4.4 SPACE DOMAIN MULTIPLE ACCESS (SDMA)

4.4.1 Basic SDMA Characteristics

While FDMA, TDMA, and CDMA use the orthogonality between the frequency, time, and code domains, SDMA takes advantage of spatial separation.

The spatial separation may be accomplished in several ways (Table 4.35).

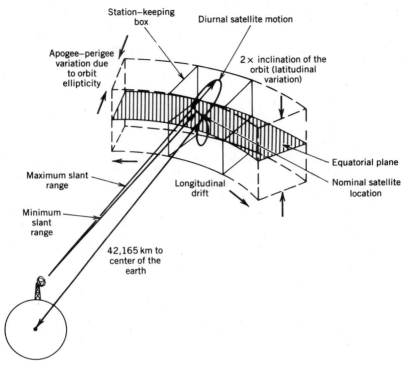

Parameter	Amount	Variation	Typical Time
Inclination	±0.1°	±73.5 km	Daily
Longitude	±0.1°	±73.5 km	Long
Apogee–perigee		30 km	Daily
Slant Range and Delay	Distance	Delay	Variation
Nominal	40,000 km	±133.43 msec	Daily

Volume of 0.2° × 0.2° × 30-km box: 648,435 km³

FIGURE 4.50 Satellite motion.

SDMA is usually accomplished in conjunction with some other form of multiple access such as shown in Table 4.36.

SDMA provides two operational advantages: frequency reuse and on-board switching. These in turn result in a great communications capacity within an assigned frequency band, increased flexibility in traffic handling, and often reduced requirements for the earth station because of the use of narrower satellite antenna beams (with their higher gains).

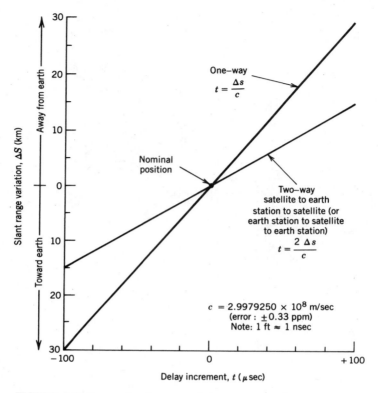

FIGURE 4.51 Propagation time uncertainties due to satellite range variation.

Traffic routing between the diverse SDMA paths may be done by the up-link frequency and on-board filters (SDMA/FDMA) (see Figure 4.54). This can be further extended by the use of remote filter switching controlled by the command link and verified via the telemetry link (Figure 4.55).

While individual antennas are shown, it is possible to have a multibeam satellite using a single reflector and many feeds (see page 351). In addition, polarization diversity can also be used to obtain additional reuse and isolation. Thus, beams 1 and 2 may even cover the same spot area with cochannel allocations, being separated by polarization only.

4.4.2 SDMA with FDMA

In the broadest of senses SDMA/SS/FDMA has been used since *Intelsat IV*, where the on-board switch permitted certain up-link frequencies to be connected to either global or spot beams. The switch(es) can be located before and/or after the transmitter. Figure 4.56 shows the connections for *Anik-B*, which used a form of satellite switching for FDMA and TDMA services. There was a single (combined W,

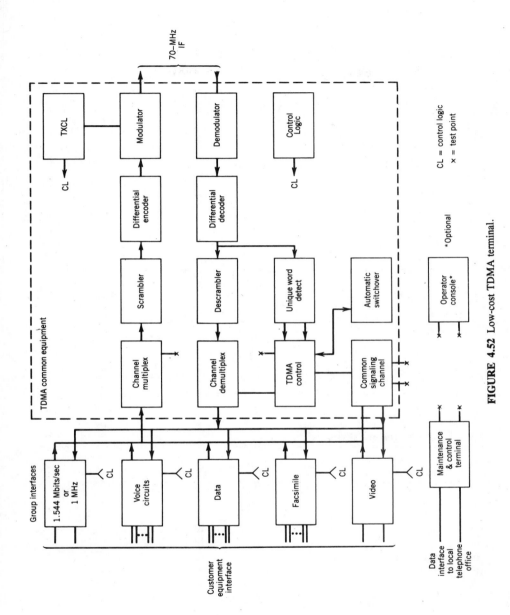

FIGURE 4.52 Low-cost TDMA terminal.

CL = control logic
x = test point

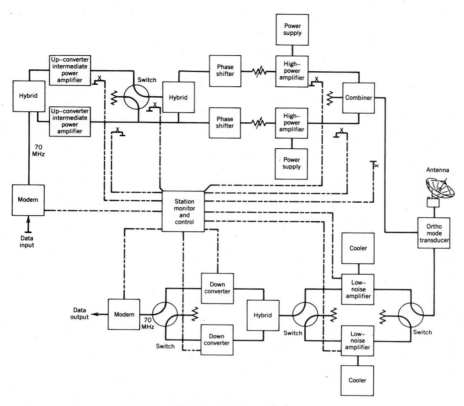

FIGURE 4.53 Major TDMA earth station diagram.

TABLE 4.35 Forms of SDMA

Spatial Separation Means	Diversity	Typical Separation	Frequency Reuses
Single Satellite			
Polarization domain	LHC & RHC	30 dB	2
	H & V	30 dB	2
Multiple-beam satellite	Beams	30 dB	Many
Multiple services	Beams and bands	Service	Many
Multiple Satellites			
Isolated satellites	Orbit longitude or latitude	Several degrees	Full
Intersatellite links	Orbit longitude, latitude, or planes	Many degrees	Full
Nongeostationary orbits	Position or phasing	Many degrees	Nearly full to full

Abbreviations: **H&V** is horizontal (or vertical) linear polarization; **LHC & RHC** is left- (or right-) hand circular polarization.

TABLE 4.36 Forms of Space Domain Multiple Access (SDMA)

Form[a]	Figure Number	Switching
SDMA/FDMA	4.54	Filters
SDMA/SS/FDMA	4.55	Switched filters
SDMA/SS/TDMA/FDMA	4.56	Switched filters
SDMA/TDMA	4.57	Time switch
SDMA/SS/TDMA	4.58, 4.59	Time switch
SDMA/SS/FDM/TDMA	Section 4.4.2	Time-switched filters
SDMA/BH/TDMA	4.60	Beam hopping
SDMA/TH/TDMA	Page 506	Transponder hopping
SDMA/FTSS/CDMA	4.70	None, full-transponder spread spectrum
SDMA/SS/CDMA	4.63	Time switch with code recognition
SDMA/SS/BH/CDMA	4.60	Beam hopping
SDMA/FH/CDMA	4.63	Frequency hopping
SDMA/hybrid MA	Section 4.4.5	See text

[a]Abbreviations: SS, satellite switched; FTSS, full-transponder spread spectrum; BH, beam hopping; FH, frequency hopping; TH, transponder hopping.

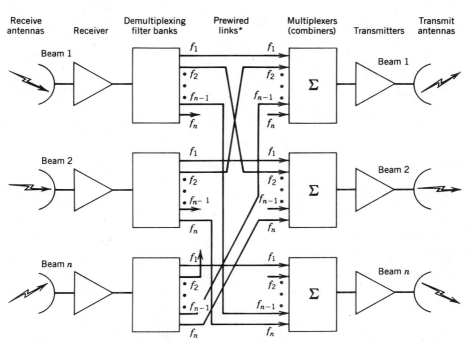

f_1, intrabeam services; all other frequencies for interbeam connections.

*In some cases, these prewired links may be established via a switch that is infrequently selected (e.g., once a year) in orbit.

FIGURE 4.54 SDMA/FDMA block diagram.

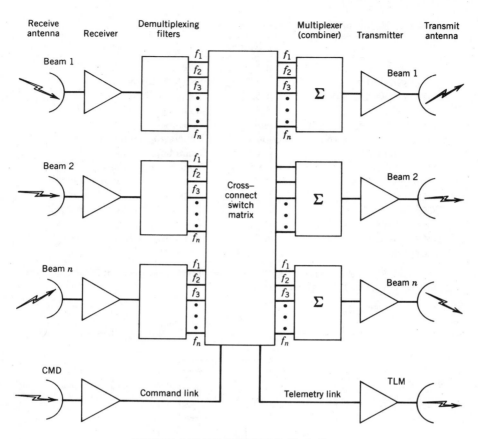

FIGURE 4.55 SDMA/SS/FDMA block diagram.

WC, EC and E) uplink beam at 14 GHz. Four TWTAs amplify six subbands and drive two to four downlink beams. This illustrates how one power amplifier output can be shared by several beams.

The SDMA methods that use filters (FDMA), time, or codes are characterized as shown in Table 4.37. The switches in the SDMA/SS/FDMA are thrown rarely, and only to reconfigure the satellite. Although this can be done on a diurnal basis, in most cases it has been done only a few times in the satellite lifetime (e.g., when a satellite is moved from one location to another) or when a different beam coverage (e.g., western United States vs. all of it) is desired by the user. A satellite operator generally makes a per-switch throw charge to discourage frequent (e.g., daily) reconfigurations. Solid-state switches (in particular, variable-power dividers) can steer predetermined or remotely selected percentages of the power amplifier between beams. These satellites must use many filters and thus tend to be heavy.

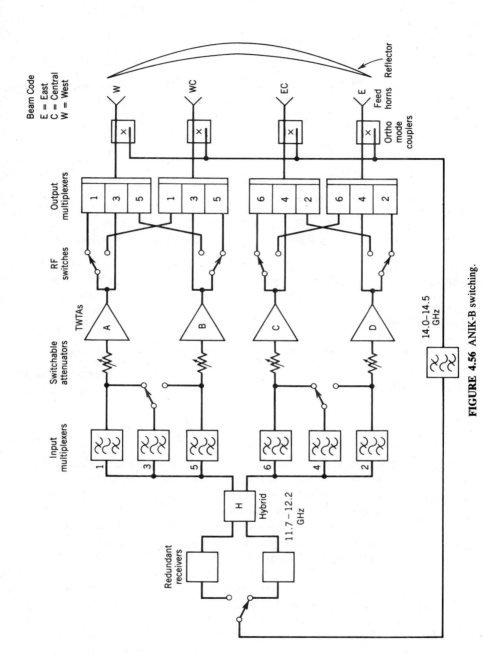

FIGURE 4.56 ANIK-B switching.

TABLE 4.37 SDMA Methods Using Satellite Switches[a]

Parameter	SDMA/SS/FDMA	SDMA/SS/TDMA	SDMA/SS/CDMA (FTSS)
Frequency assignments per band	Up to $(N_{\text{beam}})^2$	One	One
Filters			
Quantity	Up to $(N_{\text{beam}})^2$	N_{beam}	N_{beam}
Bandwidth	Narrow	Full band	Full band
Q	High	Low	Low
Mass	High	Low	Low
Total filter mass	Highest	Low	Low
Switches			
Signal type	Analog	Digital	Digital
Signal carrier frequency	IF or RF	Baseband to f_d	Baseband to f_d
Poles	few to $(N_{\text{beam}})^2$	$(N_{\text{beam}})^2$	$(N_{\text{beam}})^4$
Throw rate	Very infrequent	Per burst	Each new bit stream
Isolation	High	Low	Least
Transponder mode of operation	MCPT	SCPT	MCPT
Spectrum	Many peaks	Broad peaks	Flat, noiselike

[a]Abbreviations: N_{beam}, number of beams; SS, satellite switched; FTSS, full-transponder spread spectrum; Q, Filter quality; f_d, down-link frequency.

4.4.3 SDMA with TDMA

When SDMA and TDMA are combined, configurations such as Figures 4.57 and 4.58 are possible. The first case (SDMA/TDMA) is similar to SDMA/SS/FDMA in that a quasi-fixed switch (infrequent changes to the switch pattern for the reasons cited in the foregoing) connects one full-band (e.g., 500-MHz) TDMA receiver to one selected beam. Full interconnectivity is not provided. This is the point-to-point "cable in the sky" approach. TDMA is employed between beam pairs.

In Figure 4.58 (SDMA/SS/TDMA) time division switching is employed aboard the satellite. Each up-link beam is connected to a specified down-link beam at a certain time for a certain (the burst) duration; thus, any station in spot beam 1 may be connected to other beams at the times shown in Table 4.38. A hypothetical frame time of 125 μsec has been assumed. Beam 1 may transmit a burst to beam 2 between 23 and 28 μs after frame synchronization. The return transmission (beam 2 to 1) is not necessarily simultaneous or equal, being between 16 and 26 μsec in the burst (in this case the traffic is not instantly symmetrical). The methods of assigning the time slots and their utilization are described in the TDMA and CDMA sections. This may be done on an organized or contention basis. Alternately, the

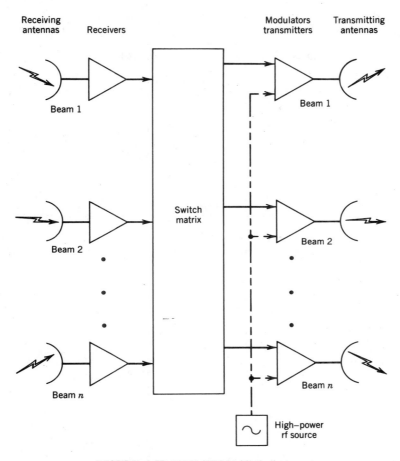

FIGURE 4.57 SDMA/TDMA block diagram.

up link may be demodulated, the destination address read, and the burst is sent to the proper down-link beam remodulator.

The satellite switch is like a high-speed, solid-state rotary switch having N_{Beam} poles. No demodulation is done in the satellite. The switches pass the TDMA bit streams at if (sometimes rf). SCPT is used.

By using filters, the satellite bandwidth can be subdivided and SS/FDM/TDMA can be done.

Figure 4.59 is a block diagram for the commercial portion of the first Tracking and Data Relay Satellite (TDRS-F1). It carried a 4 × 4 matrix to switch 225-MHz-wide, 250-Mbits/sec signals. The government purchased this portion of the satellite and never used the switch because of higher priority needs for the power amplifiers.

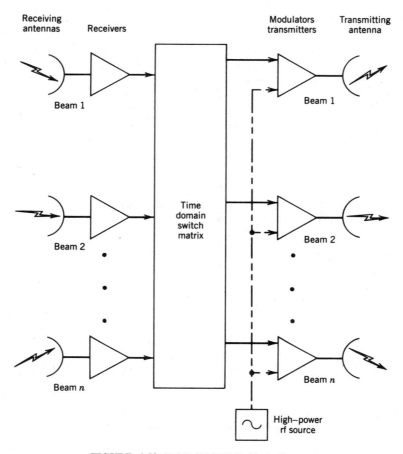

FIGURE 4.58 SDMA/SS/TDMA block diagram.

In Figures 4.57 and 4.58 an optional high-power rf source is shown. It operates at the transmit frequency (e.g., 12 GHz), and individual direct modulation modulators are provided for each beam: This has the advantage of a single high-power, high-efficiency transmitter.

Instead of having fixed beams and a moving interconnection (switch), the receive and transmit beams can be caused to move (hop), and the switch is fixed (and thus perhaps eliminated). In the case shown in Figure 4.60, the up-link beam is caused to move from one traffic source to the next. The corresponding down-link beam is steered to the desired destination earth station, and a connection can be established between the beams. Each earth station has been scheduled as to when it may transmit its burst. The receiving stations listen on the down-link frequency for any signals and decode the addresses of all of them to see if the traffic is intended for

TABLE 4.38 SS/TDMA Beam Connection Times

Beam		
From	To	Time after Frame Synchronization (μsec)
1	1	5–11
1	3	12–22
1	2	23–28
1	4	29–59
⋮	⋮	⋮
1	N	120–125
2	2	5–15
2	1	16–26
2	3	27–30 (request channel)
2	5	31–51
⋮	⋮	⋮
2	N-1	115–125
⋮	⋮	⋮

them. The dwell duration of the beam pair connection is proportional to the traffic between the stations. The burst rate must be high.

Fixed beams and moving beams can be added, with an interconnection switch. The moving or hopping beams are sometimes also called scanning or fan beams.

An antenna beam may be forced to raster scan (in a manner similar to the beam that scans the television picture tube) across the earth. (See Figure 4.61.) If Richmond, Virginia, has traffic for Memphis, Tennessee, Richmond stores its material until the beam reaches the Richmond area and then bursts the data to the satellite. The satellite stores the message until the beam is near Memphis and then transmits a burst to Memphis. The scan may cover an entire nation or only a region. The beam may be made to skip over nontraffic areas (such as an ocean). The satellite antenna gain is very high because the beam is so narrow: As a result, the satellite EIRP and G/T are high, thereby reducing the rf equipment demands on the earth station. The very high burst rate, however, makes the baseband equipment more complex, especially if there is not some way to maintain synchronism when the scan beam is pointed somewhere else. An independent (nationwide coverage) beacon with burst synchronization may be needed.

The second method is to use a wide fan-shaped beam (see Figure 4.62). In the case illustrated the beam sweeps the nation from north to south (but any orientation could be selected). Any station in the instantaneous beam (e.g., Chicago, Illinois) can transmit its traffic. Again the satellite stores the message until the beam covers

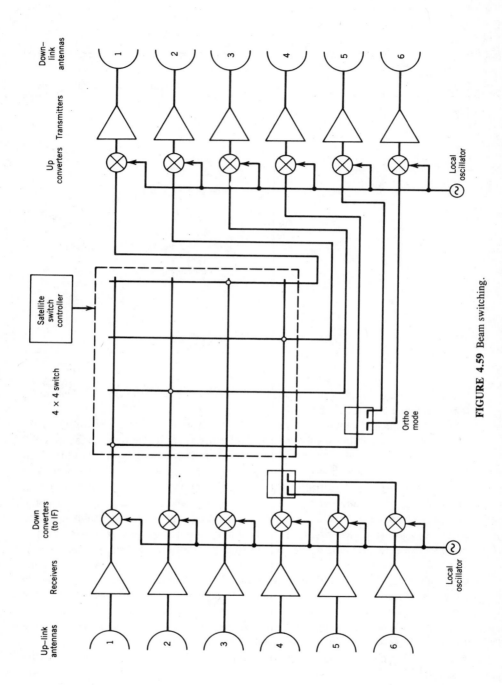

FIGURE 4.59 Beam switching.

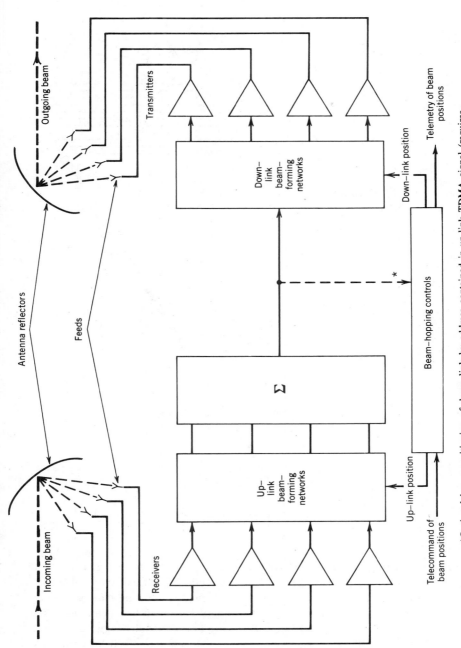

FIGURE 4.60 SDMA/BH/TDMA block diagram.

*Optional beam positioning of down link by address contained in up-link TDMA signal (requires demodulation and address detection on-board).

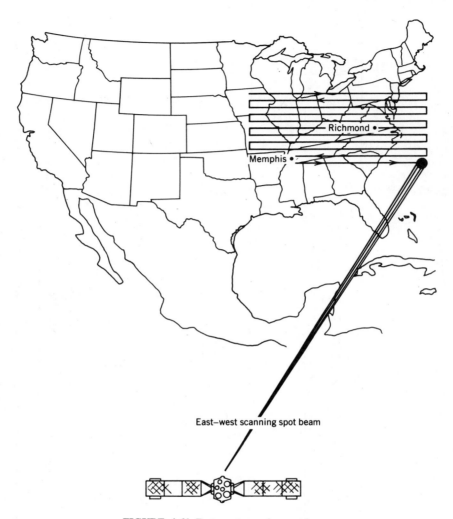

FIGURE 4.61 East–west scanning spot beam.

the destination (e.g., Jackson, Mississippi). The satellite antenna gain is lower than for the scan beam but is still much higher than for full- or half-national beams.

Extensions of this beam-hopping (BH) method include on-board store-and-forward, wherein an up-link message is held until the down-link beam arrives at the destination station. This permits the receive and transmit beams to move in orderly, raster-scan-like motions that minimize the total lost beam transmit time. On-board regeneration can take place in the satellite. As in SDMA/TDMA, the transponder is operated on a SCPT basis.

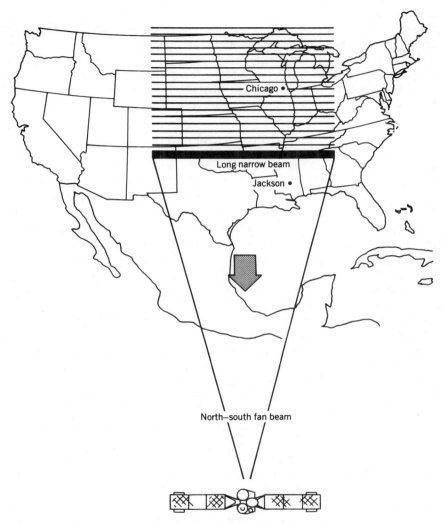

FIGURE 4.62 North–south fan beam.

The advantages of SDMA/BH/TDMA are those of TDMA plus increased down-link antenna gain while collecting hinterland traffic in the hopping beams (while the fixed beams handle the population centers using one of the other SDMA block diagrams).

The disadvantages include the larger and more complex antennas, antenna beam-steering equipment, the multiplicity of receivers (whose noises tend to add if they are clustered to form the up-link beam), the time lost in physically moving the beams (or, more likely, moving the beams electrically using lens or phased-array antennas), the high modem burst rate (e.g., 500 Mbits/sec), and the long frame times if the number of beam hops (stations) becomes large.

Instead of beam hopping, transponder hopping (TH) can be used with TDMA. Time (such as used in beam hopping) and frequency (transponder) are interchanged in going from SDMA/BH/TDMA to SDMA/TH/TDMA. The block diagram for SDMA/TH/TDMA could be Figure 4.57 or 4.58.

4.4.4 SDMA with CDMA

The combination of SDMA and CDMA permits simultaneous access to a common frequency band (see CDMA section). Up-link CDMA may be used to obtain multiple access to the satellite. Each CDMA bit stream is decoded aboard the satellite to find its destination address. Since several simultaneous addresses may occur, several code recognition equipments are needed. One method is to use direct demodulation (combining some RX and D functions of Figure 4.63). The bit stream is then switched to the proper down link, and therefore, a switch with $(N_{beam})^2 \times (N_{beam})^2$ connections is needed.

As characteristic of CDMA, simultaneous bit streams arrive at a signal summer and are handled on a MCPT basis.

A further embellishment involves the use of store-and-forward equipment placed after the beam address code recognizers: Bogus and clandestine users can be denied access to the down link. An on-board controller provides traffic management and billing information. The CDMA bit streams can be retimed, regenerated, and recorded aboard the satellite. The down-link CDMA techniques need not be the same as the up-link CDMA methods, thus optimizing each link for its particular conditions.

CDMA may be used with a SDMA spread spectrum beam-hopping satellite to permit simultaneous access by stations in a beam needing identical beam pairing (see Figure 4.60). A store-and-forward SDMA/BH/CDMA would permit greater flexibility than the SDMA/SS/BH/TDMA, which permits only one access at a time, or a SDMA/BH/FDMA, which is heavier.

A frequency-hopping CDMA (see the next section) satellite would have a block diagram similar to Figure 4.63, except that the demodulator is of a different type.

4.4.5 Other Forms of SDMA

Space domain multiple access may also be done with multiple satellites, especially if there is a link between them. The interconnected satellites need not be nearby or even in similar orbits. One may be in a geostationary orbit, the other(s) may be in a *Molniya*-type (12-hr, highly inclined elliptical), or sun synchronous orbit(s). Each satellite can be used to its unique advantage. SDMA between orbiting satellites extends the total field of view. Figure 4.64 shows the coverages of two geostationary satellites. Through the intersatellite link (ISL) any station (e.g., Brazil) in

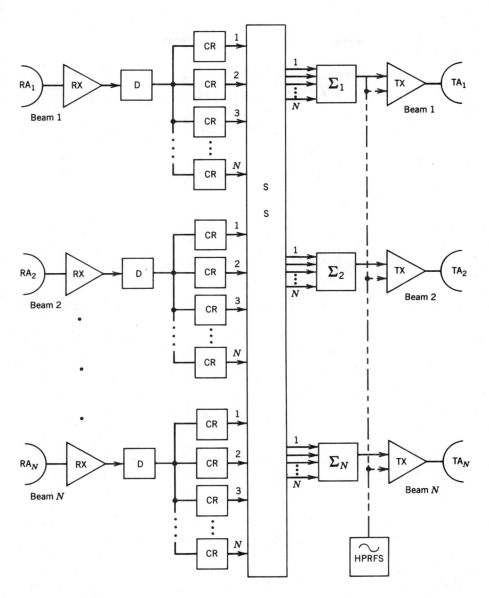

Key: RA, receive antenna; RX, receiver; D, full-band demodulator; CR, beam address code recognition; SS, satellite switch; Σ, down-link multiplexer (signal summer); TX, modulator and transmitter; TA, transmit antenna; HPRFS, high-power rf source.

FIGURE 4.63 Block diagram for SDMA/SS/CDMA and SDMA/FH/CDMA.

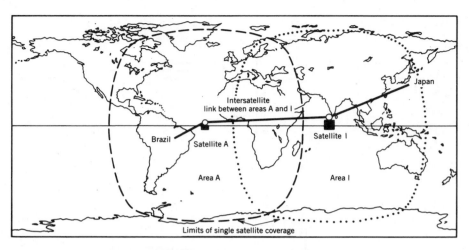

FIGURE 4.64 Coverage area by interconnected satellites.

one satellite's coverage can communicate with any station (e.g., Japan) in the other satellite's coverage area.

A multimission satellite provides services to many types of users. This form of SDMA involves the use of geographically separated beams and/or different frequency allocations with an interconnection switch aboard the satellite.

One of the difficulties with SDMA systems has been that an earth station is often forced to transmit in the "blind"; that is, the station cannot see its own transmission repeated by the satellite because the transmission is in another beam not visible to the transmit station. Therefore, the earth station cannot be absolutely certain that its signal was received or what the link time delay is. This concern is important if there are possible up-link impairments. It is also difficult to verify that the signal is entering the right transponder (earth station operators occasionally do make errors) or to verify the link quality by direct observation. Precipitation losses (at frequencies over 10 GHz), jamming (military systems), interference (all systems), and up-link power control are examples of potential problems that cannot be "seen" by the up-link station. Closed-loop control may be provided in SDMA systems if the switch matrix allows a tiny portion (e.g., 1 μsec) of each up-link burst to be transmitted back into that zone, thereby aiding the station in its burst synchronization process by measuring the round-trip time while acknowledging receipt of at least a portion of the burst.

An additional type of SDMA involves the use of on-board frequency translation. Intelsat uses an elementary example of this class (Figure 4.65). Signals arriving at 14–14.5 GHz are converted to the 3.7–4.2 GHz if used throughout the satellite. Depending on the switching, these (or other) signals may emerge in beams in the 4- or 11-GHz bands.

A second broad type of SDMA involves on-board signal processing (e.g.,

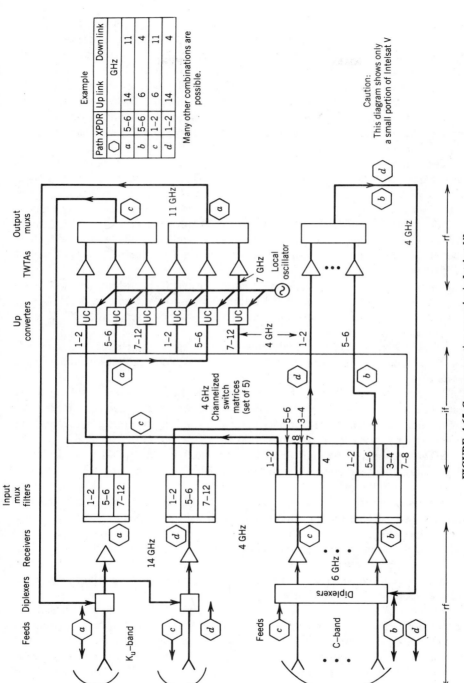

FIGURE 4.65 Cross-connection examples in Intelsat VI.

509

regeneration, retiming, store and forward, and remodulation). This signal rejuvenation process improves the earth-station-to-earth-station link by about 2–6 dB for QPSK and MSK over a conventional transponder. Store and forward permits data to enter the satellite at one burst rate (e.g., slow from a small-traffic station) and exit at another (e.g., much faster and more compatible with a large-traffic station) to permit interworking of different user and earth station classes (and equipment costs). Conventional TDMA requires all stations in a network to operate at a uniform high bit rate regardless of their traffic; thus, a small station has the rf, if, and modem costs of a large station.

The optimum up-link modulation method for a power-limited earth station may be different than the optimum method of a noise-limited down link. Remodulation allows each link to be independently optimized. At present very small aperture terminals (VSATs) are an example of this. The uplink from the power starved VSAT is demand SCPC (in various forms). The down link to the G/T_s limited VSAT is TDM to get maximum power from the satellite. A hub earth station does the remodulation, regeneration, and other functions. Eventually "smart" satellites will do these operations.

Another example involves a small earth station. The up-link and down-link conditions are quite different. A microterminal may have traffic at 56 kbits/sec. To reduce the up-link power (and cost), SCPC may be used with FEC. The up link is bandwidth limited.

After baseband processing (including determining the toll charges) on the satellite, the signal can be multiplexed with others and retransmitted to another small station.

The satellite down link to the VSAT is power limited. Often TDM is chosen (at a rate of several megahertz) for this link to keep the reception equipment costs low. This is an example of on-board signal processing where the incoming and outgoing data formats and rates are different.

SDMA using polarization diversity is limited by the polarization isolation degradation due to imperfect alignments of earth station and satellite antennas and precipitation in the path.

SDMA using multiple beams is limited by the sidelobe energy between the satellite beams. One optimization technique involves the use of several different beam sizes. Figure 4.66 shows several centers of dense population in the continental United States. A similar pattern of a few population centers and empty spaces occurs not only on land areas around the world but especially within the 71% of the earth covered by water. These are areas dotted with islands that have specialized communications needs.

Figure 4.67 shows a novel approach for handling these variations in density of population. Bandwidth, satellite power, and satellite antenna gain are the variables used to satisfy the communications needs (which may match the population). Power and antenna gain become EIRP. The fixed parameter is the channel grade of service. Each channel is uniform in data rate (e.g., 56 kbits/sec) and quality. All earth stations use the same size antenna (1.2 m) and low-noise amplifier ($T_s =$

FIGURE 4.66 Population centers (from the Census Bureau)

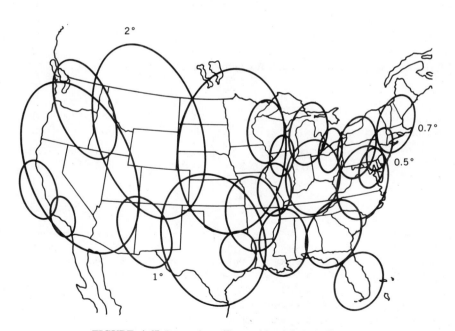

FIGURE 4.67 Domestic traffic matching using spot beams.

300 K); thus, $(G/T_s)_d$ is a constant at 15.2 dBi/K for 12.0 GHz. If the $EIRP_d$ *per channel* is held constant (at 21.4 dBW), the Q_d per channel is constant at 36.6 dBiW/K, and therefore, the service quality per channel is constant. See Figure 4.68.

The total $EIRP_d$ requirement is then proportional to the traffic in a beam. The $EIRP_d$ may be matched to the number of channels through the use of the satellite antenna gain (G_d) and sizing the power amplifier.

Table 4.39 shows a 24-beam SDMA satellite's operating levels. Multiple-power amplifiers are paralleled to reduce the number of tube types; see Figure 4.69 and Table 4.39.

The up link of Figure 4.69 uses a single coverage beam (CONUS). The up-link earth station power per 56-kbits/sec channel, P_u, is a constant. An earth station accesses any other station by operating on a SCPC basis. The carrier frequency is selected so that (after conversion in the satellite) it falls into the down-link band assigned to the distant station. The system is CONUS up link and 24-beam down link. No on-board switches are required as FDMA/SDMA is used (FDMA to select the beam) and the number of filters is small. All PAs are interchangeable. All traffic comes down on at least two (and up to eight) beams. For example, if 11.700–11.844 GHz is used in both the New York and Los Angeles beams, different polarizations may be used to get more isolation.

Variations are possible, such as additional narrower (and heavier) filters with individual beam transmitters, dual polarization, and so on.

A drawback of these down-link methods is the awkwardness of providing a "broadcast" service. To broadcast a message to all stations requires use of all the beams.

TABLE 4.39 Transponder Details for Simple Domestic Satellite

Beams		Number of Channels[a]		Total Bandwidth (MHz)	Antenna Gain Edge (dBi)	Total Actual PA Output (W)
Size	Quantity	Each	Total			
0.5°	2	1200	2400	144	46.5	25[b]
0.5°	3	600	1800	108	46.5	11.1[b,c]
0.7°	7	300	2100	126	43.5	43.8[b,d]
1.0°	8	150	1200	72	40.5	50[b]
2.0°	4	38	152	36	34.5	25[b]
	24		7652	486[e]		155[f]

[a]At 56 kbits/sec.

[b]Total power required to drive all beams in this row. 5.3 dB output backoff in all cases. EIRP per channel: 138 W or 21.4 dBW at beam edge. Value shown is saturated power rating.

[c]This is 25-W backed off 8.8 dB.

[d]This is two paralleled 25-W PAs backed off 5.9 dB.

[e]Allows full coverage (CONUS) up link frequency reuse.

[f]Total for satellite. TWTA efficiency of 50% indicates dc requirement of 350 W for transmitters, 30% SSPAs would require 580 W dc.

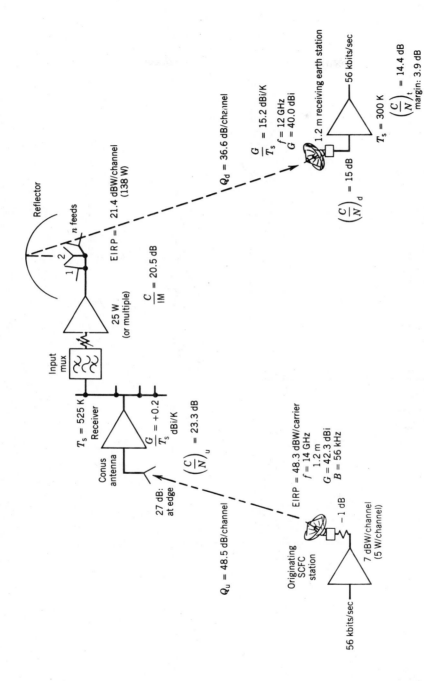

FIGURE 4.68 Link performance for a channel.

Reflector

n feeds

EIRP = 21.4 dBW/channel
(138 W)

$\dfrac{C}{IM} = 20.5$ dB

25 W
(or multiple)

Input
mux

$T_s = 525$ K
Receiver

$\dfrac{G}{T_s} = +0.2$ dBi/K

$\left(\dfrac{C}{N}\right)_u = 23.3$ dB

Conus
antenna

27 dB:
at edge

$Q_u = 48.5$ dB/channel

$Q_d = 36.6$ dB/channel

$\dfrac{G}{T_s} = 15.2$ dBi/K
$f = 12$ GHz
$G = 40.0$ dBi

1.2 m receiving earth station

$\left(\dfrac{C}{N}\right)_d = 15$ dB

$T_s = 300$ K

$\left(\dfrac{C}{N}\right)_t = 14.4$ dB
margin: 3.9 dB

56 kbits/sec

EIRP = 48.3 dBW/carrier
$f = 14$ GHz
1.2 m
$G = 42.3$ dBi
$B = 56$ kHz

Originating
SCFC
station

−1 dB

7 dBW/channel
(5 W/channel)

56 kbits/sec

513

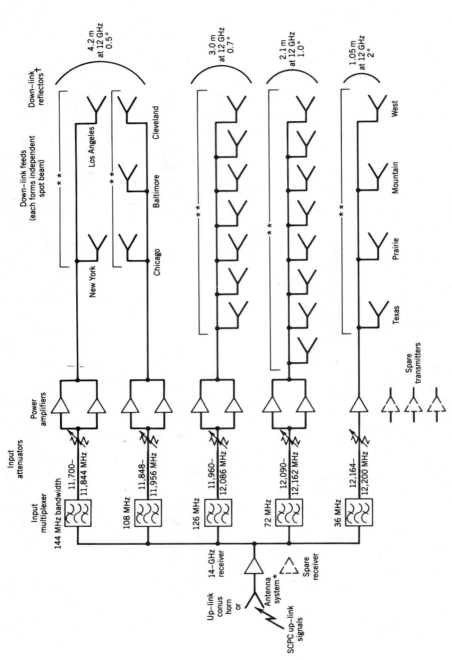

*Single polarization shown.

**All feeds in this group receive identical signals.

†Diameters shown have been oversized by 20% for multiple beams.

FIGURE 4.69 Block diagram of simple domestic satellite.

Another hybrid MA method uses a multiplicity of up-link beams (e.g., 24) with TDMA. Each up-link signal is routed to its own on-board decoder. While the bursts within a given beam might be orderly and nonoverlapping (but also see Random Multiple Access), there may be no coherence between the bursts in different beams. An on-board regenerator (with short-term buffer storage to overcome simultaneous burst problems) reassembles the burst into quasi-continuous trains of pulses for each down-link beam. In theory the guard and slack times of the up links may be eliminated. Empty down-link time slots may be stuffed with dummy bits to keep the earth station bit recovery circuits in the modems synchronized. The satellite power amplifier operates on a SCPT basis. The down link is thus SDMA/SS/TDMA.

Another example of a hybrid MA uses the unstructured random multiple access (RMA) for the up link and a highly structured method (e.g., SS/TDMA or SS/FDMA) for the down link. Hybrid MA methods lead to a few more complex satellites but permit many hundreds of simplified earth stations that use the optimum, least cost, up- and down-link MA technqiues. This is the so-called technology inversion in action. Technology inversion is the process of increasing the satellite complexity in order to simplify the earth station and thus reduce the earth station and total system costs.

4.5 CODE DOMAIN MULTIPLE ACCESS (CDMA)

4.5.1 CDMA Basics

The definition of CDMA and a comparison with other multiple-access domains are provided on page 490 (consult the index for additional CDMA cross-references).

CDMA permits simultaneous use of the spectrum and power through the use of an instantaneously orthogonal PCM spectrum (as in the case of full-band spread spectrum, or CDMA/FBSS), time varying parts of the spectrum (as in "chip," pseudorandom noise, or frequency hopping), or transponder hopping. The "hopping" may be organized but more frequently is done on a contention (random-multiple-access-link) basis.

CDMA may be combined with other multiple-access methods.

CDMA, because it uses a wide spectrum, is fairly insensitive to narrow-band interference and jamming. CDMA also tends to provide a more secure link as far as unauthorized access to the information content is concerned.

Tables 4.3 and 4.37 compare CDMA with other multiple-access methods. Table 4.36 and its figures in Section 4.4 should be consulted for basic SDMA and CDMA combinations using satellite switches (Figures 4.54–4.58), beam hopping (Figure 4.60), and narrow-band frequency hopping or CDMA/FH (also Figure 4.63).

CDMA uses transmission code spectrum orthogonality and channel contention. Each link has its own unique code assignment. This assignment may be permanent (unchanging with time) or clocked (continuously changing pseudorandom noise, or PN coding). In the latter case the code clocks at the transmitting and receiving sites must keep track of the ever-changing codes. Examples of orthogonal (or nearly orthogonal) codes include the Gold, biorthogonal, and PN codes.

Pulse address multiple-access codes may be required for a bit stream to be accepted by a satellite. If the satellite code clock malfunctions (by code errors or jitter), it may deny access to all users. Such was the case with *Courier I* of the 1960s. Soon after launch the satellite access code sequence unpredictably changed, and communications were blocked. The earth stations were unable to find the new access code stream.

Whereas FDMA uses frequency guard bands and TDMA has its guard times, CDMA uses cross-correlation gates (in time or frequency) corresponding to the instantaneous spectral lines unique to the code.

These are matched against codes contained in the receiver to determine the intelligence in the signal.

Figure 4.70 shows the time–bandwidth occupancy for CDMA. The portion (a)

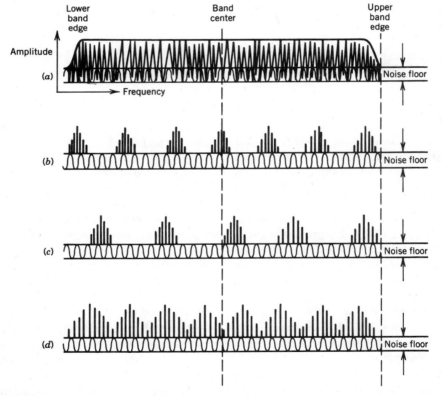

FIGURE 4.70 Code division multiple-access spectrum: (*a*) as observed on slow-scan spectrum analyzer; instantaneous spectrum at (*b*) time t_1, (*c*) t_2, and (*d*) t_3. Note: not to scale.

is that seen on an integrating spectrum analyzer; (b), (c), and (d) are what would be seen at various times if the scan time were practically instantaneous (e.g., 1 psec). What appears to be a full spectrum (a) is actually a few spectrum lines [(b)–(d)] with large gaps containing no signals. Other CDMA signals may be inserted into these gaps. This is done on a contention basis with no attempt to avoid the occasional spectral collisions between two CDMA codes. The CDMA earth station transmitters operate on a continuous SCPC basis. The sum of these many individual CDMA bit stream spectra appears to be white noise if the proper coding is used to cause the center frequency to shift (PN/FSK) with the code; otherwise, the spectra may overlap.

The satellite transponder operates on a MCPT basis. Depending on the grade of service, the probability of spectral collision and the resulting single-pulse degradation effects, the system operator may decide to operate the transponder in a backed-off (see FDMA) or saturated (see TDMA) basis. In some cases a nonlinear repeater may be desirable.

With modern microprocessors and fast Fourier transform (FFT) devices, a receiving station is able to predict the spectral signature of a mark and space (1 and 0) signal in its channel. The station can also predict the unique word signature to identify the channel's identification and synchronization. This process gates the receiver to reject other code streams.

The earth station transmitter may also be made to frequency hop (FH) within a transponder or to hop between individual transponders (TH). The situation shown in Figure 4.71 is for a single station, but it can be imagined that if there are enough randomly independent CDMA/FH signals, the result again is a white-noise-like signal. The earth station and satellite transmitters still operate on a continuous (or quasi-continuous) basis, but the earth station receiver must have a flexible local oscillator (e.g., a frequency synthesizer) to track the moving signal using a pseudo–noise code unique to that link.

Individual CDMA links need not run at a uniform bit rate; in fact, the more variation, the better the approximation to white noise. Large- and small-traffic stations can be accommodated simultaneously. In theory the low-rate stations need proportionally less transmit power (keeping the watts per bit per second constant), but small signal suppression in the transponder may reduce the advantage.

In general, CDMA is bandwidth inefficient because the bandwidth must be much greater than the information bandwidth (see Figure 4.71) but this ratio is the source of the processing gain that makes these techniques so powerful and attractive for services that otherwise would be interference or power limited.

4.5.2 Spread Spectrum

Spectrum spreading may be done in several ways, as shown in Table 4.40.

In the case of CDMA/PN each transmitter assembles a bit stream consisting of a code $[a_i(t)]$ and signal information $[S_i(t)]$. The t denotes that a_1 and S_i are functions of time, and i indicates the access number.

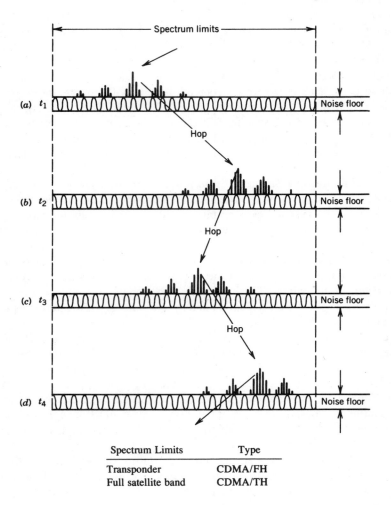

FIGURE 4.71 Signal spectra for CDMA/FH or CDMA/TH.

The receiver despreads the signal and produces a bit stream consisting of

$$\overline{[a_i^2(t)]}[S_i(t)] + \sum_{\substack{j=1 \\ j \neq i}}^{N} \overline{[a_i(t)][a_j(t)]}[S_j(t)] \qquad (4.5)$$

where j is another (unwanted) signal(s) and N is the number of accesses.

The first term, $[a_i^2(t)]S_i(t)$, is the wanted portion, whereas the second, $[a_i(t)][a_j(t)][S_j(t)]$ is unwanted (noise). The second term is zero for true

TABLE 4.40 Various Spread Spectrum Methods

Type	Technique	Advantages	Disadvantages[a]
PN	Pseudonoise code $\otimes$ PSK data carrier (direct sequence)[b]	1. Simple rf equipment 2. Compatible with coherent data modulation; best power efficiency	Difficult code acquisition 1. Complex equipment 2. Time consuming
FH/CHIRP	Pulsed rf carrier swept over frequency band during pulse interval	Simple equipment at transmitter	1. Matched filters at receiver 2. Fast receiver response needed

FH/PN	PN frequency hop $\otimes$ MFSK data carrier	1. Simpler code acquisition
		  a. Less time
		  b. Less complex
		2. No carrier acquisition: less time to acquire
			1. Complex rf generation equipment (frequency synthesizer)
			2. Less power efficient

TH	Transponder hopping	*See* FH/PH	See FH/PN
PN/TIME	PN variable time	Relatively simple receiver and modem	May be observed and tracked
TH	Time hop $\otimes$ PSK or DPSK data carrier	Compatible with packet transmission technique	Requires high-peak-power transmitter at earth terminal
Hybrid	Combination of any two (or more) of the above	Flexibility in optimization for given processing gain	Greater complexity, cost

[a]With large scale integrated circuits (LSI) some of the complexity of these techniques may be simplified from a manufacturing and operational standpoint.
[b]Exclusive OR operation, $\otimes$.

orthogonal codes. Practical codes can produce a nonzero value that consumes satellite power and produce noise. Additional noise comes from IM, interference, link, and thermal noise sources, but much of this noise is rejected by the spread spectrum aspects of CDMA, which produce a processing gain g:

$$g = \frac{P}{R_b} \quad (\text{ratio}) \qquad (4.6)$$

where P is the PN code (or "chip") rate (bits/sec) and R_b is the information transmission rate (bits/sec). In practical systems the chip rate P may be around 2.5 Mbits/sec (or 64.0 dB bits/sec) while the information rate R_b may be 19.2

kbits/sec (or 42.8 dB bits/sec). This indicates that there are 130 PN bits per information bit ($2.5 \times 10^6/19{,}200$) for a processing gain of 130 (or 21.1 dB).

Since

$$\sum_{\substack{j=1 \\ j \neq i}}^{N} \overline{[a_i(t)][a_j(t)]}[S_j(t)]$$

is spread uniformly throughout the entire available bandwidth, and since the receiver looks only at a fraction ($1/g$) of the total, CDMA "sees" only (B_{xpdr}/g)th of this term (see Figure 4.72). The CDMA bit stream bandwidth B_s is proportional to its rate R_b. With more efficient methods (higher bits per hertz), B_s is reduced and g increased.

Assuming all N accesses are the same,

$$\frac{E_b}{N_0} = \frac{S_i}{[(T-1)S_i/g]^* + \{[T + IM(R_b)]/B_{xpdr}\}^*} \qquad \text{(ratio)} \qquad (4.7)$$

where T is the total thermal noise and IM is the intermodulation noise distributed over the transponder bandwidth B_{xpdr}. The asterisk (*) desigates the use of the ratio form (*not* the dB form).

If it is assumed that there is no IM noise and $g = (B_{xpdr}/R_b)$,

$$\frac{E_b}{N_0} = \frac{[g(S_s/N_s)]^*}{[(T-1)(S_s/N_s)]^* + 1} \qquad \text{(ratio)} \qquad (4.8)$$

where S_s/N_s is the spread spectrum signal-to-noise ratio at the receiver input.

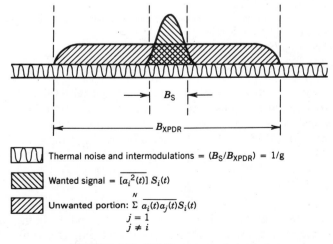

Thermal noise and intermodulations = (B_S/B_{XPDR}) = $1/g$

Wanted signal = $\overline{[a_i^2(t)]}\, S_i(t)$

Unwanted portion: $\sum_{\substack{j=1 \\ j \neq i}}^{N} \overline{a_i(t)a_j(t)} S_i(t)$

FIGURE 4.72 CDMA spectra and signals.

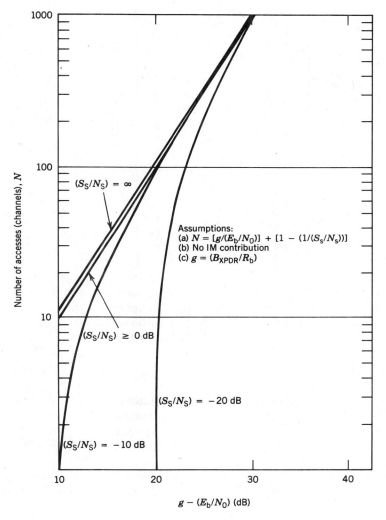

Number of accesses (channels), N

$g - (E_b/N_0)$ (dB)

$(S_S/N_S) = \infty$

Assumptions:
(a) $N = [g/(E_b/N_0)] + [1 - (1/(S_s/N_s))]$
(b) No IM contribution
(c) $g = (B_{XPDR}/R_b)$

$(S_S/N_S) \geq 0$ dB

$(S_S/N_S) = -20$ dB

$(S_S/N_S) = -10$ dB

FIGURE 4.73 CDMA multiple accesses.

The number of simultaneous accesses, N, is

$$N = \left(\frac{g}{E_b/N_0}\right)^* + \left\{1 - \frac{1}{(S_s/N_s)^*}\right\} \qquad (4.9)$$

A graph of the solution is shown in Figure 4.73. An alternative analysis is given by Maral and Bousquet (1986). If the spectrum is spread over the entire allocated band (e.g., 500 MHz), a satellite with the block diagram shown in Figure 4.74 is possible. Each up link is demodulated (D) on a per-access basis. A code recog-

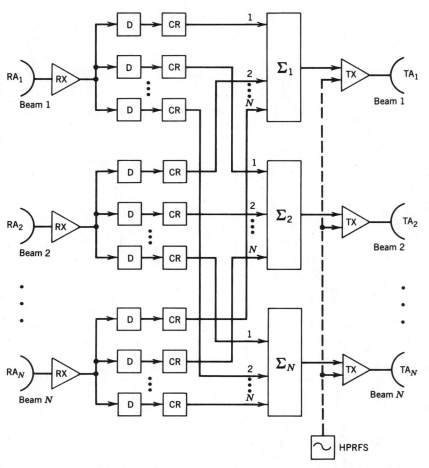

Key: RA, receive antenna; RX, receiver; D, full-band demodulator; CR, beam address code recognition; Σ, down-link multiplexer (signal summer); TX, modulator and transmitter; TA, transmit antenna; HPRFS, high-power rf source.

FIGURE 4.74 SDMA/CDMA block diagram.

nition (CR) device reads the address code and sends the signal on to the signal summer for transmission to the ground. Recoding, retiming, and access control are possible in the satellite. It may also assign priorities and resolve contention in the down link using fast fourier transform (FFT) devices.

4.5.3 CDMA Operations

The station input must be binary (or convertable to a bit stream). Table 4.41 shows a grossly simplified sample (Walsh) code using only 16 characters per 4-bit input.

TABLE 4.41 Walsh Function CDMA Codes

Generating Function (Signal)	(Decimal Equivalent)	Walsh Function Code
0 0 0 0	0	1 1 1 1 1 1 1 1 1 1 1 1 1 1 1 1
1 0 0 0	1	1 1 1 1 1 1 1 1 0 0 0 0 0 0 0 0
0 1 0 0	2	1 1 1 1 0 0 0 0 1 1 1 1 0 0 0 0
1 1 0 0	3	1 1 1 1 0 0 0 0 0 0 0 0 1 1 1 1
0 0 1 0	4	1 1 0 0 1 1 0 0 1 1 0 0 1 1 0 0
1 0 1 0	5	1 1 0 0 1 1 0 0 0 0 1 1 0 0 1 1
0 1 1 0	6	1 1 0 0 0 0 1 1 1 1 0 0 0 0 1 1
1 1 1 0	7	1 1 0 0 0 0 1 1 0 0 1 1 1 1 0 0
0 0 0 1	8	1 0 1 0 1 0 1 0 1 0 1 0 1 0 1 0
1 0 0 1	9	1 0 1 0 1 0 1 0 0 1 0 1 0 1 0 1
0 1 0 1	10	1 0 1 0 0 1 0 1 1 0 1 0 0 1 0 1
1 1 0 1	11	1 0 1 0 0 1 0 1 0 1 0 1 1 0 1 0
0 0 1 1	12	1 0 0 1 1 0 0 1 1 0 0 1 1 0 0 1
1 0 1 1	13	1 0 0 1 1 0 0 1 0 1 1 0 0 1 1 0
0 1 1 1	14	1 0 0 1 0 1 1 0 1 0 0 1 0 1 1 0
1 1 1 1	15	1 0 0 1 0 1 1 0 0 1 1 0 1 0 0 1

With the exception of 0000, all codes use an equal number of marks (1's) and spaces (0's). The information is combined (binary fashion) with a unique carrier code work $A_i(t)$, which is being continuously repeated. Figure 4.75 is a block diagram indicating the encoding process. In this process each bit in the carrier is changed if the information bit is a "zero." It is not changed if it is a "one." This process is the equivalent of an Exclusive NOR gate, which can be represented by

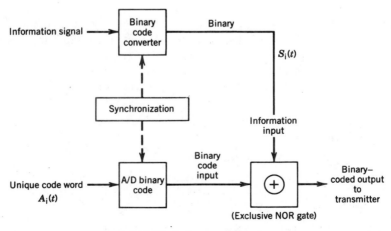

FIGURE 4.75 Block diagram of encoding process.

TABLE 4.42 Exclusive NOR Gate Truth Table

		Information Input	
		0	1
Code	0	1	0
Input	1	0	1

the truth table depicted in Table 4.42. An example of how it is used is shown in Figure 4.76.

Knowledge of the carrier code is needed for the receiver to despread the desired signal while spreading the others.

Figure 4.77 shows a sliding correlation CDMA spread spectrum receiver. When the receiver is turned on, it does not know the phasing of the PN code from the satellite. The receiver code is allowed to drift asynchronously with respect to the transmitter's code until they coincide. Then lock occurs. Let t_{CDMA}, the time to achieve lock, be defined as

$$t_{CDMA} = \frac{C_{STA}}{B_d} \quad (\text{sec}) \tag{4.10}$$

where C_{STA} is the rise time of the CDMA receiver circuitry and B_d is the postcorrelation receiver bandwidth (Hz).

Let the maximum search rate be R_s:

$$R_s = \frac{2}{t_{CDMA}} = \frac{2B_d}{C_{STA}} \quad (\text{searches/sec}) \tag{4.11}$$

The carrier-to-noise ratio at the receiver may be low (and often negative decibels). The ratio of the chip rate P to the information rate R_b is the processing gain that is added (in decibels) to C_t/N_t to obtain E_b/N_0.

Tables 4.43–4.47 trace the flow of signals in two systems. The hub stations use a large antenna and a high-powered amplifier to make up for the limitations in the small stations.

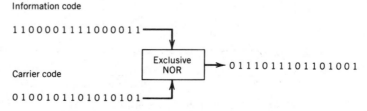

FIGURE 4.76 Example of exclusive NOR operation.

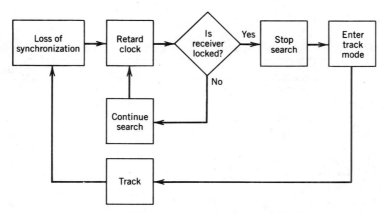

FIGURE 4.77 Spread spectrum CDMA acquisition loop.

TABLE 4.43 Typical Commercial Spread Spectrum Systems

System	Domestic	International
Small Terminal		
Operating frequencies, GHz		
Earth to space	5.925–6.425	14.0–14.5
Space to earth	3.700–4.200	10.7–12.5
Antenna		
Receive-only diameter, m (ft)	0.6 (2)	—
Transmit/receive diameter, m (ft)	1.2 (4)	1.2 (4)
Receive gain, dBi	25.5, 31.5	40.8
Transmit gain, dBi	35.0	42.5
Transmitter rating, W	1	1.3
EIRP capability (after line losses), dBW	34.0	42.4
Receiving system noise temperature, K	130	300
Figure of merit (G/T), dBi/K	+4.4, +10.4	+16.0
Price per terminal (commercial)		
Receive only (1988)	$2000	—
Transmit/receive (1988)	$6000	—
Quantity installed to date (two way)	10,000	—
Information rate (baseband) kbits/sec (dB bits/sec)	19.200 (42.8)	9.600 (39.8)
Chip rate, Mbits/sec (dBb/s)	~2.5 (64)	2.4576 (64)
Bandwidth, MHz	5	5
Processing gain, dB	21.1	24.1
Hub Terminal		
Operating frequencies, GHz		
Earth to space	5.925–6.425	14.0–14.5
Space to earth	3.700–4.200	10.7–12.5
Antenna		
Receive/transmit diameter (typical), m (ft)	10 (32)	10 (32)
Receive gain, dBi	49.8	59.4
Transmit gain, dBi	53.4	60.9
Receiving system noise temperature, K	100	350
Figure of merit (G/T), dBi/K	29.8	34

TABLE 4.44 Spread Spectrum Systems: Up Link from Microterminal to Satellite

Parameter	Domestic	International
Up-link frequency range, GHz	5.925–6.425	14.0–14.5
Transmitter power level, W	1	1.5
Spread bandwidth (2.5 Megachips), MHz, (dBHz)	5 (67)	5 (67)
Power to antenna, dBW/Hz	−67	−65.4
Antenna size, m	1.2	1.2
Antenna gain, dBi	+35	+42.5
Beamwidth, degrees	2.9	1.2
EIRP density, dBW/Hz	−32	−23
Illumination level at satellite, dBW/m^2/Hz	−195	−186
Satellite (typical)	Westar IV	Intelsat V
Satellite antenna on axis gain, dBi	+29	+39
Satellite system noise temperature		
K	1050	1000
dBK	30.2	30
Satellite G/T, dBi/K	−1.2	+9
Off-axis allowance, dB	2	2
Effective G/T, dBi/K	−3.2	+7
Up-link C/N_0, dBHz	−7	+5

TABLE 4.45 Spread Spectrum Systems: Down Link from Satellite to Hub

Parameter	Domestic	International
Down-link frequency, GHz	3.7–4.2	10.7–12.5
Satellite power amplifier, W	5	7.5
Estimated power density, dBW/Hz	−76.2	−104.7
Transmit antenna gain, dBi	+28.2	+39
EIRP density, dBW/Hz	−48	−66
Bandwidth (2.5 Megachips) MHz, (dBHz)	5 (67)	5 (67)
EIRP/carrier, dBW	+19	+1.3
Off-axis loss, dB	2	2
Illumination level, dBW/m^2/Hz	−213	−231
Earth station antenna diameter, m	10	10
Earth station antenna gain, dBi	+49.8	+59.4
Beamwidth, degrees	0.5	0.17
Earth station system noise temperature, K	100	350
dBK	20.0	25.4
Earth station G/T, dBi/K	+29.8	+34.0
Down-link C/N_0, dBHz	+12.0	−11.5
Up-link C/N_0 from microterminal, dBHz	−7	+5
Total C/N_0, dBHz	−7.1	−11.6
Information rate, bits/sec, (dBb/s)	19,200 (42.8)	9,600 (39.8)
Chip rate, Mbits/sec, (dBb/s)	2.5 (64)	2.5 (64)
Ratio of rates (processing gain), dB	21.1	24.2
C/N in information bandwidth, dB	14.0	12.6

TABLE 4.46 Spread Spectrum Systems: Up Link from Hub to Satellite

Parameter	Domestic	International
Up-link frequency range, GHz	5.925–6.425	14.0–14.5
Transmitter power level, W	0.01	0.00007
Spread bandwidth (2.5 megachips), MHz, (dBHz)	5 (67)	5 (67)
Power to antenna, dBW/Hz	−87	−99
Antenna size, m	10	10
Antenna gain, dBi	+53.4	+60.9
Beamwidth, degrees	0.35	0.15
EIRP density, dBW/Hz	−33.4	−37.7
Illumination level at satellite, dBW/m^2Hz	−197	−201
Satellite (typical)	Westar IV	Intelsat V
Satellite antenna on-axis gain, dBi	+29	+39
Satellite system noise temperature		
K	1050	1000
dBK	30.2	30
Satellite G/T, dBi/K	−1.2	+9
Off-axis allowance, dB	2	2
Effective G/T, dBi/K	−3.2	+7
Up-link C/N_0, dBHz	−8.7	−10.0

TABLE 4.47 Spread Spectrum Systems: Down Link from Satellite Hub

Parameter	Domestic	International
Down-link frequency, GHz	3.7–4.2	10.7–12.5
Satellite power amplifier, W	5	7.5
Estimated power density, dBW/Hz	−76.5	−86.2
Transmit antenna gain, dBi	+29	+39
EIRP density, dBW/Hz	−48	−47.5
Bandwidth (2.5 megachips), MHz, (dBHz)	5 (67)	5 (67)
EIRP/carrier, dBW	+19	+19.8
Off-axis loss, dB	2	2
Illumination level, dBW/m^2 Hz	−213	−212.2
Earth station antenna diameter, m	1.2	1.2
Earth station antenna gain, dBi	+31.5	+40.8
Beamwidth, degrees	4.4	1.5
Earth station system noise temperature		
K	130	300
dBK	21.1	24.8
Earth station G/T, dBi/K	10.3	16.2
Down-link C/N_0, dBHz	−7.2	−10.5
Total C/N_0 (Tables 4.45 and 4.46), dBHz	−11.2	−13.9
Total C/N_0, dBHz	−12.7	−15.5
Information rate, bits/sec, (dBb/s)	19,200 (42.8)	9,600 (39.8)
Chip rate, Mbits/sec, (dBb/s)	2.5 (64)	2.5 (64)
Ratio of rates (processing gain), dB	21.1	24.2
C/N in information bandwidth, dB	8.4	10.3

4.6 RANDOM MULTIPLE ACCESS (RMA)

4.6.1 Introduction to Random Multiple-Access Systems

Random multiple access is a demand assignment multiple-access method using little or no centralized control. Orthogonal access channels are not used, and therefore, RMA is different from CDMA, TDMA, SDMA, or FDMA.

Access to the communications link is done on a contention basis: That is, in some cases nothing is preassigned or even temporarily assigned. Terrestrial equivalents are the amateur, mobile, or U.S. citizen's band (CB) services or even a cocktail party conversation. In these cases each user wants to get his or her brief message through in spite of other cochannel users. The user may have access to several channels. If the message is "blocked" by another user, the station simply retransmits its message. There may be no master control stations to notify the blocked parties when to retransmit. If one signal is stronger than others, it may "capture" the channel during its transmissions. In a sense, this is a form of amplitude domain multiple access (ADMA).

An entire message may not be lost if the blocking lasts only a small fraction of the message duration. Some voice services may tolerate the loss of very brief speech periods (the equivalent of voice packets) and still be intelligible to the knowledgeable listener. This is a form of the so-called cocktail party effect wherein one pair of users communicate in spite of many simultaneous cofrequency communications. Forward error correction may permit concealment of errors in data and video transmissions.

A telephone subscriber might not be so tolerant unless he is used to the imperfect state of high-frequency, mobile, or CB radiotelephone services. As fiber optics is employed for regular telephony, the general public's expectations are likely to rise.

In pure RMA an earth station transmits packet bursts at random. The station usually can "see" the retransmission of its own burst to ascertain if it has collided with someone else's burst (see Figure 4.78). If a collision is detected, the burst may be retransmitted n_{RMA} times before the attempt is abandoned (n_{RMA} is between zero and infinity). Usually, there is a random delay before a station retransmits. Since all stations that collided wait random times, there is hope that there will not be a second collision, at least between the original pair of stations.

As the channel traffic increases with N_{ES} earth stations, the probability of a packet being lost (P_L) on the first attempt increases. Subsequent attempts at retransmissions only increase the congestion, further reducing the throughput or utilization U. The utilization is defined as the ratio of the time the channel is in revenue-producing use to the total time.

In Figure 4.79 the two points A and B are for the same degree of utilization. Since the traffic is on a random contention basis, it cannot be predicted from instant to instant whether the channel traffic capacity will be A or B: This is called an instability. Ideally, the traffic (G) would be directly proportional to the number of

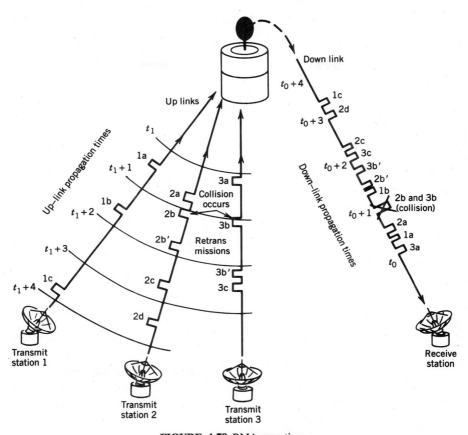

FIGURE 4.78 RMA operations.

accesses (U), as shown by the diagonal line in Figure 4.79. This implies no retransmissions.

The ALOHA system was developed by the University of Hawaii to use RMA for Pacific island communications. (Abramson, 1977)

Contention (RMA) methods are particularly attractive to data collection systems using simple transmit-only earth stations. In some instances (e.g., when seawater temperature is sampled once every 5 min), the complete loss of a measurement packet is probably tolerable because of the redundancy of the adjacent packets. However, for a seismic (earthquake) transmit-only station, all packets should be assumed to have equal value: A missing packet might contain valuable transitory data. Murphy's law almost guarantees it.

Other examples of data collection include point-of-sale cash and inventory control, meteorology, hydrology, remote telemetry, and some forms of telemail.

A wide variety of methods has been proposed to overcome ALOHA's instability, increase channel efficiency, minimize delay, and so on.

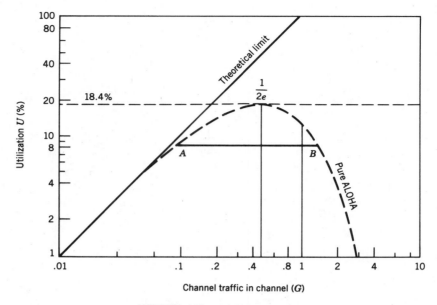

FIGURE 4.79 ALOHA performance.

Table 4.48 compares several methods.

Code domain multiple access with full transponder spread spectrum (CDMA/FTSS) is another RMA method (see Section 4.5.2). This technique may be combined with ALOHA to form CDMA/FTSS/ALOHA (or RMA/FTSS/ALOHA).

RMA and CDMA systems contend for all of the satellite's resources: frequency, time, and power. The supply of each is fixed and independent of the demand. Sharing, by the user, of frequency and time is readily apparent. Sharing of power is often overlooked, but it is important in shared systems such as ALOHA, SPADE, and any voice-activated push-to-talk system. The amount of available power per user decreases as the transponder is called upon to handle multiple carriers simultaneously (MCPT).

4.6.2 RMA Operations

The operating methods for RMA are generally flexible, and unlike FDMA, TDMA, or CDMA, individual station pairs have a greater latitude in selecting their operating rates and modes to fit their particular needs.

RMA stations tend to have few channels. In some cases these may be slow and require a narrow bandwidth. TDMA and CDMA may be uneconomical because of their high costs of equipment for only a few channels; for some services FDMA may be inefficient because of the finite width of its preassigned channels.

TABLE 4.48 ALOHA Random Multiple-Access Forms[a]

Form	Pure	Overlap	Multiple Copy	Slotted	Capture
Packet transmission protocol	Random (independent)	Random (independent)	Random	Time frame synchronous slots equal to single-packet transmission time	Random
Central coordination/control	None	None	None	None	Satellite priority ADMA
Packet collision	Retransmission	Retransmission	Multiple transmissions	Retransmissions	
Collision detection	Self-observation	Self-observation	Not required	Self-observation	
Throughput	Figure 4.79	—	—	Figure 4.80	Figure 4.80
Maximum capacity	$1/2e = 0.184$	—	$1/e = 0.368$	1	Very low
Delay (seconds)	$0.27 \times$ number of trips	$0.27 \times$ number of trips	Undetermined	$0.27 +$ delay to next slot	0.27
Retransmissions	Randomized	Randomized	Randomized	Timed	n.a.
Instabilities	Yes	Yes	—	Yes	No
Equal power for equal traffic	Yes	—	—	Yes	No
ADMA Capture (by intent)	No	n.a.	n.a.	n.a.	6 dB
Collision survival	No	No	—	No	If 6 dB stronger

[a]Abbreviation: n.a., not applicable; $e = 2.718$.

TABLE 4.49 RMA Requirements for Various Services

Service	Channel Rate, G (bits/sec)	Packet Loss Rate, P_L	Number of Reentry Attempts,[a] n (integer)	Notes
Voice (digitized)	1×10^3–16×10^3	0.05	0–2	—
Voice (mobile)	1×10^3–16×10^3	0.15	0–2	—
Facsimile (typed), telenewspaper	4.8×10^3	0.1	10	Redundancy in successive lines and text
Data collection	10×10^3	0.2	0–10	—
Meteorological	Low	0.5	0	Redundancy in successive bursts
Hydrology, oceanography	Lower	0.5	0	Redundancy in successive bursts
Seismology	1×10^3–100×10^3	0.01	10	Random data
Cash register sales	4.8×10^3	0.05	5	Accuracy more important than speed
Navigation	4×10^3–100×10^3	0.005	100	Accuracy more important than speed
Telegraph/Teletype	200	0.01	10	Telegraph code contains some redundancy, other in text
Telecomputing	64×10^3–200×10^3	0.05–0.2	5	FEC may be used to reduce P_L
Video (digital)	24×10^6–45×10^6			Generally too wideband for RMA
Tracking objects (trucks, buses, planes, bears)	50	0.8	10	Redundancy in successive bursts
Paging	300	0.1	20	Heavy use of FEC

[a]Estimated, $n \simeq (D_{AV}/T_{RT})$, where D_{AV} is the average delay and T_{RT} is the round-trip (0.27-sec) delay.

Table 4.49 evaluates potential RMA services on the basis of several important RMA aspects.

Because not all RMA packets get through the first time, some will be delayed (at least one round trip time, T_{RT}, or about 270 msec) until a successful retransmission is made. Queueing theories are applied to RMA and packet systems. For a given grade of service a reduced delay time requirement (D_{AV}/T_{RT}) is obtained at the expense of utilization and capacity. The average delay time D_{AV} is composed of three parts: delay to the next packet, packet duration (about 1 μs), and one or more round-trip delays (T_{RT} at 270 msec each). The T_{RT} round-trip time predominates.

4.6.3 ALOHA Forms of RMA

The ALOHA protocol has a series of derivatives; Slotted ALOHA, Capture ALOHA, and Round-Robin Reservation ALOHA.

The pure ALOHA transmits bursts at random times. Self-observation of the down link is used to determine the need for retransmission.

ALOHA is characterized by a low saturation capacity ($1/2e$, or 0.184) but an average delay D_{AV} of ηT_{RT}. Figures 4.80 and 4.81 show the performance of pure ALOHA.

Figure 4.82 shows the relationship between the two previous figures and the packet loss rate. For example, the pure ALOHA satellite may be operating under a nominal load at point a (left panel), which results in a given utilization indicated

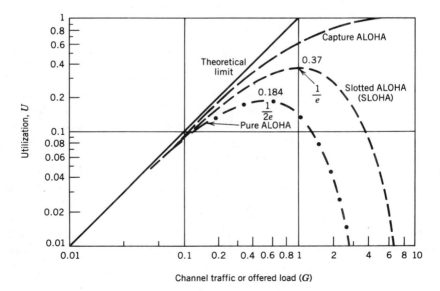

FIGURE 4.80 Utilization of three ALOHA methods.

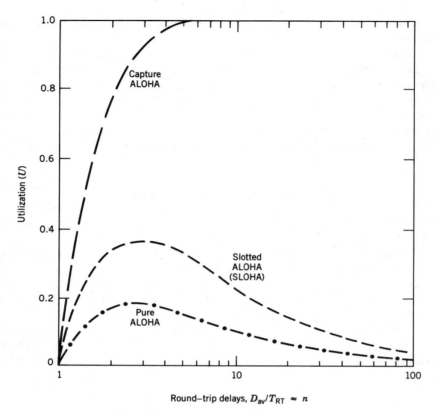

FIGURE 4.81 ALOHA delays.

by point b. This utilization results in a delay (point c, middle panel). Due to insta-
bilities (retransmissions), the real load might be at d. If this is the case, the delay
is actually at point e of the middle panel. Notice that the delays are quite different,
and one cannot predict which case (a and c or d and e) will apply at any instant.
The channel traffic may be running smoothly at the a and c points, and then several
packets may collide, requiring more retransmissions, which makes more collisions, which increases the channel rate further, which continues until the second
stable condition (d) is reached for the given value of U. Conversely, this high
number of retransmissions (and thus the longer delay) may cease once the system
catches up with itself, and the burst collisions get fewer. This will drive the system
toward the other stable points (a and c). This accounts for the flip-flop (or relaxation oscillator) type of performance common to circuits and systems with two
stable operating points and instabilities in between.

The penalty for no retransmissions is shown in the right panel. For the example

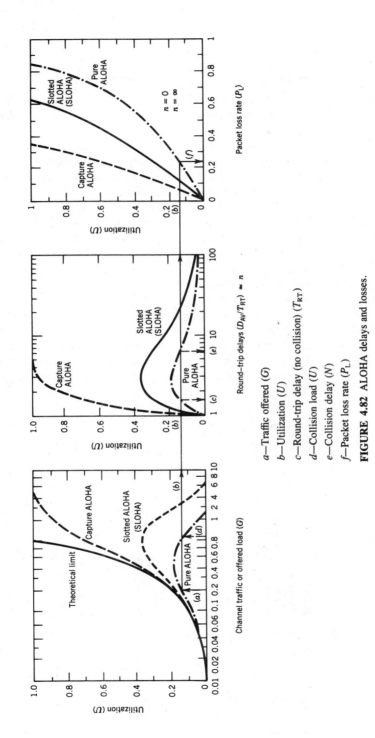

Channel traffic or offered load (G)

Round-trip delays (D_{AV}/T_{RT}) $\approx n$

Packet loss rate (P_L)

a—Traffic offered (G)
b—Utilization (U)
c—Round-trip delay (no collision) (T_{RT})
d—Collision load (U)
e—Collision delay (N)
f—Packet loss rate (P_L)

FIGURE 4.82 ALOHA delays and losses.

535

shown, about 23% of the packets (f) fail to reach their destination in the form transmitted.

In some cases, where there is a lot of redundancy in the information content, it may be possible to accept a packet in which some bits have been mutilated during a partial overlap of two (or more) packets (see Figure 4.83). If the overlap fraction Z is small, sufficient information might remain in one or both packets to be usable without retransmission.

The transmitting station makes this determination for the receiving station. This is called the Overlap ALOHA system and offers a better utilization U at a given channel rate G. During the overlap the satellite amplifier might be driven beyond saturation, causing the power drop shown in Figure 4.83.

Slotted ALOHA (SLOHA) is a more complex form of ALOHA wherein the time domain is divided into "slots" equal to a single-packet transmission time (or burst). As such, SLOHA begins to look like TDMA without assignments (see Figure 4.84). Packets cannot overlap, and either there is contention for a time slot or there is no contention.

The contention leads to the same instability (from the two stable operating points) as pure ALOHA, but the maximum channel utilization is doubled (see Figure 4.80). For a given utilization the mean delay and packet loss rate are both improved (see Figure 4.82).

Capture ALOHA is somewhat of a hybrid between an amplitude domain multiple access (ADMA) and RMA. The prior ALOHA discussion presumed that for a given amount of traffic there was one uniform value of power. Returning to our

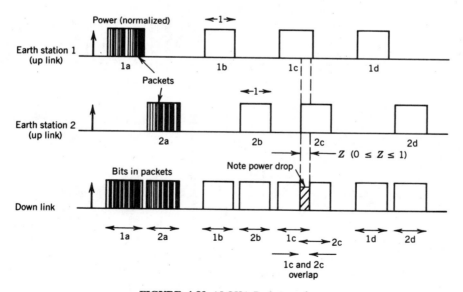

FIGURE 4.83 ALOHA Packet overlap.

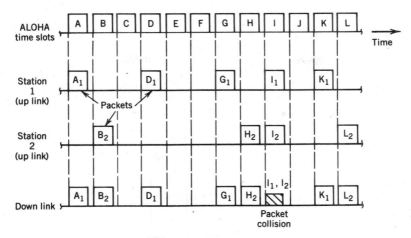

FIGURE 4.84 Slotted ALOHA (SLOHA).

cocktail party analogy, this presumption implies that all participants use equal amounts of vocal power. If the voice power of one user substantially increases (e.g., by shouting), it will ''capture'' the conversation by brute force overriding all the other users. Some of this occurs naturally because of the distance between participants. In a geostationary satellite case, the maximum range advantage (a user at the subsatellite point vs. an edge-of-earth-coverage user) is a mere 1.34 dB.

In Capture ALOHA one packet is made deliberately stronger (e.g., four times, or 6 dB) than another. This packet emerges from a packet collision unblocked. Just as in the case of the cocktail party, the other (normal level) users are interrupted by the rude shouter. The power increase may be used as a means of handling a top-priority packet (such as one that has already been retransmitted several times).

Instead of having the power increase occur at an earth station, the point of control might be in the satellite. Up-link signals arrive independently via orthogonal channels (i.e., separated by beams, polarizations, frequencies, etc.). There is a single down-link channel. An on-board signal processor/controller selects (captures) one of the up links for the down link (see Figure 4.85). The selection may be done on the basis of the retransmission number or a priority code (affixed to each transmission burst) or randomly.

The switching can be done with variable-power control (Figure 4.86) to achieve the 6-dB difference. The switching can also be done at baseband (at the individual packet basis as shown in Figure 4.84) with about a 4-dB advantage over if or rf (random-choice) switching.

Round-Robin Reservation ALOHA (RR/ALOHA) is still another method of RMA. The satellite has one-packet time slots allocated to each earth station (like TDMA). Each station is polled for traffic.

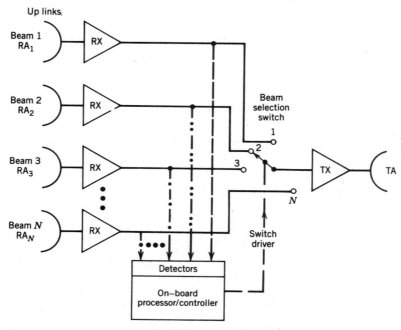

Key: RA, receive antenna; RX, receiver (frequency translator); Switch, satellite switch at rf; TX, transmitter; TA, transmit antenna.

FIGURE 4.85 Capture ALOHA satellite block diagram.

Each station's time slot occurs on a round-robin basis. Station 1's time is followed by station 2's, then stations 3's, and so on, to station N's, then back to station 1's time. In some cases a pool of unassigned time slots is also provided in the frame. This appears to be wasteful since not all stations have equal traffic. Some may have no packets, others may have to delay multipacket traffic. This problem is overcome by allowing the stations with excess packets to contend for another station's empty channel (or a pool channel) by sending a reservation request. When the assigned station wants its packet time back, it sends a "vacate my slot" signal, creating a deliberate conflict with the temporary user who now must seek some other unused channel for its excess packets.

In this case packet collisions may be used for network control if the vacate signal is sent on top of the borrower's time allocation.

There are several other ALOHA reservation methods with increasing levels of complexity. One of these is called early warning reservation. In this method stations are classified by priorities. Special time slots are set aside for reservation requests. A station transmits a reservation request 270 msec (one round trip) before it plans to transmit. It waits to see if anyone else also requests the time slot. If not, it transmits; otherwise, the stations resolve the conflict on a priority, random, or some other predetermined basis.

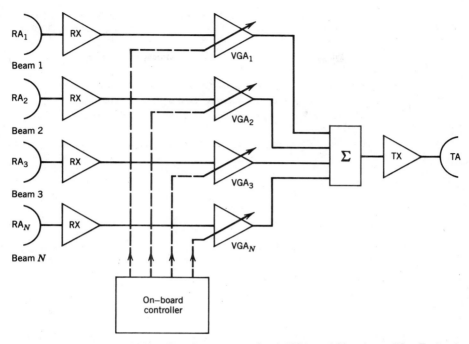

Key: RA, receive antenna; RX, receiver (frequency translator); VGA, variable-gain amplifier; Σ, signal summer; TX, transmitter; TA, transmit antenna.

FIGURE 4.86 Capture ALOHA with power controller.

Other reservation methods continuously poll or observe all the earth stations in the network for traffic. Time or frequency slot assignments are made on the basis of the poll. The poll is repeated frequently to refresh the assignments. The competition for the assignments is on a contention basis (with various schemes being used to resolve conflicts). Once an assignment is made, it is done so on an exclusive basis (until the results of the next poll are announced).

In some cases a good practice is to limit the retransmissions because it is counterproductive to continue to send retransmissions after several tries. As shown in Table 4.49, the data might become stale and dated after several attempts (e.g., two for packetized voice), and its continued transmission only serves to further congest a busy channel and delay subsequent bursts. If no retransmissions are allowed, the situation shown in the right panel of Figure 4.82 occurs, and the ALOHA protocol is stable.

REFERENCES

Abramson, N. (1977) "The throughput of packet broadcasting channels," *IEEE Transactions on Communications*, COM-25, Jan., pp. 117–128.

Dicks, J. L. and Brown, M. P., Jr. (1974), "Frequency Division Multiple Access (FDMA) for Satellite Communications Systems." *IEEE Electronics and Aerospace Systems Convention*, October 7–9, Washington, D.C.

Hilbourn, C. G. et al., (1977), "Implications of Demand Assignment for Future Satellite Communications Systems," NTIA document AD/A-043002.

Maral, M., and Bousquet, M. (1986). *Satellite Communications Systems*, Wiley, New York.

Miya, K. (1975). *Satellite Communications Engineering*, Lattice Co. Ltd., Tokyo.

Morgan, W. L. and Rouffet, D. D. (1988) *Business Earth Stations for Telecommunications*, Wiley, New York, pp. 104–119.

Stamminger, R. et al. (1976), "Communications Systems Modeling Study," NTIA document AD/A-043352.

F I V E

SPACECRAFT TECHNOLOGY

A communications satellite placed in orbit must survive and operate in an environment different from that at the earth's surface. There is little atmosphere, no apparent gravity, no means of physical support, and no easy access for maintenance. A major part of a communications satellite is the group of subsystems that support the communications payload; spacecraft technology provides the means to design and build these necessary subsystems. This chapter provides detailed discussions of spacecraft configuration, structures, and subsystems.

5.1 ENVIRONMENT OF SPACE

It will be useful to define some of the characteristics of space and discuss the effects they have on a communications satellite.

5.1.1 Zero Gravity

One operating difficulty in zero gravity is fuel management—that is, how to make sure that a liquid fuel will flow out of the tank when needed. A bladder may be used, or internal surfaces constructed so that surface tension will aid in fuel expulsion. The deployment of solar panels and antennas may be easier in a zero-gravity environment; however, it is difficult to simulate the deployment on the ground and test the equipment before launch.

5.1.2 Lack of Atmosphere

It is possible to calculate an atmospheric pressure at geostationary orbit, assuming the atmosphere obeys the perfect-gas law, hydrostatic equilibrium is maintained, and a model exists for temperature as a function of height. This leads to a pressure on the order of 10^{-17} torr; while this is not a precise measurement, there is no doubt that the pressure is quite low. At pressures below 10^{-5} torr the thermal conduction becomes negligible; as pressures go down to 10^{-10}, most adsorbed gases on surfaces are removed. Friction between surfaces increases, and lubrication of bearings requires material with low vapor pressures. Welding of contacts of mechanical relays can occur.

The electronics expert is not so much concerned with the pressure of outer space, but rather with the pressure inside an electronics package. This is a function of the internal outgassing and the venting to the outside. Pressures inside a spacecraft are considerably higher, by many orders of magnitude, than the pressure outside. In some confined spaces it is possible for the pressure to be optimum for voltage breakdown (around one torr) and for arcing to occur.

5.1.3 Intensity of Solar Radiation

The power in the sun's radiation is equal to 137 mW/cm^2 above the earth's atmosphere (air mass zero); see curve S_c in Figure 5.1. This "solar constant" is not really constant. The earth's orbit has an eccentricity of 0.0167, so there is a $\pm 1.67\%$ annual variation in distance to the sun, or a $\pm 3.34\%$ in power, with a maximum in early January. This variation is shown in the S curve of Figure 5.1.

Many communications satellites have solar arrays that are oriented toward the

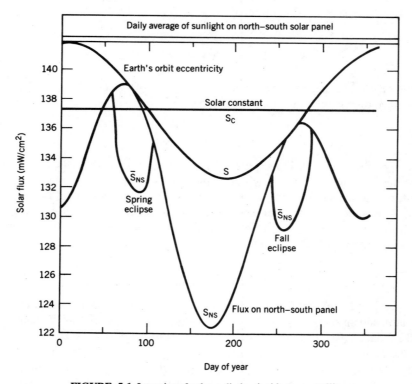

FIGURE 5.1 Intensity of solar radiation incident on satellite.

sun in terms of east–west pointing but are not oriented in terms of north–south pointing. At the vernal and autumnal equinoxes (March and September), when the sun is in the equatorial plane, the solar array is oriented directly at the sun. At the winter and summer solstices (June and December), the sun is 23.44° away from the equatorial plane, and the power on the solar array is reduced by the cosine of this angle. Since the summer solstice (sun farthest north) coincides within 2 weeks of the earth's aphelion (sun's greatest distance from earth), the lowest solar power occurs near the end of June. This flux on a north–south panel is shown as the lower smooth curve S_{NS} in Figure 5.1.

The average solar power is also reduced by eclipses. The two annual eclipse seasons are centered around the equinoxes and have a maximum duration of about 70 min. The daily average power is reduced about 5%. Fortunately, this reduction occurs at the maximum of solar power on the arrays so that it does not reduce the overall minimum, which still occurs near the summer solstice. These variations are labeled $\bar{S}_{NS}$ in Figure 5.1. Equations for these curves are in Section 5.5.2. Beyond these reductions the average solar intensity on the solar array of a spinning satellite is reduced by a factor of π.

The intensity and variations of solar intensity have a major effect on not only the design of a solar array but also the design of the battery. The direction of the sun is detected by solar sensors for purposes of satellite attitude control. Finally, the solar radiation has a major effect on the temperature of various spacecraft components.

5.1.4 Temperature of Space

On the earth the temperature of an object is often determined by the temperature of the surrounding air; in space the temperature is determined by conduction to other spacecraft parts and by both internal and external radiation. The average kinetic energy of the surrounding particles is quite high, around 10,000 K; but since there are so few particles, they have a negligible effect on a body's temperature. The equilibrium temperature with stellar radiation, the so-called temperature of outer space, is quite cold, around 4 K; however, it has practical application only for spacecraft that travel toward the outer planets and beyond. This temperature characterizes the far infrared radiation (~ 1000 GHz) from stellar radiation, and may not be the same as the noise temperature for rf radiation (see page 308).

A more practical definition of the temperature of space is the equilibrium temperature of a small black sphere. These temperatures are graphically shown in Figure 5.2. In geostationary orbit the "temperature" in sunlight is 275–280 K, while in eclipse it is 60–75 K. Deployed solar panels approach these temperatures but are somewhat higher because they are flat rather than spherical in shape. It may appear to be a happy coincidence that spacecraft tend to reach temperatures similar to temperatures on earth. The reason is that the earth is a giant spacecraft subjected to the sun's radiation, and the same laws that determine the earth's temperature also determine the spacecraft temperature.

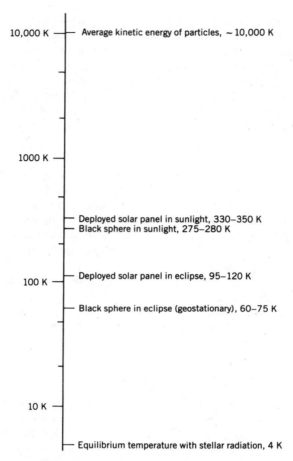

FIGURE 5.2 Various concepts of temperature in space.

5.1.5 Particles in Space

There is a variety of particles in the space environment, including cosmic rays, protons from solar flares, electrons, and meteoroids. Particle radiation has an important effect on solar cells and on some solid-state devices inside the space-craft. Particles from the sun vary depending on solar sunspot activity and come in occasional bursts. Solar cells are protected by coverslides but still suffer nonre-versible degradation. The spectrum of solar cosmic ray protons is shown in Figure 5.3; the intensity decreases for higher energy particles. Solar cell radiation degra-dation in space can be expressed in equivalent radiation by 1-MeV electrons. A well-protected solar cell may receive the equivalent of 3×10^{14} e^-/cm^2 in 7 yr at geostationary orbit.

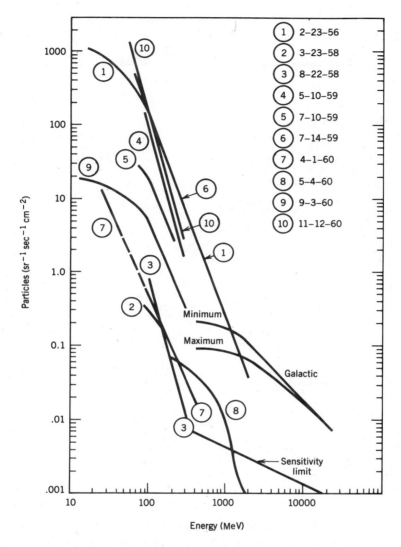

FIGURE 5.3 Solar cosmic protons (Haviland and House, 1965. Reprinted with permission). Occasional solar activity produces the higher curves, and more rapid degradation of satellite solar arrays.

The population of sporadic meteoroids is shown in Figure 5.4. In addition to these there are meteor showers where the intensity for a short time can be considerably higher. The effect of meteoroids on communications satellites is not serious; the probability of significant damage is quite small. Both the erosion of solar cell coverslides and/or the loss of an occasional solar cell are small effects that will not materially change the degradation of a solar array.

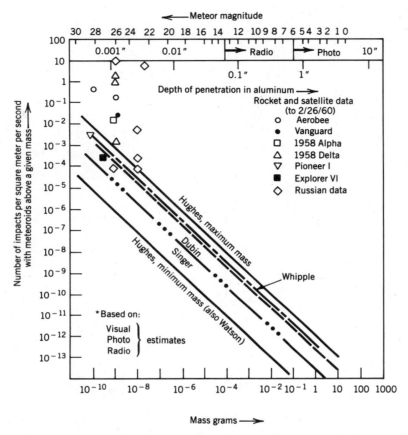

FIGURE 5.4 Population of sporadic meteoroids (Haviland and House, 1965. Reprinted with permission). In 10 yr on 1 m^2 the largest likely particle is 10^{-5} g, which would penetrate less than 1 mm.

5.1.6 Earth's Magnetic Field

The average magnetic field of the earth can be represented by the field of a magnetic dipole (see Figure 5.5). The intensity of the field at any point is

$$H = H_0 (R_e/r)^3 (1 + 3 \sin^2 L_m)^{1/2} \qquad (5.1)$$

where H_0 = strength at earth's surface
$\quad R_e$ = earth's radius
$\quad r$ = distance to point
$\quad L_m$ = latitude from magnetic equator

Since the intensity decreases as the cube of the distance, at geostationary orbit it is only one-three hundredths the value at the earth's equator. The average north–south field component is 110 Γ ($1\ \Gamma = 10^{-5}$ G). Since the geomagnetic north pole

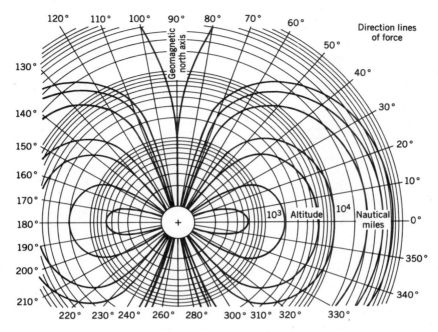

FIGURE 5.5 Contour of the earth's magnetic field (Haviland and House, 1965. Reprinted with permission). Geosynchronous orbit is at 22,767 n., mi.

is inclined to the earth's north pole by 11.5°, there is also a radial component that oscillates daily with an amplitude of 18 Γ.

The earth's magnetic field is weak at the earth's surface compared to an iron magnet and even weaker by a factor of 300 at geostationary orbit. However, if the spacecraft has a magnetic moment, a significant torque can be exerted. It is possible to control this by using a large electric coil with many turns, thus producing a desired torque on the spacecraft.

Charged particles are deflected by the earth's magnetic field, and both protons and electrons are trapped in the Van Allen radiation belts (see Figure 5.6). Fortunately, most of this radiation is considerably below the geostationary orbit, which is at 6.6 earth radii, and only a small fraction affects geostationary satellites. A spacecraft does go through it in transfer orbit, but this is of limited duration. This would become more significant if orbit raising were done with electric propulsion, since then the time spent in the Van Allen belts would be longer.

5.2 SPACECRAFT CONFIGURATION AND SUBSYSTEMS

Before discussing each of the subsystems of a communications satellite, a discussion on satellite configurations is necessary. There are many factors to consider in choosing a configuration (not all technical); the selected configuration affects many

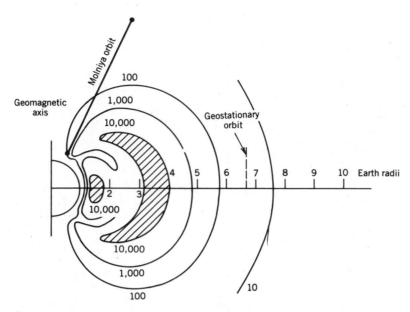

FIGURE 5.6 Van Allen radiation. Numbers are counting rates of geiger tube on *Explorer IV* and *Pioneer III* (Haviland and House, 1965. Reprinted with permission).

of the spacecraft subsystems. The fact that satellites of different configurations have been built is an indication that the decision may not be critical; many missions can be successfully implemented with a variety of configurations.

The two main configurations for communications satellites are body stabilization and spin stabilization. Body-stabilized satellites or three-axis stabilized satellites are simpler in concept but often require more hardware to implement. Body-stabilized satellites are fixed with respect to the earth; they rotate once a day as the satellite goes around the earth. Body-stabilized satellites may be either a zero-momentum system, where the reaction wheels operate around zero spin and may come to a complete stop, or a biased momentum system in which there is always a minimum angular momentum in a certain direction.

A spin-stabilized satellite has a significant part of the satellite that rotates at a rate on the order of once a second or the entire satellite rotates. Small, simple communications satellites may be single bodies that rotate around the axis with the maximum moment of inertia. They may have an omni or toroidal antenna, more complicated antennas that are despun either mechanically or electrically, or a despun communications payload. Some spin-stabilized satellites are "dual spinners." These have a rotating part and a stationary part and rotate around a minimum moment of inertia.

An example of a body-stabilized communications satellite is shown in Figure 5.7. Solar arrays are usually deployed and rotate around a north–south axis. An

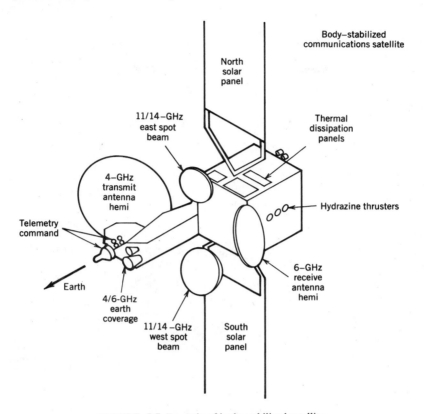

FIGURE 5.7 Example of body-stabilized satellite.

antenna farm faces the earth with the largest dishes corresponding to the narrower beams at the lower frequencies. The power amplifiers are mounted so that heat can be dissipated on the north- and south-facing panels. Thrusters are arranged so that positioning can be along both east–west and north–south directions; also pairs of thrusters are used for orientation around various axes. The external surfaces of the spacecraft are controlled so that the desired temperatures result.

An example of a spin-stabilized satellite is shown in Figure 5.8. The shape is cylindrical with the solar array mounted on the outside surface of the cylinder; the rotation axis is a north–south axis. Much of the weight is in the spinning part, but the antennas must be kept pointed at the earth. Usually this means that all the communications payload is placed on a despun shelf that is stationary with respect to the earth. The solar array, batteries, fuel tanks, and thrusters are on the spinning part. The electric power is delivered to the communications payload through slip rings. The telemetry and command functions are divided, since both parts of the spacecraft require these functions.

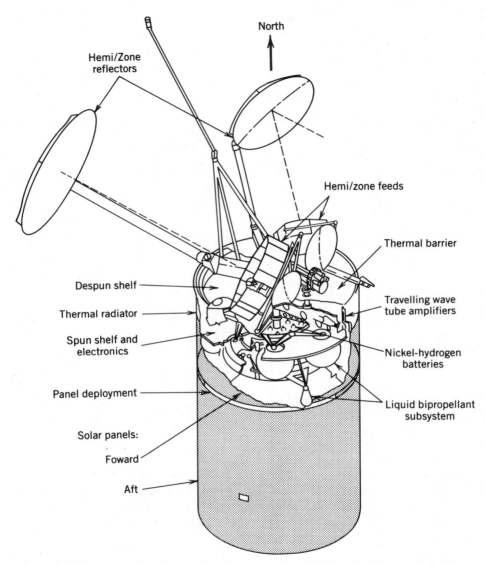

North

Hemi/Zone reflectors

Hemi/zone feeds

Thermal barrier

Despun shelf

Travelling wave tube amplifiers

Thermal radiator

Spun shelf and electronics

Nickel-hydrogen batteries

Panel deployment

Liquid bipropellant subsystem

Solar panels:

Foward

Aft

FIGURE 5.8 Example of spinner satellite (Copyright, Thompson and Johnston, 1983. Reprinted by permission of John Wiley & Sons, Ltd.).

5.2.1 Communications Subsystem

The main function of a communications satellite is to receive a radio signal from the ground, amplify it, and retransmit it (see Section 5.3). A very simple diagram is shown in Figure 5.9. The purpose here is not to explain the communications payload in detail but to relate it to the various supporting subsystems. A simple

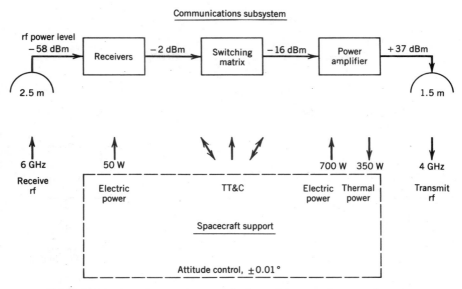

FIGURE 5.9 Block diagram of communications subsystem and spacecraft support.

satellite would be only a receive antenna, a receiver, a power amplifier, and a transmit antenna. Most communications satellites have a multiplicity of these items with various antenna beams, multiplexers, switching matrices, and redundant components.

5.2.2 Telemetry, Tracking, and Command (TT&C)

The state and health of the various subsystems is transmitted by the telemetry system. A command system transmits the desired commands to the satellite. Tracking is done by the earth station antenna and ranging by a signal sent up through the command link and returned down by the telemetry transmitter. See Section 5.4. A function diagram of the TT&C subsystem is shown in Figure 5.10.

A command starts at the satellite control center, goes via land lines to an earth station, and is then transmitted to the satellite. The command receiver in the spacecraft receives the signal, demodulates it, and processes it. The command is stored, and a verification signal is passed on to the telemetry and returned to ground. Only after verification is an execute signal sent from the control center, received, decoded, and distributed to the proper spot in the spacecraft. Commands are used to control the satellite's on-board equipment status, operational modes, position, and orientation. Discrete commands use signals that are fixed-duration pulses to change the status of equipment, such as turning it on or off. Analog signals are used to generate variable-length pulses to turn equipment on for specific time intervals. Digital data blocks can be transmitted to a satellite for reprogramming on-board processors.

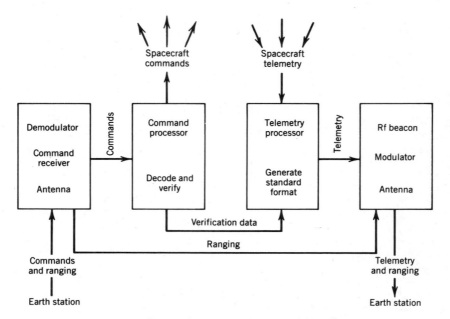

FIGURE 5.10 Telemetry, command, and ranging.

Shown on the right side, the telemetry subsystem gathers data from various subsystems, processes them into a desired form, modulates the signal of the rf beacon, and transmits it to the ground. The telemetry collects, formats, and transmits data of various types. Digital data indicate the condition of on–off equipment or the position of various switches. Many sensors monitor the performance of operational on-board systems by providing data on temperature, voltage, current, or pressure. Signals from attitude sensors and accelerometers are transmitted for interpretation on the ground. Some scientific and experimental data may be gathered (e.g., radiation, magnetic field, electrical discharges, motion of liquid fuel, etc.) to aid in the design of future communications satellites.

Tracking functions are used to: determine the satellite position, calculate future orbital motion, pointing data for earth stations, and command for thrusters to maintain position. Ranging measures the actual distance (slant range) from the earth station to the satellite as a function of time, by determining the time interval for a signal from the earth station to the satellite and back to ground. Angle tracking measures the azimuth and elevation angle from the earth station to the satellite as a function of time and requires only an rf beacon from the satellite.

5.2.3 Electric Power

The electric power subsystem (see Sections 5.5 and 5.6) furnishes power to operate the commuications payload, especially the power amplifiers that consume 70–80% of the total power requirements.

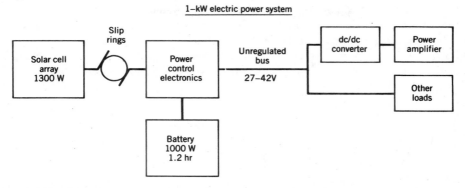

FIGURE 5.11 Electric power subsystem. Most components are duplicated for redundancy. On spinner satellites, the slip rings come after the power control electronics.

The primary power of a communications satellite is provided by silicon solar cells. During eclipse the secondary power is provided by rechargeable nickel–cadmium (NiCd) batteries or, more recently, nickel–hydrogen (NiH) batteries. A typical power subsystem is shown in Figure 5.11. On a body-stabilized satellite there are usually slip rings between the rotating solar array and the rest of the power subsystem; on a spinner the entire subsystem is on the rotating part, and slip rings deliver the needed power to the stationary communications platform.

Control electronics are required to connect the solar array and batteries to the various loads. An important function of this control is to regulate the power from the batteries and recharge them properly. In many communications satellites an unregulated bus is used. Much of the power is for the power amplifiers (PA), and it is more efficient to do the voltage regulation at the PA power supply (which often requires a dc/dc converter) rather than in the control electronics. The power requirements for a typical communications satellite are shown in Table 5.1. After

TABLE 5.1 Example of Electric Power Requirements[a]

	Autumnal Equinox	Summer Solstice	Eclipse
Communications	769	769	768
Telemetry, command, ranging	39	39	39
Attitude control	48	73	48
Electrical power control	9	9	9
Thermal control	136	86	30
I^2R harness losses	10	10	9
Battery charging	100	30	—
Total load	1111	1016	903
Power Margin	243	272	76
Total power	1354	1288	979

[a] End of life, watts.

the communications the largest requirement is for the electric heaters used to maintain a minimum temperature in various parts of the spacecraft. Because the solar array degrades due to radiation, calculations must be made for the end of life (EOL) of the satellite. It is assumed that all power amplifiers will be functioning at that time, although often this is not the case. In addition, a power margin is provided.

5.2.4 Attitude Control

The antennas require an attitude control system that will keep them pointed at the earth, frequently within $0.1°$ and $0.01°$ (see Sections 5.7 and 5.8). The parts of an attitude control system are shown in Figure 5.12. Any error in pointing is detected by a sensor and corrected by changing the speed (or axis) of a rotating wheel. If the wheels have reached their limit, thrusters are used so the wheel speed can be reduced.

The performance specifications of an attitude control system are set by the existing disturbance torques and the required pointing accuracy. The major disturbances on a geostationary satellite are torques due to solar radiation pressure and misalignment of thrusters. The main sensor used is an earth sensor or rf sensor, but other devices are used to determine attitude, including sun sensors, star trackers, and gyroscopes. Reaction wheels (which spin either way) or momentum wheels (which only spin in one direction) are used to correct the attitude of a satellite. Thrusters are used when the wheels are no longer capable of maintaining attitude. To save fuel, magnetic torquing coils, or solar flaps, can provide a weak, but steady, torque. A control system takes the information from the sensors and provides commands to the torque generators.

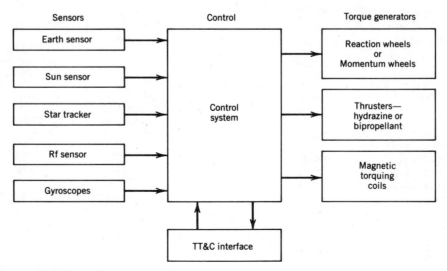

FIGURE 5.12 Attitude control subsystem. Satellites have only some of sensors shown.

5.2.5 Structure

A structure is needed to mount the components of these subsystems and bring them safely through the launch environment (see Section 5.9). A mass summary of a communications satellite is shown in Table 5.2. Other communications satellites have more or less mass, but the proportion for the different subsystems remains relatively constant. Variations occur; for example, fuel mass depends on design life. The mass of the apogee motor is about equal to that of the spacecraft after injection. The hydrazine fuel for thrusters and the electrical power system are a significant part of the spacecraft mass.

The spacecraft structure is designed to provide mechanical support for all the subsystems and components, to provide precise alignment where needed, and to aid the thermal control with thermal conduction and desired surface properties. These functions must be provided within certain constraints and trade-offs. The launch vehicle limits the shape and size of the structure. The attitude control system frequently requires the principal inertia axes to be in preferred directions and the inertia ratio to be within certain limits. It may also dictate the stiffness of the structure or parts of it. Finally, easy access during assembly, integration, and testing is important.

The structure must also be designed to sustain the loads during environmental testing, ground handling, launch into low orbit, perigee and apogee firings, and any deployment of antennas or solar arrays. After reaching final orbit position and deployment, the loads on the structure are greatly reduced in the zero-gravity environment, but the alignment requirements are more rigorous. The designer must satisfy all requirements, minimize the structure mass and cost, and still keep the probability of failure near zero.

TABLE 5.2 Example of Mass Summary

	Mass (kg)
Communications	233
Attitude, determination, and control	73
Electrical power	141
Propulsion	39
Telemetry, command, and ranging	26
Structure/thermal	184
Electrical and mechanical integration	69
Apogee motor inerts	61
Total dry spacecraft (EOL)[a]	826
Fuel/pressurant	187
Total spacecraft (BOL)[a]	1013
Adapter	21
Apogee motor consumables	871
Mass margin	24
Total in transfer orbit	1929

[a]Abbreviations: EOL, end of life; BOL, beginning of life.

The ways in which loads are imposed on the spacecraft are sketched in Figure 5.13. In the early stages of the launch the acoustic noise is highest and is transmitted from the rocket motors by the air through the fairing or housing and onto the spacecraft. Steady loads are transmitted through the structure as the rockets accelerate the spacecraft to the high velocities required by satellite orbits. Vibration of many frequencies is transmitted through the spacecraft supports from the rocket motors. The separation ring and other pyrotechnic devices send sudden shocks through the structure.

A satellite structure is shown in Figure 5.14. Surfaces are needed for support of the communications equipment, fuel tanks, batteries, and so on, and a strong structure is needed between the loads and the launch vehicle (or shuttle) interface. The basic bus structure shown is rectangular with a separate antenna support structure that holds the rf feeds and reflectors. The antenna support structure includes stationary positions for the rf feeds and deployable structures for some of the reflectors. The solar arrays consist of panels that are folded together.

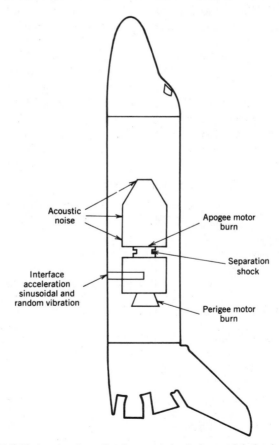

FIGURE 5.13 Acceleration, vibrations, and shocks imposed during launch.

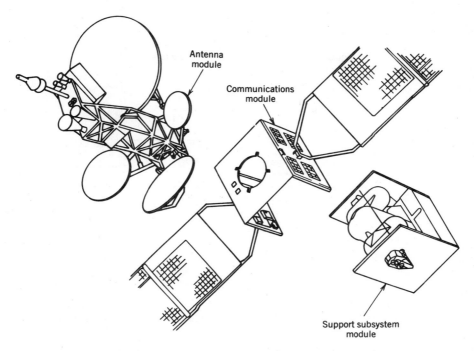

FIGURE 5.14 Structure of communications satellite.

5.2.6 Thermal Control

The thermal subsystem maintains the proper temperature (see Section 5.10); at least half the power input to the power amplifiers results in heat, which must be dissipated.

The average temperature of a satellite is based on an equilibrium between the thermal input and the power radiated (see Figure 5.15). The average temperature is controlled mainly by carefully specifying the thermal properties of different surfaces in terms of thermal emissivity and absorptivity to sunlight. Temperatures of different parts inside a satellite can be controlled by changing the thermal coupling and by the use of electric heaters. Some of the more important tasks in a communications satellite are to maintain the batteries within narrow temperature limits, keep the hydrazine fuel from freezing, and to dissipate the large amount of power generated by the power amplifiers.

5.2.7 Reliability

Table 5.3 shows the calculated 7-yr reliabilities for the different subsystems (see Section 5.12). These numbers are calculated from the reliabilities of the many components. Redundancy is used extensively so that if one part fails, the redundant unit can fulfill the mission. The number of points where a single failure would

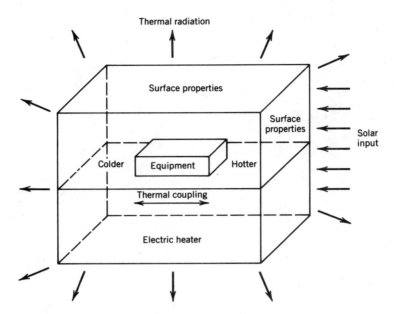

FIGURE 5.15 Major factors that determine the temperature in a satellite.

TABLE 5.3 Reliability of a Communications Satellite

	Reliability
Communications	0.90
Attitude, determination, and control	0.93
Electrical power	0.96
Propulsion	0.997
Telemetry, command, and ranging	0.95
Electrical and mechanical integration	0.986
Total spacecraft reliability[a]	0.75

[a]Usually spacecraft reliability figures exclude the launch reliability.

terminate the mission (single points of failure) is kept to a minimum. The achievement of these reliabilities requires good design, careful selection of components, quality control in the integration, and a variety of tests to detect weaknesses before launch.

5.3 COMMUNICATIONS SUBSYSTEM

5.3.1 Introduction

The major function of a communications satellite is to receive a radio signal from the ground, amplify it, and retransmit it. This basic function is illustrated in Figure 5.16. The required hardware has characteristics dictated by a number of significant

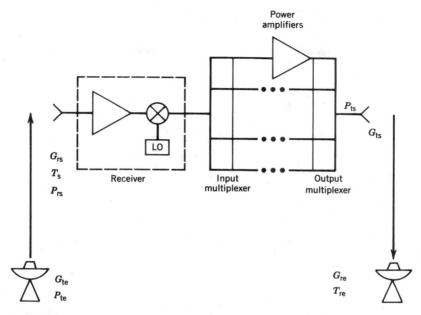

FIGURE 5.16 Major components of communications subsystem (Difonzo et al., 1980).

factors. The large distance (35,000 km) means that the space loss and the required amplification, including antenna gain, is large (200 dB). Most communications satellites have large bandwidth requirements (500 MHz or more). Finally, the use of the geostationary orbit means that mass and power are severely limited, and maintenance is out of reach.

Prior to 1979, the frequencies for international and commercial communications were the 6-GHz band (5.925–6.425 GHz) for the up link and the 4-GHz band (3.7–4.2 GHz) for the down link. Normally, the lower frequency is used for the down link because there is more attenuation of higher frequencies in the atmosphere and power is limited on the satellite. Additional frequencies were allocated in 1979, and use of these bands will grow. The 14/11-GHz band is being widely used, and its usage will continue to grow. See Tables 3.3 and 3.4 in Section 3.1.1.

The Intelsat system was already using the full 6/4-GHz band with the *Intelsat III* series of satellites (see Table 5.4). Since then, the 6/4 GHz band has been widened and a number of methods have been used or suggested to allow reuse of the same frequency band. Satellite diversity makes use of more than one satellite, at different longitudes, so that earth station antenna beams can differentiate between them. Beam diversity makes use of narrow satellite antenna beams so that different earth stations receive different signals from the same satellite. Polarization is used to double the frequency use. Reverse frequency allocation has been suggested so that the up link could reuse the lower frequency (e.g., 4 GHz) and the down link reuse the higher frequency (e.g., 6 GHz). Finally, there are higher frequency bands available at 14/12 GHz and 30/20 GHz that will grow in usage. The performance

TABLE 5.4 Communications Characteristics of Intelsat Satellites

Intelsat designation	I	II	III	IV	IV-A	V	V-A	VI	VII[a]
Transponders	2	1	2	12	20	27	—	50	36
Bandwidth, MHz	50	130	300	500	800	2,144	2,250	3,300	2,392
Voice circuits	240	240	1,500	4,000	6,000	12,000	15,000	120,000	—
Television channels	—	—	—	2	2	2	2	3	—
Antenna	Omni	Omni	Despun	Despun	Despun	Three-axis	Three-axis	Despun	Three-axis
Coverage	Hemi	Global	Global	Global, spot	Global, spot	Global, hemi, zone, spot	Global, hemi, zone, spot	Global, hemi, zone, spot	Global, hemi, zone, spot
EIRP, dBW	11.5	15.5	23	23–34	22–29	23–41	—	25–45	29–45
Year of first launch	1965	1967	1968	1971	1975	1980	1985	1989	1993
Design lifetime, yr	1.5	3	5	7	7	7	7	14	10

[a]Information available early 1988.

and the availability of current hardware for low-noise and power output applications is increasing, in spite of the problems with rain interference.

The communications payload consists of a receiving antenna, a low-noise amplifier (LNA), a down converter, a power amplifier, and a transmit antenna. The antennas are sometimes combined, so that one antenna is used both to receive and transmit. The LNA may utilize tunnel diodes or other solid-state devices. The down converter shifts the up-link frequency to the down-link frequency and consists of a local oscillator and a mixer. The main components of the power amplifier are either TWTs or solid-state amplifiers. Although TWTs are broadband devices, they cannot cover effectively the full 500-MHz band, so each one amplifies a limited width, such as 40 or 80 MHz. This requires an input multiplexer to separate the input signals for the TWTs and an output multiplexer to recombine them. In addition, the communications subsystem includes diplexers, filters, equalizers, attenuators, and many redundant components with appropriate switches.

5.3.2 Receiver

A receiver consists of several stages of amplification, a down converter, and additional amplification. The receiver circuitry shown in Figure 5.17 starts with a switch and a four-stage amplifier. The local crystal oscillator is temperature controlled and generates a 23.177083-MHz signal. This is multiplied in three stages to 2225 MHz, with suitable filtering and amplification.

The mixer is followed by a filter that rejects out-of-band signals and the second harmonic of the local oscillator. This is followed by further stages of amplification. Next, there is a PIN diode attenuator that can be commanded to insert a 7.5-dB loss. This is isolator coupled to a driver amplifier through another filter. The driver consists of additional stages of amplification using transistors.

Three types of solid-state diodes are used on communications satellites: the tunnel diode, the avalanche (IMPATT) diode, and the solar cell. All are two-terminal devices with negative resistances used to generate power, but they actually operate on very different principles. The solar cell exhibits negative resistance at the direct current, that is, its current is opposite to its voltage (see Section 5.5.3.1). The tunnel diode always has a positive resistance using direct current, but part of its $I-V$ curve exhibits a negative slope (see Figure 5.18), which can be used to generate microwave power when used with a proper dc bias. The IMPATT diode has no negative slope on its $I-V$ curve, but it exhibits a dynamic response that has the effect of a negative resistance.

Most commercial tunnel diodes are made from germanium or gallium arsenide (GaAs), since a high ratio of peak current (I_p) to valley current (I_v) is hard to obtain in silicon. Typical values of I_p/I_v are 3.5, 6, and 15 for Si, Ge, and GaAs, respectively; corresponding values are 65, 55, and 150 mV for V_p and 420, 350, and 500 mV for V_v.

A typical amplifier circuit is shown in Figure 5.19. The dc bias is provided so that the diode is at a point where it exhibits negative resistance. When microwave

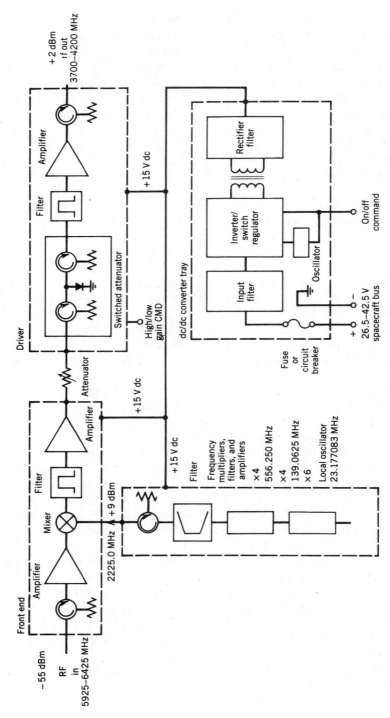

FIGURE 5.17 Block diagram of 6-GHz receiver.

562

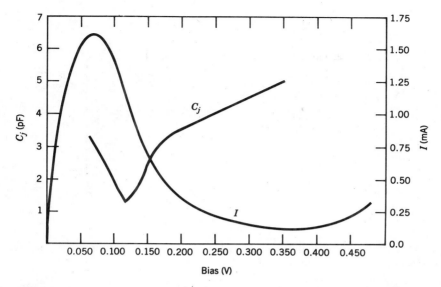

FIGURE 5.18 Current I and junction capacitance C_j as function of bias for typical 6-GHz germanium tunnel diode (Revesz and Fleming, 1978; reprinted, by permission, from *COMSAT Tech. Rev.*).

energy is transmitted toward the diode, a current is generated within the diode that reinforces the transmitted wave. Thus, rather than absorb the transmitted power, the diode enhances it, and the reflected power is greater than the incident power. A circulator is used to separate the reflected wave from the incident wave.

The other negative-resistance microwave device is the IMPATT diode (impact ionization avalanche transit time). Conditions in this diode are such that a strong

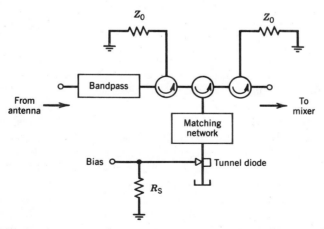

FIGURE 5.19 Schematic of tunnel diode amplifier circuit (Revesz and Fleming, 1978; reprinted by permission, from *COMSAT Tech. Rev.*).

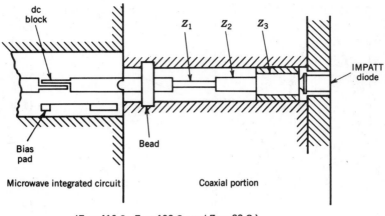

$(Z_1 = 116 \ \Omega \ , Z_2 = 102 \ \Omega \ , \text{and} \ Z_3 = 20 \ \Omega \ )$

FIGURE 5.20 Cutaway view of single-diode amplifier module using IMPATT diode (Chou, 1979; reprinted by permission, from *COMSAT Tech. Rev.*).

electric field can produce an avalanche, with free electrons gaining enough energy to produce additional free electrons. The diffusion length is long, and these electrons move through the depletion layer. However, the transit time is so long that by the time they arrive, the electric field has reversed and is a minimum. These electrons produce a current that is 180° out of phase with the voltage (an increasing current accompanying a decreasing voltage constitutes a negative resistance). A view of a single-diode amplifier is shown in Figure 5.20. The avalanching process is noisy compared to the tunneling process, so the IMPATT is not useful as a low noise amplifier (LNA). For more information on the use of receivers, see Section 3.2.2.

5.3.3 Traveling-Wave Tube

The TWT was the dominant power amplifier for communications satellites, but many communications satellites now use solid-state amplifiers. The principle of operation of a TWT may be understood by reference to Figure 5.21. An electron stream is produced by an electron gun (heated cathode and focusing electrodes), travels along the axis of the tube constrained by focusing magnets, and is finally collected by suitable collector electrodes. Spaced closely around the beam is a circuit, in this case a helix, capable of propagating a slow electromagnetic wave. The circuit is proportioned so that the phase velocity of the wave is small with respect to the velocity of light. In typical low-power tubes a value on the order of one-tenth of the velocity of light is used; for higher power tubes the phase velocity may be two or three times higher. Suitable means are provided to couple an external rf circuit to the slow-wave structure at the input and output. The velocity of the

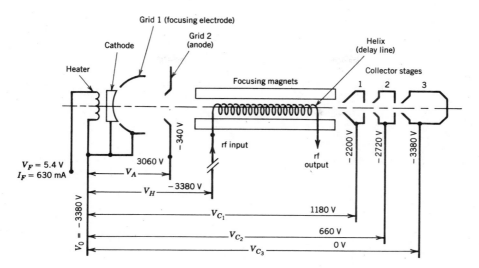

FIGURE 5.21 Schematic of travelling wave tube.

electrons is adjusted to be about the same as the axial phase velocity of the wave on the circuit.

When a wave is launched on the circuit, its longitudinal component of its field interacts with the electrons traveling in approximate synchronization with it. Some electrons will be accelerated, and some decelerated, resulting in a progressive rearrangement in phase of the electrons with respect to the wave. The electron stream thus modulated in turn induces additional waves on the helix. This process of mutual interaction continues along the length of the tube with the net result that dc energy is given up by the electron stream to the circuit as rf energy, and the wave is thus amplified.

By virtue of the continuous interaction between the wave on the helix and the electron stream, TWTs do not suffer the gain-bandwidth limitation of ordinary types of electron tubes. The helix is an extremely useful form of slow-wave circuit because the impedance that it presents to the wave is relatively high and because, when properly proportioned, its phase velocity is almost independent of frequency over a wide range. Since the electron velocity depends on the accelerating voltage, the TWT will operate correctly over only a limited range of voltage. The electrical characteristics are shown in Figure 5.22. For maximum power output and maximum efficiency, it would be desirable to work at the peak of the curve (saturation), but for linearity it is necessary to operate at a lower point.

Many TWTs for communications satellites produce 10–60 W. A 200-W direct broadcast satellite tube with multistage depressed collector is shown in Figure 5.23. The collector has electrodes at different voltages, designed so that electrons are collected at different plates, depending on their kinetic energies. The electrons are

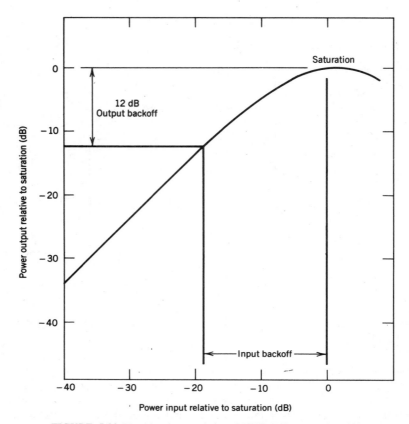

FIGURE 5.22 Transfer characteristics of TWT (Difonzo et al., 1980).

collected near the apex of their trajectories, where they have near-zero velocity. In this manner the electron kinetic energy is converted to potential energy, which reduces the power required to operate the tube.

5.3.4 Filters and Multiplexers

Filters are needed to select certain frequencies and reject others. The fundamental concept of a microwave filter is quite simple (see Figure 5.24). It consists of a cylindrical cavity, with conducting walls, slots, and irises to allow the energy to enter and exit and small tuning screws to adjust the cavity resonant frequency. The cavity resonates at a certain frequency, and this frequency and a small band of nearby frequencies are transmitted through the filter, while other frequencies are not.

The filters and resultant multiplexers have been developed considerably beyond the simple concept of a cavity with a single mode. Additional cavities are added in series, which adds to the performance of the filter. Instead of using a single-

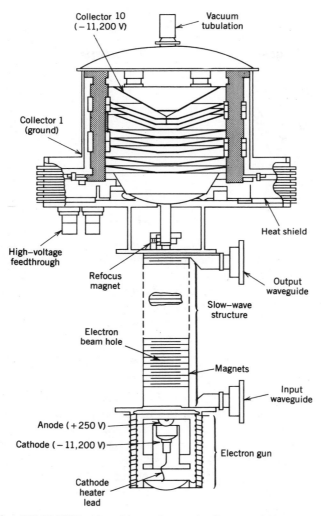

FIGURE 5.23 A 200-W TWT with multistage depressed collector used in *Hermes* satellite (CTS) (Alexovich, 1977. Reprinted by permission of the Royal Society of Canada). Slower electrons are collected by the first few electrodes, and the fastest ones by collector 10.

cavity resonance, use is made of two orthogonal modes. Folded filters have been built that not only make the packing more efficient but also allow coupling between nonadjacent cavities. Limitations on some filter designs required a minimum of one channel bandwidth as a guard band (see Figure 5.25); this meant that even and odd channels were connected to different manifolds, and two transmit antenna subsystems were required (see Figure 5.26). More recently, contiguous band multiplexers have been developed, which allow all the channels to be connected to the same antenna. An experimental four-channel waveguide multiplexer is shown

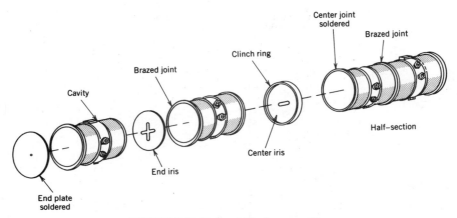

FIGURE 5.24 Lightweight microwave filter.

in Figure 5.27. Each channel has its own filter, which is then connected to the waveguide manifold. A matching network at each end of the manifold provides a necessary symmetry to all the channels. The dimensions and locations of each filter were calculated with a computer program for optimum performance. The design parameters are shown in Table 5.5. The experimental performance was compared with the computer predictions, with good agreement.

5.3.5 Antennas

Microwaves transmitted in waveguides do not produce any fields outside of the waveguide and therefore do not radiate any power. If, however, a waveguide is open at one end, fields are generated in the external space, and radiation occurs (see Figure 5.28). For most open structures a spherical wave will be radiated, but its intensity will not be the same in all directions. An open waveguide will radiate some power, but it will also reflect some power. A horn or feed is designed to look like a terminated line, that is, to radiate the maximum power, and to reflect very little.

The preceding discussion concerns transmitting antennas, but it is important to realize that if an electromagnetic wave is incident on an antenna that is connected to a transmission line, the incident wave will set up voltages and currents in the line, and power can be delivered to a load impedance at the other end of the line. Here, the antenna takes energy from space, whereas a transmitting antenna delivers energy to space. An interesting fact about antennas is that their properties are *reciprocal*. That is, they have the same impedance and radiation pattern for transmit and receive at the same frequency.

The main antennas on a communications satellite are usually parabolic reflectors (see Figure 5.29). These provide narrow beams over a wide range of frequencies. Power is transmitted (or received) through a feed horn toward the reflector. Many antennas are offset, so that the reflected beam is not intercepted by the feed horn;

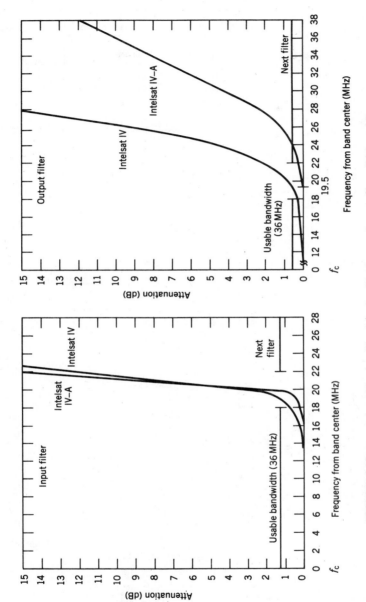

FIGURE 5.25 Attenuation of *Intelsat IV* and *Intelsat IV-A* input and output filters (Dicks and Brown, 1975; reprinted by permission from *COMSAT Tech. Rev.*).

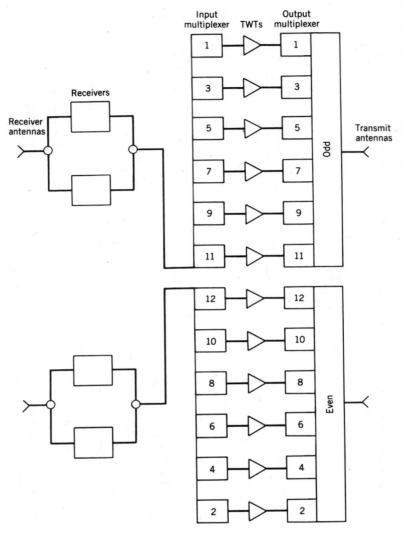

FIGURE 5.26 Even and odd channels connected to different waveguide manifolds used in *Intelsat IV* (Chen et al., 1976; reprinted by permission from *COMSAT Tech. Rev.*).

this reduces the sidelobes, especially with the large feed arrays that are sometimes required, and increases efficiency. The antenna in the figure has a gimballed reflector, so that the beam direction can be controlled.

The gain of an antenna, G, is expressed as

$$G = \frac{4\pi A_e}{\lambda^2} \qquad (5.2)$$

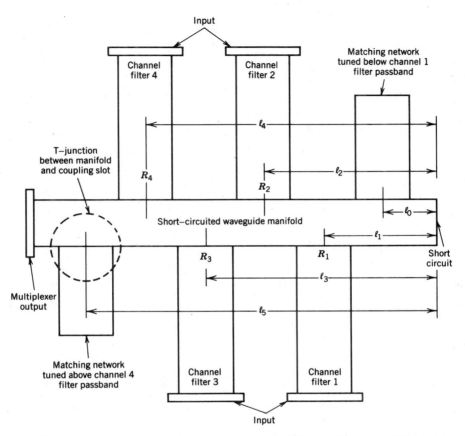

FIGURE 5.27 Mechanical configuration of contiguous band multiplexer (Chen et al., 1976; reprinted by permission from *COMSAT Tech. Rev.*).

TABLE 5.5 Design Parameters of Contiguous Band Multiplexer[a]

Parameter	Matching Network	Channel Number, k				Matching Network
		1	2	3	4	
Optimum coupling R_K	$(0.66)^b$	0.66	0.68	0.62	0.59	(0.76)
Channel center frequency, GHz	(3.66)	3.7425	3.825	3.905	3.995	(4.095)
Channel design bandwidth, MHz	(74.0)	79.0	74.0	74.0	76.0	(74.0)
Filter location after tuning (l_k of manifold of Figure 5.27), in.	1.037	3.759	5.911	8.004	9.837	10.875

[a] From Chen et al. (1976); reprinted by permission, from *COMSAT Tech. Rev.*

[b] Numbers in parentheses are for filter functions used in computer program.

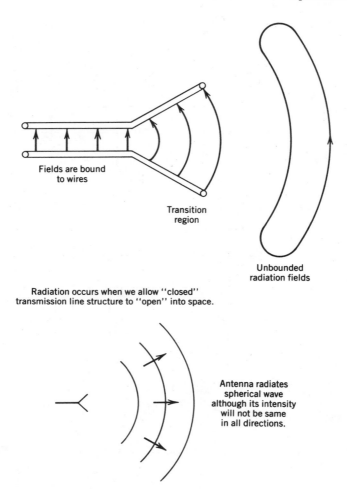

FIGURE 5.28 Radiation of microwave power into space (DiFonzo et al., 1980).

where λ is the wavelength, and A_e is the effective area of the antenna. This effective area is equal to the physical area times the antenna efficiency. The antenna gain depends on how the energy from the feed is spread over the reflector. In the simple ideal case this would be a uniform distribution, but this is difficult to attain in practice. As shown in Figure 5.30, the power decreases away from the center. If the spreading is too large, too much power is lost by not hitting the reflector; on the other hand, if the spreading is too small, the reflector is not being utilized efficiently, and the resulting beam will spread more. The actual antenna gain for various illuminations and various antenna sizes is shown in Figure 5.31. Which illustrates the decrease in antenna gain for various off-axis angles. In particular, the gain continues to increase with antenna size for symmetric (front fed) parabolic

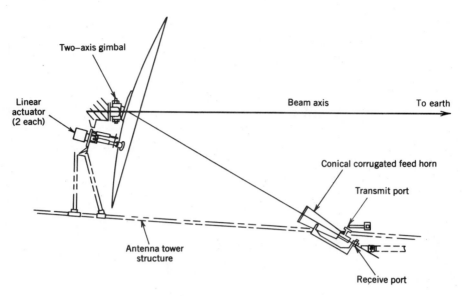

FIGURE 5.29 Offset parabolic antenna for 14/11-GHz Spot Beam.

reflectors. But with off-axis antennas the gain reaches a maximum, and then decreases with increasing reflector size.

Global beams that cover the entire earth require a circular pattern, which is simple in concept and in practice. Many communications satellite beams are designed to cover specific continents, countries, or other areas. These require beam shaping, not only to maximize the power inside the chosen area, but also to minimize the power radiated in adjacent areas. Beam shaping can be done by using multiple feeds that are fed various power levels with different phases. A simplified feed array element is shown in Figure 5.32, and a feed network is shown in Figure 5.33. The sizes, shapes, locations, power, and phase of all the feeds are varied in a computer program until an optimum feed array is selected.

5.4 TELEMETRY, TRACKING, AND COMMAND†

5.4.1 Introduction

Commands are necessary to operate most communications satellites, and in order to transmit the proper commands, information is needed on the satellite's location and condition. Telemetry, tracking, and command (TT&C) provide the means of monitoring and controlling satellite operations. These three functions are usually

†Based on a lecture and visuals by Richard S. Cooperman, whose help is gratefully acknowledged.

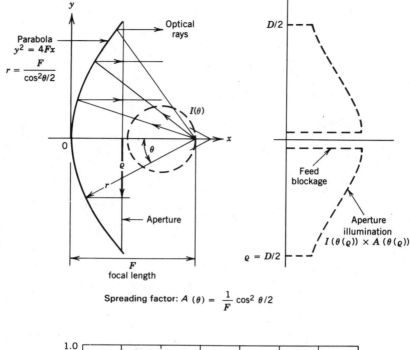

Spreading factor: $A\,(\theta) = \dfrac{1}{F}\cos^2\theta/2$

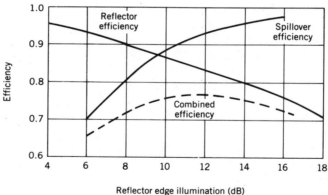

Reflector edge illumination (dB)

FIGURE 5.30 Effect of nonuniform spreading on antenna reflector (DiFonzo et al., 1980).

integrated into a single subsystem and are kept separate from the main communications. Telemetry is the means by which measurements made at a distance are transmitted to an observer. Tracking is observing and collecting data to plot the moving path of an object. And command is the means by which control is established and maintained.

A functional diagram of the TT&C subsystem is shown in Figure 5.34. A command originates at the satellite control center, is sent via land lines to an earth

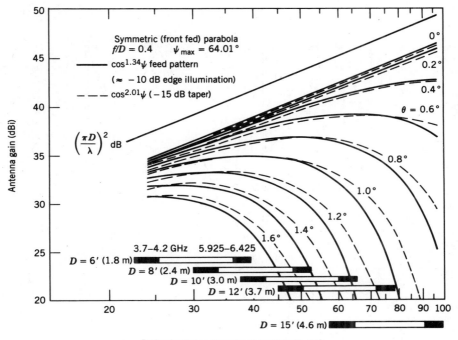

FIGURE 5.31 Antenna gain versus diameter and off-axis angle (DiFonzo et al., 1980).

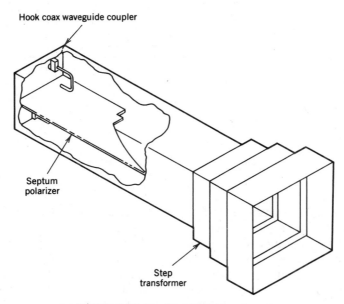

FIGURE 5.32 Simplified feed array element.

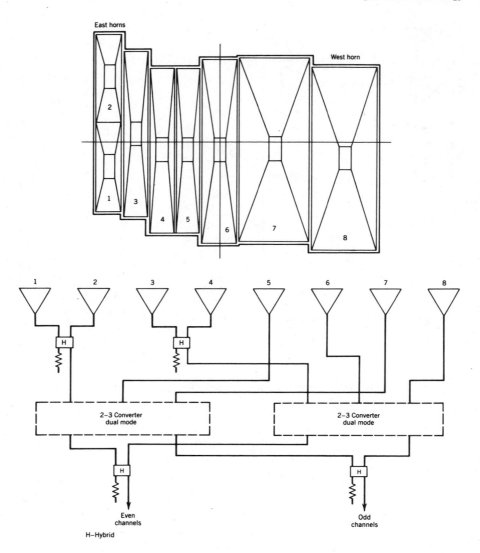

FIGURE 5.33 Transmit feed array and network.

station, and is then transmitted to the satellite. The command receiver in the spacecraft receives the signal, demodulates it, and processes it. After decoding a command, a verification signal is passed on to the telemetry and returned to the satellite control center. Only after verification is an execute signal sent from the control center, received, decoded, and distributed to the proper spot in the spacecraft. The telemetry subsystem gathers data from various other subsystems, processes the data into a desired form, modulates the signal from the rf beacon,

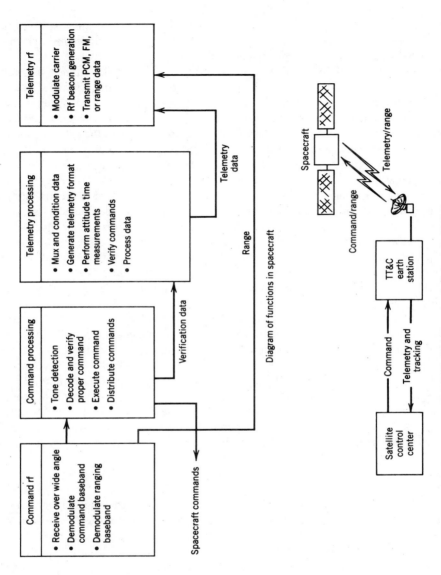

Telemetry rf

- Modulate carrier
- Rf beacon generation
- Transmit PCM, FM, or range data

Telemetry processing

- Mux and condition data
- Generate telemetry format
- Perform attitude time measurements
- Verify commands
- Process data

Command processing

- Tone detection
- Decode and verify proper command
- Execute command
- Distribute commands

Command rf

- Receive over wide angle
- Demodulate command baseband
- Demodulate ranging baseband

Telemetry data

Range

Verification data

Spacecraft commands

Diagram of functions in spacecraft

Spacecraft

Telemetry/range

Command/range

TT&C earth station

Command

Telemetry and tracking

Satellite control center

FIGURE 5.34 Functional diagram for TT&C system.

and transmits it to the ground. Tracking is done by an earth station antenna controllable in azimuth and elevation. Ranging is done by sending a signal through the command link and returning the signal via the telemetry transmitter.

The telemetry collects, formats, and transmits data of various types. Digital data indicates the condition of on–off equipment or the position of various switches. Many sensors monitor the performance of operational on-board systems by providing data on temperature, voltage, current, and pressure. Signals from attitude sensors and accelerometers are transmitted for interpretation on the ground. Some scientific and experimental data may be gathered (e.g., radiation, magnetic field, electrical discharges, motion of liquid fuel, etc.) to aid in the design of future communications satellites.

Tracking functions determine the satellite position and are used to predict future orbital position, pointing data for earth stations, and data for command of spacecraft thrusters to maintain position. Angle tracking measures the azimuth and elevation angle from earth station to satellite as a function of time and requires only an rf beacon signal from the satellite. Ranging measures the actual distance (slant range) from the earth station to the satellite as a function of time. The range is found by determining the time interval for a signal from the earth station to the satellite and back.

Commands are used to control the satellite's on-board equipment status, operational modes, position, and orientation. Discrete commands use signals that are fixed-duration pulses to change the status of equipment, such as turning it on or off. Analog signals are used to generate variable-length pulses to turn equipment on for specific time intervals. Digital data blocks may also be transmitted to a satellite for reprogramming on-board processors. Commands include a spacecraft identification code so they will not affect another spacecraft. In some satellites the command links are encrypted for added protection. Commands are stored, retransmitted to the ground, and verified before they are activated by an execution comand.

5.4.2 Telemetry

A typical telemetry system for a communications satellite is shown in Figure 5.35. Various signals are processed into the desired form, multiplexed together, and then modulated onto the telemetry beacon signal. Since telemetry is a vital part of the satellite, redundancy is used for extra reliability. At least two separate telemetry encoders and two separate beacons are cross-strapped so that each encoder can operate with either beacon. Although the points on telemetry lists are chosen so that the satellite can operate with only one encoder, the lists are often not exact duplicates. Some items may be slightly different; if both telemetry units are operating, the total information is increased.

The principal objectives for spacecraft telemetry are to provide information of operational use, failure analysis, and prediction of spacecraft performance. In routine operations the telemetry verifies commands and equipment status and alerts personnel of any unusual occurrences. In case of failures or anomalies telemetry

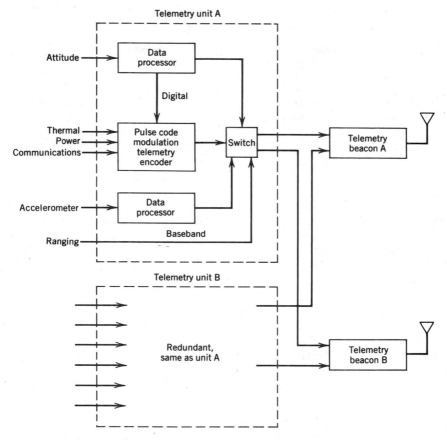

FIGURE 5.35 Typical telemetry system for communications satellite.

data is used to determine the causes, the events, and ways to counteract or alleviate the problems produced by a failure. Telemetry can also be used to analyze any degradation that might affect performance and predict its effect on the spacecraft lifetime.

A list of possible telemetry items on a communications satellite is given in Table 5.6. Whereas the list for each satellite is different, the items listed under typical data are frequently telemetered. Items listed as suggested data are usually not telemetered. A trade-off is usually necessary for each satellite to determine the benefits of the information compared to the weight and costs of including the point in the telemetry list. A few of the suggested items are difficult to measure; a number of them are curves with many points and have a considerable amount of information to be transmitted. In addition to these telemetered parameters, the communications transmissions also provide information on the on–off status of a

TABLE 5.6 Items on Telemetry Lists[a]

	Typical Data	Suggested Data
	Communications	
Receiver	Switch, S, T, I	**Output,**[b] mixer (I or V)
Switch matrix	**Switch-S,**[b] T	
Transmitter	Switch-S, T, input-I	**Output,**[b]
TWT	—	Filament-I, helix-I
Transponder	—	Out-of-band noise power
Antenna	T, actuator-S	Deployment-S, P
	Electric Power	
Solar array	T (eclipse range), **V,**[b] I[b]	I–V curve, orientation, T (high resolution)
Array drive	**Motor-I,**[b] **position,**[b] **deployment,**[b] drive-T	Brush delta-V
Battery	**V,**[b] I,[b] T,[b] ΔT, relay-S, charge rate, heater-S	**Individual cell-V,**[b] cell pressures
Power conditioning	Bus (I, V), interconnect, dissipator-T	—
	Telemetry, Tracking, and Command	
Beacon	—	**rf output**[b]
Telemetry	**Calibration,**[b] transmission-S, T, encoder-S	—
Command	**Verification-S,**[b] decoder-S, receiver-T	—
	Propulsion	
Tanks	**Pressure,**[b] T,[b] heater-S	**Fuel left,**[b] location
Valves	Valve-S, heater-S, T	**Position,**[b] I vs. time
Lines	T	Flow rate
Thrusters	**Several T,**[b] heater-S	Thrust, I–V electrothermal
Apogee motor	T, heater-S, safe/arm	Thrust
	Orientation	
Earth sensors	**Output,**[b] T	Analog output, mirror
Rate gyros	**Output**[b]	
Sun sensors	**Output**[b]	
Logic unit	**Error,**[b] **torque,**[b] modes	**Integrator,**[b] rate error
BAPTA[c]	T, I,[b] **acceleration,**[b] heater-S, voltage drop	
Wheels	**Speed,**[b] **error,**[b] T,[b] I,[b] rotation direction	Accelerometer, bearing signature

[a] Abbreviations: S status; V, voltage; I, electric current; T, temperature; P position (angle); TWT, travelling wave tube.

[b] Judged more important in determining subsystem performance.

[c] Bearing and power transfer assembly for spin stabilized satellites.

transponder, the rf power output, and the performance of the antenna pointing. On spinning satellites wobble or precession may produce variations in the rf power observed.

5.4.2.1 Transducers

Various devices are used to translate the value, or condition to be measured, into a voltage (analog or pulses). In choosing a transducer, several factors must be considered: physical quantity to be measured, possible range of values (including abnormal operation), accuracy required, form of the output, and physical principle that transforms the desired measurement into the required output.

While there are many transducers used in communications satellites, the most common items to be measured, as can be seen in Table 5.6, are voltage, current, and temperature. Voltages can often be transformed into the desired telemetry voltage, such as 5 V, with metal film resistor dividers. For current measurements resistive shunts or Hall devices[†] are used, with amplification of the voltage produced across the shunt.

For temperature measurements platinum resistors are used for large variations in temperature, such as those that can occur on a deployed solar array in eclipse. Platinum is the standard for precision temperature measurement and has a linear variation of resistance with temperature. For narrower temperature ranges (0–70°C) a thermistor may be used (the word *thermistor* means thermally sensitive resistor). The temperature coefficient for a thermistor is much greater than for platinum, and the resistance may change 6 or even 8% per degree Celsius. The temperature coefficient is negative, so the resistance decreases as the temperature increases. However, the temperature–resistance relationship is not linear, so calibration curves or computer calculations are necessary.

5.4.2.2 Telemetry Data

Many of the voltages produced by transducers are transformed into pulses by an analog-to-digital conversion. Various data codes may be used, as shown in Figure 5.36, which shows a sample data of 10110001101. In the first code, return to zero (or RZ), the first half of each time interval is either high or low; the second half of each time interval is always zero. In the next three codes there is no return to zero (NRZ). The simplest code is NRZ-L, where 1 is high and 0 is low. In the code NRZ-S the 1 is interpreted as do not change to the opposite state and the 0 as change to the opposite state. The NRZ-M is the opposite, with 1 being change and 0 being no change. As can be seen in the figure, the NRZ codes have fewer changes, have wider pulses, and would require less bandwidth to transmit. Three "biphase" codes are shown: the BIϕ-L goes down in the middle of the time interval for a 1 and goes up for a 0; at the end of a time interval it may need to change. In

---

[†]An electric current in a magnetic field produces a voltage perpendicular to both the current and the magnetic field.

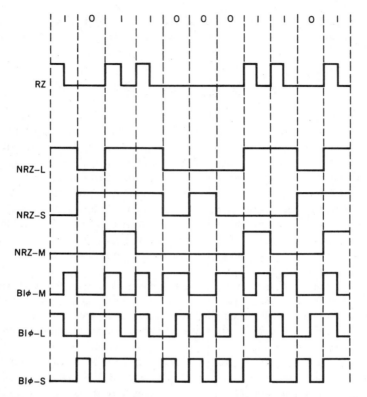

FIGURE 5.36 IRIG (InterRange Instrumentation Group) standard PCM data codes.

the BIϕ-M and BIϕ-S codes, there is always a flip at the end of each time interval; in the middle of the time interval BIϕ-M flips for a 1 and not for a 0, while BIϕ-S flips for a 0 and not for a 1. These codes are standardized by the InterRange Instrumentation Group (IRIG).

Some telemetry data consists of time intervals, as shown in Figure 5.37. These measurements are for a spinner satellite, with two sun sensors and two earth sensors on the rotating part (actually three earth sensors, but only two are measured separately). The top two lines are for the two sun sensors, where the time interval T_1 is the time for a complete rotation. The "north" earth sensor provides a signal as the earth comes into view and another signal when it disappears from view. The other earth sensor provides similar data, but since it is not mounted at the same location, the signals come at a different time in the rotation. The "platform" signal is given when the platform (nonrotating part with the earth-pointing antennas) is at a specific orientation with respect to the rotor. Finally, the last line is the frame index, a signal set by the telemetry clock. Although these time intervals are telemetered to the ground, they are also used to control the east–west pointing of the antennas by varying the motor torque.

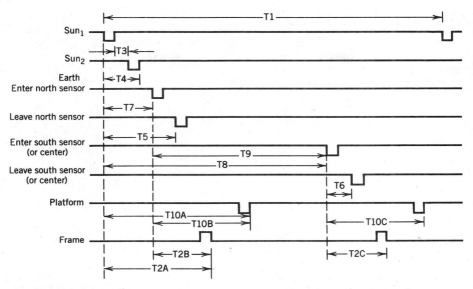

FIGURE 5.37 Time interval measurements from attitude control data. Arrow lengths indicate time intervals measured and telemetered; A, sun modes; B–north earth sensor mode; C–south or center earth sensor mode.

Some attitude data may be transmitted in real time. Figure 5.38 shows the spectrum for one particular satellite. A pilot tone standard is transmitted at 13.5 kHz, the command execute signal is sent at 13.838 kHz, earth pulses are centered at 14.222 kHz, and the sun sensor signals at 15.059 kHz. These frequencies are all chosen so that they fit into the standard IRIG channel 13, which extends from 13.412 kHz to 15.588 kHz. The limits for other channels are shown in Table 5.7.

The early satellites had a small number of channels, each fixed to a specific

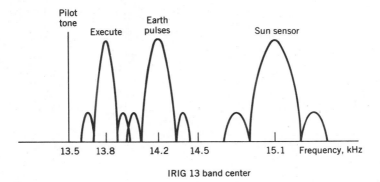

FIGURE 5.38 Real-time pulse spectrum transmitted in IRIG subcarrier channel 13 (Inter-Range Instrumentation Group). Execute signal is for verification.

TABLE 5.7 IRIG Subcarrier Channels[a]

Band	Center Frequency (Hz)	Lower Limit (Hz)	Upper Limit (Hz)	Maximum Deviation (%)	Frequency Response[b] (Hz)
1	400	370	430	±7.5	6.0
2	560	518	602	±7.5	8.4
3	730	675	785	±7.5	11
4	960	888	1,032	±7.5	14
5	1,300	1,202	1,399	±7.5	20
6	1,700	1,572	1,828	±7.5	25
7	2,300	2,127	2,473	±7.5	35
8	3,000	2,775	3,225	±7.5	45
9	3,900	3,607	4,193	±7.5	59
10	5,400	4,995	5,805	±7.5	81
11	7,350	6,799	7,901	±7.5	110
12	10,500	9,712	11,288	±7.5	160
13	14,500	13,412	15,588	±7.5	220
14	22,000	20,350	23,650	±7.5	330
15	30,000	27,750	32,250	±7.5	450
16	40,000	37,000	43,000	±7.5	600
17	52,500	48,562	56,438	±7.5	790
18	70,000	64,750	75,250	±7.5	1,050
19	93,000	86,025	99,975	±7.5	1,395
A[c]	22,000	18,700	25,300	±15	660
B	30,000	25,500	34,500	±15	900
C	40,000	34,000	46,000	±15	1,200
D	52,500	44,625	60,375	±15	1,600
E	70,000	59,500	80,500	±15	2,100
F	93,000	79,050	106,950	±15	2,790

[a] J. H. Crow, "FM Data Systems," in *AEROSPACE TELEMETRY Volume II*, Stiltz ed., (c) 1966, p. 62. Reprinted by permission of Prentice-Hall, Inc., Englewood Cliffs, New Jersey.

[b] Frequency response given based on maximum deviation and deviation ratio of 5.

[c] Bands A–E are optional and may be used by omitting adjacent bands, as shown in the following:

Bands Used	Omit Bands
A	13, 15, B
B	14, 16, A, C
C	15, 17, B, D
D	16, 18, C, E
E	17, 19, D, F
F	18, E

transducer. Each transducer was sampled in turn and the result transmitted to the ground. The sampling rate for each transducer was the same. Analysis of telemetry data shows that there are some quantities that change very slowly, so that the actual information content is quite low. Later satellites have had additional capabilities, such as using one channel for sampling additional transducers, which are thus sampled at a slower rate. Another possibility is to stop the main sampling procedure at any desired channel, dwelling on it for a period of time, and thus sampling at a much higher frequency.

With the advent of on-board microprocessors, additional flexibility is possible. Transducers can be sampled at different rates. It is also possible to change the range or the accuracy of a measurement and not be confined to a fixed calibration curve. Data compression is possible by omitting the transmission of data when the value does not change (or when it can be predicted). One example of this is shown in Figure 5.39. The solid circles are values transmitted when the values were changing rapidly. Once a data point is transmitted, it defines a range, and as long as points fall within this range (open circles) they do not need to be transmitted. Of course, if the data points are being transmitted at odd intervals, each one must be tagged so that its origin is known. This procedure of data compression is similar in nature to speech-predictive encoding (SPEC), in which voice signals are transmitted in a smaller bandwidth than would otherwise be possible.

5.4.2.3 Telemetry System

Table 5.8 shows a link budget for a telemetry system at 4 GHz. The link budget of a radio transmission link is used to determine either the power level that must be transmitted in order to provide adequate received power after the losses and gains encountered along the link or the power that will be received for a given transmitted power.

Link budgets are usually calculated in decibels because the calculations generally involve multiplication by fractions; such calculations are more easily handled by logarithmic means since values can be added instead of multiplied. Any ratio

• Significant samples

o Redundant samples

FIGURE 5.39 Data compression possible by transmitting only samples with significant changes.

TABLE 5.8 Link Budget for Typical 4 GHz Telemetry Subsystem[a]

	Absolute Value	Decibel Notation
Spacecraft beacon EIRP	1 W	0 dBW
Wavelength, λ	0.075 m	—
Path loss	$\lambda^2/(4\pi S)^2$	−196.7 dB
Range, S	40,586,000 m	—
Rain loss	0.8	−1 dB
Polarization loss	0.5	−3 dB
Earth station antenna gain	$4\pi A_e/\lambda^2$	52 dB
Effective area, A_e $= 128 \times 0.55$	70 m^2	—
Received power, C	1.3×10^{-15} W	−148.7 dBW
Earth station noise temperature, T	80 K	19.0 dBK
Boltzmann constant, k	1.38×10^{-23} W s/K	−228.6 dBWs/K
Noise density, $N_0 = kT$	1.1×10^{-21} W s	−209.6 dBW/Hz
Carrier/noise density	C/N_0	60.0 dBHz
Receiver intermediate-frequency bandwidth	100 kHz	50 dBHz
C/N in intermediate frequency		10.9 dB
Subcarrier bandwidth ratio	100,000 Hz/500 Hz	23 dB
E_b/N_0 in 0.5 kHz	—	33.9 dB
E_b/N_0 required for BER of 10^{-6}	—	16 dB
PCM margin	—	17.9 dB

[a] Abbreviations: C/N, carrier-to-noise ratio; BER, bit error rate; E_b/N_0, ratio of bit energy to noise power density; PCM, pulse code modulation.

R can be expressed in decibel form as

$$r = 10 \log_{10} R \quad (dB) \tag{5.3}$$

A similar expresison is used for absolute quantities, so that a power in, say, watts can be expressed in terms of decibels with respect to 1 W, that is, decibel watts. For example, since the logarithm of unity is zero, a power of 1 W can be expressed as 0 dBW. Similarly, a power of 100 W can be expressed as 20 dBW because the logarithm of the ratio 100:1 is 2.

The wavelength λ for any frequency is given by

$$\lambda = \frac{c}{f} \tag{5.4}$$

where f is the frequency and c is the velocity of light. Using $c = 3 \times 10^8$ m/sec and f expressed in hertz, λ will be in meters. Thus, the wavelength at 4 GHz (4 $\times 10^9$ Hz) is 0.075 m.

In the example in Table 5.8 the equivalent isotropic radiated power (EIRP) of a spacecraft beacon is assumed to be 0 dBW; that is, to the receiver, it appears that after all link losses and gains are considered, the beacon radiated 1 W in all directions. This is the equivalent, as far as the receiver is concerned, to the actual beacon transmitting power P_t multiplied by the transmitting antenna gain G_t.

The equivalent power spreads out into an equivalent sphere with a surface of $4\pi S^2$, where S is the radius of the sphere. If the receiving antenna has an effective area A_{er}, the power received, P_r, is

$$P_r = \frac{P_t G_t A_{er}}{4\pi S^2} \tag{5.5}$$

where P_t is the transmitter power and G_t the transmitter antenna gain. The relation between antenna gain and effective area is

$$G = \frac{4\pi A_e}{\lambda^2} \tag{5.6}$$

so the power received can be expressed as

$$P_r = P_t G_t G_r \frac{\lambda^2}{(4\pi S)^2} \tag{5.7}$$

This is the form normally used in link budgets, but in decibels the quantities add rather than multiply. The last part of the equation, $\lambda^2/(4\pi S)^2$, is known as the path loss or the space loss. The distance from a satellite at geostationary orbit to an earth station ranges from a minimum of 35786 km at the subsatellite point to a maximum of about 40587 km if the minimum elevation angle is 10°. At 4 GHz, with a wavelength of 0.075 m, the maximum space loss is

$$\frac{\lambda^2}{(4\pi S)^2} = 2.16 \times 10^{-20} \tag{5.8}$$

which is equivalent to -196.7 dB.

The effective area of an antenna, A_e, is the product of the efficiency η and its physical size A_p, or

$$A_e = \eta A_p \tag{5.9}$$

A 12.8-m antenna (42 ft) has a physical area of 128 m²; with an efficiency of 0.55 the effective area is 70 m², and the antenna gain is 52 dB. Starting with a beacon

EIRP of 0 dBW, with a space loss of -196.7 dB, an estimated rain loss of -1 dB, and a polarization loss of -3 dB, the result is a received power of -148.7 dBW. This is equivalent to -118.7 dBmW (generally abbreviated to dBm) or 1.3×10^{-12} mW.

Whether a certain power in the receiver is enough depends on how much noise there is. The receiver noise may be expressed as a noise temperature T, that is, the receiver noise is equivalent to that received if the antenna were looking at a surface with a particular temperature. The actual temperature of the satellite background may be considerably lower (unless the sun is behind the satellite, as seen from the earth station), and the main noise may be a contribution from the noise temperature of the earth, as seen by the antenna sidelobes. The available noise power density, N_0, at the antenna terminal is

$$N_0 = kT \qquad (5.10)$$

where k is the Boltzmann constant ($=1.38 \times 10^{-23}$ W sec/K). For a noise temperature of 80 K, the antenna noise power density is 110×10^{-23} W sec, -209.6 dBW/Hz, or -179.6 dBmW/Hz. The noise power spectral density is assumed to be flat up through the gigahertz range, and the noise power received is proportional to the bandwidth.

Continuing the example for 4 GHz in Table 5.8, the carrier-to-noise-density ratio (C/N_0) is equal to the received power (-148.7 dBW) minus the noise density (-209.6 dBW/Hz) decibel value, or 60.9 dBHz. If the receiver intermediate-frequency bandwidth is 100 kHz, the carrier-to-noise ratio (C/N) in the intermediate frequency is 10.9 dB. A pulse train with 1000-bit/sec data rate was modulated on a subcarrier with a bandwidth of 500 Hz; the ratio of bit energy to noise power density is 33.9 dB.

The error rate of pulse detection will depend on the encoding and decoding methods used. In digital phase-shift keying (PSK) digital information is transmitted in pulses of constant amplitude, angular frequency, and duration but of different relative phase. In differentially coherent detection a phase comparison is made of successive samples, and the phase transitions between carrier pulses are detected. The theoretical probability of error, P_e, for phase reversal PSK is

$$P_e = \tfrac{1}{2} e^{-E_b/N_0} \qquad (5.11)$$

for a given E_b/N_0. This is plotted in the center curve of Figure 5.40 along with two other methods of encoding and detection. If a BER of 10^{-6} is desired (one error in every million pulses), an E_b/N_0 of at least 11 dB is required; for a more practical system a value of 16 dB is used. Since the actual ratio is 33.9 dB, there is still a margin of 17.9 dB in the link budget. For telemetry there is a high level of redundancy; often there is little change between samples of the same point, and some errors can, and often are, tolerated. For command, accuracy requirements are much higher, and error rates must be lower.

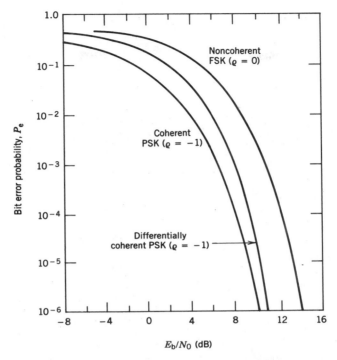

FIGURE 5.40 Error rates for different ratios of energy per bit to noise density (Lawton, 1958). Exact values depend on method used in digitizing information. See also Figure 4.49 on p. 490.

5.4.2.4 Antenna Patterns

The patterns used for a telemetry antenna must consider both normal and abnormal operation. During normal operation a communications satellite has a group of antennas oriented toward the earth; when the spacecraft antenna for the rf beacon is focused on the earth station, it is most efficient. However, during transfer orbit, injection, or some abnormal period, it is possible that the antennas are not oriented toward the earth. At such times it is critically important that telemetered information be available. Antenna patterns for both a spin-stabilized and a three-axis-stabilized satellite are shown in Figure 5.41. In each case there is an earth coverage horn for telemetry during normal operation. There is also a bicone toroidal pattern antenna for abnormal operation. In the spin-stabilized satellite the toroidal pattern is such that coverage would tend to be maintained even if the despun platform started spinning as long as the body axis remained in the north–south direction. For the three-axis-stabilized satellite tumbling motion would probably interrupt continuous telemetry coverage, but the toroidal pattern would usually intersect the earth during part of each rotation.

To ensure the telemetry reliability, there is redundancy in the telemetry beacons

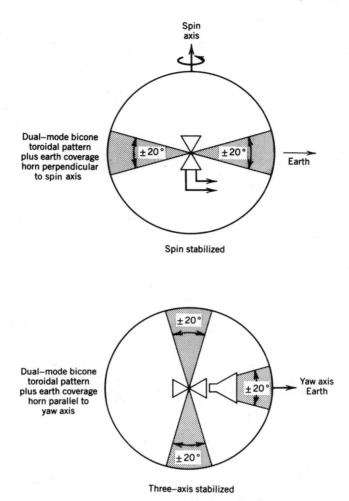

FIGURE 5.41 Radiation patterns for typical telemetry antennas.

and the connections to the antennas, as shown in the example in Figure 5.42. There are two separate beacons. In normal mode, the signal from each beacon goes directly into global and spot beams that are beamed at the earth. If attitude control is lost, the signal from each beacon can be amplified by a TWT and fed to the bicone antenna. This is based on the reasonable assumption that if the bicone antenna is needed due to loss of attitude control, the power amplifiers are not useful for communications. The narrow-beam communications antennas are not pointed at the earth, and the power amplifier might as well be used to amplify the telemetry signal (the bicone antenna does not have the antenna gain of a spot beam).

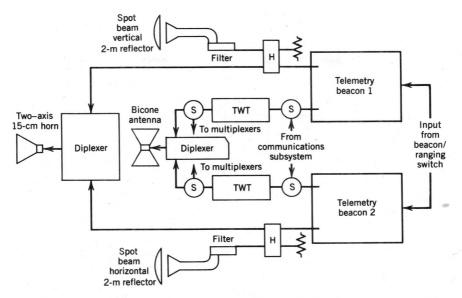

FIGURE 5.42 Telemetry beacons and antennas. Normal mode: global horn and spot beams for telemetry, and TWTs for communications. Abnormal mode: bicone antenna and TWTs for telemetry.

5.4.2.5 Distributed Telemetry System

Early satellites had wires running from all transducers to a central multiplexer and encoder. With the advent of integrated circuits and digital circuits, it became more advantageous to do some of the processing in remote encoders. The main features are shown in Figure 5.43. A number of remote encoders are located in different parts of the satellite, usually one to each subsystem. The desired sensor data are gathered at each location, encoded, and transmitted over a data bus to a central encoder; the latter does the final processing and then passes the final signal on to a telemetry transmitter. The use of redundancy is shown in the figure, since there is a duplicate remote encoder at each location, plus duplicate buses, duplicate central encoders, and duplicate transmitters. Furthermore, there is cross-strapping after the central encoders so that either transmitter can transmit information from either central encoder. Notice that the data bus is not a large bundle of wires, one for each transducer, but rather a single channel over which signals from different transducers are transmitted in turn.

There are several advantages of the distributed telemetry system. The harness has a lower weight and is easier to manage; separate wires are not needed from each transducer to the central encoder. The modularity provides flexibility and added test benefits; it is easier to add transducers when needed, and the remote encoder can be used for data collection during subsystem tests and is itself tested.

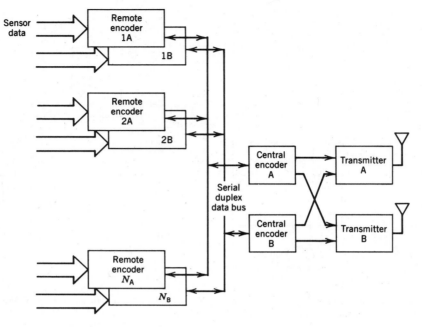

FIGURE 5.43 Distributed telemetry system with full redundancy.

By using programming control in the remote encoder, changes in the processing of data can be made in the software without any hardware changes, a real benefit in the last stages of satellite construction. Finally, the distributed telemetry gains in reliability by having many autonomous remote encoders; trouble in one is unlikely to affect the operation of the others.

5.4.3 Command

In some respects the command system is similar to the telemetry system. Instead of collecting information from various points in a satellite and transmitting it to the ground, the command system receives information from the earth station and distributes it to various points in the satellite. The number of different commands is usually smaller than the telemetry list, but the reliability must be higher. In the telemetry a small but measureable error rate can be tolerated; but in the command system special pains are taken to prevent an incorrect command from being executed.

A block diagram of a typical communications satellite command system is shown in Figure 5.44. For redundancy there are two active receivers and two active command decoders with cross-strapping. The decoded outputs may be put into a voting circuit and are accepted only if both decoders have the same answer. The ranging signal goes directly to the telemetry beacon for retransmission. After each

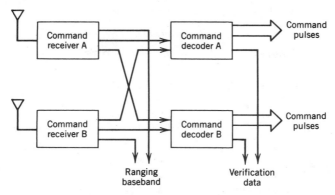

FIGURE 5.44 Typical command system for communications satellite.

command is decoded, it is sent back to the earth station for verification; only after it has been verified on the ground is the "execute" signal transmitted and the command actually carried out.

A list of typical items on a command list is given in Table 5.9. Each satellite will have a different list depending on its own construction. For example, spin-stabilized satellites have BAPTA (bearing and power transfer assembly), whereas body-stabilized satellites have solar array drives. The location of heaters will depend on the spacecraft thermal design.

The performance for a typical command system is given in Table 5.10. Compared to the telemetry system, the frequency is usually higher, the bit rate is lower, and the word length is longer. The data modulation is RZ rather than NRZ (see Figure 5.36).

The link budget for a typical command system (shown in Table 5.11) is similar to the budget for the telemetry system. The ground station transmitter has considerably greater power than the spacecraft beacon transmitter. In addition, because of the higher frequency used for the up link and because larger antennas can be used on the ground, the EIRP from the earth station is greater. However, these advantages are partially balanced by the increased path losses at the higher up-link frequency. The input power at the spacecraft receiver is considerably greater (than at the ground station telemetry receiver), but the noise density is also greater. Not only does the command receiver have to look at the earth (which is a 290-K noise source), but also the receiver has greater constraints on power and mass than the restraints applicable to the ground station receiver. The result is a noise temperature of 2000 K used for the command receiver, compared to 80 K for the telemetry receiver. It doesn't have to be this high, but this does allow also for possible terrestrial interference sources. The final result is a C/N for the command link of 7.2 dB, fairly similar to the telemetry C/N ratio of 10.8 dB.

Typical antenna patterns are shown in Figure 5.45. For a spin-stabilized satellite a toroidal pattern includes the earth as long as the spin axis is in the north–

TABLE 5.9 Typical Items on Command Lists

Communications

Receiver	On–off
Transponders	On–off
Switch matrices	Switch status
Spot beams	Pointing

Electric Power

Array drive	Normal, fast forward, reverse, heater
Battery	Charge, recondition, heater
Power conditioning	Control

Telemetry, Tracking, and Command

Beacon	On–off, modes
Encoder	On–off, modes
Command	Execute

Propulsion

Valves	Interconnect, cut off
Line	Heater
Thrusters	On–off
Apogee motor	Fire, heater

Orientation

Earth sensors	On–off, mode
Sun sensors	On–off, mode
Wheels	On–off
BAPTA[a]	Torque, heater
Solar array[b]	Torque motor

[a]Bearing and power transfer assembly, for spin stabilized satellites.

[b]Body stabilized satellites.

south direction. The pattern shown for a three-axis-stabilized satellite is a cardioid of revolution and is an omnidirectional antenna except for a cone of silence where the spacecraft shadows the antenna.

The format of a command is illustrated in Figure 5.46. In this example there is a preamble, thirteen 0's and a 1, to reduce the chances of a stray signal being interpreted as a command. Each command has a spacecraft address and the actual command; the memory load command is followed by the data to be loaded; any other command may be followed by another command. Commands are stored in

TABLE 5.10 Performance of Typical C-Band Command System

Carrier frequency	6423 MHz
Intermediate-frequency bandwidth	1.5 MHz nominal
Receiver threshold	−105 dBm
Receiver noise figure	9 dB
Carrier modulation	FM
FM deviation	400 kHz peak
Data modulation	Pulse code modulation/ return to zero (PCM/RZ)
Bit rate	≤ 100 bits/sec
Baseband frequency	200 Hz to 30 kHz
Word length	25 bits, typical
Verification	Full verification via telemetry
Command baseband	One tone, zero tone, execute tone in range 5 kHz to 15 kHz
Execute output	Equal to duration of execute tone

TABLE 5.11 Link Budget for Typical 6-GHz Command System

	Absolute Value	Decibel Notation
Transmitter power, P	245 W	23.9 dBW
Wavelength, λ	0.05 m	
Earth station antenna gain	$4\pi A_e/\lambda^2$	55.5 dB
Effective area, A_e	70 m^2	
Earth station EIRP	—	79.4 dBW
Rain loss	0.63	−2 dB
Path loss	$\lambda^2/(4\pi S)^2$	−200 dB
Range, S	40,586,000 m	
Satellite antenna gain	0.63	−2 dBi
Coaxial loss	0.63	−2 dB
Receiver input power	2×10^{-10} mW	−126.6 dBW
Noise temperature, T	2000 K	33.0 dBK
Boltzmann constant, k	1.38×10^{-23} W s/K	−228.6 dB Ws/K
Noise density, kT	2.76×10^{-20} W s	−195.6 dBW/Hz
Receiver intermediate-frequency bandwidth	1.5 MHz	61.8 dBHz
Receiver noise power	4×10^{-11} mW	−133.8 dBW
Carrier to noise		7.2 dB

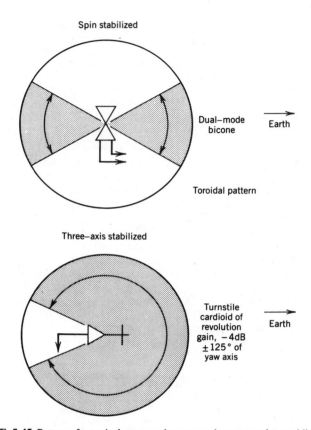

FIGURE 5.45 Patterns for typical command antennas that are nearly omnidirectional.

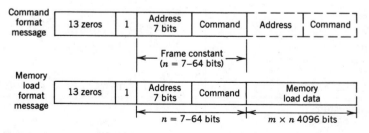

The symbol n is assigned a single value for any given spacecraft.

The symbol m is an integer and is assigned not more than four values for any given spacecraft.

*Memory load instruction command.

**There will be a break of subcarrier between messages.

FIGURE 5.46 Command format and memory load format.

memory, retransmitted to the ground, and only executed when an execution signal is received from the earth station.

Just as was true of the telemetry system, a distributed command system has a number of advantages. As shown in Figure 5.47, a central decoder sends commands over a data bus to remote decoders located in different parts of the spacecraft. The remote decoder recognizes the commands intended for that location by the address and then executes the necessary instruction. For reliability, there is redundancy of command receivers, central decoders, data buses, and remote decoders. The advantages include the lower weight of harness, ease of management, flexibility of modules, better subsystem testing, and added reliability.

A central command decoder is shown in Figure 5.48. Digital data is sent directly to its destination. Commands are checked for the correct preamble, stored in registers, and transmitted to the telemetry system for verification. They are then sent to the remote decoders shown in Figure 5.49. The remote decoder stores the command temporarily in a buffer, decodes it, and sends it on to the proper location.

5.4.4 Tracking and Ranging

There are two routine methods to find the position of a satellite: the angular coordinates to a satellite can be measured from an earth station antenna (angle tracking), and the distance to the satellite can be measured by finding how long it takes a signal to go to a satellite and return (ranging). Other methods are possible, but they are not used routinely by operators of communications satellite systems. Radar can detect a satellite in geostationary orbit. Communications satellites with earth-pointing reflectors have a large radar cross section; the radar signal is collected by

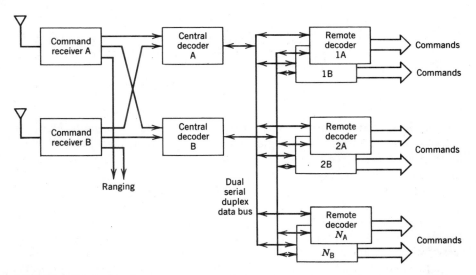

FIGURE 5.47 Distributed command system with full redundancy. Central decoders may include a decryption unit.

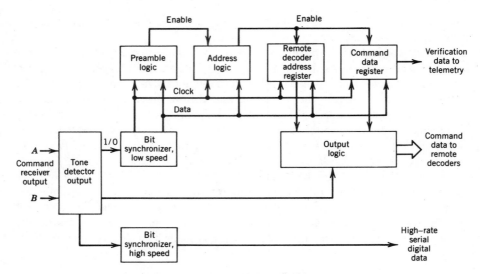

FIGURE 5.48 Central decoder for distributed command system.

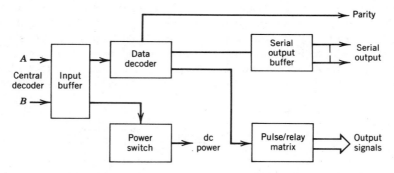

FIGURE 5.49 Remote decoder for distributed command system.

the antenna, reflected by the feed, and returned toward the earth. Powerful telescopes and sensitive photoimaging tubes used in a ground-based electrooptical deep-space surveillance (GEODSS) system can detect sunlit objects as small as one meter in diameter in geostationary orbit.

5.4.4.1 Angle Tracking

Angles to a satellite are measured at fixed time intervals by TT&C stations. The elevation is the angle from the horizon to the satellite; the azimuth is the angle from geographic north to the projection of the satellite's direction on the horizontal plane measured clockwise. The simplest measurement technique would be to monitor the signal from the satellite, point the antenna for maximum intensity, and read the antenna's azimuth and elevation.

For increased accuracy, multiple-feed horns are used that effectively create two antennas with patterns in slightly different directions. The principle for a single plane is shown in the top of Figure 5.50. Two signals (one from each antenna) enter a beam-forming network, which produces both the sum and the difference. An automatic gain control (AGC) is used to determine the ratio of difference to sum, thus making the error signal independent of the signal strength. This error signal is then used to drive the antenna until the error signal is zero.

To measure both the azimuth and elevation, it is necessary to determine the errors in two different planes. This could be done with two dual-channel systems or, more simply, with the triple-channel system shown in the bottom of Figure 5.50. Four different signals come from four separate horns (A, B, D, and C); one error signal determines the elevation ($A + B - D - C$) and another determines the azimuth ($A + D - B - C$). The sum of all four signals is used with an AGC for normalization of the error signals.

Angle measurements from a single earth station taken over a day or more are sufficient to determine a satellite's orbit. Astronomers have done this for centuries. Measurements from additional earth stations can be used for increased accuracy. Typical angular resolution for a 12.8-m antenna (42 ft) is $\pm 0.01°$.

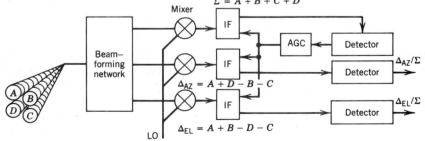

FIGURE 5.50 Simplified monopulse tracking system. Equipment used to generate error signals in azimuth and elevation for tracking of rf transmitter. Abbreviations: AGC, automatic gain control; IF, intermediate frequency; LO, local oscillator.

5.4.4.2 Ranging

One method of measuring range is to transmit a single frequency and measure the phase of the returned signal. A high frequency is needed for accuracy, but lower frequencies are needed to resolve ambiguities, so a number of different frequencies are needed. Another method is to transmit pseudorandom digital pulses and measure the correlation of the transmitted and returned signal. Communications satellites that use TDMA are more dependent on the ranges to earth stations than those that use frequency division multiple access (FDMA). The switching part of the communications package sometimes has need for ranging data and can also be used to provide ranging data.

The Intelsat/Comsat ranging system uses four frequencies: 27,777, 3968.25, 283.447, and 35.431 Hz; these are generated from 1 MHz divided successively by 36, 7, 14, and 8. The highest frequency f is 27.777 kHz, which has a wavelength λ of 10.8 km. If the phase is measured to an accuracy ($\Delta\phi$) of $\pm 1°$, the range accuracy is

$$\Delta S = \lambda \frac{\Delta\phi}{360°} = \pm 30 \text{ m} \tag{5.12}$$

This accuracy exceeds any requirements for antenna pointing data. Additional accuracy could be obtained using higher frequencies, but fluctuations of transit time through the ionosphere become a limiting factor. Range data from a single earth station does not completely determine a satellite orbit (it cannot separate daily oscillations due to inclination and those due to eccentricity), so the data must be combined with range data from other earth stations or with angle-tracking data.

Any phase measurement of a sine wave has an ambiguity equal to the wavelength. For example, if a phase angle of $225° \pm 1°$ is measured on a tone frequency of 27,777 Hz, as shown in Figure 5.51, the phase and distance can be written as

$$\text{Phase} = 360°j + 225° \pm 1° \tag{5.13}$$

$$\text{Distance} = 10j + 6.25 \pm 0.03 \text{ km} \tag{5.14}$$

where j is any integer. The first term shows the ambiguity, the second term is the actual measurement, and the third term is the accuracy. In the figure the distance might be 6.25, 16.25, 26.25, etc. The ambiguity can be resolved by a measurement on the lower tone frequency of 3968.25 Hz, also shown in Figure 5.51. If the phase measurement is $135° \pm 1°$, the phase and distance are

$$\text{Phase} = 360°i + 135° \pm 1° \tag{5.15}$$

$$\text{Distance} = 70i + 26.25 \pm 0.2 \text{ km} \tag{5.16}$$

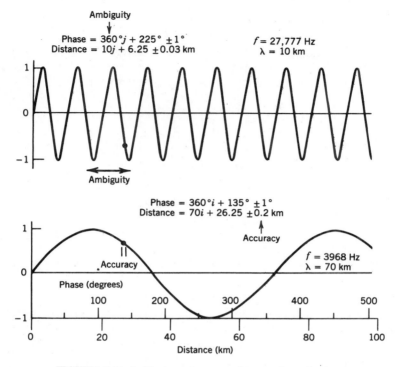

FIGURE 5.51 Ambiguity and accuracy in range determination.

where i is another integer. Comparison of these two measurements shows that j cannot be 0 or 1, but j can be 2, 9, 16, and so on; instead of 10 km, the ambiguity is now up to 70 km, the wavelength of the lower frequency. In the figure, the distance can only be 26.25 or 96.25 km.

To resolve an ambiguity and to be useful, the phase measurement, $\Delta\phi$, on the lower frequency must be accurate to

$$\Delta\phi = \pm \frac{180°}{a} \qquad (5.17)$$

where a is the ratio of frequencies. For a ratio of 7 the phase accuracy must be at least $\pm26°$. Actually, the phase is usually measured to an accuracy of a few degrees or better, so the ambiguity is easily resolved.

A block diagram of a multitone ranging system is shown in Figure 5.52. A stable crystal oscillator generates a 1-MHz frequency, which is then divided to obtain the desired frequencies. These signals are then filtered to obtain pure sine waves and improve the accuracy in phase measurement. The signals are then sent to the satellite, and upon return they are filtered once more to eliminate any distor-

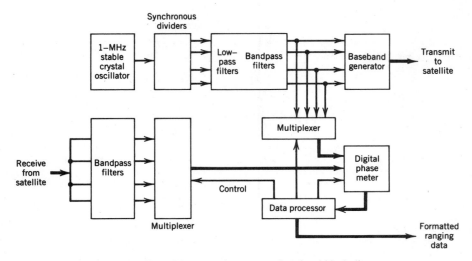

FIGURE 5.52 Multitone ranging system, functional block diagram.

tion they may have suffered in transit. A digital phase meter measures the phase change, and multiplexers are used so that one phase meter can be used for all four frequencies. The results go into a data processor that can then provide the range and range rate (velocity toward or away from the station).

Other ranging methods are used. The NASA standard system, GRARR (Goddard Range and Range Rate), uses a hybrid harmonic/pseudorandom baseband signal system. The tone frequencies used are 500, 100, 20, and 4 kHz and 800, 160, 32, and 8 Hz. The harmonic ambiguity is $\pm 18{,}740$ km based on the lowest frequency of 8 Hz, but by use of the pseudorandom signals, the ambiguity is increased to 1,213,600 km; this may be useful for a spacecraft going to Jupiter but is unnecessary for geostationary communications satellites.

5.5 SOLAR ARRAYS†

5.5.1 Introduction

Communications satellites require electric power to function. Even the passive ECHO balloon had solar cells to power a small rf beacon. Communications satellites use silicon solar cells as a primary energy source that transforms solar radiation into electric power. The amount of power generated depends on the intensity and direction of the solar radiation and the characteristics of the solar cells.

---

†Thanks to W. Billerbeck and D. Curtin for help in preparing Section 5.5.

5.5.2 Solar Radiation

The average power in the sun's radiation above the earth's atmosphere (air mass zero) is equal to

$$S_c = 137 \text{ mW/cm}^2 \qquad (5.18)$$

This power can be expressed in a variety of units, some of which are listed in Table 5.12.

Although the preceding number is often referred to as the "solar constant," it is not really constant (see Figure 5.1). The maximum power occurs at perihelion (closest approach to the sun) in early January, about 2 weeks after the winter solstice (December 23); the latter is when the sun has its most southern declination. The intensity of solar radiation in the vicinity of the planet Earth is

$$S = S_c \left(\frac{a}{r_s}\right)^2 \qquad (5.19)$$

where r_s is the radial distance from the earth to the sun and a is the mean distance (149,597,870 km).

Many communications satellites have solar arrays that are not oriented toward the sun in terms of north–south pointing. When the sun is shining on the array, the solar power is

$$S_{NS} = S_c \left(\frac{a}{r_s}\right)^2 \cos \delta_s \qquad (5.20)$$

where, for geostationary satellites, δ_s is the sun's declination. In general, it would be the angle between the normal to the solar array and the sun's rays.

The average intensity on satellite solar arrays is also reduced by eclipses, when the satellite is in the earth's shadow for part of its orbit. The daily average of

TABLE 5.12 Solar Constant[a]

Units	Value
Centimeter-gram-second	137 mW/cm^2
International System (SI)	1370 W/m^2
Heat	1.96 cal/min cm^2
English	434 Btu/hr ft^2

[a]Total radiant power per unit area due to sun at average distance to earth, outside the earth's atmosphere.

sunlight on a north–south solar panel is then

$$\overline{S}_{NS} = S_c \left(\frac{a}{r_s}\right)^2 \cos \delta_s \left[1 - \frac{1}{\pi} \cos^{-1} \frac{\left(1 - R^2/r^2\right)^{1/2}}{\cos \delta_s} \right] \qquad (5.21)$$

where R/r is the ratio of the earth's radius R to the geostationary orbit radius r
(0.1513). Note that if the argument of the arc cosine is greater than 1, there is no
eclipse; in a computer program choose between 1 and the argument shown, which-
ever is less. The average power is reduced about 5% during maximum eclipses;
fortunately, this reduction occurs when the sun's declination δ_s is near zero, so it
does not reduce the overall minimum. Equations 5.18 through 5.21 are plotted in
Figure 5.1 on page 542.

The spectral distribution for the sun's radiation is shown in Figure 5.53. The
energy of each photon in light is given by

$$E \text{ (eV)} = \frac{1.242}{\lambda(\mu m)} \qquad (5.22)$$

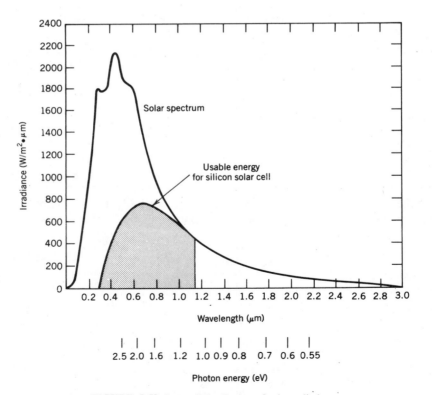

FIGURE 5.53 Spectral distribution of solar radiation.

The photon energy in electron volts is also shown in Figure 5.53. The radiation varies from 0.1 μm in the ultraviolet to 2 or 3 μm in the infrared and includes the visual spectrum.

5.5.3 Solar Cells

5.5.3.1 Physical Characteristics

A cross section of a typical silicon solar cell is shown in Figure 5.54. The cell is a single crystal of pure silicon doped so that most of it is p-type, in which positive electron holes carry most of the electricity. The top layer is doped differently, so it is n-type, where free electrons are the majority carriers. Free electrons are collected by a grid on top of the cell, which forms the negative electrode, and positive current is collected by conducting metal on the bottom of the cell (positive electrode). A cover slide is placed over the cell to protect it from particle radiation in space; an antireflective coating reduces the amount of light lost by reflection.

The principle on which a solar cell works is illustrated in Figure 5.55, where energy is shown as a function of depth in the solar cell. The crystalline nature of silicon allows electrons in the valence band and conduction band but not for intermediate energies in the energy gap. The valence band is almost full of electrons, but in p-type materials it contains electron holes that move around as a positive carrier. The conduction band is mostly empty of electrons, but in n-type materials there are a few free electrons that can conduct electric current. An intrinsic electric field keeps the free electrons in the n-type material from recombining with the electron holes in the p-type material. When sunlight is incident near the junction, electrons are knocked out of the valence band into the conduction band, creating free electrons and electron holes. Free electrons then flow into the n-type material, electron holes flow into the p-type material, and an electric current is thereby generated. The current is proportional to the number of photons absorbed, and the voltage is a function of the band gap of the material (Lindmayer and Wrigley, 1965).

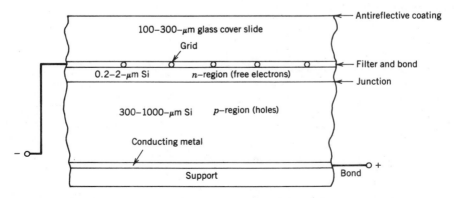

FIGURE 5.54 Schematic of typical single-crystal n-on-p silicon solar cell.

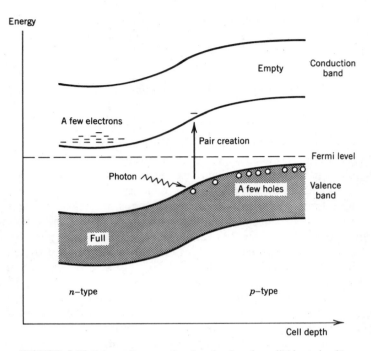

FIGURE 5.55 Energy diagram of *pn* junction in solar cell (short circuit).

5.5.3.2 Electrical Characteristics

The current–voltage characteristics of typical solar cells are shown in Figure 5.56 for 2 × 2-cm cells along with curves of constant power. The nonreflective cell has a short-circuit current of 182 mA and an open-circuit voltage of 590 mV. The maximum power output occurs at the knee of the curve and is about 85 mW for the nonreflective cell; based on a 4-cm² area, this is equivalent to 15.5% efficiency.

A simple equation for the current I of a solar cell is

$$I = I_L - I_0(e^{qV/akT} - 1) \tag{5.23}$$

where I_L = current generated by solar radiation
I_0 = reverse saturation current
q = charge of electron
k = Boltzmann constant
T = absolute temperature
a = adjustable constant on order of unity

The same equation can be written in terms of the voltage as

$$V = \frac{akT}{q} \ln \frac{I_L + I_0 - I}{I_0} \tag{5.24}$$

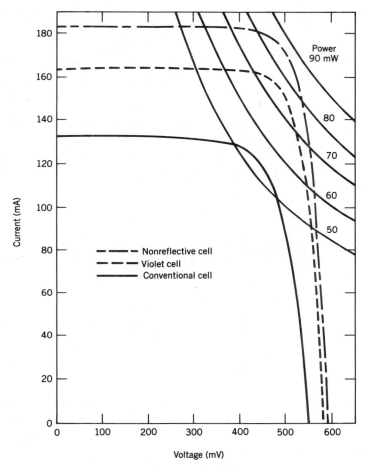

FIGURE 5.56 Current–voltage characteristics of nonreflective, violet, and conventional cells (Allison et al., 1975; reprinted by permission from *COMSAT Tech. Rev.*).

Note that I_0 is a strong function of temperature, so that, in general, the voltage decreases with increasing temperature. The actual *I–V* characteristics of a nonreflective cell at temperatures from 10 to 70°C is shown in Figure 5.57. The short-circuit current increases slightly with temperature, but because of the voltage decrease, the net effect on power is a sharp decrease of maximum power with increasing temperatures.

The points on the *I–V* curve most commonly measured are the short-circuit current I_{sc}, the open-circuit voltage V_{oc}, and the current and voltage at the maximum power point, I_{mp} and V_{mp}. The parameters for the simple equation can then be calculated by

$$I_L = I_{sc} \tag{5.25}$$

$$\frac{akT}{q} = \frac{V_{oc} - V_{mp}}{\ln\left[I_{sc}/(I_{sc} - I_{mp})\right]} \qquad (5.26)$$

$$I_0 = \frac{I_{sc}}{\left(e^{qV_{oc}/akT} - 1\right)} \qquad (5.27)$$

These equations do not require that I_{mp} and V_{mp} be the maximum power point, but the derivation does assume that $I_{sc} - I_{mp}$ is large compared to I_0.

There are many factors that cause deviations from the simple I–V equation shown in the preceding. Many alternatives have been suggested in the literature. One form

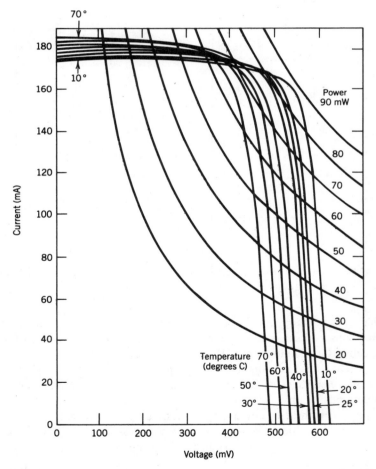

FIGURE 5.57 Current–voltage characteristics of nonreflective cell at temperatures from 10 to 70°C (Allison et al., 1975; reprinted by permission from *COMSAT Tech. Rev.*).

is

$$I = I_{\text{L}} - I_0 \left[\exp \frac{q}{akT} (V + R_{\text{s}}I) - 1 \right] + \frac{V}{R_{\text{sh}}} \tag{5.28}$$

where R_{s} is series resistance, and R_{sh} is shunt resistance. With five adjustable parameters, a better fit can be made to a measured I-V curve, but the equation becomes harder to handle.

5.5.4 Reliability and Lifetime

Damaging space particulate radiation consists primarily of penetrating electrons and solar flare protons. The maximum power output is shown in Figure 5.58 as a function of 1-MeV electron irradiation fluence. A cell that is well protected (spinner, and a coverslide some 300 μm thick) might receive the equivalent of 3 $\times$ 10^{14} e^-/cm^2 in 7 yr at geostationary orbit. A cell on a lightweight deployed solar array, with a thin coverslide and receiving radiation from both sides, may receive the same amount of radiation in 1 yr. The intensity of solar flare protons also varies from year to year, depending on the solar activity. While it is customary to plot radiation degradation with the electron fluence on a logarithmic scale, it can be misleading to an individual not used to this graph. If it takes 7 yr to reach 3 $\times$ 10^{14} e^-/cm^2, it takes 70 yr to reach 3 $\times$ 10^{15} e^-/cm^2. As time goes on cells

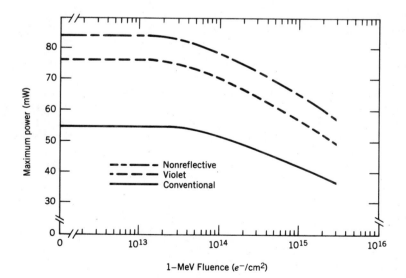

FIGURE 5.58 Maximum power output of nonreflective, violet, and conventional cells as function of 1-MeV electron irradiation fluence (Allison et al., 1975; reprinted by permission from *COMSAT Tech. Rev.*).

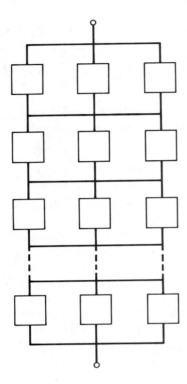

FIGURE 5.59 Typical method of connecting string of solar cells in parallel and series.

become more resistant to particle radiation, and the changes in later years are much less than the first year.

A complete solar array requires many thousands of solar cells. The number of cells in series is set by the desired array voltage. For reliability reasons, strings of cells are often configured as shown in Figure 5.59. A few cells are connected in parallel, and then these sets are connected in series. For maximum power the cells are matched so that I-V characteristics are compatible. A short circuit in a cell will cause a power loss from all the cells in parallel, but the string will still deliver power at a somewhat reduced current. The current I for a string with a shorted cell would be

$$I = N_{par} f\left[\frac{V_{bus}}{N_{ser} - 1}\right] \tag{5.29}$$

where N_{par} is the number of cells in parallel, N_{ser} is the number in series, V_{bus} is the bus voltage, and $f(\ )$ represents the I-V curve. If operation had been near the maximum power point, this is approximately

$$I \approx N_{par} I_{mp} \frac{N_{ser}}{N_{ser} - 1} \tag{5.30}$$

The power is lost from the shorted cell and the other cells in parallel that are also shorted, but the remaining cells deliver approximately the same power.

A less likely occurrence is for the cell to fail open, but this has more serious consequences. All the current must go through the other cells in parallel. If there are three cells in parallel and one fails open, the string current is reduced from three times the maximum power current to approximately twice the short-circuit current. The string tries to force more current through the two remaining cells. However, the *I–V* curve in Figure 5.57 shows that the current through the two remaining cells can only increase slightly. The voltage across these two cells drops and may even go below zero (reverse voltage). The power generated by the entire string depends on the current and drops to roughly two-thirds (65–75%). The power loss in such a string might be 25% even though only one cell in several hundred is defective.

5.5.5 Mass

The mass of solar arrays is shown in Figure 5.60. It is approximately proportional to the electric power delivered. The cylindrical solar arrays on spinning satellites have about three times the mass, primarily because they need over three times as many solar cells. Lighter deployed solar arrays are technically possible, and a few demonstrations have been flown.

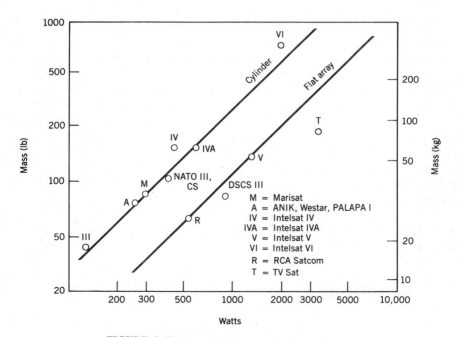

FIGURE 5.60 Mass of solar array (including substrate).

5.6 ELECTRIC POWER†

5.6.1 Introduction

Most communications satellites use silicon solar cells as a primary energy source and rechargeable batteries for eclipse operations. In spite of much research and development, nuclear power is not suitable for commercial communications satellites. Batteries are either nickel–cadmium (NiCd) or nickel–hydrogen (NiH). If a solar array, battery, and loads all operated at the same constant voltage, there would be no need for any power control; all the equipment could simply be wired to the same bus. However, the solar array delivers no power during eclipse, has a higher than usual voltage right after eclipse, and has a rather sharp maximum power point of operation. A battery requires different voltages for charging and discharging, has a declining voltage during discharge, and requires control to prevent overcharging or too deep a discharge. The loads on a communications satellite may vary as a function of time and require a variety of voltages, each maintained within specified limits. The function of the power system is to provide the power to all the loads, drawing power from the solar array and battery as required.

A simplified diagram of an electric power subsystem for a body-stabilized communications satellite is shown in Figure 5.61. It contains a solar array, battery, provision for recharging the battery, a bus with some voltage regulation, distribution to a variety of individual loads, electronic power conditioning (EPC) for each load, and power control through automatic devices and ground control. On body-stabilized satellites the solar arrays are usually on a sun-oriented flat panel that turns daily on a north–south axis. This requires a bearing, slip rings to transfer power, and a motor to turn the panel. The main array is made up of a number of strings of solar cells, and each string may have a blocking diode in case a cell in the string shorts to ground. The solar array was described in the previous section; the other power subsystem components are described in the following pages.

One or more buses transmit the electricity from the solar arrays to the individual loads. This bus may be regulated with the voltage maintained close to a single value, or it may be completely unregulated so that no attempt is made to control the voltage. A common compromise is to limit the upswings of the solar array with a voltage limiter but allow a voltage drop with battery operation because power is at a premium.

The battery operates in three modes: delivering power to the bus, recharging with power from the solar array, and reconditioning when it is discharged into a resistive load. It requires a higher voltage for charging, which may be provided by a boost regulator or by a secondary charging array (as shown). Provision must be made to control the charging current and to limit the duration of the charge. Reconditioning a battery, which discharges it beyond its normal depth of discharge, tends to rejuvenate a battery and extend its life.

The electric power is distributed throughout the satellite through the bus to

†Thanks to W. Billerbeck, N. Jacobus, and J. Stockel for help in preparing Section 5.6.

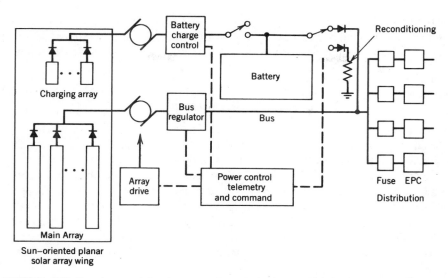

FIGURE 5.61 Components of electric power subsystem for body-stabilized satellite. Most loads have their own electric power conditioners (EPC), and a fuse or circuit breaker to protect the distribution bus.

individual loads, many with their own electric power control. While these controls may be shown as part of the electric power subsystem, they are physically located adjacent to their respective loads. The distribution includes a fuse or circuit breaker to limit the potential damage of a failure. A power control system monitors and telemeters voltages and currents to fully characterize the power subsystem. Some regulation is done automatically, and some actions are initiated by ground command. While Figure 5.61 shows only a single array, battery, and bus, many satellite power subsystems have two separate arrays, two batteries, and two buses, so that the satellite can operate even if a bus is shorted.

5.6.2 Differences on Spinner and Body-Stabilized Satellite

A communications satellite with a spinner configuration requires some differences in the electric power subsystem. The solar cells on a spinner are mounted on the outside of a cylinder, rotating around a north–south axis. At any one time, only a fraction of the cells are in sunlight, so that for a given electric power the number of cells required is increased by a factor of π (the ratio of the curved surface of a right cylinder to the projected area of the cylinder). Some advantages of a spinner are that deployment of solar arrays is frequently not necessary, and the solar cells tend to be at a lower temperature.

On a body-stabilized satellite, most of the electric power subsystem is on the main spacecraft frame, and only the solar arrays (with maybe some diodes and shunt regulators) are on a panel that rotates once a day. On a spinner, much of the electric power subsystem, including the solar array and battery, is on the spinning part that rotates about once a second. The despin part contains the communications payload, with the major part of power consumption, so the power flows through slip rings from the spinning side to the despun side.

5.6.3 Power Budget

Typical power requirements for a communications satellite of about 1 kW is shown in Table 5.13. Three columns show three critical times during the year. These are shown for the end-of-life (EOL) conditions, after the maximum expected degradation of the solar array by radiation. During the summer solstice, the power delivered by a north–south oriented array is at a minimum, but there is no eclipse. During autumnal equinox, there is an eclipse season; the power available is slightly greater when the array is sunlit, but extra power is needed for recharging the batteries each orbit. When in eclipse, power is only available from the batteries; this is less than that available from the solar arrays but is still sufficient to operate the satellite.

The communications payload requires the bulk of the electric power, often 70–80% of the total load (see Figure 5.62). The maximum eclipse is 5% of the orbit period, and battery recharging will require a greater percentage of the total load, on the order of 10%. The thermal control may or may not consume large amounts of electric power, depending on the use of electric heaters. The other subsystems of the satellite require lesser amounts of power.

In addition to the various satellite loads, a power margin is usually maintained. Frequently this is a 10% margin at launch plus another 10% that is used in the original design and may be absorbed by modifications during the construction and testing of the satellite. This power margin has proved adequate. Future television broadcast satellites will have higher powers, and the designers may not be able to afford the luxury of such a power margin.

TABLE 5.13 Electric Power Average Requirements (W)^a

	Autumnal Equinox	Summer Solstice	Eclipse
Communications	769	769	768
Telemetry, command, and ranging	39	39	39
Attitude control	48	73	48
Electric power	9	9	9
Thermal control	136	86	30
I^2R harness losses	10	10	9
Battery charging	100	30	—
Total load	1111	1016	903
Power margin	243	272	76
Total power	1354	1288	979

^{a} Power at end of life (EOL).

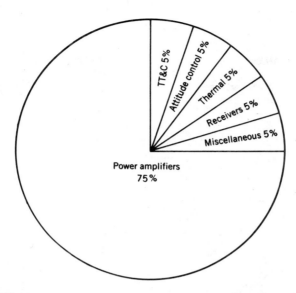

FIGURE 5.62 Primary power distribution in typical communications satellite.

5.6.4 Bus and Bus Regulation

The selection of bus voltage depends sometimes on nontechnical factors. Frequently it is based on existing equipment, already proven on another satellite. The main power-switching semiconductors are limited to approximately 100 V by present technology, although a few operating to 400 V exist. The minimum bus voltage is determined by the distribution losses that can be tolerated; low voltages require high currents and corresponding high I^2R losses. Small systems with short leads and low power can operate as low as 5 V; larger systems require higher voltages. In the 1960s, bus voltages of 20–30 V were common; present bus voltages are on the order of 40 V; in the 1990s bus voltages of 50–100 V are expected.

Once the system voltage is set, the number of solar cells and battery cells in series is determined. The number of solar cells in series is equal to the minimum bus voltage divided by the voltage per solar cell; the latter is the voltage at maximum power point (around 0.45 V) or close to it. The number of battery cells in series is equal to the minimum bus voltage divided by the minimum cell voltage (about 1 V).

If there is no regulation of bus voltage, there will be considerable fluctuations due to various causes, as pictured in Figure 5.63. During discharge a battery voltage decreases so that at the start of an eclipse the voltage would be higher than at the end of the eclipse (e.g., from 42 to 27 V). The solar array provides different voltages due to annual variations of solar intensity and long-term degradation due to radiation. In addition, there is a voltage peak when a satellite comes out of eclipse due to the lower temperature of the solar cells. This peak is especially high

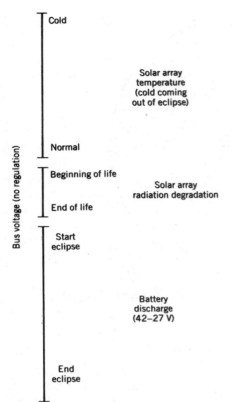

FIGURE 5.63 Possible bus voltage fluctuations due to solar array and battery.

for body-stabilized satellites because the lightweight, deployed arrays fall to lower temperatures during eclipse.

These inherent bus voltage variations can be controlled by various methods of voltage regulation, or they can be ignored, allowing the bus voltage to fluctuate. Many satellite components require regulated voltage, but this regulation either can be provided on the bus or provided in the individual electric power conditioner. In some electric power subsystems, the choice is made to limit the voltage due to solar array changes but allow the voltage to decrease when battery voltage falls. Excess array voltage usually means excess power so there is no great loss in dissipating or limiting the power when solar array voltages are high. On the other hand, power is at a premium during battery operation in eclipse, and dissipating power to reduce voltage would be undesirable. There are a variety of voltage regulation techniques; a few will be described here to illustrate possible methods.

The principle of a series regulator is shown in Figure 5.64. This is not practical for bus regulation, but is useful in understanding PWM regulators. A higher unregulated voltage is stepped down to a lower regulated voltage through a

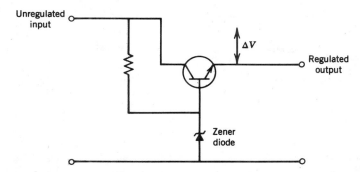

FIGURE 5.64 Series regulator that steps down to regulated voltage from higher unregulated voltage.

transistor. The transistor can be thought of as a variable resistance, that is, adjusted so that the voltage drop is equal to the difference between the unregulated input and the regulated output. Since there is a current I through a voltage difference ΔV, the series regulator does require power dissipation equal to $I \Delta V$.

To reduce energy losses, power conditioning systems make extensive use of switching methods. A pulse width modulation (PWM) regulator is outlined in Figure 5.65. This is similar to a series regulator, but instead of opening the valve (transistor) part way all the time, the valve (switch) is opened all the way part of the time. A small energy storage circuit is used to maintain the voltage when the switch is open. An inductance keeps the current flowing. When the switch is closed, the current tends to increase because the input voltage is higher; when the switch is open, the current comes through the diode from ground but decreases because the voltage is lower. By controlling the relative amount of time that the switch is open versus the time it is closed (the width of the pulse), the desired voltage can be maintained.

A simple shunt regulator, shown in Figure 5.66, is similar to the series regulator, but the control is in parallel to the load. This also is not practical for bus regulation,

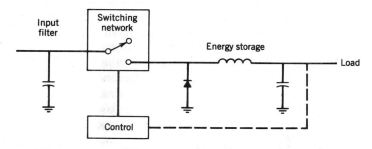

FIGURE 5.65 Main components of pulse width modulation (PWM) regulator.

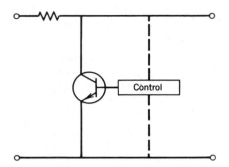

Reduces voltage by shunting
current

FIGURE 5.66 Basic principle of shunt regulator.

but is useful in understanding solar array shunts. Instead of producing a voltage drop from the bus to the load, it shunts the current to ground; with most power sources this produces the desired voltage drop. The power loss is equal to the voltage times the current going through the shunt. An interesting variation of a shunt regulator is shown in Figure 5.67. A shunt is used across half of a solar cell string; this reduces the voltage on one-half of the string and increases the voltage on the other half. In both cases the operating point is moved away from the voltage at which maximum power is produced, so both halves of the string produce less power, and hence the bus voltage is decreased.

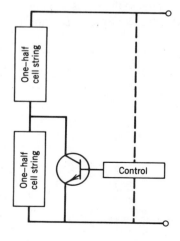

Reduces voltage by short–
circuiting one–half string

Used separately on different
strings

FIGURE 5.67 Special application of shunt regulator to solar array (repeated on many strings).

5.6.5 Rechargeable Batteries

5.6.5.1 Electrochemical Reaction

The usual nickel–cadmium battery is a sealed prismatic structure composed of alternating flat plates of sintered nickel electrodes and cadmium electrodes (see Figure 5.68). Between the plates is an inert separator and an electrolyte consisting of an aqueous solution of 35% potassium hydroxide. The electrochemical equation for the main reaction at the positive electrode (Ni) is

$$2NiOOH + 2H_2O + 2e^- \underset{\text{charge}}{\overset{\text{discharge}}{\rightleftharpoons}} 2Ni(OH)_2 + 2OH^-$$

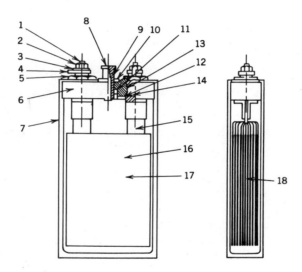

Parts list

Item no.	Description	Number required	Item no.	Description	Number required
1	Post	2	10	Vent washer	1
2	Post nut, top	2	11	Vent nut	1
3	Connector washer	2	12	Nut seal	1
4	Connector		13	Gland	2
5	Post nut, bottom	2	14	Post ring	4
6	Jar cover	1	15	Plate tab	
7	Jar	1	16	Plate, positive	
8	Vent plug	1	17	Plate, negative	
9	Vent ring	1	18	Separator	

FIGURE 5.68 Construction of typical wound-separator prismatic nickel–cadmium cell (Bauer, 1968).

where, on discharge, nickelic hydroxide and water are then changed to nickelous hydroxide and a hydroxyl ion. At the negative cadmium electrode the equation is

$$Cd + 2OH^- \; \underset{\text{charge}}{\overset{\text{discharge}}{\rightleftharpoons}} \; Cd(OH)_2 + 2e^-$$

where during discharge cadmium and a hydroxyl ion form cadmium hydroxide. The net reaction for the cell is

$$Cd + 2NiOOH + 2H_2O \; \underset{\text{charge}}{\overset{\text{discharge}}{\rightleftharpoons}} \; Cd(OH)_2 + 2Ni(OH)_2$$

As can be seen, the negative electrode is not pure nickel but a carefully prepared surface of nickel hydroxide.

An alternative to the nickel–cadmium cell is a nickel–hydrogen cell that is lighter and/or more reliable. The electrochemical equation for the main reaction at the positive electrode (Ni) is

$$NiOOH + H_2O + e^- \; \underset{\text{charge}}{\overset{\text{discharge}}{\rightleftharpoons}} \; Ni(OH)_2 + (OH)^-$$

which is exactly the same as for the nickel–cadmium cell. At the negative electrode the equation is

$$\tfrac{1}{2} H_2 + (OH)^- \; \underset{\text{charge}}{\overset{\text{discharge}}{\rightleftharpoons}} \; H_2O + e^-$$

Since hydrogen is a gas, the actual electrode is made from an inert material (Pt), and the gas diffuses to the electrode for the necessary reaction. The net reaction for the cell is

$$\tfrac{1}{2} H_2 + NiOOH \; \underset{\text{charge}}{\overset{\text{discharge}}{\rightleftharpoons}} \; Ni(OH)_2$$

Since hydrogen is needed for discharge and evolved during charging, the gas must be contained. A sketch of a nickel–hydrogen cell is shown in Figure 5.69, with the surrounding pressure vessel. The pressure is an accurate indication of the state of charge and reaches a maximum on the order of 30 atm (several hundred pounds per square inches) when the cell is fully charged.

5.6.5.2 Electrical Characteristics

The *I–V* characteristics of a battery can be represented by an electromotive force and a series resistance. The terminal voltage is a linear function of the discharge current, as shown in Figure 5.70. The voltage is a function of temperature, so the

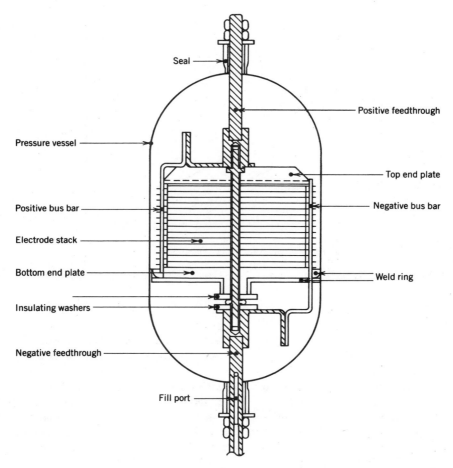

FIGURE 5.69 Typical physical arrangement of nickel–hydrogen cell (Esch et al., 1976; paper presented at the 27th IAF Congress, Anaheim, CA. Reprinted with permission).

terminal voltage V can be written as

$$V = V_G + V_T - IR \qquad (5.31)$$

where V_G is a voltage that is a function of the depth of discharge, V_T is a correction for temperature, I is the discharge current, and R the internal resistance.

The linear characteristic of the I–V curve applies only to a limited portion of the curve. In particular, it cannot be extrapolated to zero current because of effects of activation polarization and concentration polarization. The actual open-circuit voltage is higher, and so is the portion to the left of the axis (charging current).

To compare the performances of different batteries, it is customary to divide the current (amperes) by the battery capacity (ampere-hours), so that the perfor-

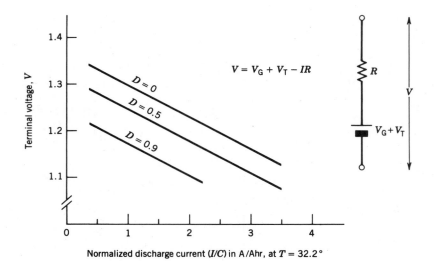

FIGURE 5.70 Ohmic losses in typical sealed nickel–cadmium cell for different depths of discharge *D*.

mance is independent of the size of the battery. For example, a current of 15 A for a 30 Ahr cell has a ratio of $1/2$; this current is often referred to as a $C/2$ current.

The general dependence of terminal voltage on depth of discharge D is shown in Figure 5.71. There is a sharp drop in voltage at the start of discharge, a much slower decline during the major part of the discharge, and then a gradually increasing voltage drop. The curve shown is for a typical sealed nickel–cadmium cell, and other cells would have different numerical values. These curves were fitted to a fifth-order polynomial, and the result was

$$V_G = 1.3597 - 0.83986D + 3.6972D^2 - 8.6476D^3$$
$$+ 9.9853D^4 - 4.4239D^5 \tag{5.32}$$

where D is the depth of discharge as a fraction. This curve should not be used in the deep-discharge region below 1 V per cell. The curve for V_G is shown as the top curve; as was noted before, this is not the open-circuit voltage because of polarization effects.

Similar curves are shown by Esch, Billerbeck, and Curtin, (1976) in Figure 5.72 for different temperatures. Most of this can be modeled by the temperature correction

$$V_T = -0.00072(32.2 - T_c) \tag{5.33}$$

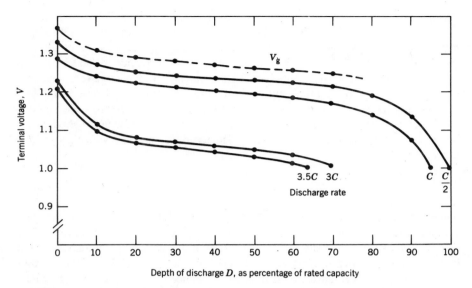

FIGURE 5.71 Discharge characteristics of typical Ni–Cd cell at 32.2 °C (100 °F) for different discharge rates (Billerbeck, 1979). A discharge rate of $C/2$ means 15 A for a 30 Ahr cell.

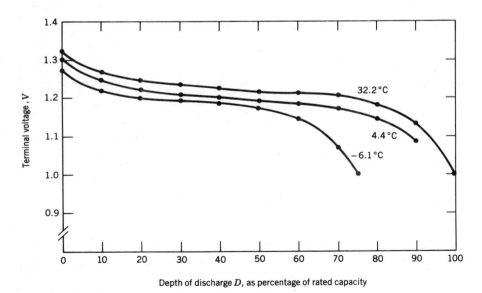

FIGURE 5.72 Discharge characteristics of typical sealed Ni–Cd cell at $I/C = \frac{1}{2}$ for various temperatures (Billerbeck, 1979). At lower temperatures there is less voltage and power, especially at higher depth of discharge.

where T_c is the end of discharge temperature in degrees celsius. The rate of change of voltage with temperature is 0.00072 V / °C. This is approximate, and the experimental points show that at low temperatures and at a high depth of discharge, the temperature drops far more than that predicted by the linear model. During discharge a battery cell dissipates heat, so that the end of discharge temperature is a function of the surrounding temperature and the thermal dissipation available.

The internal resistance R is approximately constant and has been measured as 7 mΩ for a typical 20-Ahr cell. For other capacities, the resistance should vary inversely with capacity. The internal resistance does increase with age and has been modeled as

$$R = R_0 + Kt^2 \qquad (5.34)$$

where $R_0 = 7 \pm 0.8$ mΩ, and $K = 0.6 \pm 0.2$ mΩ/yr². Figure 5.73 shows how the end-of-discharge voltage varies with time (Esch, Billerbeck, and Curtin, 1976). The flight data shows considerable variation from one battery to another. Note that batteries on a regular cycling schedule do not suddenly go bad. Rather, their capacity gradually decreases, their internal resistance increases, and the voltage at the end of a discharge cycle becomes smaller.

The nickel–hydrogen cell has different construction with different materials. It might be expected that the general features would be the same but that performance

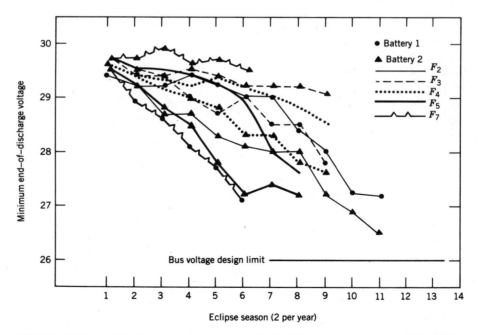

FIGURE 5.73 End-of-discharge voltages for some 21-Ah (15-Ah rated capacity) satellite Ni–Cd batteries operated at 7A discharge current (Billerbeck, 1979).

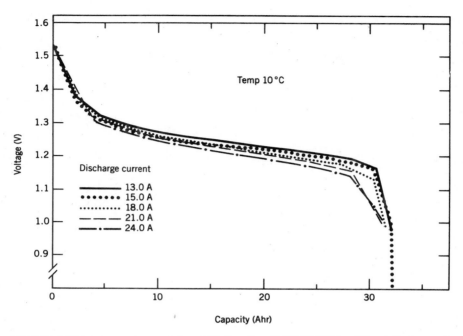

FIGURE 5.74 Discharge characteristics of Ni–H$_2$ cell at 10 °C (50 °F) for different discharge rates (Stockel, 1980).

would differ in some details. Curves for the voltage during discharge of a nickel–hydrogen cell are shown in Figure 5.74 for different discharge currents. Unlike the nickel–cadmium cell, the internal resistance appears to increase with depth of discharge. The discharge voltage for a nickel–cadmium and a nickel–hydrogen cell with about the same capacity and similar date of construction is shown in Figure 5.75. Comparison of these curves indicates that (at least for these cells) the voltage of nickel–cadmium tends to be slightly flatter than for nickel–hydrogen; that is, there is less drop in voltage during discharge. Also, the nickel–cadmium cell shows an increase of capacity at higher temperatures, while the nickel–hydrogen cells shows a decrease.

5.6.5.3 Charge Control

The battery cell is sensitive to charging rates and times, so control of the charging current is necessary. The input source may be the main bus or it may be a booster solar array. There are several charging methods. A constant current charging requires a cutoff when the charge is completed; constant potential charging can also be used. After the battery is fully charged, trickle charging is often used to maintain full charge. Sensing the state of charge of a nickel–cadmium cell can be done by sensing the voltage, the internal cell pressure, the auxiliary electrode, a

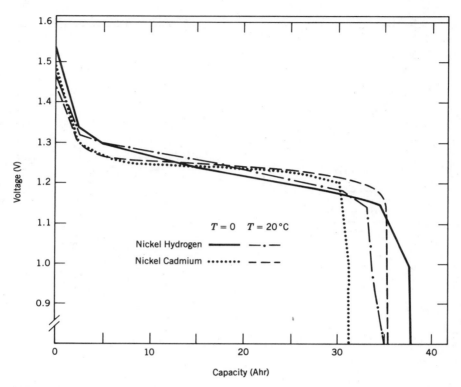

FIGURE 5.75 Comparison of Ni–Cd and Ni–H$_2$ cells at 0 °C (32 °F) and 20 °C (68 °F) (Stockel, 1980).

coulometer or ampere-hour integrator, or just measuring time. If the cell voltage is used, temperature compensation is needed.

5.6.5.4 Reliability and Lifetime

The number of years that a battery can perform its mission depends on a number of factors: the planned depth of discharge, the temperature control of the batteries, the method of storage during the noneclipse season, and any other variations in charging and discharging. Sometimes a battery can be "reconditioned" by discharging it to a lower voltage than normal operation. This is a shock treatment that can have both good and bad effects; it often raises the end-of-discharge voltage and increases the useful battery life, and it could produce a short that decreases the life.

Extra life can also be built into a battery by how it is built up from single cells. The number of cells in series is set by the desired battery voltage. If one or more extra cells are added to the series string, this protects the battery against a short-circuited cell (the most likely failure). More than one cell has to fail before the battery fails to deliver the required voltage. A numerical example is given in Section 5.12.7. Protection against an open-circuited cell (less likely failure) can be provided by a diode bypass that will carry the current around any cell that is open.

If a communications satellite is to have full operation during eclipse, the mass of the batteries is a significant part of the electric power subsystem. The exact mass depends on the desired lifetime and reliability. Figure 5.76 shows the capacity per mass of nickel–cadmium batteries for different depths of discharge and different amounts of redundancy. The total measured capacity of nickel–cadmium cell may be 30 or 40 Whr/kg, but operation to total capacity would require operation at very low voltages and also would severely curtail the cell life. A typical operating regime is operating to 40% of measured capacity and less than 20 Whr/kg. A nickel–hydrogen cell has slightly higher measured capacity per mass (50 Whr/kg) and can probably operate to a higher depth of discharge. Nevertheless, providing 1000 W for a 70-min eclipse still requires 40–100 kg for the battery, which is a significant fraction of the mass of a spacecraft.

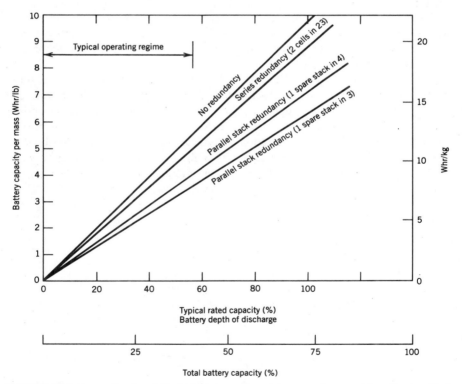

FIGURE 5.76 Energy density of Ni–Cd batteries packaged for geostationary orbit application (Esch et al., 1976. Paper presented at the 27th IAF Congress, Anaheim, CA, USA. Reprinted with permission).

5.6.6 Distribution

The distribution system has a number of features that affect the reliability of the electric power subsystem. The bus should be protected from the various subsystems, with fuses or resettable circuit breakers, so that a load will disconnect rather than overload the bus. Each fuse location may have two fuses in parallel, with a small resistance in one, to decrease the chance that the fuse fails with an insignificant overload. Multiple buses can be used to eliminate a single point of failure if a short occurs (the satellite SEASAT failed by a short on a single bus). A small essential bus can be provided for critical services, such as command, some telemetry, and attitude control. If the voltage drops below normal, the loads should disconnect automatically, in reverse order to their importance, command being last. Careful consideration should be made of control functions and whether they should be made on the satellite, on the ground, or by a combination.

Many electronic subsystems have their own power supply to generate the required voltages and to provide an additional regulation of voltage. A typical dc/dc converter is shown in Figure 5.77. A chopper on one side of a transformer sends direct current first through coils in one direction and then in the opposite direction. This produces alternating current on the other side of the transformer, with any desired voltage, which is then rectified by a network of diodes. A capacitor is used to smooth the voltage variations. A series of output coils can be used for a variety of desired voltages.

Some communications satellites use TWTs as the power amplifiers in the transmitters. In a typical communications satellite, over 80% of the power is used in power amplifiers. Because of special voltage needs, a TWT requires its own power supply. This tends to favor an unregulated bus, since a TWT power supply can also do the required voltage regulation. A typical TWT power supply is shown in Figure 5.78. The collector requires a very high voltage, the filament a very low voltage, and other components intermediate voltages. With multicollectors, the required power may vary with the impressed rf signal.

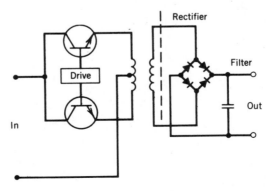

FIGURE 5.77 A dc-to-dc converter.

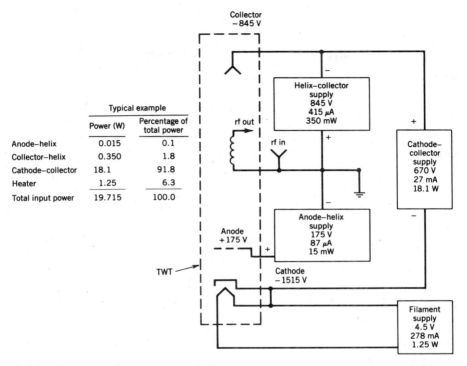

The typical example table within the figure:

	Typical example	
	Power (W)	Percentage of total power
Anode–helix	0.015	0.1
Collector–helix	0.350	1.8
Cathode–collector	18.1	91.8
Heater	1.25	6.3
Total input power	19.715	100.0

FIGURE 5.78 Typical power supply for travelling-wave tube. Most of the power is in the cathode-collector circuit, where the voltage must be accurately controlled. (Billerbeck, 1973; © 1973 IEEE).

5.6.7 Conclusion

A power control system monitors and controls the various components of an electric power subsystem. Some of this is done by control on the satellite. Other functions are measured on the satellite, telemetered to the ground, and controlled by ground commands. The batteries require special handling, both in controlling the daily charging during eclipse season and in the reconditioning done prior to an eclipse season. The Air Force has tended toward putting most of the control of satellite functions on the spacecraft, while some commercial satellite operations tend to put more control in the control center.

For additional reliability, many communications satellites have a dual redundant electric power subsystem. They are not truly redundant, for each half only provides half the power of the satellite. But each system has its own solar array, battery, bus, and distribution to different parts of the spacecraft. Thus, a failure in either half reduces the power and capability of the satellite but does not result in complete mission failure. Some cross-strapping is also provided. While the two systems are independent, they can be connected at various points by ground command; thus, a single failure in each system would not necessarily result in total loss of power.

The lifetime of a power subsystem depends on many factors. The solar array degrades with time; while this can be predicted fairly well, variations in solar activity can modify the expected degradation. The greatest source of a catastrophic failure of a solar array is the array drive (for oriented arrays) and slip rings. The batteries are the most serious constraint for long life of an electric power system and possibly of the entire communications satellite. The nickel–cadmium battery has had many years of development and failure analysis but continues to exhibit serious degradation with time; the nickel–hydrogen battery has longer lifetimes of 10–20 yr. The power conditioning is mostly susceptible to random failures; these can be handled with quality control, derating, redundancy, and testing.

5.7 SPACECRAFT ATTITUDE†

5.7.1 Introduction

The earliest satellites radiated signals in all directions with omnidirectional antennas. Nowadays, communications satellites require antennas that are pointed at the earth and that focus the radiated power into a narrow beam. The orientation of the satellite is maintained by the attitude control subsystem.

The concept and parts of an attitude control system are shown in Figure 5.79. Any error in pointing is detected by a sensor and corrected by changing the speed (or axis) of a rotating wheel. If the wheels have reached their limit, thrusters are used so the wheel speed can be reduced.

The required performance of an attitude control system is set by the existing disturbance torques and the required pointing accuracy. The major disturbances on a geostationary satellite are torques due to imbalances in solar radiation and misalignment of thrusters. The principal sensor used is an earth sensor or rf sensor, but other devices are used to determine attitude, including sun sensors, star trackers, and gyroscopes. Reaction wheels (which can spin either way) and momentum wheels (which only spin in one direction) are used to correct the attitude of a satellite. Hydrazine thrusters are usually used when the wheels are no longer capable of maintaining attitude. Finally, a control system takes the information from the sensors and provides appropriate commands to the wheels and thrusters.

The fundamental physical concepts used in attitude control are listed in Table 5.14. These concepts apply either to the entire spacecraft or to a wheel in the spacecraft. Each of these quantities is measured around an axis of rotation and would be different for different axes. The angular position of a body around an axis is measured by an *angle*, either in radians or degrees. The rate of change of this angle with time is the *angular velocity* of an object. When a force is applied to change the angular velocity, its effectiveness is proportional to the perpendicular distance to the axis of rotation (lever arm); the product of the force and the lever

†Thanks to Alberto Ramos for help in preparing Section 5.7.

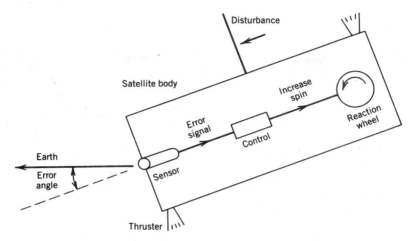

FIGURE 5.79 Concept and parts of attitude control system in body-stabilized satellite.

arm is the *torque*. Applying a torque during a time interval generates an *angular momentum*, which is a product of torque and time. The angular momentum of a body is proportional to its angular velocity; the ratio of the angular momentum to the angular velocity (in radians per second) is the *moment of inertia*. Finally, the *radius of gyration* is the square root of the ratio of moment of inertia to mass. That is, the moment of inertia is equal to the mass of the body times the square of the radius of gyration. If all the mass of a body was at this distance from the axis of rotation, the moment of inertia would be the same. It is useful to think of angular quantities as analogous to the more familiar linear quantities. In the last column of Table 5.14 is listed the linear analog, which should help in remembering the equations and understanding these angular quantities.

The coordinate system used to define the attitude of a satellite is illustrated in Figure 5.80. The coordinate system is fixed with respect to the spacecraft structure. The origin of this coordinate system is the center of mass of the satellite. In

TABLE 5.14 Fundamental Physical Concepts Used in Attitude Control

Quantity	Formula	Metric	English	Linear analog
		Dimensions		
Angle	θ	rad	rad	Distance
Angular velocity	$\omega = \dot{\theta}$	rad/sec	rad/sec	Speed
Torque[a]	$N = Fd$	N-m	ft-lb	Force
Angular momentum	$h = Nt$	N-m-sec	ft-lb-sec	Momentum
Moment of inertia	$I = h/\omega$	N-m-sec^2	ft-lb-sec^2	Mass
		kg-m^2	slug ft^2	
Radius of gyration	$r_g = \sqrt{I/m}$	m	ft	—

[a]Torque N equal to force F times lever arm d, the perpendicular distance to center of rotation.

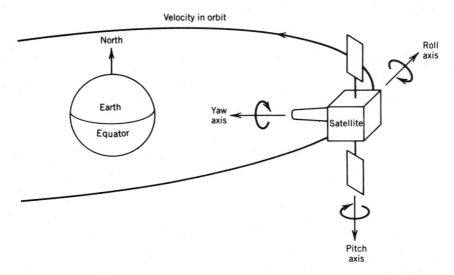

FIGURE 5.80 Coordinate system used for geostationary satellite. A few use a different system, with pitch and yaw interchanged.

the usual satellite attitude, the yaw axis is the line through the origin and the earth's center of mass and is positive in the direction to the earth; the pitch axis is through the origin and is practically perpendicular to the orbit plane, and its positive direction is to the south with a normal satellite; the roll axis is perpendicular to the other two axes and therefore lies practically along the velocity vector, with the positive direction in the direction of motion. Rotations about these axes are defined by the "right-hand rule"; that is, if the right-hand fingers are curled around an axis, with the thumb pointing in the positive direction, the fingers point in the positive direction of rotation. Attitude changes of the satellite are rotations about these axes. Figures 5.81–5.83 show how these attitude changes cause movement of the satellite beams on the earth. A positive pitch rotation moves a beam eastward, a positive roll rotation moves a beam northward, and a positive yaw rotation moves a beam clockwise around the subsatellite point.

If a point on the earth is separated from the subsatellite point by a central angle β_0 (see Figure 5.84), the angular displacement α_0 as seen by the satellite is

$$\tan \alpha_0 = \frac{R_e \sin \beta_0}{r - R_e \cos \beta_0} \tag{5.35}$$

where R_e is the earth's radius (6378 km), and r is the satellite orbit radius (42,164 km for geostationary orbit). The elevation angle at the earth station h is

$$\tan h = \frac{r \cos \beta_0 - R_e}{r \sin \beta_0} \tag{5.36}$$

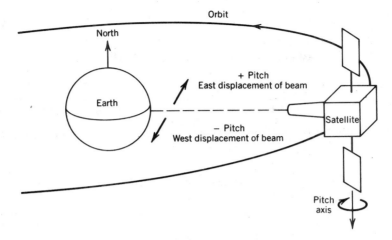

FIGURE 5.81 Eastward displacement of beam due to positive pitch.

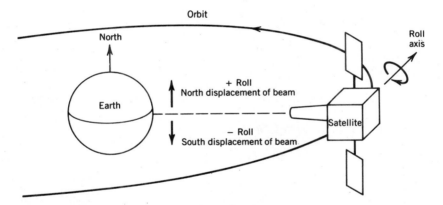

FIGURE 5.82 Northward displacement of beam due to positive roll.

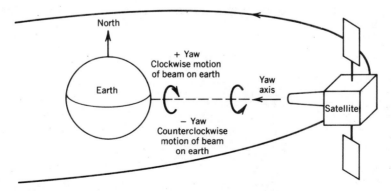

FIGURE 5.83 Clockwise displacement of beam around subsatellite point due to positive yaw.

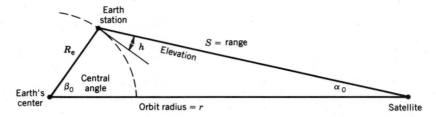

FIGURE 5.84 Geometric relation between satellite angular displacement, earth station elevation angle, and central angle from earth's center.

and the range S from the earth station to the satellite is

$$S^2 = r^2 + R_e^2 - 2rR_e \cos \beta_0 \qquad (5.37)$$

These quantities can be written in terms of the elevation angle h as

$$\sin \alpha_0 = \frac{R_e}{r} \cos h \qquad (5.38a)$$

$$\beta_0 = \cos^{-1}\left(\frac{R_e}{r} \cos h\right) - h \qquad (5.38b)$$

$$S = \sqrt{r^2 - R_e^2 \cos^2 h} - R_e \sin h \qquad (5.38c)$$

In terms of the range S, the equations are

$$\cos \alpha_0 = (r^2 - R_e^2)/(2Sr) + S/(2r) \qquad (5.39a)$$

$$\cos \beta_0 = (r^2 + R_e^2 - S^2)/(2rR_e) \qquad (5.39b)$$

$$\sin h = (r^2 - R_e^2)/(2SR_e) - S/(2R_e) \qquad (5.39c)$$

And in terms of the angular displacement α_0, as seen from the satellite, the equations are:

$$\beta_0 = \sin^{-1}\left(\frac{r}{R_e} \sin \alpha_0\right) - \alpha_0 \qquad (5.40a)$$

$$\cos h = (r/R_e) \sin \alpha_0 \qquad (5.40b)$$

$$S = r \cos \alpha_0 - \sqrt{R_e^2 - r^2 \sin^2 \alpha_0} \qquad (5.40c)$$

Of the four quantities β_0, h, S, α_0, if two are known, the other two can be found with the following simple equations

$$90° = h + \alpha_0 + \beta_0 \tag{5.41a}$$

$$S \cos h = r \sin \beta_0 \tag{5.41b}$$

$$S \sin \alpha_0 = R_e \sin \beta_0 \tag{5.41c}$$

These equations are listed with numerical constants in Figures 3.23 and 3.24. The relation between these various quantities can be seen in Table 5.15, which lists them for different elevation angles. The distance from earth station to subsatellite point is simply $R_e \beta_0$, where β_0 must be in radians.

For an earth station at a latitude of L and a difference in longitude of $\Delta\lambda$ from the subsatellite point (see Figure 5.85), the central angle is given by

$$\cos \beta_0 = \cos L \cos \Delta\lambda \tag{5.42}$$

This can be used in the previous equations. The line from the satellite to the subsatellite point (or to the center of the earth) is called the bore axis. The pitch angle θ_p and the roll angle θ_r are

$$\tan \theta_p = \frac{\sin \Delta\lambda}{(r/R_e) \sec L - \cos \Delta\lambda} \tag{5.43a}$$

TABLE 5.15 Geometric Relationships between Earth Station and Satellite

Elevation, h (degrees)	Off Axis, α_0 (degrees)	Central, β_0 (degrees)	Range, S (km)	Time (sec)	Loss (dB)	Elevation, h (degrees)
0	8.700	81.30	41,679	0.139	1.3	0
5	8.667	76.33	41,127	0.137	1.2	5
10	8.567	71.43	40,586	0.135	1.1	10
15	8.402	66.60	40,061	0.134	1.0	15
20	8.172	61.83	39,554	0.132	0.9	20
25	7.880	57.12	39,070	0.130	0.8	25
30	7.527	52.47	38,612	0.129	0.7	30
35	7.118	47.88	38,181	0.127	0.6	35
40	6.654	43.35	37,780	0.126	0.5	40
45	6.140	38.86	37,412	0.125	0.4	45
50	5.580	34.42	37,078	0.124	0.3	50
55	4.977	30.02	36,780	0.123	0.2	55
60	4.338	25.66	36,520	0.122	0.2	60
65	3.665	21.33	36,297	0.121	0.1	65
70	2.966	17.03	36,114	0.120	0.1	70
75	2.244	12.76	35,971	0.120	0.0	75
80	1.505	8.49	35,868	0.120	0.0	80
85	0.755	4.24	35,807	0.119	0.0	85
90	0.000	0.00	35,786	0.119	0.0	90

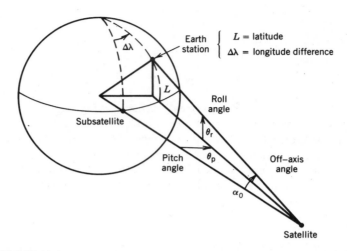

FIGURE 5.85 Pitch and roll angles for earth stations at different latitudes and longitudes.

$$\sin \theta_r = \frac{R_e \sin L}{\left(r^2 + R_e^2 - 2rR_e \cos L \cos \Delta\lambda\right)^{1/2}} \qquad (5.43b)$$

These angles are used in pointing various antenna beams at different locations on the earth's surface.

5.7.2 Body-Stabilized Satellites

The two basic configurations for communications satellites are body stabilized and spin stabilized. Body-stabilized satellites are simpler in concept but often require more hardware to implement. An example is shown in Figure 5.7. Body-stabilized satellites are fixed with respect to the coordinates shown in Figure 5.80; that means they rotate once a day, as the satellite goes around the earth. Body-stabilized satellites may be zero-momentum systems, where the reactions wheels operate around zero spin, and may come to a complete stop; or they may be a biased momentum system in which there is always a minimum angular momentum in a certain direction.

The system components required for a zero-momentum system are shown in Figure 5.86. A horizon sensor (also called an infrared sensor) or an rf sensor is used to determine pitch and roll with respect to the earth. A yaw sensor is needed and can either be a star tracker or a gyro. At least three reaction wheels are needed to control rotation around the three axes; additional wheels may be used for redundancy. A processor takes the information from the sensors and generates the necessary information for the wheels. Not shown in the figure are thrusters that would be used to dump angular momentum from the wheels.

A biased momentum system is shown in Figure 5.87. In this system there is a

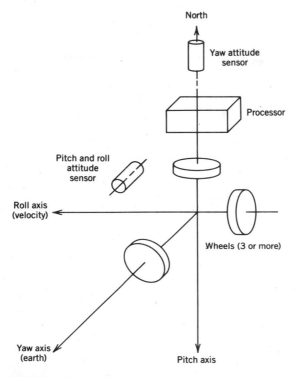

FIGURE 5.86 System components of zero-momentum system.

momentum wheel that always rotates in one direction and provides a significant amount of angular momentum aligned primarily along the pitch axis (north–south). Pitch errors can be detected by the pitch and roll sensor and corrected by speeding up or slowing down the momentum wheel. Because of the gyroscopic stiffness of the spinning wheel, the system is less sensitive to torques around the roll and yaw axis; instead of producing a constant rotation around the axis, a torque causes only a change in the spin axis by a much smaller angle. Errors in roll can be detected by the sensor and corrected. Errors in yaw are not detected directly by the earth sensor, but a yaw error is converted in 6 hr to a roll error, which can then be detected and corrected. Roll errors can be corrected with thrusters, with gimbals on the momentum wheel, or with a secondary momentum wheel.

5.7.3 Simple Spinner Satellites

A spin-stabilized satellite has a significant part of the satellite that rotates at a rapid rate (on the order of 1 sec^{-1}) or the entire satellite rotates. An example is shown in Figure 5.8. Small, simple communications satellites may be single bodies that rotate around the axis with the maximum moment of inertia. They may have an

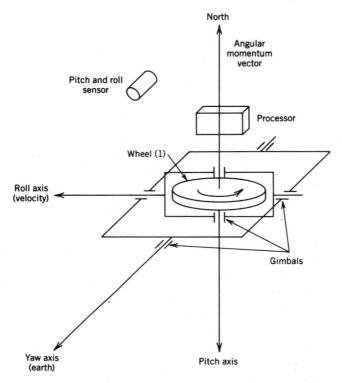

North

Angular
momentum
vector

Pitch and roll
sensor

Processor

Wheel (1)

Roll axis
(velocity)

Gimbals

Yaw axis
(earth)

Pitch axis

FIGURE 5.87 System components of biased momentum system using double-gimballed momentum wheel.

omni or toroidal antenna or more complicated antennas that are despun either mechanically or electrically. Some spin-stabilized satellites are "dual spinners." These have a rotating part and a stationary part and rotate around a minimum moment of inertia.

Most bodies in space cannot spin steadily around an arbitrary axis. In general, rigid bodies can rotate smoothly near two axes; one of these is the axis of maximum moment of inertia, the other is the axis of minimum moment of inertia, and these are always 90° apart. Communications satellites that spin around the maximum moment of inertia are easier to design and operate. The kinetic energy is equal to $h^2/2I$, and the angular momentum h is constant in the absence of external torques. Hence, the largest moment of inertia I corresponds to the smallest energy, and the satellite is in a stable state; energy dissipation on the spacecraft tends to drive it toward this state and cannot disturb it when it is in this lowest energy state.

A spinning communications satellite may have either a toroidal antenna (all directions perpendicular to the spin axis) or a despun antenna mounted on bearings so it does not partake of the satellite spin. If something is amiss, a spinning satellite may either wobble or precess.† A wobble means that the principal axis does not coincide with the bearing axis. The satellite spins around the principal axis, but the bearing axis "cones" (precesses) around this axis. The antenna beam will exhibit a north–south oscillation at the rotation frequency, which usually will result in a power variation in the communications link. A wobble usually results from a mass redistribution on the spacecraft, producing the change in direction of the principal axis. Note that in a wobble, the spin axis remains fixed in the spacecraft, and there is no variation in acceleration at any point in the spacecraft.

If a spinning satellite is disturbed, such as by firing a thruster, it frequently exhibits precession. The principal axis is still along the symmetry axis or bearing axis, but the instantaneous spin axis is displaced slightly from this axis. The spin axis does not remain fixed but precesses around the principal axis. The frequency of this precession, or coning, is the frequency λ:

$$\lambda = \left[\frac{(I_1 - I_2)(I_1 - I_3)}{I_2 I_3} \right]^{1/2} \omega_0 \qquad (5.44)$$

where ω_0 is the rotation frequency, I_1 is moment of inertia around the principal axis close to the spin axis, and I_2 and I_3 are the moments of inertia around the other two axes. In many satellites these last two are equal, and the precession frequency λ can be written as

$$\lambda = \frac{I_1 - I_t}{I_t} \omega_0 = (\sigma - 1)\omega_0 \qquad (5.45)$$

where σ is the ratio of the moment of inertia around the desired spin axis I_1 to the transverse moment of inertia I_t. This precession is a force-free motion of a rigid body; it is relative to the body axes, which are themselves rotating in space with the larger frequency ω_0.

Precession can be, and sometimes is, detected with an accelerometer on the spacecraft. Since the spin axis changes with respect to the spacecraft, the centrifugal accelerations observed also vary as a function of time. Any spacecraft parts not rigidly fixed may move as a result of this varying acceleration; this may include liquids, flexible antennas, or other components. Since most motion involves some viscosity or friction, energy will be absorbed, which will decrease or dampen the precession for a satellite rotating around its maximum moment of inertia. Instead of the coning angle θ between the spin axis and principal axis remaining fixed, it will gradually decrease with time t. This relation can be written approximately as

$$\theta = \theta_0 e^{-t/\tau} \qquad (5.46)$$

---

†This precession is often called nutation, but this is not the nutation defined in physics.

where θ_0 is the initial coning angle, and τ is the time constant. The time constant τ can be written in terms of the energy dissipation $\dot{E}$ as

$$\frac{1}{\tau} = \frac{\dot{E}/\theta^2}{\sigma(\sigma - 1) I_t \omega_0^2} \tag{5.47}$$

For many damping mechanisms, the energy dissipation is proportional to the nutation angle θ^2, so the time constant is independent of the magnitude of the coning angle. If the spacecraft is spinning around the maximum moment of inertia, the ratio σ is greater than 1, the time constant is positive, and the angle decreases. If the spacecraft is spinning around the minimum moment of inertia, the ratio σ would be less than 1, the time constant would be negative, and any energy dissipation makes the coning angle increase.

In most spinning spacecraft, a damper (often called a nutation damper) is included that exhibits some motion under precession, dissipates energy, and reduces the coning angle. Different devices have been used, including mechanical pendulums, half-filled annular rings, and rings completely filled with liquid. For the damper to function at very small angles, it must be sensitive to small accelerations and not be locked by static friction below a certain angle. An annular ring as a damper is shown in Figure 5.88; the liquid is driven to one side of the ring, and as precession continues, the liquid is driven around the ring.

5.7.4 Dual-Spinner Satellites

Satellites designed to spin around the minimum moment of inertia are referred to as gyrostats or dual spinners (see Figure 5.89). The need for such satellites arose as larger communications satellites were built, but rotor diameters were limited due to launch vehicle shroud constraints. The feasibility was proven by showing that energy dissipation on the despun portion of a dual spinner would offset the destabilizing effect of dissipation in the rotor. According to the energy sink approx-

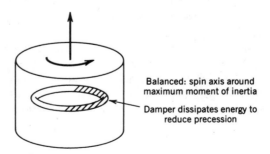

Balanced: spin axis around maximum moment of inertia

Damper dissipates energy to reduce precession

FIGURE 5.88 Spinning satellite, turning around its maximum moment of inertia, unconditionally stable.

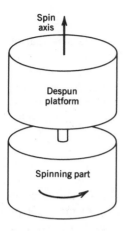

FIGURE 5.89 Dual spinner, turning around minimum moment of inertia, conditionally stable.

imations for an axisymmetric dual spinner, the time constant τ can be expressed as

$$\frac{1}{\tau} = \frac{\dot{E}_r/\theta^2}{\sigma(\sigma - 1)\,I_t\omega_0^2} + \frac{\dot{E}_p/\theta^2}{\sigma^2 I_t\omega_0^2} \qquad (5.48)$$

where σ is the ratio of the rotor moment of inertia around the spin axis to the transverse moment of inertia I_t of the entire spacecraft around the center of mass. For a dual spinner, this ratio is less than 1, so the first term is negative, and any energy dissipation on the rotor, $\dot{E}_r$, is destabilizing. However, the second term is always positive, so any energy dissipation on the despun platform, $\dot{E}_p$, is stabilizing; if there is enough platform dissipation, the second term dominates, and the satellite is stable.

The preceding oversimplified picture is presented to show the fundamental concepts of a dual spinner. More detailed analyses are in the literature (Likins, 1967; Kaplan, 1976). The system is not closed with respect to energy, since there is a motor to maintain rotation between the rotor and despun platform. Furthermore, differences in the two transverse moments of inertia and any imbalances in the platform may affect spacecraft stability. In an actual communications satellite, the largest energy dissipation in the rotor is often sloshing in the hydrazine tanks; the magnitude of this energy dissipation is difficult to calculate and changes as the hydrazine is used during the life of the spacecraft.

A significant problem with dual spinners is that to achieve stability, the platform must be despun. Problems with the control system, the motor, or the bearings may cause the platform to start to spin. With a spinning platform, the satellite is unstable, the coning angle increases, and the satellite eventually ends up spinning around the maximum moment of inertia, referred to as a flat spin. This has happened with several satellites. Methods have been developed to recover from a flat spin, and these have been executed on a few occasions.

5.7.5 External Torques

External torques affect the attitude of a satellite, and these can be disturbance torques or can be used to control the attitude. The largest disturbance torque at geostationary orbit is usually solar pressure. Another source of error (not really external) is misalignment of the north–south stationkeeping thrusters. Magnetic torques and gravity gradient have been used for attitude control, although both are quite weak at geostationary altitude. Disturbance torques can also be generated by micrometeoroids, but fortunately these usually are negligible.

5.7.5.1 Solar Pressure

The largest disturbance torque on a geostationary communications satellite from external sources is the pressure from solar radiation. An estimate of the order of magnitude of this torque is easy; a precise calculation may be difficult. The average solar flux S is 1370 W/m², although it varies by ±3.34% throughout the year, with a minimum in early July. If this power is absorbed completely by a surface, the effective pressure is equal to S/c, where c is the velocity of light, or 4570 × 10^{-9} N/m² (95 × 10^{-9} lb/ft²).

As shown in Figure 5.90, the force F exerted by absorbed sunlight is

$$F = \frac{S}{c} A \cos i \qquad (5.49)$$

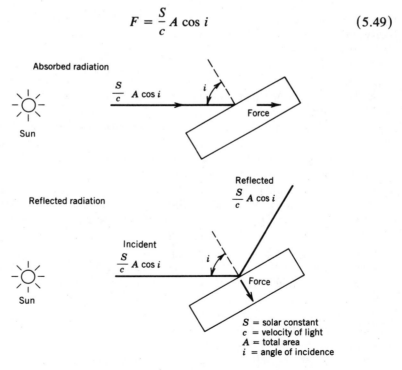

FIGURE 5.90 Forces produced by solar pressure from absorbed radiation and reflected radiation.

where A is the surface area and i is the angle of incidence, the angle that the sun's rays make with a perpendicular to the surface. The force exerted by absorbed sunlight is in the direction of the sun's rays. The torque depends on the distance from the center of solar pressure to the satellite center of mass. As an example, absorbed radiation at normal incidence on a 1-m^2 surface, an average of 1 m from the center of mass for 6 hr, will generate an angular momentum of about 0.1 N-m-sec. Reflected radiation also exerts a force so that the effective force is twice the normal component of the radiation, or

$$F = 2 \frac{S}{c} A \cos^2 i \qquad (5.50)$$

Note that this force is exerted in a direction normal to the surface.

On satellites with large solar arrays the effect of solar torques is significant. Added areas (flaps or vanes) have been used to cancel undesirable solar torques. Magnetic coils have also been used.

5.7.5.2 Thrusters

Various thrusters are placed on a spacecraft to control both its position and orientation. The use of thrusters for control will be discussed in detail in Section 5.8. There is a side effect that merits inclusion in a compilation of disturbance torques. The major use of a thruster on a communications satellite is for north–south stationkeeping. Ideally, such a thruster has a thrust vector through the center of mass of the satellite and therefore should not generate a torque. But any misalignment θ of the thruster (see Figure 5.91) will produce a torque N equal to $N = Fd$

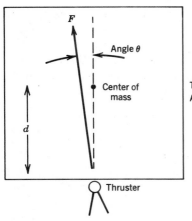

Torque: $N = Fd \sin \theta$
Angular momentum: $L = Nt$

FIGURE 5.91 Torque produced by thrust-misalignment during north–south stationkeeping maneuver.

sin θ, where F is the force of the thruster, and d its distance from the center of mass. A typical misalignment is $\pm 0.25°$. If a 0.5-N thruster is 0.5 m from the center of mass, it will generate a torque of 0.001 N-m; if it fires for 1000 sec, the angular momentum will be 1 N-m-sec. This torque is usually canceled by firing other attitude control thrusters. With electric propulsion thrusters, it is possible to change the direction of a thruster electrically, but with hydrazine thrusters such an alignment in space is usually not done.

5.7.5.3 Magnetic Field

The earth's magnetic field B at geostationary orbit has a magnitude of about 0.00110 G, or 110×10^{-9} N/A-m (or tesla). The torque N on a coil with n_t turns of wire carrying a current I is

$$N = n_t BAI \cos \alpha \qquad (5.51)$$

where A is the area of the coil and α the angle of the coil plane and the direction of B. As an example, a coil of 100 turns, with a current of 1 A and an area of 1 m^2, perpendicular to the earth's magnetic field, will experience a torque of 1.1×10^{-5} N-m. Spacecraft torques can occur due to residual magnetic dipole moments in the spacecraft, but these tend to be quite small at geostationary orbit. It is also possible to use a coil in the spacecraft to develop useful torques for attitude control. The earth's magnetic field is not constant at geostationary orbit, especially in periods of high solar flare activity, but the magnetic torques are so weak that usually long-term averages are the only important quantities.

5.7.5.4 Gravity Gradient

The gravitational field of the earth can produce a significant torque on a satellite. Different points on the spacecraft are at different distances to the earth's center and therefore have different amounts of gravitational acceleration. One stable position is to have the maximum moment of inertia along the pitch axis (orbit normal) and the minimum moment of inertia along the yaw axis (toward the earth). This usually means the longest dimension along the local vertical (toward and away from the earth) and the shortest dimension along the pitch axis (see Figure 5.92). For small variations from this stable position, the gravity gradient torque (Mueller and Spangler, 1964) around the pitch axis (orbit normal) is

$$N_p = -\frac{3\mu}{a^3} (I_r - I_y) \theta_p \qquad (5.52)$$

where μ is the gravitational parameter (GM), a is the orbit radius, θ_p is the angle between the principal axis and the pitch axis, I_r is the moment of inertia around the roll axis, and I_y is the moment of inertia around the yaw axis. Similarly, the

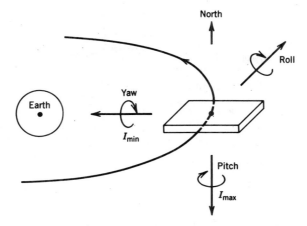

FIGURE 5.92 Stable configuration for satellite subject to gravity gradient forces.

gravity gradient torque around the roll axis is

$$N_r = -\frac{3\mu}{a^3}(I_p - I_y)\,\theta_r \qquad (5.53)$$

where θ_r is the angle between the principal axis and the roll axis and I_p the moment of inertia around the pitch axis.

While the gravity gradients have always appeared to be an attractive method of attitude control, the torques are small and the control is weak. The quantity μ/a^3 is equal to the square of the angular velocity of the orbit, n, and the frequency of oscillation about the pitch axis is given by

$$\omega_p^2 = 3n^2\frac{I_r - I_y}{I_p} \qquad (5.54)$$

Similarly, the frequency for oscillation around the roll axis is given by

$$\omega_r^2 = 3n^2\frac{I_p - I_y}{I_r} \qquad (5.55)$$

It is possible to make I_y small compared to the other two moments of inertia by extending booms along the local vertical. But the period of oscillation around one axis is bound to be at least on the order of 10 hr. In addition, it necessary to introduce a mechanism to damp these oscillations. For communications satellites at geostationary orbit, gravity gradient is not usually an important external torque, either as a disturbance or as a control mechanism.

5.8 ATTITUDE CONTROL†

5.8.1 Introduction

In order for a geostationary communications satellite to retain its effectiveness, it must be maintained—within narrow limits—not only at the geostationary position but also at its proper attitude with respect to the earth.

5.8.2 Attitude Sensors

To determine the orientation of a communications satellite, its attitude is measured with respect to the earth, to the sun, or to a star. The earth is the most important reference; however, its infrared radiation, which is usually used by sensors, is affected by the atmosphere and causes the earth's edge to appear fuzzy. In addition, earth sensors are sometimes confused by the sun or moon. In spite of these limitations, the earth sensor is still the most important attitude sensor for communications satellites. In addition to these earth measurements, it is useful to measure rotation around the earth–satellite line (i.e., the yaw axis). The sun is easy to detect, is a fairly small point, but unfortunately sometimes is along the earth–satellite line and is then not useful as a yaw sensor. A star, such as the north star, is a sharp reference point and is located in a good position, but its intensity makes it harder to sense. Star trackers are heavier than other sensors, and most communications satellites get along without them.

5.8.2.1 Earth Sensors

Horizon sensors can detect the disk of the earth against the background of space by measuring the infrared radiation, usually at wavelengths near 15 μm. Three types of horizon sensors are commonly used: the static horizon sensor generates an error signal until it is aligned precisely with the earth. The scanning horizon sensor uses an oscillating mirror to pass the earth's image over a sensor and generates a signal for the leading edge and the trailing edge. A spin infrared sensor uses the same principle but is used in a spinning satellite, which generates its own scan as it rotates. Many other variations and combinations are possible.

The basic components of a static horizon sensor are shown in Figure 5.93. The earth is imaged by a germanium lens, which transmits the infrared radiation. Four detectors are located around the edge of the image. If the image moves toward a detector, made of a thermopile, the additional radiation heats the detector and generates a stronger signal. By taking the difference between opposite detectors, sensitive error signals can be generated. For maximum precision, corrections must be made for the noncircularity of the earth's disk.

A horizon sensor used on spinning satellites is sketched in Figure 5.94. A single bolometer measures the presence or the absence of the earth's image. As the satel-

†Thanks to George Huson and Alberto Ramos for help in preparing Section 5.8.

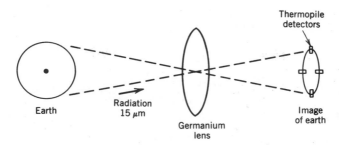

FIGURE 5.93 Static infrared horizon scanner (balanced radiation).

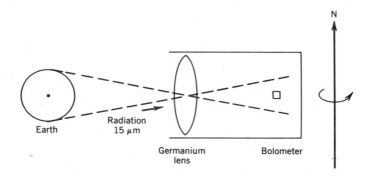

FIGURE 5.94 Spin earth infrared horizon scanner (edge tracking).

lite rotates, the earth's image sweeps across the detector. The midpoint of the signal provides pitch information, when the detector is centered on the earth's meridian going through the subsatellite point. The length of the signal provides roll information; the nominal scan is offset from the equator. If the pulse length is larger than nominal, the scan is closer to the equator, and if it smaller, it is farther from the equator. A scanning horizon sensor operates in the same general way for a body-stabilized satellite; a special oscillating mirror moves the earth's image across a detector.

Another method of determining the direction of the earth is to detect a radio frequency from the earth with an rf sensor. The principle for a single plane is shown in Figure 5.95 and is the same as tracking satellites with earth station antennas (see Section 5.4.4.1). Two feeds are used with a single reflector to produce two antenna patterns in slightly different directions. If a source of radiation is more in the A pattern, the amplitude of the signal received in feed A is larger than that in feed B. By changing the satellite attitude until the two amplitudes are equal, the boresight axis will then be pointed at the rf signal. To measure both pitch and roll, four feeds are normally used, and two error signals are generated. One advantage of rf sensors is that they tend to increase in accuracy as the need for increased accuracy develops. If a satellite has very narrow beams, it needs

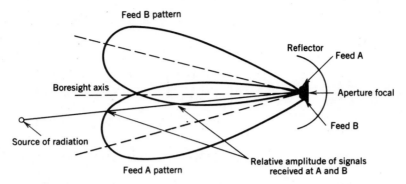

FIGURE 5.95 Amplitude sensing antenna pattern of rf sensor (Lebsock, 1970; reprinted by permission from Lockheed Missiles and Space Co., Inc., Sunnyvale, CA).

more accurate pointing; but it will also have larger reflectors, so that the patterns for the rf sensor can also be narrower. In contrast, horizon sensors are approaching their limit in accuracy (approximately $\pm 0.1°$), and it is more difficult to increase the accuracy by an order of magnitude.

5.8.2.2 Sun and Star Sensors

One type of sun sensor of shown in Figure 5.96. Half of the sensor consists of two solar cells located behind a meridional slit. When the two solar cells are generating the same signal, the sun shines through the slit on the boundary between them. On a spinning satellite, the slit can be north–south and provides information when the sun is aligned on that meridian. The other part of the sun sensor has a slit that may be inclined at an angle of 30°. This sensor provides information on whether the sun is farther north or farther south, since it will be advanced or delayed by the inclined slit.

A sun sensor for a body-stabilized satellite, shown in Figure 5.97 (Lebsock,

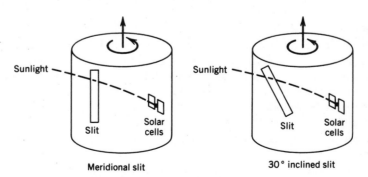

FIGURE 5.96 V-slit sun sensor for spinning satellite.

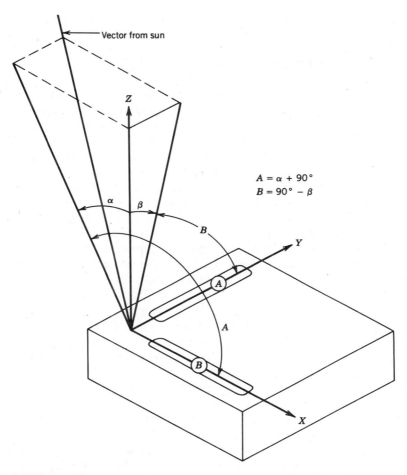

FIGURE 5.97 Sun sensor with two slits that measure sun's position over wide field of view and provides digital information (Lebsock, 1970; reprinted by permission from Lockheed Missiles and Space Co., Inc., Sunnyvale, CA).

1970), uses a similar principle. Two slits A and B are set at $90°$. Detectors under the A slit determine angle A, and detectors under the B slit determine angle B. Thus, the direction of the sun can be determined over a wide field of view. Frequently several detectors under each slit are coded with masks, so a different combination of detectors is illuminated for each angle. The information can thus be provided directly in digital form.

In a star tracker, a starfield image is focused on an image dissector tube through a suitable lens system. A televisionlike field raster scans the field, searching for the desired star. To determine the yaw attitude of a geostationary satellite, Polaris is preferred since it is $89° 06'$ from the orbit plane, and the yaw information can be obtained directly from its attitude with respect to the satellite. A red filter may be employed to differentiate Polaris from a few other bright stars.

5.8.3 Gyroscopes and Wheels

The gyroscope is essentially an attitude detector. Although the gyroscope cannot provide absolute attitude information, it can sense any change in the inertial-space orientation of its spin axis. Therefore, the gyro can be used to provide error information pertaining to perturbations in spacecraft orientation. Many past applications did not use gyros in communications satellites because satellite accuracy requirements were low, adequate lifetimes were not available, power consumed was excessive, and the gyro's use required some computational capacity. Some of these factors have changed, and gyros are now being used.

A conventional gyroscope has a rapidly rotating and delicately balanced mass (usually in the form of a wheel), a motor to maintain rotation, gimbals to support the rotating mass and provide the desired attitude information, and a sealed housing to protect the gyro. The main bearings are usually either ball bearings or gas bearings; the gimbals can be supported on jewel bearings.

Since the gyroscope wheel tends to maintain its spin axis orientation in inertial space (and therefore tends to maintain the orientation of its gimbals in inertial space), any rotation of the satellite body about any axis other than the gyroscope spin axis causes initiation of relative motion between the spacecraft structure and the gimbals. This relative motion is detected and used for both the attitude control system and, if required, to determine the appropriate corrective precession for the gyroscope. With programmed control of precession, desired rotation rates of the structure can be established, and deviations therefrom detected.

Reaction wheels are essentially devices to store momentum for accumulation or for release. The reaction wheel is a high-inertial rotor of a motor. When the motor exerts a torque on the wheel (either accelerating or decelerating it), an equal and opposite torque is transmitted through the motor mounting pads to the spacecraft structure. The moment of inertia of the wheel is sufficiently large, so that a change in the wheel's angular momentum has a significant effect on the spacecraft attitude.

Some reaction wheels can go either way and are used in zero-momentum systems where the total angular momentum could be zero. Other wheels, often called momentum wheels, are designed to turn in only one direction and always store a certain minimum angular momentum. Some momentum wheels have a fixed axis with respect to the spacecraft; others are gimballed, so the axis can be moved with respect to the spacecraft. Wheels are mounted inside a sealed case and are partially evacuated to reduce air friction (in ground tests). Gimballed wheels may have the gimbals inside or outside the case.

Conventional bearings limit both the angular speed and lifetime of most wheels. Some wheels have been developed that use a magnetic bearing suspension; an optic sensor detects the position of the wheel, and electromagnets can push or pull the wheel into position. This reduces the friction and increases the maximum angular speed and lifetime. The angular speed is constrained mainly by the strength of the rotor, so that centrifugal forces do not tear it apart.

5.8.4 Propulsion System

Both attitude control and stationkeeping (maintaining the position of a satellite in the desired orbit) require a propulsion system. For attitude control, a thruster should be mounted far from the center of mass, with the thrust vector at right angles to the direction of center of mass; this provides the largest torque. For stationkeeping, the thruster is mounted so that the thrust vector goes through the center of mass, or a pair of thrusters is used so that the torque is approximately zero. For attitude control, the propulsion system is used whenever a wheel approaches its maximum speed to reduce the total angular momentum of the spacecraft by "dumping" some into space.

The most widely used propulsion system on communications satellites is a catalytic monopropellant system using hydrazine fuel. Unlike many fuels, hydrazine does not require an oxidizer but decomposes with heat and/or the presence of a suitable catalyst. The chemical reaction produces heat and a higher thrust than a cold gas system with an equivalent mass. The main components of a hydrazine propulsion system are shown in Figure 5.98. The hydrazine (N_2H_4) is stored in propellant tanks and flows through a fuel line with filters and valves to a thruster. The latter includes the catalyst bed, a combustion chamber, and an expansion nozzle. A complete propulsion system often includes several tanks and many thrusters mounted in different places on the spacecraft.

Some communications satellites use a bipropellant system for the apogee kick motor, to get from transfer orbit to geostationary orbit. This means a fuel and an oxidizer in separate tanks. When this is done, the bipropellant system can also be used for other purposes, especially north–south stationkeeping.

The force F exerted by a thruster can be written as

$$ F = \dot{m}v_e = \frac{\dot{w}}{g} v_e = \dot{w}I_{sp} \tag{5.56} $$

where $\dot{m}$ is the mass flow rate, $\dot{w}$ the weight flow rate, v_e the effective exhaust velocity, and g the acceleration of gravity on earth (9.8 m/sec^2). The specific impulse I_{sp} is the most commonly used figure of merit for thrusters, and (for chemical propulsion) efforts are made to increase the specific impulse. As seen from the preceding equation, the specific impulse is the ratio of thrust to weight flow rate:

$$ I_{sp} = \frac{F}{\dot{w}} \tag{5.57} $$

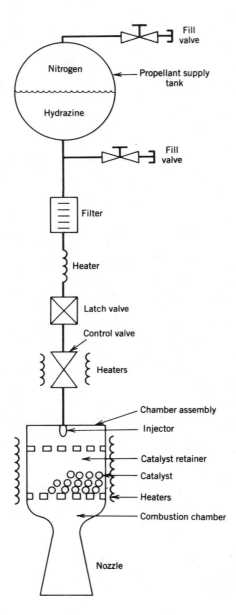

FIGURE 5.98 Catalytic monopropellant propulsion system (Rogers, 1977). Actual system has several tanks, many thrusters, and a complex network of pipes and valves.

The use of specific impulse originated in English units. For example, a thrust of 0.46 lbf and a weight flow rate of 0.002 lbm/sec means a specific impulse of 230 sec. The concept of a specific impulse is more natural in English units, while the effective exhaust velocity is more natural in SI units. Typical values of specific impulse are hydrazine, 230 sec; solid propellants, 290 sec; and liquid hydrogen and oxygen, 445 sec.

The product of the average value of thrust and the time during which it acts is the total impulse

$$I_t = \int F \, dt = \int I_{sp} \dot{w} \, dt \tag{5.58}$$

If the specific impulse I_s is constant with time, then the total impulse I_t is

$$I_t = I_{sp} w = I_{sp} mg \tag{5.59}$$

where w is the fuel weight, and m is the fuel mass. (Sutton and Ross, 1976).

5.8.4.1 Hydrazine

Anhydrous hydrazine is a colorless, toxic, corrosive liquid with a distinct ammonialike odor. It is insensitive to shock or friction and is a stable chemical that can be stored for long periods of time. Properties of hydrazine are listed in Table 5.16. For more information see Schmidt (1984). Many physical properties of hydrazine are similar to water.

In the presence of a catalyst, liquid hydrazine fuel may decompose according to either of the two following reactions:

$$3N_2H_4 \xrightarrow{cat} 4NH_3 + N_2 + 3.484 \text{ J/kg}$$

$$3N_2H_4 \xrightarrow{cat} 6H_2 + 3N_2 + 1.574 \text{ J/kg}$$

In the first step, liquid hydrazine is broken catalytically into gaseous ammonia (NH_3) and nitrogen; in the second step, the liquid hydrazine is disassociated into nitrogen and hydrogen. If the fraction of molecules obeying the second equation is represented by X, the two reactions can be combined into

$$3N_2H_4 \xrightarrow{cat} 4(1 - X)NH_3 + 6XH_2 + (2X + 1)N_2$$
$$+ (3.484 - 1.91X)\text{J/kg}$$

The catalyst, temperature, flow variables, and combustion chamber control the mix of final products, or the value of X. The second reaction, producing hydrogen, produces less thermal energy. However, it also lowers the mean molecular weight, thus increasing the velocity and partially offsetting the loss in energy. Optimum I_{sp} is obtained at low ammonia dissociations and high chamber temperatures.

TABLE 5.16 Physiochemical Properties of Anhydrous Hydrazine[a]

Property	Value	Metric	English
Molecular formula, N_2H_4			
Molecular weight	32.04	—	—
Boiling point	—	114.2 °C	237.6 °F
Critical properties	$P_c = 145$ atm, $T_c = 380$ °C (716 °F), $d_c = 0.231$ g/cm³		
Density	—	1.0259 g/cm³, 0 °C	8.562 lb/gal, 32 °F;
		1.0083 g/cm³, 20 °C	8.415 lb/gal, 68 °F;
		1.0040 g/cm³, 25 °C	8.379 lb/gal , 77 °F;
		0.981 g/cm³, 50 °C	8.188 lb/gal, 122 °F
Dielectric constant	51.7 at 25 °C (77 °F)	—	—
Electrical conductivity	2.3×10^{-6}–2.8×10^{-6} Ω^{-1}, 25 °C (77 °F)	—	—
Explosive limits (in air, 1 atm.)	4.7% lower, 100% upper	—	—
Fire point (tag open cup)	—	52 °C	125.6 °F
Flash point (tag open cup)	—	52 °C	125.6 °F
Freezing point	—	2.0 °C	35.6 °F
Heat capacity (liquid)	23.62 cal/mol °C, 25 °C	0.737 cal/g °C, 25 °C	0.737 Btu/lb °F, 77 °F
Heat of combustion, N_2H_4 (liq) + O_2 = N_2 + $2H_2O$ (liq)	—	−148.6 kcal/mol, 25 °C	−8.359 Btu/lb, 77 °F

654

Heats of formation at 25 °C (77 °F)

$N_2 + 2H_2 = N_2H_4$ (g)	22.434 kcal/mol	1260 Btu/lb
$N_2 + 2H_2 = N_2H_4$ (liq)	12.054 kcal/mol	677 Btu/lb
$N_2 + 2H_2 + H_2O = N_2H_4 \cdot H_2O$	10.300 kcal/mol	579 Btu/lb
$N_2 + 2H_2 + aq = N_2H_4$ aq	8.140 kcal/mol	457 Btu/lb
Heat of fusion	3.025 kcal/mol, 2.0 °C	170 Btu/lb, 35.6 °F
Heat of solution (liquid)	−3.9 kcal/mol, 25 °C	−219.6 Btu/lb, 77 °F
Heat of vaporization	9.34 kcal/mol, 114.2 °C	539 Btu/lb, 237.6 °F
Index of refraction, n_D	1.4683 at 25 °C (77 °F)	—
Ionization constants		
K_1	8.5×10^{-7} at 25 °C (77 °F)	—
K_2 (approximately)	1×10^{-16} at 25 °C (77 °F)	—
Surface tension	66.5 dyn/cm, 25 °C	0.00456 lb/ft, 77 °F
	64.0 dyn/cm, 35 °C	0.00439 lb/ft, 95 °F
Vapor pressure	10.4 mm, 20 °C;	10.4 mm, 68 °F;
	14.2 mm, 25 °C;	0.0187 atm, 77 °F;
	20 mm, 30.7 °C;	0.0263 atm, 87.3 °F;
	100 mm, 62.5 °C	0.131 atm, 144.5 °F
Viscosity	1.23 cP, 5 °C;	0.000830 lb/ft sec, 41 °F
	1.06 cP, 15 °C;	0.000712 lb/ft sec, 59 °F
	0.91 cP, 25 °C	0.000615 lb/ft sec, 77 °F

[a]Data from the RRC Monopropellant Hydrazine Design Data booklet, courtesy of Olin Defense Products Group, Rocket Research Corp. (1966). Updated by private communication; a second edition is being prepared. For more information on hydrazine, see Schmidt (1984).

5.8.4.2 Propellant Tanks

An inert and strong material is needed for the tanks, and stainless steel or titanium may be used. Most hydrazine systems use a "blowdown" system to expel the fuel from the tank. Nitrogen gas is stored in part of the tank, and the pressure forces hydrazine out of the tank. As the fuel is used, the gas expands to a larger volume, the pressure decreases, the flow is slower, and the resulting thrust is smaller.

Propellant management is needed to make sure that the fuel is expelled from the tank and not the gas. In a spinning satellite, the outlet is placed at the greatest distance to the spin axis, and centrifugal force pushes the fuel toward the outlet. In some spinners, a sphere-conical tank is used that puts the outlet both at the maximum radius for space use and also the lowest point for ground tests. In a three-axis stabilized satellite, a bladder may be used for propellant management or various barriers or screens used so that the fuel flows to the outlet aided by surface tension.

5.8.4.3 Distribution

Fuel lines are needed to take the hydrazine from the tanks to the thrusters located on various spacecraft surfaces. Latching isolation valves are used in addition to the thruster valves to ensure that a leak will not result in total fuel loss. Filters in the line prevent small particles from interfering with the closing of valves. Fill and drain valves enable the hydrazine and nitrogen gas to be loaded into the system.

Hydrazine freezes at 2 °C, and in some spacecraft heaters and insulation are necessary to make sure the fuel in the lines remains liquid. While hydrazine is extremely stable, bubbles have formed in the fuel lines over a period of months. Usually such a bubble merely produces a transient in the thrust when it leaves the system. In some cases, the bubble has interfered with fuel flow, and special maneuvers were required to expel the bubble (Gordon et al., 1974).

5.8.4.4 Thrusters

The catalytic thrusters (Huzel and Huang, 1971) usually contain a propellant valve, a decomposition chamber, and a nozzle (see Figure 5.99). The valves often have

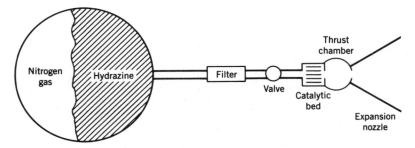

FIGURE 5.99 Catalytic hydrazine thruster, usually 0.5–20 N thurst.

dual-actuation coils and dual-series seats. The decomposition chamber includes
the catalyst bed, which may be heated electrically before the hydrazine valves are
opened. Beyond the catalytic bed is a mixing chamber and a convergent section.
Most spacecraft thrusters used in geostationary orbit vary from 0.5 to 20 N (0.1–
4 lbf), with the smaller ones used exclusively for attitude control and the larger
ones for north–south stationkeeping.

5.8.4.5 Alternatives to Hydrazine

While the hydrazine thruster has been widely used in communications satellites,
other thrusters have flown in space. A cold gas is the utmost in simplicity, but the
low specific impulse means its primary application is for low values of thrust. A
bipropellant thruster increases the complexity, requiring two tanks and two distribution systems; the specific impulse is higher. An electrothermal thruster is
basically a hydrazine thruster but uses electrical energy to increase the specific
impulse (from 230 to 290 sec in one case). A pulsed plasma thruster creates a
plasma by producing a spark over a Teflon surface; the actual thrust is quite low,
but the specific impulse is high.

5.8.5 Control System

As noted in the introduction, a control system is needed that takes information
from the sensors and provides appropriate commands to the wheels and thrusters.
These commands should reduce any errors in attitude. In its simplest form, an
error angle θ is measured, and a voltage is provided to a wheel that results in a
torque N. The result of this torque is an angular acceleration $\ddot{\theta}$ of the spacecraft.
A simple control law can be written as

$$N = I\ddot{\theta} = -K_\theta \theta - K_\theta \tau_D \dot{\theta} \tag{5.60}$$

where $\dot{\theta}$ is the angular velocity, K_θ the proportional gain, τ_D the ratio of two gains,
and I the moment of inertia of the spacecraft around the axis being considered.
The necessity for the last term in the equation will be clearer later.

Once the control equation is written, the variation of the angle with time is
determined. In understanding the solution of the equation, it is useful to discuss
two parameters. The undamped natural frequency ω_n is

$$\omega_n = \sqrt{\frac{K_\theta}{I}} \tag{5.61}$$

which is the square root of the ratio of the proportional gain to the moment of
inertia. If the last term in the first equation is zero, or near zero, the system will
oscillate at the undamped natural frequency. The damping ratio ξ is given as

$$\xi = \tfrac{1}{2}\omega_n \tau_D \tag{5.62}$$

which is one-half the product of the undamped natural frequency ω_n and the ratio of gains τ_D. This constant determines the behavior of the angle as a function of time.

Several solutions are shown in Figure 5.100. The first one is for no damping, ($\xi = 0$), where the correcting torque is only a function of angle. What happens is shown in the figure. The torque does indeed get the angle back to zero, but the spacecraft has angular momentum and keeps going past the desired zero angle. Thus, zero damping results in oscillations around the desired angle. The critically damped case ($\xi = 1$) is the usually desired control. With a large error angle and no angular velocity, a torque is applied to correct the angle; but as the angular velocity builds up and the desired angle is approached, the torque is decreased. As it nears the axis, an opposite torque is applied, so that the angular velocity is slowed down, and the angle can reach and remain at the desired zero value. If the damping ratio is too large, the system is overdamped ($\xi = 4$). There is no oscillation, but the system takes longer to reach the desired value.

The foregoing example just touches on the problems in designing a control system. In a satellite there are three axes to control. In a zero-momentum system the three control systems are fairly independent; in other systems there must be adjustments for the coupling between different axes, especially roll and yaw. In most satellite systems there are redundant components that must be switched in either automatically or by ground command. Systems are also designed to maintain pointing during momentary outages. Powerful communications satellites have large antennas and solar arrays, with low vibrational frequencies for which compensation is necessary. Techniques of adaptive control can be used so that actual frequencies do not have to be predicted. Finally, attitude control systems should make maximum use of on-board processors and of the information on the satellite's attitude, both present and past.

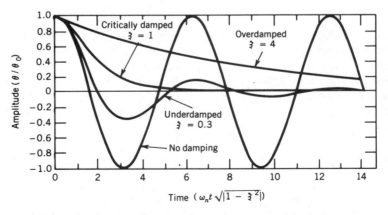

FIGURE 5.100 Effect of damping on satellite attitude for simple control system. Initial angular velocity $\dot{\theta}_0$ assumed zero.

Many modes of operation must be included in the design of the overall stabilization system, as described well by Lebsock:

Possible modes for a synchronous communications satellite are as follows:

Ascent Mode—which controls the spacecraft along the transfer ellipse from booster separation to initiation of acquisition mode.

Acquisition Mode—which establishes the spacecraft in a known orientation starting from an arbitrary attitude and possibly some suitably small values of initial rates.

Beam Pointing Mode—which is the long-term mode of operation during which precise beam pointing is needed.

Stationkeeping Mode—which maintains precise beam pointing during periods of linear velocity adjustment which are required to maintain proper orbit parameters. The spacecraft returns to beam pointing mode after conclusion of this mode.

Reacquisition Mode—reestablishes the spacecraft to a known orientation after some unpredicted phenomena causes the loss of attitude reference. The spacecraft returns to beam pointing mode after conclusion of this mode.

Back-Up Mode—maintains attitude reference after malfunction of some component of the beam pointing mode. Back-up mode may result in reduced performance requirements.

Consideration of the use of ground based control modes is important, particularly for increasing the reliability during beam pointing mode of operation. During this mode, the control system need only counteract environmental torques; therefore, rapid system response is not needed. The low response requirement allows a low control command bit rate. Rapid system response may be required in certain modes, requiring some onboard logic not dependent on a ground link. In addition, some onboard control logic is needed to maintain pointing during periods of data link dropout. While ground control modes are desirable from a reliability standpoint, overall mission aspects must be evaluated before deciding its applicability (Reprinted by permission, from Lebsock, 1970).

5.8.6 Errors and Accuracy

Many factors enter into the accuracy of an attitude control system. Usually only a rough estimate can be made of some of these factors. Representative values are listed in Table 5.17. The effect of stationkeeping error is minimized by the use of earth sensors. If two star trackers were used for attitude control, a $0.1°$ error in longitude would be equivalent to a $0.1°$ error in pointing. With earth sensors, a $0.1°$ error in longitude produces an error in pitch of about one-fifth that amount. Misalignment errors can be corrected after launch if there is provision for introducing a bias by ground command.

The accuracy for several satellite systems is listed in Table 5.18. While there are variations, most communications satellites have accuracies of a few tenths of

TABLE 5.17 Typical Accuracy of Attitude Control Components During Normal Beam-Pointing Mode

Error Sources	RSS Values[a]	Pitch (degrees)	Roll (degrees)	Yaw (degrees)
Orbit inclination, ($\pm 0.1°$)	—	—	± 0.019	± 0.10
Orbit longitude, ($\pm 0.1°$)	—	± 0.019	—	—
Infrared sensor accuracy	X	± 0.05	± 0.05	—
Infrared sensor alignment	X	± 0.03	± 0.03	—
Wheel torque disturbance	X	± 0.05	—	—
Wheel wobble	X	—	± 0.01	± 0.01
Wheel alignment	X	—	—	± 0.05
Control electronics	X	± 0.05	± 0.05	—
Wheel control loop deadband	—	± 0.02	—	—
Roll/yaw control loop deadband	—	—	± 0.05	± 0.23
Total errors		± 0.131	± 0.146	± 0.381

[a] In combining errors, the square root of the sum of the squares of the items marked with an X is added to the unmarked items.

TABLE 5.18 Performance of Various Attitude Control Systems

Satellite	Pitch East–West (degrees)	Roll North–South (degrees)	Yaw (degrees)
ANIK-A	± 0.169	± 0.134	—
ANIK-B	± 0.12	± 0.14	± 0.25
AUSSAT	± 0.05	± 0.05	± 0.209
BS (YURI-1)	± 0.1	± 0.1	± 0.6
CS (SAKURA-1)	± 0.222	± 0.143	—
ETS II (KIKU-2)	± 0.346	± 0.40	—
FLTSATCOM	± 0.17	± 0.17	—
INTELSAT IV	± 0.35	± 0.3	—
INTELSAT V	± 0.14	± 0.14	± 0.41
INTELSAT VI	± 0.15	± 0.15	—
NATO III	± 0.246	± 0.247	—
OTS	± 0.2	± 0.2	± 0.5
SATCOM, C-Band	± 0.21	± 0.19	—
SATCOM, K_u Band	± 0.1	± 0.1	—
SBS	± 0.032	± 0.04	± 0.153
TDRS	± 0.1	± 0.1	± 0.25

a degree. The pitch and roll errors are usually similar. The allowable yaw error is larger due to the geometry. Even for an earth station at the rim of the earth's disk, a yaw error only produces less than one-sixth the effect of a roll or pitch error. The effect of an error in attitude is usually a decrease in the intensity of transmitted and received signals. So the accuracy is dictated by the width of the narrowest antenna beam and the allowable decrease in intensity. As spot beams get narrower, attitude control requirements increase. Examples are AUSSAT (0.05°), SBS (0.04°), and GE Series 4000 (0.03°).

5.9 STRUCTURE†

5.9.1 Introduction

The function of the spacecraft structure is to provide mechanical support for all the subsystems and components, to provide precise alignment where needed, and to aid the thermal control with thermal conduction and desired surface properties. These functions must be provided within certain constraints and trade-offs. The volume provided by the launch vehicle is limited. The attitude control system frequently requires the principal inertia axes to be in preferred directions and the inertia ratio to be within certain limits and may also dictate the stiffness of the structure or parts of it. Finally, there should be easy access during assembly, integration, and testing.

The structure must be designed to sustain the loads during environment testing, ground handling, launch into low orbit, perigee and apogee firings, and any deployment of antennas or solar arrays. After reaching its final orbit position and deployment, the loads on the structure are greatly reduced in the zero-gravity environment, but the alignment requirements are more rigorous. The designer must satisfy all requirements, minimize the structure mass and cost, and still keep the probability of failure near zero.

The acoustic noise is highest in the early stages of the launch and is transmitted from the rocket motors by the air through the fairing or housing and onto the spacecraft. Steady loads are transmitted through the structure as the rockets accelerate the spacecraft to the high velocities required by satellite orbits. Vibration at many frequencies is transmitted through the spacecraft supports from the rocket motors. The separation ring and other pyrotechnic devices send sudden shocks through the structure.

The volume for payloads inside the shuttle orbiter is 18.3 m long and has a diameter of 4.6 m. Usually, this space must also be used for the perigee and apogee motors, as well as other spacecraft, so the volume available for a single spacecraft is much smaller. As shown in Figure 5.101, the larger satellites have their symmetry axis along the shuttle axis, and the satellite has the full 4.6 m diameter, while the smaller satellites are turned 90°, so their length is constrained to 4.6 m.

†Based on a lecture and visuals by Brij N. Agrawal, whose help is gratefully acknowledged.

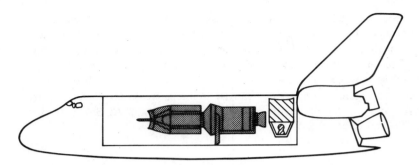

FIGURE 5.101 Configuration of spacecraft inside shuttle orbiter bay.

The shuttle fees are strongly dependent on the length of the payload, so there is incentive to decrease the payload length.

For other launch vehicles (such as large expendable ones), the shape of the physical volume tends to be more like that available in an Atlas Centaur fairing shown in Figure 5.102. For such fairings, the length is often ample; the available length shown is 8.23 m, and in one type of extended fairing the available length is 10.3 m. However, the diameter shown is only 2.39 m. Other launch vehicles are larger, but still constraining. Ariane IV and Titan III/PAM-D2 are only 3.65 m. These dimensional differences mean that a satellite optimized for the shuttle will not be optimized for launching in an expendable vehicle.

The way a satellite is stowed for launch is shown in Figure 5.103, which shows

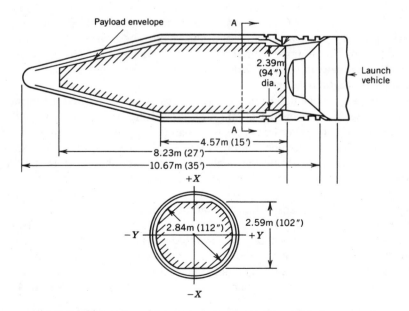

FIGURE 5.102 Atlas Centaur fairing, showing space available for spacecraft.

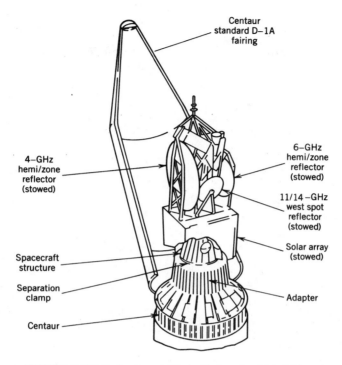

4–GHz
hemi/zone
reflector
(stowed)

Centaur
standard D–1A
fairing

6–GHz
hemi/zone
reflector
(stowed)

11/14–GHz
west spot
reflector
(stowed)

Solar array
(stowed)

Spacecraft
structure

Separation
clamp

Centaur

Adapter

FIGURE 5.103 Packaging satellite inside launch vehicle fairing.

an *Intelsat V* satellite configured for an Atlas Centaur launch. The rf feeds are fixed, and only the reflectors are moved into a stowage position. The solar arrays consist of panels that are folded together. The adapters carry the loads from the large diameter of the Centaur to a smaller diameter separation ring. Figure 5.104 shows the actual elements of a structure. The basic bus structure is rectangular and supported on a cylinder, and a separate antenna support structure holds the rf feeds and reflectors. The structure of a different satellite is shown in Figure 5.105. This shows cones to support the basic load, platforms to hold equipment, and the deployment mechanism for the solar array.

5.9.2 Basic Elements in Structure Design

To analyze a spacecraft structure, it is divided into simple elements, for which analytical equations are available to predict distortions (strains) and stresses under given forces (loads). These simple elements simplify both the construction and the analysis of a spacecraft structure. In the first steps of a design, a few major elements are analyzed that will carry the main load of a satellite. Later, a computer model with hundreds or thousands of these elements provides a more detailed analysis of the proposed structure.

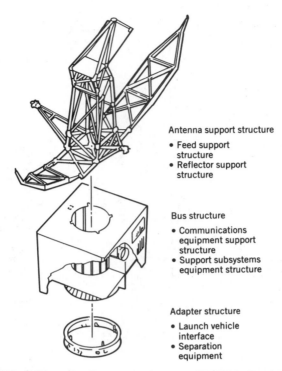

Antenna support structure

- Feed support
  structure
- Reflector support
  structure

Bus structure

- Communications
  equipment support
  structure
- Support subsystems
  equipment structure

Adapter structure

- Launch vehicle
  interface
- Separation
  equipment

FIGURE 5.104 *Intelsat V* structure. (Courtesy of INTELSAT and FACC.)

The elements discussed here include struts, cylindrical or conical shells, and rectangular panels. Struts are a simple length of material, with axial loading, either in extension or compression; they may also be called bars or columns. Cylindrical or conical shells have been used as the central support of some spacecraft and support both axial and transverse loads. Rectangular panels are frequently used for mounting equipment. Many other elements have been used and analyzed, including beams with transverse loadings, disks or circular plates, and spherical shells. The ends (or edges) of these elements can be free, simply supported (end is fixed but can rotate), or clamped.

The fundamental physical concepts used in structural analysis are shown in Figure 5.106. In part (a), a simple strut has a length L, a width W, and a cross-sectional area A. When an axial force P is applied, the length changes from L to $L + \delta$. The ratio of force to area is called stress, and the ratio of the deformation, δ, to the length L is called a strain ϵ. The ratio of stress to strain is a property of the material and is the modulus of elasticity E. These quantities are summarized in Table 5.19. In addition to the elongation, a transverse contraction also occurs when the axial force P is applied. The transverse strain ϵ_T is the ratio of the contraction to the width. The ratio of the transverse strain ϵ_T to the axial strain ϵ is known as Poisson's ratio.

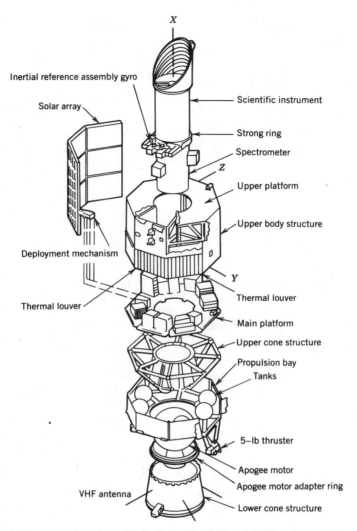

FIGURE 5.105 Structural configuration for International Ultraviolet Experiment spacecraft (IUE).

The shear quantities are defined in terms of part (b) of Figure 5.106. Tangential forces are applied to opposite sides of a material in opposite directions. The force per unit cross-sectional area is the shear stress, and the deformation per unit length is the shear strain. The ratio of shear stress to shear strain is a property of the material, called the modulus of rigidity, or shear modulus. It can be shown that the modulus of rigidity G and the modulus of elasticity E are related by the equation

$$E = 2G(1 + \mu) \tag{5.63}$$

where μ is Poisson's ratio.

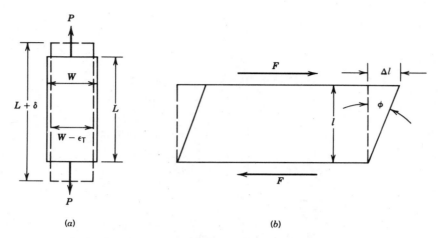

FIGURE 5.106 Cross sections of solid, showing elongation and shear.

TABLE 5.19 Fundamental Physical Concepts Used In Structural Analysis

Quantity	Ratio	Form	Symbol	SI	English
				Units	
Elongation					
Stress	Force/area	P/A	σ	N/mm^2	psi
Strain	Deformation/length	δ/L	ϵ	—	—
Modulus of elasticity	Stress/strain	σ/ϵ	E	N/mm^2	psi
Shear					
Shear stress	Force/area	F/A	σ_s	N/mm^2	psi
Shear strain	Deformation/length	$\Delta l/l$	ϵ_s	—	—
Modulus of rigidity	Shear stress/strain	σ_s/ϵ_s	G	N/mm^2	psi
Relation					
Poisson's ratio	$\dfrac{\text{Transverse contraction}}{\text{elongation}}$	ϵ_T/ϵ	μ	—	—

For small values, the stress is proportional to the strain, and the ratio of these two (the modulus of elasticity) is indeed a constant. For higher loads, a point will ultimately be reached where the ratio is no longer constant (see Figure 5.107). When the "yield stress" is reached, the material starts to yield, the deformation increases more than expected, and the deformation tends to be permanent. If the "yield stress" is exceeded, the object will not return to its original dimensions when the stress is removed and will have had inelastic deformation. If the "ultimate stress" is exceeded, the material breaks.

For a structure to be safe, the stresses encountered should be well below the yield stress. However, the need to minimize the mass of the structure tends to increase the actual stresses within safe limits. For use in structure design, the limit

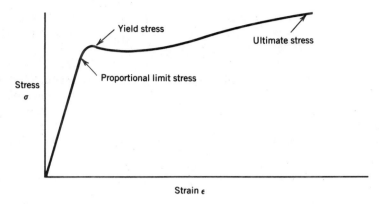

FIGURE 5.107 Elastic region, yield point, and ultimate failure of typical material.

load is taken as the 2-sigma point of flight loads, that is, on 98% of the flights the loads will be below the limit loads. The ultimate loads are taken arbitrarily as 1.5 times the limit loads. The margin of safety is then defined as

$$\text{Margin of safety} = \frac{\text{allowable stress}}{\text{actual stress}} - 1 \qquad (5.64)$$

where the actual stress is that calculated to occur under an ultimate load, and the allowable stress is the maximum stress the structure can sustain. Under ultimate loads, the spacecraft structure should not have inelastic deformation, and the margin of safety should be greater than zero.

5.9.2.1 Strut

Under tension, the allowable stress on a strut is limited to the linear portion of the curve. Under compression, a strut can bend and break for a much lower stress. The allowable stress f_c under compression is given by Euler's formula as

$$f_c = \frac{C\pi^2 E}{(L/\rho)^2} = \frac{\pi^2 E}{(L'/\rho)^2} \qquad (5.65)$$

where E is the modulus of elasticity, L' is the effective length ($L' = L/\sqrt{C}$), and ρ is the radius of gyration of the cross section. The radius of gyration is the distance from the axis at which the total mass might be concentrated without changing its moment of inertia; for a rod of radius r it is just $r/2$, and for a square cross section it is $\rho = w/\sqrt{12}$ for a width w. Notice that failure of a strut under tension depends on how strong it is (yield stress F), but failure under compensation depends on how stiff it is (Young's modulus E).

The constant C depends on the end conditions, that is, how the strut is held.

Part (a) of Figure 5.108 illustrates a few possibilities. If the ends are simply supported, $C = 1$; this is the case when the ends are held at fixed points but are free to rotate. If the ends are simply supported and the center of the strut is constrained, the effective length is cut in half, the allowable stress is increased by 4, and $C = 4$. If the ends are clamped and not allowed to rotate, $C = 4$. The fourth case shows one end clamped and the other end free; the value of C is $\frac{1}{4}$. The last case shows one end clamped and the other end simply supported; the resultant C is 2.05.

Euler's formula applies only to the case of a strut bending. As the strut gets

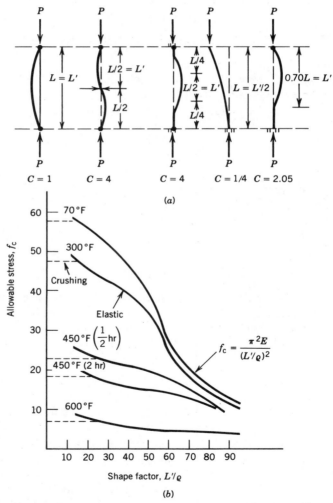

FIGURE 5.108 Allowable stress for strut under axial compression (Bruhn, 1965). Notice that the theoretical formula (Equation 5.65) is just a rough approximation to the experimental curve.

thicker or the length gets shorter, other failure mechanisms take over. Part (b) of Figure 5.108 shows a transition range of shape factors where failure is due to inelastic instability of the column. The range for shorter columns is where failure is due to the plastic crushing of the column.

5.9.2.2 Shell

The thin-walled cylinder has become important in the design of launch vehicles and spacecraft structures. The strength design of such structures is based primarily on best-fit curves to test results, using theoretical parameters and curves when they appear applicable. The term *monocoque cylinder* means a thin-walled cylinder without any additional stiffeners. The geometry is shown in Figure 5.109. For the case of local buckling, the critical stress σ_{cr} can be written as

$$\sigma_{cr} = K_c \frac{Et/r}{\sqrt{3(1 - \mu^2)}} \tag{5.66}$$

where E is the modulus of elasticity, t the shell thickness, r the cylinder radius, μ Poisson's ratio, and K_c is a buckling coefficient of the order of unity that depends

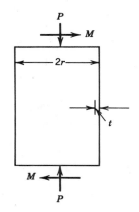

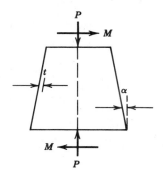

FIGURE 5.109 Thin-walled cylinder and cone (monocoque).

on edge boundary conditions. The critical axial force or load P_{cr} is

$$P_{cr} = K_c \frac{2\pi E t^2}{\sqrt{3(1 - \mu^2)}} \qquad (5.67)$$

and is obtained by multiplying the cylinder cross section $2\pi rt$ by the critical stress.

Lateral loads also occur on monocoque cylinders, and it is necessary to find the bending strength of the cylinders. The equations take the same form as for axial compression but with a bending coefficient K_b substituted for the buckling coefficient K_c. Some designers have assumed that the two coefficients are equal, while others have assumed that K_b is 30% higher than K_c; the truth is probably between these two limits. Bruhn (1965) points out that under compression all parts of the cylinder are under stress, whereas with bending the stress varies from zero at the neutral axis to a maximum at the most remote element. The lower probability of an imperfection occurring within the smaller highest stressed region would lead one to conclude that higher buckling stresses in bending should be expected.

A thin-walled (monocoque) conical shell is similar to the cylinder. By including an extra term, the critical force P_{cr}, or load, can be written as

$$P_{cr} = K_c \frac{2\pi E t^2}{\sqrt{3(1 - \mu^2)}} \cos^2 \alpha \qquad (5.68)$$

where α is the semi-vertex angle of the cone.

5.9.2.3 Panel

A flat surface is often used to mount the equipment on a spacecraft; the panel may be rectangular (see Figure 5.110). The maximum stress σ_{max} of a uniformly loaded

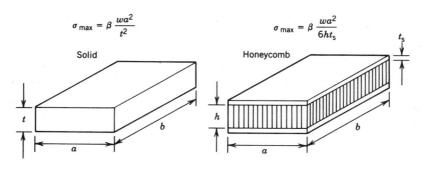

All edges supported

b/a	1	1.4	1.6	2	3	4	5	∞
β	0.2874	0.4530	0.5172	0.6102	0.7134	0.7410	0.7476	0.75

FIGURE 5.110 Coefficient for solid and honeycomb panels.

panel can be written as

$$\sigma_{max} = \beta \, \frac{wa^2 C}{6I} \tag{5.69}$$

where w is the uniform load per area, a is the smaller side dimension, C is the distance from the center of the section to the top skin, and I is the cross-sectional moment per unit length. The coefficient β is a function of the plate aspect ratio (length–width) and the support on the edges. If all edges are simply supported, the values of the coefficient for a number of aspect ratios are given in Figure 5.110. For a square the coefficient is 0.2874, and for an infinitely long panel the coefficient is 0.75.

For a solid homogeneous panel of thickness t the formula for the maximum stress σ_{max} reduces to

$$\sigma_{max} = \beta \, \frac{wa^2}{t^2} \tag{5.70}$$

since $C = t/2$ and $I = t^3/12$. The formula for a honeycomb panel is more complicated, but it is important because of the reduction in mass that is possible. If the faces of the honeycomb each have a thickness t_s and the distance between the two faces is h, the cross-sectional moment per unit length is $I = (h + t_s)^2 \, t_s/2$, and the distance from the honeycomb center to the center of the skin is $C = (h + t_s)/2$. The maximum stress of a uniformly loaded panel is then

$$\sigma_{max} = \beta \, \frac{wa^2}{6ht_s} \tag{5.71}$$

where the assumption is made that the skin thickness t_s is small compared to the distance h between the two skins. In all these formulas, it should be remembered that a specific failure mode is assumed; other failure modes require different formulas, which in some cases may take precedence.

5.9.3 Resonance Effects

A structure may be perfectly safe under static loads and yet fail because of resonance effects. A small force applied periodically at just the resonant frequency may produce an oscillation that increases gradually in amplitude and that can then cause damage. The resonant frequency ω_n of a rectangular panel can be expressed as

$$\omega_n = \beta \, \sqrt{\frac{D}{\rho a^4}} \tag{5.72}$$

where D is the stiffness, ρ the mass per area, and a the smaller side dimension. The coefficient β depends on the aspect ratio (width–length) and how the edges are supported; a few representative values are given in Table 5.20. More detailed tables are in the literature (Pilkey and Chang, 1978). For a homogeneous panel, the stiffness D is

$$D = \frac{Et^2}{12(1 - \mu^2)} \qquad (5.73)$$

where E is the modulus of elasticity, t the thickness of the panel, and μ the Poisson ratio. For a honeycomb panel, the stiffness D is equal to

$$D = \frac{Eht_s^2}{2(1 - \mu^2)} \qquad (5.74)$$

where t_s is the thickness of the honeycomb skin, and h is the distance between the two skins.

TABLE 5.20 Coefficient To Determine Natural Frequencies of Panels[a]

Edges	Ratio						
All edges S	b/a	1.0	1.5	2.0	2.5	3.0	∞
	β	19.74	14.26	12.34	11.45	10.97	9.87
C left; S top, S right, S bottom	b/a	1.0	1.5	2.0	2.5	3.0	∞
	β	23.65	18.90	17.33	16.63	16.26	15.43
	a/b	1.0	1.5	2.0	2.5	3.0	∞
	β	23.65	15.57	12.92	11.75	11.14	9.87
C left, C right; S top, S bottom	b/a	1.0	1.5	2.0	2.5	3.0	∞
	β	28.95	25.05	23.82	23.27	22.99	22.37
	a/b	1.0	1.5	2.0	2.5	3.0	∞
	β	28.95	17.37	13.69	12.13	11.36	9.87
All edges C	b/a	1.0	1.5	2.0	2.5	3.0	∞
	β	35.98	27.00	24.57	23.77	23.19	22.37

[a] Harris and Crede (1976). Copyright McGraw-Hill 1976; reproduced with permission.
[b] Abbreviations: S, simply supported; C, clamped edge.

5.9.3.1 System with One Degree of Freedom

To understand the effect of resonances in a spacecraft structure, consider a few simple cases. In Figure 5.111 a mass M is suspended by a spring that has an elastic constant K; the mass moves in a single direction. The displacement x then is governed by the differential equation

$$M\ddot{x} + Kx = 0 \qquad (5.75)$$

which says that the restoring force Kx must be equal and opposite to the acceleration $M\ddot{x}$. The solution is

$$x = A \sin (\omega_n t + \phi) \qquad (5.76)$$

The displacement x varies sinusoidally with an arbitrary amplitude A, an arbitrary phase ϕ, and an undamped natural frequency ω_n equal to

$$\omega_n = \sqrt{\frac{K}{M}} \qquad (5.77)$$

determined by the elastic constant K and the mass M.

If a damping force proportional to $\dot{x}$ is introduced, the equation becomes

$$M\ddot{x} + C\dot{x} + Kx = 0 \qquad (5.78)$$

where C is a constant of proportionality. This can be written as

$$\ddot{x} + 2\xi\omega_n\dot{x} + \omega_n^2 x = 0 \qquad (5.79)$$

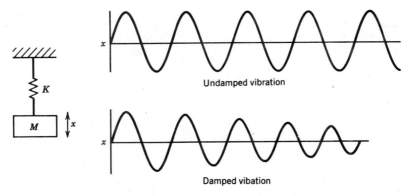

Undamped vibration

Damped vibration

FIGURE 5.111 Oscillations of system with single degree of freedom.

where the damping ratio $\xi = C/C_c$, and the critical damping constant $C_c = 2\sqrt{KM}$. The solution to this equation for damping ratios less than unity can be written as

$$x = Be^{-\xi\omega_n t} \sin\left(\omega_n \sqrt{1 - \xi^2}\, t + \phi\right) \qquad (5.80)$$

which is the equation for damped oscillations. Any oscillations generated by the initial conditions will gradually decay in time with a time constant of $1/\xi\omega_n$.

An additional change of the problem is to apply a force of $P_0 \cos \omega t$ with an amplitude P_0 and a driving frequency ω. The differential equation can then be written as

$$M\ddot{x} + C\dot{x} + Kx = P_0 \cos \omega t \qquad (5.81)$$

Solutions for different damping ratios are plotted in Figure 5.112 as a function of driving frequency ω. Note that the acceleration amplitude is greatest when the driving frequency ω matches the natural frequency ω_n. Also, the smaller the amount of damping, ξ, the larger the amplitude. In a spacecraft, the natural frequencies can be predicted from the structural analysis and can also be measured in tests. It is more difficult to predict how much damping will be present in an actual structure, and this damping influences the resulting amplitudes.

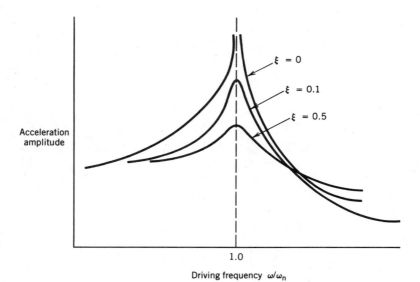

FIGURE 5.112 Amplitude of systems with one degree of freedom for different damping ratios.

5.9.3.2 System with Two Degrees of Freedom

A more complex resonant system is shown in Figure 5.113, where a second mass is attached to the first mass with another spring. It is assumed the masses are equal and the spring constants are equal. Equating the forces, the two differential equations are

$$M\ddot{x}_1 + Kx_1 + K(x_1 - x_2) = 0 \tag{5.82}$$

$$M\ddot{x}_2 + K(x_2 - x_1) = 0 \tag{5.83}$$

where x_1 is the displacement of the first mass, and x_2 the displacement of the second mass. These two equations can be written as a single equation in matrix notation as

$$\begin{bmatrix} M & 0 \\ 0 & M \end{bmatrix} \begin{bmatrix} \ddot{x}_1 \\ \ddot{x}_2 \end{bmatrix} + \begin{bmatrix} 2K & -K \\ -K & K \end{bmatrix} \begin{bmatrix} x_1 \\ x_2 \end{bmatrix} = 0 \tag{5.84}$$

To solve these equations, a solution is assumed in the form of the two masses oscillating at the same frequency ω but with different amplitudes C_1 and C_2:

$$\begin{bmatrix} x_1 \\ x_2 \end{bmatrix} = \begin{bmatrix} C_1 \\ C_2 \end{bmatrix} e^{j\omega t} \tag{5.85}$$

Substituting this assumed solution into the differential equation,

$$\begin{bmatrix} 2K - \omega^2 M & -K \\ -K & K - \omega^2 M \end{bmatrix} \begin{bmatrix} C_1 \\ C_2 \end{bmatrix} e^{j\omega t} = 0 \tag{5.86}$$

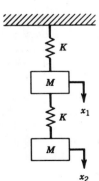

x_2 **FIGURE 5.113** System with two degrees of freedom.

This equation can only be satisfied if the determinant of the matrix is equal to zero. This can be set to zero and solved for the angular frequency ω. For each frequency, the solution also requires a certain ratio of the two amplitudes C_1 and C_2, that is, one mass may oscillate at an arbitrary amplitude, but then the other one has an oscillating amplitude set by the equations.

The solution with the lower frequency is

$$\omega_1 = 0.61 \sqrt{\frac{K}{M}} \qquad C_1 = 0.61\, C_2$$

In this solution, the two masses are in phase and oscillating at a frequency of 0.61 of the frequency of the single mass by itself. The amplitude of the second mass is greater than the amplitude of the first mass. The solution with the higher frequency is

$$\omega_2 = 1.6 \sqrt{\frac{K}{M}} \qquad C_1 = -1.6\, C_2$$

In this solution, the two masses are out of phase and oscillating at a frequency higher than the single mass frequency. When the first mass goes down the second mass goes up, and vice versa. The amplitude for the first mass is greater than the amplitude for the second mass. Since the two masses are out of phase, the second spring tends to oppose the motion of the first mass, so the resonant frequency is higher than the original frequency.

The general solution can be written as the sum of the two preceding solutions; in matrix form,

$$\begin{bmatrix} x_1 \\ x_2 \end{bmatrix} = K_1 \begin{bmatrix} 0.61 \\ 1 \end{bmatrix} \sin\left(\omega_1 t + \phi_1\right) + K_2 \begin{bmatrix} -1.6 \\ 1 \end{bmatrix} \sin\left(\omega_2 t + \phi_2\right) \quad (5.87)$$

The constants K_1, K_2, ϕ_1, and ϕ_2, depend on the initial conditions. If $K_2 = 0$, only the first mode is excited, and the two masses oscillate at a frequency ω_1; conversely, if $K_1 = 0$, the two masses oscillate at a frequency ω_2. When both modes of oscillation are present, the actual motion is a sum of sinusoidal motion of the two frequencies.

In actual structures some damping is always present. If damping terms are introduced into the equation, exponential terms that decrease with time will be in the solution. When the damping of one oscillatory mode is greater than the other, that oscillation will decrease faster; the result is that even if both modes are excited initially, eventually only oscillation at one frequency is observed. Often higher frequency modes decay first, and lower frequencies are observed the longest.

Of more practical interest for spacecraft structures is the case of forced vibration. For the two masses, an oscillating force on the first mass would introduce a sinusoidal term in the first equation; a force on the second mass would correspond to a forcing term in the second equation. In either case the solution, after transients decay, is a steady oscillation of the two masses, with the amplitude depending on the strength of the forcing function, the driving frequency, and the damping present. As the driving frequency is changed, the amplitudes will be greater as the frequency approaches either of the two resonant frequencies ω_1 and ω_2. The amplitudes will also be larger if the damping is smaller. Theoretically, if the driving frequency equals a resonant frequency and if there is no damping, the amplitude will increase to infinity (i.e., there is no steady-state value).

5.9.3.3 System with Many Degrees of Freedom

An actual spacecraft structure consists of many masses connected by structural members. Each mass has six degrees of freedom, but some are not modeled, so each mass may have one to six degrees of freedom in the model. There are many natural resonant frequencies each with its own mode of vibration. At a given frequency, some masses may oscillate at a greater amplitude than others, but usually all masses oscillate to some extent. During launches, these modes are driven by the vibrations caused by the rocket motors. The lower frequencies are driven by oscillations of the mounting surface, where the spacecraft is attached to the launch vehicle. The higher frequencies may be driven by acoustic coupling, where the larger spacecraft surfaces are driven by the sound vibrations in the air (before the vehicle has completely left the earth's atmosphere).

A detailed analysis of a spacecraft structure is usually done with a computer program. A program widely used and well known is NASTRAN, although other excellent programs are available. NASTRAN uses a lumped-element approach, in which the distributed physical properties of a structure are represented by a model consisting of a finite number of idealized elements interconnected at a finite number of grid points. Static or dynamic loads can be applied and displacements, accelerations, and stresses calculated. Line drawings can also be generated showing the intersections (nodes), connecting lines (bars), and sections of surface (panels); an example is shown in Figure 5.114.

Computer models are usually generated before the structure is constructed. If the maximum stresses are too large, struts can be made stronger or additional members added. If the maximum stresses are quite small, weight savings can be made by reducing the mass or the number of elements. After the structure is built, test results can be compared with the computer model results. Additional modifications can be made to bring the model into closer correspondence with the actual structure.

5.9.4 Material Properties

The mechanical properties for a number of materials used on spacecraft are listed in Table 5.21. The ultimate tensile strength for extension and for shear are listed in columns 4 and 5. The longitudinal tensile yield shows the stress at which perma-

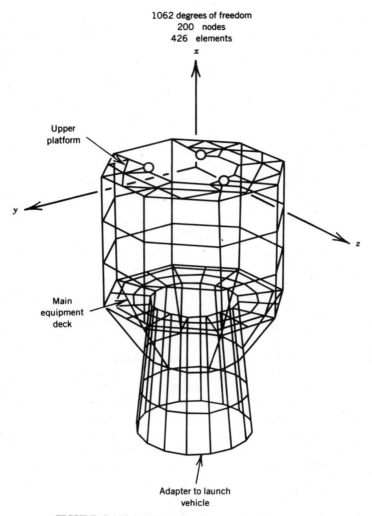

1062 degrees of freedom
200 nodes
426 elements

Upper platform

Main equipment deck

Adapter to launch vehicle

FIGURE 5.114 NASTRAN model of satellite structure.

nent deformation starts to occur. The modulus of elasticity (Young's modulus) and the modulus of rigidity (shear modulus) are listed in columns 7 and 8. The last two columns are figures of merit for spacecraft materials, since it is important to save weight. The lightest materials are desired, so the important criterion is the ultimate tensile strength and the modulus of elasticity divided by the density of the material. Some of the better values of these two figures of merit are plotted in Figure 5.115.

The thermal properties of the same materials are listed in Table 5.22. Higher values of specific heat decrease temperature variations since it enables more heat to be stored. Low values of thermal expansion are desirable, especially where

TABLE 5.21 Mechanical Properties of Spacecraft Materials

Material Type	Density, ρ (kg/m³)	Longitudinal Ultimate Tensile Strength, F (N/mm²)	Transverse Ultimate Tensile Strength (N/mm²)	Longitudinal Tensile Yield (N/mm²)	Youngs Modulus, E (N/mm²)	Shear Modulus, G (N/mm²)	Specific Strength, F/ρ (km/sec)²	Specific Stiffness, E/ρ (km/sec)²
Aluminum alloy								
Sheet 2014–T6	2,800	440	—	390	72,000	28,000	0.16	26
Sheet 2024–T36	2,770	480	—	410	72,000	28,000	0.17	26
Sheet 6061–T6	2,710	290	—	240	68,000	26,000	0.11	25
Sheet 7075–T6	2,800	520	—	450	71,000	27,000	0.19	25
Beryllium								
Extrusion —	1,850	620	—	410	290,000	140,000	0.33	158
Lockalloy BE–38% AL	2,100	427	—	430	190,000	—	0.20	88
Sheet Cross rolled	1,850	450	—	290	290,000	140,000	0.24	158
Wrought Hot pressed	1,830	280	—	180	290,000	140,000	0.15	160
Boron–epoxy Avco 5505/4								
[0] v_f 50%	2,000	1320	72	—	200,000	5,000	0.66	100
[0,/±45] v_f 55%	2,000	720	108	—	120,000	—	0.36	60
Graphite-epoxy								
[0] HTS	1,490	1,340	67	—	150,000	5,900	0.89	100
[0,/±45] HTS	1,490	640	290	—	83,000	—	0.43	56
[0] HM	1,600	680	29	—	190,000	5,900	0.42	120
[0,/±45] UHM	1,690	620	20	—	290,000	4,100	0.37	170
Invar 36 Annealed	8,080	490	—	280	145,000	56,000	0.06	18
Magnesium								
Extrusion tubes AZ31B	1,770	220	—	110	45,000	16,000	0.12	25
Sheet AZ31B–H24	1,770	270	280	200	45,000	16,000	0.15	25
Steel								
PH15–7 Mo RH1050	7,670	1,310	—	1,200	200,000	76,000	0.17	26
4130 Cr Mo 1350°F	7,830	860	—	710	200,000	76,000	0.11	25
Ti6AL–4V								
Sheet	4,430	1,100	—	1,000	110,000	43,000	0.25	25
Forgings and bar	4,430	1,030	—	960	110,000	43,000	0.23	25
Kevlar 49, [0] v_f 60%	1,380	1,380	30	—	76,000	2,000	1.00	55
"S" glass v_f 60%	2,020	1,620	—	—	54,000	—	0.80	27

TABLE 5.22 Thermal Properties of Spacecraft Materials

	Material Type	Specific Heat c J/kg K	Thermal Expansion α 10^{-6}/K	Thermal Conductivity κ W/m K	Thermal Parameters κ/α 10^6 W/m	$\kappa/\rho c$ mm^2/sec
Aluminum alloy						
Sheet	2014-T6	960	22.5	160	6.9	69
Sheet	2024-T36	880	22.5	120	5.4	60
Sheet	6061-T6	960	23.4	170	7.1	76
Sheet	7075-T6	840	22.9	140	5.9	69
Beryllium						
Extrusion	—	1860	11.5	180	15.6	62
Lockalloy	BE-38% AL	—	16.9	210	12.6	—
Sheet	Cross rolled	1860	11.5	180	15.6	62
Wrought	Hot pressed	1860	11.5	180	15.6	63
Boron–epoxy	Avco, 5505/4					
[0]	v_f 50%	920	4.18	1.9	0.45	1
[0,/±45]	v_f 55%	—	4.64	0.3	0.08	—
Graphite–epoxy						
[0]	HTS	—	-0.4	—	—	—
[0,/±45]	HTS	—	—	—	—	—
[0]	HM	—	—	—	—	—
[0,/±45]	UHM	—	-1.04	—	—	—
Invar 36	Annealed	510	1.3	13.5	10.7	3
Magnesium						
Extrusion tubes	AZ31B	1050	25.2	97	3.8	63
Sheet	AZ31B-H24	1050	25.2	97	3.8	63
Steel						
PH15-7 Mo	RH1050	—	11.0	15	1.40	—
4130 Cr Mo	1950°F	480	11.3	38	3.35	12
Ti6AL-4V						
Sheet		500	8.8	7.4	0.84	4
Forgings and bar		500	8.8	7.4	0.84	4
Kevlar 49, [0]	v_f 60%	—	-4.0	1.7	-0.44	—
"S" glass	v_f 60%	—	—	—	—	—

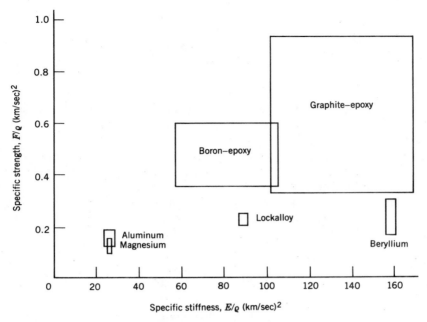

FIGURE 5.115 Figures of merit for materials used in spacecraft structures.

precise alignment is needed, such as for antennas. High values of thermal conductivity are useful in conducting heat away from concentrated heat sources, such as power amplifiers. Column 6 lists the ratio of thermal conductivity to thermal expansion; a high value of this ratio is desirable. The last column lists the thermal diffusivity.

Many spacecraft structures are metals such as aluminum and magnesium that are readily available, low cost, and easily fabricated. Other materials are chosen for specific applications. Table 5.23 summarizes a few of the principal elements

TABLE 5.23 Materials for Structural Elements

Element	Design Criterion	Desired Properties	Examples of Advanced Materials
Struts	Prevent buckling	High modulus	Beryllium, unidirectional composites
Shells	Prevent buckling	High modulus	Beryllium, advanced composites
Panels	Increase frequency, conduct heat	Stiffness, high thermal conductivity	Metal honeycomb with metal faces; graphite–epoxy and boron–epoxy honeycomb need metal inserts
Rings	Strength, conduct heat	High tensile strength, high thermal conductivity	For Al, Mg, lock alloy face sheets use same material; for Be, graphite–epoxy face sheets use titanium

in a structure, the design criterion used in choosing materials, the desired properties, and some examples of more advanced materials used.

Beryllium has a specific stiffness six times higher than conventional materials.
It is brittle and tends to fracture under low-impact loads. There is a significant
difference in the mechanical properties in the direction of rolling and transverse to
it. Etching is necessary to remove microcracks due to machining and drilling. The
chief disadvantages of beryllium are the dust particles in the air that can cause
berylliosis and the high cost for both the material and for manufacture.

Graphite–epoxy consists of graphite filaments imbedded in a matrix of epoxy.
It has a high specific strength and stiffness. Its low thermal expansion makes it
ideal for antennas and antenna support. It can be contoured into intricate shapes
without filament breakage. In contrast, boron–epoxy, which uses boron filaments,
is more difficult to contour into intricate shapes. Boron filaments require diamond
or carbide cutters because of their extreme hardness. Both graphite–epoxy and
boron–epoxy can be joined only by bonding or with mechanical fasteners.

5.10 THERMAL CONTROL

5.10.1 Introduction

There is no air in the geostationary orbit, so heat is transferred only by conduction
through solids and radiation through space. The temperature of a spacecraft is set
by incident solar energy, internal electrical dissipation, and thermal radiation into
space. The heat radiated by a black body, q, is proportional to the fourth power
of the absolute temperature T and is given by the Stefan–Boltzmann law of radiation as

$$q = A\sigma T^4 \qquad (5.88)$$

where A is the surface area and σ the Stefan–Boltzmann constant ($=5.6703 \pm$
0.007×10^{-8} W/m^2 K^4). For nonblack bodies the radiated power is less and is
written as

$$q = \epsilon A\sigma T^4 \qquad (5.89)$$

where the emissivity ϵ is always between zero and unity.

The radiated power of a black body per unit area is given in Table 5.24 for
various temperatures. Thus, for a temperature of 294 K (70 °F) the radiation is
423 W/m^2. If 42 W are to be radiated, a radiator of at least 0.1 m^2 is required
(black, with no incident radiation). If there is 310 W/m^2 incident (from an environment of 272 K), the net heat transfer is 113 W/m^2, and 0.1 m^2 radiator would
only radiate a net power of 11.3 W. The value of σT^4 always appears in radiation
problems, and the spacecraft thermal designer uses it as an alternate temperature
scale.

The temperature of space can be defined as the equilibrium temperature of a small black sphere (see Section 5.1.4). For the spacecraft thermal designer, this definition is more useful than alternate definitions. The temperature of space is shown in Figure 5.116 for geostationary orbit. The input thermal power is the flux S times the projected area, πr^2, and the heat radiated by the black sphere of radius r is $4\pi r^2 \sigma T^4$; in equilibrium these two must be equal. In sunlight, the solar flux averages 1370 W/m², but varies annually from 1325 to 1415 W/m²; the sunlight temperature is then 275–280 K. In eclipse, the major input is thermal radiation from the earth; the earth's equivalent temperature is roughly from 220 to 275 K. The radiation near the earth's surface is 140–320 W/m²; at the geostationary orbit it is 3–7 W/m², and the temperature is 60–75 K there.

The spectral distribution of sunlight and of the earth's radiation is shown in Figure 5.117. Sunlight is mostly in the visible spectrum and in the near infrared; thermal radiation from the earth is at longer wavelengths, on the order of 10 μm. The absorptivity of any spacecraft surface to the earth's radiation is usually very close to the emissivity since the wavelengths are similar. The absorptivity to sunlight may be quite different and is denoted by α. Typical values of emissivity ϵ and absorptivity α are shown in Table 5.25. These are typical values, but actual values vary depending on the preparation of the surface. The ratio of absorptivity α to emissivity ϵ tends to be between 0.1 and 10; special care is often taken to get a high or low ratio. It is also important to know the values at the end of the spacecraft lifetime since some white surfaces darken with age.

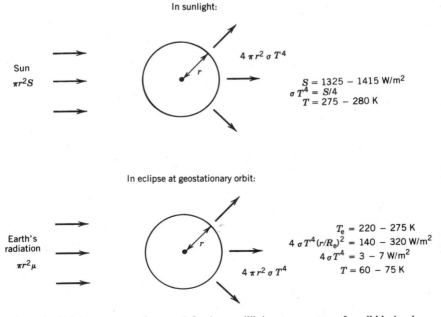

FIGURE 5.116 Temperature of space, defined as equilibrium temperature of small black sphere.

TABLE 5.24 Power Radiated by Black Body[a]

°C	K	W/m²	°C	K	W/m²	°C	K	W/m²	°C	K	W/m²
-271	2	0.00	-171	102	6.14	-71	202	94.41	29	302	471.67
-269	4	0.00	-169	104	6.63	-69	204	98.20	31	304	484.28
-267	6	0.00	-167	106	7.16	-67	206	102.11	33	306	497.15
-265	8	0.00	-165	108	7.71	-65	208	106.14	35	308	510.28
-263	10	0.00	-163	110	8.30	-63	210	110.28	37	310	523.66
-261	12	0.00	-161	112	8.92	-61	212	114.54	39	312	537.31
-259	14	0.00	-159	114	9.58	-59	214	118.92	41	314	551.22
-257	16	0.00	-157	116	10.27	-57	216	123.43	43	316	565.40
-255	18	0.01	-155	118	10.99	-55	218	128.07	45	318	579.85
-253	20	0.01	-153	120	11.76	-53	220	132.83	47	320	594.57
-251	22	0.01	-151	122	12.56	-51	222	137.73	49	322	609.58
-249	24	0.02	-149	124	13.41	-49	224	142.76	51	324	624.86
-247	26	0.03	-147	126	14.29	-47	226	147.92	53	326	640.44
-245	28	0.03	-145	128	15.22	-45	228	153.23	55	328	656.30
-243	30	0.05	-143	130	16.19	-43	230	158.68	57	330	672.45
-241	32	0.06	-141	132	17.21	-41	232	164.27	59	332	688.90
-239	34	0.08	-139	134	18.28	-39	234	170.01	61	334	705.65
-237	36	0.10	-137	136	19.40	-37	236	175.90	63	336	722.71
-235	38	0.12	-135	138	20.56	-35	238	181.93	65	338	740.07
-233	40	0.15	-133	140	21.78	-33	240	188.13	67	340	757.74
-231	42	0.18	-131	142	23.05	-31	242	194.48	69	342	775.73
-229	44	0.21	-129	144	24.38	-29	244	200.99	71	344	794.04
-227	46	0.25	-127	146	25.76	-27	246	207.66	73	346	812.66
-225	48	0.30	-125	148	27.21	-25	248	214.49	75	348	831.62
-223	50	0.35	-123	150	28.71	-23	250	221.50	77	350	850.90
-221	52	0.41	-121	152	30.27	-21	252	228.67	79	352	870.52
-219	54	0.48	-119	154	31.89	-19	254	236.02	81	354	890.47
-217	56	0.56	-117	156	33.58	-17	256	243.54	83	356	910.76

684

°C	K	σT_K^4	°C	K	σT_K^4	°C	K	σT_K^4	°C	K	σT_K^4
−215	58	0.64	−115	158	35.34	−15	258	251.24	85	358	931.40
−213	60	0.73	−113	160	37.16	−13	260	259.12	87	360	952.39
−211	62	0.84	−111	162	39.05	−11	262	267.18	89	362	973.73
−209	64	0.95	−109	164	41.02	−9	264	275.44	91	364	995.43
−207	66	1.08	−107	166	43.06	−7	266	283.88	93	366	1017.49
−205	68	1.21	−105	168	45.17	−5	268	292.51	95	368	1039.91
−203	70	1.36	−103	170	47.36	−3	270	301.34	97	370	1062.71
−201	72	1.52	−101	172	49.63	−1	272	310.37	99	372	1085.87
−199	74	1.70	−99	174	51.98	1	274	319.60	101	374	1109.41
−197	76	1.89	−97	176	54.41	3	276	329.04	103	376	1133.33
−195	78	2.10	−95	178	56.92	5	278	338.68	105	378	1157.64
−193	80	2.32	−93	180	59.52	7	280	348.53	107	380	1182.33
−191	82	2.56	−91	182	62.21	9	282	358.59	109	382	1207.42
−189	84	2.82	−89	184	64.99	11	284	368.88	111	384	1232.91
−187	86	3.10	−87	186	67.87	13	286	379.38	113	386	1258.80
−185	88	3.40	−85	188	70.83	15	288	390.10	115	388	1285.09
−183	90	3.72	−83	190	73.90	17	290	401.05	117	390	1311.79
−181	92	4.06	−81	192	77.06	19	292	412.23	119	392	1338.91
−179	94	4.43	−79	194	80.32	21	294	423.64	121	394	1366.44
−177	96	4.82	−77	196	83.68	23	296	435.28	123	396	1394.40
−175	98	5.23	−75	198	87.15	25	298	447.17	125	398	1422.78
−173	100	5.67	−73	200	90.72	27	300	459.29	127	400	1451.60

[a]The first two columns are the temperature in Celsius and Kelvin. The third column is σT_K^4.

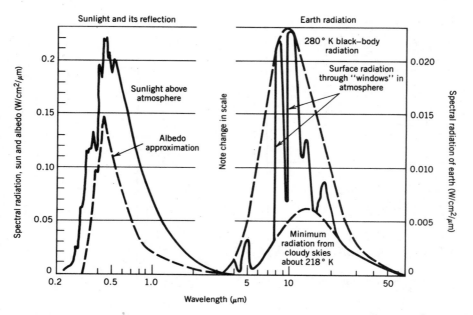

FIGURE 5.117 Spectral distribution in solar radiation, reflected solar radiation (albedo), and earth radiation (Van Vliet, 1965). Reprinted with permission of Macmillan Publishing Company from *Passive Temperature Control in the Space Environment* by Robert M. Van Vliet. Copyright © 1965 by Robert M. Van Vliet.

The average satellite temperature is found in Figure 5.118 and is a function of the input radiation (usually solar, S); any internal dissipation Q; the ratio of absorptivity to emissivity, α/ϵ; and the ratio of projected area to total surface area, a/A. By equating the input to the output, the average temperature is found to be

$$\sigma T^4 = \frac{\alpha}{\epsilon} \frac{a}{A} S + \frac{Q}{\epsilon A} \qquad (5.90)$$

where σ is the Stefan–Boltzmann constant. In thermal calculations the input electrical power that is converted to rf energy is usually excluded from the energy balance equation.

For simplicity, an isothermal spacecraft has been assumed in the previous discussion. The calculations are, however, applicable to any spacecraft if averages are suitably defined by the equation

$$\int_A \epsilon\sigma T^4 \, dA = \int_a \alpha S \, da + Q \qquad (5.91)$$

The average absorptance is averaged over the projected area a. The average temperature is the fourth-power average taken over the total spacecraft surface A

TABLE 5.25 Thermal Properties of Typical Surfaces, Solar Absorptivity and Thermal Emissivity[a]

	α	ϵ	α/ϵ
Flat Reflectors (low α, low ϵ)			
Al acrylic paint; nonleafing (Sherwin Williams)	0.42	0.47	0.89
Al–silicone paint, W. P. Fuller, 171–A–152, leafing Al	0.42	0.45	0.93
Cat-a-Lac Al paint (Finch Paint and Chemical)	0.25	0.25	1.0
Flat Absorbers (high α, high ϵ)			
Cat-a-Lac black 433–1–8 epoxy (Finch Paint and Chemical)	0.96	0.85	1.10
Dow 17 on HM21 Mg (Dow Chemical)	0.78	0.70	1.11
Stainless steel, sand-blasted	0.75	0.85	0.88
Solar Absorbers (high α, low ϵ)			
Polished Al	0.20	0.05	4.0
Al foil, plain, dry, annealed	0.12	0.05	2.4
$CeO_{2x}/Mo_x/MgF_2$ (1000–200 A) (500–1000 A)	0.85	0.11	7.7
Solar Reflectors, Inorganic (low α, high ϵ)			
SP–500 ZnO in PS–7 K_2SiO_3 (Z–93)	0.16	0.86	0.19
$ZrO_2 \cdot SiO_2–K_2SiO_3$ (synthetic)	0.07	0.85	0.08
$LiAlSiO_4$ (synthetic)–K_2SiO_3	0.12	0.90	0.13
Solar Reflectors, Organic (low α, high ϵ)			
ZnO (SP–500) in methyl silicone (GE RTV–602) (IITRI S–13)	0.18	0.87	0.21
Solar Reflectors, Mirrors (low α, high ϵ)			
Optical solar reflector; Ag–coated fused silica, overcoated with vapor-deposited Inconel	0.05	0.80	0.062

[a]Excerpts from a larger table (Rittenhouse and Singletary, 1968). Reprinted by permission from Lockheed Missiles and Space Co. Inc., Sunnyvale, CA.

but weighted by the emittance at each surface element. Similarly, if there are cyclic variations of orbits or equipment cycles, the quantities are averaged over time.

More important as a practical matter is the notion that the average temperature represents a number that will be the temperature of some point within the spacecraft. Thus, by quick and simple calculations the temperature of the spacecraft has been determined even if the location of this particular temperature may still be unknown. It also represents a goal that can often be achieved for any desired

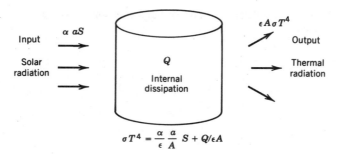

$$\sigma T^4 = \frac{\alpha}{\epsilon}\frac{a}{A}\, S + Q/\epsilon A$$

FIGURE 5.118 Average temperature of satellite depends on ratio of surface absorptivity α to emissivity ϵ and ratio of projected area a to total surface area A.

internal point if suitable insulation or heat transfer paths are provided. Finally, any fluctuations calculated for the average spacecraft temperature, such as those due to a change in sun angle, will be reflected in similar temperature fluctuations for every point in the spacecraft. The calculation of the average spacecraft temperature is a valuable tool even though it will be supplanted by more detailed calculations of the temperature at many points in the spacecraft and at many different times.

5.10.2 Transients

Often the major interest in a thermal analysis is the steady-state solution, when all inputs are constant and the temperature of each body has stabilized so it no longer changes with time. The steady-state solution is important, not only because the temperatures are often stabilized but because it can be a first approximation to a transient solution and sometimes the first step in finding the solution of a time-varying system.

When there is more heat going into a body than coming out, its temperature rises. The net thermal power input is equal to power in from radiant, thermal, or electrical sources minus the thermal power dissipated. The rate at which the temperature rises (dT/dt) is equal to the net thermal power input q divided by the thermal mass of the body. The thermal mass is equal to the product of the mass m and the specific heat c. The specific heat is a property of a solid and can be defined by the equation

$$q = mc\,\frac{dT}{dt} \tag{5.92}$$

If the thermal power input q is measured in watts, and the mass m is in kilograms, the specific heat c is given in J/kg K. The specific heat of water is equal to 4180 J/kg K, that of aluminum is 900 J/kg K; other values are in Table 5.22 on page 680. While spacecraft are made up of many different materials, an average specific heat value close to that of aluminum, or one-fifth that of water, is usually a good approximation if more accurate numbers are not available.

A thermal equation for an isothermal spacecraft in sunlight can be written as

$$mc\frac{dT}{dt} = \alpha a S + Q - \epsilon A \sigma T^4 \tag{5.93}$$

where the rate of change of the temperature (dT/dt) depends on the solar input, $\alpha a S$, the internal heat generated, Q, and the heat radiated, $\epsilon A \sigma T^4$. This equation can be written in terms of the equilibrium temperature T_E:

$$mc\frac{dT}{dt} = \epsilon A (\sigma T_E^4 - \sigma T^4) \tag{5.94}$$

where the equilibrium temperature is the same as before:

$$\sigma T_E^4 = \frac{\alpha}{\epsilon}\frac{a}{A} S + \frac{Q}{\epsilon A} \tag{5.95}$$

5.10.2.1 No Incident Radiation

If the thermal input to the body is negligible, S and Q are zero, and then the differential equation can be written as

$$\frac{dT}{dt} = -\frac{\epsilon A \sigma T^4}{mc} \tag{5.96}$$

This is equivalent to stating that the equilibrium temperature T_E is close to absolute zero. This means the cooling body will continue to approach absolute zero. This is often a useful approximation for spacecraft entering the earth's shadow, when the earth-emitted radiation may be negligible.

The differential equation can be solved by multiplying through by dt/T^4 and integrating. The result is

$$\frac{1}{T^3} = \frac{3\epsilon A \sigma}{mc} t + \frac{1}{T_i^3} \tag{5.97}$$

where T_i is the initial temperature. The solution is shown in Figure 5.119 for various values of $m/\epsilon A$, the ratio of mass to effective area. The area A is the total surface area, so for solar panels both sides must be included. As the temperature drops, the rate of decrease becomes smaller since the thermal power radiated decreases according to the Stefan–Boltzmann law. The graph is plotted for a specific heat of 1000 J/kg K (which is one-fifth of the value for water) but can be used for other values by multiplying $m/\epsilon A$ by the ratio of the actual specific heat to 1000.

As an example of the use of the graph, consider a body with 5 kg/m^2 starting at 80 °C. The constant of integration, C, is first evaluated by finding the point

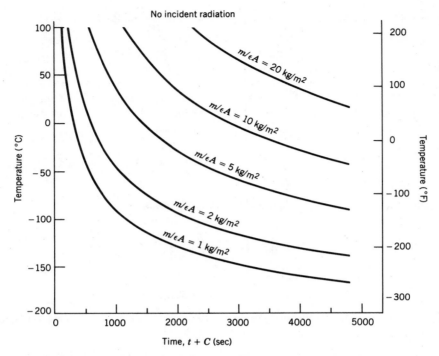

FIGURE 5.119 Radiant cooling curve, with no incident radiation. Assuming the specific heat is 1000 J/kg K.

where the curve for 5 kg/m^2 crosses the temperature of 80 °C; the constant C is found to equal 600 sec. If the temperature after half an hour (1800 sec) is desired, $t + C$ is equal to 2400 sec; the final temperature is read from the graph and is equal to −45 °C.

5.10.2.2 Linearized Solution

When the temperature T is close to the equilibrium temperature T_E, the Newtonian approximation can be used. The basic differential equation 5.94 then becomes

$$mc \frac{dT}{dt} = \epsilon A(\sigma T_E^4 - \sigma T^4) \approx \epsilon A[4\sigma T_E^3(T_E - T)] \qquad (5.98)$$

It is convenient to write this equation in the form

$$\tau \frac{dT}{dt} = T_E - T \qquad (5.99)$$

where a time constant τ is defined by

$$\tau \equiv \frac{mc}{4\epsilon A\sigma T_E^3}$$

The time constant is a measure of how rapidly the temperature of the body will approach its equilibrium value.

The solution of Equation (5.99) for $T_E - T$ can be obtained by separating the variables and integrating. The result is

$$T = T_E + (T_i - T_E)e^{-t/\tau} \tag{5.100}$$

The constant of integration has been evaluated by specifying the initial temperature T_i at the initial time $t = 0$. When the time is zero, the exponential term is unity, and the temperature indeed is equal to T_i; as the time gets larger, the exponential term gets smaller, and the temperature approaches the equilibrium temperature T_E, as expected.

The temperature difference between the body and its equilibrium value decreases by a factor of $1/e$ for each time constant elapsed. This temperature difference usually becomes negligible when it is smaller than the accuracy of temperature measurement, or smaller than any variation in T_E; the body can then be considered to have "reached" equilibrium. This may take from one to perhaps five time constants, depending on whether a factor of $1/e$ or a factor of $1/100$ is required. Thus, if the temperature is desired for an elapsed time longer than five time constants, only the steady-state solution is necessary.

5.10.2.3 Mathematical Solution of General Case

The differential equation can be written in terms of the time constant τ as

$$\frac{dT}{dt} = \frac{T_E^4 - T^4}{4\tau T_E^3} \tag{5.101}$$

This equation can be solved by separating variables and partial fractions. The solution can be written in terms of inverse trigonometric functions and either logarithmic or inverse hyperbolic functions. The solution also varies in form depending on whether the initial temperature is above or below the equilibrium temperature.

For the case of radiative cooling, the solution can be written in the form

$$\frac{t}{\tau} + C = \ln\frac{T/T_E + 1}{T/T_E - 1} - 2\tan^{-1}\frac{1}{T/T_E} \tag{5.102}$$

where C is the constant of integration. The functions are natural logarithm and arc tangent in radians. Note that while time can be written explicitly as a function of temperature, it is not possible to write temperature as a function of time. It is useful to plot the equation as a graph so that the temperature can be obtained if the time is known; this is shown in Figure 5.120. In practice, the initial temperature is usually known, and the graph is first used to determine the constant of integration. Then the time increment (t/τ) is added, and the graph is used again to determine the final temperature.

For the case of radiative heating, the solution is

$$\frac{t}{\tau} + C = \ln \frac{1 + T/T_E}{1 - T/T_E} + 2 \tan^{-1} \frac{T}{T_E} \tag{5.103}$$

This is similar in form and use to the equation for cooling, except that the initial temperature is lower than the equilibrium temperature. The curve is shown in Figure 5.121. Useful approximations to the exact solutions are also shown in Figures 5.120 and 5.121 for the ends of the curves.

5.10.3 Radiative Coupling in Simple Two-Body Cases

In the previous section, the average temperature of various spacecraft was determined. However, spacecraft often have thermal gradients of $20°$ or more, and they may be as high as $100°$. These gradients will now be discussed, mainly in terms

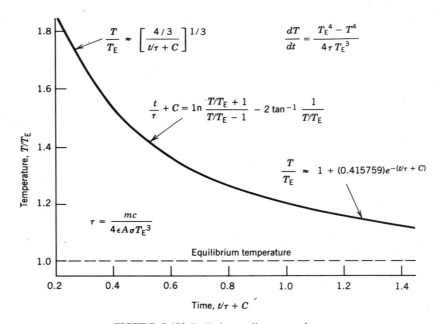

FIGURE 5.120 Radiative cooling, general case.

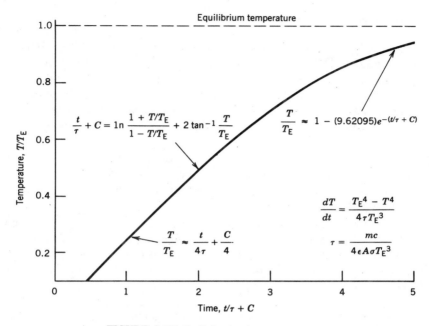

FIGURE 5.121 Radiative heating, general case.

of radiation coupling between different parts of the spacecraft and then with some comments on coupling by thermal conduction through the solid structure.

5.10.3.1 Basic Principles

To determine the temperature of a specific component in a spacecraft, the heat balance equation for that component is needed. It is therefore necessary to determine the heat that will flow between one component and another. Each component will radiate some heat, and a fraction of this radiation will be incident on another component. Using the surface properties and the geometry involved, the heat balance equation for each component is written. This set of equations can then be solved for the temperature of each component.

While the preceding procedure is satisfactory if only one solution is desired, it is complicated and time consuming. Often more than one solution is desired for a variety of different times and seasons; also, the temperature of one component may be more important than that of another component. For this reason, procedures have been formulated that essentially solve the set of equations for the temperature of one component, eliminating the temperature of other components from the equations. This solution can then be used for any set of radiation inputs, that is, the temperature of a single point on the spacecraft can then be easily obtained for any desired orbit conditions. A few special cases will first be considered, dealing only with two bodies; these have practical applications and also illustrate the basic principles that will be discussed in detail later.

Certain approximations are usually made:

1. It is assumed that the spacecraft is made up of a number of isothermal bodies. No matter how complete an analysis on a body is done, there are always a few neighboring points that are assumed to be at the same temperature. Whether the whole spacecraft is assumed isothermal or whether the spacecraft is divided into hundreds of isothermal bodies, this isothermal approximation must be made. Even in discussing shape factors from one surface to another, it is assumed that each surface is at a single temperature.

2. Each surface is assumed to be a Lambert surface, that is, both the emitted and the reflected radiation intensity are assumed to follow Lambert's law (see page 699). This is a good approximation for most surfaces, although there are important exceptions.

3. The last approximation is the "grey approximation"; that is, the emissivity as a function of wavelength is fairly flat. A different emissivity and absorptivity are assumed; in other words, there is one absorptance at the visual wavelengths and another at the infrared, and this difference is taken into account; but within a band the surface is assumed "grey."

Often the significance of these approximations is not appreciated. If there are two surfaces facing each other with an emissivity 1 for surface 1 and emissivity 2 for surface 2, it is necessary to calculate the radiation from surface 1 to surface 2, and various reflections back and forth. It could happen that the only radiation that surface 1 emits is at angles less than $45°$, and the only radiation that surface 2 emits is at angles greater than $45°$; then each radiation will all be reflected back. There would then be no coupling between surfaces 1 and 2, even though each has a certain average emissivity. A similar action could occur for different wavelengths. If surface 1 emits within a certain range of wavelengths and surface 2 emits within an entirely different range, neither surface will absorb any of the radiation emitted by the other surface, and the two surfaces would not be coupled.

5.10.3.2 Plane Surfaces

The heat flowing between two plane surfaces at temperatures T_1 and T_2 will first be calculated. The two surfaces are parallel, and the spacing between them is small compared to other dimensions. Since the heat flow will be proportional to the area, the net heat flow per area will be calculated, and this is the net irradiance in watts per square meters. Each surface has a certain emittance, 1 and 2, respectively. By Kirchhoff's law, the absorptivity must equal the emissivity for each surface since the same infrared spectrum is involved. Surface 1 will emit a certain irradiance, $\epsilon_1 \sigma T_1^4$, some of which is absorbed by surface 2 and some of which is reflected back to surface 1. This reflected radiation is then partly absorbed by surface 1 and partly reflected back to surface 2, and so on. See Figure 5.122.

To determine the net heat flow, consider the total irradiance from surface 1 to surface 2, which is made up partly of radiation emitted by surface 1 and partly by

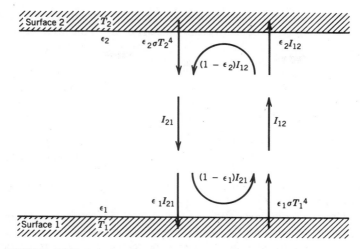

FIGURE 5.122 Heat flow between two plane surfaces, with radiation coupling.

radiation reflected by surface 1. This total irradiance, I_{12}, is equal to

$$I_{12} = (1 - \epsilon_1)I_{21} + \epsilon_1 \sigma T_1^4 \qquad (5.104)$$

where I_{21} is total irradiance from surface 2 to surface 1. When radiation I_{12} strikes surface 2, $\epsilon_2 I_{12}$ will be absorbed, and $(1 - \epsilon_2)I_{12}$ will be reflected. The total irradiance from surface 2 back to surface 1 will then be made up of this reflected radiation plus the emitted radiation, $\epsilon_2 \sigma T_2^4$:

$$I_{21} = (1 - \epsilon_2)I_{12} + \epsilon_2 \sigma T_2^4 \qquad (5.105)$$

These two equations can be solved explicitly for irradiances I_{12} and I_{21} in terms of the surface temperatures.

The net power flow per area from surface 1 to surface 2 is then equal to the difference between the two irradiances:

$$\frac{q}{A} = I_{12} - I_{21} = \frac{\sigma T_1^4 - \sigma T_2^4}{1/\epsilon_1 + 1/\epsilon_2 - 1} \qquad (5.106)$$

This equation can be put into the simple form

$$q = \epsilon_e A(\sigma T_1^4 - \sigma T_2^4) \qquad (5.107)$$

if we define an effective emittance ϵ_e as

$$\frac{1}{\epsilon_e} \equiv \frac{1}{\epsilon_1} + \frac{1}{\epsilon_2} - 1 \qquad (5.108)$$

This equation can also be considered a law of addition for emittances. Note that if one emittance is unity, the effective emittance is equal to the other emittance. If one emittance is very small, the effective emittance is equal to that particular emittance; and if both emittances are small and equal, the effective emittance is equal to one-half of that value.

5.10.3.3 Effective Emittance and Absorptance of Covered Surface

Consider the case of two plane surfaces in which one surface is a thin foil covering the surface of principal interest. The concept of an effective emittance is especially useful. Assume that the surface is at temperature T, it is covered by some foil at temperature T_w, and there are various emissivities and absorptivities. The temperature that this extra surface will attain depends on how much radiation is incident and on the value of T, the lower surface temperature. The surface then reaches its own equilibrium temperature T_w, again in the steady state. In many of these problems there is no interest in T_w; whatever temperature the foil reaches is immaterial because the foil can stand a large range of temperatures. Of real interest, however, is what values of heat flow in and out of the body at temperature T. To simplify, the problem can be considered equivalent to another body at temperature T with a different emissivity; this is denoted by the equivalent emittance, or the effective emittance of the composite body. Once this is done, the foil can be ignored, and this different emittance is used to continue with the problem solution.

The surface properties are defined in Figure 5.123, and it is assumed that there is a certain incident infrared radiation μ and an incident solar radiation of S. The net irradiance from the foil to the surface was derived previously and is equal to

$$\frac{q}{A} = \frac{1}{1/\epsilon + 1/\epsilon_1 - 1}(\sigma T_w^4 - \sigma T^4) \tag{5.109}$$

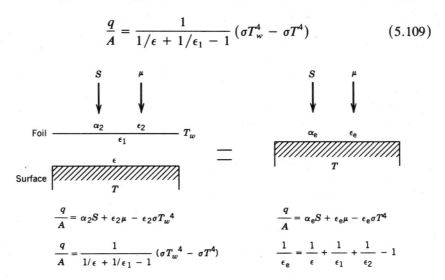

FIGURE 5.123 Heat flow through foil to external environment.

The net irradiance from space into the foil is equal to

$$\frac{q}{A} = \alpha_2 S + \epsilon_2 \mu - \epsilon_2 \sigma T_w^4 \qquad (5.110)$$

In the steady-state case, these two net irradiances must be equal and will be positive if there is a heat flow into the main body and negative if the net flow is out.

Eliminating σT_w^4 from the two equations, the heat flow is

$$\frac{q}{A} = \alpha_e S + \epsilon_e \mu - \epsilon_e \sigma T^4 \qquad (5.111)$$

where effective emittance ϵ_e and effective absorptance α_e are defined as

$$\frac{1}{\epsilon_e} \equiv \frac{1}{\epsilon} + \frac{1}{\epsilon_1} + \frac{1}{\epsilon_2} - 1 \qquad (5.112)$$

$$\alpha_e \equiv \frac{\alpha_2/\epsilon_2}{1/\epsilon + 1/\epsilon_1 + 1/\epsilon_2 - 1} \qquad (5.113)$$

Note that the ratio α/ϵ has not changed; that is, $\alpha_e/\epsilon_e = \alpha_2/\epsilon_2$, which could be proved by general physics principles.

While the concept of an effective emittance may appear artificial at first sight, it has been useful in many calculations. It is independent of the main body surface temperature, of the actual amount of incident infrared radiation, and of the incident solar radiation. One interesting point is that if all surfaces are black, the shields still have an effect; in fact, the highest effective emissivity is only $\frac{1}{2}$, even though all the surfaces have unity emissivity.

Several layers of foils may be placed on a surface. The effective emittance can be found by repeated applications of the previous formula. The effective emittance ϵ_e for n foils is shown in Figure 5.124 and is equal to

$$\frac{1}{\epsilon_e} = \frac{1}{\epsilon_T} + \frac{2n}{\epsilon} - n \qquad (5.114)$$

where ϵ_T is the emissivity of the top surface, and ϵ the emissivity of internal surfaces. Layers of aluminized mylar are often used for thermal insulation in satellites. Equation (5.114) provides an estimate of the insulation provided, but beyond 10 or 20 layers, the formula is not accurate because of trapped vapors between the layers, leakage through pinholes made by sewing, leakage around edges and through supports, or where a blanket is fastened.

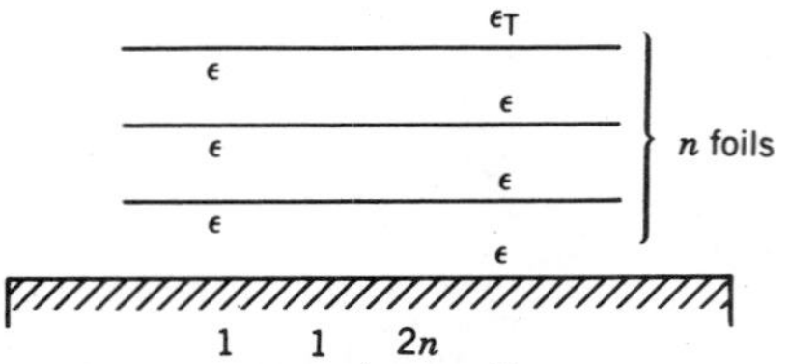

$$\frac{1}{\epsilon_e} = \frac{1}{\epsilon_T} + \frac{2n}{\epsilon} - n$$

ϵ_e = effective emissivity
ϵ = emissivity of layers $q = \epsilon_e A \sigma T^4$
n = number of layers
ϵ_T = emissivity of top

FIGURE 5.124 Effective emissivity of surface covered with several layers of foils.

5.10.4.3 Two Concentric Spheres

The heat flow between two concentric spheres can be calculated by a method similar to that used for two plane surfaces. The major difference between the two cases is that not all the radiation that leaves the outer sphere is incident on the inner sphere; some of this radiation will miss the inner sphere and be incident on the outer sphere. The effective emittance of the inner sphere is then

$$\frac{1}{\epsilon_e} = \frac{1}{\epsilon_1} + \left(\frac{1}{\epsilon_2} - 1\right)\frac{A_1}{A_2} \tag{5.115}$$

where A_1 is the area of inside sphere, and A_2 the area of the outer sphere. This equation is known as Christiansen's equation and is strictly true for two concentric spheres. It can be applied to a large number of cases of a solid inside a cavity and is approximately true for many cases. Thus, for a spacecraft inside a vacuum chamber, the second term will be small if either (a) the emittance of the walls, ϵ_2, is close to unity or (b) the ratio of the areas, A_1/A_2, is small; that is, a small spacecraft is inside a large chamber. If neither of these conditions are satisfied, this equation can be used to estimate the error incurred.

5.10.4 Configuration Factors and Radiation Coupling Factors

In the previous section the radiant heat transfer between two bodies was calculated. In this section a more formal approach will be followed with a solution in a general form applicable to any number of bodies. In the following section effects of conduction coupling will be discussed. The radiation coupling solution can be applied to simple problems as well as to complicated problems to be solved on a computer.

First, the form factors F_{ij} are determined, which are the fraction of the radiation from one body (i) that is directly incident on another body (j). To determine the form factors, it is necessary to know the angular distribution of emitted radiation, and for black bodies this is the Lambert distribution, or the cosine law. Second, the radiation coupling factors R_{ij} are determined, which are the actual thermal coupling between body i and body j, including radiation reflected by a third body. Finally, the temperature of each body is determined in terms of the internal heat dissipated in each body plus the effects of the external environment.

5.10.4.1 Lambert's Law

The angular distribution of radiation from a black-body plane surface varies as the cosine of the angle to the surface normal. That is, the radiation in any direction from a plane is proportional to the projected area, as seen from that direction. The intensity of radiation I is

$$I = \frac{\sigma T^4}{\pi} \cos \theta \qquad (5.116)$$

which is the thermal radiation emitted by a black body at temperature T per unit area per unit solid angle and in a direction θ. The total thermal power emitted by an area dA in a solid angle $d\omega$, shown in Figure 5.125, is

$$I \, dA_1 \, d\omega = \frac{\sigma T^4}{\pi} \cos \theta \, dA_1 \, d\omega \qquad (5.117)$$

This thermal power may be incident on another area dA_2, which will have a projected area of $\cos \theta_2 \, dA_2$, where θ_2 is the angle between the surface normal and the line connecting the two areas. The solid angle intercepted will then be the projected area divided by the square of the distance:

$$d\omega = \frac{\cos \theta_2 \, dA_2}{r^2} \qquad (5.118)$$

Hence, the total thermal power incident on dA_2 from dA_1 is given by

$$q_{12} = I \, dA_1 \, d\omega = \frac{\sigma T_1^4}{\pi} \frac{\cos \theta_1 \cos \theta_2}{r^2} dA_1 \, dA_2 \qquad (5.119)$$

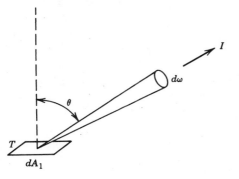

$$I \, dA_1 \, d\omega = \frac{1}{\pi} \sigma T^4 \cos \theta \, dA_1 d\omega$$

FIGURE 5.125 Angular distribution of radiation by Lambert's law.

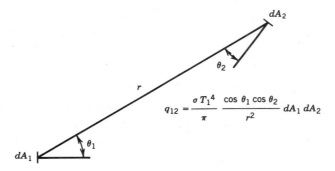

FIGURE 5.126 Radiant heat transfer between two infinitesimal areas.

where subscripts have been inserted to identify the emitting black body (first body), and the receiving area (second body). The geometry is shown in Figure 5.126. If both areas are black bodies at the same temperature, it can be seen from symmetry that the thermal power from dA_1 to dA_2 is equal to the thermal power from dA_2 to dA_1; that is, the two areas are in thermal equilibrium with each other, and the net heat transfer is zero.

If the surfaces are not black bodies, the portion of the thermal power from any surface that is absorbed by a second surface is given by

$$q_{12} = \frac{\epsilon_1 \epsilon_2 \sigma T_1^4}{\pi} \frac{\cos \theta_1 \cos \theta_2}{r^2} dA_1 \, dA_2 \qquad (5.120)$$

where dA_1 and dA_2 are the areas, ϵ_1 and ϵ_2 are the respective emissivities, and θ_1 and θ_2 are the respective angles from the surface normal to the connecting line (see Figure 5.126). This equation assumes a cosine angular distribution, which is often a good assumption, but not necessarily true for nonblack surfaces.

5.10.4.2 Configuration Factors

The next step is to consider the heat transfer between extended areas rather than between infinitesimal areas. The configuration factor has also been called form factor, shape factor, and angle factor. The configuration factor F_{12} is defined as the fraction of the total radiant flux leaving the area A_1 that is incident on the area A_2. Thus, by definition the value of the configuration lies between 0 (no radiation) and 1 (all the radiation). The configuration factor is a function of the geometry of the two surfaces A_1 and A_2, their relative position, and the directional distribution of the radiation from the source. It will be assumed that the angular distribution will follow Lambert's law, in which case the configuration factor is only a function of the two areas, their shape, and relative position.

In practice, the configuration factor is needed for areas that are not infinitesimal but extend over a finite area and are often curved surfaces. The form factor for these areas is obtained by a double integration over both areas:

$$A_1 F_{12} = \frac{1}{\pi} \int_{A_1} \int_{A_2} \frac{\cos \theta_1 \cos \theta_2}{r^2} \, dA_2 \, dA_1 \tag{5.121}$$

The length r is the distance between dA_1 and dA_2; θ_1 is the angle between r and the normal to dA_1, while θ_2 is the angle between r and the normal to dA_2.

The evaluation of the configuration factor, at best, will involve laborious integration. In many cases numerical methods, with the aid of a computer, are necessary. To avoid duplication of effort, many results have been recorded in the literature. As an example, Figure 5.127 shows the configuration factors for two rectangular areas with a common side, with planes at 90° to each other. These results can be extended to other cases by means of "configuration factor laws" and "configuration factor algebra," which will now be explained.

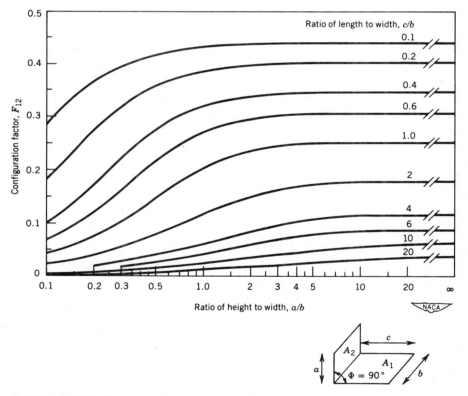

FIGURE 5.127 Radiant heat configuration factors between two perpendicular rectangular plates with common side (Hamilton and Morgan, 1952).

5.10.4.3 Basic Reciprocity Law

The product of an area A_1 and the shape factor of A_1 relative to another area A_2 is equal to the product of the area A_2 and the shape factor of A_2 relative to the A_1:

$$A_1 F_{12} = A_2 F_{21} \qquad (5.122)$$

This law is useful in cases where the evaluation of one configuration factor is easier than the other. This law follows directly from the symmetry of the defining equation.

5.10.4.4 Summation Law

If the interior surface of a completely enclosed space is subdivided into parts having areas $A_1, A_2, A_3, \cdots, A_n$ and each area is uniformly irradiated, the configuration factors for one area will sum to unity:

$$\sum_{j=1}^{n} F_{ij} = F_{i1} + F_{i2} + F_{i3} + \cdots + F_{in} = 1 \qquad (5.123)$$

5.10.4.5 Decomposition Law

Given two surfaces A_1 and A_2, if surface A_1 is subdivided into A_3 and A_4, the total shape factor F_{21} is related to the two subsidiary configuration factors F_{23} and F_{24} according to

$$A_2 F_{21} = A_2 F_{23} + A_2 F_{24} \qquad (5.124)$$

By use of the reciprocity law, this can be transformed to

$$A_1 F_{12} = A_3 F_{32} + A_4 F_{42} \qquad (5.125)$$

5.10.4.6 Configuration Factor Algebra

By use of the preceding laws, it is possible to derive configuration factors for more complicated geometries from simpler cases. Figure 5.128 shows cases of two perpendicular rectangles with a common axis but where the common sides do not coincide. Various configurations factors are

$$2A_1 F_{1(2,4)} = (A_1 + A_3) F_{(1,3)(2,4)} + A_1 F_{12} - A_3 F_{34} \qquad (5.126a)$$

$$2A_1 F_{14} = 2A_3 F_{32} = (A_1 + A_3) F_{(1,3)(2,4)} - A_1 F_{12} - A_3 F_{34} \qquad (5.126b)$$

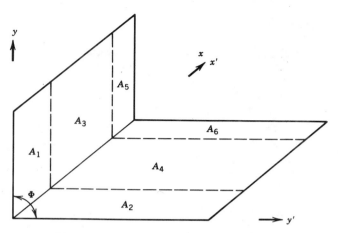

FIGURE 5.128 Example of configuration factor algebra.

$$2A_1F_{16} = (A_1 + A_3 + A_5)F_{(1,3,5)(2,4,6)} - (A_1 + A_3)F_{(1,3)(2,4)}$$
$$- (A_3 + A_5)F_{(3,5)(4,6)} + A_3F_{34} \qquad (5.126c)$$

$$2A_3F_{3(2,4,6)} = (A_1 + A_3)F_{(1,3)(2,4)} + (A_3 + A_5)F_{(3,5)(4,6)}$$
$$- A_1F_{12} - A_5F_{56} \qquad (5.126d)$$

Note that the right sides are all configuration factors of rectangles with a common side, so the values are available from Figure 5.127. More complicated geometries can be handled in a similar fashion. Other equations are in the literature (Hamilton and Morgan, 1952).

5.10.4.7 Radiation Coupling Factors

The form factor F_{ij} is an excellent measure of how much radiation is emitted by one body and directly incident on another body. In general, however, there is radiation that undergoes a number of reflections before it is finally absorbed. Even in the simple two-body problems discussed in the previous chapter, some of the radiation was reflected back and forth several times between the two bodies before it was absorbed. In this section a radiation coupling factor R_{ij} will be derived that is a measure of the total amount of radiation emitted by the ith body and absorbed by the jth body regardless of the path taken by the radiation.

It is assumed that each surface has a uniform temperature and a uniform emissivity and the incident radiation on each surface is uniformly distributed over the surface (the last assumption is the one most often violated). These assumptions can always be checked by dividing the surface into a number of smaller surfaces and repeating the problem; however, this is easier said than done. It is also assumed that the surfaces form a completely closed system; if not, the opening to "space" can be replaced by an equivalent wall at a fixed calculated temperature.

The net heat transfer from the kth body to the ith body is given by

$$q_{ki} - q_{ik} = R_{ki}(\sigma T_k^4 - \sigma T_i^4) \tag{5.127}$$

where the radiation coupling factor R_{ki} is a measure of the total coupling between the ith body and the kth body. If the kth body is at temperature T_k and all other bodies are at zero temperature, $R_{ki}\sigma T_k^4$ is the total radiation absorbed by the ith body.

The radiation coupling factors can be calculated by the equation

$$R_{ki} \equiv \epsilon_k A_k \left\{ \sum_{j=1}^{n} F_{kj} \left[\delta_{ij} - (1 - \epsilon_i) F_{ij} \right]^{-1} \right\} \epsilon_i \tag{5.128}$$

where the ϵ's are emissivities, F_{ij}'s the configuration factors, A_k the surface area, and δ_{ij} the identity matrix.

For the special case of all black bodies, where all the emissivities are unity, the equation can be simplified to

$$R_{ki} = A_k F_{ki} \tag{5.129}$$

which is simply the fraction of radiation emitted by the kth body and absorbed by the ith body. Since the emissivities are unity, there are no reflections to complicate the analysis.

5.10.4.8 Temperature of Each Body

Once the radiation coupling factors R_{ki} are known, it is possible to write down a heat balance equation for each body in an n-body problem. This set of n equations can then be solved for the n temperatures by standard methods. Consider a spacecraft made up of n isothermal bodies placed in a vacuum chamber with walls maintained at various temperatures T_w as shown in Figure 5.129. This is equivalent to a spacecraft in space if the wall temperatures are properly chosen.

A distinction is made between the wall temperatures T_w and the spacecraft body temperatures T_j. Essentially, the wall temperatures are known temperatures, although the heat flux through them is unknown. In contrast, the temperatures of the spacecraft components T_j are unknown, although the heat flux into them, Q_j, is known (frequently zero).

Consider the jth body in the satellite, which is at a temperature T_j. The total heat radiated by this jth body is

$$\epsilon_j A_j \sigma T_j^4 \tag{5.130}$$

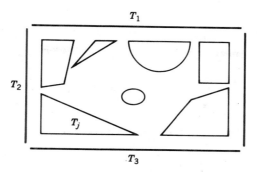

FIGURE 5.129 Generalized spacecraft with n isothermal bodies enclosed in vacuum chamber with three walls at three different temperatures.

The total heat received by the jth body is equal to

$$\sum_w R_{jw}\sigma T_w^4 + \sum_i R_{ji}\sigma T_i^4 + Q_j \qquad (5.131)$$

which includes the heat from the chamber walls from the other parts of the spacecraft ($i = 1, 2, 3, \cdots, n$) and any heat generated internally, Q_j. Note that R_{jw} and R_{ji} are coupling coefficients that depend on the emissivities of the various bodies and on the geometry. The preceding equations not only include the heat directly from i to j, but also radiation that starts from the ith body is reflected by other bodies and eventually is absorbed by the jth body. The heat balance equation for the jth body is

$$m_j c_j \frac{dT_j}{dt} = -\sum_i \left(\delta_{ij}\epsilon_j A_j - R_{ij}\right)\sigma T_i^4 + \sum_w R_{jw}\sigma T_w^4 + Q_j \qquad (5.132)$$

where δ_{ij} is the identity matrix.

For a transient analysis, the heat balance equations can be solved by numerical methods of solving differential equations. A set of initial temperatures is used, and the temperatures are calculated as a function of time. For an equilibrium solution, two methods are in use. The first is to treat the problem as a transient problem and simply continue the numerical solution for several time constants, when the temperatures are no longer varying. The second solution is to consider the problem as a set of simultaneous linear equations and solve for the unknown (T_j^4).

The equilibrium temperatures can be written as

$$\sigma T_i^4 = \sum_w c_{iw}\sigma T_w^4 + \sum_j b_{ij}Q_j \qquad (5.133)$$

where the coefficient for the internal dissipation terms, b_{ij}, is

$$b_{ij} = \left(\delta_{ij}\epsilon_j A_j - R_{ji}\right)^{-1} \qquad (5.134)$$

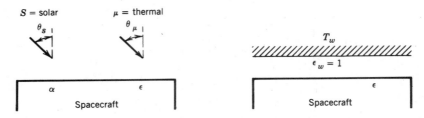

FIGURE 5.130 Wall temperatures equivalent to radiation input.

and the coefficients for the external inputs are

$$c_{iw} = \sum_j b_{ij} R_{jw} \qquad (5.135)$$

The external inputs can be expressed in terms of the incident radiation. Note that the number of bodies in a thermal analysis may be quite large, but the solution in terms of wall temperatures T_w and internal heat dissipators Q_j usually involves a smaller number of terms. With this equation the temperatures of important spacecraft components can easily be calculated for several different conditions.

For a spacecraft in space (see Figure 5.130), it is convenient to use an equivalent thermal radiation input I_w, where

$$I_w = \sigma T_w^4 = \frac{\alpha}{\epsilon} (S\psi \cos \theta_s) + \mu \cos \theta_\mu \qquad (5.136)$$

which is a function of only the incident radiation and the absorptivity–emissivity ratio α/ϵ. This equation is expressed for a solar input of intensity S and angle to the normal θ_s and a thermal input radiation of intensity μ and angle θ_μ. The fraction ψ may be unity for equilibrium temperatures in the sun, or it may be the fractional suntime if the average temperatures during an orbit with eclipse are desired.

5.10.5 Example of Radiatively Coupled Spacecraft

To illustrate the equations used in the previous section, a numerical example will now be considered. The spacecraft is a hollow cylinder with a height and radius of approximately 0.56 m (actually $1/\sqrt{\pi}$). The external and internal surface properties are given in Table 5.26. There are 20 W dissipated in the bottom surface. The cylinder axis makes a 60° angle with the sun's rays, as shown in Figure 5.131. The assumption will be made that each of the three surfaces is isothermal. The problem is to determine the temperatures of each surface.

The solution will follow the same order as the general equations derived previously. First, the configuration factors F_{ij} are determined; second, the radiation

TABLE 5.26 Spacecraft Characteristics for Sample Problem

	Internal Emissivity, ϵ	External Emissivity, ϵ	External Absorptivity, α	Surface Area, $A\,(\mathrm{m}^2)$	Heat Dissipation, $Q\,(\mathrm{W})$
1. Top	0.6	0.88	0.88	1	0
2. Side	0.7	0.88	0.88	2	0
3. Bottom	0.8	0.35	0.35	1	20

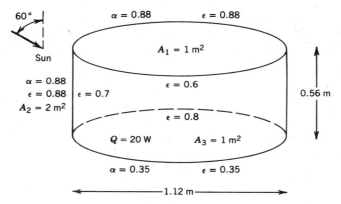

FIGURE 5.131 Example of radiatively coupled satellite in form of hollow cylinder.

coupling factors R_{ij} are calculated; and finally, the temperature of each body is determined in terms of the internal heat and the external radiation.

5.10.5.1 Configuration Factors

The first configuration factor to be obtained is that from one cylinder end to the other; when this is known, all the others can be obtained from configuration factor laws. For the two parallel, directly opposed, equal plane circular ends, the config-uration factor is

$$A_1 F_{13} = \frac{\pi}{2}\left[2r^2 + d^2 - d\sqrt{4r^2 + d^2}\right] \tag{5.137}$$

as given by Hamilton and Morgan (1952). For this problem, where the cylinder radius r is equal to the cylinder height d, the configuration factor from the top (1) to the bottom (3) is

$$F_{13} = 0.38$$

All the configuration factors from a particular area must add to unity. Hence, the configuration factor from the top (1) to the side (2) is

$$F_{12} = 1 - F_{13} = 0.62$$

By the reciprocity law, the configuration factor from the side (2) to the top (1) is

$$F_{21} = \frac{A_1}{A_2} F_{12} = 0.31$$

and by symmetry the configuration factor from the side to the bottom, F_{23}, must also be 0.31. The summation law is then used to obtain the configuration factor from the side (2) to the side (2):

$$F_{22} = 1 - F_{21} - F_{23} = 0.38$$

which means that 38% of the radiation that leaves the side of the cylinder will be incident on the side at another spot.

Summarizing the preceding results, all the configuration factors are shown in Figure 5.132 and equal to

$$F_{11} = 0 \qquad F_{12} = 0.62 \qquad F_{13} = 0.38$$
$$F_{21} = 0.31 \qquad F_{22} = 0.38 \qquad F_{23} = 0.31$$
$$F_{31} = 0.38 \qquad F_{32} = 0.62 \qquad F_{33} = 0$$

$$A_1 F_{13} = \frac{\pi}{2}\left[2r^2 + d^2 - d\sqrt{4r^2 + d^2}\right]$$

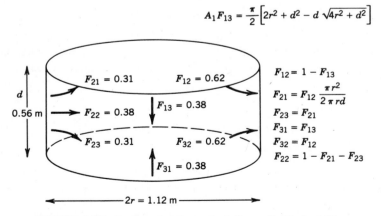

FIGURE 5.132 Configuration factors for hollow cylindrical satellite.

5.10.5.2 Radiation Coupling Factors

If the internal surfaces had unity emissivity, the radiation coupling factors could be obtained by multiplying the form factors by the respective areas. Since the emissivities are not unity, the radiation coupling factors R_{ki} for the inside of the hollow cylindrical spacecraft are determined by

$$
R_{ki} =
\overset{\epsilon_k A_k}{
\begin{bmatrix} 0.6 \text{ m}^2 & & \\ & 1.4 \text{ m}^2 & \\ & & 0.8 \text{ m}^2 \end{bmatrix}}
\overset{F_{kj}}{
\begin{bmatrix} & 0.62 & 0.38 \\ 0.31 & 0.38 & 0.31 \\ 0.38 & 0.62 & \end{bmatrix}}
$$

$$
\left[\delta_{ij} - (1 - \epsilon_i) F_{ji}\right]^{-1}
$$

$$
\cdot \begin{bmatrix} 1 & -0.4(0.62) & -0.4(0.38) \\ -0.3(0.31) & 1 - 0.3(0.38) & -0.3(0.31) \\ -0.2(0.38) & -0.2(0.62) & 1 \end{bmatrix}^{-1}
$$

$$
\cdot \overset{\epsilon_i}{\begin{bmatrix} 0.6 & & \\ & 0.7 & \\ & & 0.8 \end{bmatrix}} = \begin{bmatrix} 0.04 \text{ m}^2 & 0.33 \text{ m}^2 & 0.23 \text{ m}^2 \\ 0.33 \text{ m}^2 & 0.59 \text{ m}^2 & 0.48 \text{ m}^2 \\ 0.23 \text{ m}^2 & 0.48 \text{ m}^2 & 0.10 \text{ m}^2 \end{bmatrix} \quad (5.138)
$$

This R_{ki} matrix is symmetrical; that is, the radiation coupling factor from the top to the side is equal to that from the side to the top. Of major physical significance are the off-diagonal terms pictured in Figure 5.133. The diagonal terms represent radiation emitted and absorbed by the same body and do not really affect temperatures; they are useful for checking since the sum of each row or column is equal to the product of the area times the emissivity.

There are also external radiation coupling factors, R_{jw}, through which heat flows between the spacecraft and outer space. Since this particular example is a convex spacecraft, all the configuration factors are unity. Hence, the radiation coupling factors, R_{jw}, are simply the product of the external emissivity and the area:

$$
R_{1w} = 0.88(1 \text{ m}^2) = 0.88 \text{ m}^2
$$

$$
R_{2w} = 0.88(2 \text{ m}^2) = 1.76 \text{ m}^2
$$

$$
R_{3w} = 0.35(1 \text{ m}^2) = 0.35 \text{ m}^2
$$

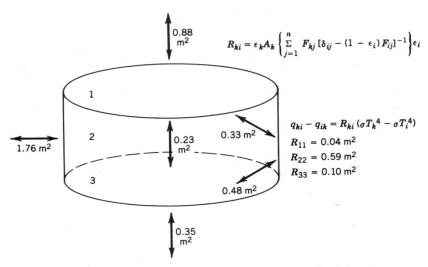

$$R_{ki} = \epsilon_k A_k \left\{ \sum_{j=1}^{n} F_{kj} [\delta_{ij} - (1 - \epsilon_i) F_{ij}]^{-1} \right\} \epsilon_i$$

$$q_{ki} - q_{ik} = R_{ki} (\sigma T_k{}^4 - \sigma T_i{}^4)$$

$$R_{11} = 0.04 \text{ m}^2$$

$$R_{22} = 0.59 \text{ m}^2$$

$$R_{33} = 0.10 \text{ m}^2$$

FIGURE 5.133 Radiation coupling factors R_{ki} for hollow cylindrical satellite.

5.10.5.3 Temperature of Each Body

The transient equations for each of the spacecraft parts can be written from the radiation coupling terms:

$$m_1 c_1 \frac{dT_1}{dt} = 0.88 \text{ m}^2 (I_T - \sigma T_1^4) + 0.33 \text{ m}^2 (\sigma T_2^4 - \sigma T_1^4)$$

$$+ 0.23 \text{ m}^2 (\sigma T_3^4 - \sigma T_1^4) + Q_1 \qquad (5.139a)$$

$$m_2 c_2 \frac{dT_2}{dt} = 0.33 \text{ m}^2 (\sigma T_1^4 - \sigma T_2^4) + 1.76 \text{ m}^2 (I_S - \sigma T_2^4)$$

$$+ 0.48 \text{ m}^2 (\sigma T_3^4 - \sigma T_2^4) + Q_2 \qquad (5.139b)$$

$$m_3 c_3 \frac{dT_3}{dt} = 0.23 \text{ m}^2 (\sigma T_1^4 - \sigma T_3^4) + 0.48 \text{ m}^2 (\sigma T_2^4 - \sigma T_3^4)$$

$$+ 0.35 \text{ m}^2 (I_B - \sigma T_3^4) + Q_3 \qquad (5.139c)$$

In each equation, the self-coupling term has been omitted since it consists of heat radiated by a body and absorbed by the same body.

The steady-state solutions can be obtained by setting the left side of the equations to zero and solving for the three temperatures. The numerical result is

$$\sigma T_1^4 = 0.67 I_T + 0.25 I_S + 0.08 I_B + 0.76 \text{ m}^{-2} (Q_1)$$

$$+ 0.14 \text{ m}^{-2} (Q_2) + 0.23 \text{ m}^{-2} (Q_3)$$

$$\sigma T_2^4 = 0.12 I_T + 0.80 I_S + 0.08 I_B + 0.14 \text{ m}^{-2}(Q_1)$$
$$+ 0.45 \text{ m}^{-2}(Q_2) + 0.23 \text{ m}^{-2}(Q_3)$$

$$\sigma T_3^4 = 0.20 I_T + 0.41 I_S + 0.39 I_B + 0.23 \text{ m}^{-2}(Q_1)$$
$$+ 0.23 \text{ m}^{-2}(Q_2) + 1.10 \text{ m}^{-2}(Q_3)$$

where I_T, I_S, and I_B are the equivalent thermal radiation incident on the top, sides, and bottom, respectively, and Q_1, Q_2, Q_3, are the internal heat dissipations. Once the solution is obtained in this form, it is simple to calculate the three spacecraft temperatures for any radiation input and any internal heat dissipation. Notice that for each equation, the equivalent thermal radiation coefficients are pure numbers (no units) and must add to unity. The coefficients for the internal heat dissipation terms are in reciprocal area units. Also, there is a reciprocity: if 1 W in body 2 increases σT_3^4 by 0.23 W/m^2, 1 W in body 3 must increase σT_2^4 by the same amount.

For a sun shining at a 60° angle to the cylinder axis and with absorptivity-emissivity ratios of unity, the equivalent radiation inputs can be calculated as

$$I_T = 1370 \text{ W/m}^2 \,(\cos 60°) = 685 \text{ W/m}^2$$
$$I_S = 1370 \text{ W/m}^2 \,(\sin 60°)/\pi = 378 \text{ W/m}^2$$
$$I_B = 0$$

The internal power dissipations were given as

$$Q_1 = 0 \qquad Q_2 = 0 \qquad Q_3 = 20 \text{ W}$$

The temperature of the bottom surface, T_3, can be found by substituting these input values into the equation for σT_3^4. The result is

$$\sigma T_3^4 = 0.20(685 \text{ W/m}^2) + 0.41(378 \text{ W/m}^2) + 1.10 \text{ m}^{-2}(20 \text{ W})$$
$$\sigma T_3^4 = 314 \text{ W/m}^2 \qquad T_3 = 273 \text{ K}$$

Similar calculations will yield the temperatures of the top and sides. The heat flows can be calculated from the temperatures and the radiant coupling factors. The temperatures and heat flows are shown in Figure 5.134. Knowing the heat flows, it is possible to predict the effect of changing the couplings between two bodies. For example, the heat flow from the top to the sides is 55 W; if the coupling were decreased, the heat flow would be reduced, the temperature of the top would increase, and the temperature of the sides would decrease.

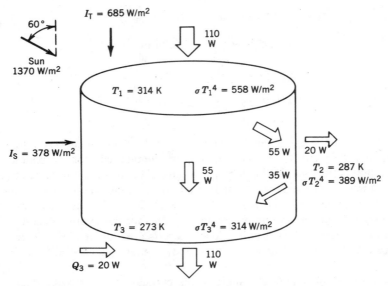

FIGURE 5.134 Calculated temperatures and heat flows for satellite example.

5.10.6 Heat Transfer by Conduction and Radiation

Inside a spacecraft heat is often transferred by conduction through the solid parts of the spacecraft. For localized heat sources, conduction often must be used to spread the heat over a large area before this heat can be radiated. Both thermal conduction and thermal radiation are important in determining the internal temperature distribution of a spacecraft.

5.10.6.1 Thermal Conductivity: Definition

When different parts of a solid body are at different temperatures, heat flows from the hotter to the colder portions by thermal conduction. The heat flowing in any small volume in the body will be proportional to the cross-sectional area A and to temperature gradient dT/dx. The thermal conductivity κ can be defined by

$$q = -\kappa A \frac{dT}{dx} \qquad (5.140)$$

where q is the power (in watts) flowing; thus, the thermal conductivity can be expressed in watts per meter per degree. The negative sign means that if the temperature gradient dT/dx is positive in a certain direction (temperature increasing), the heat flow will be in the opposite direction (from hot to cold). The physical significance can better be appreciated by considering linear heat flow and other simple geometries. It will be assumed here that the thermal conductivity does not change with temperature, which is not rigorously true but is usually a good approximation.

For uniform heat flow, the cross-sectional area A is constant. The heat flow can be written as

$$q = \kappa A \frac{T_1 - T_2}{x_2 - x_1} \tag{5.141}$$

where temperature T_1 is at position x_1 and temperature T_2 is at position x_2. Thus, the heat flow is directly proportional to the thermal conductivity κ, the cross-sectional area A, and the temperature difference $T_1 - T_2$ and inversely proportional to the length, $x_2 - x_1$. The geometry of linear flow is shown in Figure 5.135.

The heat flow by conduction is always expressed in terms of the first power of the temperatures. For radiation, the heat transferred between two bodies is proportional to the difference between the fourth power of each temperature. This is a basic difference between conduction and radiation. Physically, it means that radiation becomes dominant at high temperatures and conduction dominates at low temperatures. Mathematically, any mixture of conduction and radiation terms becomes difficult to handle. In some cases this can be done. In many cases it is better to use some approximation to simplify the terms. For many years the Newtonian approximation has been used to linearize the terms in radiation, making them similar to conduction terms. For spacecraft an approximation to convert the conduction terms into a difference between the fourth powers of the temperatures is sometimes used.

5.10.6.2 Approximate Relations

If the two temperatures are not too different (on the absolute scale), radiation coupling can be written as

$$q = R_{12}(\sigma T_1^4 - \sigma T_2^4) \approx R_{12} 4\sigma T_0^3 (T_1 - T_2) \tag{5.142}$$

where the temperature T_0 is a rough average of the two temperatures. To be mathematically accurate, $4T_0^3$ should be equal to $T_1^3 + T_1^2 T_2 + T_1 T_2^2 + T_2^3$, but the value

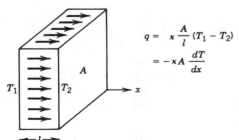

$$q = \kappa \frac{A}{l}(T_1 - T_2)$$
$$= -\kappa A \frac{dT}{dx}$$

FIGURE 5.135 Thermal conductivity definition in terms of linear heat flow.

of the approximation is lost if too much time is spent calculating T_0. The best procedure is to pick T_0, solve the problem, and then check to see whether the calculated T_0 is close enough for the desired accuracy. Occasionally, one temperature, T_2, may be close to absolute zero so that the approximation is no longer valid; in these cases, it is better to omit the 4 and have T_0 approximately equal to T_1. In either case, Newton's law of cooling states that the heat radiated is proportional to the excess of temperature over that of the surrounding medium, which will hold true if the excess is not too large. (The same type of approximation is used for convection cooling.)

For spacecraft calculations, it may be better to take the few conduction terms and by an approximation transform each of them into a radiation term. The justification is the same as before, but the direction in which the approximation is used is reversed. The temperature difference $T_1 - T_2$ is approximated by

$$q \propto \kappa(T_1 - T_2) \approx \kappa \frac{\sigma T_1^4 - \sigma T_2^4}{4\sigma T_0^3} \qquad (5.143)$$

where T_0 is again an average temperature, intermediate between T_1 and T_2. To be accurate, $4T_0^3$ should be equal to $T_1^3 + T_1^2 T_2 + T_1 T_2^2 + T_2^3$, but in general, any value in the general vicinity of the temperatures involved is sufficient.

The concept of a conductivity length is useful. If two parallel plates (emissivity of unity) are at two temperatures, T_1 and T_2, the net heat flow radiated per area is equal to

$$\frac{q}{A} = \sigma T_1^4 - \sigma T_2^4 \qquad (5.144)$$

If the space between the two plates is filled with a substance with a thermal conductivity κ, the heat flow is approximately

$$\frac{q}{A} \approx \frac{\kappa/4\sigma T_0^3}{x_2 - x_1}(\sigma T_1^4 - \sigma T_2^4) \qquad (5.145)$$

Comparing these two equations, the relative magnitude of the heat flows depends on whether the distance between the plates, $x_2 - x_1$, is greater or less than the conductivity length, $\kappa/4\sigma T_0^3$. While it is somewhat unusual to consider the thermal conductivity as a length, this is a useful concept in problems that have both conduction and radiation.

Remember that the conductivity length is a function of the temperature. It may be tabulated for various temperatures for different materials. These values are often accurate enough for a rough estimate; for greater accuracy, a calculation from the fundamental value of the thermal conductivity may be necessary.

5.10.6.3 Conduction through Thin Surface

Suppose there is a plate with one surface held at temperature T_1, as shown in Figure 5.136, and that the heat is conducted through the plate and is then radiated into space (assume no radiation inputs). The heat conducted through the plate can be approximated by a radiation coupling term, with T_2 the unknown temperature of the outer surface. The heat radiated is then equal to

$$\frac{q}{A} = \epsilon \sigma T_2^4 \tag{5.146}$$

where ϵ is the emittance of the outer surface. The unknown temperature, T_2, can be eliminated from the two equations and solved in terms of the heat flow. The effective emittance ϵ_e is then given by

$$\frac{1}{\epsilon_e} = \frac{1}{\epsilon} + \frac{x_2 - x_1}{\kappa/4\sigma T_0^3} \tag{5.147}$$

The effect of the conduction through the plate is to reduce the emittance of a surface. Similar calculations can be made with radiation inputs. If the plate thickness is small compared to the conductivity length, the change in emittance is small; that is, the temperature gradient through the plate can be neglected.

5.10.6.4 Radiating Rod of Infinite Length

The long rod with constant cross section is a practical problem that can be solved and is also a good illustration of conduction and radiation. One end of the rod is raised to some known temperature T_0; heat from this end is conducted down the rod and then radiated from the rod. As shown in Figure 5.137, the rod has an

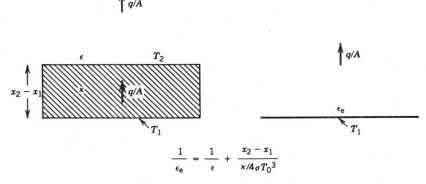

FIGURE 5.136 Effect of conductance in reducing effective emittance.

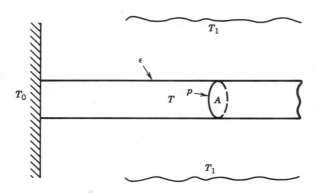

FIGURE 5.137 Radiating rod of infinite length.

emissivity ϵ and a perimeter p ($2\pi r$ for a circular rod) and is in an environment at a temperature T_1. The rod temperature decreases along the rod, finally reaching the environment temperature at great distances from the end. Such a rod might be an antenna or boom attached to a spacecraft or a large heat source fastened to an internal spacecraft strut.

An effective length for the rod can be defined such that if the rod were all at the temperature T_0 and if the rod were only of this length, the heat radiated would be the same. The effective length l_e can therefore, be defined by stating that the heat radiated is

$$q = \epsilon p l_e \sigma (T_0^4 - T_1^4) \tag{5.148}$$

The exact mathematical solution shows the effective length to be

$$l_e^2 \approx \frac{2\kappa A (T_0^5 - 5T_1^4 T_0 + 4T_1^5)}{5 p \epsilon \sigma (T_0^4 - T_1^4)^2} \tag{5.149}$$

The equation

$$l_e^2 \approx \frac{A}{p} \frac{\kappa / 4\sigma T_0^3}{\epsilon} \tag{5.150}$$

can be derived for $T_1 \approx T_0$. Even for cases where $T_1 \approx 0$, this equation is only off by a factor of $\frac{8}{5}$.

5.11 SPACECRAFT TESTING

5.11.1 Introduction

Any complex and expensive system requires testing to ensure adequate performance. Communications satellites require extensive testing for two additional reasons: they operate in an environment considerably different from that in which they were built and, after launch, they are inaccessible to routine maintenance and repair. The objective of the testing is not necessarily to duplicate the space environment but to approach it sufficiently so that any spacecraft that passes the tests will operate successfully.

The major features of the space environment that are difficult to simulate exactly are zero gravity, high vacuum, solar radiation, particle radiation, lack of nearby obstructions, and a full design lifetime (see Table 5.27). Except for zero gravity, approximate simulation of these parameters is possible for most spacecraft components. Gravity affects the deployment of solar arrays and antennas, and liquid motion in components such as fuel tanks, heat pipes, and batteries.

Major testing of a spacecraft occurs between integration and shipment to the launch site. This testing takes a considerable length of time (e.g., 5 months). A typical sequence of tests is shown in Table 5.28. The main backbone of testing comprises systems performance tests, mechanical vibration tests, and thermal vacuum tests. Alignment of various components is needed, including sensors, antennas, thrusters, and wheels. The slant rf test measures the electrical performance of the antennas; of interest is not only the intensity distribution in an antenna beam but also the strength of the sidelobes and possible interference between beams. The mechanical vibration tests and thermal vacuum tests are described in detail in later sections. The mass properties includes the total mass, the position of the center of mass, and various moments of inertia. Since the performance of the apogee motor is fixed (solid rocket), the spacecraft must have a calculated mass; often small differences just before launch are offset by adding extra fuel to the propellant system.

Most subsystems are tested separately to detect problems before spacecraft integration. Each subsystem has performance measurements, vibration tests, and thermal vacuum tests. Specifications for these tests reflect the requirements for

TABLE 5.27 Difficulties in Simulating Space

Space Condition	Critical Components	Simulation
Zero gravity	Array deployment, antenna deployment, liquid motion	Counterweights, balloons, horizontal position
Vacuum (10^{-10}–10^{-20} torr)	Sliding friction, thermal control, mechanical relay contacts	Use solid lubricant, thermal vacuum tests, reduce current
Solar radiation	Solar array, component temperature	Solar simulator, thermal vacuum tests
Particle radiation	Solar array, semiconductor components	1-MeV electrons
Lack of nearby obstructions	Antenna patterns	Anechoic chamber, outdoor range
Design life of many years	Bearings, batteries, power amplifiers	Accelerated testing

TABLE 5.28 Tests of Integrated Spacecraft (Before Launch)

Test	Parameters or Units Measured
System performance	All units
Alignment 1	Antennas, sensors, thrusters, wheels
Rf range 1	Antennas
Vibration tests	Static, sine, acoustic, shock
Functional	Performance of all units
Thermal vacuum	Temperatures, performance
Alignment 2	Antennas, sensors, thrusters, wheels
Rf range 2	Antennas
Mass properties	Mass, center of mass, moments of inertia

each subsystem, the results of computer models of structure and thermal design, and experience with previous spacecraft component testing. Many of these subsystem tests are similar to the spacecraft tests described later. The solar cells comprise a subsystem that requires a unique set of tests, and these will be described in some detail.

5.11.2 Solar Cell Testing

The performance of an illuminated solar cell is obtained by measuring its electrical characteristics. The most common solar simulation technique is the use of a xenon arc lamp, with filters, to remove undesired line spectra in the near infrared. A close spectral match to the solar spectrum is obtained, as shown in Figure 5.138. Incandescent tungsten sources are not suitable because their radiation peaks in the red and near infrared, which is the wavelength region causing the greatest rate of degradation in the cell's response to radiation. Tests using tungsten sources show greater cell degradation than tests using a suitable solar simulator; therefore, the tests using the solar simulator provide a much better indication of the cell's expected life in space.

The electrical characteristics are determined by using load resistances from zero to infinity and measuring the cell voltage and current. The load resistance may be varied manually or electronically. Solar cell response is a strong function of temperature, so the temperature must be determined. Cells measured under one sun irradiance at room temperature can be stabilized at 29 °C with adequate heat sinking and cooling. To ensure that the voltages measured are representative of those developed on the cell contacts, separate probes are employed to measure cell voltage and current. The current–voltage data are usually plotted with an X–Y recorder. Often the graph paper has suitable hyperbolas plotted, so the maximum power point can be read directly. The short-circuit current, I_{sc}, and open-circuit voltage, V_{oc}, can be read directly with digital meters.

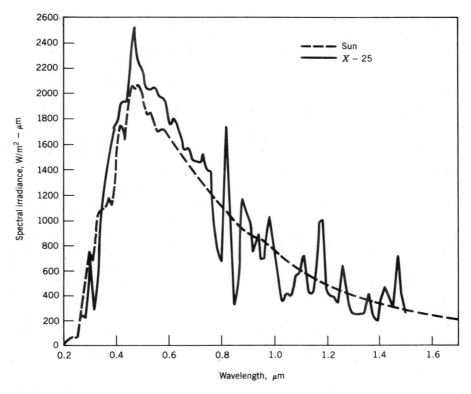

FIGURE 5.138 Spectral output. Spectrolab X-25 Spectrosun (Carter and Tada, 1973).

During component testing, the solar cell is subjected to a number of environments, as shown in Table 5.29. The humidity, peel, and thermal cycling tests are primarily to test the adhering electrodes and interconnects. The radiation test is for degradation of the basic solar cell. The humidity test is used because moisture penetration under the electrode (before launch) can weaken the bond. The peel test

TABLE 5.29 Solar Cell Testing

Test	Typical Parameter	Observation or Failure[a]
I–V test	Filtered xenon lamp, 1 sun	I_{sc}, V_{oc}, I_{mp}, V_{mp}, temperature coefficient
Humidity	95%, many hours	Electrode separation
Peel	Sticky tape, two directions	Electrode separation
Thermal cycling	70 to 300 K, 1000 cycles	Interconnect failure
Radiation	1-MeV electron beam	Power degradation

[a]Abbreviations: I_{mp}, current at maximum power; V_{mp}, voltage at maximum power; I_{sc}, short circuit current; V_{oc}, open circuit voltage.

consists of putting some sticky tape on a cell and peeling it off; it does not simulate any actual environment but is an easy method to test the adhesion of a cell electrode. The thermal cycling test is often done by dipping the cell in liquid nitrogen and then warming it up to room temperature (or above); while this is more severe than what cells experience going into eclipse, it is also a simple test that many cells can survive.

Determining the expected degradation of solar cells in the space radiation environment is done in two parts. The cells are exposed to a 1-MeV electron beam at normal incidence, and performance degradation is measured. The equivalent 1-MeV electron fluence is calculated based on the known radiation environment and the mission parameters.

Charged-particle accelerators are the primary sources for space radiation simulation. Electron energies of 1 MeV are usually obtained with Van de Graaff electrostatic or Dynamatron accelerators. Accelerators produce irradiation rates many orders of magnitude greater than space environments; test times are only a few minutes. A successful experiment includes accurate knowledge of the particle energy and the beam intensity. Control samples of solar cells with known properties are used to check that beam intensity is correct. Mechanical motion is used to change the positions of the cells, so that any variation in cross-sectional beam intensity is canceled.

In the space environment, there are electrons and protons trapped in the earth's magnetic field. In addition, there are protons from solar flares that occasionally produce significant degradation in a short time. Finding the equivalent 1-MeV electron fluence depends on the shielding on the solar cell, the activity of the sun (in the 11-yr sunspot cycle), whether the prediction is for the short-circuit current I_{sc} or for the open-circuit voltage V_{oc} and maximum power P_{max} and the design lifetime of the satellite.

The predicted annual equivalent 1-MeV electron fluence for short-circuit current is shown in Table 5.30. With no shielding, the effect of protons trapped in the earth's magnetic field is overwhelming; but with only a little shielding, their effect is negligible. The effects of trapped electrons and solar flare protons are comparable in magnitude. For a given mission the effects of both for each year can be added, and the result is the equivalent 1-MeV electron fluence. The degradation of I_{sc} measured for that amount of radiation is the degradation to be expected in the mission. Similar numbers for the degradation in open-circuit voltage V_{oc} and maximum power P_{max} are listed in Table 5.31. The estimated flare environments were calculated by Weidner (1969) and are commonly used in connection with military satellite systems. They lead to a worst case degradation estimate, useful in designing the size of a solar array.

5.11.3 Vibration Testing

The mechanical launch environment for the payload of a three-stage launch vehicle is shown in Figure 5.139. The first stage starts at lift-off, later jettisons the solid rocket boosters, and ends with the main engine cutoff (MECO). The second stage

TABLE 5.30 Short-Circuit Current I_{sc} Predicted Annual Equivalent 1-MeV Electron Fluence at Geostationary Orbit[a]

	Shielding[b] (g/cm^2)						
	0 (0)	0.0168 (3)	0.0335 (6)	0.0671 (12)	0.1006 (18)	0.1675 (30)	0.3350 (60)
Trapped electrons	6.35×10^{13}	3.67×10^{13}	2.55×10^{13}	1.35×10^{13}	6.46×10^{12}	2.81×10^{12}	8.61×10^{11}
Trapped protons	8.22×10^{18}	6.06×10^{5}	1.12×10^{2}	2.68×10^{-2}	0	0	0
Solar flare protons, Minimum							
1975–1977, 1985–1987	—	2.9×10^{13}	1.7×10^{13}	8.7×10^{12}	5.3×10^{12}	3.7×10^{12}	1.9×10^{12}
1978–1979, 1983–1984	—	1.5×10^{14}	1.3×10^{14}	4.4×10^{13}	2.6×10^{13}	1.8×10^{13}	9.7×10^{12}
Solar flare protons, Maximum,							
1980–1982	—	2.9×10^{14}	1.7×10^{14}	8.7×10^{13}	5.3×10^{13}	3.7×10^{13}	1.9×10^{13}

[a]Infinite backshielding assumed. Density of fused silica, 2.2 g/cm^3. Predictions for trapped electrons for 35,900-km orbit, 0° circular orbit, for trapped protons at 33,300 km, and for solar flare protons at 1 AU. From Weidner (1969).

[b]Number in parentheses are thicknesses of fused silica in mils.

TABLE 5.31 Open-Circuit Voltage V_{oc} and Maximum Power P_{max} Predicted Annual Equivalent 1-MeV Electron Fluence at Geostationary Orbit[a]

	Shielding[b] (g/cm²)						
	0 (0)	0.0168 (3)	0.0335 (6)	0.0671 (12)	0.1006 (18)	0.1675 (30)	0.3350 (60)
Trapped electrons	6.35×10^{13}	3.67×10^{13}	2.55×10^{13}	1.35×10^{13}	6.46×10^{12}	2.81×10^{12}	8.61×10^{11}
Trapped protons	1.43×10^{25}	2.02×10^{6}	5.32×10^{2}	9.15×10^{-2}	1.22×10^{-5}	0	0
Solar flare protons, minimum							
1975–1977, 1985–1987	—	6.7×10^{13}	3.4×10^{13}	1.6×10^{13}	8.6×10^{12}	5.5×10^{12}	2.6×10^{12}
1978–1979, 1983–1984	—	3.3×10^{14}	1.7×10^{14}	8.0×10^{13}	4.3×10^{13}	2.7×10^{13}	1.3×10^{13}
Solar flare protons, maximum,							
1980–1982	—	6.7×10^{14}	3.4×10^{14}	1.6×10^{14}	8.6×10^{13}	5.5×10^{13}	2.6×10^{13}

[a]Infinite backshielding assumed. Density of fused silica, 2.2 g/cm³. Predictions for trapped electrons for 35,900-km orbit, 0° circular orbit, for trapped protons at 33,300 km, and for solar flare protons at 1 AU. From Weidner (1969).

[b]Numbers in parentheses are thickness of fused silica in mils.

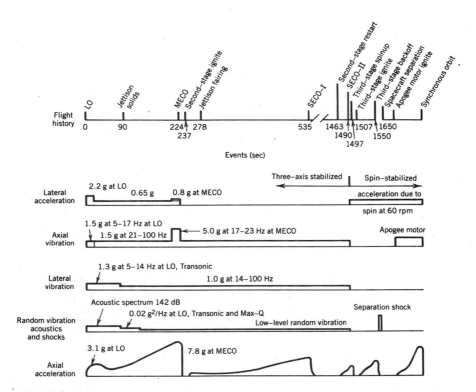

FIGURE 5.139 Mechanical launch environment for three-stage launch vehicle: LO, lift-off; MECO, main engine cutoff; SECO, secondary engine cutoff.

is ignited, the fairing is jettisoned (not needed when there is no air), and then the engine is turned off for a coast period—secondary engine cutoff-I (SECO-I). There is a quarter of an hour coast period in low circular orbit until the equator is reached. The second stage is then restarted, for the perigee burn, and ends with the secondary engine cutoff-2 (SECO-2). The third stage is spun up (it has no guidance), ignited, and continues until burnout (BO). The spacecraft then separates from the launch vehicle and is in transfer orbit. After a coast of a few days, the apogee motor is fired, and the satellite is in near-geostationary orbit.

The accelerations and vibrations for the different launch events are shown in Figure 5.139. The axial acceleration increases during burn since the thrust of a rocket motor tends to remain constant and the mass is constantly decreasing. Lateral accelerations occur near launch (especially on a windy day) and when the spacecraft is spin stabilized. The vibration levels can be high whenever rockets are firing. The acoustic vibrations are high near liftoff, when there is air to conduct the sound from the motor to the spacecraft. The shocks occur at different points, with one strong shock when the spacecraft separates from the launch vehicle.

To test a spacecraft structure, it is not necessary to duplicate the entire launch

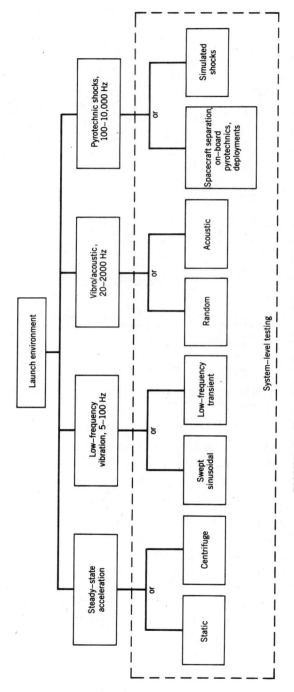

FIGURE 5.140 Simulation of launch environment at system level.

environment. Tests are designed so that if the spacecraft passes the test, it has a high probability of surviving the launch environment. In Figure 5.140, the four main components of the launch environment are shown: steady-state acceleration, low-frequency vibration (5–100 Hz), vibro/acoustic (20–2000 Hz), and pyrotechnic shocks with methods of simulating each component. The choice of method may depend on the spacecraft being tested and on the facilities available. Many test specifications are set by NASA or the Air Force. Several of these tests will now be considered.

5.11.3.1 Static Testing

The test for steady-state acceleration determines whether the spacecraft can survive the accelerations of the launch vehicles. For some launch vehicles this may be 7 g or higher. Accelerations for the shuttle are shown in Figure 5.141. The maximum along the shuttle axis (X axis) is 3.2 g for ascent and 2 g for landing. The maximum along the Z axis (up–down during landing) is 2.5 g for ascent and as high as 4.2 for landing. The maximum for the Y axis, along the wings, is 1.0 for launch and 1.25 for landing. While there is no plan for the spacecraft to be in the shuttle during landing, it should be designed to survive if the mission is aborted.

One way to do static testing is to attach cables to various parts of the structure using a harness to distribute the loads and gradually increase the forces on the structure with hydraulic pumps. Strain gages are used to measure the structure response. An alternative is to use a centrifuge and subject the entire structure to

Flight Event	Load Factor			Angular Acceleration (rad/sec²)		
	N_x	N_y	N_z	$\ddot{\phi}$	$\ddot{\theta}$	$\ddot{\psi}$
Lift-off	−0.2 −3.2	1.0 −1.0	2.5 −2.5	0.1 −0.1	0.15 −0.15	0.15 −0.15
Post-SRB staging	−1.10	±0.12	−0.59			
High-Q boost envelope	−1.9	±0.40	0.25 −0.50	±0.10	±0.15	±0.15
Integrated vehicle boost maximum N_x	−2.9	±0.06	−0.10	±0.20	±0.25	±0.25
Orbiter boost maximum N_x	−3.17	0.0	−0.60	±0.20	±0.25	±0.25

NASA will supply updated numbers for any spacecraft testing.

FIGURE 5.141 Limit loads for payloads on shuffle orbiter during lift-off and ascent.

the desired acceleration. For the centrifuge test, it is necessary to load the structure with the equipment or with dummy weights.

The advantages of a static test is that it requires only an inexpensive testing facility, there are no constraints on the number or type of transducers, and if critical deformation occurs, the test can be stopped to avoid complete structural failure. The disadvantage of a static test is that concentrated loads are applied only at relatively few selected points. The centrifuge test, on the other hand, has the advantage that every item in the spacecraft experiences acceleration. The disadvantages of the centrifuge are the existence of an acceleration gradient, the need of an expensive test facility, the limitations on the number of channels of data, the difficulty in measuring deflection, and the extra hazard of the test.

5.11.3.2 Sinusoidal Vibrations

Vibration tests are done on a shaker, which creates vibration by the electromagnetic action of coils, similar to a huge loudspeaker. The main components are shown in Figure 5.142. The vibration starts in an oscillator and goes through a gain control, then to a power amplifier, and finally into the coils of the shaker. For control, an accelerometer on the shaker measures the amplitude, and the accelerometer response returns through a feedback loop to control the gain and thus the vibration amplitude.

The oscillator is swept through the desired frequencies at a certain rate. The sweep rate cannot be too fast or there will not be a chance for any resonant effects to build up to large amplitudes. On the other hand, the sweep rate cannot be too slow or some of the structure may suffer from fatigue by being cycled too many times. The acceleration magnitude as a function of frequency for a few actual tests at a particular frequency-sweep rate is shown in Figure 5.143. The swept frequency as a function of time can be written as

$$f = f_0 \, 2^{Kt/60} \tag{5.151}$$

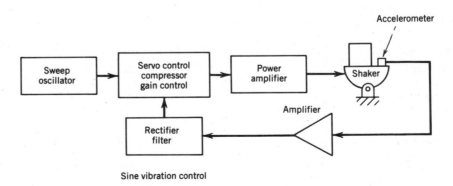

FIGURE 5.142 System for swept sinusoidal vibration test.

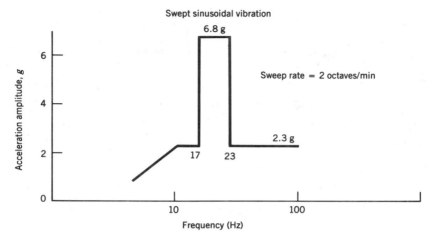

FIGURE 5.143 Levels for sinusoidal vibration test ($1 \text{ g} = 9.8 \text{ m/sec}^2$).

where f_0 is the starting frequency at t_0, t is the time in seconds, and K is the sweep rate in octaves per minute. In order to hold acceleration amplitude constant during specific time intervals while frequency is, say, increasing, the amplitude of the vibration is decreased gradually as a function of time during that interval.

For larger spacecraft, a modification of the sinusoidal test is sometimes required, referred to as "notching." Small spacecraft have no effect on the launch vehicle; for these it is only necessary to determine how much the launch vehicle vibrates the mounting surface and then measure the effect of this vibration on the spacecraft. However, with larger spacecraft, when a resonance is reached, the spacecraft absorbs vibration energy, and the amplitude of mounting surface vibration is decreased. Holding constant the oscillation amplitude of the original mounting surface would overtest the spacecraft structure.

To explain this phenomenon, consider the oscillation of two masses, as shown in Figure 5.144. This is similar to the example studied earlier, except that the second mass is small compared to the first mass. The system is driven at a frequency ω, and the amplitude is plotted in the figure. The amplitude of the large mass A_1 drops in the vicinity of the resonant frequency ω_n because the small mass starts to oscillate with a larger amplitude and absorbs more energy. Figure 5.145 depicts the similar situation for a launch vehicle and a spacecraft; the latter has a single resonant frequency ω_n. At this frequency the spacecraft oscillation increases, but the vibration amplitude of the launch vehicle mounting surface decreases. In running a shaker test, it is necessary to reduce the amplitude of the mounting surface to simulate the actual launch.

To determine the amount of notching, a computer model of the spacecraft is combined with a computer model of the launch vehicle (or shuttle). The maximum amplitude of oscillation of various components of the spacecraft can then be determined. In subsequent shaker tests, these components are monitored. If these

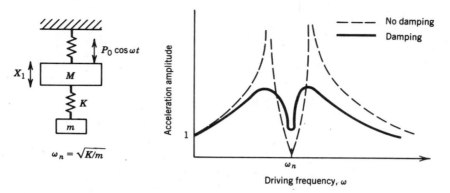

FIGURE 5.144 Acceleration amplitude of large mass when driving frquency is near resonance of coupled small mass.

maximum values are reached, the shaker excitation is reduced in such a way that these maxima are never exceeded.

Notching is often required for several different resonant modes, each one having a maximum amplitude on some particular spacecraft component. A method of implementation is sketched in Figure 5.146. The spacecraft structure has controls on several points where the response is to be limited. Signals from these controls are normalized and fed into a multilevel selector, which finds the control that is most restrictive; this is then used by the servo control to determine the signal driving the shaker. Figure 5.147 shows an actual graph of the acceleration of the base. This is nominally being run at an acceleration of 1.5 g. The dips in the curve show frequencies at which the different controls took over, limiting the excitation to a much lower acceleration. The result for a particular spacecraft component is similar to that shown in Figure 5.148. The response shows a normal resonance peak, but the top has been chopped off by the notching of the base excitation.

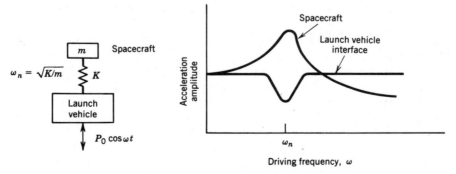

FIGURE 5.145 Decrease of amplitude of launch vehicle interface vibration near spacecraft resonance.

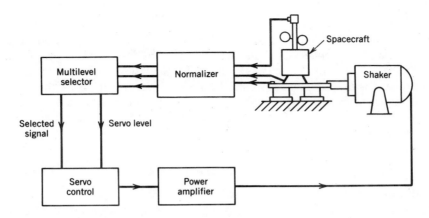

FIGURE 5.146 Spacecraft monitor controls used in notching.

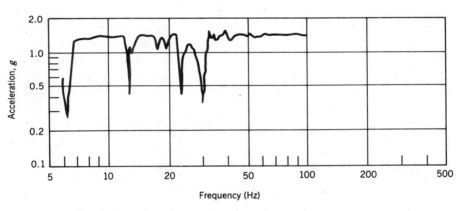

FIGURE 5.147 Acceleration level of shaker during spacecraft test with notching.

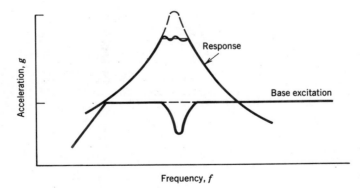

FIGURE 5.148 Truncation of spacecraft response due to notching.

5.11.3.3 Random Vibration

Instead of oscillating the shaker at a single frequency at a time, it is possible to excite a variety of frequencies simultaneously. In some respects this is more analogous to actual launches, when a variety of frequencies is present. The spectral density curves used in some test specifications are shown in Figure 5.149. The curve increases from a low frequency, has a flat portion over a middle band of frequencies, and then tails off at higher frequencies. This frequency spectrum is maintained for the duration of the test.

5.11.3.4 Acoustic Vibration

A different type of test is conducted in an acoustic chamber, where the vibrations are transmitted through the gas from giant horns rather than through the spacecraft support (see Figure 5.150). The acoustic levels used in this test are shown in Figure 5.151, and the overall sound level is quite high.

The acoustic and random vibration tests are to some extent complementary, and usually one or the other is done but not both. The acoustic test has the advantage of a good simulation for larger spacecraft with complex shapes and provides adequate excitation to the subsystems on the spacecraft exterior. Its disadvantages

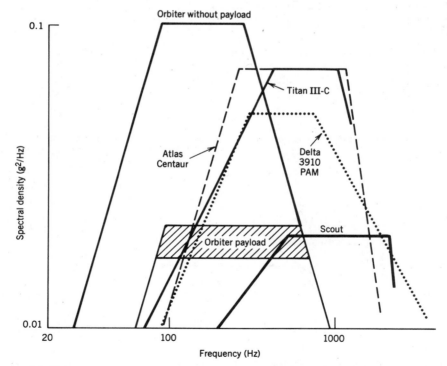

FIGURE 5.149 Test levels for random vibration tests. Get updated levels for any spacecraft testing.

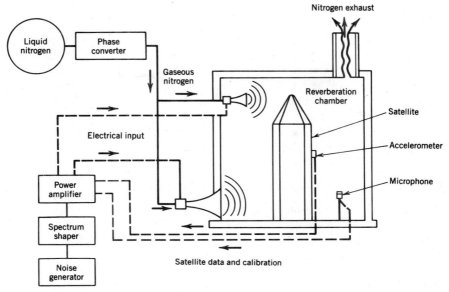

FIGURE 5.150 Acoustic chamber for spacecraft tests.

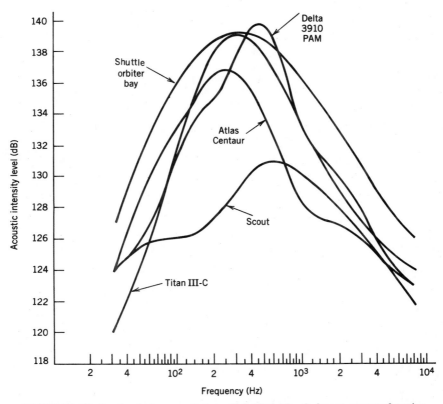

FIGURE 5.151 Test levels for acoustic tests. Get updated levels for any spacecraft testing.

are that it is expensive, is unable to excite subsystems near the mounting surfaces, and is unable to provide enough force in the low-frequency region. In contrast, the random vibration tests have the advantages of good simulation for smaller, compact spacecraft and provide adequate excitation to subsystems near the mounting surface. The random vibration test has the disadvantage that it is unlikely to provide sufficient excitation at the forward end of the spacecraft (far from the mounting surface) and is unlikely to provide sufficient excitation to subsystems on the exterior of the spacecraft (such as solar panels and antennas).

5.11.4 Thermal Vacuum Tests

The importance of environmental tests for spacecraft cannot be minimized. The closest simulation of the actual space environment is the thermal vacuum test. A typical thermal vacuum test cycle is shown in Figure 5.152. Cold tests alternate with hot tests for a 2-week test. There are two purposes in a thermal vacuum test.

One purpose originates from the difficulties in the accurate prediction of spacecraft temperatures. Some of these uncertainties may arise from thermal resistance between contacting spacecraft parts, inaccuracies in calculating the heat radiated from one part to another, and the exclusion of many small parts and wires in the

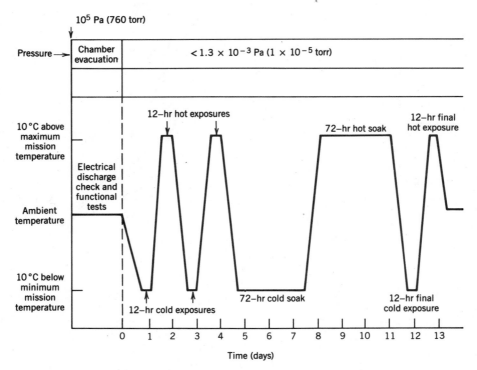

FIGURE 5.152 Typical thermal vacuum test cycle (NASA/GSFC, 1977).

mathematical model. Therefore, a test is needed to verify the basic thermal design, prove the approximations used in the analysis are valid, and check temperature predictions. This test may be called a thermal balance test and often is done only on the first of a group of satellites.

A second purpose of a thermal vacuum test is to provide an environment similar to that encountered in space—and see if the spacecraft works. It is usually not possible to simulate space exactly, with the extreme vacuum and the complex sources of radiation (thermal, visual, and particles), but it is possible to simulate space sufficiently so that most potential troublespots show up. Also, it is difficult to simulate the exact time dependence of the environment; it is easier to choose two extreme conditions. Thus, there would be a "hot" test when maximum temperatures are encountered and a "cold" test with minimum temperatures. This test must be done on every flight spacecraft.

A naive goal for a thermal vacuum test is to duplicate the environment encountered in space. Such an idealistic aim quickly degenerates in the face of practical difficulties. In light of the high costs, a more realistic goal is to design a simple test with a high probability of detecting any inherent defects in the spacecraft. For thermal tests, the criterion of any compromise can usually be estimated in the resulting temperature change in the spacecraft. Any change in the test that produces only a fraction of a degree change in spacecraft temperature can usually be tolerated. Such a criterion can be used in setting tolerance limits on chamber wall emissivities, wall temperatures, vacuum achieved, supports for spacecraft, and so on.

5.11.4.1 Supports

Fixtures used to support the spacecraft should not significantly change the temperature distribution. In particular, the heat conducted through the supports should be small compared to the heat flow by radiation. The best supports would probably be suspension by wires or nylon cords, thus providing the maximum strength with the minimum heat flow path. Stainless steel rods can also be used because of their low thermal conductivity. For many satellites solid supports of other metals are adequate.

5.11.4.2 Wall Emittance

If the walls of the vacuum chamber are not perfectly black, some of the thermal radiation from the spacecraft will be reflected by the walls back to the spacecraft. This will decrease the net heat flow out from the spacecraft and, if the electrical heating in the spacecraft is significant, raise the spacecraft temperature. An estimate of this error can be obtained from Christiansen's equation:

$$\frac{1}{\epsilon_e} = \frac{1}{\epsilon_1} + \left(\frac{1}{\epsilon_2} - 1\right)\frac{A_1}{A_2} \tag{5.152}$$

where ϵ_e is the effective spacecraft emittance, ϵ_1 the actual emittance, ϵ_2 the chamber wall emittance, and A_1/A_2 the ratio of spacecraft surface area to vacuum chamber wall area. Christiansen's equation is discussed on page 698, and can be derived using the method in Section 5.10.5.

If the major effect on spacecraft temperature by the wall emittance is due to the electrical heat of the spacecraft, q, it can be shown that the fractional change in spacecraft temperature $(\Delta T/T)$ is equal to

$$\frac{\Delta T}{T} \approx \frac{q}{4\sigma T^4} \left(\frac{1}{\epsilon_2} - 1 \right) \frac{1}{A_2} \tag{5.153}$$

where ϵ_2 is the chamber wall emittance, and A_2 is the area of the vacuum chamber wall. Usually, if the wall emittance is greater than 0.9, and the chamber wall area is 3 or 10 times greater than the spacecraft surface area, the change in spacecraft temperature is negligible; these specifications are relatively easy to meet.

5.11.4.3 Vacuum Requirements

The vacuum required to eliminate conduction through the gas is usually overestimated. The requirement can be estimated by calculating the number of molecules ν that impinge on a surface per unit time and per unit area. Assuming the average energy transfer cannot be greater than $3k(T - T_0)/2$, the fractional error E incurred in the heat transferred is

$$E \leq \frac{3}{2} \frac{k(T - T_0) \nu}{\epsilon\sigma(T^4 - T_0^4)} < \frac{3}{2} \frac{k\nu}{\epsilon\sigma T^3} = \frac{pc}{\sqrt{6\pi\epsilon\sigma T^3}} \tag{5.154}$$

where k is the Boltzmann constant, p is the gas pressure, and c is the rms molecular speed. For a temperature $T = 300$ K, this simplifies for air to

$$E < \frac{35p_{\text{torr}}}{\epsilon} \tag{5.155}$$

where p is measured in torr (or mm Hg). Unless the emittances are abnormally low, the error is usually negligible for pressures below 10^{-4} torr.

5.11.4.4 External Thermal Inputs

Various methods are used to simulate the solar radiation or to provide a proper thermal input to a surface. In general, the more complex the external surfaces, the more difficult it is to achieve this simulation. The ultimate is to simulate the sun in intensity, direction, and spectrum; while this can be done, it is costly and not used in the majority of tests. Electrical heaters can be used imbedded in the spacecraft skin or suspended between the spacecraft and a cold wall. Alternatively, the chamber wall can be maintained at an appropriate temperature so that its radiation is the right amount. Different walls may be at different temperatures so that the thermal power absorbed on each surface is duplicated. Finally, there are isothermal tests, where all the chamber walls are at one temperature; also, components are often mounted on a plate at a fixed temperature.

5.11.4.5 Thermocouples

Besides using thermistors coupled to the satellite telemetry system, during test additional temperatures are often measured with thermocouples. These should be easily installed and removed and unaffected by incident radiation or by the rate of change of surface temperature. In some locations it makes little difference how the thermocouple is made or attached; at other locations, however, the measured temperature should be unaffected by incident radiation.

To test the effect of incident radiation, a thermocouple was placed on each side of a thin copper plate, which was in a vacuum between two walls. With one wall at 100 °C and the other at −185 °C, the measured temperature difference between the two test thermocouples was only 0.5 °C. To test the transient response, a reference thermocouple was made by spot welding copper and constantan wires to a thin aluminum plate. The test thermocouple was then attached on top and measured the correct temperature for temperature changes up to 0.5 °C/sec.

5.11.4.6 Monitoring a Test

A useful monitoring tool is to plot the temperatures as a function of $1 - e^{-t/\tau}$, as shown in the sample graph in Figure 5.153. While the equilibrium temperatures of several points on a spacecraft may be different, the time constants are often the same, so that a number of temperatures may be plotted on the same graph. Also, as the terms with shorter time constants disappear, the linearized solution becomes applicable, and the points follow a straight line. As time goes on during the test, additional points can be plotted; the equilibrium values can always be determined by a linear extrapolation to the right-hand side (where $t = \infty$) and become more accurately known as closer points are plotted. The graph also has the advantage that all the data can be used in determining an equilibrium value rather than selecting a few points that may be in error.

Frequently, after faster transients have disappeared, a temperature is following an exponential approach to an equilibrium temperature. If three temperature readings, T_1, T_2, and T_3, are taken evenly spaced in time, the equilibrium temperature T_E can be calculated as

$$T_E = T_2 + \frac{(T_2 - T_1)(T_3 - T_2)}{(T_2 - T_1) - (T_3 - T_2)} \tag{5.156}$$

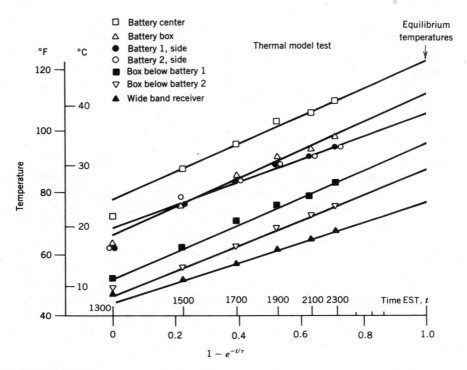

FIGURE 5.153 Test temperatures in thermal vacuum test extrapolated to equilibrium. Time constant is 8 hours.

and the time constant τ is given by

$$\tau = \frac{h}{\ln\left[(T_2 - T_1)/(T_3 - T_2)\right]} \qquad (5.157)$$

where h is the time interval between temperature readings.

5.11.5 In-Orbit Tests

After the satellite is launched, there is usually an additional period of in-orbit testing. These tests take several weeks and can usually be made while the satellite is drifting to its operational longitude. The main test objectives are:

Verify deployment of arrays and antennas.

Verify that satellite system is operating properly.

Verify operation of redundant units.

Provide baseline as reference for future tests.

Measure those properties that can be evaluated more accurately in space, such as actual antenna patterns.

TABLE 5.32 Representative In-Orbit Tests

Attitude	Normal mode and stationkeeping mode, operations with different gain settings, wheel unload, thruster calibration, magnetic torque operation, loss of earth pointing
Power	Unit power consumption, solar array power, solar array orientation
Thermal	Temperatures for different power dissipations, heater function, automatic heater operation
Telemetry, tracking, and command	Command sensitivity, command bandwidth, beacon EIRP, modulation index, beacon frequency
Communications	EIRP at saturation, flux density for saturation, gain transfer, G/T; Antenna pattern: $\pm 12°$ east–west (pitch) at 4–6 cuts, copolarization and crosspolarization pattern, isolation between beams; local oscillator frequency; in-band and out-of-band frequency response; group delay

Baseline measurements are especially useful for units that may suffer degradation with time, such as solar cells, bearings, cathode coatings, and so on. Some properties, such as antenna patterns and thermal couplings, can never be measured completely on the ground, and the last word is never obtained until measurements are made in space.

The variety of possible in-orbit tests is shown in Table 5.32. Many of these are a final measurement of a parameter, more accurate than can be done on the ground, that will provide a baseline for future reference. Any unexpected results may cause modifications of operations procedures to compensate for an unknown factor.

5.12 RELIABILITY

5.12.1 Introduction

Reliability is the probability that no failure will occur in a given time interval of operation. A reliability must be between 0 and 1. The reliability may be very close to 1 for a highly reliable component, such as 0.99999, but as the time interval is increased, the reliability will decrease. Some satellites have a specified reliability of 0.7 of operating for 10 yr.

More precisely, reliability is the probability of a device performing its purpose adequately for the period of time intended under the operating conditions encountered. For a precise definition, there must be a clear understanding of what constitutes adequate performance of a device and under what conditions it is supposed to operate.

The simplest way to visualize reliability is to have N_0 components and to put them all in operation at the same time. As time goes on, some components fail

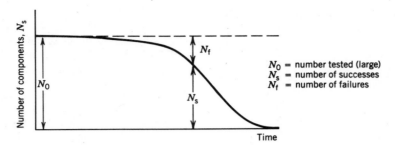

FIGURE 5.154 Number of components operating as function of time.

and others continue operating. The number of components still operating, N_s, can be plotted as a function of time, as shown in Figure 5.154. The number of failures N_f is the difference between the initial number N_0 and the number still operating, N_s.

If the number of initial components, N_0, is large, an estimate of the reliability is

$$R = \frac{N_s}{N_0} = \frac{N_s}{N_s + N_f} \qquad (5.158)$$

that is, the fraction of components that are still operating is an estimate of the reliability. It is also useful to define an unreliability Q as the probability that a failure has occurred in a certain time interval. Then

$$Q = \frac{N_f}{N_0} = \frac{N_f}{N_s + N_f} = 1 - R \qquad (5.159)$$

A reliability curve is shown in Figure 5.155, which is the same as Figure 5.154 but with a normalized axis.

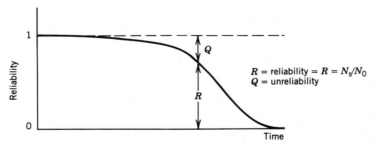

FIGURE 5.155 Reliability for various times.

The failure density function is obtained by taking the derivative of the reliability curve:

$$f = -\frac{dR}{dt} = \frac{1}{N_0}\frac{dN_f}{dt} \qquad (5.160)$$

and this is shown in Figure 5.156. This typical curve shows the distribution of failures in time on a per-component basis (the numbers are actually taken from the 1958 Standard Mortality Table). The peak of the failure density curve occurs at 75 yr (3% of all people born die at age 75). While this age has more individuals dying than any other age, there are many more that die before the age of 75 than after, which is shown by the area under the curve being larger to the left of age 75.

The failure rate λ is the number of failures per hour, dN_f/dt, divided by the number of units still operating, N_s:

$$\lambda = \frac{1}{N_s}\frac{dN_f}{dt} = -\frac{1}{N_s}\frac{dN_s}{dt} = -\frac{1}{R}\frac{dR}{dt} \qquad (5.161)$$

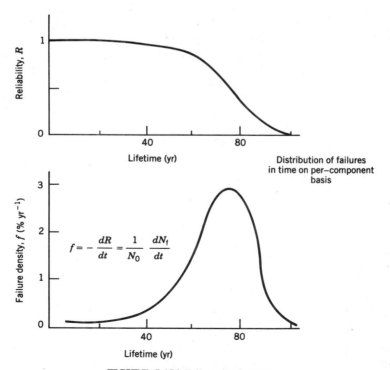

FIGURE 5.156 Failure density curve.

The failure rate that corresponds to the previous curves is shown in Figure 5.157. The reliability can be written in terms of the failure rate λ as

$$R = \exp\left(-\int_0^t \lambda\, dt\right) \qquad (5.162)$$

This is a general equation, and no assumption has been made of how the failure rate varies with time.

Failure rates for satellite components are often specified in FITS, as the number of failures in 10^9 hours. If a component has a probability equal to 10^{-9} of failing in 1 hr, its failure rate is 1 FIT. This is extremely small, but if a satellite with thousands of components is designed to operate for several years, some components must have reliabilities as low as a few FITS. While specifications in terms of 10^9 hr are common in satellite designs, failures in 10^6 hr are often used in the military; unfortunately, the terms FITS has been used for the latter unit.

Many items have a failure rate curve similar to the one in Figure 5.157, where the failure rates at the start and end are higher than in the middle. This has led to the classification of failures into three types: early failures, random failures, and wear-out failures. A fourth type is design failures. Many failures in communications satellites tend to be design failures or random failures.

Fundamental mistakes in design are called *design* failures; when they are detected, the design is changed, if possible, to eliminate this source of failures. Often errors in quality control are also included in design failures, although they are not due to the actual engineering design. Regardless of the source, a design error occurs in a number of units; when it is discovered, it means that the failure rate is much higher than originally planned.

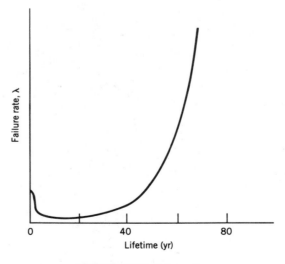

FIGURE 5.157 Failure rate.

Early failures, also known as infant mortality, are any other failures that occur early in the life of the equipment. It is usually considered that most early component failures will be eliminated either by the standard satellite testing or by additional burn-in periods.

Chance failures occur at random throughout the life of the equipment. They cannot be eliminated, but they can be reduced. Reliability engineering is mainly focused on chance failures because it affects equipment in actual operational use—after the equipment has been debugged and put in service and before it begins to wear out.

Finally, *wear-out* failures are a symptom of components aging. It may be a predictable occurrence, such as fuel depletion, or one less predictable, such as failure of a bearing or an electric battery. Whatever may be the cause, the occurrence of failures increases after a certain operational period, and these failures are classified as wear-out failures.

5.12.2 Combining Probabilities

If A and B are two independent events with the probabilities $P(A)$ and $P(B)$, the probability that both events will occur, $P(AB)$, is equal to the product of the probabilities that each will occur;

$$P(AB) = P(A) \cdot P(B) \tag{5.163}$$

This equation is only true if the probabilities are independent, which means that the probability of one event is absolutely independent of whether or not the other event occurs.

Two events are mutually exclusive if the occurrence of either one absolutely excludes the occurrence of the other. In this case, the probability that either event A or event B occurs, $P(A \text{ or } B)$ is equal to the sum of the two probabilities:

$$P(A \text{ or } B) = P(A) + P(B) \tag{5.164}$$

Two events are complementary if one or the other must occur but not both. If event A does not occur, B must occur and vice versa. In this case

$$P(A) + P(B) = 1 \tag{5.165}$$

and the probabilities add to unity. In reliability calculations, the unreliability Q is the probability that the unit has failed in a given time. This is related to the reliability R by

$$R + Q = 1 \tag{5.166}$$

since, by definition, the unit must either be still working or have failed.

The binomial distribution (Burington and May, 1958) is used in finding the probabilities when there are n identical components each with a reliability R. The probability that all n components work is R^n, and the probability that exactly k components work is

$$P_k = {}_nC_k R^{n-k} Q^k = \frac{n!}{(n-k)!\,k!}\, R^{n-k}(1-R)^k \qquad (5.167)$$

To obtain the probability that k components or more are working, it is necessary to sum all the terms from P_k to P_n.

An example of the binomial distribution is shown in Table 5.33 for eight launch vehicles, each with a success probability estimated at 0.75. Since the success probability is $\frac{3}{4}$, an average of two failures out of eight is expected. However, the probability of exactly two failures is only 0.3; the probability of only one failure is almost 0.3, and the probability of three failures is 0.2. To obtain the probability of six successful launches or more, it is necessary to add the probabilities of six, seven, and eight successes; the result is 0.679, or about two-thirds.

5.12.3 Constant Failure Rates

For many satellite components, it is common to assume a constant failure rate (see Table 5.34). The failure rate is not decreasing with time (as would be true for early failures), nor is it increasing with time (as usually happens with wear-out failures). A constant failure rate λ leads to an exponential curve for the reliability function

$$R = e^{-\lambda t} \qquad (5.168)$$

The reciprocal of the failure rate is called the mean time between failures (MTBF) and is denoted by m. Then

$$R = e^{-t/m} \qquad (5.169)$$

TABLE 5.33 Binomial Distribution for Eight Launches, Each with 75% Probability of Success

Successes	Failures	Calculation	Probability
8	0	$(0.75)^8$	0.10011
7	1	$8(0.75)^7(0.25)$	0.26697
6	2	$28(0.75)^6(0.25)^2$	0.31146
5	3	$56(0.75)^5(0.25)^3$	0.20764
4	4	$70(0.75)^4(0.25)^4$	0.08652
3	5	$56(0.75)^3(0.25)^5$	0.02307
2	6	$28(0.75)^2(0.25)^6$	0.00384
1	7	$8(0.75)(0.25)^7$	0.00037
0	8	$(0.25)^8$	0.00002

TABLE 5.34 Estimated Failure Rates[a]

Part Type	Failure Rate	Part Type	Failure Rate
Resistor		Bearing	50
Composition	0.1	Slip ring	40/Brush
Metal film	0.1	Motor (despin)	100
Wire wound	2		
Potentiometer	200	Tunnel diodes	200
		(germanium)	
Capacitors			
Ceramic		Transistors	
General purpose	0.1	(silicon planar)	
Feed through	0.5	General purpose	2
Glass	0.2	Switch	5
Mylar	0.1	rf	8
Paper	0.5	Power	10
Tantalum, solid	1	Field effect	15
Trimmer	0.2		
High voltage (>500 V peak)	10	Integrated circuits,	
		monolithc	
Diodes (silicon)		Digital (bipolar)	20
Switching	0.6	Analog (op amps)	30
Power	1		
Zener	2	Hybrid Circuits	
Detector/mixer	10	Simple (<10	60
		monolithic chips)	
Filters (per section)		Complex (10–20 chips)	
Hybrid	1		100
Bandpass	1	Traveling-wave tubes	
		4 GHz, 4–12 W	1,000
		12 GHz, 10–20 W	1,500
Couplers	1		
Circulators	10	Transformers	
Relays (SPST[b])	50	Power	20
		Signal	10
rf switches			
Ferrite (SPDT[b])	20		
Mechanical (SPDT)	25	Inductors	
		Power	20
		Signal	10
Solder and weld connections	0.2		
Plug and receptable connectors	0.4	Crystals	40

[a]Failures in 10^9 operating hours, at 20% stress and 25 °C (where applicable).
[b]Abbreviations: SPDT, single pole double throw switch; SPST, single pole single throw switch.

The MTBF may not have any connection with the actual life, wear-out life, or mean time to failure but is the average life if the failure rate is indeed constant.

The reliability curve R is shown in Figure 5.158. This shows that the number of surviving components decreases exponentially. The failure density curve ($f = -dR/dt$) is also shown in Figure 5.158. The frequency of failures is equal to the failure rate (assumed constant) multiplied by the population; it is also a decreasing exponential curve.

Complete systems usually consist of many components or subsystems. When there is no redundancy, if one component fails, the system fails. On a reliability diagram, all these components are in series. The reliability of the system is equal to the product of the reliabilities of the components:

$$R_s = R_1 \cdot R_2 \cdot R_3 \cdot R_4 \cdots R_n \tag{5.170}$$

If all the components have constant failure rates, the system reliability can be written as

$$R_s = e^{-\lambda_1 t} \cdot e^{-\lambda_2 t} \cdot e^{-\lambda_3 t} \cdots e^{-\lambda_n t}$$
$$= e^{-(\lambda_1 + \lambda_2 + \lambda_3 + \ldots + \lambda_n)t} \tag{5.171}$$

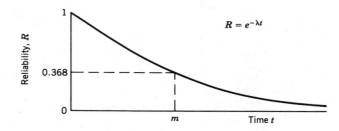

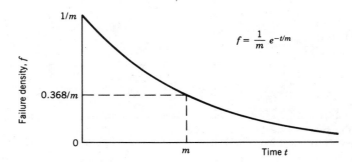

FIGURE 5.158 Constant failure rate; reliability curve R and its negative derivative, failure density curve f.

This shows that the failure rate of the system is equal to the sum of the failure rates of its components. When all components are needed for successful operation of the system and if each unit has an exponential reliability curve, the reliability of the system is also an exponential curve.

A simple example typical of calculations of electronic circuits is shown in Table 5.35. The circuit is made of several types of components, and the number of each component and the failure rate for each type is calculated. If the failure rate for a resistor is 0.1 failures per 10^9 operating hours and there are 20 resistors, there would be an average of 2 resistor failures in 10^9 operating hours. Combining all the failure rates for all the components, the net result is 30 failures in 10^9 operating hours for this simple example.

5.12.4 Wear-Out Distribution

In studying wear-out failures, the normal distribution is used; however, the reliability function R is not normally distributed but rather the derivative of the reliability function. The negative of the slope, $-dR/dt$, is called the failure density function. If we start with a large number of components, N_0, this curve shows where the actual failures occur. The failure density function f is simply the failure rate (failures per hour) divided by the initial number of components, N_0.

The failure density function f for wear-out failures is a normal curve and is shown in Figure 5.159. The normal curve is defined by two parameters: the mean life M and the standard deviation σ.

In a wear-out distribution, the initial time is important. When a random failure curve is plotted as a function of the operating time t, it can start at any particular time; that is, if we know that at a certain time the component is operating adequately, this time can be used as $t = 0$. The wear-out curve is plotted in terms of the operating life, or age, of the component; this always must be measured from the time the component was first operated.

The reliability curve R for wear-out failures is also shown in Figure 5.159. It is the negative integral of the density function. It is useful to label the reliability curve as R_W to define it explicitly as the wear-out reliability.

In many systems there are combined effects of chance and wear-out failures.

TABLE 5.35 Example of Reliability in Simple Circuit

Part	Number	Failure Rate[a]	
		Individual	Total
Resistors	20	0.1	2
Capacitors	10	0.2	2
Diodes	6	1	6
Transistors	4	5	20
Total failure rate, λ			30

[a] In FITS (failures in 10^9 hr).

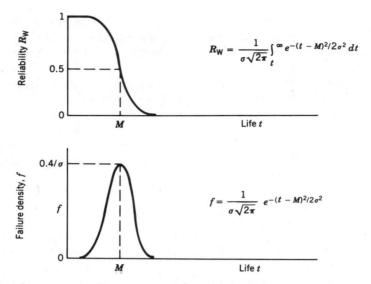

$$R_W = \frac{1}{\sigma\sqrt{2\pi}} \int_t^\infty e^{-(t-M)^2/2\sigma^2} dt$$

$$f = \frac{1}{\sigma\sqrt{2\pi}} e^{-(t-M)^2/2\sigma^2}$$

FIGURE 5.159 Wear-out distribution; reliability curve R and its negative derivative, failure density curve f.

The chance failures can be characterized by a mean time between failures of m and the wear-out by a mean life M and a standard deviation σ. Since reliability is equal to a probability of success, the two reliability curves can be multiplied together to obtain the system reliability curve. The first part of the curve will follow the exponential curve but then be quickly brought to zero by the effect of wear-out failures.

5.12.5 Redundancy

When high reliabilities are required, duplicate units may be needed. These are connected so that if either unit works, the system will function; from a reliability view, these units are considered to be in parallel (this may or may not correspond to the electrical connections).

If the reliabilities of $n + 1$ units are independent, the probability that all of them fail is equal to the product of the individual unreliabilities Q_i:

$$Q_p = Q_1 \cdot Q_2 \cdot Q_3 \cdots Q_{n+1} = \prod_{i=1}^{n+1} Q_i \tag{5.172}$$

Frequently, these are all equal so that

$$Q_p = Q^{n+1} \qquad R_p = 1 - Q^{n+1} = 1 - (1 - R)^{n+1} \tag{5.173}$$

where R is the reliability of a single unit, and Q is the unreliability.

Parallel redundancy is not particularly useful when the unit reliability is poor, and it is not needed if the unit reliability is excellent. For example, if the unit reliability is 0.3, three redundant units will have a reliability of only 0.65, which is probably still not sufficient. However, if the unit reliability is fairly good, such as 0.995, a twofold redundancy will increase it to 0.999975, and the redundancy is quite useful.

If each unit has a random failure rate λ, the probability that at least one unit out of $n + 1$ is working is

$$R_p = 1 - (1 - e^{-\lambda t})^{n+1} \qquad \lambda_{\text{spare}} = \lambda_{\text{op}} \qquad (5.174)$$

where it is assumed that a nonoperating unit fails at the same rate as an operating unit. Reliability numbers are shown in Table 5.36 for 5-, 7-, and 10-yr satellites. Since the equation is in terms of λt, a 10-FITS component in a 5-yr satellite has the same reliability as a 5-FITS component in a 10-yr satellite. The unreliabilities $1 - R$ have been rounded to two significant figures. As an example, a 1000-FITS components has a 7-yr reliability of 0.941 for zero spares; if there is one spare and if nonoperating units have the same failure rate as operating units, the probability is 0.9965 that one out of two is still working.

Alternatively, it can be assumed that a standby unit has a zero failure rate until it is turned on and that the switching device also has a zero failure rate. Then when n spare units are standing by to replace *one* that operates, the reliability of the system (Bazovsky, 1961) is

$$R_p = e^{-\lambda t}\left[1 + \lambda t + \frac{\lambda^2 t^2}{2!} + \cdots + \frac{\lambda^n t^n}{n!}\right] \qquad \lambda_{\text{spare}} = 0 \qquad (5.175)$$

Reliability numbers for different component failure rates are shown in Table 5.37. This differs from Table 5.36 in that the spare units are assumed to have zero failure rates until they are needed. Thus, the calculated reliabilities are always higher.

Table 5.37 can also be used for several operating units. For example, it there are five 1000-FITS operating units, each with a 7-yr reliability of 0.941 and three spares, the row for 5000 FITS shows that the reliability is 0.74 for no spares and 0.99971 for three spares. For k operating units and n spare units, the reliability of the whole system is

$$R_p = e^{-k\lambda t}\left[1 + k\lambda t + \frac{k^2\lambda^2 t^2}{2!} + \cdots + \frac{k^n\lambda^n t^n}{n!}\right] \qquad (5.176)$$

where the failure rate of a spare λ_{spare} is zero and that of operating units is λ.

In actual practice, the failure rate of units on stand by is usually between the two preceding cases; the failure rate is less than that of an operating unit but greater than zero. Consider a system of k operating units and n spare units. The operating unit has a failure rate of λ, and the spares have a failure rate of $s\lambda$ when on stand by; when they are switched on, their failure rate is assumed to be λ. A study of

TABLE 5.36 Reliability of a Unit with N Operating Spares[a]

FITS (failures in 10^9 hr)			Spares				
5 yr	7 yr	10 yr	0	1	2	3	4
1	0.7	0.5	0.999956				
1.4	1	0.7	0.999939				
2	1.4	1	0.999912				
2.8	2	1.4	0.99988				
4	2.9	2	0.99982				
5	3.6	2.5	0.99978				
7	5	3.5	0.99969				
10	7.1	5	0.99956				
14	10	7	0.99939				
20	14.3	10	0.99912	0.999999			
28	20	14	0.9988	0.99998			
40	28.6	20	0.9982	0.99997			
50	35.7	25	0.9978	0.99995			
70	50	35	0.9969	0.99991	1		
100	71.4	50	0.9956	0.99981	1		
140	100	70	0.9939	0.999963	1		
200	142.9	100	0.9913	0.999924	0.999999		
280	200	140	0.988	0.99985	0.999998		
400	285.7	200	0.983	0.99970	0.999995	1	
500	357.1	250	0.978	0.99953	0.999990	1	
700	500	350	0.970	0.99909	0.999972	0.999999	1
1000	714.3	500	0.957	0.9982	0.999921	0.999997	1
1400	1000	700	0.941	0.9965	0.99979	0.999987	0.999999
2000	1428.6	1000	0.916	0.9930	0.99941	0.999951	0.999996
2800	2000	1400	0.88	0.987	0.9985	0.99982	0.99998
4000	2857.1	2000	0.84	0.974	0.9958	0.99933	0.99989
5000	3571.4	2500	0.80	0.961	0.9924	0.9985	0.99971
7000	5000	3500	0.74	0.93	0.982	0.9951	0.9987
10000	7142.9	5000	0.65	0.87	0.955	0.984	0.9944
14000	10000	7000	0.54	0.79	0.904	0.956	0.980
20000	14285.7	10000	0.42	0.66	0.80	0.88	0.932
28000	20000	14000	0.29	0.50	0.65	0.75	0.82
40000	28571.4	20000	0.17	0.32	0.44	0.53	0.61
50000	35714.3	25000	0.11	0.21	0.30	0.38	0.45
70000	50000	35000	0.05	0.09	0.13	0.17	0.21

$\lambda_{spare} = \lambda$

[a]Assuming spares have *same* failure rate as operating unit.

TABLE 5.37 Reliability of a Unit with *N* Zero-Failure Spares[a]

| FITS (failures in 10^9 hr) | | | Spares | | | | |
5 yr	7 yr	10 yr	0	1	2	3	4
1	0.7	0.5	0.999956				
1.4	1	0.7	0.999939				
2	1.4	1	0.999912				
2.8	2	1.4	0.99988				
4	2.9	2	0.99982	1			
5	3.6	2.5	0.99978	1			
7	5	3.5	0.99969	1			
10	7.1	5	0.99956	1			
14	10	7	0.99939	1			
20	14.3	10	0.99912	1			
28	20	14	0.9988	0.999999			
40	28.6	20	0.9982	0.999998			
50	35.7	25	0.9978	0.999998			
70	50	35	0.9969	0.999995			
100	71.4	50	0.9956	0.999990			
140	100	70	0.9939	0.999981	1		
200	142.9	100	0.9913	0.999962	1		
280	200	140	0.988	0.999925	1		
400	285.7	200	0.983	0.99985	0.999999		
500	357.1	250	0.978	0.99976	0.999998		
700	500	350	0.970	0.99954	0.999995	1	
1000	714.3	500	0.957	0.99907	0.999986	1	
1400	1000	700	0.941	0.9982	0.999963	0.999999	
2000	1428.6	1000	0.916	0.9964	0.99990	0.999998	1
2800	2000	1400	0.88	0.9931	0.99972	0.999991	1
4000	2857.1	2000	0.84	0.986	0.99921	0.999966	0.999999
5000	3571.4	2500	0.80	0.979	0.9985	0.999920	0.999996
7000	5000	3500	0.74	0.962	0.9962	0.99971	0.999982
10000	7142.9	5000	0.65	0.928	0.990	0.9989	0.999907
14000	10000	7000	0.54	0.87	0.976	0.9964	0.99956
20000	14285.7	10000	0.42	0.78	0.941	0.988	0.9979
28000	20000	14000	0.29	0.65	0.87	0.964	0.9915
40000	28571.4	20000	0.17	0.48	0.74	0.90	0.967
50000	35714.3	25000	0.11	0.36	0.63	0.82	0.929
70000	50000	35000	0.05	0.19	0.41	0.63	0.80

$\lambda_{\text{spare}} = 0$

[a]Assuming spares have *zero* failure rate.

flight data suggests that $s = \lambda_{spare}/\lambda$ is on the order of 0.1, but this ratio depends on the type of unit. The system reliability R_{pn} for 0, 1, 2, and 3 spares is

$$R_{p0} = R^k \qquad 10\,\lambda_{spare} = s\lambda$$

$$R_{p1} = R^k[1 + F]$$

$$R_{p2} = R^k[1 + F + (1 + s/k)\,F^2/2] \tag{5.177}$$

$$R_{p3} = R^k\big[1 + F + (1 + s/k)\,F^2/2$$
$$+ (1 + s/k)\,(1 + 2s/k)\,F^3/6\big]$$

where

$$R^k = e^{-k\lambda t} \qquad F = k(1 - R^s)/s \tag{5.178}$$

As s approaches zero, the expression for F becomes indeterminate, and it may be more accurate to use the series approximation

$$F \approx -k\ln R - ks(\ln R)^2/2 \tag{5.179}$$
$$- ks^2(\ln R)^3/6 - \cdots$$

The preceding equations can be generalized for any number of spare units, and put into a form for computer use. Succeeding terms T_n are calculated and then summed.

$$T_0 = R^k$$

$$R_{p0} = R^k \tag{5.180}$$

$$T_n = T_{n-1}\big[1 + (n-1)\,s/k\big]\,F/n$$

$$R_{pn} = R_{p,n-1} + T_n$$

The first two equations set the initial values, and the last two equations are repeated for different values of n. Numerical values for one operating unit and several spares, assuming λ_{spare}/λ is 10% are shown in Table 5.38. As expected, the system reliability values are intermediate between the two preceding tables.

The preceding formulas assume that failures are independent; that is, a failure in one unit does not affect the failure rate of other units. Frequently this is not true for satellite components. Either a design defect or lack of quality control during manufacture causes a number of similar units to fail. This is portrayed graphically in Figure 5.160. If there is no redundancy, the reliability of the system is equal to the component reliability $(R_p = R)$. If there is twofold redundancy, the system reliability has increased to $R_p = 1 - (1 - R)^2$. If there is fourfold redundancy, the system reliability of $R_p = 1 - (1 - R)^4$ appears to be quite high. The assumption is that any failure is independent of any other failure.

TABLE 5.38 Reliability of a Unit with N Realistic Spares

$$\lambda_{spare} = 0.1\lambda$$

FITS (failures in 10^9 hr)			Spares				
5 yr	7 yr	10 yr	0	1	2	3	4
1	0.7	0.5	0.999956				
1.4	1	0.7	0.999939				
2	1.4	1	0.999912				
2.8	2	1.4	0.99988				
4	2.9	2	0.99982	1			
5	3.6	2.5	0.99978	1			
7	5	3.5	0.99969	1			
10	7.1	5	0.99956	1			
14	10	7	0.99939	1			
20	14.3	10	0.99912	0.999999			
28	20	14	0.9988	0.99998			
40	28.6	20	0.9982	0.99997			
50	35.7	25	0.9978	0.99995			
70	50	35	0.9969	0.99989			
100	71.4	50	0.9956	0.999979	1		
140	100	70	0.9939	0.999958	1		
200	142.9	100	0.9913	0.999918	1		
280	200	140	0.988	0.99983	1		
400	285.7	200	0.983	0.99974	0.999999		
500	357.1	250	0.978	0.99949	0.999998		
700	500	350	0.970	0.9990	0.999994	1	
1000	714.3	500	0.957	0.9980	0.999982	1	
1400	1000	700	0.941	0.9960	0.999952	0.999999	
2000	1428.6	1000	0.916	0.9923	0.99986	0.999996	1
2800	2000	1400	0.88	0.985	0.99963	0.999986	1
4000	2857.1	2000	0.84	0.977	0.99897	0.999943	0.999997
5000	3571.4	2500	0.80	0.958	0.9981	0.99986	0.999992
7000	5000	3500	0.74	0.922	0.9951	0.99952	0.999960
10000	7142.9	5000	0.65	0.86	0.987	0.9982	0.99979
14000	10000	7000	0.54	0.77	0.969	0.9942	0.99905
20000	14285.7	10000	0.42	0.63	0.927	0.981	0.9956
28000	20000	14000	0.29	0.45	0.85	0.946	0.983
40000	28571.4	20000	0.17	0.33	0.70	0.86	0.940
50000	35714.3	25000	0.11	0.17	0.57	0.76	0.88
70000	50000	35000	0.05		0.35	0.54	0.70

[a]Assuming failure of spares is 0.1 of operating unit until it replaces a failed unit.

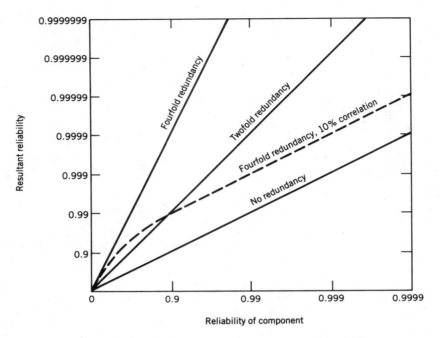

FIGURE 5.160 Reliability of redundant system with correlated failures.

But suppose that if a unit fails, 10% of the time all other similar units will fail. The system reliability is then a combination of "no redundancy" 10% of the time and "fourfold redundancy" 90% of the time. This combination is shown in Figure 5.160 as a dashed line. Redundancy does improve reliability but not nearly as much as predicted by the simple equations. This effect, called correlated failures, is frequently observed in actual spacecraft. For more than twofold redundancy, it is better to ensure independency by procuring units from different sources.

5.12.6 Reliability Model

As an example of the analysis of a satellite, a mathematical model for the *Intelsat IV* will be presented (Binckes, 1970). The basic criterion for success was that nine of the twelve transponders operate for seven years from the time of launch.

The satellite was divided into six major subsystems, as shown in Figure 5.161. Each subsystem is necessary for the functioning of the satellite, so the satellite

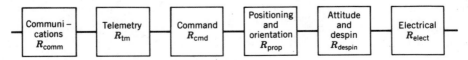

FIGURE 5.161 Overall satellite reliability model.

reliability is

$$R_{sys} = R_{comm} R_{tm} R_{cmd} R_{prop} R_{despin} R_{elect} \qquad (5.181)$$

The command subsystem includes components necessary to receive and decode commands sent from the ground to the satellite. The telemetry subsystem encodes information from the satellite and transmits it to the ground. The electric power subsystem receives power from the sun, stores it, and switches it to the two main power buses. The propulsion subsystem includes the tanks, thrusters, and valves necessary to develop thrust for positioning and orienting the satellite. The despin control subsystem despins the antenna and keeps it pointed at the earth. Finally, the communications subsystem provides the satellite payload that receives the signals from the earth, amplifies them, and transmits them back to earth.

The command subsystem (see Figure 5.162) is fully redundant, with the exception of the receiving antenna, which has a high reliability. Not shown are the transformer windings that carry the signals from the despun to the spinning side; they are part of the BAPTA and are included in the despin control subsystem.

To find the reliability of the command subsystem, the reliability of each component is first calculated by $R_i = \exp(-\lambda_i t)$. The two command receivers are in parallel, so unreliabilities must be multiplied; the unreliability of one receiver is $1 - R_2$, and the unreliability of the pair of receivers is then $(1 - R_2)^2$; the reliability is then $1 - (1 - R_2)^2$. The reliabilities of the despun decoders and of the spinning decoders are found in a similar manner. The four sections are in series, so the reliabilities must be multiplied; the final result is

$$R_{cmd} = R_1 \left[1 - (1 - R_2)^2 \right] \left[1 - (1 - R_3)^2 \right] \left[1 - (1 - R_4)^2 \right] \qquad (5.182)$$

When the numbers are put in the equations for a time interval of 61,320 hr (7 yr), a reliability of 0.97 is found for the command subsystem.

The telemetry subsystem shown in Figure 5.163 is fully redundant, but unlike the command subsystem, it does not have any cross-strapping; it consists of two

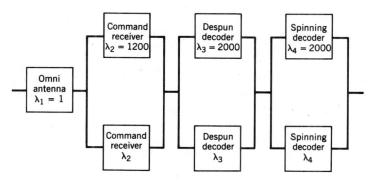

FIGURE 5.162 Command subsystem logic.

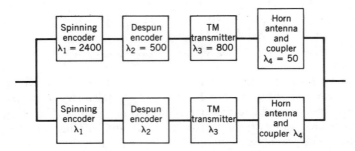

FIGURE 5.163 Telemetry subsystem logic.

separate chains, and every component in one of the chains must operate. Like the command subsystem, the two pairs of rotary transformer windings are included in the despin control subsystem.

In analyzing any system, the criterion for success must be carefully defined. In the telemetry subsystem there is not full redundancy, and some signals in one telemetry chain are not duplicated in the other. Nevertheless, a judgment was made that the satellite can function with either telemetry chain working.

To find the reliability of the telemetry subsystem, first the reliability of each component is calculated. In each chain the components are in series, so these reliabilities are multiplied, and the reliability of the chain is $R_1 R_2 R_3 R_4$. The two chains are in parallel, so the unreliabilities must be multiplied. The result is

$$R_{tm} = 1 - (1 - R_1 R_2 R_3 R_4)^2 \qquad (5.183)$$

and the reliability of 0.96 is found for the telemetry subsystem.

The electric power subsystem (see Figure 5.164) has two chains similar to the telemetry subsystem. However, in this case each chain delivers only half the power so both chains are needed for the nine transponders that define a successful satellite; hence, the chains are in series.

Note that if one battery failed, it would affect operation only during eclipses of maximum duration. One battery could carry seven, or maybe eight, transponders through the longest eclipse and operate nine transponders through shorter eclipses. If that were classified as a "success," the two batteries would be shown in parallel.

To find the reliability of the electric power subsystem, the reliabilities of each component are first calculated. The two load relays are shown in parallel, so their unreliabilities are found; the unreliability of one is $1 - R_4$, the unreliability of both is $(1 - R_4)^2$, and the reliability is $1 - (1 - R_4)^2$. The rest of the components of a chain are in series, so the reliabilities are multiplied; the reliability of a chain is $R_1 R_2 R_3 [1 - (1 - R_4)^2]$, and the reliability of the subsystem is

$$R_{elect} = \left\{ R_1 R_2 R_3 [1 - (1 - R_4)^2] \right\}^2 \qquad (5.184)$$

Numerically, it is found to be 0.91 for 7 yr.

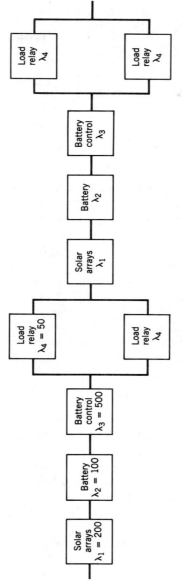

FIGURE 5.164 Electric power subsystem logic.

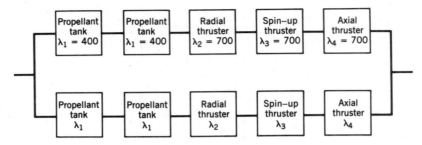

FIGURE 5.165 Propulsion subsystem logic.

The propulsion subsystem shown in Figure 5.165 also consists of two chains. These are completely redundant except that the propellant tanks in one chain do not contain enough fuel for the full 7-yr mission. Also, the axial thruster requires the largest amount of fuel, and if one axial thruster fails, the fuel available for the other may not be sufficient. A complete reliability analysis would have to account for failures at different times of the mission; the simplification shown in the preceding is representative of many assumptions that must be made in order to obtain a useful mathematical model.

Since the components of a chain are in series, the reliabilities are multiplied. Another method of calculating would be to add the failure rates for the chain and then find the chain reliability from the total failure rate. Finally, the two chains are in parallel, so the unreliabilities are multiplied. The equation is

$$R_{\text{prop}} = 1 - \left(1 - R_1^2 R_2 R_3 R_4\right)^2 \qquad (5.185)$$

The resultant 7-yr reliability is 0.97.

The attitude and despin control subsystem (see Figure 5.166) contains a variety of sensors to determine the attitude of the antennas with respect to the earth. The sun sensors will not work during eclipse, but signals from the control center can substitute for the sun sensors. The bearing and power transfer assembly (BAPTA) includes the bearings, rotary transformers, motors, nutation damper, and master index pulse generator. Determination of the failure rate for the BAPTA is difficult, but it may be higher than the number shown in Fig. 5.166.

Calculation of the reliability is divided into two steps. First, the reliability of the chain labeled R_{chain} is

$$R_{\text{chain}} = \left[1 - (1 - R_1)(1 - R_2)\right] R_3 R_4 R_5 R_6 \qquad (5.186)$$

which includes the sensors and the despin control electronics in one chain. The sun sensor and the earth sensors are in parallel, so the unreliability of the sun sensor, $1 - R_1$, is multiplied by the unreliability of the earth sensor, $1 - R_2$, to obtain the unreliability of the combination. The rest of the calculations are similar

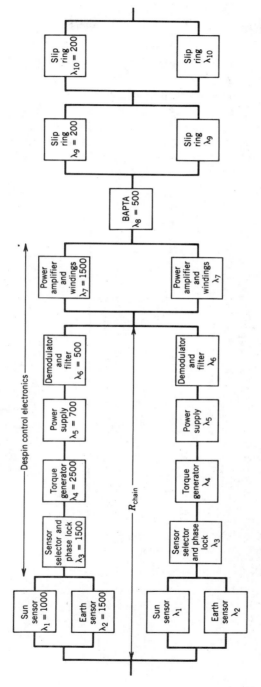

FIGURE 5.166 Attitude and design control subsystem logic. Abbreviation: BAPTA, bearing and power transfer assembly.

to previous ones. The reliability of the entire subsystem is

$$R_{\text{despin}} = \left\{1 - (1 - R_{\text{chain}})^2\right\}\left[1 - (1 - R_7)^2\right] \\ \cdot R_8\left[1 - (1 - R_9)^2\right]\left[1 - (1 - R_{10})^2\right] \tag{5.187}$$

Numerically, it is 0.89 and lower than any other subsystem.

The communications subsystem shown in Figure 5.167 actually has two global receive and two global transmit antennas, but for simplicity each pair is lumped together; with their high reliabilities, they could even be omitted. The TWTA failure rate (λ_4) includes the power supply; the common transponder elements

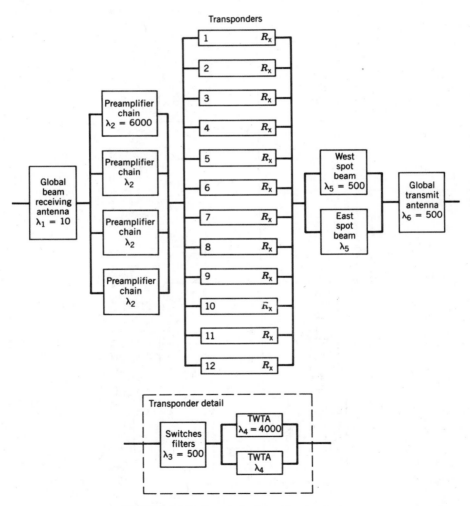

FIGURE 5.167 Communications subsystem logic.

(λ_3) includes equalizers and attenuators as well as switches and filters. The criterion for success was defined as one spot beam operating, although if both spot beams failed, the satellite would hardly be called a total failure—in the Indian Ocean they might not be used anyway.

The reliability of the set of transponders is

$$R_{\text{set}} = R_x^{12} + 12R_x^{11}(1 - R_x) + 66R_x^{10}(1 - R_x)^2 + 220R_x^9(1 - R_x)^3 \quad (5.188)$$

where $R_x = R_3[1 - (1 - R_4)^2]$. Each transponder has an operating TWT, a spare, and various common elements. The equation is for 9 out of 12 transponders required for success. Calculations can also be made for 10, 11, or 12 transponders required by omitting terms in the equation for R_{set}. The first term is for all 12 working, the second term is for exactly 11 operating, and so on; this is an application of the binomial distribution. The reliability of the entire communications subsystem is

$$R_{\text{comm}} = R_1[1 - (1 - R_2)^4]R_{\text{set}}[1 - (1 - R_5)^2]R_6 \quad (5.189)$$

The resultant 7-yr reliability is 0.95.

The reliability of each subsystem can be calculated for various times and the results combined to obtain the reliability of the entire satellite. The result is shown in Figure 5.168 for 9 out of 12 transponders and also for a higher number of operating transponders. This reliability model was formulated in 1970, almost a year before the first satellite launch. Figure 5.168 also shows the actual reliability curve for the seven satellites that were successfully launched. In Table 5.39 is a summary table of the predicted number of failures of various components, and the actual number observed. While these comparisons are difficult to make, there is a correlation between the reliability model and the results obtained.

5.12.7 Improving Reliability

If the overall system reliability is too low, the reliability of one or more units that make up the system will have to be improved. There are two general methods of increasing reliability: improve the unit reliability or introduce redundant units.

To improve unit reliability, the stress on components can be reduced mechanically, electrically, or thermally. Reducing the mechanical tension, working voltages, or temperature extremes tends to increase component reliability. Also, if equipment can be designed with fewer components, there will be fewer components to fail.

Little can be done about unit reliability once the unit has been designed and produced. It may turn out then that the only way to improve system reliability is to put in redundant units. This may be simpler from a design viewpoint but often entails extra penalties in weight. The closer to the failure point that the redundancy can be introduced, the less weight penalty is involved.

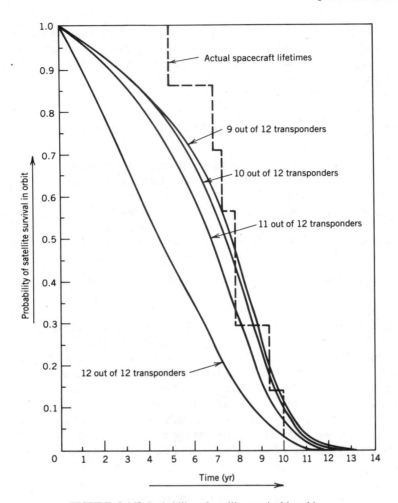

FIGURE 5.168 Probability of satellite survival in orbit.

As an example, consider a battery of 23 cells in series. The reliability of a cell depends on many factors. Assume a reasonable number such as 100 FITS. Then the 7-yr reliability of a cell, R, is 0.994. If all 23 cells must operate, the battery reliability, R^{23}, is only 0.87, which is too low. By including a redundant battery of 23 cells and doubling the weight, the reliability can be increased to 0.98, which is still not satisfactory.

A better way to improve reliability is to introduce redundant cells rather than a redundant battery. Since cells usually fail in a shorted condition, 25 cells can be put in series, and the circuits arranged so that adequate voltage is maintained if 23 of the 25 cells are functioning. The reliability is increased to

$$R_B = R^{25} + 25R^{24}(1 - R) + 300R^{23}(1 - R)^2 = 0.9995 \quad (5.190)$$

TABLE 5.39 Failures of Spacecraft Components[a]

Number per Satellite	Component	Failure Rate, FITS	Failures Predicted	Actual
24	Travelling-wave tube amplifier	4000	58.3	36
4	Receiver, pre-amplifier	6000	13.8	34
2	Attitude control electronics	5200	6.1	3
2	Command decoder	4000	5.4	0
12	Transponder switches and filters	500	4.1	0
2	Telemetry encoder	2900	3.9	2
2	Thruster (set of three)	2100	2.8	4
2	Power amplifier (attitude control)	1500	2.0	0
2	Earth sensor	1500	2.0	3
2	Command receiver	1200	1.6	0
2	Sun sensor	1000	1.3	0
2	Hydrazine tank (pair)	800	1.1	0
2	Telemetry transmitter	800	1.1	1
2	Battery and control	600	0.8	6
2	Spot beam	500	0.7	0

[a] Based on 78.5 operating years

which is much better. With less than a 10% increase in weight, the battery reliability has increased from 0.87 to 0.9995. If the cell failure rate can be made less than 100 FITS, the battery reliability is even higher.

If 23 out of 25 cells are required, a battery reliability of 0.9995 was calculated. Whenever calculated reliabilities are high, it is time to look for other failure modes. Individual cells usually fail short, but occasionally (about 1 out of 100) they fail open.

As described so far, an open cell would stop current flowing through all the cells and make the battery useless. If the failure rate for open cells is 1 FIT, the 7-yr reliability is 0.99994. The reliability of the battery is then $(0.99994)^{25} = 0.9985$. Note that the extra two cells that help reduce the problem of shorted cells actually increase the likelihood of the battery failing open.

The chance of this entire battery failing open (0.0015) is actually greater than of the battery shorted close (0.0005). The overall battery reliability is now 0.998. Further improvements would be to introduce diode bypass circuits that would conduct the current around a cell that has failed open.

5.12.8 System Trade-Offs

Most spacecraft designs include a reliability chart that lists the component reliabilities, the redundancy provided, and a reliability figure calculated for the entire spacecraft. Frequently the reliability chart is used only to determine the final system reliability. The approaches used to determine redundancy are very simple, such as: twofold redundancy everywhere, equalize the subsystem redundancies, or just look for weak spots and add redundancy.

In this section a systematic method is used to measure the value of each redundant part. The effect of each redundant component on system reliability is calculated, and a figure of merit based on mass is derived. A series of optimum configurations with different masses is determined, which can be compared with the initial configuration. The trade-off between reliability and mass can be shown on a graph.

As an example, one subsystem with four basic components will be analyzed, but this analysis can be extended to an entire spacecraft, with hundreds of components. An attitude control subsystem has four components: an earth sensor, an rf sensor, control electronics, and a momentum wheel. It is assumed that one of each component is required for successful operation. The assumed probabilities of failure after seven years are 0.09, 0.09, 0.04, and 0.04 for components with a mass of 5, 1, 2.5, and 15 kg, respectively.

A possible configuration shown in Figure 5.169 includes three earth sensors and three rf sensors for redundancy. The control electronics and the momentum wheel have no redundancy. The problem is to determine whether this is an optimum configuration; that is, a good trade-off between reliability and mass. The overall subsystem reliability is 0.92 for the "3 3 1 1" configuration shown, and the mass is 35.5 kg. An answer could be obtained by trial and error, but a systematic method is needed that can be applied to an entire spacecraft with hundreds of components.

The systematic approach is possible because the spacecraft reliability is equal to the product of the reliabilities of many small units. If a unit reliability is increased by a certain ratio, then the system reliability increases by the same ratio. In the example, the improvement from one earth sensor reliability, R, to a two earth sensor reliability, R', is given by

$$r = \frac{R'}{R} = \frac{1 - Q^2}{1 - Q} = \frac{0.9919}{0.91} = 1.09 \qquad (5.191)$$

since the unit unreliability Q is 0.09.

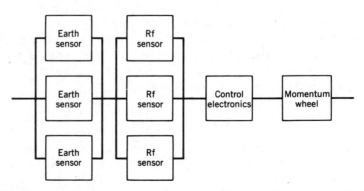

FIGURE 5.169 Unoptimized attitude control subsystem with reliability of 0.92 and a 35.5 kg mass.

This increase in reliability by a factor of 1.09 can be expressed as a 0.374 dB increase, with the advantage that increases can then be added, rather than multiplied. The increase in reliability achieved by adding redundancy also adds to the mass and to the cost. In this analysis, we will concentrate on the mass, M, although a similar analysis could be made in terms of cost. The key factor is the reliability increase per kilogram. In the example, the earth sensor is 5 kg, so the improvement is

$$\text{Figure of merit} = \frac{10 \log r}{M} = \frac{0.374 \text{ dB}}{5 \text{ kg}} = 0.075 \text{ dB/kg} \quad (5.192)$$

This figure of merit is a direct measurement of the value of the first earth sensor spare to the spacecraft reliability.

The figure of merit is shown in Table 5.40 for the first three spares for each component. Listed for each case is the reliability, the factor by which the system of reliability is increased, and the calculated figure of merit. For the earth sensor, the figure of merit for the first spare is 0.075 dB/kg, for the second is 0.0064 dB/kg, and for the third only 0.0006 dB/kg. Each additional spare has less effect on system reliability than the preceeding unit.

The second earth sensor increased the system reliability by a factor of $r = 1.09$. A second rf sensor would improve the reliability by the same factor, since the unit reliabilities and unreliabilities are the same. However, the mass of the rf sensor is only 1 kg, while the earth sensor is 5 kg; hence the figure of merit for the rf sensor is higher.

A series of optimum configurations can be obtained by adding in turn always the component with the highest figure of merit. These are listed in Table 5.41 which starts with a "1 1 1 1" configuration, that is, one earth sensor, one rf

TABLE 5.40 Reliability Improvements for Redundant Components

Number of units	Equation Mass, M, kg Unreliability, Q	Earth sensor 5.0 0.09	rf sensor 1.0 0.09	Control electronics 2.5 0.04	Momentum wheel 15.0 0.04
1	$R = 1 - Q$	0.91	0.91	0.96	0.96
1–2	$r = (1 - Q^2)/(1 - Q)$	1.09	1.09	1.04	1.04
1–2	$(10 \log r)/M$	0.075	0.374	0.068	0.011
2	$R = 1 - Q^2$	0.9919	0.9919	0.9984	0.9984
2–3	$r = (1 - Q^3)/(1 - Q^2)$	1.0074	1.0074	1.0015	1.0015
2–3	$(10 \log r)/M$	0.0064	0.032	0.0026	0.0004
3	$R = 1 - Q^3$	0.999271	0.999271	0.999936	0.999936
3–4	$r = (1 - Q^4)/(1 - Q^3)$	1.0007	1.0007	1.00006	1.00006
3–4	$(10 \log r)/M$	0.0006	0.0030	0.0001	0.0000
4	$R = 1 - Q^4$	0.999934	0.999934	0.999997	0.999997

TABLE 5.41 Optimum Configurations for Attitude Control Subsystem

	Figure of merit, dB/kg	Configuration[a]				Mass, kg	Subsystem reliability
No redundancy		1	1	1	1	23.5	0.763
Add rf sensor	0.374	1	2	1	1	24.5	0.832
Add earth sensor	0.075	2	2	1	1	29.5	0.907
Add control electronics	0.068	2	2	2	1	32	0.943
Add rf sensor	0.032	2	3	2	1	33	0.950
Add momentum wheel	0.011	2	3	2	2	48	0.988
Add earth sensor	0.0064	3	3	2	2	53	0.995

[a]Number of earth sensors, rf sensors, control electronics, and momentum wheels, respectively.

sensor, one electronics control, and one momentum wheel. This has a reliability of 0.763 and a total mass of 23.5 kg. The highest figure of merit is a second rf sensor, which yields a "1 2 1 1" configuration. The optimum configurations are plotted in Figure 5.170 where the 7-yr reliability is plotted as a function of subsystem mass.

When a massive component, such as the momentum wheel is added, there is a large jump in the total mass. In the example, when the second momentum wheel is added, the subsystem mass jumps from 33 kg to 48 kg. Additional configurations (labeled as "good configurations") can be obtained by skipping the massive unit, and adding lighter components still in the order of figure of merit. These configurations are still the best at a given total mass, but would rarely be chosen in practice. For example, for a mass of 46.5 kg, the good configuration is "4 4 3 1" and the reliability is 0.96. But with the addition of 1 kg a reliability of 0.98 is possible, a substantial improvement. These "good configurations" are also plotted in Figure 5.170. In dealing with a large number of units, such as an entire spacecraft, the fine detail is less important, and the "optimum configurations" are usually sufficient.

Two optimum configurations can be seen in the figure that are definitely preferable to the "3 3 1 1" configuration originally shown. Both the "2 2 2 1" configuration and the "2 3 2 1" configuration have greater reliabilities and smaller masses than the "3 3 1 1" configuration. The "2 3 2 1" configuration (i.e., two earth sensors, three rf sensors, two control electronics, and one momentum wheel) is shown in Figure 5.171; it is 2.5 kg lighter and has a reliability of 0.95 instead of 0.92 for the "3 3 1 1" configuration.

This same analysis was applied to an entire spacecraft (except for the communications subsystem) by Henry Meyerhoff, and is shown in Figure 5.172. Many more components are included, and the curve of reliability versus mass tends to be smooth, without any sharp bends. During the design and initial fabrication of a spacecraft, there is a total mass that defines a point on the curve. The slope, or figure of merit at that point, determines a critical reliability improvement (in

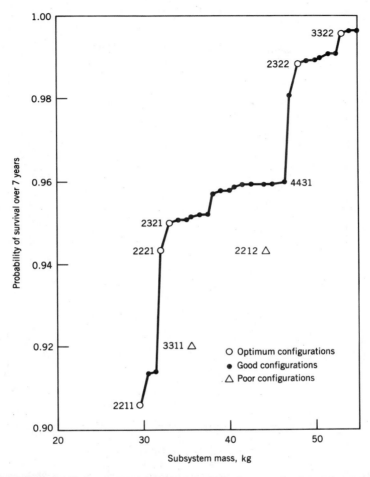

FIGURE 5.170 Reliability versus mass for attitude control subsystem showing optimum redundancy configurations.

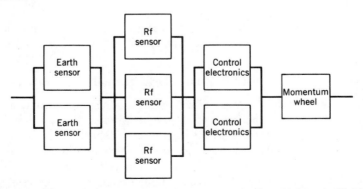

FIGURE 5.171 Optimized attitude control subsystem with reliability of 0.95 and a 33.0 kg mass.

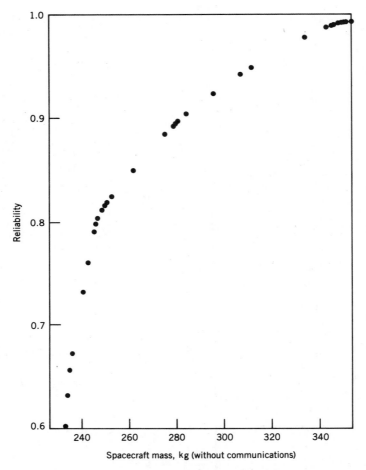

FIGURE 5.172 Reliability versus mass for a proposed spacecraft excluding the communications subsystem.

dB/kg). This threshold can be applied to all subsystems, adding redundancy where the figure of merit is above the critical value, and removing it when it is below.

5.12.9 Monte Carlo Method

The Monte Carlo method consists of generating randomness by some method and actually performing an experiment to observe the results. While it was developed before the age of the computer, it has been used widely since the computer came on the scene. The programs are usually relatively easy to write, and the computer is capable of repeating an experiment a large number of times.

A Monte Carlo simulation of a system of eight satellites is discussed to provide insight as to the importance of different parameters and to show how the reliability of a system of satellites is found. The initial impetus for this study was concern over the replacement time for a satellite raised to geostationary orbit with electric propulsion.

A standard system was chosen as a baseline and consisted of eight satellites purchased; the first four satellites were launched at 4-month intervals, other satellites were launched 6 months after a failure (either in orbit or a launch failure), launch reliability was assumed to be 75%, and satellite reliability was based on a reliability model. A "successful" satellite is one that has 9 (out of 12) or more transponders working.

The flowchart for one system of satellites is in Figure 5.173. The reliability curve (top curve in Figure 5.168 for 9 out of 12 surviving transponders) for the

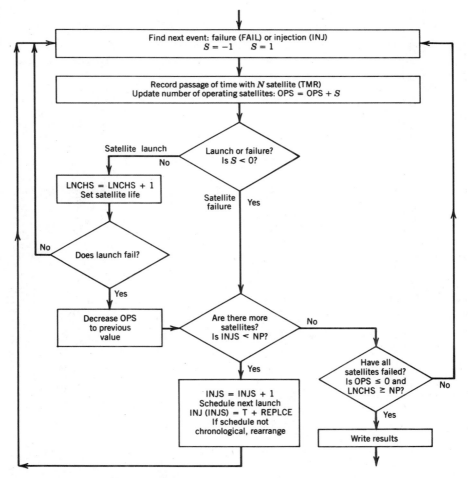

FIGURE 5.173 Computer flowchart for system of satellites.

TABLE 5.42 Simulation of Eight Satellites

Years	Satellite Number and Event	Status[a] of Each Satellite	Number of Satellites			Time Satellites Operating				
			Scheduled	Launched	Operating	0	1	2	3	4
0.00	1 Launched	CRRRGGGG	4	1	1	0.00	0.00	0.00	0.00	0.00
0.33	2 Launched	CCRRGGGG	4	2	2	0.00	0.33	0.00	0.00	0.00
0.67	3 Launched	CCCRGGGG	4	3	3	0.00	0.33	0.33	0.00	0.00
1.00	4 Launched	CCCCGGGG	4	4	4	0.00	0.33	0.33	0.33	0.00
2.38	3 Fail	CCFCRGGG	4	4	3	0.00	0.33	0.33	0.33	1.38
2.88	5 Launched	CCFCCGGG	5	5	4	0.00	0.33	0.33	0.83	1.38
7.25	5 Fail	CCFCFRGG	5	5	3	0.00	0.33	0.33	0.83	5.75
7.75 7.75	6 Launched 6 Fail	CCFCFFRG	6	6	3	0.00	0.33	0.33	1.33	5.75
8.25 8.25	7 Launched 7 Fail	CCFCFFFR	7	7	3	0.00	0.33	0.33	1.83	5.75
8.45	1 Fail	FCFCFFFR	8	7	2	0.00	0.33	0.33	2.03	5.75
8.75	8 Launched	FCFCFFFC	8	8	3	0.00	0.33	0.64	2.03	5.75
8.88	4 Fail	FCFFFFFC	8	8	2	0.00	0.33	0.64	2.16	5.75
12.13	2 Fail	FFFFFFFC	8	8	1	0.00	0.33	3.88	2.16	5.75
16.64	8 Fail	FFFFFFFC	8	8	1	0.00	4.85	3.88	2.16	5.75

[a] Abbreviations: C, operating; F, failed; G, ground; R, ready for launch

satellite was used to determine the lifetime of each satellite. A random number from 0 to 1 was generated, this number used for the ordinate (equivalent to R), and the lifetime read on the abscissa. For example, if the random number was 0.70, the satellite lifetime is 6 yr.

A summary of a single run is shown in Table 5.42. The first four launches took place as scheduled; satellite 3 failed at 2.38 yr, so number 5 was launched at 2.88 yr. Satellites 6 and 7 failed during launch, and satellite 1 failed before 8 was launched, so for awhile (0.3 yr), the system was down to two operating satellites.

Column 3 shows the system status, where G stands for satellite on the ground, R for one scheduled for launch (ready), C for one operating (communicating), and F for one that has failed. The next three columns show the number of satellites scheduled for launch, the number that have been launched, and the number operating. The last five columns show the accumulated time during which any particular number of satellites were operating.

The program was run 200 times; a summary of 17 runs, shown as 17 lines in Table 5.43, lists the time of injection and time of failure for each satellite. Satellite 1 was always injected at time 0.0, and satellite 2 was injected at 0.3 yr. Satellite 3 was scheduled for 0.7 yr, unless the first launch had failed, in which case it was 0.5 yr.

An interesting result of this study is the rather large variability from one system to another based purely on statistical fluctuations. For example, line 2 is a disaster with six launch failures. Notice that the failure time of satellites 1, 3, 4, 5, 7, and 8 are the same as their injection time. On the other hand, line 3 is terrific with the first four satellites working for 6.9 yr. The variability can also be seen by looking at the time of injection of the eighth satellite: when this is over 7 yr, the system is in fair shape. But in 5 of the 17 runs, this number is less than 2.4 yr, which indicates the system is in serious trouble.

Two sets of 100 runs were made, and the fraction of time that at least N satellites were operating is plotted in Figure 5.174. As expected, the two runs agree within a percent or two for most points. The most significant curve is the one for $N = 3$, which applies for a system of three satellites with one spare in orbit.

For the first 5 yr, the reliability of the system was between 95 and 99%, but after that there is a steady decline. This is due to cases in which the eight satellites are used up and there are no more to be launched. In actual practice, more satellites would be ordered, but the lead time could be long so that some of this drop could happen.

To observe the effect of various assumptions, each of the inputs was varied, and 100 systems were run for each variation. The results are shown in Figure 5.175. Increasing the number of spares or decreasing the replacement time did not significantly change the reliability during the first 5 yr. Some improvement was observed by increasing the number of satellites purchased or improving launch reliability. One run was made (not plotted) where all factors were improved (16 satellites, two spares, 1-month replacement, and 90% launch reliability), and then the reliability was over 99% for about 14 yr.

TABLE 5.43 Injection and Failure Times for Many Runs

Time of Injection (yr)[a]						Time of Failure (yr)[b]							
3	4	5	6	7	8	1	2	3	4	5	6	7	8
0.7	1.0	1.5	2.0	6.4	6.9	13.0	9.2	5.9	1.0	1.5	6.7	6.4	9.7
0.5	0.7	1.0	1.0	1.2	1.5	0.0	11.8	0.5	0.7	1.0	7.4	1.2	1.5
0.7	1.0	7.4	10.3	10.7	10.8	9.8	6.9	10.2	10.3	15.4	18.1	16.7	19.6
0.7	1.0	1.2	6.8	10.6	10.9	10.4	6.3	0.7	10.1	10.8	12.6	16.9	10.9
0.7[c]	1.0	2.9	7.8	8.3	8.8	8.5	12.1	2.4	8.9	7.3	7.8	8.3	16.6
0.7	1.0	7.4	8.2	11.6	13.5	13.0	6.9	7.7	11.1	15.1	16.7	17.0	22.3
0.7	0.8	1.0	1.3	6.5	7.0	7.8	0.3	6.0	0.8	7.8	10.2	6.5	10.4
0.7	1.0	1.2	5.1	9.0	9.8	8.5	4.6	0.7	10.0	9.3	15.5	16.2	18.5
0.7	0.8	1.0	1.3	1.5	2.0	5.7	0.3	8.0	0.8	1.0	6.0	1.5	2.2
0.7	0.8	1.0	3.7	4.2	7.9	7.5	0.3	9.2	7.4	3.2	3.7	13.0	15.9
0.7	0.8	1.0	3.7	8.2	8.5	9.9	0.3	8.0	3.2	7.7	11.0	18.5	18.5
0.7	1.0	7.3	7.4	7.8	7.9	6.8	6.9	8.6	8.7	7.3	7.4	7.8	16.5
0.7	1.0	5.5	6.0	6.2	8.6	5.7	5.0	8.1	10.5	5.5	17.7	14.3	15.6
0.7	1.0	1.2	1.5	1.7	2.0	7.3	9.0	0.7	1.0	1.2	1.5	11.0	8.2
0.7	0.7	1.0	1.0	1.5	2.0	0.0	0.5	10.0	5.8	1.0	3.0	1.5	2.0
0.7	0.8	1.0	2.5	6.7	7.9	7.4	0.3	6.2	2.0	8.1	13.0	15.2	8.1
0.5	0.7	0.8	1.0	1.3	2.4	0.0	0.3	5.2	8.0	0.8	8.0	1.9	7.2

[a] Injection times for first two satellites are always 0.0 and 0.3 yr.
[b] Satellite lifetime is year of failure minus the year of injection.
[c] Line 5 corresponds to run shown in Table 5.42.

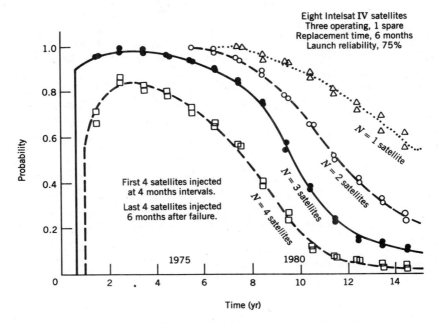

FIGURE 5.174 Probability of at least *n* satellites in operation.

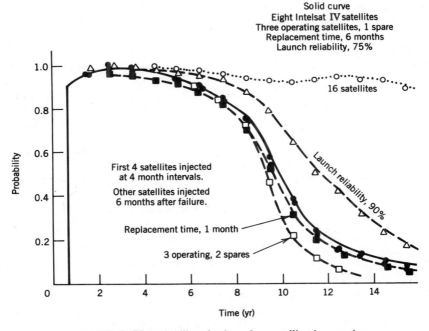

FIGURE 5.175 Probability of at least three satellites in operation.

One factor is not included in this study, the graceful degradation of satellites, that is, the warning that some satellites give of impending failure; this warning might be the running out of fuel or loss of redundant components. The criterion for launch decision becomes more difficult, but there is no doubt that an increase in system reliability would result.

This study shows that these factors are related and that making a vast improvement in one factor will not often make a substantial change in system reliability. Careful study is required to isolate the factor (or factors) that is limiting the system reliability and then to look for improvements in this area.

REFERENCES

Alexovich, R. E. (1977). Performance of the 12 GHz, 200 watt transmitter experiment package for the Hermes satellite, in *Hermes* (I. Paghis, ed.), Vol. 2. Royal Society of Canada, Ottawa, Ontario, Canada.

Allison, J. F., Arndt, R., and Meulenberg, A. (1975). A comparison of the COMSAT violet and non-reflective solar cells. *COMSAT Tech. Rev.* **5**(2), 211–223.

Bauer, P. (1968), *Batteries for Space Power Systems*, NASA SP-172, U.S. Govt. Printing Office, Washington, DC.

Bazovsky, I. (1961). *Reliability Theory and Practice*, Prentice-Hall, Englewood Cliffs, NJ.

Billerbeck, W. J. (1973). Electric power for state-of-the-art communications satellites. Paper presented at National Telecommunications Conference, Atlanta, GA. Sponsored by IEEE, New York, NY.

Billerbeck, W. J. (1979). Long-term prediction of power system performance for geosynchronous spacecraft. *Proc. Intersoc. Energy Convers. Eng. Conf.*, **14** Boston, MA.

Binckes, J. B. (1970). Personal communication.

Bruhn, E. F. (1965). *Analysis and Design of Flight Vehicle Structures*, Tri-State Offset Company, Cincinnati, OH.

Burington, R. S., and May, D. C. (1958). *Handbook of Probability and Statistics with Tables*, Handbook Publishers, Sandusky, OH.

Carter, J. R., and Tada, H. Y. (1973), *Solar Cell Radiation Handbook*, TRW, 21945-6001-RU-00, Contract NAS 7-100. U.S. Govt. Printing Office, Washington, DC.

Chen, M. H., Assal, F. T., and Mahle, C. E. (1976). A contiguous band multiplexer. *COMSAT Tech. Rev.* **6**(2), 285–307.

Chou, S. (1979), 12-GHz 10-W amplifier using GaAs IMPATT diodes. *COMSAT Tech. Rev.* **9**(2B), 617–628.

Crow, J. H. (1966), FM Data Systems, in *Aerospace Telemetry*, Vol. 2, Stiltz ed., Prentice-Hall, Englewood Cliffs, NJ, p. 62.

Dicks, J. L., and Brown, M. P. (1975). INTELSAT IV-A transmission system design. *COMSAT Tech. Rev.* **5**(1), 73–104.

DiFonzo, D., Gruner, R., Carpenter, E., and Persinger, R. (1980). "Earth Station Antenna Technology Seminar," COMSAT Laboratories, Clarksburg, MD.

Esch, F., Billerbeck, W. J., and Curtin, D. (1976). Electric power systems for future communications satellites. *Proc. Int. Astronaut. Fed.*, 27th IAF Congress, Anaheim, CA, Paper 76–237.

Gordon, G. D., Huson, G. R., and Slabinski, V. J. (1974), Blocking bubbles in the INTELSAT IV fuel lines. *COMSAT Tech. Rev.* **4**(2), 499–506.

Hamilton, D. C., and Morgan, W. R. (1952). Radiant-interchange configuration factors. *Natl. Advis. Comm. Aeronaut.*, *Tech. Notes* **NACA TN 2836**.

Harris, C. M., and Crede, C. E. (1976). *Shock and Vibration Handbook*, McGraw-Hill, New York.

Haviland, R. P., and House, C. M. (1965). *Handbook of Satellites and Space Vehicles*, Van Nostrand, New York.

Huzel, D. K., and Huang, D. H. (1971). Design of liquid propellant rocket engines. [*NASA Spec. Publ.*] NASA SP-125, N71-29405-416.

Kaplan, M. H. (1976), *Modern Spacecraft Dynamics and Control*, Wiley, New York.

Lawton, J. G. (1958), Comparison of binary data transmission systems. *Proc. Nat. Conf. Mil. Electron.*, 2nd.

Lebsock, K. L. (1970). "Wheel Attitude Control System Comparison Study," Final Rep. A965660, Lockheed Missiles and Space Company, Inc., Sunnyvale, CA.

Likins, P. W. (1967). Attitude stability criteria for dual spin spacecraft. *J. Spacecr. Rockets* 4(12), 1638–1643.

Lindmayer, J., and Wrigley, C. Y. (1965), *Fundamentals of Semiconductor Devices*, Van Nostrand-Reinhold, New York.

Mueller, G. E., and Spangler, E. R. (1964), *Communications Satellites*, Wiley, New York, p. 105.

NASA/Goddard Space Flight Center (1977), *General Environmental Test Specification for Spacecraft and Components*, GETS (ELV)-1. Natl. Aeronaut. Space Admin., Washington, DC.

Pilkey, W. D., and Chang, P. Y. (1978), *Modern Formulas for Statics and Dynamics*, McGraw-Hill, New York.

Revesz, A. G., and Fleming, P. L. (1978), Tunnel diodes in satellite communications. *COMSAT Tech. Rev.* 8(2), 257–271.

Rittenhouse, J. B., and Singletary, J. B. (1968), "Space Materials Handbook," Tech. Rep. AFML-TR-68-205, Lockheed Missiles and Space Company, Inc., Sunnyvale, CA.

Rocket Research Corp. (1966), *Monopropellant Hydrazine Design Data*, Seattle, WA.

Rogers, M. V. (1977), *Reliability Estimating Procedures for Electric and Thermochemical Propulsion Systems*, AFRPL-TR-76-99, No. 9006-085-1, Booz Allen Applied Research.

Schmidt, E. W. (1984), *Hydrazine and Its Derivatives*, Wiley, New York.

Stockel, J. (1980), Personal communications.

Sutton, G. P., and Ross, D. M. (1976), *Rocket Propulsion Elements*, Wiley, New York.

Thompson, P. T., and Johnston, E. C. (1983). INTELSAT VI. A new satellite generation for 1986-2000. *Int. J. Satellite Commun.*, 1(1), 3–14.

Van Vliet, R. M. (1965), *Passive Temperature Control in the Space Environment*, Macmillan, New York.

Weidner, D. K. (1969), Natural space environment criteria for 1975-1986, NASA space stations. *NASA Tech. Memo.* NASA TM-X-53865.

SATELLITE ORBITS

6.1 FUNDAMENTALS OF CIRCULAR ORBITS

6.1.1 Geocentric Equatorial Coordinate Systems

Orbits of communications satellites are generally specified in terms of a geocentric equatorial coordinate system using either rectangular or spherical coordinates and with axes fixed with relation to the stars. The basic coordinates are shown in Figure 6.1. The origin is at the center of the earth. The z axis is the earth's spin axis and is approximately the direction to the North Star. The x axis is defined by the intersection of the earth's equatorial plane and the ecliptic; the positive x-axis direction is the direction of the sun when it crosses the equator in March. This intersection is called the vernal equinox, and its projection on the celestial sphere is called the first point of Aries, ϒ.

The geocentric *rectangular* equatorial coordinate system is right-handed, with the positive y axis in the equatorial plane and 90° to the east of the positive x axis. Distances will be specified in kilometers.

A geocentric *spherical* equatorial coordinate system is also used: In this latter system, right ascension α is the angle in the equatorial plane measured increasingly positive eastward from zero at the x axis; declination δ is the angle from the equatorial plane toward the z axis—measured increasingly positive toward the north pole from zero at the equator and increasingly negative toward the south pole. The radius r is the distance to the origin at the center of the earth.

The relation between the rectangular coordinate system (x, y, z) and the spherical coordinate system (α, δ, r) is shown in Figure 6.2. The equations to convert from spherical coordinates to rectangular coordinates are

$$x = r \cos \delta \cos \alpha$$

$$y = r \cos \delta \sin \alpha$$

$$z = r \sin \delta \tag{6.1}$$

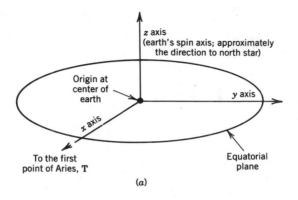

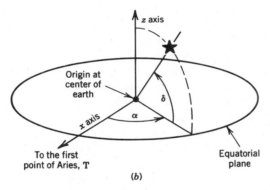

FIGURE 6.1 Inertial geocentric coordinate systems: (*a*) rectangular; (*b*) spherical. Main spherical coordinates are right ascension α and declination δ.

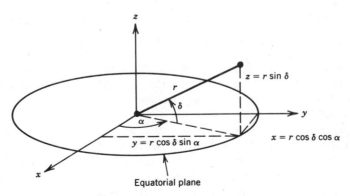

FIGURE 6.2 Relation between geocentric rectangular and spherical coordinate systems.

The equations to convert from rectangular coordinates to spherical coordinates are

$$\tan \alpha = \frac{y}{x}$$

$$\tan \delta = \frac{z}{\left(x^2 + y^2\right)^{1/2}}$$

$$r = \left(x^2 + y^2 + z^2\right)^{1/2} \tag{6.2}$$

The latter equations are written in a form suitable for computer calculations, and the angles will be in the correct quadrant if the two-argument arctangent function is used.

The earth's spin axis does not remain pointed in the same direction: The axis precesses and makes a complete revolution around the ecliptic pole every 26,000 yr. To be precise, the preceding coordinate system is defined at a specific time (or epoch), such as at the start of 2000. Then it is given in terms of the "true equator of data" and the "mean equinox of date."

Astronomers frequently use spherical coordinates since the direction of celestial objects is often better known, and of more interest, than the distance. Spherical coordinates are often used to specify the positions of the moon and the sun, which infuence communications satellites. For calculations of satellite positions and velocities, the rectangular coordinate system is often more convenient.

Two other coordinate systems will be also be used. For one of them, a coordinate system in the orbit plane (with origin still at the earth's center) is useful in exploring the details of an elliptical satellite orbit. Another system, used in calculating the direction of a satellite from a given earth station, is a topocentric coordinate system. This coordinate system has the origin at the earth station; the satellite's azimuth angle is in the horizontal plane and is measured eastward from geographic north, and the satellite elevation angle is measured from the horizontal toward the local vertical.

6.1.2 Planar Projection

The right ascension varies from $0°$ to $360°$, and the declination varies from $-90°$ to $+90°$. The declination is directly related to the latitude, that is, the vector from the center of the earth to a location on the surface has a declination equal to the latitude of that location.

The right ascension is similar to the longitude, but there is a time variation. The celestial sphere is fixed in space, whereas the earth rotates once a day. Longitudes are measured from the Greenwich meridian, and the right ascension is measured from the first point of Aries; these two usually coincide once a day.

When orientation of an orbit plane is measured, it is easier to refer to the orbit normal. The ecliptic plane's orientation can be specified by the ecliptic pole, which is on the normal to the earth's orbit plane. The angle between the equatorial plane and the ecliptic plane, known as the obliquity of the ecliptic, is 23° 26.5'. To plot these quantities on a flat surface, some distortion is required, such as that present on the planar projection shown in Figure 6.3, which is similar to the projection used on many world maps. Hence, the position of the ecliptic pole is specified by a right ascension of 270° and a declination of 66° 33.5', as indicated by the cross in Figure 6.3. Also plotted is declination of the ecliptic as a function of right ascension.

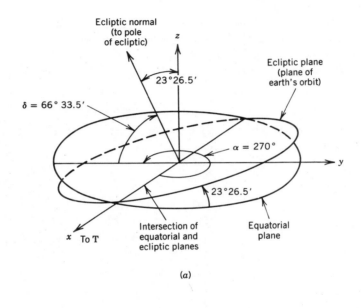

(a)

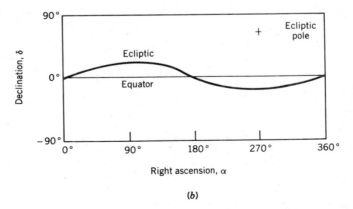

(b)

FIGURE 6.3 Equatorial plane and earth's orbit plane (ecliptic): (a) three dimensional view; (b) planar projection of right ascension and declination.

6.1.3 Polar Projection

A polar projection of spherical coordinates is often useful for accurate representations of directions near the north pole. Geostationary satellites have orbit normals (the line perpendicular to the orbit plane) closely parallel to the earth's spin axis. Also, the angular momentum vector of a spinning satellite, or of some satellites with momentum wheels, will be close to the earth's spin axis and in the direction of the north pole.

In the projection shown in Figure 6.4, lines of equal declination, δ, are circles, and lines of equal right ascension, α, are straight lines going through the center. In the figure, a right ascension of zero is to the right, and other values of right ascension are measured from this line in a counterclockwise direction.

If two directions are specified by their declination δ and their right ascension α, the central angle β_0, between these two vectors can be calculated by

$$\cos \beta_0 = \sin \delta_1 \sin \delta_2 + \cos \delta_1 \cos \delta_2 \cos (\alpha_1 - \alpha_2) \qquad (6.3)$$

This equation is obtained by forming a spherical triangle of great-circle arcs using these two points and the north pole.

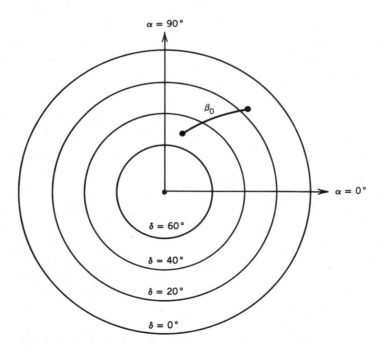

FIGURE 6.4 Polar projection of spherical coordinates. Circles are lines of constant declination.

6.1.4 Spherical Trigonometry

A spherical triangle is formed by three points on a sphere. The three sides of the triangle are formed by arcs of great circles, and the angles between sides, at the vertices, are denoted in Figure 6.5 by A, B, and C. The sides of a spherical triangle are not measured as a distance but in terms of the central angle from the center of the sphere (the size of the sphere is usually not significant). In Figure 6.5 the sides a, b, and c are opposite the angles A, B, and C, respectively.

A spherical triangle can usually be specified by any three of the six parameters, although in some cases there are two different solutions. The most commonly used equations are the law of sines,

$$\frac{\sin A}{\sin a} = \frac{\sin B}{\sin b} = \frac{\sin C}{\sin c} \tag{6.4}$$

the law of cosines for three sides,

$$\cos a = \cos b \cos c + \sin b \sin c \cos A \tag{6.5}$$

and the law of cosines for three angles,

$$\cos A = -\cos B \cos C + \sin B \sin C \cos a \tag{6.6}$$

Note that by permuting the letters, each law of cosines can be written in three different ways. Other equations are available in mathematical references.

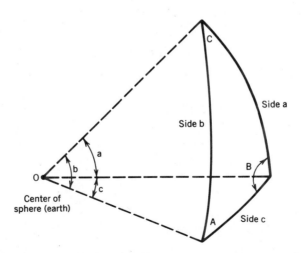

FIGURE 6.5 Spherical triangle. Point O, earth's center; A, B, and C, points on surface of sphere; a, b, and c are called sides, but are actually angles measured from the sphere's center.

The law of sines is the simplest equation and is usually applied whenever possible; sometimes it is sufficient. If two sides and an included angle are known, the law of cosines (three sides) provides the third side. If three sides are known, the same equations can be solved for the angle A. In either case the law of sines can then be used to get the other angles. The law of cosines (three angles) is used in a similar fashion for the case of two angles and an included side and for the case of three known angles.

Frequently, one of the angles of the spherical triangle is equal to 90°. As is true of plane right triangles, many of the equations are considerably simplified. There are then two other angles and three sides; any two of these five parameters specify the triangle. There are actually 10 simple equations relating any three parameters. A right spherical triangle is shown on the left side of Figure 6.6, where $C = 90°$, and the angle A is omitted. The four equations relating a, b, c, and B are

$$\cos c = \cos a \cos b \qquad \sin b = \sin c \sin B$$

$$\cos B = \tan a \cot c \qquad \sin a = \tan b \cot B \qquad (6.7)$$

If the angle A (opposite the side a) is needed, it can be found from $\cos A = \cos a \sin B$. Other relations can be found from the preceding equations by interchanging a and b and interchanging B and A as is evident from the symmetry of the triangle.

A right spherical triangle often used with satellite orbits is shown on the right of Figure 6.6. The satellite's direction from the earth's center is given by the right ascension α and the declination δ. The orbit plane has an inclination i with respect to the equatorial plane and a right ascension of the ascending node, Ω. The angle from node to the satellite is measured by the sum of the argument of perigee ω and the true anomaly v. The four equations for the right spherical triangle then become

$$\cos (\omega + v) = \cos (\alpha - \Omega) \cos \delta$$

$$\sin \delta = \sin (\omega + v) \sin i \qquad (6.8)$$

$$\cos i = \tan (\alpha - \Omega) \cot (\omega + v)$$

$$\sin (\alpha - \Omega) = \tan \delta \cot i \qquad (6.9)$$

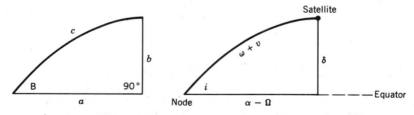

FIGURE 6.6 Right spherical triangle as applied to satellite orbit parameters.

This set is redundant in that any equation can be derived from two others; however, such derivations are not trivial.

6.1.5 Circular Orbit Parameters

Astronomy is an old field, and the symbols have a long history; these symbols are often different from the corresponding ones in other fields. In physics the law of gravitation is

$$F = \frac{GMm}{r^2} \tag{6.10}$$

where F is force, G the fundamental gravitational constant, M the earth's mass, m the satellite mass, and r the distance from the center of the earth to the satellite. In astronomy GM always occurs together and is given a separate symbol. In this book, the symbol μ is used for the product GM; the symbols k^2 and γ have also been used in the literature for GM and sometimes for G. Acceleration (F/m) is of more practical important than forces, and $\ddot{r}$ will denote acceleration. The law of gravitation then becomes

$$\ddot{r} = \frac{\mu}{r^2} \tag{6.11}$$

The subscript e will be used on μ (as in μ_e) to denote the earth's mass when needed to differentiate it from other celestial bodies.

The most important constant for satellites around the earth is the gravitational parameter μ ($=398600.5$ km^3/sec^2). This is equal to the gravitational constant G multiplied by the mass of the earth, M. The product is known more accurately than either of the two factors; in fact, the best way to determine the earth's mass is by dividing the parameter μ by the gravitational constant.

In a circular orbit, the apparent centrifugal acceleration is equal to $\dot{s}^2/r$ (see Figure 6.7). Equating the two accelerations, it can be shown that the velocity $\dot{s}$ is

$$\dot{s} = \left(\frac{\mu}{r}\right)^{1/2} \tag{6.12}$$

In physics the radius is usually denoted by the symbol r; however, in most of this chapter, a will be used to denote the orbit radius, so that it will be in harmony with the equations on elliptical orbits, where a denotes the semimajor axis. In physics the velocity is usually denoted by v, however astronomers use v for an angle (true anomaly), so $\dot{s}$ will be used here for velocity.

The physical and astronomical constants for the earth, sun, and moon are shown

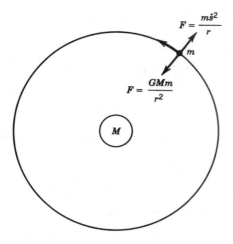

$$F = \frac{m\dot{s}^2}{r}$$

m

$$F = \frac{GMm}{r^2}$$

M

FIGURE 6.7 Circular satellite orbit showing gravitational force and apparent centrifugal force away from earth's center.

in Table 6.1. The gravitational parameters for the sun and moon are dominant in calculations for satellites going around those bodies. For communications satellites, these two parameters determine perturbation effects, which affect slightly the satellite's orbit around the earth.

The equatorial radius of the earth is also listed in Table 6.1. This is often used to determine the slant range, although for high accuracy the exact distance from an earth station to the center of the earth must be calculated. For infrared earth sensors, the apparent radius of the earth is larger, due to the earth's atmosphere, and is equal to 6408 km.

Listed in Table 6.2 are some equations and numerical values for circular orbits around the earth. The radius a is the distance from the center of the earth to the satellite. The orbit period P is the time it takes the satellite to go around the orbit, in inertial space (i.e., with reference to the stars). The acceleration g is the rate of change of the vectorial velocity; since this acceleration is exerted by a gravitational force, any mass in the satellite is in "free fall" and acts as if there is no gravity present.

Three sets of equations are shown in Table 6.2. The first is in terms of the orbit radius a and the gravitational parameter μ. The second is in terms of the velocity $\dot{s}$ and the radius; the third is in terms of the velocity $\dot{s}$ and the gravitational parameter.

Two sets of numerical values are given: for an orbit at the earth's radius and one at geostationary orbit. The first is only a theoretical limit; in practice, an orbit must be a few hundred kilometers greater than the earth's radius because of the earth's atmosphere. The geostationary orbit is chosen so that the orbit period is equal to one sidereal day, that is, the time for one earth rotation with respect to a fixed star (about 23 hr, 56 min).

The equations and values in Table 6.2 are based on a spherical earth and two-body equations. The actual geostationary orbit is affected by the bulge of the earth's

TABLE 6.1 Astronomical Constants[a]

Constant	Symbol	Earth	Sun	Moon	Units
Gravitational parameter	$\mu = GM$	398,600.5	$132{,}712.438 \times 10^6$	4902.79	km^3/sec^2
	$\sqrt{\mu} = \sqrt{GM}$	631.3482	364297.2	70.0199	$km^{3/2}/sec$
Mass	M	5.9733×10^{24}	1.9888×10^{30}	7.3472×10^{22}	kg
Radius	R	6,378.140	696000	1738.	km

[a]Physical constants: Gravitational constant, $G = 6.673 \times 10^{-20}$ $km^3/kg\ sec^2$; Velocity of light, $c = 299792.458$ km/sec.

783

TABLE 6.2 Circular Orbit Parameters

Quantity	Equations In Terms Of			At Earth's Radius	In Geostationary Orbit	Unit
	a and μ	a and $\dot{s}$	$\dot{s}$ and μ			
Radius	—	—	$a = \mu/\dot{s}^2$	6378.140	42164.570	km
Velocity	$\dot{s} = \sqrt{\mu/a}$	—	—	7.905364	3.074689	km/sec
Angular velocity	$n = \sqrt{\mu}/a^{3/2}$	$n = \dot{s}/a$	$n = \dot{s}^3/\mu$	0.00123944 7	72.92115×10^{-6}	rad/sec
Orbit period	$P = 2\pi a^{3/2}/\sqrt{\mu}$	$P = 2\pi a/\dot{s}$	$P = 2\pi\mu/\dot{s}^3$	5069.347	86164.091	sec
Acceleration[a]	$g = \mu/a^2$	$g = \dot{s}^2/a$	$g = \dot{s}^4/\mu$	9.79828×10^{-3}	0.2242099×10^{-3}	km/sec²

[a]Calculated according to simplified two-body theory. On earth, acceleration of gravity is affected by the earth's shape and rotation; standard $g = 9.80665 \times 10^{-3}$ km/sec². In orbit, all the objects in the satellite have almost identical acceleration, and there appears to be zero gravity.

equator and the sun and moon perturbations. The average radius of a geostationary orbit is 42164.570 km, and the earth's attraction μ seems to be slightly greater.

6.1.6 Orbits of Different Radii

Circular orbits of different sizes have various parameters. Listed in Table 6.3 are the orbit period, the radius, the height above the earth's surface, and the velocity for various circular orbits. The table is for increments of 0.2 km/sec in velocity, except for the first entry of zero height and the last entry for a geostationary orbit. The orbit period is shown in terms of total seconds and in hours and minutes.

Table 6.3 is presented for circular orbits, but it is also useful for elliptical orbits. The a (in Table 6.2) is then the semimajor axis, or the average of the maximum and minimum radii. The tabulated velocity is then the geometric mean (square root of the product) of the maximum and minimum velocities.

TABLE 6.3 Period and Velocity for Orbits of Different Heights

Period (sec)	Period hr	Period min	Radius (km)	Height (km)	Velocity (km/sec)
5,069	1	24	6,378	0	7.9054
5,278	1	28	6,552	173	7.8000
5,705	1	35	6,901	523	7.6000
6,180	1	43	7,279	901	7.4000
6,710	1	52	7,689	1,311	7.2000
7,302	2	2	8,135	1,757	7.0000
7,965	2	13	8,620	2,242	6.8000
8,711	2	25	9,151	2,772	6.6000
9,554	2	39	9,731	3,353	6.4000
10,509	2	55	10,369	3,991	6.2000
11,595	3	13	11,072	4,694	6.0000
12,836	3	34	11,849	5,471	5.8000
14,261	3	58	12,710	6,332	5.6000
15,905	4	25	13,669	7,291	5.4000
17,812	4	57	14,741	8.363	5.2000
20,036	5	34	15,944	9,566	5.0000
22,646	6	17	17,300	10,922	4.8000
25,730	7	9	18,837	12,459	4.6000
29,401	8	10	20,589	14,211	4.4000
33,804	9	23	22,596	16,218	4.2000
39,133	10	52	24,913	18,534	4.0000
45,642	12	41	27,604	21,226	3.8000
53,680	14	55	30,756	24,378	3.6000
63,721	17	42	34,481	28,103	3.4000
76,431	21	14	38,926	32,548	3.2000
86,164	23	56	42,164	35,786	3.0747

The most important element of a geostationary orbit is the radius for a circular orbit (or the semimajor axis for an elliptical orbit) since this determines the orbit period. If the radius or semimajor axis is not controlled, the satellite will gradually drift until it no longer is over the desired area. The required orbit period is one sidereal day, 86164.091 sec, or 23 hr, 56 minutes, and 4.091 sec, which is not quite 24.00 hr. The 24.00-hr day is the average time between consecutive crossings of any particular meridian by the sun. The sidereal day is the time between consecutive crossings of any particular meridian by any one star. The sidereal day is shorter than the mean solar day by one part in 365.25 because of the apparent motion of the sun against a stellar background.

6.1.7 Near-Geostationary Orbits

A circular orbit is shown in Figure 6.8. Looking down on the north pole, the earth rotates counterclockwise at an angular rate of 72.92115×10^{-6} rad/sec, or 360.9856° per solar day. A solar day, with 86400 seconds, is longer than a sidereal day. The earth rotates 360° per sidereal day, and 360.9856° per solar day. If an equatorial orbit is chosen with the same angular rate, the satellite will keep in step with the earth. If the radius is slightly smaller, the velocity is larger, and the orbit period is smaller; the satellite will drift to the east with respect to the earth. Conversely, for a slightly larger radius, the satellite will drift westward.

By taking the derivative of the equations relating period, radius, and velocity to the angular rate, equations can be derived showing the effect of eastward drift on various quantities. In terms of the difference in angular velocity, Δn, the relative change in orbit period, ΔP, is

$$\frac{\Delta P}{P} = -\frac{\Delta n}{n} \tag{6.13}$$

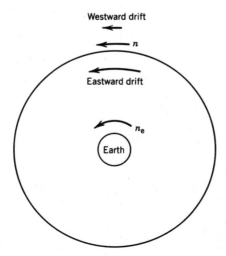

FIGURE 6.8 Satellite orbits close to geostationary orbit showing relative angular velocities.

The fractional change in n is equal in magnitude to the fractional change in P but is opposite in sign; that is, when the drift is eastward, the angular velocity is larger, the period decreases, and vice versa.

A similar relation can be found for Δa, the change in orbit radius a:

$$\frac{\Delta a}{a} = -\frac{(2/3)\,\Delta n}{n} \tag{6.14}$$

A decrease in radius of 0.78 km is a fractional change of 0.78 km/42,164.57 km, or 19 ppm; this corresponds to a 2×10^{-9} rad/sec change in angular rotation, or an eastward drift of $0.01°$ per day.

The effect of an eastward drift on velocity is

$$\frac{\Delta \dot{s}}{\dot{s}} = \frac{(1/3)\,\Delta n}{n} \tag{6.15}$$

or

$$\Delta \dot{s} = \tfrac{1}{3}\,a\Delta n$$

This indicates the actual velocity changes needed to correct for eastward or westward drift. A common mistake is to assume that since $\dot{s} = an$, it must follow that $\Delta \dot{s} = a\,\Delta n$, forgetting that the semimajor axis a will change when the velocity is changed.

The effect of small deviations for a geostationary orbit is shown in Table 6.4. For an eastward drift of $0.01°$ per day the radius (or semimajor axis) is 0.78 km smaller, the period is 2.39 sec smaller, and the velocity is 0.028 m/sec larger, or 28×10^{-6} km/sec larger. The spread of values is typical of communications satellites; some satellites are held to closer tolerances than others.

In the range of the table, the variations are close to linear, so the difference equations just shown could be used. Exact values can be calculated for any drift using the basic equations for the radius, period, and velocity in terms of the angular rotation (exact equations were used in calculating the table).

Circular orbits are assumed, although some concepts apply also to near-circular orbits. A change from one circular orbit to another cannot be done with a single velocity change. Two-body equations have also been used in calculating the values in Table 6.4; effects of the sun and moon increase the average radius slightly to 42164.57 km.

The simple two-body analysis indicates that a satellite would remain forever in the geostationary orbit once it is injected in the right location with the right velocity. In practice, there are many secondary influences that change the orbit slowly over a long period of time. The major perturbations for communications satellites in geostationary orbit are listed in Table 6.5. The sun and moon's effect is to change the plane of the orbit so that it forms an increasing angle with the equatorial plane;

TABLE 6.4 Effects of Eastward Drift[a]

Δn (degrees/day)	a (km)	Δa (km)	P (sec)	ΔP (sec)	$\dot{s}$ (km/sec)	$\Delta \dot{s}$ (m/sec)	Δn (degrees/day)
-0.05	42168.09	3.89	86176.03	11.94	3.074520	-0.142	-0.05
-0.04	42167.31	3.12	86173.64	9.55	3.074548	-0.114	-0.04
-0.03	42166.53	2.34	68171.25	7.16	3.074577	-0.085	-0.03
-0.02	42165.75	1.56	86168.87	4.77	3.074605	-0.057	-0.02
-0.01	42164.98	0.78	86166.48	2.39	3.074634	-0.028	-0.01
0.00	42164.20	0.00	86164.09	0	3.074662	0.000	0.00
0.01	42163.42	-0.78	86161.70	-2.39	3.074690	0.028	0.01
0.02	42162.64	-1.56	86159.32	-4.77	3.074719	0.057	0.02
0.03	42161.86	-2.34	86156.93	-7.16	3.074747	0.085	0.03
0.04	42161.08	-3.11	86154.54	-9.55	3.074776	0.114	0.04
0.05	42160.30	-3.89	86152.16	-11.93	3.074804	0.142	0.05

[a]Negative Δn is westward drift. Calculated according to simplified two-body theory; actually a is about 0.37 km more, and $\dot{s}$ about 0.000027 km/sec more.

TABLE 6.5 Main Perturbations of Geostationary Orbit

Cause	Effect	Orbit Element	Magnitude	Varies with	7-yr ΔV Requirement	Mass of Hydrazine (%)
Lunar–solar	North–south oscillation	i	0.75°–0.95° per year	Year	320 m/sec	13
Earth	East–west drift	a	±0.002° per day^2	Longitude	15 m/sec	1
Solar pressure	East–west oscillation	e	—	Area per mass	—	—

this increase is $0.75°-0.95°\ \text{yr}^{-1}$ and varies from year to year. When the orbit plane is no longer in the equatorial plane, the spacecraft performs a daily north–south oscillation (see Figure 3.62).

The earth is slightly nonspherical and, as a result, produces an acceleration in longitudinal position. There are two stable points, at 79° E and 252.4° E; at other longitudes there is an acceleration toward one of these two points, which may be as high as $0.002°$ per day^2. A third perturbation is due to the solar pressure on the area of a spacecraft; this accelerates the spacecraft when it is going away from the sun and retards it when it is approaching the sun. The solar pressure effect increases the eccentricity of the orbit if it starts at zero eccentricity (circle).

All geostationary communications satellites must maintain their longitude within certain limits. Thrusters are used periodically to maintain the satellite at the desired location. If hydrazine thrusters are used, fuel is needed, up to 1% of the spacecraft mass, for a 7-yr mission; how much is needed depends on the longitude. Many communications satellites also perform north–south stationkeeping, which requires about 13% of the spacecraft mass for hydrazine.

6.2 DIRECTION OF ORBIT NORMALS AND OF SUN

6.2.1 Orientation of Orbit Plane

The orientation of the satellite's orbit plane (see Figure 6.9) is traditionally specified by the plane's inclination and the location of the ascending node. The inclination i is the angle between the orbit plane and the earth's equatorial plane. During each orbit the satellite crosses the equatorial plane twice; the point where the satel-

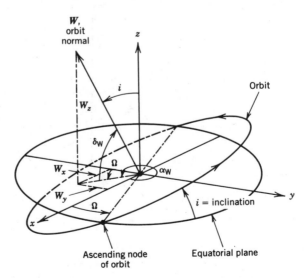

FIGURE 6.9 Orientation of satellite orbit plane with respect to inertial geocentric coordinate system.

lite crosses the equatorial plane from south to north is called the ascending node. The location of the ascending node is specified by Ω, the right ascension of the ascending node (in a geocentric system).

The orbit plane can also be defined by the orbit normal **W**, the line perpendicular to the orbit plane. The direction of the orbit normal follows the right-hand rule and is positive in the northerly direction for satellites traveling eastward. Just as the directions of celestial bodies are defined by the right ascension and declination of the geocentric celestial sphere, the orbit plane can be defined by the right ascension α_W and the declination δ_W of the orbit normal, or pole of the orbit. The orbit normal may also be defined by the three components of a unit vector, W_x, W_y, and W_z.

Most satellites move in an eastward direction, the orbit normal is then in the northern hemisphere, and the inclination angle is less than 90°. A satellite in a retrograde orbit moves westward and has an orbit normal pointed into the Southern Hemisphere and an inclination angle between 90° and 180°.

The orientation of the orbit plane can be specified by (1) the components of the orbit normal, (2) the declination and right ascension of the orbit normal, or (3) the inclination and right ascension of the ascending node. The relationship between these quantities can be seen in the geometric drawings in Figure 6.10.

The components of the orbit normal **W** are

$$W_x = \sin \Omega \sin i \qquad W_x = \cos \alpha_W \cos \delta_W$$

$$W_y = -\cos \Omega \sin i \qquad W_y = \sin \alpha_W \cos \delta_W$$

$$W_z = \cos i \qquad W_z = \sin \delta_W \qquad (6.16)$$

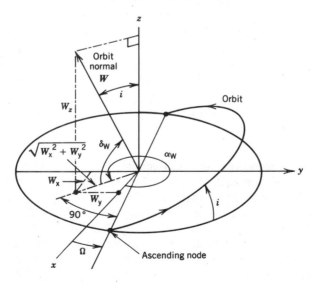

FIGURE 6.10 Relation between components of orbit normal **W**, direction of orbit normal (α_w, δ_w), and classical orbital elements (i, Ω).

The three equations on the left show the components of the orbit normal in terms of the right ascension of the ascending node, Ω, and the inclination, i. The second set shows them in terms of the orbit normal declination δ_W and the orbit normal right ascension α_W.

The declination and right ascension of the orbit normal are

$$\delta_W = 90° - i \qquad \alpha_W = \Omega + 270° \tag{6.17}$$

in terms of the inclination and right ascension of the ascending node. The orbit normal declination and right ascension are

$$\delta_W = \sin^{-1} W_z$$

$$= \tan^{-1} \frac{W_z}{\left(W_x^2 + W_y^2\right)^{1/2}}$$

$$\alpha_W = \tan^{-1} \frac{W_y}{W_x} \tag{6.18}$$

in terms of the components of the orbit normal.

Finally, the inclination and the right ascension of the ascending node can be written as

$$i = 90° - \delta_W \qquad \Omega = \alpha_W - 270° \tag{6.19}$$

in terms of the declination and right ascension of the orbit normal. The inclination and right ascension of the ascending node are

$$i = \cos^{-1} W_z$$

$$= \tan^{-1} \frac{\left(W_x^2 + W_y^2\right)^{1/2}}{W_z}$$

$$\Omega = \tan^{-1} \left(\frac{-W_x}{W_y}\right) \tag{6.20}$$

in terms of the components of the orbit normal **W**.

If the declination of a satellite, δ, is plotted as a function of the right ascension α, the result is a planar projection of the satellite orbit (see Figure 6.11). If the satellite orbit is close to the equatorial plane, the inclination i is small, and the tangent of the declination is approximately equal to the declination (in radians); with this approximation, the planar projection plot of the orbit is a sine curve.

The orbit normal can be plotted on the planar projection, and is shown in Figure 6.11. Its declination is $90° - i$, and its right ascension is $90°$ less than the ascending node (or $270°$ greater). Thus, the right ascension of the orbit normal is the same as the satellite's right ascension when it has its most southern declination.

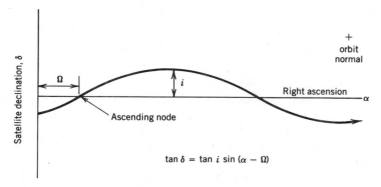

FIGURE 6.11 North–south oscillation of satellite in inclined orbit.

The most obvious result of an inclined orbit is the north–south oscillation of the satellite. The declination has a maximum positive value of $+i$ and a maximum negative value of $-i$ during each revolution. The geosynchronous orbit has a period of 1 day, so this north–south oscillation occurs once a day. The point at which the orbit crosses the equatorial plane, from south to north is called the ascending node; this node can have any right ascension, as the oscillation can have any phase.

While the main effect of orbit inclination is to produce a north–south oscillation, there are other effects. A relatively small effect is an east–west oscillation. Most communications satellites have an inclination of less than $1°$, and the east–west oscillation is less than $0.005°$. This oscillation is a well-known effect, but many individuals overemphasize its importance when they discuss the figure-eight.

A satellite in a circular orbit proceeds along the orbit at a uniform angular rate; this is indicated in Figure 6.12 by labeling the distance along the orbit as nt_i. By

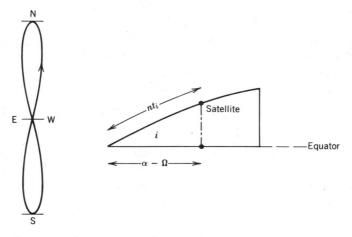

FIGURE 6.12 Twelve-hour east–west oscillation of satellite in inclined orbit due to 24-hr north–south oscillation.

spherical trigonometry, the satellite's right ascension α is then given by

$$\tan (\alpha - \Omega) = \cos i \tan nt_i \qquad (6.21)$$

where i is the inclination, n the angular rate of rotation, and t_i time.[a] When the circular orbit is inclined, the rate of change of satellite right ascension varies during each quarter of the orbit with respect to the average rate for the entire orbit. During the first quarter of the orbit from the ascending node, the rate of change of α is slowest at the beginning and fastest at the end. Therefore, during the first quarter the satellite appears to first drift westward, then remain essentially motionless, and then drift eastward during the last part of the quarter. During the second quarter the satellite appears to drift eastward during the early part and westward during the latter part. The process is repeated during the third and fourth quarters. The result is an east–west oscillation twice daily, that is, a period of half a day.

The ground track performs a figure-eight as it moves north and south; but in most cases this figure-eight is practically a straight line moving north and south. The maximum amplitude of the east–west oscillation can be shown to be

$$(nt - \alpha)_{max} = \sin^{-1} \left(\tan^2 \frac{i}{2} \right) \qquad (6.22)$$

for a circular orbit with an inclination i. The approximate amplitude of the east–west oscillation is

$$(nt - \alpha)_{max} \cong \tfrac{1}{4} i^2 \quad \text{(angles in radians)}$$

$$\cong \tfrac{1}{229} i^2 \quad \text{(angles in degrees)} \qquad (6.23)$$

where i is the orbit inclination. The amplitude of the east–west oscillation of a circular orbit is plotted in Figure 6.13 for various orbit inclinations. With many communications satellites, the east–west oscillation due to orbit eccentricity is dominant, and the effect due to orbit inclination can be neglected.

6.2.2 Solar Eclipses

The longest eclipse occurs when the sun is in the orbit plane. The earth's shadow is approximated by a long cylinder with radius R (the earth's radius). The angular distance traversed by the satellite from the eclipse midpoint until it emerges from eclipse is denoted by η (see Figure 6.14). For a geostationary circular orbit this

---

[a]Assuming time t_i starts at ascending mode.

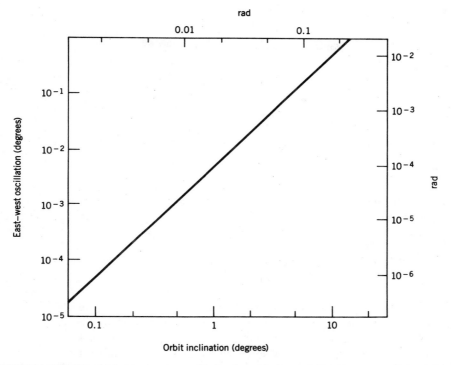

FIGURE 6.13 Magnitude of east–west oscillation due to inclination. For many geostationary satellites, this is small compared to east–west oscillation due to orbit eccentricity.

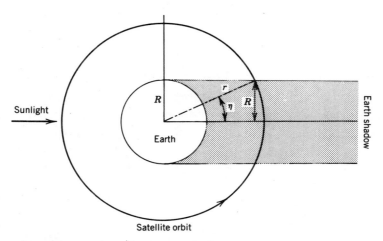

FIGURE 6.14 Maximum eclipse. Part of satellite orbit in earth's shadow when sunlight is parallel to equatorial plane. Not to scale for geostationary orbit.

angle is

$$\eta = \sin^{-1} \frac{R}{r} = \sin^{-1} \left(\frac{6378 \text{ km}}{42164 \text{ km}} \right)$$

$$= 0.151 \text{ rad} = 8.67° \qquad (6.24)$$

The eclipse duration is obtained by multiplying by P/π, so

$$\text{Eclipse}_{\max} = \frac{P\eta}{\pi} = \frac{P}{\pi} \sin^{-1} \frac{R}{r} = 4150 \text{ sec} \qquad (6.25)$$

where P is the orbit period (86,164 sec for geostationary orbit).

For a zero-inclination satellite orbit, the sun is in the orbit plane at the vernal and autumnal equinoxes, when the sun's declination is zero. This occurs around March 21 or September 23 and produces the longest eclipse, equal to 4150 sec, or about 70 min. The preceding assumes a point source for the sun and a spherical earth with no atmosphere. Since the sun subtends half a degree, it takes about 2 min from total sunlight to total darkness. The earth's atmosphere bends the sunlight around the earth. Combining the two effects, the geometric eclipse is roughly the time from when the sunlight starts to decrease to the time it is fully restored.

When the sun is not in the orbit plane, it casts an elliptical shadow in the orbit plane, shown in Figure 6.15. The semiminor axis is the earth's radius R, and the semimajor axis is $R/\sin \delta_e$, where δ_e is the angle between the sun's rays and the orbit plane. (For an equatorial orbit, δ_e is the sun's declination.) An eclipse occurs only if this semimajor axis of the earth's shadow is greater than the satellite orbit

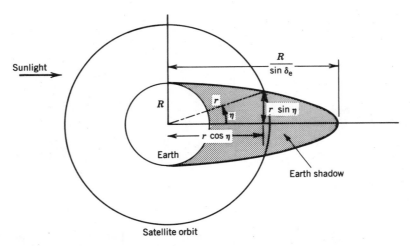

FIGURE 6.15 Shorter eclipse. Earth's shadow intercepted by equatorial plane when sunlight is not parallel to plane.

radius, that is, if $\sin \delta_e < R/r$. For a geostationary orbit this means the sun's declination must be less than $8.7°$.

The equation of the ellipse can be written as

$$\left(\frac{r \cos \eta}{R/\sin \delta_e}\right)^2 + \left(\frac{r \sin \eta}{R}\right)^2 = 1 \qquad (6.26)$$

where the point at which the satellite emerges from the eclipse is given by $x = r \cos \eta$ and $y = r \sin \eta$. Solving the preceding equation for the half-angle η, the result is

$$\eta = \cos^{-1} \frac{(1 - R^2/r^2)^{1/2}}{\cos \delta_e} \qquad (6.27)$$

The fractional nighttime is the half-angle η divided by π radians. So the total duration of eclipse is

$$\frac{P\eta}{\pi} = \frac{P}{\pi} \cos^{-1} \frac{(1 - R^2/r^2)^{1/2}}{\cos \delta_e} = \frac{P}{\pi} \tan^{-1} \frac{(R^2 - r^2 \sin^2 \delta_e)^{1/2}}{(r^2 - R^2)^{1/2}} \qquad (6.28)$$

For zero-inclination orbits, the angle δ_e is the declination of the sun. For near-zero inclination orbits, the sequence of eclipse times is approximately the same, but the maximum may be shifted a few days.

Eclipses are of shorter duration when the sun is not at equinox. When the sun has a certain declination, the earth's shadow is displaced from the equator, and the synchronous satellite moves through a shorter dark path. The eclipse durations for various sun declinations and various dates are shown in Figure 6.16.

For geostationary satellites, eclipses can be expected near local midnight of the subsatellite point for a total of about 90 evenings per year in the spring and fall. Spring eclipses begin in late February or early March and end about mid-April. Fall events begin about September 1 and end mid-October. Eclipses lasting about 70 min occur around March 21 and September 23; those lasting longer than 1 hr occur about 50 days per year. Any slight inclination of the satellite orbit may shift the eclipse season by a few days.

6.2.3 Direction of Sun

The fundamental time to which the positions—with respect to the earth—of the sun and moon are referred is 2000 January 0 at noontime, ephemeris time, which is Julian day 2451545.0 (Julian days start at 4713 B.C.). Listed in Tables 6.6 and 6.7 are the number of days from the fundamental time to the year and month listed. This number is negative for years before 2000. January "0" is actually the last day of December; it is convenient to list "zero" days because the day of the month

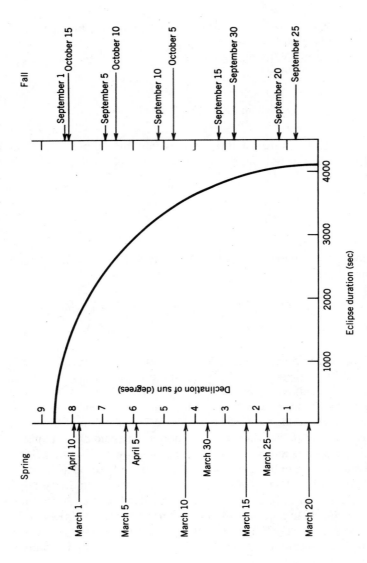

FIGURE 6.16 Duration of eclipse on different dates for satellite in geostationary orbit.

TABLE 6.6 Julian Day Minus 2,451,545 (January–June)

Year	January 0	February 0	March 0	April 0	May 0	June 0	Year
1985	−5,479.5	−5,448.5	−5,420.5	−5,389.5	−5,359.5	−5,328.5	1985
1986	−5,114.5	−5,083.5	−5,055.5	−5,024.5	−4,994.5	−4,963.5	1986
1987	−4,749.5	−4,718.5	−4,690.5	−4,659.5	−4,629.5	−4,598.5	1987
1988	−4,384.5	−4,353.5	−4,324.5	−4,293.5	−4,263.5	−4,232.5	1988
1989	−4,018.5	−3,987.5	−3,959.5	−3,928.5	−3,898.5	−3,867.5	1989
1990	−3,653.5	−3,622.5	−3,594.5	−3,563.5	−3,533.5	−3,502.5	1990
1991	−3,288.5	−3,257.5	−3,229.5	−3,198.5	−3,168.5	−3,137.5	1991
1992	−2,923.5	−2,892.5	−2,863.5	−2,832.5	−2,802.5	−2,771.5	1992
1993	−2,557.5	−2,526.5	−2,498.5	−2,467.5	−2,437.5	−2,406.5	1993
1994	−2,192.5	−2,161.5	−2,133.5	−2,102.5	−2,072.5	−2,041.5	1994
1995	−1,827.5	−1,796.5	−1,768.5	−1,737.5	−1,707.5	−1,676.5	1995
1996	−1,462.5	−1,431.5	−1,402.5	−1,371.5	−1,341.5	−1,310.5	1996
1997	−1,096.5	−1,065.5	−1,037.5	−1,006.5	−976.5	−945.5	1997
1998	−731.5	−700.5	−672.5	−641.5	−611.5	−580.5	1998
1999	−366.5	−335.5	−307.5	−276.5	−246.5	−215.5	1999
2000	−1.5	29.5	58.5	89.5	119.5	150.5	2000
2001	364.5	395.5	423.5	454.5	484.5	515.5	2001
2002	729.5	760.5	788.5	819.5	849.5	880.5	2002
2003	1,094.5	1,125.5	1,153.5	1,184.5	1,214.5	1,245.5	2003
2004	1,459.5	1,490.5	1,519.5	1,550.5	1,580.5	1,611.5	2004
2005	1,825.5	1,856.5	1,884.5	1,915.5	1,945.5	1,976.5	2005
2006	2,190.5	2,221.5	2,249.5	2,280.5	2,310.5	2,341.5	2006
2007	2,555.5	2,586.5	2,614.5	2,645.5	2,675.5	2,706.5	2007
2008	2,920.5	2,951.5	2,980.5	3,011.5	3,041.5	3,072.5	2008
2009	3,286.5	3,317.5	3,345.5	3,376.5	3,406.5	3,437.5	2009
2010	3,651.5	3,682.5	3,710.5	3,741.5	3,771.5	3,802.5	2010

can be easily added. For example, for July 4, 1990, the −3,472.5 from the table is added to 4, and −3,468.5 is the start of July 4.

Strictly speaking, the time used should be ephemeris time (ET), a uniform measure of time. However, this differs only by seconds from universal time (UT), which is used as the basis for all civil time keeping, and this difference will be ignored here. Universal time may be identified with Greenwich mean time (GMT), and that will be used in the examples.

An example will illustrate the use of the tables. The time d, for 3 P.M. EDT on July 4, 1990 is:

1990, July 4,   3 P.M. eastern daylight time

3 P.M. EDT = 2 P.M. EST = 1900 GMT

$$d = -3,472.5 + 4 + 1900/2400 = -3,467.708 \qquad (6.29)$$

TABLE 6.7 Julian Day Minus 2,451,545 (July–December)

Year	July 0	August 0	September 0	October 0	November 0	December 0	Year
1985	−5,298.5	−5,267.5	−5,236.5	−5,206.5	−5,175.5	−5,145.5	1985
1986	−4,933.5	−4,902.5	−4,871.5	−4,841.5	−4,810.5	−4,780.5	1986
1987	−4,568.5	−4,537.5	−4,506.5	−4,476.5	−4,445.5	−4,415.5	1987
1988	−4,202.5	−4,171.5	−4,140.5	−4,110.5	−4,079.5	−4,049.5	1988
1989	−3,837.5	−3,806.5	−3,775.5	−3,745.5	−3,714.5	−3,684.5	1989
1990	−3,472.5	−3,441.5	−3,410.5	−3,380.5	−3,349.5	−3,319.5	1990
1991	−3,107.5	−3,076.5	−3,045.5	−3,015.5	−2,984.5	−2,954.5	1991
1992	−2,741.5	−2,710.5	−2,679.5	−2,649.5	−2,618.5	−2,588.5	1992
1993	−2,376.5	−2,345.5	−2,314.5	−2,284.5	−2,253.5	−2,223.5	1993
1994	−2,011.5	−1,980.5	−1,949.5	−1,919.5	−1,888.5	−1,858.5	1994
1995	−1,646.5	−1,615.5	−1,584.5	−1,554.5	−1,523.5	−1,493.5	1995
1996	−1,280.5	−1,249.5	−1,218.5	−1,188.5	−1,157.5	−1,127.5	1996
1997	−915.5	−884.5	−853.5	−823.5	−792.5	−762.5	1997
1998	−550.5	−519.5	−488.5	−458.5	−427.5	−397.5	1998
1999	−185.5	−154.5	−123.5	−93.5	−62.5	−32.5	1999
2000	180.5	211.5	242.5	272.5	303.5	333.5	2000
2001	545.5	576.5	607.5	637.5	668.5	698.5	2001
2002	910.5	941.5	972.5	1,002.5	1,033.5	1,063.5	2002
2003	1,275.5	1,306.5	1,337.5	1,367.5	1,398.5	1,428.5	2003
2004	1,641.5	1,672.5	1,703.5	1,733.5	1,764.5	1,794.5	2004
2005	2,006.5	2,037.5	2,068.5	2,098.5	2,129.5	2,159.5	2005
2006	2,371.5	2,402.5	2,433.5	2,463.5	2,494.5	2,524.5	2006
2007	2,736.5	2,767.5	2,798.5	2,828.5	2,859.5	2,889.5	2007
2008	3,102.5	3,133.5	3,164.5	3,194.5	3,225.5	3,255.5	2008
2009	3,467.5	3,498.5	3,529.5	3,559.5	3,590.5	3,620.5	2009
2010	3,832.5	3,863.5	3,894.5	3,924.5	3,955.5	3,985.5	2010

The time of day must be converted to GMT and then into fractions of a day. In the example, 3 P.M. eastern daylight time (EDT) is equivalent to 1900 GMT, so that 1900/2400 is the fraction of day to be added. The result is −3,467.708; a date before the year 2000 will be negative.

The longitude of the sun is the angle between the direction of the sun and the x axis (First Point of Aries). The sun's longitude, L_{sun}, can be calculated as

$$L_{mean} = 280.460° + 0.9856474°d$$

$$M_E = 357.528° + 0.9856003°d$$

$$L_{sun} = L_{mean} + 1.915° \sin M_E + 0.020° \sin 2\,M_E \qquad (6.30)$$

where L_{mean} is the mean longitude of the sun (the longitude if the sun moved uniformly), and M_E is the earth's mean anomaly. These equations have an accuracy on the order of $0.01°$, which may be sufficient for many purposes.

The angle between the ecliptic plane (earth's orbit) and the equatorial plane, ϵ, is approximately constant and is given by

$$\epsilon = 23.439° - 0.000\ 0004\ d \tag{6.31}$$

This angle equals 0.409 rad.

If the celestial longitude of the sun, L_{sun}, and the angle of the ecliptic, ϵ, are known, the right ascension α and the declination δ can be found from the properties of a right spherical triangle (see Figure 6.17). The equations are

$$\tan \alpha = \frac{\cos \epsilon \sin L_{sun}}{\cos L_{sun}} \qquad \tan \delta = \tan \epsilon \sin \alpha \tag{6.32}$$

where L_{sun} is the sun's mean longitude and ϵ the obliquity of the ecliptic.

The sun's right ascension α is always in the same quadrant as the sun's longitude L_{sun}. For manual calculations this can be done by inspection. For computer calculations it is useful to use the two-argument arctangent functions. By using $\cos \epsilon \sin L_{sun}$ as the first argument and $\cos L_{sun}$ as the second, the result will be in the correct quadrant. The declination δ can only be in the first or fourth quadrant, so a single-argument arctangent function can be used. If an arcsine function is available, the declination can be calculated directly from the longitude with $\sin \delta = \sin \epsilon \sin L_{sun}$ without using the right ascension.

The radial distance from the earth to the sun r_s can be determined by

$$\frac{r_s}{a_s} = 1.00014 - 0.01671 \cos M_E - 0.00014 \cos 2M_E \tag{6.33}$$

where the mean anomaly M_E was given previously. The mean distance to the sun, a_s, is equal to 149,597,870 km.

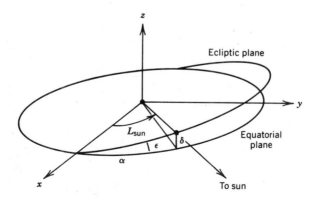

FIGURE 6.17 Spherical coordinates of sun and relation to celestial longitude, L_{sun}.

6.3 ELLIPTICAL ORBITS IN PLANE

For a system of two point masses, such as a satellite around the earth, the satellite remains in a fixed plane. The position vector and velocity vector determine this orbit plane; since the gravitational acceleration is along the position vector, there is no acceleration or velocity to get the satellite out of the plane.

In the orbit plane, the position of the moving mass (satellite) can be expressed in rectangular coordinates as

$$x_\omega = r \cos v \qquad y_\omega = r \sin v \qquad (6.34)$$

where r is the radius vector, and the angle v is usually referred to as the true anomaly (see Figure 6.18). The converse equations can also be written as

$$r = \left(x_\omega^2 + y_\omega^2\right)^{1/2} \qquad v = \tan^{-1}\frac{y_\omega}{x_\omega} \qquad (6.35)$$

It can be shown that the path of the satellite is a conic in the orbit plane. Our primary concern will be with elliptical orbits. In the next few pages are described some features of conics in general and some features of ellipses.

6.3.1 Properties of Ellipses

A conic is made of all the points whose distances r from a fixed point (focus F) bear a constant ratio (eccentricity e) to their perpendicular distances from a straight

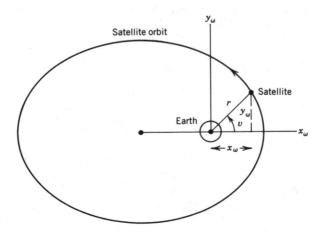

FIGURE 6.18 Rectangular coordinates (x_ω, y_ω) and polar coordinates (r, v) in plane of satellite orbit.

line (directrix). The eccentricity e is a function of the type of curve:

$$e = 0 \quad \text{(circles)}$$

$$0 < e < 1 \quad \text{(ellipses)}$$

$$e = 1 \quad \text{(parabola or straight line)}$$

$$e > 1 \quad \text{(hyperbola)} \tag{6.36}$$

In the equation for the conic, the distance from the focus to the directrix is p/e (see Figure 6.19). When the point is on the y_ω axis, the radius is equal to the parameter p, (i.e., $p = r$). In general, for any point on the conic, this distance (p/e) is equal to the distance from the point to the directrix (r/e) plus the abscissa of the point, x_ω, or $p/e = r/e + x_\omega$. Substituting in $x_\omega = r \cos v$, the result is the conic equation

$$r = \frac{p}{1 + e \cos v} \tag{6.37}$$

where p is the parameter, e the eccentricity, r the radius vector, and v the true anomaly (true angle).

Another useful angle in the analysis of orbits is the eccentric anomaly E. Anomaly is an old-fashioned word for angle. An auxiliary circle is drawn using the major axis of the ellipse as a diameter (see Figure 6.20). The position of the satellite on the ellipse is then projected onto the circle. The eccentric anomaly is

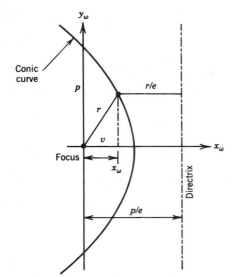

FIGURE 6.19 Coordinates used in defining parameters for polar equation of any conic.

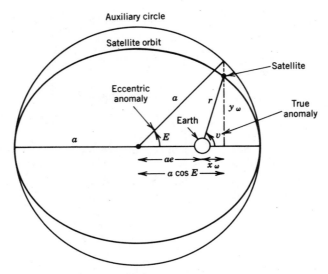

FIGURE 6.20 Relation between elliptical orbit, with its true anomaly v, and auxiliary circle, with its eccentric anomaly E.

the angle measured from the center of the ellipse between the major axis and the projected point.

A simple relation for the radius vector and eccentric anomaly E is

$$r = a(1 - e \cos E) \tag{6.38}$$

where a is the semimajor axis and e the orbit eccentricity.

The relations between the eccentric anomaly E and the true anomaly v are

$$\sin E = \frac{(1 - e^2)^{1/2} \sin v}{1 + e \cos v} \qquad \sin v = \frac{(1 - e^2)^{1/2} \sin E}{1 - e \cos E}$$

$$\cos E = \frac{\cos v + e}{1 + e \cos v} \qquad \cos v = \frac{\cos E - e}{1 - e \cos E} \tag{6.39}$$

where the only parameter is the orbit eccentricity e that determines the shape of the orbit.

A relation between velocity and radius based on the conservation of energy is

$$\dot{s}^2 = \mu \left(\frac{2}{r} - \frac{1}{a} \right) \tag{6.40}$$

where $\dot{s}^2$ is the square of the velocity. From the preceding, the maximum velocity (at perifocus) and the minimum velocity (at apofocus) can be found:†

$$\dot{s}_p^2 = \frac{\mu r_A}{a r_p} \qquad \dot{s}_A^2 = \frac{\mu r_p}{a r_A} \tag{6.41}$$

It is sometimes useful to refer to a characteristic velocity in the orbit of $(\mu/a)^{1/2}$, which is the geometric mean of the maximum and minimum velocities.

6.3.2 Kepler's Equation

The mean anomaly M is a fictitious angle measured from perigee through which a body would have traveled if it were moving uniformly at its mean angular velocity n. This can be written as

$$M = M_0 + n(t - t_0) \tag{6.42}$$

where

$$n = \frac{\mu^{1/2}}{a^{3/2}}$$

Notice in Figures 6.20 and 6.21 that at perigee the mean anomaly M, the eccentric anomaly E, and the true anomaly v are all zero (or multiples of 360°); at apogee they are all 180°, or π radians. Usually the value of M is given between 0° and 360°, and multiples of 360° are subtracted (or added) if necessary. At time t_0 (or epoch) the mean anomaly is M_0; frequently M_0 is taken as zero, and then $t_0 = T$; that is, the initial time is taken as the time the satellite passes through perigee.

Calculating the mean anomaly M from the preceding equations is usually the first step in determining the position r and velocity $\dot{r}$ of the satellite at some speci-

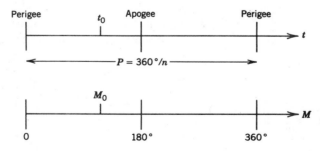

FIGURE 6.21 Relation between mean anomaly M and time t from one perigee to the next.

†The terms *perifocus* and *apofocus* apply to any orbit. For orbits around the earth, the equivalent terms are *perigee* and *apogee*.

fied time. The next step is to determine the eccentric anomaly by solving Kepler's equation:[†]

$$M = E - e \sin E \tag{6.43}$$

where the mean anomaly M is $M_0 + n(t - t_0)$, and the mean angular rotation rate n is equal to $\mu^{1/2}/a^{3/2}$.

To obtain the eccentric anomaly E from the mean anomaly M, Kepler's equation must be solved the "hard way." There are many ways to solve a transcendental equation. The orbit eccentricity e for most communications satellites is quite small so a series approximation is quite accurate:

$$E = M + e \sin M + \tfrac{1}{2}e^2 \sin 2M + \tfrac{1}{8}e^3(3 \sin 3M - \sin M) + \cdots \tag{6.44}$$

If, for example, the eccentricity is less than 0.001, three terms of the series provide about nine significant figures.

For eccentricity values not close to zero, an iterative calculation can be used; a good one is the Newton–Raphson method. To start the solution, a trial value of E, labeled E^*, is found from the preceding series approximation. Substituting this value in Kepler's equation, the value of the mean anomaly M^* can be calculated and compared with the true value M. A change in E is then calculated as

$$\Delta E^* = \frac{M - M^*}{1 - e \cos E^*} \qquad E^*_{i+1} = E^*_i + \Delta E^* \tag{6.45}$$

A better value of E^* is obtained by adding ΔE^* to the old value. As shown in Figure 6.22, this is equivalent to taking a line at the calculated point (E^*, M^*) with the same slope as the actual curve $(1 - e \cos E^*)$ and extrapolating it until it intersects the desired value of M.

The correction is made on E^*, and the process is repeated as many times as needed. In rough calculations this might be until ΔE^* is 10^{-3} rad, while in more accurate computations it might be 10^{-6} rad. While the convergence may be slow at first, after two or three significant figures are obtained, others come rapidly.

6.3.3 Summary of Equations for Elliptical Orbits

The semimajor axis a and the eccentricity e define the size and shape of an elliptical orbit. If the initial mean anomaly M_0 is known at an initial time t_0, then for some later time t we can find the mean anomaly M and the eccentric anomaly E from the equations

$$M = M_0 + n(t - t_0) \qquad M = E - e \sin E \tag{6.46}$$

[†]Radians must be used for M and E in Kepler's equation.

Mean anomaly

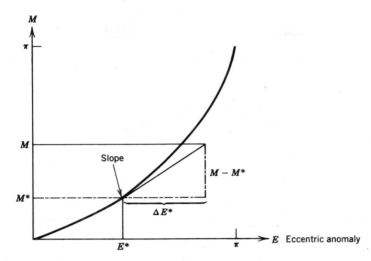

FIGURE 6.22 Iterative solution of Kepler's equation using Newton–Raphson method.

The first equation is the definition of the mean anomaly. The second equation is Kepler's equation, which must be solved for E, either by an iterative procedure or from the series expansion.

The position of the satellite in terms of the rectangular coordinates in the orbit plane is given by

$$x_\omega = a(\cos E - e) \qquad y_\omega = a(1 - e^2)^{1/2} \sin E \qquad (6.47)$$

These equations come directly from the geometry of the eccentric anomaly (see Figure 6.20).

The equations for the satellite velocity are

$$\dot{x}_\omega = -\frac{(\mu a)^{1/2} \sin E}{r} \qquad \dot{y}_\omega = \frac{[\mu a(1 - e^2)]^{1/2} \cos E}{r} \qquad (6.48)$$

Once the position in the orbit is determined, it is possible to shift coordinate systems and determine the satellite position in a celestial coordinate system or its direction from a desired earth station.

The properties of elliptical orbits are summarized in Figure 6.23. Various quantities are shown on the left mainly in terms of the semimajor axis a and the eccentricity e. On the right they are expressed principally in terms of the maximum radius r_A and the minimum radius r_p. The equations for the satellite position and velocity are summarized in Figure 6.24; on the left they are expressed in terms of the true anomaly v and on the right as a function the eccentric anomaly E.

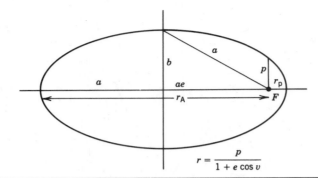

$$r = \frac{p}{1 + e \cos v}$$

Perigee distance:	**Eccentricity:**
$$r_p = a(1 - e)$$	$$e = \frac{r_A - r_p}{r_A + r_p}$$
Apogee distance:	
$$r_A = a(1 + e)$$	**Semimajor axis:**
	$$a = \tfrac{1}{2}(r_A + r_p)$$
Parameter:	**Parameter:**
$$p = a(1 - e^2)$$	
	$$p = \frac{2 r_A r_p}{r_A + r_p}$$
Semiminor axis:	
$$b = a\sqrt{1 - e^2}$$	**Semiminor axis:**
	$$b = \sqrt{r_A r_p}$$

Orbit period:	**Perigee velocity:**
$$P = 2\pi a^{3/2}/\sqrt{\mu}$$	$$\dot{s}_p = \dot{s}_c \sqrt{r_A/r_p}$$
Mean velocity:	**Apogee velocity:**
$$\dot{s}_c = \sqrt{\mu/a}$$	$$\dot{s}_A = \dot{s}_c \sqrt{r_p/r_A}$$
Eccentric anomaly (E):	**Mean anomaly:**
$$r = a(1 - e \cos E)$$	$$M = M_0 + n(t - t_0)$$
Angular momentum:	**Kepler's equation:**
$$\mathbf{h} = \mathbf{r} \times \dot{\mathbf{r}} = \sqrt{\mu p}\,\mathbf{W}$$	$$M = E - e \sin E$$
Velocity:	**Mean angular velocity:**
	$$n = \sqrt{\mu}/a^{3/2}$$
$$\dot{s}^2 = \mu\left(\frac{2}{r} - \frac{1}{a}\right)$$	

FIGURE 6.23 Parameters of elliptical orbit and relations between orbit parameters.

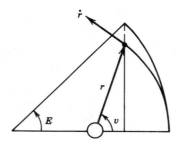

As function of true anomaly v	As function of eccentric anomaly E
Radius:	Radius:
$$r = \frac{p}{1 + e \cos v}$$	$$r = a(1 - e \cos E)$$
Position:	Position:
$$x_\omega = \frac{p \cos v}{1 + e \cos v}$$ $$y_\omega = \frac{p \sin v}{1 + e \cos v}$$	$$x_\omega = a(\cos E - e)$$ $$y_\omega = a\sqrt{1 - e^2} \sin E$$
Velocity:	Velocity:
$$\dot{x}_\omega = -\sqrt{\mu/p} \sin v$$ $$\dot{y}_\omega = \sqrt{\mu/p}(\cos v + e)$$	$$\dot{x}_\omega = -\sqrt{\frac{\mu}{a}} \frac{\sin E}{1 - e \cos E}$$ $$\dot{y}_\omega = \sqrt{\frac{\mu}{a}} \frac{\sqrt{1 - e^2} \cos E}{1 - e \cos E}$$
Eccentric anomaly:	True anomaly:
$$\sin E = \frac{\sqrt{1 - e^2} \sin v}{1 + e \cos v}$$ $$\cos E = \frac{\cos v + e}{1 + e \cos v}$$ $$\tan E = \frac{\sqrt{1 - e^2} \sin v}{\cos v + e}$$	$$\sin v = \frac{\sqrt{1 - e^2} \sin E}{1 - e \cos E}$$ $$\cos v = \frac{\cos E - e}{1 - e \cos E}$$ $$\tan v = \frac{\sqrt{1 - e^2} \sin E}{\cos E - e}$$

FIGURE 6.24 Satellite position and velocity as function of true anomaly v or eccentric anomaly E.

6.3.4 East–West Oscillation due to Eccentricity

An orbit with the correct period and zero inclination may still exhibit east–west oscillations due to orbit eccentricity. From perigee to apogee it will be east of its average position, and from apogee to perigee, it will be on the west side. To determine the magnitude of the swing, a calculation will be made for $E = 90°$, which is about the maximum eastward point (see Figure 6.25). The mean anomaly M is equal to $E - e \sin E = \pi/2 - e$, so that it takes less than a quarter of a

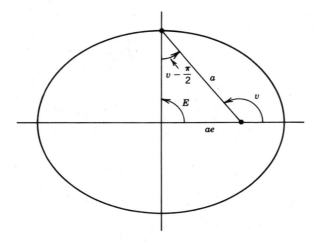

FIGURE 6.25 East–west oscillation for eccentric geosynchronous orbit. Mean anomaly (not shown) is less than 90°, eccentric anomaly E = 90°, and true anomaly v is greater than 90°.

period to reach this point. However, the true anomaly v is greater than one-fourth of a revolution; from the figure, $\sin(v - \pi/2) = ae/a = e$; hence, $v \cong \pi/2 + e$. Since v is close to $\pi/2$, the angle $v - \pi/2$ is small, and the sine is approximately equal to the angle.

The mean anomaly M is an angle that turns at a constant rate (the rate of the earth), so it is equivalent to a point at a constant longitude. The true anomaly is the actual satellite angle, so that $v - M$ is the difference in longitude:

$$\Delta\lambda = v - M = 2e \tag{6.49}$$

A similar calculation for $E = 270°$ would show a westward swing of $2e$. The effect of an orbit eccentricity is to produce an east–west oscillation with an amplitude of $2e$. An orbit eccentricity of about 0.01 would produce an amplitude of 0.02 radians or about 1°.

6.4 POSITION FROM ORBITAL ELEMENTS

Six elements are necessary to define an orbit and the satellite position. The classical elements include three dimensional elements and three orientational elements. The three dimensional elements are the semimajor axis a, the eccentricity e, and the mean anomaly at t_0 (M_0): The first two are shown in Figure 6.26. The three orientational elements are the orbit inclination i, the right ascension of the ascending node (Ω), and the argument of perigee (ω). Four of these elements are also used for circular orbits. The dimensional elements were covered in the last section, and the orientational elements are covered in this section.

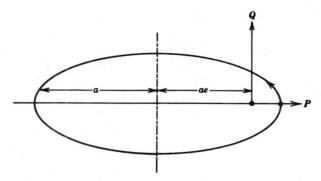

FIGURE 6.26 Elliptical orbit. Size and shape are measured by semimajor axis a and eccentricity e.

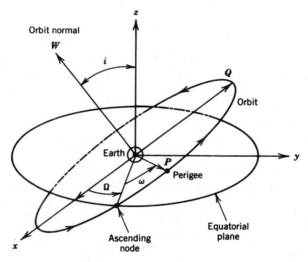

FIGURE 6.27 Orientation of elliptical orbit measured by orbit inclination i, right ascension of ascending node Ω, and argument of perigee ω.

6.4.1 Orientation of Elliptical Orbit

The orientational elements relate the orbit plane rectangular coordinate system (**P**, **Q**, **W**)† to the geocentric equatorial system (**x**, **y**, **z**) described in Section 6.1. (See Figures 6.1 and 6.27). The right ascension of the ascending node, Ω, is the angle measured in the equatorial plane from the vernal equinox to the ascending node. The inclination of the orbit, i, is the angle between the orbit normal and the north celestial pole. The argument of the perigee, ω, is the angle between the

†In the orbit plane, the earth's center is the coordinate system origin, the positive **P** axis is the direction toward perigee, the positive **Q** axis is in the orbit plane and 90° E of the **P** axis, and **W** is the orbit normal. When the orbit is perfectly circular, the directions of **P** and **Q** are indeterminate.

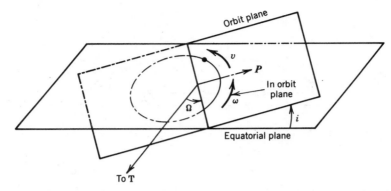

FIGURE 6.28 Combination of satellite position in orbit plane and orientation of orbit plane in space.

ascending node and the perigee. Once the elements are determined, it is relatively easy to calculate position and velocity at any future time.

The orientation of the orbit is shown in Figure 6.28. The orbit plane is inclined to the equatorial plane by the inclination angle i. The line of intersection is defined by the right ascension of the ascending node, Ω. The angle between the line of nodes and the perigee is the argument of perigee, ω. In a classical two-body problem these elements are constant and do not change with time. In reality, these elements do change slowly with time.

For geostationary orbits the inclination and the eccentricity are close to zero, and these classical elements may not be the best choice of parameters. The right ascension of the ascending node, the argument of perigee, and the mean anomaly, may have very different values for successive orbit determinations. They are indeterminate when i and e equal exactly zero.

With satellites in geostationary orbit, the longitude and rate of change of the mean longitude are frequently included with the orbit elements. They are important parameters, and not easily calculated from the orbit elements.

6.4.2 Satellite Position and Velocity

Determining the satellite position from the orbit elements is done in two parts. The unit vectors $\mathbf{P}$ and $\mathbf{Q}$ are expressed in terms of the orbit elements (referred to the basic xyz coordinate system), and the satellite position in terms of these unit vectors and the orbit plane coordinates (x_ω, y_ω). The latter relation is shown in Figure 6.29.

The components of the unit vector $\mathbf{P}$, in the direction of perigee, can be written in terms of the three orientational orbit elements: orbit inclination i, right ascension of the ascending node Ω, and the argument of perigee ω:

$$\mathbf{P}_x = \cos \omega \cos \Omega - \sin \omega \sin \Omega \cos i$$

$$\mathbf{P}_y = \cos \omega \sin \Omega + \sin \omega \cos \Omega \cos i$$

$$\mathbf{P}_z = \sin \omega \sin i \tag{6.50}$$

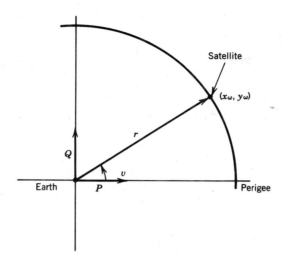

FIGURE 6.29 Satellite position in orbit plane in rectangular coordinates (x_ω, y_ω) and polar coordinates (r, v).

Similarly, the components of the parameter unit vector $\mathbf{Q}$, at right angles to both the perigee vector and the orbit normal are:

$$\mathbf{Q}_x = -\sin \omega \cos \Omega - \cos \omega \sin \Omega \cos i$$

$$\mathbf{Q}_y = -\sin \omega \sin \Omega + \cos \omega \cos \Omega \cos i$$

$$\mathbf{Q}_z = \cos \omega \sin i \qquad (6.51)$$

The components of the unit vector along the orbit normal $\mathbf{W}$ (perpendicular to the orbit plane) are not used in this section, but were given in Section 6.2.1, and are repeated here for completeness:

$$\mathbf{W}_x = \sin \Omega \sin i$$

$$\mathbf{W}_y = -\cos \Omega \sin i$$

$$\mathbf{W}_z = \cos i \qquad (6.52)$$

The vectors $\mathbf{P}$, $\mathbf{Q}$, and $\mathbf{W}$ form a rectangular coordinate system.

The satellite position and velocity, in terms of the geocentric equatorial coordinate system (xyz) are then

$$\mathbf{r} = x_\omega \mathbf{P} + y_\omega \mathbf{Q} \qquad \dot{\mathbf{r}} = \dot{x}_\omega \mathbf{P} + \dot{y}_\omega \mathbf{Q} \qquad (6.53)$$

All the equations needed to determine satellite position and velocity are summarized in Figure 6.30. For each equation there is a reference section number, the name of the quantity being calculated, the pertinent equation, and the numerical

Section Number	Name	Equation	Example[a]
6.1.5	Angular velocity	$n = \sqrt{\mu}/a^{3/2}$	0.004178 deg/sec 7.2921×10^{-5} rad/sec
6.3.2	Mean anomaly	$M = n(t - t_0) + M_0$	75°, 1.309 rad
6.3.2	First guess	$E \simeq M + e \sin M$ $+ \frac{1}{2}e^2 \sin 2M$	75.001°, 1.3092 rad
6.3.2	Kepler's equation	$M^* = E - e \sin E$	75°, 1.309 rad
6.3.2	Newton's equation	$\Delta E = \dfrac{M - M^*}{1 - e \cos E}$	-1.77×10^{-10} deg -3.1×10^{-12} rad
6.3.2	Eccentric anomaly	$E = E + \Delta E$, repeat last 3 equations until $\Delta E \simeq 0$	75.001°, 1.3092 rad
6.3.3	Position (in orbit)	$x_\omega = a(\cos E - e)$ $y_\omega = a\sqrt{1 - e^2} \sin E$	10,897 km 40,730 km
6.3.0	Radius	$r = \sqrt{x_\omega^2 + y_\omega^2}$	42,162 km
6.3.3	Velocity (in orbit)	$\dot{x}_\omega = -\sqrt{\mu a}(\sin E)/r$ $\dot{y}_\omega = \dfrac{\sqrt{\mu a(1 - e^2)}(\cos E)}{r}$	-2.9702 km/sec 0.79525 km/sec
6.4.2	Perigee unit vector	$P_x = \cos \omega \cos \Omega$ $\quad - \sin \omega \sin \Omega \cos i$ $P_y = \cos \omega \sin \Omega$ $\quad + \sin \omega \cos \Omega \cos i$ $P_z = \sin \omega \sin i$	-0.64222 -0.76637 0.01513
6.4.2	Parameter unit vector	$Q_x = -\sin \omega \cos \Omega$ $\quad - \cos \omega \sin \Omega \cos i$ $Q_y = -\sin \omega \sin \Omega$ $\quad + \cos \omega \cos \Omega \cos i$ $Q_z = \cos \omega \sin i$	0.76280 -0.64091 -0.08582
6.4.2	Position	$\mathbf{r} = x_\omega \mathbf{P} + y_\omega \mathbf{Q}$	24,071; $-34,455$; $-3,331$ km
6.4.2	Velocity	$\dot{\mathbf{r}} = \dot{x}_\omega \mathbf{P} + \dot{y}_\omega \mathbf{Q}$	2.5141, 1.7666, -0.11321 km/sec
6.1.1	Right ascension	$\alpha = \tan^{-1}(y/x)$	$-55.06°$, -0.96101 rad
6.1.1	Declination	$\delta = \tan^{-1}(z/\sqrt{x^2 + y^2})$	$-4.53°$, -0.079087 rad

[a]Example input: $\mu = 398,600.5$ km^2/sec^3, $a = 42,164.57$ km, $e = 0.002$, $i = 5°$, $\Omega = 60°$, $\omega = 170°$, $M_0 = 75°$, $t_0 = 0$, $t = 0$.

FIGURE 6.30 Summary of equations to obtain satellite position and velocity from orbit elements.

value of a sample calculation. Kepler's equation must be solved "the hard way," with a series expansion or an iterative process; Newton's approximation is outlined and converges rapidly for small eccentricities.

The equations to calculate the satellite position and velocity from the classical orbital elements are also shown as a FORTRAN program in Figure 6.31. The required input are the orbital elements: mean anomaly (MA), semimajor axis (SMA), eccentricity (ECC), inclination (INC), right ascension of the ascending node (NODE), and argument of perigee (PER). The outputs include the satellite's

```
      SUBROUTINE ORBIT(SMA,ECC,INC,NODE,PER,MA,X,Y,Z,XDOT,YDOT,ZDOT)
C SEP 79 - POSITION AND VELOCITY FROM ORBIT ELEMENTS, G. D. GORDON
      REAL INC, NODE, MA, MU, N, MSTAR
      D = ATAN(1.)/45.
      MU = 398600.5
C                                               SOLUTION OF KEPLER'S EQUATION
      RD = 57.2957795130823
      EA = MA +(ECC*SIN(D*MA) + (ECC*ECC/2.)*SIN(D*2.*MA))*RD
         DO 10 K = 1, 10
         MSTAR = EA - ECC*SIN(D*EA)*RD
         DELTA = (MA - MSTAR)/(1. - ECC*COS(D*EA))
         EA = EA + DELTA
         IF(ABS(DELTA) .LT. 1.E-8) GO TO 20
   10    CONTINUE
C                                         POSITION AND VELOCITY IN THE ELLIPSE
   20 XW = SMA*(COS(D*EA) - ECC)
      YW = SMA*(SQRT(1. - ECC*ECC))*SIN(D*EA)
      R = SQRT(XW**2 + YW**2)
      XWDOT = - SQRT(MU*SMA)*SIN(D*EA)/R
      YWDOT = SQRT(MU*SMA*(1. - ECC**2))*COS(D*EA)/R
C ORBIT ORIENTATION: P = PERIGEE UNIT VECT ; Q = PARAMETER UNIT VECT
      PX = COS(D*PER)*COS(D*NODE) - SIN(D*PER)*SIN(D*NODE)*COS(D*INC)
      PY = COS(D*PER)*SIN(D*NODE) + SIN(D*PER)*COS(D*NODE)*COS(D*INC)
      PZ = SIN(D*PER)*SIN(D*INC)
      QX =-SIN(D*PER)*COS(D*NODE) - COS(D*PER)*SIN(D*NODE)*COS(D*INC)
      QY =-SIN(D*PER)*SIN(D*NODE) + COS(D*PER)*COS(D*NODE)*COS(D*INC)
      QZ = COS(D*PER)*SIN(D*INC)
C            SATELLITE POSITION AND VELOCITY
      X = XW*PX + YW*QX
      Y = XW*PY + YW*QY
      Z = XW*PZ + YW*QZ
      XDOT = XWDOT*PX + YWDOT*QX
      YDOT = XWDOT*PY + YWDOT*QY
      ZDOT = XWDOT*PZ + YWDOT*QZ
C FOR RIGHT ASCENSION AND DECLINATION, INCLUDE IN SUBROUTINE CALL
      ALPHA = ATAN2(Y,X)/D
      DECL = ATAN(Z/SQRT(X*X + Y*Y))/D
      RETURN
      END
```

FIGURE 6.31 FORTRAN program to calculate satellite position and velocity from orbit elements.

position in rectangular coordinates (X, Y, Z), velocity (XDOT, YDOT, ZDOT), right ascension (ALPHA), declination (DECL), and radius (R).

In the FORTRAN program, the four orbit elements (inclination INC, right ascension of the ascending node NODE, argument of perigee PER, and mean anomaly MA) are always in degrees. This means that Kepler's equation must have the conversion factor $RD = 180/\pi$. In addition to that, the program is written so the computer trigonometric functions can be either in degrees or in radians. If they are in degrees, the first executable statement results in $D = 1$, which has no effect in the rest of the program (and could be deleted). If the computer trigonometric functions are in radians, the value of D will be $\pi/180$ and angles will be converted to radians before they are used in trigonometric functions.

6.5 ORBITAL ELEMENTS FROM POSITION AND VELOCITY

Given the position and velocity of a satellite at a specified time, the orbital elements can be determined. This problem in celestial mechanics has been solved in a variety of ways: The set of equations given here has been optimized for computer solutions of general problems and follows closely the set given by Herrick (1971). Note that if the eccentricity or inclination are identically zero, some elements become indeterminate; however, this is a deficiency with the definition of the classical elements and would occur with any method of solution. For orbits close to the geostationary orbit, other methods of solution might be better, but this set has wide applicability.

Previously a set of equations was presented to determine position and velocity from the orbital elements. Note that there are six components of position and velocity as well as six orbital elements. Either set of six numbers determines the orbit. But it is not a simple matter to reverse one set of equations to determine the second set. The two sets of equations shown here use different intermediate quantities.

In this section the orbital elements are determined from the position and velocity at a single time. In most practical cases, orbit determinations are made from many observations that do not necessarily include both position and velocity. Astronomical observations frequently have only the direction of the object (right ascension and declination), and the orbit must be determined from a series of these observations. For a communications satellite both the range and direction can be measured (with varying degrees of accuracy) and the orbit determined from a series of observations. The set of equations presented here can be used to determine a new orbit after a planned change in velocity (thruster maneuver).

6.5.1 Mathematical Formulation of Equations

To calculate the satellite orbit elements it is useful to define a quantity D and an eccentricity vector $\mathbf{e}$. The quantity D is defined $\mathbf{r} \cdot \dot{\mathbf{r}}/\mu$ and originated in parabolic orbits. The D defined here differs by $\mu^{1/2}$ from the parabolic anomaly (D) defined in the literature. In terms of scalars, $D = r \cdot \dot{r}/\mu$, where $\dot{r}$ is the range rate and not the magnitude of the velocity, $\dot{s}$. The quantity D can also be written as

$$D = \frac{a^{1/2}}{\mu^{1/2}} e \sin E = \frac{ey_\omega}{(p\mu)^{1/2}} \qquad (6.54)$$

in terms of various orbit elements and the variables of eccentric anomaly E and the distance of the satellite from the major axis of the ellipse, y_ω. As defined here, the units of D are in reciprocal velocity units, or seconds per kilometer.

The rate of change of the quantity D can be written as

$$\dot{D} = \frac{1}{r} e \cos E = \frac{\dot{s}^2}{\mu} - \frac{1}{r} = \frac{1}{r} - \frac{1}{a} \qquad (6.55)$$

where the time-varying quantities are the eccentric anomaly E, the radius r, and the satellite velocity $\dot{s}$. It may be noted that the quantity D becomes zero at perigee and apogee, when the satellite crosses the major axis of the ellipse. The time derivative $\dot{D}$ becomes zero when the satellite crosses the minor axis.

The eccentricity vector $\mathbf{e}$ shown in Figure 6.32 is a vector that points from the earth's center to the perigee (the minimum radius) and has a magnitude equal to the eccentricity e. It is a constant of the motion and points in the same direction as the unit vector $\mathbf{P}$. The eccentricity vector is

$$\mathbf{e} = \dot{D}\mathbf{r} - D\dot{\mathbf{r}} \qquad (6.56)$$

and even though the quantity D and the radius vector $\mathbf{r}$ vary with time, the eccentricity vector remains constant.

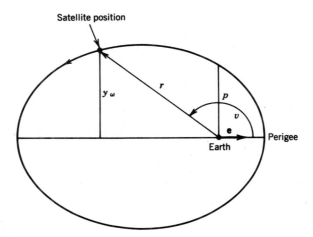

FIGURE 6.32 Eccentricity vector $\mathbf{e}$ in direction of perigee, radius vector $\mathbf{r}$, and satellite position.

6.5.2 Classical Orbital Elements

The set of equations shown in Figure 6.33 is sufficient to calculate the six classical orbital elements, starting with the position (x, y, z) and the velocity $(\dot{x}, \dot{y}, \dot{z})$ of the satellite. With each equation is included the name of the quantity being calculated, a reference to where the equation is described, and the calculated result for a numerical example. The latter is the same orbit as used in the set of equations to calculate the position and velocity from the orbit elements. The first half of the equations make use of the equations for energy and angular momentum per unit

Section Number	Name	Equation	Example[a]
	Radius, magnitude	$r_0 = \|\mathbf{r}_0\| = \sqrt{x_0^2 + y_0^2 + z_0^2}$	42162 km
	Velocity, magnitude	$\dot{s}_0 = \|\dot{\mathbf{r}}\| = \sqrt{\dot{x}_0^2 + \dot{y}_0^2 + \dot{z}_0^2}$	3.0748 km/sec
6.3.1	Semimajor axis	$\dfrac{1}{a} = \dfrac{2}{r_0} - \dfrac{\dot{s}_0^2}{\mu}$	2.3717×10^{-5} km^{-1}
6.5.2	Angular momentum	$\mathbf{h} = \mathbf{r}_0 \times \dot{\mathbf{r}}_0$	$9{,}785;\ -5{,}650;\ 129{,}148$ km^2/sec
	Angular momentum, magnitude	$h = \|\mathbf{h}\|$	129,641 km^2/sec
6.2.1	Inclination	$i = \tan^{-1}(\sqrt{h_x^2 + h_y^2}/h_z)$	$5°;\ 0.087266$ rad
6.2.1	Right ascension of ascending node	$\Omega = \tan^{-1}(h_x/-h_y)$	$60°;\ 1.0472$ rad
6.5.1	Parabolic anomaly[b]	$D_0 = (\mathbf{r}_0 \cdot \dot{\mathbf{r}}_0)/\mu$	6.2835×10^{-5} sec/km
6.5.1	Parabolic anomaly time derivative	$\dot{D}_0 = 1/r_0 - 1/a$	1.2268×10^{-9} km^{-1}
6.5.1	Eccentricity vector	$\mathbf{e} = \dot{D}_0 \mathbf{r}_0 - D_0 \dot{\mathbf{r}}_0$	$-1.2844 \times 10^{-4},$ $-1.5327 \times 10^{-4},$ 0.0303×10^{-4}
6.5.1		$e \sin E_0 = D_0 \sqrt{\mu/a}$	1.9320×10^{-4}
6.3.1		$e \cos E_0 = 1 - r_0/a$	0.5172×10^{-4}
	Eccentricity	$e = \sqrt{(e \sin E_0)^2 + (e \cos E_0)^2}$	0.0002
	Eccentric anomaly	$E_0 = \tan^{-1}(e \sin E_0/e \cos E_0)$	$75.012°;\ 1.3092$ rad
6.3.2	Mean anomaly	$M_0 = E_0 - e \sin E_0$	$75°;\ 1.3090$ rad
6.4.2	Argument of perigee	$\omega = \tan^{-1}[e_z h/(e_y h_x - e_x h_y)]$	$170°;\ 2.9670$ rad

[a]Input for example: $\mathbf{r}_0 = 24{,}070.7459,\ -34{,}455.3428,\ -3{,}331.0039$ km; $\dot{\mathbf{r}}_0 = 2.5141112,\ 1.76659356,\ -0.113209229$ km/sec, $\mu = 398{,}600.5$ km^3/sec^2.

[b]Quantity D is parabolic anomaly (as defined in the literature) divided by $\sqrt{\mu}$.

FIGURE 6.33 Summary of equations to obtain orbit elements from satellite position and velocity.

mass; the second half of the equations uses the parabolic anomaly and the eccentricity vector. In the equations, a zero subscript denotes the instantaneous value of a quantity that changes as the satellite moves in the orbit. A FORTRAN program to calculate the classical orbital elements is shown in Figure 6.34.

The semimajor axis can be calculated from the magnitude of the radius and velocity using the energy equation (vis viva equation). While in most cases the semimajor axis could be calculated from its reciprocal ($1/a$), some experts advise using the reciprocal in computer calculations; for a parabolic orbit, the reciprocal semimajor axis is infinite, and using the reciprocal eliminates a potential problem.

The angular momentum per unit mass, **h**, is obtained from the cross product of the radius and the velocity. This vector is in the direction of the orbit normal and also defines the orbit plane. From this vector the inclination i and the right ascension of the ascending node, Ω, are calculated.

To find the other three elements, we first determine the parabolic anomaly D

```
      SUBROUTINE ELEMENT(SMA,ECC,INC,NODE,PER,MA,X,Y,Z,XDOT,YDOT,ZDOT)
C     MAR 81, ORBIT ELEMENTS FROM POSITION AND VELOCITY, G.D. GORDON
      REAL INC, INVA, MA, MU, NODE
      RD = 57.2957795130823
      DG = ATAN(1.)/45.
      MU = 398600.5
C     SAT RADIUS (KM), SAT VELOCITY (KM/SEC), 1/SEMI-MAJ AXIS (1/KM)
      R = SQRT(X*X + Y*Y + Z*Z)
      SDOT = SQRT(XDOT**2 + YDOT**2 + ZDOT**2)
      INVA = 2./R - SDOT**2/MU
C     ANGULAR MOMENTUM (KM**2/SEC)
      HX = Y*ZDOT - Z*YDOT
      HY = Z*XDOT - X*ZDOT
      HZ = X*YDOT - Y*XDOT
      H = SQRT(HX*HX + HY*HY + HZ*HZ)
C     RIGHT ASC. OF ASC. NODE, INCLINAT., PARABOLIC ANOMALY (SEC/KM)
      INC = ATAN(SQRT(HX*HX + HY*HY)/HZ)/DG
      NODE = ATAN2(HX, -HY)/DG
      D = (X*XDOT + Y*YDOT + Z*ZDOT)/MU
      DDOT = 1./R - INVA
C     ECCENTRICITY VECTOR AND MAGNITUDE
      EX = DDOT*X - D*XDOT
      EY = DDOT*Y - D*YDOT
      EZ = DDOT*Z - D*ZDOT
      ESINEA = D*SQRT(MU*INVA)
      ECOSEA = 1. - R*INVA
      ECC = SQRT(ESINEA**2 + ECOSEA**2)
C     ECCENTRIC ANOMALY, MEAN ANOMALY, ARG. OF PERIGEE, SEMI-MAJ AXIS (KM)
      EA = ATAN2(ESINEA, ECOSEA)/DG
      MA = EA - ECC*SIN(DG*EA)*RD
      PER = ATAN2(EZ*H, EY*HX - EX*HY)/DG
      SMA = 1./INVA
      RETURN
      END
```

FIGURE 6.34 FORTRAN program to calculate orbit elements from satellite position and velocity.

and the eccentricity vector **e**. The parabolic anomaly defined here differs from that in the literature by a factor of $\mu^{1/2}$, and this simplifies some equations. The quantity defined here is in units of inverse velocity (sec/km), whereas the quantity in the literature is in units of the square root of a length ($km^{1/2}$). The first equation for D_0 is simply the original definition. The time derivative is the difference between the inverse radius and the inverse semimajor axis. Knowing these last two quantities, it is possible to calculate the eccentricity vector **e** and then the argument of perigee, ω. The eccentric anomaly E is found by first calculating $e \sin E_0$ and $e \cos E_0$ and taking the ratio.[†] The orbit eccentricity e can also be found from these two quantities. The mean anomaly M_0 is found from the eccentric anomaly by solving Kepler's equation the easy way.

The necessary inputs to the FORTRAN program are the satellite position (X, Y, Z) and the satellite velocity (XDOT, YDOT, ZDOT). The orbit inclination is found using an arctangent function, but if an arccosine function is available, it can be found as $\cos^{-1}(h_z/h)$. The calculated elements consist of the semimajor axis, SMA, the orbit inclination, INC, the right ascension of the ascending node, NODE, the orbit eccentricity, ECC, the mean anomaly, MA, and the argument of perigee, PER.

6.6 EARTH STATION: AZIMUTH, ELEVATION, AND RANGE

As the earth rotates in space, there is a changing relation between an earth station and the inertial coordinate system normally used in defining the position of the satellite. One method of determining the direction of a satellite from an earth station is to first determine the location of the satellite in earth coordinates. Most important is the determination of the satellite's longitude: Once the satellite's longitude, latitude, and radius are known, its relation to the earth station can be determined from geometry.

6.6.1 Longitude of Satellite

The inertial coordinate system used for satellite orbits and the earth's coordinate system are quite similar (see Figure 6.35). Both use the earth's center as the origin and the earth's spin axis as the z axis. The difference is in the x-axis locations: The inertial coordinate system uses the first point of Aries (the vernal equinox) for zero right ascension, and in the earth's coordinate system the Greenwich meridian is used for zero longitude. The basic relation between the two is the angle between the two axes, that is, the angle between the first point of Aries and the Greenwich meridian. This angle is called the Greenwich hour angle or the Greenwich sidereal time and is frequently given in hours or in seconds.

The Greenwich hour angle (GHA) is a constantly changing angle as the earth

---

†Note that **e** is the eccentricity vector, and e is the scalar value of eccentricity.

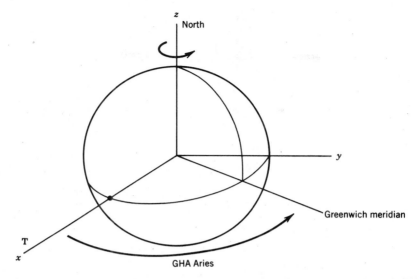

FIGURE 6.35 Greenwich hour angle (GHA) is angle from first point of Aries T to Greenwich meridian, or time since Greenwich meridian passed first point of Aries.

rotates in inertial space and can be written as

$$\text{GHA} = 100.4602346 + 0.985647348\,(\text{JDO} - 2451545) + 15.041068\,\text{UT}$$

$$(6.58)$$

where JDO is the Julian day at midnight; that is, it must be a half integer. Values of JDO − 2,451,545 are tabulated in Tables 6.6 and 6.7 on pages 799–800. The universal time (UT) equivalent to Greenwich mean time (GMT) must be in decimal hours. The result of the equation will be the desired Greenwich hour angle in degrees.

As an example, take midnight at the start of September 28, 1989. The angle between the first point of Aries and the Greenwich meridian is 100.4602346 + 0.985647348 (−3775.5 + 28) + 0, which is 6.7468°. It is usually necessary to add or subtract multiples of 360° to obtain the angle in a convenient range. Accuracy of the preceding formula is about ±0.01°.†

Longitude is the angle from the Greenwich meridian to the meridian of an object or place (see Figure 6.36). East longitudes are positive, and west longitudes are negative. To find the longitude (λ) of an object such as a satellite, from the right ascension α subtract the Greenwich hour angle:

$$\lambda = \alpha - \text{GHA} \qquad (6.59)$$

†For communications satellites, right ascension of the satellite (and the corresponding longitude of the subsatellite point) is usually expressed in degrees. It should be noted that the right ascension of an astronomical object is expressed in hours—usually in hour–minute–second format and sometimes in decimal hours, where 1 hr corresponds to 15°.

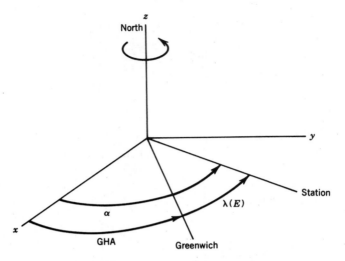

FIGURE 6.36 Relation between Greenwich hour angle (GHA), east longitude of earth station λ, and the right ascension of earth station α.

Note that when α is less than GHA, the longitude of the satellite is negative, and therefore λ is west longitude. When α is greater than GHA, the satellite λ is east longitude. Conversely, if the longitude of a place (earth station) is known, its right ascension can be found by adding the longitude (east) to the GHA. It may be necessary to add or subtract multiples of 360° to obtain a result in a proper range.

The relation between right ascension and longitude cannot be determined exactly because the rotation of the earth is not a mathematical constant. There is a nutation that introduces a periodic variation and also a slowing down of the earth's rotation due primarily to tidal friction. Fortunately, these effects are small and do not concern most users of communications satellites. For more accurate values, see the Astronomical Almanac.

The right ascension of a satellite can be obtained from the orbital elements with the equations shown in Figure 6.30 in Section 6.4.2. Alternatively, it can be obtained from

$$E = M + e \sin M + \frac{e^2}{2} \sin 2M + \cdots$$

$$v = \tan^{-1} \frac{(1 - e^2)^{1/2} \sin E}{\cos E - e}$$

$$\alpha = \Omega + \tan^{-1} \frac{\cos i \sin (\omega + v)}{\cos (\omega + v)} \qquad (6.60)$$

The first equation is a series solution of Kepler's equation for small eccentricities, which provides the eccentric anomaly E from the mean anomaly M. The true anomaly v is calculated in the second equation. Finally, the right ascension of the satellite, α, is calculated from the right ascension of the ascending node, Ω, the orbit inclination i, and the argument of perigee, ω.

In geostationary orbits the inclination and the eccentricity are small. The right ascension of the satellite is approximately the sum of the right ascension of the ascending node, the argument of perigee, and the true anomaly (approximately the mean anomaly), as shown in Figure 6.37. Including a few more terms in a series expansion, the satellite longitude can be written as

$$\lambda(E) = \left[\Omega + \omega + M + 2e \sin M + \tfrac{5}{4} e^2 \sin 2M \right.$$

$$\left. - \tfrac{1}{4}i^2 \sin 2(\omega + M) + \cdots \right] - \text{GHA} \qquad (6.61)$$

Note that angles must be in radians. There are daily fluctuations in the satellite longitude due to eccentricity and other variables. The mean longitude for a 24-hr period is a more significant parameter for some purposes. The change of this mean longitude from one day to the next provides a measure of the satellite drift rate.

As an example, suppose a satellite orbit has an eccentricity of 0.0001231, an inclination of 0.07281°, a right ascension of the ascending node of 247.0636°, an argument of perigee of 353.833°, and a mean anomaly of 357.4509°. These elements are for the start of the day (GMT) of September 28, 1988. Then the GHA is 6.9855°, the right ascension is 238.35°, and the east longitude is 231.36°E.

Locations on earth are usually specified by their latitude (degrees from the

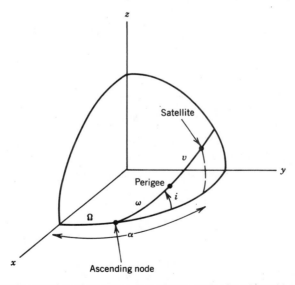

FIGURE 6.37 Right ascension of satellite as sum of three angles for orbits with small inclination.

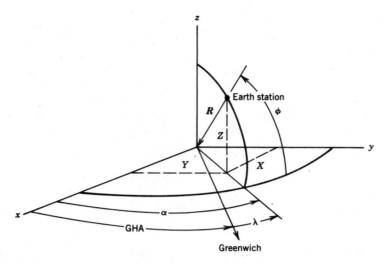

FIGURE 6.38 Location of earth station in rectangular coordinates (X, Y, Z) and in spherical coordinates (R, λ, φ).

equator) and their longitude (degrees from Greenwich meridian). North latitudes and east longitudes are customarily designated as positive, and south latitudes and west longitudes are negative. To relate earth stations to satellites, it is necessary to specify their locations in the same coordinate system. While in most of this section the satellite's location is calculated in the earth's coordinate system, the earth station's location in an inertial system is described here.

The earth station position (which is moving in the inertial coordinate system) is defined by the vector **R** from the station to the center of the earth (See Figure 6.38). The instantaneous coordinates of the earth station (X, Y, Z) are given in terms of the latitude φ and longitude λ as

$$X = -(C + H) \cos \phi \cos \alpha$$
$$Y = -(C + H) \cos \phi \sin \alpha$$
$$Z = -(S + H) \sin \phi \qquad (6.62)$$

where $\alpha = \lambda + \text{GHA} \times 15° \text{ hr}^{-1}$ (λ is the station longitude) and GHA is the Greenwich hour angle.† The angle α changes at the earth's rotation rate.

The earth station velocity coordinates (Ẋ, Ẏ, Ż) are

$$\dot{X} = \dot{\theta}(C + H) \cos \phi \sin \alpha$$
$$\dot{Y} = -\dot{\theta}(C + H) \cos \phi \cos \alpha$$
$$\dot{Z} = 0 \qquad (6.63)$$

†Since α and φ (and λ) are generally expressed in degrees for these calculations, GHA must also be expressed in degrees. The term $15° \text{ hr}^{-1}$ is used only when GHA is expressed in hour-angle units, since there are 15° of earth rotation per hour of hour-angle measurement.

where ϕ is the geodetic latitude and $\dot{\theta}$ the rotational rate of the earth ($2\pi/$ 86164.091 sec^{-1}). The quantities in parentheses are approximately the earth radius R_e, or more accurately

$$C \cong R_e\left[1 - (2f - f^2)\sin^2\phi\right]^{-1/2}$$
$$S \cong C(1 - f)^2 \quad f = 1/298.25$$
$$H = \text{height above sea level} \qquad (6.64)$$

and the equatorial radius R_e is 6378.140 km.

6.6.2 Azimuth and Elevation

The elevation h is the angle between an extraterrestrial object (e.g., star or satellite) and the observer's horizon plane. The azimuth A is the direction of the object measured in the horizontal plane from geographic north in a clockwise direction (eastward). In the case of a star, by plotting the north pole, the star's position, and the observer's zenith (point directly overhead), the desired quantities can be calculated by spherical trigonometry (see Figure 6.39).

The difference in longitude, $\Delta\lambda$, is

$$\Delta\lambda = \alpha - \theta = \lambda(E)_{\text{obj}} - \lambda(E)_{\text{sta}} \qquad (6.65)$$

where α is the right ascension of the star and θ is the right ascension of the station. The difference can also be the difference in longitude (i.e., the east longitude of the station minus the east longitude of the object).

The elevation of a star, h_∞, can be found from the law of cosines applied to the spherical triangle in Figure 6.39:

$$\sin h_\infty = \sin\delta\sin\phi + \cos\delta\cos\phi\cos\Delta\lambda \qquad (6.66)$$

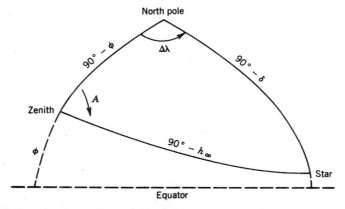

FIGURE 6.39 Azimuth A and elevation of star h_∞ as seen from earth station.

where δ is the star's declination and ϕ the observer's latitude. The subscript on h_∞ emphasizes that this equation applies to objects at infinity.

The azimuth A can then be calculated as

$$\tan A = \frac{\sin \Delta\lambda}{\cos \phi \tan \delta - \sin \phi \cos \Delta\lambda} \tag{6.67}$$

If the denominator is negative, the angle is in the second or third quadrant ($90°$ $< A < 270°$), that is, southward. The use of the ATAN2 function in computer programs will yield the correct quadrant.

6.6.2.1 Parallax

For objects that are not infinitely far away, such as communications satellites, the size of the earth causes a shift in the apparent position of objects; this shift is called parallax. This does not affect the azimuth, but it decreases the apparent elevation. That is, the elevation from the surface of the earth is less than the elevation calculated for the center of the earth (see Figure 6.40). Additional equations for this figure can be found in Figures 3.23, 3.24 and in Section 5.7.1. The change in angle is smallest for objects near the zenith (straight up) and increases to a maximum at the horizon.

The observed elevation, h (i.e., above the observer's horizontal) can be written as

$$\tan h = \frac{\sin h_\infty - R/r}{\cos h_\infty} \tag{6.68}$$

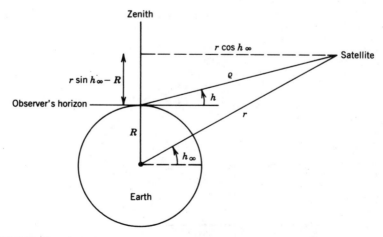

FIGURE 6.40 Elevation of satellite, h, from earth station and relation to theoretical elevation h_∞ from earth's center.

TABLE 6.8 Parallax Correction for Elevation of Geostationary Satellite

Elevation from earth's center h_∞ (deg)	True elevation h (deg)	Difference $h_\infty - h$ (deg)
90	90.00	0.00
80	78.23	1.77
70	66.55	3.45
60	55.03	4.97
50	43.72	6.28
40	32.69	7.31
30	21.93	8.07
20	11.47	8.53
10	1.30	8.70
0	−8.60	8.60

where h_∞ is the calculated elevation (from earth's center), r is the distance to the object (from earth's center), and R is the radius of the earth. For communications satellites in geostationary orbit, $R/r = 0.15127$, and the relation between the two angles is shown in Table 6.8. Around 60° there is a 5° difference, and at the horizon it is 8.7°.

If the observed elevation h is known, the theoretical elevation from the center of the earth h_∞ is more easily calculated with the equation

$$\sin\left(h_\infty - h\right) = \frac{R}{r}\cos h \qquad (6.69)$$

For a user at the north pole, the preceding equations will show the communications satellite in geostationary orbit to be 8.6° below the horizon and therefore not usable. Even at 80° latitude, the maximum elevation is only 1.3°, normally not considered a usable elevation. This shows that from 70° to 80° latitude use of geostationary satellites is difficult, and from 80° to 90° it is impossible.

6.6.2.2 Summary

A summary of equations is shown in Figure 6.41 to calculate the longitudal difference $\Delta\lambda$, the elevation without parallax correction (h_∞), the actual elevation h, and the azimuth A. The general equations are on the left. Many communications satellites in geostationary orbit have a very small declination δ; if the declination is assumed zero, the equations simplify to the expressions on the right.

A graph of the azimuth and elevation for geostationary satellites is shown in Figure 6.42. The azimuth is shown as a function of elevation angle, and lines of

Any Satellite	Geostationary Satellite ($\delta = 0$)
Longitude difference	Longitude difference
$$\Delta\lambda = \lambda(E)_{\text{sat}} - \lambda(E)_{\text{sta}}$$	$$\Delta\lambda = \lambda(E)_{\text{sat}} - \lambda(E)_{\text{sta}}$$
Star elevation	Star elevation
$$\sin h_\infty = \sin\delta\sin\phi + \cos\delta\cos\phi\cos\Delta\lambda$$	$$\sin h_\infty = \cos\phi\cos\Delta\lambda$$
True elevation	True elevation
$$\tan h = \frac{\sin h_\infty - R/r}{\cos h_\infty}$$	$$\tan h = \frac{\sin h_\infty - R/r}{\cos h_\infty}$$
Azimuth	Azimuth
$$\tan A = \frac{\sin\Delta\lambda}{\cos\phi\tan\delta - \sin\phi\cos\Delta\lambda}$$	$$\tan A = \frac{\sin\Delta\lambda}{-\sin\phi\cos\Delta\lambda}$$

FIGURE 6.41 Equations for calculating azimuth and elevation of satellite. Simplified equations for geostationary satellite (0° latitude).

contour mark the longitudinal difference (difference in longitudes between the earth station and the satellite) and the earth station latitude.

To use Figure 6.42, find the intersection of the two appropriate contour lines and read off the elevation and azimuth from the proper scales. For azimuth there are four different scales, labeled with the direction from the station location to the satellite. As an example, assume a satellite is at 263° E, and a station is at Washington, DC (39° N, 283° E). The longitudinal difference is 20°, and the latitude is 39°. The elevation read from the figure is 40°. The direction from Washington to the satellite is southwest, so the azimuth angle is 210° (measured clockwise from geographic north).

6.6.3 Range and Range Rate

The range is the distance from the satellite to some point on earth. From Figure 6.40, the range ρ can be written as

$$\rho = \sqrt{r^2 - R^2\cos^2 h} - R\sin h \tag{6.70}$$

in terms of the distance from satellite to earth's center r, the radius of the earth R, and the actual elevation h. The time delay τ for signals relayed through a satellite is equal to $(\rho_1 + \rho_2)/c$, the sum of the two ranges divided by the velocity of light ($c = 299{,}792$ km/sec), plus any delay because of equipment circuitry. The Doppler shift of frequencies relayed through the satellite is equal to $f_2 - f_1 = -(\dot\rho_1 + \dot\rho_2)/c$, the sum of the range rates divided by the velocity of light.

If the satellite was in a theoretical geostationary orbit, the range would be

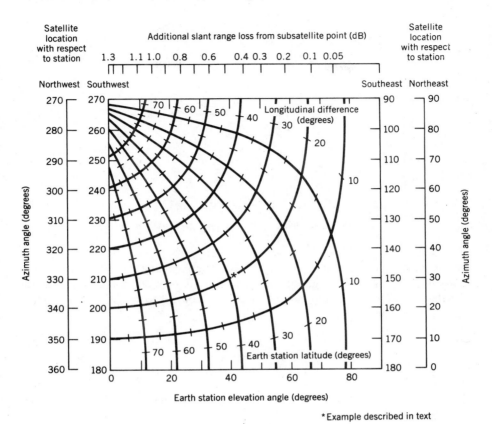

FIGURE 6.42 Graph of azimuth and elevation of geostationary satellite (0° latitude).

constant, and the range rate would be zero. Such an orbit would require constant thrusting, which is never done. If a satellite is in an orbit close to geostationary, the eccentricity, inclination, and drift rate are small. It is useful to determine the effect of small values of these parameters so that estimates of range variation and range rate may be made. This is done in the following paragraphs.

6.6.3.1 Eccentricity

A satellite in an elliptical orbit has a radius variation of $\pm ea$, where e is the eccentricty and a the semimajor axis (see Figure 6.43). When the earth station is not in the orbit plane, the range is

$$\rho = \rho_m - \frac{ea^2}{\rho_m} \cos nt_p \qquad (6.71)$$

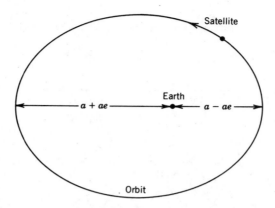

FIGURE 6.43 Effect of orbit eccentricity on satellite range.

where ρ_m is the mean range, t_p is the time from perigee, and n is the orbital angular velocity. The range for a geostationary satellite is between 35,786 km (satellite overhead) and 41,679 km (satellite at the horizon). The range rate is the time derivative of the range, or

$$\dot{\rho} = \frac{ea^2 n}{\rho_m} \sin nt_p \qquad (6.72)$$

For an eccentricity of 0.001, the cyclical range variation is about ± 42 km, and the cyclical range rate is about ± 0.003 km/sec.

The orbit eccentricity is related to motion toward and away from the earth. The orbit inclination defines a north–south motion and the drift rate an east–west motion. For an earth station located at a subsatellite point, the effect of inclination and drift rate on the range is negligible. However, for an earth station not on the equator, the orbit inclination will produce a daily oscillation in the range. And if the earth station is not at the satellite's longitude, the orbit drift rate will produce a steady drift in range.†

6.6.3.2 Inclination

The orbit inclination defines a motion with an amplitude of $\pm ia$, where i is the inclination in radians and a is the semimajor axis (see Figure 6.44). Taking the component of this motion along the range results in a range of

$$\rho = \rho_m - \frac{iaR}{\rho_m} \sin \phi \sin nt_i \qquad (6.73)$$

---

†In practice, inclination and orbit drift rate are usually corrected by satellite maneuvers.

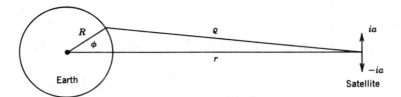

FIGURE 6.44 Effect of orbit inclination on satellite range.

where ρ_m is the mean range, i the inclination, a the semimajor axis, R the earth's radius, ϕ the latitude of the earth station, and n the orbital angular velocity. The time t_i is measured starting with the satellite at the ascending node. The range rate is obtained by differentiating the equation for the range and is

$$\dot{\rho} = -\frac{iaRn}{\rho_m} \sin \phi \cos nt_i \tag{6.74}$$

An orbit inclination of $1°$ results in a range variation of ± 110 km and a range rate of ± 0.008 km/sec.

The drift rate defines a motion of the satellite of Dat_d, where D is the drift rate in radians per second, a the semimajor axis, and t_d the time from some arbitrary starting value. Taking the component along the range results in a range variation

$$\rho = \rho_m - \frac{DaRt_d}{\rho_m} \cos \phi \sin \Delta\lambda \tag{6.75}$$

where $\Delta\lambda$ is the difference in longitude between the satellite and the earth station. The range rate due to a small drift rate is

$$\dot{\rho} = \frac{DaR}{\rho_m} \cos \phi \sin \Delta\lambda \tag{6.76}$$

obtained by differentiation. A drift rate of $1°$ per day, or 2×10^{-7} rad/sec results in a range rate of 0.0013 km/sec.

The preceding equations are approximate and are presented to provide a physical understanding and for those interested in order-of-magnitude estimates of time delay and Doppler shift. More accurate equations for the range and range rate can be obtained from Slabinski (1974).

6.7 LUNAR AND SOLAR PERTURBATIONS

According to the simple two-body theory, a satellite moves in a perfect, unchanging ellipse. In practice, various forces produce perturbations. For a satellite in geostationary orbit, the principal causes of perturbations, in order of their importance, are the moon's mass, the sun's mass, the nonspherical earth, and the sun's radiation (solar pressure). The first two are quite similar, and the sun's effect will be considered first because in some ways the analysis is simpler than the corresponding analysis for the moon.

6.7.1 Perturbations due to Sun

The major effect of the sun on a satellite is often forgotten because it is so obvious. The sun's gravity keeps the satellite in an orbit around the sun, just as it keeps the earth in orbit. In any day a satellite in a geostationary orbit travels 260,000 km around the earth, but it travels 2,600,000 km as part of its annual trip around the sun. But when an earth-centered coordinate system is used, we expect the satellite to travel with the earth around the sun. Our only concern is that the sun produces a different acceleration on the satellite than it does on the earth.

When the satellite is closer to the sun (i.e., on the sun side of the earth), the gravitational attraction is greater, and hence the sun causes an extra acceleration toward the sun. Conversely, when the satellite is farther from the sun (i.e., on the night side of earth), the latter causes an acceleration away from the sun (compared to the earth). The magnitude of this net acceleration is about $3\mu_s r/r_s^3$, where μ_s is the sun's gravitational parameter ($132,712.438 \times 10^6$ km^3/sec^2), r_s is the distance from the sun to the earth center, and r is the component of the distance from the earth center to the satellite measured along the earth–sun line.

The same effect is used in gravity gradient satellites, where an object at the end of a boom toward the earth receives an extra attraction toward the earth. This gravity gradient of the earth's gravitational field also acts on the moon and keeps the same face pointed toward the earth.

The solar perturbation produces two effects on a satellite's orbit. It distorts the orbit from a true ellipse, shortening it along the earth–sun axis. And it produces a torque that produces a secular drift of the direction of angular momentum or the orbit normal. For orbits that start with 0° inclination, this latter effect causes an approximate 0.9°-yr^{-1} increase in inclination (due to sun and moon). Correction of this inclination is often the major need for propellant aboard communications satellites.

The daily average precession due to the sun on a geostationary orbit is

$$\dot{W}_p = \frac{3\mu_s a^2}{4hr_s^3} 2 \sin \delta_s \cos \delta_s \tag{6.77}$$

where μ_s is the sun's gravitational parameter, δ_s the sun's declination, a the orbit semimajor axis, h the satellite angular momentum per unit mass [$h^2 = \mu a(1 - e^2)$], and r_s the distance from the earth to the sun. This represents a precession of the orbit normal at right angles to the sun's direction and this precession is proportional to $2 \sin \delta_s \cos \delta_s$, or $\sin 2\delta_s$ (see Figure 6.45).

As the sun goes from vernal equinox ($\alpha_s = 0$ in March) to autumnal equinox

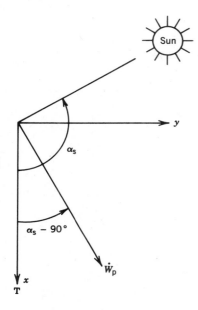

FIGURE 6.45 Precession of orbit normal due to sun's action. Precession $\dot{W}_p$ is at right angles to direction of sun.

($\alpha_s = 180°$ in September), the declination δ_s is positive, and there will always be a component of orbit precession toward Aries (x axis). From autumnal equinox to vernal equinox, the sun's declination is negative, the direction of the precession in the orbit plane perpendicular to the sun's direction is reversed, and there is still a component of precession toward the first point of Aries.

The approximate annual average motion of the orbit normal is

$$\dot{W}_x(\text{annual}) = \frac{3\,\mu_s a^2}{4\,hr_s^3} \sin \epsilon \cos \epsilon$$

$$\dot{W}_y(\text{annual}) = 0 \tag{6.78}$$

where ϵ is the angle between the ecliptic and the equator ($23.44°$). It is assumed that the sun's celestial longitude changes uniformly with time; that is, that the eccentricity of the earth's orbit is zero (actually it is 0.0167). The net change in the satellite orbit normal due to the sun is toward the first point of Aries (positive x axis), proportional to $\sin \epsilon \cos \epsilon$, or $\sin 2\epsilon$. For a solar gravitational parameter of $132,712 \times 10^6 \text{ km}^3/\text{sec}^2$, a satellite orbit radius of 42,164 km, an orbital angular momentum of 129,641 km²/sec, and a distance to the sun of 149,597,870 km, the numerical coefficient is 4×10^{-10} rad/sec, or $0.74°$ yr^{-1}, and the average annual precession due to the sun is $0.269°$ yr^{-1}.

The precession of the orbit normal due to the sun does not proceed at a uniform rate but is in a series of arcs, as shown in Figure 6.46. Each arc represents 6 months travel. The change in the orbit normal is greatest at the midpoint, when

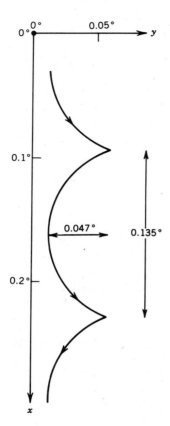

FIGURE 6.46 Precession of orbit normal, over a year, due only to solar perturbations.

the sun's declination is maximum, and slows to near zero at the ends of the arc, when the sun's declination is close to zero. The actual change in the orbit normal is tabulated in Table 6.9 at 2-week intervals throughout an average year.

The length of each arc is one-half the annual precession, or 0.135°. The width of each arc is 0.047°. This means that along the y axis there is a 6-month cycle of ±0.024° due to solar perturbation.

6.7.2 Perturbations due to Moon

The moon's gravitational attraction also creates a gravitational gradient field and a precession of the satellite orbit normal, just as the sun did. As is well known with tides, the moon's effect is larger than the sun. It is also more complicated because the moon's orbit normal changes with time, whereas the sun's (the ecliptic pole) remains fixed (within the accuracy of concern here).

The location of the lunar orbit normal is shown in relation to the ecliptic pole and the north pole in Figure 6.47. While the angle between the lunar orbit normal and the ecliptic pole is fixed, the lunar orbit normal rotates around the ecliptic pole

TABLE 6.9 Solar Effect on Orbit Normal of Geostationary Satellite, in 2-week Increments

Days		ΔW_x (deg)	ΔW_y (deg)
1	14	0.0188	0.0064
15	28	0.0150	0.0100
29	42	0.0101	0.0113
43	56	0.0052	0.0098
57	70	0.0016	0.0060
71	84	0.0000	0.0008
85	98	0.0009	−0.0045
99	112	0.0038	−0.0088
113	126	0.0083	−0.0110
127	140	0.0131	−0.0109
141	154	0.0173	−0.0083
155	168	0.0200	−0.0040
169	182	0.0206	0.0012
183	196	0.0190	0.0061
197	210	0.0156	0.0097
211	224	0.0110	0.0112
225	238	0.0063	0.0104
239	252	0.0024	0.0072
253	266	0.0003	0.0025
267	280	0.0003	−0.0029
281	294	0.0027	−0.0076
295	308	0.0068	−0.0106
309	322	0.0118	−0.0111
323	336	0.0165	−0.0090
337	350	0.0197	−0.0048
351	364	0.0207	0.0007

every 18.6 yr. The right ascension α_W and declination δ_W of the lunar orbit normal are

$$i_e = 5.1453964°$$

$$\epsilon = 23.439° - 0.0000004\,d$$

$$\Omega_m = 125.044484° - 0.05295378\,d$$

$$\alpha_W - 270° = \tan^{-1} \frac{2 \sin i_e \tan (\Omega_m/2)}{\sin (\epsilon + i_e) + \sin (\epsilon - i_e) \tan^2 (\Omega_m/2)}$$

$$\delta_W = \sin^{-1} (\cos i_e \cos \epsilon - \sin i_e \sin \epsilon \cos \Omega_m) \qquad (6.79)$$

where i_e is the angle between the lunar orbit normal and the ecliptic pole (or between the lunar orbit plane and the earth's orbit plane), ϵ is the obliquity of the ecliptic

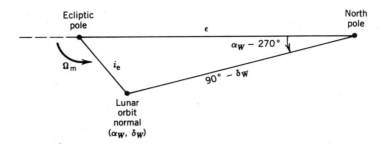

FIGURE 6.47 Position of lunar orbit normal with respect to ecliptic pole and north pole.

(angle between the earth's orbit plane and the equatorial plane), and Ω_m is the right ascension of the ascending node for the moon with respect to the ecliptic. These are written in terms of the time d in days with respect to the start of the year 2000 (see page 799).

The location of the lunar orbit normal for the start of various years is given in Table 6.10 and is illustrated in Figure 6.48. The angle between the lunar orbit normal and the north pole reached a maximum in November 1987 when the angle between the lunar orbit normal and the north pole was $23.45° + 5.15°$, or an angle of $28.6°$. On this year the lunar perturbations were maximum, and so were the satellite fuel requirements. A minimum occurs in February 1997. The period of this precession is 6798 days, or 18.6 yr, so another maximum does not occur until June 2006.

The angle between the moon's orbit normal and the ecliptic pole remains fixed, but the orbit normal precesses around the ecliptic pole. The reason for this precession is exactly the effect under discussion; the sun's gravitational gradient field produces a torque on the moon's orbital angular momentum. The precession of the satellite's orbit normal averaged over a lunar month is in the direction of the moon's ascending node. While this is in the general direction of the sun's ascending node (the x axis or First Point of Aries), it does vary by a number of degrees. This precession can be divided into two components along the x axis and the y axis

$$\dot{W}_x(\text{month}) = \frac{3\mu_m a^2}{4hr_m^3}\sin \delta_w \cos \delta_w(-\sin \alpha_w)$$

$$\dot{W}_y(\text{month}) = \frac{3\mu_m a^2}{4hr_m^3}\sin \delta_w \cos \delta_w \cos \alpha_w \qquad (6.80)$$

For a lunar gravitational constant μ_m of 4902.78 km^3/sec^2, a satellite orbit radius a of $42,164$ km, an orbital angular momentum h of $129,641$ km^2/sec, and a distance to the moon, r_m, of $384,400$ km, the numerical coefficient is 8.9×10^{10} rad/sec, or $1.6°$ yr^{-1}. This effect is added to the precession due to the solar perturbation of $0.269°$ yr^{-1}. Since δ_w varies from $71.7°$ to $61.4°$, the total precession in a year varies from $0.75°$ to $0.94°$.

TABLE 6.10 Lunar Orbit Normal Coordinates

Start of Year	Right Ascension (deg)	Declination (deg)
1985	279.43	63.30
1986	276.47	62.23
1987	273.07	61.59
1988	269.47	61.42
1989	265.89	61.73
1990	262.60	62.51
1991	259.82	63.69
1992	257.84	65.19
1993	256.98	66.88
1994	257.58	68.60
1995	259.88	70.12
1996	263.90	71.22
1997	269.13	71.70
1998	274.53	71.45
1999	279.01	70.53
2000	281.88	69.11
2001	283.00	67.43
2002	282.57	65.72
2003	280.91	64.14
2004	278.36	62.85
2005	275.19	61.93
2006	271.68	61.47
2007	268.07	61.49
2008	264.58	61.98
2009	261.44	62.93
2010	258.94	64.24

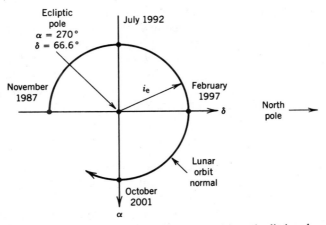

FIGURE 6.48 Precession of lunar orbit normal around ecliptic pole.

TABLE 6.11 Annual Lunar Effect on Orbit Normal of Geostationary Satellite

Year	ΔW_x (deg)	ΔW_y (deg)
1985	0.6466	0.0910
1986	0.6644	0.0560
1987	0.6729	0.0153
1988	0.6715	−.0270
1989	0.6603	−.0666
1990	0.6401	−.0993
1991	0.6129	−.1217
1992	0.5812	−.1308
1993	0.5483	−.1252
1994	0.5183	−.1050
1995	0.4949	−.0721
1996	0.4813	−.0303
1997	0.4792	0.0153
1998	0.4892	0.0589
1999	0.5096	0.0953
2000	0.5378	0.1200
2001	0.5702	0.1306
2002	0.6026	0.1263
2003	0.6317	0.1081
2004	0.6544	0.0784
2005	0.6688	0.0406
2006	0.6735	−.0012
2007	0.6683	−.0429
2008	0.6534	−.0802
2009	0.6302	−.1094
2010	0.6009	−.1269

The yearly change of a satellite orbit normal due to lunar perturbations is tabulated in Table 6.11 in terms of the rate toward the x axis and the y axis. The precession toward the y axis can be either positive or negative and averages zero over the 18.6 yr of a lunar cycle. This precession is also plotted in Figure 6.49. The $0.269° \ \mathrm{yr}^{-1}$ for the solar perturbation has been added to the graph. Thus, the lunar effect starts from the origin (0, 0), and the total lunar and solar effect is pictured as starting at the top point ($x = -0.269°$, $y = 0$).

6.7.3 Perturbations due to the Sun and Moon

An example of the combined effect of the sun and moon on the orbit normal is shown in Figure 6.50. North–south stationkeeping within 0.1° (the circle shown in the figure) ceased in November 1985 for the satellite. During 1986 the inclination increased 0.93°, and the satellite had daily north–south oscillations of this magnitude. Points in the figure are plotted at 1-week intervals. In March and

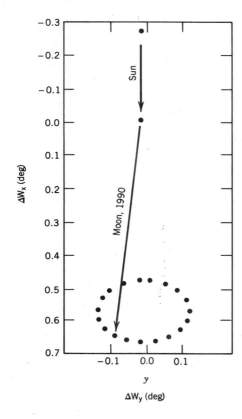

FIGURE 6.49 Annual effect on geostationary satellite's orbit normal of sun (constant) and moon (variable).

September, the sun's effect is almost negligible, and the points are closer together. The 6-month waviness is also due to the sun. The main trend in the figure is downwards, toward the First Point of Aries.

The theoretical precession due to lunar and solar perturbations is shown in Table 6.12 for 4-week intervals. From one year to the next there are changes due to the change in lunar orbit normal. During one year the greatest change is due to the position of the sun along the ecliptic. In any one 4-week interval the total precession can vary from $0.037°$ to $0.092°$. This table is approximate, but it shows the variations that can be expected.

During a long time period the precessions along the y axis cancel. There are 2-week oscillations of $\pm 0.004°$ as the moon goes around its orbit; there are 6-month oscillations of $\pm 0.05°$ as the sun moves on the ecliptic; and there are 18-yr oscillations of $\pm 0.4°$ due to the motion of the lunar orbit normal. For most communications satellites some of these precessions along the y axis are negligible. Other precessions can be cancelled by stationkeeping maneuvers. The dominant stationkeeping maneuver is to cancel the precession along the x axis; by changes in angle, some of the precession along the y axis can be canceled without greatly affecting the fuel required.

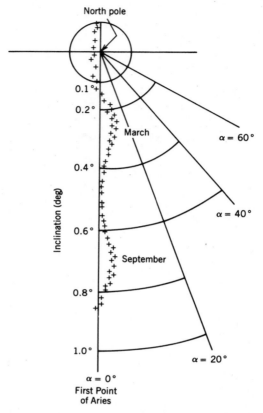

FIGURE 6.50 Orbit normal of geostationary satellite plotted at weekly intervals, starting in November of 1985.

A summary of the lunar and solar perturbations is shown in Table 6.13. Listed first are the basic physical constants for a satellite in geostationary orbit. These lead to the basic coefficient $3a^2\mu/4\ hr^3$, listed in radians per second and degrees per year. The rest of the table lists the secular and periodic terms of the precession.

6.8 PERTURBATIONS FROM NONSPHERICAL EARTH

The fact that the earth is not a sphere affects satellite orbits. Several effects will be discussed: The flattening of the poles affects the orbit normal, the flattening also changes slightly the radius of a geostationary orbit, and the noncircularity of the equator causes an acceleration of the satellite longitude.

6.8.1 Effect on Orbit Normal

The earth is "flattened" at the poles and "bulges" at the equator. The polar radius (6356.77 km) is 21 km less than the equatorial radius (6378.14 km). A satellite in space experiences an extra attraction toward the "bulge" on the equator, so that

TABLE 6.12 Lunar and Solar Effects on Orbit Normal of Geostationary Satellite, 4-week Intervals

Year	4-week ΔWx (thousandths of degrees)	Annual ΔWx (deg)	4-week ΔWy (thousandths of degrees)	Annual ΔWy (deg)
1985	83, 64, 50, 54, 71, 88, 90, 76, 58, 50, 60, 79, 92	0.915	25, 30, 14, −6, −15, −5, 15, 28, 25, 6, −13, −15, 2	0.091
1986	85, 66, 52, 55, 73, 89, 91, 78, 60, 51, 61, 80, 93	0.933	23, 27, 12, −9, −18, −8, 12, 26, 22, 3, −15, −18, −1	0.055
1987	86, 67, 63, 56, 73, 90, 92, 79, 60, 52, 61, 80, 93	0.941	20, 24, 9, −12, −21, −11, 9, 22, 19, 0, −19, −21, −5	0.015
1988	86, 67, 53, 56, 73, 89, 92, 79, 60, 52, 61, 80, 92	0.940	16, 21, 6, −15, −24, −15, 5, 19, 16, −3, −22, −24, −8	−0.027
1989	85, 66, 42, 56, 73, 89, 91, 77, 59, 51, 60, 79, 91	0.929	13, 18, 2, −18, −27, −17, 3, 16, 12, −7, −25, −27, −10	−0.067
1990	84, 65, 51, 54, 71, 87, 89, 76, 57, 49, 58, 77, 89	0.909	10, 15, −0, −21, −30, −20, −0, 14, 10, −9, −27, −29, −13	−0.100
1991	83, 63, 49, 52, 69, 85, 87, 74, 55, 47, 56, 75, 87	0.881	8, 13, −2, −23, −32, −22, −2, 12, 9, −10, −29, −31, −14	−0.122
1992	80, 61, 47, 49, 66, 83, 85, 72, 53, 44, 53, 72, 85	0.849	7, 12, −3, −24, −33, −23, −3, 11, 8, −10, −29, −31, −15	−0.131
1993	77, 58, 44, 47, 64, 80, 82, 68, 50, 42, 51, 70, 82	0.817	7, 12, −3, −24, −32, −22, −2, 12, 8, −10, −28, −30, −13	−0.125
1994	75, 56, 41, 45, 62, 78, 80, 66, 48, 39, 49, 68, 80	0.787	8, 13, −2, −22, −31, −21, −0, 14, 10, −8, −26, −28, −11	−0.105
1995	73, 54, 40, 43, 60, 76, 78, 65, 46, 38, 47, 66, 79	0.763	10, 15, 1, −20, −29, −18, 2, 16, 13, −5, −23, −25, −8	−0.072
1996	72, 53, 38, 41, 58, 75, 77, 64, 45, 37, 46, 65, 78	0.750	13, 18, 4, −17, −25, −15, 5, 19, 16, −2, −20, −22, −5	−0.030
1997	71, 52, 38, 41, 59, 75, 77, 63, 45, 37, 46, 66, 78	0.748	17, 21, 7, −14, −22, −11, 9, 23, 20, 1, −17, −18, −1	0.015
1998	71, 52, 38, 42, 59, 75, 78, 64, 46, 38, 47, 67, 79	0.758	20, 25, 10, −10, −18, −8, 12, 26, 23, 5, −13, −15, 2	0.059
1999	73, 54, 40, 43, 60, 77, 79, 66, 48, 40, 49, 69, 81	0.778	23, 28, 14, −7, −16, −5, 15, 29, 26, 8, −11, −13, 4	0.095
2000	75, 56, 42, 45, 62, 79, 82, 69, 50, 42, 51, 71, 84	0.806	25, 30, 16, −5, −14, −4, 17, 31, 28, 9, −9, −11, 5	0.120
2001	77, 58, 44, 48, 65, 82, 84, 71, 52, 44, 54, 74, 86	0.839	27, 32, 17, −4, −13, −2, 18, 32, 28, 9, −9, −11, 6	0.130
2002	79, 60, 47, 50, 68, 84, 87, 73, 55, 47, 56, 76, 89	0.871	27, 32, 17, −4, −13, −3, 17, 31, 28, 9, −9, −11, 5	0.126
2003	82, 63, 49, 52, 70, 86, 89, 76, 57, 49, 58, 78, 91	0.900	26, 31, 16, −5, −14, −4, 16, 30, 26, 7, −11, −13, 3	0.108
2004	84, 65, 51, 54, 71, 88, 91, 77, 59, 51, 60, 79, 92	0.923	24, 29, 14, −7, −16, −7, 13, 27, 24, 5, −13, −16, 0	0.078
2005	85, 66, 52, 56, 73, 90, 92, 78, 60, 52, 61, 81, 93	0.937	22, 26, 11, −10, −19, −9, 11, 25, 21, 2, −17, −19, −2	0.040
2006	86, 67, 53, 56, 74, 90, 92, 78, 60, 52, 61, 81, 93	0.942	19, 23, 8, −13, −22, −12, 8, 21, 17, −2, −20, −22, −6	−0.002
2007	86, 67, 53, 56, 73, 89, 92, 78, 60, 51, 60, 80, 92	0.937	15, 20, 5, −16, −25, −16, 4, 18, 14, −4, −23, −25, −9	−0.043
2008	85, 66, 52, 55, 72, 88, 91, 77, 59, 50, 59, 78, 91	0.922	12, 17, 2, −19, −28, −19, 1, 15, 12, −7, −26, −28, −12	−0.081
2009	83, 64, 50, 53, 71, 87, 89, 75, 56, 48, 57, 77, 89	0.899	10, 14, −1, −22, −31, −21, −1, 13, 9, −10, −28, −30, −13	−0.110
2010	82, 62, 48, 51, 68, 84, 86, 73, 54, 46, 55, 74, 86	0.869	8, 12, −3, −24, −32, −22, −2, 12, 8, −11, −29, −31, −14	−0.127

TABLE 6.13 Lunar and Solar Perturbations: Summary

Quantity	Symbol	Moon	Sun	Units
Orbit constants[a]				
Semimajor axis	a	42,164.57	42.164.57	km
Orbit angular momentum	h	129,641	129,641	km²/sec
Perturbation				
Gravity parameter	μ	4,902.78	$132,712.4 \times 10^6$	km³/sec²
Distance	r	384,400	149,504,200	km
Coefficient $3a^2\mu/4hr^3$	—	888×10^{-12}	408×10^{-12}	rad/sec
		1.61	0.738	degrees/yr
Secular term precession				
minimum	$\dot{W}_x$	0.48	0.269	degrees/yr
maximum	$\dot{W}_x$	0.67	0.269	degrees/yr
Periodic term				
Period	—	13.6	182.6	days
Amplitude	$\dot{W}_y$	±0.004	±0.05	degrees
Period	—	6,798		days
Amplitude	$\dot{W}_y$	±0.4		degrees

[a]These orbital constants are for a geostationary satellite, and not for the moon or sun.

in general the gravitational force will not be directed exactly toward the earth's center (see Figure 6.51). The earth's gravitational potential can be expressed as a series of harmonic terms. A simple form is

$$\phi = \frac{\mu}{r}\left[1 + \frac{J_2 R_e^2}{2r^2}(1 - 3\sin^2 \delta) + \cdots\right] \qquad (6.81)$$

The first term in brackets, 1, leads to the principal law of gravitation, while the next term, with the coefficient J_2, expresses the oblateness of the earth. The coefficient J_2 has a value of 1082.63×10^{-6}.

The flattening of the earth has two effects on a geostationary orbit. One effect is that the gravitational acceleration in a geostationary orbit is increased; for the orbit period to equal exactly one sidereal day, the orbit radius must be slightly greater. The other effect is that if the orbit has any inclination, the satellite will experience a force toward the equatorial plane that will cause a precession of the orbit normal around the earth's spin axis. While there is no effect if the orbit inclination is exactly zero, it becomes more important as the inclination angle increases.

From the gravitational potential it is possible to determine the precession of the orbit normal around the North Pole. The rate of change of the right ascension of the ascending node is

$$\dot{\Omega} = -\frac{3J_2\mu^{1/2}R_e^2}{2a^{7/2}(1-e^2)^2}\cos i = -(4.9° \text{ yr}^{-1})\frac{\cos i}{(1-e^2)^2} \qquad (6.82)$$

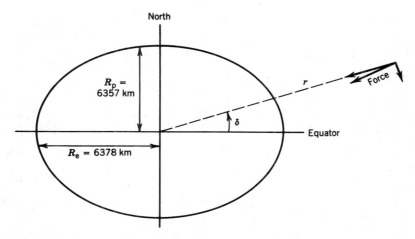

FIGURE 6.51 Oblateness of earth. Gravitational attraction is not exactly toward center of earth.

The value of J_2 is 1082.63×10^{-6}, and so the value of the coefficient, $\frac{3}{2} J_2 \mu^{1/2} R_e^2 / a^{7/2}$, is 2.71×10^{-9} rad/sec, or $4.90°$ yr^{-1} for a geostationary orbit. Hence, the oblateness of the earth causes a precession of the ascending node by $4.9°$ yr^{-1}, or a complete revolution in 73 yr.

The combined effect on the orbit normal of the earth's oblateness and the lunar and solar perturbations will now be considered. For equatorial orbits the influence of the earth is negligible. But as soon as the lunar and solar perturbations have produced some inclination, the earth exerts an influence. Instead of the orbit normal precessing around the ecliptic pole, the earth forces the precession into a tighter circle. The center of this circle is approximately one-third of the way from the north pole to the ecliptic pole. The exact location depends on the location of the lunar orbit normal and is a vector sum of the three perturbing influences.

If the satellite normal is at a point about $7.5°$ from the north pole, the various effects just cancel (see Figure 6.52). The moon and the sun will try to precess the

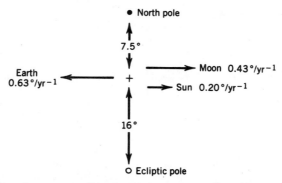

FIGURE 6.52 Point where average effect of sun and moon on satellite orbit normal would be canceled by effect of earth.

orbit normal toward the First Point of Aries, and the flattening of the earth will precess the orbit normal in the opposite direction. Some periodic terms will produce oscillations about this point.

For other orbits, the combined effect is a precession around this stable point. A satellite that starts off in an equatorial geosynchronous orbit will have an inclination angle that increases at the rate of slightly less than $1° \text{ yr}^{-1}$. This angle will increase up to a maximum of $15°$ and then will decrease back to zero. The total cycle would take about 53 yr. A few typical curves are shown in Figure 6.53 for different starting dates.

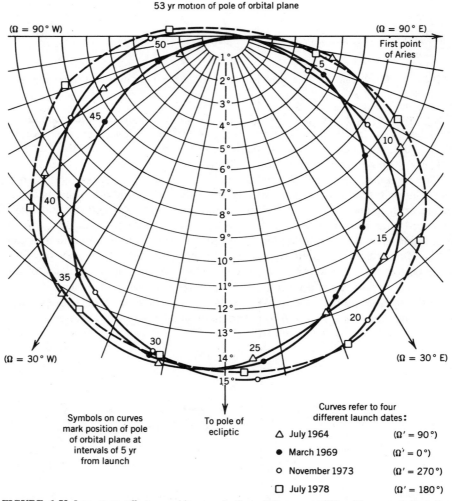

FIGURE 6.53 Long-term effect on orbit normal of geostationary satellite with no stationkeeping (Allan, 1963). Copyright 1963. Reproduced with permission of the Controller of Her Brittanic Majesty's Stationery Office.

6.8.2 Effect on Satellite Longitude

Considered now are perturbations that affect the satellite longitude. The major effect can be visualized as an equatorial cross section that is not circular. The earth can better be described as a triaxial ellipsoid rather than an ellipsoid of revolution about the polar axis.

Because of the earth's triaxiality, the force of gravitational attraction is toward the nearest equatorial bulge and not exactly toward the earth's center (or spin axis). This creates a component of force acting along or opposite to the satellite's velocity.

A force parallel to the satellite's velocity produces an average acceleration *opposite* to the force. A force in the same direction as the velocity increases the energy, increases the semimajor axis, increases the period, and results in a decrease in the average velocity. Therefore, an eastward force will increase a westward drift (see Figure 6.54). Conversely, a westward force will increase an eastward drift or decrease a westward drift. The components of gravitational force are toward the major axis of the ellipse (the equatorial bulge), but the satellite drift acceleration is toward the nearest minor axis.

The longitudinal acceleration is toward two points at 79° E and 252.4° E (107.6° W). These points are approximately, but not quite, opposite each other. The first is over a point close to Sri Lanka in the Indian Ocean; the second is over the Pacific Ocean, off the coast of Ecuador. This effect is caused by very small irregularities in the mass distribution of the earth. The equatorial bulge is quite small, equivalent to less than 100 m difference between different points on the equator. For most satellites, any asymmetry around the earth's spin axis averages to zero. But a geostationary satellite maintains a constant relation with the mass asymmetry, and the effect accumulates over a period of days or weeks.

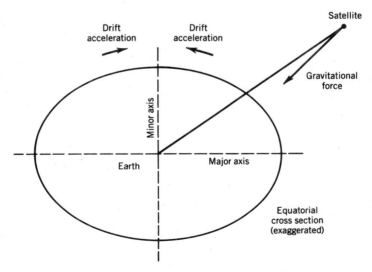

FIGURE 6.54 Effect on satellite longitude of non-circular equatorial cross section of earth.

The actual acceleration at different longitudes is shown in Figure 6.55, adapted from a graph by Wagner (1970). Where the acceleration is positive, the longitude tends to increase, and the satellite eventually moves to the right. So the zero crossings at 79° E and 252.4° E are stable points; a satellite at that longitude will stay there, and a satellite near that longitude will oscillate about that longitude. The other two zero crossings are unstable; satellites near that longitude are accelerated away from that point.

The curve shown in Figure 6.55 is not symmetrical. The first low point near 120° E has more acceleration than the second point near 300° E. The curve was obtained from observations on a number of satellites and shows that even a triaxial symmetrical model will not completely describe the earth's gravitational potential.

To picture the satellite motion, Figure 6.56 shows the geostationary arc as two valleys and two hills. If left alone, a satellite tends to roll "downhill." To maintain its longitude, it is necessary to push it back up hill every so often, that is, to fire a thruster. If this is not done, a satellite placed at 310° E will "roll downhill" to the valley at 252.4° and continue up the "next hill" until it reaches a point at the same height, which corresponds to about 200° E.

Mathematically, Figure 6.56 is the integral of Figure 6.55. The vertical scale is similar to kinetic energy ($\frac{1}{2} mv^2$) per unit mass. The numbers correspond to changes in the square of the drift rate. If a satellite with zero drift rate started near the top of the graph (i.e., in the Pacific Ocean region), when it reached the bottom

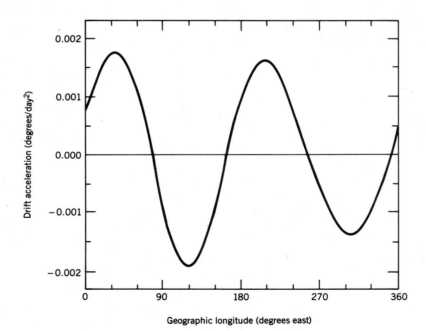

FIGURE 6.55 Acceleration in longitude of geostationary satellite.

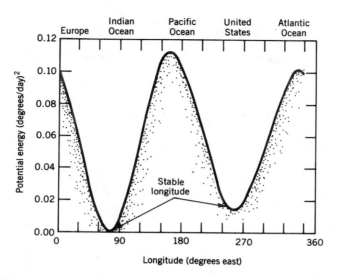

FIGURE 6.56 Longitudinal acceleration pictured as hills and valleys. Satellite tends to "roll downhill" and needs thrusters to keep it in place.

the difference would be about 0.09 (degrees/day)^2, and its drift rate would be 0.3° per day.

To illustrate the effect of longitudinal acceleration, the longitude of *Early Bird* (the first Intelsat satellite) is shown in Figure 6.57 for several years. This shows a period when it was no longer maintained at a constant longitude. Originally it had been stationed at 310° E, but after it ran out of fuel, the longitudinal acceleration was no longer nullified. It then accelerated toward the stable longitude of 252.4° E. When it reached that longitude, it had a considerable drift rate, so it went past that longitude, and its drift rate started to decrease. Its drift rate reached zero at about 200° E longitude, when it turned around to go back to 310° E.

The period of this oscillation is about 3 yr. For small longitudinal amplitudes (less than 20°) the period would be independent of amplitude. For large amplitudes, however, the period decreases with amplitude, as the acceleration is smaller at the extreme longitudes.

A typical drift correction cycle is shown in Figure 6.58 for a satellite maintained at 6° W longitude with limits of ±0.1°. At day 0, a westward drift rate is imparted of such a magnitude that the satellite arrives at its western limit with zero drift rate. Drift reversal occurs automatically, without any thrusting maneuver. When the satellite returns to its eastern limit, another maneuver is necessary, and the cycle is repeated. During the 60 days between maneuvers, the drift rate increases from about −0.013° to 0.012° per day, or an average acceleration of 0.0004° per day^2. Note that there is superimposed a periodic variation (caused by the moon) with a period of 13.6 days.

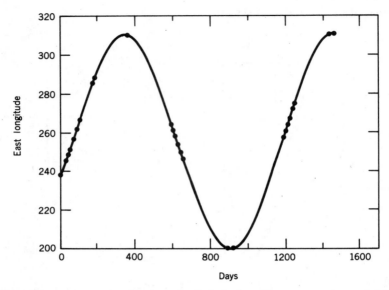

FIGURE 6.57 Free oscillation in longitude for *Early Bird* satellite over several years from January 1, 1972.

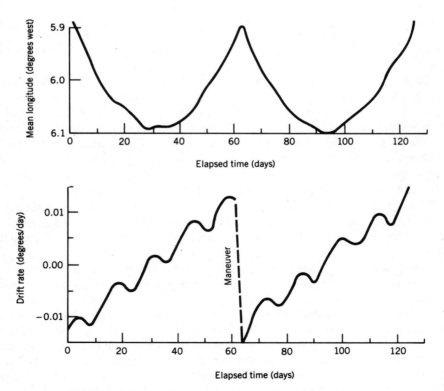

FIGURE 6.58 Control of mean longitude to keep satellite at 6° W.

If the orbit is eccentric, there is also a variation with a daily period. This variation disappears by defining the *mean longitude* as the arithmetic average of the instantaneous longitude of the subsatellite point taken over a mean solar day. The *drift rate* is defined as the difference between the mean longitudes on two successive days.

6.9 STATIONKEEPING IN GEOSTATIONARY ORBIT

A truly geostationary orbit would have zero inclination and zero eccentricity and the satellite would be at a fixed longitude and not move with respect to the earth. In actual practice, the elements are allowed to deviate from the desired values. The process of maintaining these elements within specified limits is called stationkeeping.

The orbit element that requires the most thrusting to maintain is the inclination; the required fuel for north-south stationkeeping is usually an order of magnitude greater than for the other elements. The most important element to control is the longitude, for in most cases lack of control would result in the greatest deviation from the desired station. The easiest element to control is the eccentricity. There is no secular perturbation that produces a steady increase in eccentricity (for periods of a year or longer). The control of eccentricity can usually be combined with east-west stationkeeping, with little or no additional fuel requirements.

The control of each of the three orbit elements is discussed in the following pages, with estimates for the first two of velocity changes, fuel requirements, and frequency of stationkeeping maneuvers.

The required velocity changes for stationkeeping are obtained with various types of thrusters. If a thruster ejects a certain mass of fuel (dm) with an effective velocity v_e, the spacecraft mass m experiences a velocity change dV:

$$m \, dV = -v_e \, dm \tag{6.83}$$

This follows from the conservation of linear momentum and is illustrated in Figure 6.59.

The exhaust velocity v_e is usually given in terms of a specific impulse I_{sp} measured in seconds. This is the ratio of thrust to rate of fuel usage, which is equivalent to the effective exhaust velocity divided by the acceleration of gravity ($I_{sp} = v_e/g$). A common thruster for communications satellites uses a monopropellant called hydrazine. A typical specific impulse is 230 sec, although this will vary with the thruster and will be lower for short pulses. Electric thrusters will have higher specific impulses.

Combining the preceding equations to obtain a desired velocity difference (dV), the fuel required is

$$dm = -\frac{m \, dV}{g I_{sp}} \tag{6.84}$$

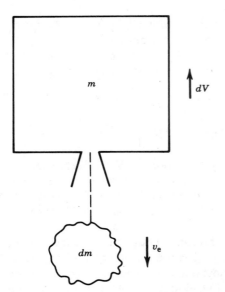

FIGURE 6.59 Satellite velocity change produced by small mass of fuel expended.

The negative sign indicates the spacecraft gets lighter as the fuel is used up. This equation can be integrated, but for stationkeeping it will be assumed that the fuel used is small compared to the spacecraft mass.

6.9.1 North–South Stationkeeping: Inclination

North–south stationkeeping requires that the orbit normal be maintained within a prescribed angle, Δi, of the north pole of the earth. Shown in Figure 6.60 is a circle with a typical radius of $0.1°$ that defines an allowed region for the orbit normal. While maintaining the orbit normal within these limits, two secondary objectives may be followed: (1) to minimize the total fuel consumption and (2) to maximize the spacing between maneuvers.

Lunar and solar perturbations tend to move the orbit normal toward Aries ($\alpha = 0°$). Therefore, inclination maneuvers move the orbit normal generally in the opposite direction ($\alpha = 180°$). To maximize the number of days between maneuvers, the stationkeeping may be done so that the future trace of the orbit normal passes through, or very close to, the north pole. In the example in Figure 6.60, the orbit normal precessed from A to B; a maneuver then moved the orbit normal from B to C, after which the orbit normal precessed from C to D.

To minimize the effect of daily fluctuations in the orbit normal position, the instantaneous inclination can be averaged over a mean solar day. This mean inclination can then be monitored and controlled rather than an instantaneous inclination.

The orbit normal is perpendicular to both the radius vector and the velocity vector. An inclination maneuver is done when the radius vector is perpendicular

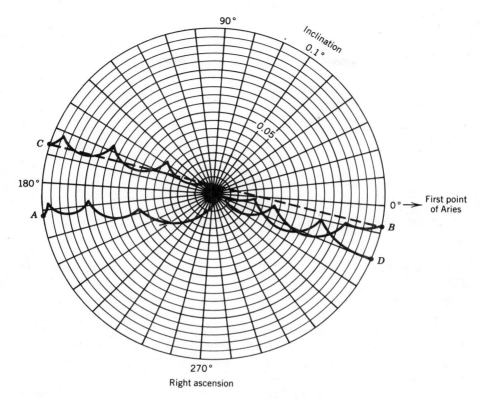

FIGURE 6.60 Typical motion of orbit normal through north pole of geostationary satellite. Thruster firing moves orbit normal from *B* to *C*.

to both the old and new orbit normals (this usually means when the satellite is close to the equatorial plane). A typical north–south stationkeeping maneuver occurs at a descending node around $\alpha = 270°$. A northward velocity is produced by a thruster on the south end of the satellite (see Figure 6.61). The descending node becomes an ascending node as the orbit plane is rotated an angle θ. Typically, the angle θ is twice the inclination limit Δi. Conversely, an equivalent change of the orbit normal can be obtained at the ascending node by using a thruster on the north end of the satellite to produce a southward velocity change.

The magnitude of the required velocity change ΔV is a function of the plane change angle θ. For a geostationary satellite, the velocity is 3074 m/sec, so a ΔV of 5.4 m/sec is needed for each 0.1° of plane change (i.e., 5.4/3074 = 0.0018 rad $\cong 0.1°$).

The maximum velocity difference ΔV for an inclination maneuver occurs when the angle of plane change is twice the inclination limit. If the inclination limit Δi is 0.1°, the maneuver goes from 0.1° on one side to 0.1° on the other. The inclination does not actually change, but the orbit normal shifts by 0.2°, and the right

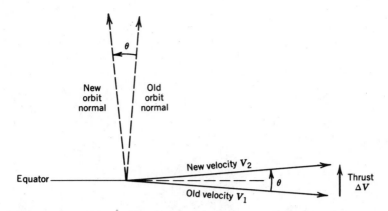

FIGURE 6.61 Effect on orbit normal of velocity change produced by thruster maneuver.

ascension of the ascending node shifts from the neighborhood of 90° to around 270°.

As described previously, the direction of the precession will vary, depending on the time of year for the sun and the location of the lunar orbit normal. In general, each maneuver will be designed mainly to cancel the previous motion of the orbit normal but will also optimize the initial position for the next precession. The required velocity change and the fuel required to produce this velocity change are shown in Figure 6.62. The fuel used is calculated for a 1000-kg satellite, assuming a specific impulse of 230 sec, which is typical for a hydrazine thruster. The fuel used is proportional to the mass of the spacecraft, so that the graph can be interpreted as simply the fraction of spacecraft mass used in the maneuver in parts per thousand.

The time interval between maneuvers is shown in Figure 6.63. The velocity difference and fuel used per maneuver increase for larger inclination limits, but so does the time between maneuvers. The precession per year is a little less than 1° yr^{-1}, so if the limits of inclination are 0.5°, each maneuver will be about 1°, and inclination maneuvers have to be done only once a year.

The precession rate varies from day to day and from year to year, as seen in Table 6.12. The top curve in Figure 6.63 is for a good year, such as 1997, when the lunar orbit normal will be closest to the north pole. The bottom curve applies when the lunar orbit normal is farthest from the north pole, such as 1987 or 2001. In addition to these variations, for small inclination limits (intervals much less than a year) there will be variations due to the sun depending on the time of year.

To the accuracy used in the previous graphs, both the fuel per maneuver and the time between maneuvers are proportional to the inclination limit. Therefore, over a long period of time the total fuel consumption is independent of the inclination limit and of how frequently the stationkeeping maneuvers are done. However, there is one free traverse of the circle; that is, there is no need for the orbit normal at the end of the mission to be where it started. In particular, if the

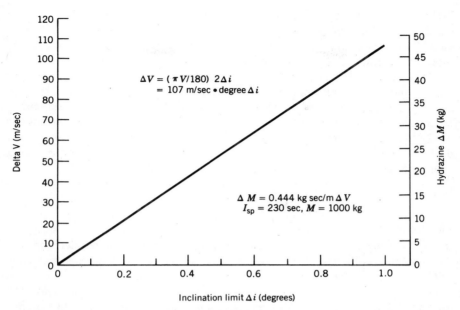

$$\Delta V = (\pi V/180)\ 2\Delta i$$
$$= 107\ \text{m/sec} \cdot \text{degree}\ \Delta i$$

$$\Delta M = 0.444\ \text{kg sec/m}\ \Delta V$$
$$I_{sp} = 230\ \text{sec},\ M = 1000\ \text{kg}$$

FIGURE 6.62 Velocity change and hydrazine required for each inclination maneuver (1000 kg satellite).

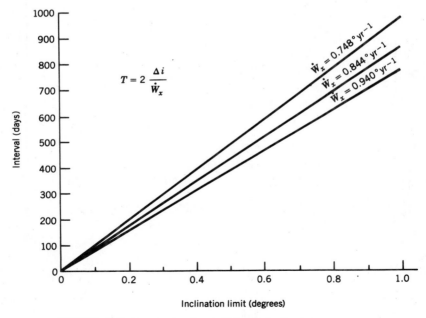

$$T = 2\ \frac{\Delta i}{\dot{W}_x}$$

$$\dot{W}_x = 0.748^\circ\ \text{yr}^{-1}$$
$$\dot{W}_x = 0.844^\circ\ \text{yr}^{-1}$$
$$\dot{W}_x = 0.940^\circ\ \text{yr}^{-1}$$

FIGURE 6.63 Average time interval between inclination maneuvers.

limits are large enough, there may be no need for north–south stationkeeping. For example, for limits of $\pm 3°$ it takes almost 7 yr to cross the circle once. There is an additional bonus in that for most launch sites it is easier to reach an inclined orbit than to reach an equatorial orbit that has $0°$ inclination.

The total ΔV and fuel requirements are shown in Figure 6.64. For small inclination limits the maneuvers must be started soon and done frequently. For inclination limits on the order of $0.01°–0.1°$ the fuel requirements are essentially independent of the actual limits. In the curves in Figure 6.64 an average precession of $0.844° \text{ yr}^{-1}$ is assumed. There will be some variation from year to year depending on the lunar orbit normal, with fuel requirements being larger for smaller values of the lunar orbit normal declination (see Table 6.10).

6.9.2 East–West Stationkeeping: Longitude

The satellite longitude must be maintained within certain specified limits, $\pm \Delta \lambda$. In most locations in geostationary orbit, there is a longitudinal acceleration $\ddot{\lambda}$ in one direction. If the satellite starts with zero drift rate at one limit, a maximum drift rate of $\dot{\lambda}_m$ will be reached at the other limit. A thruster can be used to change the velocity and reverse the drift rate; the satellite will then go back and just reach the first limit. A plot of longitude as a function of time is shown in Figure 6.65.

Reversal of drift rates will be required at intervals of time T. The time interval

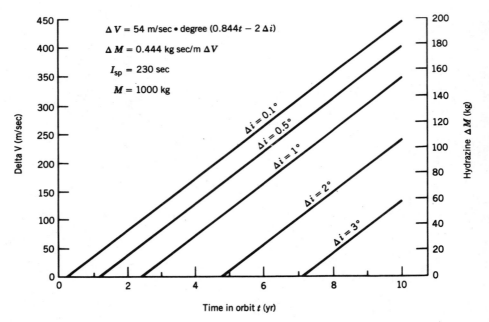

FIGURE 6.64 Velocity change and hydrazine required for all inclination maneuvers (1000 kg satellite).

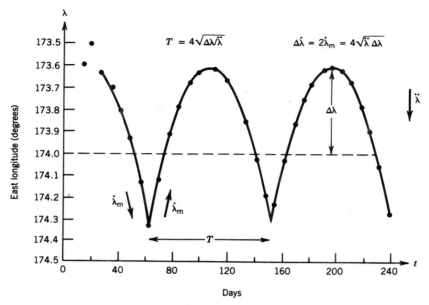

FIGURE 6.65 Longitudinal motion of geostationary satellite, Intelsat IV F-8, maintained at 174° E ± 0.5°.

is

$$T = 4\left(\frac{\Delta\lambda}{\ddot{\lambda}}\right)^{1/2} \tag{6.85}$$

and the required velocity change at each time interval is

$$\Delta\dot{\lambda} = 2\dot{\lambda}_m = 4(\ddot{\lambda}\,\Delta\lambda)^{1/2} \tag{6.86}$$

These equations are similar to the more familiar equations for gravity acceleration of $s = \frac{1}{2}gt^2$ and $v^2 = 2gs$ except for various factors of 2. Note that the average drift rate change per unit time, $\Delta\dot{\lambda}/T$, is the longitudinal acceleration $\ddot{\lambda}$. Values of $\ddot{\lambda}$ for different longitudes are given in Figure 6.55 on page 846.

To change the drift rate of a satellite, the velocity must be changed. However, changing the velocity at one point in the orbit does not produce a similar velocity change in the entire orbit. For east–west stationkeeping it is necessary to determine the relation between a velocity change at one point to the change in average drift rate, or the change in average angular velocity.

For satellite orbits the equation for conservation of energy is expressed in the *vis viva* equation

$$\dot{s}^2 = \mu\left(\frac{2}{r} - \frac{1}{a}\right) \tag{6.87}$$

The relation between the mean angular velocity n and the semimajor axis a is

$$n^2 = \frac{\mu}{a^3} \qquad (6.88)$$

Differentiate the last two equations, and eliminate the change in semimajor axis, da, and the result is the velocity change

$$d\dot{s} = -\frac{a^2 n}{3\dot{s}} \, dn \qquad (6.89)$$

Note that the velocity change is being made at one point, so that the radius r is treated as a constant. The velocity change $d\dot{s}$ is proportional to the change in the mean angular velocity dn. The change in mean angular velocity dn, is equal to the change in drift rate, $\Delta\dot{\lambda}$, in radians per second. For near circular orbits the proportionality coefficient is approximately equal to $-a/3$, so

$$\Delta V = -\frac{a}{3} \, \Delta\dot{\lambda} = -\left(2.839 \, \frac{\text{m/sec}}{\text{degrees/day}} \right) \Delta\dot{\lambda} \qquad (6.90)$$

A desired drift rate change of $1°$ per day will require a velocity change ΔV of 2.839 m/sec. Since the drift rate change $\Delta\dot{\lambda}$ is equal to $4(\ddot{\lambda} \, \Delta\lambda)^{1/2}$, the velocity change can be written as

$$\Delta V = \left(2.839 \, \frac{\text{m/sec}}{\text{degrees/day}} \right) 4(\ddot{\lambda} \, \Delta\lambda)^{1/2} \qquad (6.91)$$

The necessary velocity change, and the fuel needed are plotted in Figure 6.66 for various limits on longitude and for three longitudinal accelerations. An acceleration of $0.002°$ per day^2 is about the maximum acceleration anywhere in geostationary orbit; a value of $0.001°$ per day^2 is more typical, and at a few longitudes the acceleration approaches zero.

To calculate the fuel requirements, a specific impulse of 230 sec and a spacecraft mass of 1000 kg were assumed. For longitudinal limits of $\pm 0.1°$ and an acceleration of $0.001°$ per day^2, the velocity difference per maneuver is 0.11 m/sec, and the fuel used is 0.05 kg.

The interval between the east–west stationkeeping maneuver is shown in Figure 6.67 for various longitudinal accelerations. If the longitude tolerance is roughly comparable to the inclination limit, the longitudinal maneuvers (east–west) are done more frequently than inclination maneuvers (north–south). However, if the limits are very small, it is possible that inclination maneuvers would have to be done more frequently.

For a typical longitude limit of $\pm 0.1°$ and an acceleration of $0.001°$ per day^2,

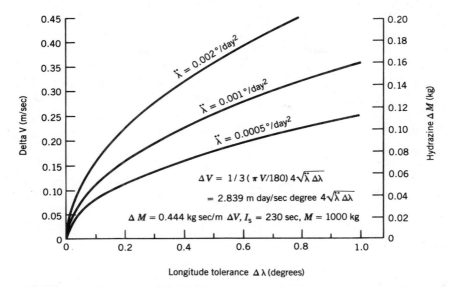

FIGURE 6.66 Velocity change and hydrazine required for each longitude maneuver (1000 kg satellite).

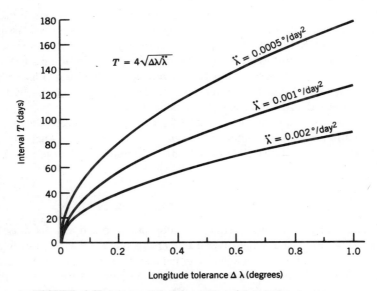

FIGURE 6.67 Average time interval between longitude maneuvers.

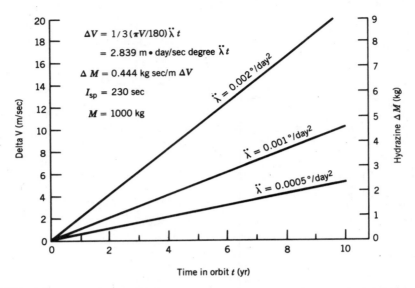

FIGURE 6.68 Velocity change and hydrazine required for all longitude maneuvers (1000 kg satellite).

the time between maneuvers would be about 40 days. This means 20 days from one limit to the opposite limit and 20 days back. Because of the parabolic effect of the constant acceleration, it would take 6 days from the maneuver to the longitudinal midpoint and then 14 days to get to the other limit.

The total velocity difference and fuel requirements are shown in Figure 6.68. A spacecraft mass of 1000 kg and a specific impulse of 230 sec are assumed. A longitudinal acceleration of $0.002°$ per day^2 is the maximum in geostationary orbit; $0.001°$ per day^2 is more typical. A comparison with the requirements for north–south stationkeeping shows that most of the fuel is required for inclination control. Typical values for a 7-yr-lifetime satellite would be 100 kg for north–south stationkeeping but less than 5 kg for east–west stationkeeping.

While some longitudes require less fuel for east–west stationkeeping than others, this is usually not a major factor in the selection of longitude. Visibility of earth stations and availability of slots usually are the dominant factors in choosing a longitude.

6.9.3 Control of Eccentricity

Orbit eccentricity is caused by solar radiation pressure and also by most satellite maneuvers. A north–south maneuver usually has a small sideways force that affects eccentricity; an east–west maneuver can have a major effect on eccentricity. The timing of east–west maneuvers is usually used to control orbit eccentricity.

Looking down on the earth's north pole, the satellite moves counterclockwise. At dawn, satellite local time, the force is opposite to the satellite velocity, so it slows the satellite. The main effect is seen 12 hours later, when the satellite altitude is decreased. Conversely at dusk, the force is in the same direction as the satellite velocity, and increases the speed; this increases the satellite altitude at dawn. The effect depends on the ratio of the satellite area to the satellite mass. Many communications satellites have large solar arrays which increase the area to mass ratio, and increase the effect of solar radiation pressure.

When an east–west stationkeeping maneuver is made, there will be a change in eccentricity. Satellites in regions with an eastward longitudinal acceleration, such as the region from 0 to 60° E, require an eastward force to control the mean longitude. If the force is applied at satellite local dawn (0600) it will tend to decrease the effect of solar radiation. Conversely, satellites in regions with a westward longitudinal acceleration need a maneuver at local dusk (1800) to decrease the effect of solar radiation.

To illustrate the velocity differences, determine the requirement to circularize an orbit at perigee. The velocity at perigee, $\dot{s}_p$, is given by

$$\dot{s}_p^2 = \frac{\mu(1 + e)}{r_p} \tag{6.92}$$

and the velocity needed for a circular orbit at that radius, $\dot{s}_n$, is given by

$$\dot{s}_n^2 = \frac{\mu}{r_p} \tag{6.93}$$

The velocity difference is then

$$\Delta V = \dot{s}_p - \dot{s}_n = \frac{\dot{s}_n^2 e}{\dot{s}_p + \dot{s}_n} \cong \frac{\dot{s}e}{2} = (1537 \text{ m/sec})e \tag{6.94}$$

An eccentricity of 0.0001 can be eliminated with a ΔV of about 0.15 m/sec. This will also produce a change in drift rate of 0.05° per day, which may or may not be desired.

One can do half a maneuver at perigee and half a maneuver at apogee. If the thrusts are in the same direction (both eastward or both westward), the drift rate changes and not the eccentricity. If the thrusts are in opposite directions, the eccentricity changes but not the drift rate. For many communications satellites, fuel is used only to change the drift rate. By thrusting east at apogee or west at perigee, the eccentricity is minimized, and the resulting values usually result in negligible east–west oscillations.

6.10 LAUNCHING INTO GEOSTATIONARY ORBIT

To inject a satellite into a desired orbit, it is necessary first for the satellite to reach a point in the orbit and then to modify the velocity so that the satellite has the required velocity for that point in the desired orbit. Changes of velocity are made with rocket motors, and the thrust durations are usually short enough so for many

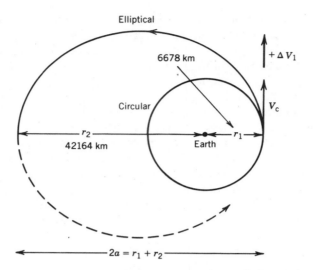

FIGURE 6.69 Transfer from circular orbit to elliptical orbit by increasing velocity.

launches a velocity change can be considered to have occurred at one point in space (for approximate calculations).

A change from a circular orbit to an elliptical orbit is shown in Figure 6.69. A satellite is in a circular orbit with radius r_1, and its velocity is increased by ΔV_1 with no change in direction; the result is the elliptical orbit shown, with perigee r_1 and apogee r_2. The velocity for the circular orbit, V_c, and the perigee velocity V_p are given by

$$V_c = \left(\frac{\mu}{r_1}\right)^{1/2} \qquad V_p = \left(\frac{\mu r_2}{ar_1}\right)^{1/2} \tag{6.95}$$

where a is the semimajor axis of the elliptical orbit $[a = \frac{1}{2}(r_1 + r_2)]$, and μ is the gravitational parameter. The velocity difference is then

$$\Delta V_1 = V_p - V_c = \frac{\mu^{1/2}}{r_1^{1/2}}\left(\frac{r_2^{1/2}}{a^{1/2}} - 1\right) \tag{6.96}$$

and this velocity difference must be furnished by the rocket. As an example, a circular orbit 300 km above the earth's surface has a velocity of 7.7 km/sec. To reach an apogee of 42,164 km (for geostationary orbit), a perigee velocity of 10.1 km/sec is needed, so the required velocity change (ΔV_1) is 2.4 km/sec.

To study injection into stationary orbits, Hohmann transfers will first be considered. A Hohmann transfer is a minimum-energy method to transfer between two coplanar circular orbits (see Figure 6.70). For geostationary satellites launched at

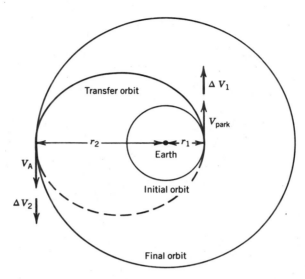

FIGURE 6.70 Hohmann transfer. Transfer between two coplanar circular orbits with two velocity changes.

Cape Kennedy, the transfer is further complicated by a necessary plane change that is normally done as part of the last maneuver (velocity change at apogee).

In a Hohmann transfer, an initial velocity change ΔV_1 puts the satellite in an elliptical transfer orbit, and a second velocity change ΔV_2 then puts the satellite in the final orbit. The necessary velocity changes are

$$\Delta V_1 = \frac{\mu^{1/2}}{r_1^{1/2}} \left(\frac{r_2^{1/2}}{a^{1/2}} - 1 \right) \tag{6.97}$$

and

$$\Delta V_2 = \frac{\mu^{1/2}}{r_2^{1/2}} \left(1 - \frac{r_1^{1/2}}{a^{1/2}} \right) \tag{6.98}$$

where a is the semimajor axis and μ the gravitational parameter.

As an example, suppose the initial orbit is at 300 km height ($r_1 = 6678$ km), and the final orbit is at geostationary height ($r_2 = 42,164$ km). The velocity in the initial orbit is 7.7 km/sec. An additional velocity of 2.4 km/sec will put the satellite into transfer orbit, and a final velocity change of 1.5 km/sec at apogee will circularize the orbit. It will be seen later that an additional 0.3 km/sec will be required to change the orbit plane for satellites launched from Cape Kennedy.

A graph showing velocity changes required for final orbits of different radii (starting from an initial orbit of 300 km height) is shown in Figure 6.71. It is interesting to note that the sum of the two velocity changes required to attain

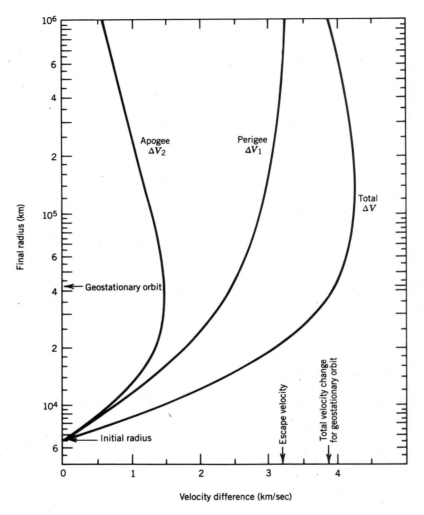

FIGURE 6.71 Velocity differences required for Hohmann transfer from low parking orbit to higher circular orbit.

geostationary orbit (42,164 km) is greater than the escape velocity; this means it takes less fuel to escape the earth's gravity completely (i.e., from parking orbit) than it does to insert into geostationary orbit from the parking orbit.

The Hohmann transfer is the minimum energy maneuver for transfer between a low circular earth orbit and a high circular orbit with a radius less than 76,000 km. For very high circular orbits there are more complicated maneuvers that actually require less energy and less fuel. The graph shows that for very high orbits the velocity requirements for Hohmann transfers actually decrease with altitude.

6.10.1 Steps to Geostationary Orbit

For launching a satellite into geostationary orbit, two intermediate orbits are used (see Figure 6.72): a low circular orbit to reach the equator, called a parking orbit or a coast period, and a transfer orbit that connects the parking orbit (perigee of transfer orbit) to the geostationary orbit (apogee of transfer orbit). For launch from any location not on the equator, the sequence from low orbit to geostationary orbit is similar to a Hohmann transfer but requires a plane change; most of the change in inclination is done at apogee. The magnitudes of the velocity differences required are shown in Figure 6.72. Although the major velocity change is needed to get into space (parking orbit), the additional changes to get to geostationary orbit are not insignificant.

Many constraints dictate the exact orbits used. The latitude of the launch site generates an inclined parking orbit. Injection into transfer orbit with traditional launch vehicles is done at the first crossing of the equator, but with the NASA space shuttle it is done at later crossings. Usually, the satellite is left in transfer orbit for several orbit periods for determination of orbit and attitude accuracy. The desired position of the sun with respect to the transfer orbit may favor a desired time of day for launch. Choices of which equatorial crossing for perigee and which apogee for final injection will determine the initial longitude at geostationary orbit.

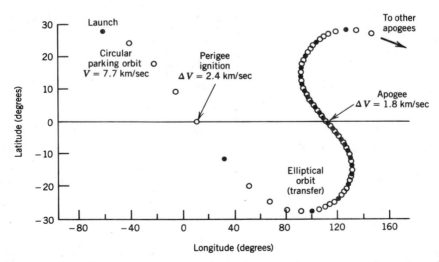

FIGURE 6.72 Typical launch sequence. Launch shown from Kennedy Space Center by expendable launch vehicle. Open circles are every 5 min (if possible), and solid circles are every 25 min.

Table 6.14 shows a summary of the parameters for a typical parking orbit at 300 km height, a geostationary orbit, and the intermediate transfer orbit that would connect the two. The parking orbit is approximately circular and has a low altitude. Since apogee must be on the equator, the perigee must also be on the equator, which is why the parking orbit is needed (for nonequatorial launch sites).

At launch a ΔV of at least 7.3 km/sec is needed to get from 0.4 km/sec of Cape Kennedy (due to earth's rotation) to 7.7 km/sec of parking orbit. At perigee a ΔV of 2.4 km/sec is needed to attain the 10.1 km/sec for transfer orbit. Finally, at apogee the velocity must be changed from 1.6 to 3.07 km/sec: Because of the plane change (to be discussed shortly), an additional ΔV of 1.8 km/sec is required.

Because the orbits are fixed in inertial space, the earth's rotation causes a 23° W change in longitude for each successive parking orbit and a 160° W change in longitude for each successive transfer orbit apogee. These numbers are based on the typical orbits given in Table 6.14. The parking orbit period is 1.5 hr, and during this period the earth rotates through 23°. For the transfer orbit starting with this parking orbit perigee and an apogee height for stationary orbit, the orbit period is 10.5 hr, and the earth rotates through 160° during this period.

Thus, the final longitude depends on the equatorial crossing at which the perigee kick is made and also on when the apogee motor is fired. There is usually some residual drift rate, and the satellite is drifted to the desired location before the final orbit trim maneuvers are made.

TABLE 6.14 Example of Three Orbits Used in Launch from Kennedy Space Center[a]

Parking Orbit (Coast)	Transfer Orbit	Geostationary Orbit[b]
$a - R_e = 300$ km	$r_p - R_e = 300$ km	$a - R_e = 35786$ km
$a = 6678$ km	$r_a = 42150$ km	$a = 42164$ km
$e \simeq 0$	$e = 0.73$	$e = 0$
$i = 28°$	$i = 28°$	$i = 0°$
$P \doteq 5400$ sec	$P = 38000$ sec	$P = 86164$ sec
$= 1$ hr 30 min	$= 10$ hr 33 min	$= 23$ hr 56 min
$V = 7.7$ km/sec	$V_p = 10.1$ km/sec	$V_s = 3.07$ km/sec
	$V_a = 1.6$ km/sec	
$\Delta\lambda = 23°$ West per orbit	$\Delta\lambda = 200°$ East per orbit	$\Delta\lambda = 0$

[a]Kennedy Space Center: latitude = 28.5°, longitude = 279.5° E, V = 0.4 km/sec
[b]Approximate longitude of successive apogees: 90° E, 290° E, 130° E, 330° E, 170° E, . . .

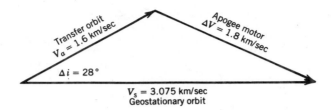

FIGURE 6.73 Change of orbit plane when firing apogee rocket motor.

An example of typical velocities in the plane change maneuver is shown in Figure 6.73. For a minimum total ΔV, most of the plane change should be done at apogee, but a little may be done at perigee. As a rule of thumb, any change in angular momentum should be done as far as possible from the center of mass (earth) in order to have the maximum radial arm. This is in contrast to changes in satellite energy, where the maneuver is done as close as possible to the center of mass in order to have the maximum velocity.

If there was no plane change, the ΔV required at apogee to change from 1.6 km/sec to 3.075 km/sec would be only about 1.5 km/sec. Because of the plane change, about 1.8 km/sec is needed. The exact value can be determined from the law of cosines for plane triangles:

$$\Delta V^2 = V_a^1 + V_s^2 - 2 V_a V_s \cos \Delta i$$

$$1.8^2 = 1.6^2 + 3.075^2 - 2 \times 1.6 \times 3.075 \times \cos 28° \qquad (6.99)$$

where V_a and V_s are the apogee velocity and the geostationary velocities, respectively, and Δi is the plane change angle (normally the inclination of the transfer orbit). The orbit velocity at apogee depends on the perigee radius. The required ΔV depends both on the value of perigee and the angle of plane change, as shown in Table 6.15. The required angle may be less than the launch site latitude, either because some plane change is done at perigee or because an inclined initial orbit is desired.

Most apogee motors are solid-propellant rockets, which deliver a fixed impulse. During the fabrication it is possible to omit some fuel (off-load) and reduce the impulse. A way to change the ΔV just before launch is to increase the spacecraft mass. But the solid rocket is rather inflexible. Liquid fuel rockets have been used at apogee, and these can be turned off after any desired value of ΔV is achieved.

6.10.2 Repositioning A Satellite

A geostationary satellite at one longitude may have to be moved to a different longitude. This can be done by changing its velocity at one point, leaving it in the elliptical orbit a number of days, and then making an equal and opposite velocity

TABLE 6.15 Required Velocity Change at Apogee

Angle (degrees)	$r_p = 300$ km (km/sec)	$r_p = 500$ km (km/sec)
28	1.819	1.807
27	1.797	1.784
26	1.775	1.762
25	1.754	1.741
24	1.734	1.720
23	1.714	1.699

change to put it back into a geostationary orbit: Since the elliptical orbit has a different period, this produces a change in longitude. As illustrated in Figure 6.74, an increase in eastward velocity produces an orbit with a longer period that will produce a westward shift in longitude.

The longitude change $\Delta\lambda$ produced in one orbit period (one sidereal day) is equal to $-360°(\Delta P/P)$. In terms of the velocity change required, the change in longitude for one revolution is

$$\Delta\lambda = -(360°)\frac{3}{2}\frac{2\,\Delta V}{V} = \frac{-1080°}{3075 \text{ m/sec}}\Delta V$$

This is the longitude change produced per day by an initial ΔV. To get back into geostationary orbit, an equal and opposite ΔV is needed, so a total of $2\,\Delta V$ is required, which is

$$\frac{2\,\Delta V}{\Delta\lambda/P} = \frac{2PV}{3(360°)} \equiv 5.7\frac{\text{m/sec}}{\text{degrees/day}} \tag{6.100}$$

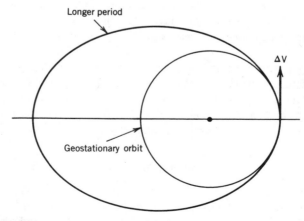

FIGURE 6.74 Change of orbit for repositioning geostationary satellite to different longitude (scale exaggerated).

The fuel required for longitude changes depends on how fast such a change is required. If a drift of $0.1°$ per day is sufficient, a total of 0.57 m/sec velocity change (half at the start, half at the finish) is all that is needed. If a drift of $1°$ per day is needed, 10 times the velocity change and 10 times as much fuel will be required.

6.10.3 Rockets and Launch Vehicles

To achieve required velocities, chemical rockets are presently used. If a small mass, dm, is ejected from a rocket with a velocity v_e (with respect to the rocket), the change in rocket momentum, $m\, dV$, is equal to the momentum of the ejected mass, $-v_e\, dm$ (see Figure 6.75):

$$m\, dV = -v_e\, dm \tag{6.101}$$

If the exhaust velocity v_e is constant, the equation can be integrated, and the result is

$$m = m_0 \exp\left(-\frac{\Delta V}{v_e}\right) \tag{6.102}$$

The exhaust velocity is usually given in terms of the specific impulse I_{sp} (measured in seconds), where I_{sp} is equal to the effective exhaust velocity v_e divided by the acceleration of gravity ($g = 0.0098$ km/sec^2); that is, $I_{sp} = v_e/g$, or $v_e = I_{sp}g$:

$$m = m_0 \exp\left(-\frac{\Delta V}{I_{sp}g}\right) \tag{6.103}$$

The specific impulse may be 290 sec for a rocket using solid propellants, higher for liquid bipropellant combinations (up to 445 sec for liquid hydrogen and liquid oxygen), and lower for monopropellants (230 sec for hydrazine).

The reduction in mass is illustrated in Table 6.16 by specific payloads for various orbits. Mass ratios of 5:1 or 10:1 are obtainable in a single stage, but ratios like 100:1 need multiple stages. From the table, a composite average v_e can be obtained by calculating the ratio of V to $\ln(m_0/m)$. The average exhaust velocity v_e is a

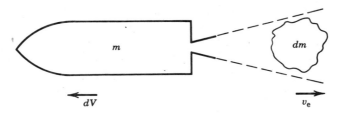

FIGURE 6.75 Satellite velocity change produced by small expulsion of fuel mass.

TABLE 6.16 Launch Vehicle Capabilities[a]

Launcher	Initial Mass	Low Orbit	Transfer to Geostationary Orbit[b]	Geostationary	Units
Velocity change	0.0	7.7	10.1	11.9[c]	km/sec
Ariane-4	200,000	–	4,000	2,400	kg
Atlas Centaur	150,000	–	3,000	1,500	kg
China, CZ[d]	–	–	4,000	–	kg
Delta	–	–	2,000	1,000	kg
Japan, H-II	–	–	–	1,500	kg
USSR-Proton	–	–	4,600	2,000	kg
STS-Centaur	2,000,000	30,000	–	4,000	kg
Titan, 34D-IUS	500,000	15,000	5,000	1,500	kg

[a]Launch vehicles come in many sizes, and are continually being upgraded (Delta capabilities have grown at least by a factor of 10). This table comes from various sources; some numbers are not comparable. The largest mass ratio is from initial weight to low orbit, but there is still a significant ratio (6:1 to 10:1) to get to geostationary orbit.

[b]Often called the Geostationary Transfer Orbit (GTO).

[c]For Kennedy Space Center launches. Launches near equator (e.g., Ariane) only need a total velocity change of 11.5 km/sec.

[d]Also called the Long March launch vehicle.

little over 2 km/sec (I_{sp} = 200 sec), with some values up to 2.6 km/sec (I_{sp} = 250 sec). This includes effects of air drag, dropping empty stages, and nonoptimum solid rocket stages, so the actual exhaust velocity and specific impulse of some stages are considerably higher.

The best mass ratio from low orbit to transfer orbit is about 3:1, and from transfer orbit to geostationary orbit about 2:1. While these ratios are smaller than the ratio from launch to low orbit (e.g., 30:1), they are not insignificant. To get a 1000 kg satellite into geostationary orbit requires 2000 kg in transfer orbit, and 6000 kg in low orbit. If retrieval was desired, 5/6 of the 1000 kg would have to be fuel, so it could be brought back to low orbit, and rendezvous with the shuttle. Instead of a 1000 kg satellite, there would really only be a 150 kg satellite. Such impossibilities are known by launch experts, but are often not understood by nontechnical individuals.

A few satellites are launched with the Space Shuttle. It consists of a non-expendable orbiter vehicle, an expendable external tank, and a pair of recoverable solid rocket boosters launched from the Kennedy Space Center. After completing its mission of launching a satellite from a parking orbit, and other missions, the orbiter reenters the atmosphere and lands like an airplane without propulsion.

For reaching geostationary orbit, the satellite launched from the orbiter needs additional propulsion. Low-cost spinning solid upper stages are single stages without active guidance for getting payloads into transfer orbit. The satellite also uses an apogee motor for insertion into geostationary orbit. For larger payloads other propulsion units are required.

The shuttle orbit is low and inclined. The satellite, with propulsion for perigee and apogee stages, is deployed from the shuttle and separated by a safe distance. Perigee ignition can occur either on an ascending node or a descending node. The satellite reaches geostationary orbit altitude after 5.3 hr, and the apogee motor may be ignited then or on a subsequent apogee.

FURTHER READING

Celestial mechanics is concerned with the motion of objects in astronomical space. It has been studied for centuries, and there is a vast number of books. One excellent reference is by Brouwer and Clemence (1961).

Astrodynamics includes both celestial mechanics and space navigation. It includes not only predicting future positions of objects, but also actions to change this motion. The literature on astrodynamics is much smaller. Good introductions include Berman (1961), Baker and Makemson (1961), and Kaplan (1976). The definitive book on astrodynamics is Herrick (1971); it is quite mathematical, and also computer-oriented.

The geostationary orbit is a unique orbit, and merits special treatment. There are hundreds of satellites in almost identical orbits, and the number is growing. A few satellites are maintained close together (colocation), nominally at the same longitude. Satellites in geostationary orbit are particularly sensitive to the asymmetries in the earth's gravitational field. After their use is over, satellites should be moved out of the geostationary orbit. The orbits have very small inclinations and eccentricities, so standard celestial mechanics calculations are sometimes not appropriate. In particular, a geostationary coordinate system (rather than inertial) is useful for some calculations. Many of these features have been covered in technical journals, but there are few books that provide a comprehensive study of the geostationary orbit. One is by Pocha (1987).

Acknowledgments

Much of the author's knowledge of geostationary satellite orbits has come from many members of the COMSAT orbital mechanics group. In particular, thanks to William Kinney, who headed the group for many years, and to Victor Slabinski, the resident astronomer expert.

REFERENCES

Allan, R. R. (1963). *Perturbations of a Geostationary Satellite, Lunisolar Effects*, Tech. Note No. SPACE 47, Royal Aircraft Establishment, England.

Astronomical Almanac for 1987 (1986). U.S. Naval Observatory and Royal Greenwich Observatory, U.S. Govt. Printing Office, Washington, D.C. Issued every year.

Baker, R. M. L., Jr., and Makemson, M. W. (1961). *Astronautics*, Wiley, New York.

Berman, A. I. (1961). *Astronautics*, Wiley, New York.

Brouwer, D., and Clemence, G. M. (1961). *Methods of Celestial Mechanics*, Academic Press, New York.

Herrick, S. (1971). *Astrodynamics*, Vols. 1 and 2, Van Nostrand, New York.

Kaplan, M. H. (1976). *Modern Spacecraft Dynamics and Control*, Wiley, New York.

Pocha, J. J. (1987). *Mission Design for Geostationary Satellites*, Reidel, Dordrecht, Holland.

Slabinski, V. J. (1974). Variations in range, range-rate, propagation time delay, and Doppler shift for a nearly geostationary satellite. *Prog. Astronaut. Aeronaut.*, 33, 3.

Wagner, C. A. (1970). *Low Degree Resonant Geopotential Coefficients from Eight 24-Hour Satellites*, Rep. No. X-552-70-402, Goddard Space Flight Center, Greenbelt, MD.

INDEX

Absorption bands, 164–166, 254–256, 309
Absorptivity:
  effective, 696–697
  to sunlight, values, 683, 687
  temperature, effect on, 686–688
Accelerated testing, 717
Acceleration:
  centrifugal, 781–782
  launch environment, 556, 723–724
  in longitude, 845–848
  of test mass, 673–677
Accelerator, charged particle, 720
Accelerometer:
  in ground test, 727
  in satellite, 639
  signal in telemetry, 579
Access to satellite, 2, 8
Accuracy in range determination, 601
Acoustic:
  chamber test, 730–731
  noise, Shuttle, 556, 731
  vibration:
    launch, 661, 723
    simulation, 677
Acquisition:
  attitude, spacecraft, 659
  CDMA, 524–525
Acronym, see List of Acronyms in back of book
  carrier description, 414
Activation, voice, 422
Adapter, to launch vehicle, 663–665
Adaptive control for attitude, 658
Adaptive network, TDMA, 471–472
Address:
  bits, command, 596
  general, 100, 412, 416, 480
  packet, 415
  spacecraft, 578, 594, 596
Adjacent satellite:
  interference, 287, 292–293, 435
    TDMA, 486

  polarization, 246
ADMA (amplitude domain multiple access), 408, 528
Advantages, of satellites, 10
Aeronautical mobile satellite, 127
  earth station radome, 359
  frequency allocations, 146–151
Aging of HPA, 358
Air mass zero, solar radiation, 542, 603–604, 686
Alarm, circuit out of specification, 97
Albedo, reflected earth radiation, 686
Alignment of spacecraft component, 555, 717
Allocation:
  dynamic, 97, 125
  frequency, 134–166
ALOHA, 528–539
  delay, 531–539
  early warning reservation, 528, 538
  forms, 531
  instability, 528–530
  performance, 530
  reservation, 531
  RMA:
    example of, 401, 406, 416
    forms of, 533
  slotted, 531–537
  system, 529
  types, 531–539
Altitude of satellite, 220, 265, 271, 785
Aluminized mylar, 697
Aluminum properties, mechanical and thermal, 679–681
Amateur radio, 528
Amateur satellite frequency allocation, 152–160
Amplifiers:
  parametric, cooled, 273–274, 318
  type, 321
Amplitude domain multiple access, see ADMA (amplitude domain multiple access)
Amplitude modulation (AM):
  broadcast, multiple point, 87

Amplitude modulation (AM) (*Continued*)
  distortion in TDMA, 487
  intersatellite link, use of, 418
  nonlinearity, 432
Amplitude shift keying, *see* ASK (amplitude shift
  keying)
Analog:
  bandwidth, 14–17
  to digital conversion, 18, 22, 29–30, 78, 339
  services, 76
Anechoic chamber, 717
Angle:
  factor, thermal, 698–703
  modulation, *see* PSK
  tracking, 259, 264, 597–599, 648
Angular:
  acceleration, attitude control, 657
  attitude of spacecraft, 632–636
  momentum, definition, 631
  momentum control, 636–638, 650
  momentum per unit mass, 808, 818–819
  separation between satellites, 218–220, 230–231
  velocity, 631, 657, 784, 808
Anhydrous hydrazine, 653–655
Anik-B switching, 492, 497
Annual effect on orbit normal, 841
Anomaly, *see* Eccentric anomaly; Mean anomaly;
  Parabolic, anomaly; True anomaly
Antenna, 276–307, 568–575
  area, actual, 351
  beam, 268–308. *See also* Beam, antenna
  beamwidth, 275, 280–284, 391
  bicone, 589–591
  bifocal aplanatic, 295
  blockage elements, 292
  boresight gain, 276–277
  Cassegrain, 295
  CCIR and FCC requirements, 277, 284, 286, 291
  command subsystem, 596
  coverage, 220, 223–225, 287–288, 296, 300–306,
    371. *See also* Coverage
  cross-section area, 281–282
  diameter, equations, 284, 290, 388–399
  earth coverage (global), 285, 288, 549
  earth station, 276–291. *See also* Earth station
  effective area, 278, 351, 587
  efficiency, 278, 279, 572–575
  elevation angle, 192, 219–232, 825–829
  fan beam, 501, 505
  feed, 288, 359, 573–576
  flat plate, 278
  flexible, 639
  focal length, 574
  gain, 226, 278, 279, 284, 572, 575, 587
    coverage, 281, 285, 294, 371
    vs. diameter, 278–279, 284, 572, 575, 587
    limitation, 284, 289
    off-axis, 277, 286, 291
    vs. population density, 510, 511
    vs. power, 226

global beam, 285, 288
Gregorian, 295
G/T, 291–326. *See also* Figure of merit (G/T)
half-power beamwidth, 280–285
horn, 294, 568–576
ice on, 299
ideal, 290–291, 307
under illumination, 351
in-orbit test, 736–737
lens, 295, 506
mass, 295, 351–352
module, 557
mount, 287
multibeam service, 279, 294, 323–326, 347–352,
  508
noise, 291–315. *See also* Noise
offset-feed, 295, 568, 573
omnidirectional, 548, 596
parabolic, 226, 568, 573–575
pattern, 276–277, 310, 589–590
peak (boresight) gain, 276
phased array, 295, 506
pointing, *see* Pointing
  earth station, 218–226, 825–829
  satellite, 299–305, 632–636
polarization misalignment, 243
power received, 234
rain, effects on noise temperature, 297, 299,
  311–312
range test, 717–718
reflector, 226, 295, 568, 570, 573–575
scanning beam, 501, 504
shaped beam, 281–283, 287–290, 294, 296, 371,
  573–576
sidelobe, 219, 276–277, 286–287, 291–293, 310,
  435. *See also* Sidelobe
spacecraft, 279, 351–352, 568–575, 589–590,
  596
spot beam, 290, 300–305, 469–470, 511, 661
support structure, 550, 664
telemetry, 549, 591
test, 717, 736–737
tolerance, 284, 289
toroidal, 548, 588–590, 596
types, 295
Antireflective coating, solar cell, 605
Aperture, *see* Antenna, diameter, equations
Aphelion, earth's, 543, 603, 801
Apofocus, 805
Apogee:
  longitudes at, 864
  Molniya, 9, 265
  velocity, 808
Apogee kick motor (AKM):
  launch environment, 556, 723
  mass, 555, 717
  specific impulse, 651–652
  velocity change, 860–866
Arc of great circle, 41, 778
Arcing, electrical (in space), 542

Area code, telephone, 39, 471
Argument of perigee:
definition, 811–813
from r and ṙ, 818–819
Ariane launch vehicle, 868
Aries, first point of, 774–776
ARR (automatic repeat request), 273
Array, solar, 602–611. See also Solar, array
Array drive, solar, 543, 593–594, 630
Ascending node, 790–791
Ascent mode, attitude control, 659
ASK (amplitude shift keying), 22–23, 339, 473
Assembly, easy access, 661
Assignment:
channel, 404, 462, 477
demand, 97, 402–406, 425. See also DAMA (demand assignment multiple access)
methods, TDMA, 462
Astronomical constants, 783, 801
Astronomy, radio frequencies for, 145, 164–166
Asynchronous terminal, 102
Atlas Centaur, 723, 730–731, 868
Atmospheric path loss, 254–256, 309
Atmospheric pressure in space, 541
A-to-D conversion, 18, 22, 30, 78, 339
Attenuation, 327, 338–340
atmosphere, 254–256, 262
frequency dependence, 252–256
noise, 297
rain, 254–256, 261–262, 311, 426–428, 487
waveguide, 309, 316
Attenuator in spacecraft, 438, 444, 497, 561–562
Attitude, 630–636
Attitude control, 646–661
accuracy, 659–661
adaptive control, 658
biased momentum system, 636–638
coordinates, 632–633
error, 657–660
perturbations, 642–645
physical concepts, 630–631
reliability, 756–757, 762–765
sensors, 646–650
summary, 554, 630–631
telemetry data, 577–583
wheels, use of, 636–638
zero momentum system, 636–637, 650, 658
Attitude and design control subsystem, 646–661, 756–757
Audio:
in TV signal, 453–457
weighting network, 438–441
Automatic gain control (AGC), 122, 599
Automatic repeat request (ARQ), 273
Automation, 83
Automobile population growth, 65
Autumnal Equinox, 542–543, 614, 798–801
Auxiliary circle, 803–804
Availability, 101, 262–263, 771
Avalanche (IMPATT) diode, 561, 563

Average calling rate, 50
Axial ratio, polarization, 248, 250
Axial strain, 664
Axis of moment of inertia, 638–641
Azimuth angle, 210–211, 228, 776, 825–829
equations, 193, 225, 828
vs. latitude and longitude, 218, 222, 226
tracking, 598–599

Backoff in TWT operation:
earth station, 326, 335, 340
energy dispersion, 239
equation, link, 427, 436–437, 478
FDMA earth station, 426–431
input-output curve, 337, 362–363, 432–433, 450, 463, 566
input power, 362–364, 436–437, 445–446, 478
intermodulation, 438, 445–446, 450
modulation, various, 428
multicarrier, 349, 361–363, 369
multiple access, 410–411
multiple video, 433
output selection, 358
satellite, 336, 361–364, 566
SCPC, 450
TDMA, 478
up-link fade, 463
Back-up mode, attitude control, 659
Backyard earth station, 87
Band designations, 181, 212–213
Bandpass, see Bandwidth
Bandwidth:
bit rate reduction techniques, 82
Carson's rule, 330, 438, 448
CCIR, 234, 236
compression, 457, 459
digital, 452
down-link equations, 395–397
expansion in SSMA, 519–524
facsimile, 12–13
filter, 349, 571
information handling capacity, 14–17
limited case, TDMA, 474, 481–482, 484
modulation, 452
necessary (codes), 182, 214
noise, 329–330, 394, 397
spread spectrum usage, 520
television, 386
transponder, 365
typical, 12–17
up-link equations, 389–390
Banking, electronic, 76, 81
Baseband processor, 506–510
Basic group and supergroup, 18–19
Battery, 612, 619–627
charging, 612, 625–626
discharge, 616, 620–626
failure, 760–761
lifetime, 624, 626
mass, 347, 627

Battery (*Continued*)
  nickel-cadmium, 619–627
  nickel-hydrogen, 620–627
  operation, 612–613, 625
  power budget, 614
  recondition, 612–613, 626
  reliability, 626–627, 754–755, 760–761
  thermal requirements, 557
BBC (British Broadcast Company), 6
Beacon, telemetry, 577–579
  block diagram, 591
  in-orbit test, 737
  link budget, 586
  reliability, 754
  synchronization, with beam scan, 501
Beam:
  antenna, 268–308. *See also* Antenna
  cluster, 283, 288, 294
  coverage area *vs.* gain, 281, 285
  diversity, 559
  elliptical, 287, 294
  fan, 501, 505
  hopping or scanning, 412, 500–504
  isolation, 242, 737
  multiple, 279, 294, 323–326, 347–352, 494, 508
  overlap, 251–253
  pointing, 283, 287, 633–636
  vs. population, 510
  projection onto map, 268–269, 283, 287
  rf sensor, 554, 598–599, 647–648, 762
  shape, coverage, 281–283, 287–288, 294, 296, 371, 573–576
  spot, 251–253, 290–294, 300–305, 469–470, 511, 661
  switching (TDRS), 499, 502
  United States coverage, 34, 41, 290, 297–306
Beamwidth, antenna, 275, 280–284, 391
Bearer (B) channels in ISDN, 17
Bearing, wheel, 650
Bearing and power transfer assembly (BAPTA), 593
Bending coefficient K_b, 670
BER, *see* Bit error rate (BER)
Beryllium properties, mechanical and thermal, 679–682
Bias, attitude control, 659
Biased momentum attitude system, 636–638
Bicone antenna, 589–591
Bifocal aplanatic antenna, 295
Billing, TDMA, 471–477
Binary phase shift keying, *see* BPSK (binary PSK, 2PSK)
Binomial distribution, launches, 742
Biorthogonal code, 516
Biphase data code, 581–582
Bipropellant propulsion, 651, 657, 867
Bit, 57, 474
  communications, total per year, 75–76
  energy (E_b), 331, 337, 465
  rate, 57, 474
    vs. bandwidth, 452

reduction technique, 13, 82, 585
  TDMA, 474–475, 480–481
    reduction technique, 82
Bit error rate (BER):
  vs. bit energy per noise density, 337, 452, 465, 471–473, 484, 489–490, 588–589
  vs. figure of quality, 485
  TDMA, 484, 489–490
  terrestrial connection, 101
Black-body radiation, 682–686
Bladder, propellant tank, 541, 656
Blind transmission, 508
Blockage:
  antenna beam, 292
  communications capacity, 410, 528
  grade of service, 53–54
  intersatellite link, 272
Blowdown system, fuel, 656,
B-MAC (multiplexed analog component), 13, 18, 454
Body-stabilized satellite:
  configuration, 548–549
  control, 636–638
  power, 613
Boltzmann constant (k), 329, 586, 595, 606, 734
Booster separation, 663, 723
Boost regulator, power, 612
Bootlace lens antenna, 295, 506
Bore axis, satellite, 635
Boron-epoxy properties, mechanical and thermal, 679–683
Box, stationkeeping, 262, 265, 488, 491, 787–790
Box beam, pointing, 300–305
BPSK (binary PSK, 2PSK):
  bandwidth, 452
  bit error rate (BER), 452, 465, 472, 490
  capacity, TDMA, 484
  coherent, 473
  power spectral density, 488
Broadcasting satellite service (BSS), 87, 127, 500
  assignments, 145–209
  channels and longitude, 166–171, 177–209
  coverage beams, 512–514
  EIRP limit, WARC-77, 326–334
  feeder link frequency, 137–138
  frequency assignments, 145–209
  link budget, 380–381
  location maps, 180–193
  traveling wave tube, 200-W, 567
Buckling coefficient, structure, 669–670
Budget:
  link, 355–382
  mass, 555
  noise, 449
  power, 614
  reliability, 558
Building blocks:
  analog, 13, 19
  digital, 13, 24, 25
Bulge of earth's equator, 845

Bulk transponder business, 4
Burst, 413, 415, 463
  bit rate, TDMA, 481
  color television, 454
  communications, 415
  digital bit stream, 458
  hopping beam, 500–501
  lost, 485
  packet, 100
Bus:
  electric power, 344, 613, 615–618, 628
  structure, 663, 664
Bush traffic, Canada, 271
Business network, 32
Busy:
  blockage, 53–54, 407
  grade of service, 50, 52–54
  hour traffic, 49, 79
  reroute (bypass), 97
  signaling, 112–115
  traffic nomogram, 51

c (speed of light), 180, 273, 783
C, rate of battery discharge, 622–623
Cable in the sky (satellite), 86
Cable television (CATV):
  frequency plan, 107
  growth, 6, 66, 82–83
  plant, 87, 97–99, 101, 106
CAD (computer-aided design), 79
CAE (computer-aided engineering), 79
Calculator, pocket, 65
Calendar, Julian, 799–800
Calls, telephone:
  characteristics, 47, 53, 57, 64
  by mileage, 45
  signaling tones, 114–118
  by time, 48–50
  traffic, United States, 76
CAM (computer-aided manufacturing), 79
Canada bush traffic, 271
Capacity, see Traffic
Cape Kennedy launch, 859–869
Capture ALOHA, 533–539
Capture of channel, 528
Carbon dioxide (CO_2) absorption, 256
Carrier:
  and bit timing, 479
  to interference ratio (C/I), 287, 435, 438
  to intermodulation noise ratio (C/IM), 358, 438,
      445–446, 450
  modulation, 8, 181–183, 214–216, 414, 479
  to noise density ratio (C/N$_o$), 329–333, 389–397,
      438, 465, 471
  to noise ratio, 329–333, 386–397, 436–438, 465,
      471. See also C/N (= C/kTB)
  per transponder, 430–431
  power allocation, 365
  to thermal noise ratio, 328, 373–378, 389–397.
      See also C/T or C/N$_o$ (= C/kT)

Carson's rule, 330, 443, 448
Cartesian coordinates, 774–776
Cassegrain antenna, 295
Catalytic monopropellant, hydrazine, 651–655
Catalytic thruster, 656-657
CATV, see Cable television (CATV)
C-band, 181, 212–213
CCIR, 124
  antenna requirements, 277, 284, 286, 291
  bandwidth, 234, 236
  video quality, 385
CCITT:
  data service, 441
  network hierarchy, 88, 91
  offices, 92
  rain region, 256–258
  switching centers, 88, 92
  tones, 119–120
  weighting factor, 384
CCS (hundred call seconds), 50
CDMA, 515–527
  code locking, 524
  definition, 402, 515
  link budget, 526
  multiple-access techniques, comparison,
      409–412
  operations, 522
  shared transponder, 467
  spread spectrum, 417–418
Celestial mechanics, 774–849. See also Orbit
Cell, see Battery; Solar, cell
Center of mass, spacecraft, 643–644
Central angle, 220, 223, 225, 632–635
Central office, 88, 101–105
Centrifugal acceleration, 781–782
Centrifuge test, 724–726
CEPT (European Posts and Telecommunications),
      24
Chamber:
  acoustic, 730–731
  anechoic, 717
  vacuum, 704–705, 732–733
Chance failure, 741
Channel, 95
  assignment, TDMA, 404, 462, 477
  bank (multiplexer), 78
  BSS assignment, 145–209
  capacity:
    CDMA, 521
    FDMA, 425–431
    TDMA, 477, 482–483
  group, 19
  noise, adjacent, 451
  requests (DAMA), 402–403, 405
  signaling (CSC), 448, 451
  television, 145. See also Broadcasting satellite
      service (BSS)
Charge control, battery, 612, 625–626
Charged-particle accelerator, 720
Charging array, solar, 612–613
Chart, see Nomogram

Chip rate, 519
Christiansen's equation, thermal, 698, 733
Chromanance, television, 13, 18, 456
C/I, 287, 435, 438
C/IM, 358, 438, 445–446, 450
Circuit, 91
 breaker, 613, 628
 duplex, full or half, 91, 95–96
 individual, retail, 4
 loading (FDMA), 363, 369, 429
 switching, 90, 100–101
Circulator, microwave, 563
Cities, 39, 46
Citizen's band (CB), 528
C/kT, see C/N$_o$ (= C/kT)
Classical elements, 810–812
Classification:
 earth and spare stations, 217–218
 multiple access, 401–403
 telecommunications, 12, 14–17
 telecommunications offices, 88
 telephone offices, 90
Clear sky antenna temperature, 312
C-line conditioning, 109–110
Clip level, limiter, 432, 433
Clock synchronization (TDMA), 479–480, 485
Cluster beams, 283, 288, 294
C-MAC, 13, 18, 454. See also Multiplexed analog
 component (MAC)
C-message weighting, 441
CMSA (metropolitan statistical area), 39
C/N (= C/kTB), 329–333, 465, 471
 cross-link, 393–394
 down-link, 395–397, 435, 437
 in link budget, 384
 optimization, 445, 450
 total, 331–337, 438–439
 up-link, 389–390, 435–436
C/N$_o$ (= C/kT), 329–333, 465, 471
 cross-link, 393, 394
 down-link, 395–397
 in link budget, 382
 total, 438–439
 up-link, 389–390
Coast period, launch, 863–864
Cocktail party effect, 528
Code:
 area and zip, 39, 471
 biorthogonal, 516
 biphase data, 581–582
 country symbols, 128–134
 emission designation (ITU), 181–182, 214–216
 encryption (DES), 457
 error correcting (FEC), 100, 416, 465, 479–481.
  See also FEC (forward error correction)
 Golay, 472
 Gold, 516
 orthogonal, 414, 519
 parity, 412
 rate 1/2, 416, 472–473

 spread spectrum, 417
 station class (ITU), 183, 217–218
 Walsh, 522–523
Code division multiple access, see CDMA
Code domain multiple access, see CDMA
Coherent phase-shift keying (PSK), 468, 473, 589
Coil, magnetic, 554, 643–644, 737
Cold gas propulsion, 657
Collision between accesses, 404, 528
Color burst (television), 454
Combinations of SDMA, 505
Combiner, loss, 359
Combining C/Ns, 331–332, 341
Command subsystem, 592–597
 antenna, 596
 decoder, 593, 597–598
 execute signal, 576–577, 583, 597
 format of data, 596
 functional diagram, 552–577
 in-orbit test, 737
 link budget, 595
 list of items, 594
 redundancy, 593, 597
 reliability, 753
 summary, 577–578
Common carrier, 4, 96–98
Common signaling channel (CSC), 429, 448
Communications history, 20
Communications subsystem, 343, 558–576
 antennas, 568–576
 filters and multiplexers, 556–571
 power amplifier, 465–567
 receiver, 561–564
 reliability, 758
 telemetry items, 580
Community reception, broadcasting, 127
Compander, 457, 459
Compression, 22, 457, 459, 585
Computer, 31
 aided applications (CAD, CAE, CAM), 79
 calls, duration, 48
 erlang program, 56–57
 filter optimization, 568
 orbit elements program, 819
 personal, 78
 population growth, 64
 satellite position program, 815
 structure model (NASTRAN), 677–678
 thermal model, 704–712
Concentration of traffic:
 area, 30
 time, 51
Conduction band, solar cell, 605–606
Conductivity, thermal, 541, 680–681, 712–716
Conductivity length, 714
Configuration factor, thermal, 698–706
Congestion of calls, 53
Conical shell, 669–670
Conic curve, 803
Coning, nutation, 639

Constant(s):
  astronomical, 783, 801, 842
  link, *see List of Constants in back of book*
  structure, 679
  thermal, 680
Constant failure rate, 742–745
Consultive International Committee on Radio
    Communications, 124. *See also* CCIR
Contention methods (RMA), 402–407, 515,
    528–529
Contiguous band multiplexer, 567, 571
Continuously variable-slope, delta modulation, 416
Continuous-phase FSK (CPFSK), 473
Control:
  attitude, 646–661. *See also* Attitude control
  center, satellite, 551, 574, 577, 629
  echo, 458, 471
  orbit, 849–859. *See also* Stationkeeping
  power, 553, 615–618, 625
  satellite, 592–597. *See also* Command subsystem
CONUS (contiguous 48 United States), 282
Convection cooling approximation, 714
Conversion:
  analog to digital, 18, 22, 30, 78, 339
  dB to ratio, 438
  frequency, in satellite, 435, 561
  frequency to wavelength, 180, 211, 277, 586
  ratio to dB, 184–185, 586–587
  units, *see List of Constants in back of book*
Convolution noise, 441
Cooled parametric amplifier, 273–274, 318
Cooling, 688–693
Coordinates:
  geocentric equatorial, 774–778
  orbit plane, 802
  satellite antenna pointing, 299, 632–636
  topocentric (az-el), 228, 825
Coordination (interference), 128, 287
Copper-based telecommunications, 84
Correlated failures, 750, 752
Cosatellite interference, 486
Cosines, law of, 778–780
Cosmic:
  background rf noise temperature, 292, 296, 299,
    308
  protons, 544–545, 609, 717, 720–722
Cost, 10
  economy of scale, 11, 84
  vs. growth, 97
  reduction, 83, 493
  vs. traffic potential, 85
  user satisfaction, 97
Country symbols (letter codes), 128–134
Courier I satellite, 8, 516
Coverage:
  angle, 220, 223, 225, 632–635
  cluster, 288
  elliptical, 287
  European, 296
  vs. gain, 281, 285

  from geostationary satellite, 34
  optimized beam, 280–281
  United States, 306, 371
Cover slide, solar cell, 605, 720–722
CPSK or CPFSK, 473
Crane rain model, 256–261
Cross band connection, 509
Crossbar switch, 469
Cross-connect switch, 492–509
Cross-correlation gate (CDMA), 516
Cross-link, *see* Intersatellite, link
Crossover distance, tradeoff, 84–85
Cross polarization isolation, 242–251, 403
Cross-strapped frequencies, 509
Cross-strapping (TDMA), 591
Crosstalk ratio (XTR), 441, 447
Cryogenic LNA, 318
Crystal oscillator drift, 435, 601
CSC (common signaling channel), 448, 451
CSSB/AM audio, 457
C/T, 443
  cross-link, 393, 394
  down-link, 395–397
  in link budget, 379–380
  nomogram, 385
  total, 373–378
  up-link, 388–390
CT (Centre du Transit) telephone offices, 92
CTI switching nodes, 88
CTS satellite, 567
Current-voltage curve, 606–608, 622
Customer premises earth stations, 97, 104, 510
CVSD, 416
Cylinder shell, 664, 669–670

D-to-A conversion, 18, 22–23, 30, 78, 339
DAMA, *see* Demand assignment multiple access
    (DAMA)
Damping of nutation, 639–641
Damping ratio u, 657–658, 673–674
Darken with age, surface, 685
Data:
  baseband, 13–17, 25
  burst, 413, 415, 463. *See also* Burst
  codes, telemetry, 581–583
  collection, 4, 529
  compression, 22, 457, 459, 585
  delay, *see* Delay
  encryption standard (DES), 457
  error rate, *see* Bit error rate (BER)
  hierarchy, 88
  processing, 31
  rate, 57
  services, 76
  terminal, 78, 81
  traffic, Europe, 80, 82, 296
  transmission, *see* BPSK (binary PSK, 2PSK);
      CDMA; QPSK (quadrature phase shift key-
      ing)
Data base sharing, 80

Date, Julian, 797–800, 821
Day, solar, 786
Daylight time, 799
Days between maneuvers, 853, 857
dB, 184–185, 585–586. *See also* Decibel
dBi, ratio to isotropic, 276–278. *See also* Antenna, gain
dBK, temperature, 317, 319
dBm0, mean noise power, 121
dBNF, noise figure, 316–319
DBS, 87, 127. *See also* Broadcasting satellite service (BSS)
dBW, power, 185
dBW/m², power flux density, 186–191
dc/dc converter, 553, 628
DCPSK (differentially coded PSK), 473, 490
Decibel, 184–185, 585–586
    convert to ratio, 438
    ratio to isotropic (dBi), 276–278. *See also* Antenna, gain
    reliability increase, 763
Declination, 774–780
    of lunar orbit normal, 835
    from orbit elements, 814–815
    of sun, 603–604, 797–801
Decoder (satellite), 593, 597–598
Decommissioning satellites, 870
Decomposition law, thermal, 702
Dedicated leased line, 8, 52, 101
Deemphasis (inverse of preemphasis), 384
Deep-space mission, 192
Definitions:
    multiple access, 402
    saturation, 432
    users (ITU), 126
Degradation:
    FDMA, 434–435
    solar cell, 545, 554, 609, 616
    TDMA, 486–487
Degree, from radian, 447, 816
Degrees of freedom, structure, 677
Dehopping, frequency, 417
Delay:
    ALOHA, 531–539
    equations, 391
    increment, 488, 492
    intersatellite, 272–274, 506–508
    satellite range, 225, 270–273, 828
    SDMA, 506
    tolerant protocol, 273, 481
    units, 271, 272
    variation in TDMA, 261, 488–492
Delta (D or data) channel in ISDN, 17
Delta launch vehicle, 730–731, 868
Delta V (velocity difference):
    inclination maneuver, 851–853
    launch, 863–869
    longitude maneuver, 854–858
Demand assignment multiple access (DAMA), 402, 406

FDMA, 425
    SPADE, 406
    TDMA, 462
    telephone circuits, 8
Demodulation, 418, 461
Demultiplexing, 12
Density of material, 679
Depolarization, 243, 249
Deployment:
    in-orbit tests, 736
    mechanism, 661, 663
    zero gravity, 541, 717
Depth of discharge, battery, 622–626
Depth of penetration, meteoroid, 546
Deregulation, 83
DES (Digital Encryption Standard), 457
Descrambler, 459
Design:
    failure, 740
    life, 737
    margin, power, 614
    tradeoff, 343–354
Designations for IFRB filings, 181–183, 214–216
Despreading spectrum, 518
Despun shelf, 548, 550
Destination of call, 64
Destruction of satellite, 9, 869
Diagnostics, 100, 580
Dial tones and pulses, 114
Dial-up telephone facility, 8
Diameter, antenna, 284, 290, 388–399. *See also* Antenna
Dielectric lens, 295
Differentially coherent detected (DCPSK), 473, 490
Differentially coherent PSK, 589
Diffusivity, thermal, 680–681
Digital:
    to analog conversion, 18, 22–23, 30, 78, 339
    audio, 454–457
    baseband, 76
    building blocks, 13, 24
    burst, 413, 415, 458, 463. *See also* Burst
    conversion from analog, 29
    data traffic, 80–81
    Encryption Standard (DES), 457
    hierarchy, 27
    ISDN, 17, 78, 101
    link performance, 386–388
    modulation terms, 465, 468
    radio (microwave), 25, 105
    source rates, 13, 26
    speech interpolation (DSI), 97, 421–422, 425, 458, 477
    sun sensor, 649
    telephony, 419
    television, 13, 338, 388
    traffic, 412
    transmissions, 13, 25, 473
    under voice (DUV), 18

units, 468
voice, 29–30, 78
Diode bypass, battery, 761
Diplexer, 359, 509, 561
Dipole, magnetic, 644
Direct broadcast satellite (DBS), 87, 127, 500. *See also* Broadcasting satellite service (BSS)
Directional coupler, 359
Direction of sun, 797–801
Direction of traffic, 12
Direct modulation, 500
Direct reception, 127. *See also* Broadcasting satellite service (BSS)
Directrix of orbit, 803
Direct sequence CDMA, 417, 516
Discharge, battery, 616, 620–626
Dissemination of information, 4
Dissipation:
    nutation, 639–641
    thermal, 549, 557, 682–686
Distance:
    between points, 40, 41, 778
    between rate centers, 42
    vs. delay nomogram, 273
    to geostationary satellite, 784. *See also* Range
    insensitivity of satellite, 2, 82–83, 84
    intersatellite, 218–220, 230–231
    to moon, 836, 842
    to sun, 801
Distortion, 57
    dual-path, 433–434
    group delay, 441, 448
    intersymbol, 465, 485
    TDMA, 486–487
Distributed command system, 597
Distributed telemetry system, 591
Distribution:
    electric, 344, 613, 615–618, 628
    fuel, 656
Disturbance torque, 630, 642–645. *See also* Perturbations
Diurnal satellite motion, 491, 793–795, 809–810
Diurnal traffic variation, 48, 52, 108
Diversity, antenna beams, 559
D-MAC, D2-MAC, 13, 18, 456
Domain, analog, 78
Domain or division, 402
Domestic satellite, 5, 512
D1 channel bank, 78
Doppler effect, 264–265, 828–831
Double-gimballed momentum wheel, 638
Double hop, 271–272
Double reflector antenna, 295
Double sideband suppressed carrier (DSSC), 473
Down converter, 419, 561
Down-link (space-to-earth link):
    access, 506
    broadcasting satellite (DBS), 380–381
    budget calculation, 361–368
    equations (summary), 395–399

figure of quality, 363, 485
frequency allocations, 140–143
impairment, in TDMA, 487
Intelsat V, 379, 383
maritime satellite, 381–382
Downloaded TDMA software, 471
DPSK, DBSK, 473
Drift, oscillator, 435, 601
Drift rate, 845–858
    acceleration, 845–848
    control of, 854–858
    and range, 831
    repositioning, 865–867
    vs. semi-major axis, 786–788
Driver amplifier, 561–562
Drop, *see* Rain
DS, 25, 104–105
DS-CDMA, *see* CDMA
DSI, 97, 421–422, 425, 458, 477
Dual hop, 271–272
Dual-path distortion, 433–434
Dual polarization, 242–246
Dual-spin satellite, 548, 640
Dual video per transponder, 240, 434, 451, 453
Duplex circuit, 91, 95–96
Duration, call, 47
Dwell duration of hopping beam, 501
Dynamatron accelerator, 720
Dynamic allocation, 97
Dynamic range, reduction, 457, 729

Early Bird satellite, 73, 408, 847–848
Early failure, 740–741
Early warning reservation ALOHA, 528, 538
Earth:
    aphelion, 543, 603, 801, 842
    blockage of ISL link, 272
    bulge, 845
    constants, 783, 842
    coverage beam, 285, 288, 549. *See also* Global beam
    distance to sun, 543, 603, 801, 842
    escape, 862
    flattened, 840, 842
    gravitational potential, 842–848
    magnetic field B at geostationary orbit, 546–547, 552, 644
    mean anomaly, 800
    noise in sidelobe, 292, 588
    nonspherical, 262, 840–848
    orbit, 542, 800
    radiation, 686
    radius, 223, 271, 782–784, 825, 843, 845
    rotation, 259, 784–786, 821
    to satellite link, *see* Up-link (earth-to-space link)
    sensor, 554, 583, 646–648, 660, 757, 762
    services by satellite, 3
    shadow, 689, 794–798
    spin axis, 776
    station, *see* Earth Station

Earth (*Continued*)
  synchronous, 259
Earthquake warning, 4, 529
Earth segment:
  noise budgets, 449
  resource, 273–275
Earth station, 86–87
  antenna, 276–291
  antenna noise, 309
  block diagram, TDMA, 494
  CATV, 106. *See also* Cable television (CATV)
  control via downloaded software, 471
  coordinates, 824–825
  customer premises, 97, 104, 510
  definition, 126
  down-link, 275, 395–397
  EIRP, 335
  FDMA, 418
  figure of merit (G/T), 291–325. *See also* Figure
    of merit (G/T)
  heavy route, 494
  impairments, TDMA, 486
  Intelsat standard, 273–274
  location from satellite, 223, 372, 634
  maritime, 381–382
  monitoring, 471
  path length variation, 270–273, 491–492, 830–831
  pointing, 218, 226, 282, 286, 825–829
    error, 340, 359, 364, 737
  population by longitude, 36
  rf power, 478
  shielding, 275–276
  size, 37
  TDMA, 459, 493–494, 565
  tracking, 259, 264, 597–599
  TV receive only (TVRO), 87, 455
  unmanned, 489
  up-link, 354–361, 388–392
  very small aperture terminal (VSAT), 11, 97,
    271, 359, 510
Ease of use of network, 85
East-west:
  drift, 845–858. *See also* Drift rate
  maneuver, 854–858
  oscillation:
    eccentricity, 809–810
    inclination, 793–795
  stationkeeping, 854–858
EBHC (equated busy hour call), 52
Eccentric anomaly, 803–804
  equations, 808–809
  from r and ṙ, 818–819
  series approximation, 806
Eccentricity, 802, 804
  control of, 858–859
  earth orbit, 542
  east-west oscillation, 809–810
  effect on range, 265, 829–830
  equations, 808
  in Kepler's equation, 806
  from r and ṙ, 818–819

vector, 817–819
Echo:
  cancellation, 458
  control, 458
  control defeat, 471
  single and double hop, 272
  supression disable tones, 117
  time delay, 271, 828
ECHO balloon, 8, 602
Eclipse, 794–798
  bus voltage, 616
  power requirements, 614
  solar radiation, average, 542, 604
  temperature, 543–544, 683, 688–690
Ecliptic, 776–777
  plane, 777, 801
  relation to orbit normal, 836–837, 843–844
Economy of scale, 11, 84
Effective absorptance to sunlight, 696–697
Effective area (aperture), 278, 351, 587
Effective emittance, thermal, 694–697, 715
Effective input temperature, 706, 711
Effective isotropic radiated power, *see* EIRP
Effective length, 667, 716
Efficiency:
  antenna, 278, 279, 572–575
  power, 344–347, 361, 364
  reliability, 759–766
  solar cell, 604, 606
  TDMA, 483
Eight phase PSK (8PSK), 452
EIRP, 226, 274
  budgets and losses, 326–336
  cross-link, 393
  density limit, 237–243
  down-link equations, 395–397
  earth station, 335
  for FDMA, 430–431
  in footprint, 287, 371
  in-orbit tests, 737
  nomogram, 233
  satellite budgets, 336
  for scan, fan or hopping beam, 501
  up-link equations, 390
Electric:
  arcing, 542
  field strength, *see* Field strength (intensity)
  heater, 557, 614
  power, telemetry items, 580
  power conditioner (EPC), 612–618, 625
  power subsystem, 612–630, 754
  propulsion thruster, 644, 657
Electromagnetic spectrum, 210
Electron:
  beam, 720
  fluence, 544, 721–722
  free, 605–606
  volt, 604
Electronic funds transfer (EFT), 76, 81
Elements, orbit, 810–812, 818–819
Elevation angle, 192, 219–232, 825–829

equations, 223–225, 634
  vs. latitude and longitude, 193, 226, 228, 635, 828–829
  vs. range, 223–225, 229–230, 632–635
  vs. sky noise, 313
  tables, 232, 635
  tracking, 598–599
Ellipsoid:
  earth, 842–848
  inertia, 638
Elliptical:
  antenna gain, 280–282
  beams, 287, 294
  orbit, 9, 263, 802–810
Elongation, structural, 666
Emergency position indicating radiobeacon, 127
Emission(s):
  ITU designation, 181–182, 214–216
  spurious, 434
Emission line, 145, 164–166
Emissivity:
  effective, 694–697, 715
  thermal, 682–687
E_b/N_0 (energy per bit to noise density ratio):
  vs. bit error rate (BER), 489, 589
  CDMA, 417
  cross-link, 393–394
  down-link, 395–397
  equation, 471
  with FEC, 481
  threshold, 337
  up-link, 389, 390
Encoder, 523, 579, 754
Encryption, 457, 578
  of audio, 457
End-of-discharge voltage, battery, 623–626
End of fuel lifetime, 259, 741, 854–858
End of life (EOL):
  power requirements, 614–616
  reliability, 760
  surface absorptivity, 683
End office, 86, 88, 101, 111
Energy:
  per bit, 331, 337, 465
  conservation equation, 804, 808, 855
  density, battery, 627
  diagram, solar cell, 606
  dispersion waveform (EDW), 239
  dissipation, nutation, 638, 641
  to noise ratio (E_b/N_0), 331, 471
  photon, 604–605
  sink approximation, 640
  solar protons, 545
  storage, battery, 344–347, 619–626
Environmental torques, 642–645, 659
Environment of space, 541–548, 717
Engineering service channels, 122
E-1 bandwidth, 452
Ephemeris, 799–801
EPIRB (emergency position indicating radio beacon), 127

Epoch, 797–800, 805
Epoxy, 679–683
Equalization filter, 103, 441
Equated busy hour call (EBHC), 52
Equations:
  earth station and satellite geometry, 223–225, 634
  elliptical orbit, 808–809
  link budget, 388–399
  orbit elements, 814, 818
Equator, magnetic, 546
Equatorial plane, 4, 774–777
Equatorial radius of earth, 782–783, 825, 845
Equilibrium temperature, 689–693, 735
Equinox:
  mean, 776
  spring and fall, 542–543, 614, 798–801
Equipment:
  in networks, 101–106
  prices, 83
Equivalent isotropically radiated power,
    see EIRP
Erlang B, traffic unit, 50–64, 75
Error:
  angle in attitude, 657–660
  bit, *see* Bit error rate (BER)
  box, beam pointing, 300–305
  data, 588
  detection coding, 100. *See also* FEC (forward error correction)
  performance, bit error rate, 472, 589
  in pitch, roll, and yaw, 632–637
  rates, TDMA, 489
Escape velocity, 862
Euler's formula, structure, 667–668
Europe, 80, 82, 296
Exclusive NOR gate, 523
Execute signal, command, 576–577, 583, 597
Exhaust velocity, 651, 849, 867
Expansion, thermal, 680–681
Expendable launch vehicles, 662, 868
External tank of shuttle, 869
Extremely high frequency (EHF), 390

Facilities underestimation, 11
Facsimile, 76
  advanced, 22
  analog, 21
  bandwidth, 12–13
  transmission time, 13, 22
  white paper compression, 13, 22
Fading, 485
  FDMA, 425
  margin, 426–428, 471
  TDMA, 486
Failure, 737–746
  analysis, 578. *See also* Telemetry
  density function, 739, 744–746
  mean time between, 108, 742, 744
  modes, structure, 669, 671

Failure (*Continued*)
  probability (unreliability), 738
  rate, 739–745
  single point, 557–558, 746
  solar cell, 610–611, 618, 628
  of spacecraft component, 759–761
  in time, 739
  type of, 741
Fairing, launch vehicle, 662, 723
Fan beam, 501, 505
Faraday rotation, 255
Fast fourier transform (FFT), 517
Fatigue, structure, 726
FBSS/CDMA, 467
FCC (Federal Communications Commission),
  128
FDM, 12, 103
  format, 122
  spectrum, 420
FDMA, 418–457
  backoff, 349, 358, 409–415, 426–431
  in CATV system, 101
  characteristics, basic, 418–445
  definition, 402
  forms, 421–422
  formulas, 435–445
  impairments, 432–435
  management, 425
  operation, 419
  operations, 464
  power amplifier, 425–426
FDMA/SDMA, without switches, 512
FDM/FM/FDMA, 420–421, 429
FEC (forward error correction), 100, 416, 465,
  479–481
  use of, 273, 451, 510, 532
Federal Communications Commission (FCC), 128
Feed, antenna, 288, 295, 359, 573–576
Feedback, 657–658, 728–729
FFT, fast Fourier transform, 517
FH (frequency hop), 414, 417
Fiber optics:
  communications, 2, 11, 83
  DS3 and DS4, 25
  local, 97
  network, 5
  standards, 13, 26
  and terrestrial rf interference, 252, 254
Field strength (intensity):
  formula, 371
  vs. illumination, 220, 232
  nomogram, 224, 233
Figure eight, due to inclination, 265, 491, 793–795
Figure of merit (G/T), 291
  vs. antenna size, 319
  earth station, 317–320
  with EIRP, 274–275
  gain slope (tilt), 466
  graphs, 322–325
  Intelsat Standard A, 273

  reliability increase, 763
  for scan, fan, and hopping beams, 501
  structure, 667
Figure of quality, Q_d, 363, 446, 485
Filter:
  bandpass, use of, 349, 571
  entrance for digital service, 489
  fuel line, 656
  and multiplexer, 556–571
  power loss *vs.* bandwidth output, 475
  receiver, 561
  SDMA, 498
  TDMA, 460
  tilt in gain spectrum, 466
Final Acts of WARC (or RARC), 125
First point of Aries, 774–776
FITS, failure rate, 740, 745
Fixed satellite service (FSS):
  definition, 126
  frequency allocations, 136–143
Flap, solar torque, 554, 643
Flare, solar, 544–545, 609, 720–722
Flat:
  absorber, 687
  plate antenna, 278
  reflector, 687
  spin, 641
  weighting network, 441
Flattened earth, 840, 842
Fluence, *see* Illumination level
  particle, 609, 720–722
  solar radiation, 542, 603–604, 686
  thermal, 694–695, 699, 712
Flux density (PFD), 183–243
Flywheel (momentum wheel), 630, 650–651,
  762–764
FM:
  bandwidth, 330
  radio, 224
  radio signal strength, 233
Focal length, antenna, 574
Focus of ellipse, 802–803
Fog, 256, 299
F1 weighting network, 441
Footnotes in ITU and FCC documents, 134, 145
Footprint, 287, 371. *See also* Coverage
Force on satellite, 262, 642–645, 781, 849
Foreign exchange (FX) office, 99
Form factor, 700–703
Fortran programs:
  Erlang B, 56–63
  orbital elements, 819
  satellite position, 815
Forward error correction (or control), *see* FEC (forward error correction)
Four-wire telephone connection, 90–95
Frame, 413
  command, 594, 596
  synchronization, 415
  TDMA, 479

time, 506
timing, 485
Free space path loss, 226, 254–256, 309, 338, 360, 587
Freeze-frame video, 82
Frequency:
  aeronautical satellite, 146–151
  allocations (assignments), 134–166
  amateur satellite, 152–160
  band designations, 181, 212–213
  broadcasting satellite, 161–163
  cable television (CATV), 107
  channel assignments, 196–209
  conversion in satellite, 435, 508, 561
  coordination, 128
  deviation, television, 386
  division (or domain) multiple access, 402, 418–457
  fixed satellite:
    earth to space, 136–139
    space to earth, 140–143
  hop (FH), 241, 414, 417, 515, 517
  intersatellite, 144–145, 254
  meteorological satellite, 153–154
  mobile satellite, 146–151
  modulation, *see* FM
  natural, 657–658, 671–677
  navigation satellite, 146–151
  of nutation (precession), 639
  of orbit maneuvers, 853, 857
  orthogonal polarization, 242, 245, 494
  planning, 402
  radio astronomy, 145, 164–166
  reuse, 134, 242–253, 494, 559
  signaling, telephone, 109, 111–118, 477
  space operations, 152–160
  space research, 152–160
  vs. spatial separation, 490, 494
  standard time signal satellite, 152–160
  Table of Frequency Allocations, ITU, 125, 128
  touch-tone, 118–120
  to wavelength conversion, 180, 211, 277, 586
Friction, 542
FSS, *see* Fixed satellite service (FSS)
FT (fiber transmission standards), 26
Fuel, 651–657, 849–858
  depletion, 259, 741, 844, 847–848
  distribution and tanks, 541, 652, 656
  management, 541, 656, 717
  mass, 555, 653
  pressurant, 555, 656
  properties (hydrazine), 651–655
  requirements, 852–858
  slosh, 639–641, 717
  tanks, 541, 555, 641, 651–652, 656
Full duplex, 91, 95–96
Fully variable demand access (FVDA), 419
Fuse or circuit breaker, 613, 628
FVDA (fully variable demand access), 419

Gain:
  amplifier, 314
  antenna, 226, 276–285, 570. *See also* Antenna, gain
  attitude control, 657
  automatic control (AGC), 122, 599
  contours, 371
  vs. coverage, 281, 285
  footprint, 287, 371
  LNA, LNB, or LNC, 445
  maximum due to off-axis angle, 291, 572, 575
  maximum due to surface tolerance, 283–284, 289
  processing, 417, 519
  switched steps, 438, 444, 497, 561–562
  vibration testing, 727–729
Galactic noise temperature, 292, 296, 299, 307–312
Gallium arsenide tunnel diode, 561
Gas pressure, 541, 656, 734
GDP *vs.* telephones, 30, 35
General WARC, 124
Geocentric rectangular equatorial coordinates, 774–776
Geographic longitude of satellite, 820–823. *See also* Longitude
Geometry:
  antenna beam, 287
  satellite and earth station, 223, 634
  spherical, 778–780
Geostationary orbit, *see* Orbit
  calculation, 818–819
  growth, number of satellites, 66
  launching, 859–869
  parameters, 784–788
  perturbations, 789, 831–849
  radius, 220–223, 784
  satellite, definition, 127, 258–260, 782
  stationkeeping, 849–859. *See also* Stationkeeping
  transfer orbit (GTO), 659, 723, 860–865, 869
Geosynchronous orbit, 127, 258–260, 264–265
Germanium tunnel diode, 561
Gimballed:
  momentum wheel, 650
  reflector, 570, 573
Glass properties, 679–680
Global beam:
  antenna, 549
  beamwidth and gain, 282, 285
  cluster, 288
  options, 294
  telemetry, 590
Goddard range and range rate (GRARR), 602
Golay code, 472
Gold code, 516
Grade of service, 50–63
Graphite-epoxy, 679–682
Gravitational:
  constant, 781, 783
  field of earth, 644, 842–848
  parameter (GM), 781, 783
Gravity, zero, 541, 656, 717, 784

Gravity gradient, 644–645, 832–840
Great circle distance, 778
Great Depression, 70
Greenwich:
  hour angle (GHA), 820–822
  mean time (GMT), 799–800, 821
  meridian, 776, 820–822
Gregorian date, 799
Grey approximation, thermal, 694
Gross domestic product (GDP), 30, 35
Gross national product (GNP), 32, 72
Ground-based electrooptical deep-space surveil-
    lance (GEODSS), 598
Ground command, 592–597, 629
Ground track, 863
Group of circuits, 12, 18–19
Group delay, 441, 448, 466
  TDMA, 486, 487
Growth in telecommunications services, 9, 64–82,
    101, 108
G/T, see Figure of merit (G/T)
Guard:
  band, frequency, 365
  time, TDMA, 415, 479
Gyration, radius of, 631
Gyroscope, 554, 636, 650
Gyrostat, 640–641

Hail, 299
Half-circuit, 9, 86, 91, 96
Half-duplex, 91, 95–96
Half-hop, 271–272
Half-power beamwidth, 280–285
Hall device, 581
Harmonic oscillation, 671–677
Harmonics, 145, 171, 195
Harness, 344, 597, 613–618
HDTV (high definition television), 82
Header information, 101
Heat, see Thermal
  balance equation, 683, 704–705
  from communications equipment, 361, 364, 551
  conducted, 712–715
  flow, 694–695, 699, 712
  radiated, 684–685, 704
Heater, use in satellite, 614, 656, 737
Heating, 688–693
Height of satellite, 220, 265, 785
Hermes satellite, 567
Hertz per bit, modulation methods, 452
Hierarchy of networks, 88–91
High definition television, see HDTV (high defini-
    tion television)
High power amplifier (HPA), 358, 383, 459, 486,
    564–567
High vacuum, 541, 717, 734
History of communications, 20
Hodocentric map, 268–269, 297
Hohmann transfer, 860–862
Holding time, 47–48, 50–52, 73

Home banking, 118
Honeycomb panel, 670–671, 681
Hop, satellite, 271–272
Hopping, 413–414
  beam, 412, 500–501, 503
  frequency, 241, 414, 417, 515, 517
  SSMA, 519
  transponder, 414, 506, 507
Horizon sensor, 583, 636, 646–648, 660, 757
Horizontal retrace, video, 454
Horn antenna, 294, 568–576
Hour angle, Greenwich, 820–822
Housekeeping power, 344, 614
HPA, 383, 459, 486, 564–567. See also Traveling-
    wave tube
Hub in network, 88, 272, 510
Humidity test, solar cell, 719
Hybrid:
  in feed array, 576
  multiple access method, 515
  satellite, 508
  telephone elements, 90
Hydrazine, 651–657, 849–858. See also Fuel
  properties, 651–655
  requirements, 852–858
Hydrogen lines, 164
Hydrogen and oxygen I_{sp}, 652
Hyperbolic orbit, 803

Ice on antenna, 299
Ideal antenna (1 m²), 290–291, 307
Ideal BER performance, 490
IF (intermediate frequency), 421, 424
IFRB, 124–128, 183, 214–216
Illumination level:
  definition, 183–192
  at earth station, 395
  vs. field strength, 220, 232–233
  illustrations, 221–222
  intersatellite link, 371, 391
  at satellite, 388, 476, 478
IM, see Intermodulation (IM)
Image, see Facsimile; Television
Impairment, 432–435, 484–489, 586
IMPATT diode, 561, 563
Implementation margin, 337
Impulse, specific, 651–652, 657, 849, 867–869
Impulsive thrust, 653
In-band signaling, 109, 114–117
Inclination, 265, 790–794
  control, 850–854
  vs. declination, 780
  definition, 790
  earth's orbit, 777
  east-west oscillation, 793–794
  effect on range, 830–831
  equations, 792
  figure eight, 265, 491, 793–795
  from r and ṙ, 818–819
  maximum over years, 844

north-south oscillation, 488, 491, 793–795
perturbations, 832–842
Individual reception, 127
Inelastic deformation, 666
Inertia, moment of, 631, 638–639, 661
Inertial coordinate system, 774–776
Infant mortality, 740–741
Inflation, 83
Influences on satellite services, 33
Information transfer in U.S., 75
Infrared, 210
  from earth, 686
  earth sensor, 583, 636, 646–648, 660, 757
  radius of earth, 782
  in sunlight, 604
Initial acquisition, CDMA, 524
Injection into orbit, 860–865
Inmarsat, 4
In-orbit tests, 736–737
Input backoff, see Backoff in TWT operation
Input multiplexer, 561, 566–571
Input temperature, 706, 711
Insulation, thermal, 550, 656, 696–698
Integrated services digital network (ISDN), 17, 78, 101
Integrated spacecraft tests, 718
Intelsat:
  capability, 449
  control, 453
  cross-band connection, 509
  earth station standards, 273–274
  frequency reuse, 253
  frequency stacking, 122
  half circuits, growth, 9, 66
  noise budget, 445–449
  power levels in link, 383
  ranging system, 600–602
  satellites, 560
    Intelsat I (Early Bird), 73, 408, 847–848
    Intelsat II, 408
    Intelsat III use of 500-MHz band, 559
    Intelsat IV, 492, 752–761
    Intelsat V link budget, 379, 383
    Intelsat V structure, 663, 664
  television service by region, 38
  traffic, 30, 36–38, 66
Intensity, field, 233, 371. See also Field strength (intensity)
Interactive terminal, 31
Interexchange:
  call, 111
  facilities, 93–94, 101, 105
Interface:
  launch vehicle, 556
  message processor (IMP), 100, 102
  satellite and terrestrial, 23
Interference:
  adjacent satellite, 287, 292–293, 435
  cancellation, 276
  cross polarization, 245, 250

dual path, 434–435
immunity (CDMA), 417, 515
intermodulation, 425. See also Intermodulation (IM)
intersymbol, 465
power, 287, 331, 374–387
reduction, 287
sources, 219, 292
spurious emissions, 434
sun noise, 311, 314
between TDMA users, 485
terrestrial, 134, 252, 254, 275–276
Inter-LATA carrier, 31, 33
Intermediate frequency (IF), 421, 424
Intermodulation (IM):
  amplitude/phase nonlinearity, 432
  backoff, 438, 445–446, 450
  C/IM, 361–363
  interference, 425
  models, 438
  noise, 450
  products, 420
  in TDMA, 466
  and transponder backoff, 347, 349, 363, 445–446
Internal resistance, battery, 624
International:
  switching center (CT-1), 92
  traffic growth, 9, 72
  treaty agreement, 124
International Frequency Registration Board, 124–128
International Telecommunications Satellite Organization, see Intelsat
International Telecommunications Union, 124–128. See also CCIR; CCITT
InterRange Instrumentation Group (IRIG), 582–584
Intersatellite, 192, 272
  delay, 272–274, 506–508
  distance, 218–220, 230–231
  frequency allocations, 144–145, 254
  link, 418, 458, 506–507
    access, 401
    budget, 369–378, 391–394
    losses, 327, 339
    noise, 315
    power levels, 383–384
    service, definition, 127
Intersymbol:
  distortion, 465, 485
  interference, 465
Interval between maneuvers, 853, 857
Intraexchange calls, 111
Intrasatellite switching, 401
Intrasystem signaling, 109
Invar, 679–680
Inventory control, 529
Investment in space segment, 10
Ionosphere, 70, 252, 256, 600

Irradiance, 604, 683, 695
ISDN (integrated services digital network), 17, 78, 101
ISL, 192, 272. *See also* Intersatellite, link
Isolation, beams and polarization, 242, 246–251, 403, 737
Isothermal approximation, 683, 694
Isotropic antenna definition, 276, 587
Iterative solution, 56, 806–807, 814–815
ITU, 124–128. *See also* CCIR; CCITT
    emissions designation, 181–182, 214–216
    station class codes, 183, 217–218
I-V curve, 606–608, 622

Jamming, 486
Jansky, flux density, 234
Julian date, 797–800, 821
Jumbogroup, 18
Junction, p-n, 606
Junk, space, 9, 844, 848

K-band, 181, 212
Kelvin temperature, 291, 544, 684–685
Kennedy Space Center launch, 859–869
Kepler's equation, 805–807
Kevlar, 679–680
Kinetic energy, 544, 734, 808, 847
Kirchoff's law, 694

Lack of facilities, 11
Lambert's law, 694, 699–700
LAN (local area network), 23, 31, 78, 88
Land mobile satellite, 127
    frequency allocations, 146–151
Large-scale integrated circuits, 417, 519
Laser communications, 458
LATA (local area transport arrangement), 23
Latitude:
    of earth station, 193, 225
    geographical, 268, 823
    of Kennedy Space Center, 863
    and longitude, 41, 823
    and pitch angle, 636
    of satellite, 780, 814–815
    Stefan-Boltzmann, 682–688
Launch, 859–869
    Hohmann transfer, 860–862
    Monte Carlo simulation, 766–772
    reliability, 10, 742
    sequence, 863–865
    vehicle adapter, 663–665
    vehicle capability, 867–868
    vehicle fairing, 662, 723
    vibration, 555–556, 661, 720–732
Law:
    addition for emittances, 696
    configuration factor, 701–702
    cosines, 778–780
    gravitation, fundamental, 781, 842
    Kirchoff, 694

Lambert, 694, 699–700
    sines, 779–780
    Stefan-Boltzmann, 682–688
Layers of foils, insulation, 550, 656, 696–698
L-band, 212
Leased line, 8, 52, 101
Length, effective, 667, 716
Lens antenna, 295, 506
Letter designations, bands, 181, 212–213
License, FCC, 128
Lifetime:
    component, 609, 624, 717, 745, 852
    satellite, 10, 259, 741, 760–761, 766–772, 848
Light, *see* Solar
    electromagnetic spectrum, 210
    scattering, 171
    velocity of, 180, 273, 783
Limitations of satellites, 10, 760–761
Limit of coverage, 4
Limited:
    assigned TDMA, 462
    motion video, 83
Limiter, 411, 432–433, 463
Line:
    leased, 8, 52, 101
    losses, 309, 311, 359, 614
    noise temperature, 317
    number needed (telephone), 53
    radio astronomy absorption, 164–166
Linearized TWTAs, 433, 450
Linear nomogram, *see* Nomogram, linear
Linear polarization, 251
Link:
    definition, 126
    intersatellite, 126, 391–394, 418, 458, 506–507.
        See also Intersatellite
    losses, 220, 326–327, 340
    performance, 513
    requirements, 365
Link budget, 355–382
    command subsystem, 595
    direct broadcast satellite, 380–381
    equations, 388–399
    figures, 383–384
    Intelsat V, 379
    maritime service satellite, 381–382
    telemetry subsystem, 585–588
    total, 331–332, 341–342, 438–439
Liquid bipropellant, 550, 865, 867
Liquid motion, 639–641, 717
LNA, 312, 318, 561–563
LNA, LNB or LNC, 320, 445
Load, structure, 556, 661–671, 723
Lobe, 292. *See also* Sidelobe
Local:
    access loops, 81
    area network (LAN), 23, 31, 78, 88
    area transport arrangement (LATA), 23
    earth stations, 11, 87, 97, 271, 359, 510
    exchange, 88

exchange carrier, 23, 31, 33
fiber optics, 97
loop, 23, 31, 90, 93, 97, 101
oscillator, 435, 561, 601
outside plant, 101, 104
Location, satellite, 266–267, 810–816. *See also*
Longitude
Lockalloy, 681
Logarithmic units, 185. *See also* Decibel
Longitude:
acceleration, 845–848
BSS orbit location, 166–171, 177–209
change of, 865–866
control (stationkeeping), 854–858
difference for two satellites, 218–220, 230–231
of earth station, 824–825
from elements, 820–823
elevation angle, 193, 226, 228, 635, 828–829
great circle distance, 41, 778
after launch, 864
perturbations, 845–849
pitch angle, 635
vs. population, 34, 36
satellite motion, 491
of satellites, 266–267, 820–823
of sun, 800
vs. traffic, 36, 194
variation (E/W), 491
view from satellite, 4, 268
zero (Greenwich), 776, 820–822
Long March, 868
Look angles, 219–232, 825–829. *See also* Azimuth
angle; Elevation angle
Loss:
additional link, 220, 326–327, 340
earth pointing, 737
vs. noise temperature, 297
packet, 528, 532
path, 226, 254–256, 327, 338, 587
Lost burst, 485
Lost packet, 528, 532
Low noise amplifier (LNA), 312, 318, 561–563
Low orbit, 8, 265, 785, 863–864, 869
LSI (large scale integrated circuit), 417, 519
Lubricant, solid, 542, 717
Lunar:
constants, 222, 783, 836, 842
orbit normal, 835–837
perturbations, 834–838, 841–842
rf noise, 299

MAC (multiplexed analog component), 13, 18,
454
Magnesium, 679–681
Magnetic:
bearing suspension, 650
coil on satellite, 554, 643–644, 737
field of earth, 546–547, 552, 644
Main engine cutoff (MECO), 720, 723
Main lobe, 292. *See also* Antenna

Maintenance, *see* Stationkeeping
antenna misalignment, 243
battery charging, 612, 625–626
losses due to, 340
satellite replacement, 766–771
MAMSK (multiamplitude minimum shift keying),
473
Maneuver, 849–859. *See also* Stationkeeping
Man-made interference and noise, 134, 252
Maps, 4, 180–193, 268–269, 296–305
Margin:
power, 614
precipitation, 261–262, 426–428. *See also* Rain
of safety, structure, 667
system, 471
Maritime, 2, 7, 127
frequency allocations, 146–151
link budget, 381–382
pointing losses, 359
population (ships), 44
time delay, 271
Markets for satellite services, 5–7
Mark signal, 517
M-ary, 452, 473
Mass, *see specific items*
center of, 643–644
expulsion, 849–850, 867
modeling, 337–354
saving, 567, 677, 682, 761–766
summary for satellite, 555
Mastergroup, 18
Matched filter PSK, 473
Matching spot beams to the demand, 511
Material:
properties, 654–655, 679–680, 687
for structural elements, 681
Matrix switch, 349–350, 495–507. *See also* Switch
(in satellite)
Maximum:
moment of inertia, 637–641
power point, solar cell, 607, 610
stress, 667–671
MBPC (multiple burst per carrier), 414
MBPT (multiple burst per transponder), 415
MCPC (multiple channel per carrier), 414, 418
MCPT (multiple channel per transponder), 415,
418
TDMA, 458
video, 451
Mean anomaly, 805–806
equinox of date, 776
Kepler's equation, 805–807
longitude of satellite, 848–849
longitude of sun, 800
from r and ṙ, 818–819
Mean solar day, 786
Mean time between failures (MTBF), 108, 742, 744
Mechanical:
deployment, 661, 663
properties of materials, 679

Mechanical (*Continued*)
    vibration, 673–678, 720–732. *See also* Vibration
MECO (main engine cutoff), 720, 723
Memory load data, command, 596
Mercator projection, 268
Meridian, Greenwich, 776, 820–822
Message telephone service (MTS), 77
Meteoroids, 544–546, 642
Meteorological satellite frequencies, 153–154
Metropolitan areas, 39, 46
MFC Bern Region R2 tones, 120
Microprocessor, 585, 658
Microterminal earth station, 33, 403, 406. *See also* VSAT (very small aperture terminal)
Microwave filter, 566–568
Microwave radio relay:
    and central office, 105
    cost, 84
    growth, 67
    interference, 219
    terrestrial bit rate, 25
Midocean communications, 4
Milky Way (noise source), 296
Minimum moment of inertia, 637–641
Misalignment:
    of antenna, 243
    of sensors, 659
    of thrusters, 643
Mixed use transponders, 467
Mixer in receiver, 318, 561
Mobile, 2, 7, 127. *See also* Maritime
    frequency allocations, 146–151
Model:
    Crane rain, 225–261
    reliability, 752–759
    satellite, 337–354
    structure, 677–678
    system, 337–343
    thermal, 704–712
Modem, 103, 425, 486
Modulation, 8, 181, 214–216, 452, 459
Modulus of elasticity (Young's modulus), 664–667, 678–679
Modulus of rigidity (shear modulus), 665, 678–679
Molniya orbit, 9, 263, 265, 506, 548
Moment of inertia, 631, 638–639, 661
    change, 639
    gravity gradient, 644–645
    maximum and minimum, 637–641
    test, 717
    torque, 657
    transverse, 641
    wheel, 650
Momentum exchange, 849–850, 867
Momentum wheel, 554, 630, 650–651, 762–764
M1-2 or M2-3 multiplexer, 78
Monitor and alarm (TDMA), 468, 471–472
Monocoque cylinder, 669–670
Monopropellant propulsion, 580, 644, 651–661, 756. *See also* Fuel; Thruster

Monopulse tracking system, 599
Monte Carlo method, 766–772
Moon:
    constants, 222, 783, 836, 842
    effect on satellite orbit, 834–838, 841–842
    orbit normal, 835–837
    rf noise, 299
Mortality Table, 739–740
Motion of satellite, 488, 491, 632–633
Moving satellite's location, 865–867
MPSK (M-ary phase shift keying), 452, 473
MSA (metropolitan statistical area), 39, 46
MSK spectrum, 488
MTS (message telephone service), 77
Multiband satellite, 253
Multibeam service, 279, 294, 323–326, 347–352, 494, 508
Multicarrier operation, 361–363, 370. *See also* Backoff in TWT operation
Multihop, 271–272
Multimission satellite, 401, 418, 508
Multipath, 434, 466, 487
Multiple access, 401–540
    backoff, 428
    options, 408
Multiple beams, 279, 294, 323–326, 347–352, 494, 508
Multiple satellite interference, 293
Multiplexed analog component (MAC), 13, 18, 454
Multiplexer, 103
    digital, 78, 101–105
    satellite, 561, 566–571
    U.S. telephone, 12, 18
Multipoint, 87
Multisatellite:
    link, definition, 126
    SDMA, 506, 507
Multiservice satellite, 401, 418, 508
Multitone ranging system, 600–602
Mutual interference, 9
Mux, *see* Multiplexer
Mux/Mod/Mac codes, 422
Mylar, aluminized, 697

Nadir angle, 220, 223–227, 634
Narrow-band TV, 451
NASTRAN, lumped element program, 677–678
National dial code, 88
Natural frequency, 657–658, 671–677
Natural interference, 134
Navigation satellite frequency allocations, 146–151
Near-geostationary orbit, 786–788
Negative resistance diode, 561–564, 606–609
Net heat flow in satellite, 712
Network, 85–123
    adaptive, TDMA, 471–472
    audio weighting, 438–441
    business, 32
    four-wire, 93
    half-duplex, 91, 95–96

manager, 97
Operations Center, 407
packet switched, 97, 100, 102, 416
partially interconnected, 89
private line, 98
ring, 89
simplex, 95
star, 88
telecommunications, 85
two-wire, 94
type, 12
New services, 11, 76
News gathering, 6
Newtonian approximation, thermal, 690, 714
Newton-Raphson iterative method, 56, 806–807, 814–815
Newton's:
gravitational equation, 781, 842
law of cooling, 714
N_2H_4 hydrazine fuel, 651–656. *See also* Fuel
Nickel-cadmium (NiCd) battery, 619–627
lifetime, 630
mass, 346–347, 627
reliability, 755, 760–761
use, 553, 612
Nickel-hydrogen (NiH) battery, 620–627
lifetime, 630
location in spacecraft, 550
mass, 346–347
use, 553, 612
Nitrogen, 656, 731
NOC (Network Operations Center), 407
Nodes:
communications network, 88, 100
orbit, 790–791
precession, 776, 832–844
structure, 677–678
thermal, 704–712
Noise, 291–315
ambient, 252
amplifiers and mixers, 318
budgets, 449
convolution, 441
cosmic, galatic, tropospheric, 292, 296, 299, 307–312
density (N_0), 329, 331
vs. elevation, 308–309, 313
figure nomogram, 319
from losses, 297
power, 443, 588
pseudorandom, 417, 516
rain effect, 297, 299, 311–312
shuttle, 556, 731
sources for antenna, 292, 384
telemetry and command system, 586–588, 593–595
temperature:
sources of, 308–309, 312, 320
of sun, moon, earth, 299, 307–309, 311, 314–315
weighting network, 438–441

Nominal satellite location, 488, 491, 787. *See also* Longitude
Nomogram:
linear:
carrier to noise *vs.* data rate, 386–388
field strength, illumination, EIRP, 233
S/N *vs.* C/T, 385
TDMA spectra, 474
telephone traffic, 51
spiral:
combining carrier to noise ratios, 332, 342
frequency *vs.* wavelength, 180, 211
noise figure *vs.* noise temperature, 319
range, range loss, elevation, 229
time delay *vs.* distance, 273
Noncoherent FSK, 589
Nongeostationary orbit, 3, 265, 785–787, 864
Noninterference basis, 128
Nonlinear:
channel impairment, 486
devices, 432–433, 561
Nonreflective solar cell, 607–609
Nonreturn to zero code, 581–582
Nonspherical earth, 262, 840–848
Nonswitched service, 57, 86
No return to zero (NRZ), 581
NOR gate, 523–524
Normal distribution, 745–746
North American hierarchy, 90
analog, 18, 88–90, 92
digital, 13, 27, 105
North Atlantic telephone traffic, 73
North celestial pole, 775–776
North-south:
maneuver, 651, 789, 849–854
oscillation, 790–793
roll accuracy, 637, 660
solar array, 549, 604
Notching in vibration tests, 727–729
Nozzle, thruster, 656
NRZ-L, no-return-to-zero code, 581–582
NTSC television, 82–83, 385. *See also* Television
Nuclear power, 612
Nutation of spinner satellite, 581, 637–641

Oblateness of earth, 262, 840–848
Obliquity of ecliptic, 777, 801
Off-axis:
antenna gain, 291, 572, 575
loss, 364. *See also* Pointing, satellite, control, error
Off-hook telephone, 109, 113
Office:
central, 88, 101–105
class (telephone), 90–92
communications, 87
of the future, 81
symbol, 88, 92
Offset:
antenna feed, 295, 568, 573

Offset (*Continued*)
  keying (OK/QPSK), 488
OK/QPSK spectrum, 488
OMA, definition, 402
Omnidirectional (omni) antenna, 548, 596
On-axis fed antenna, 295
On-board:
  frequency band translation, 435, 508, 561
  microprocessor, 585, 658
  processing, 401, 416, 506, 510
  recoding, 506
  regeneration, 416, 479
  remodulation, 350, 416
  retiming, 416, 506
  switching, 418, 458. *See also* Switch (in satellite)
On-demand services, 8
On-hook telephone, 109
Operating point, 438, 443. *See also* Backoff in
  TWT operation
Operations:
  Network Center (NOC), 407
  Spacecraft Control Center (SCC), 551, 574, 577,
    629
  TDMA, 464, 468
Operator signaling, 109. *See also* Signaling
Optical link, 2, 145, 171. *See also* Fiber optics; Intersatellite
Orbit, 774–870. *See also* Geostationary orbit
  angular momentum, 808, 818–819
  change:
    drift rate, 786–788
    repositioning, 865–867
  circular, 781–788
  conic, 803
  constants for geostationary satellite, 784, 842
  of earth, 542, 800
  eccentricity, 802, 804. *See also* Eccentricity
  elements:
    from r and ṙ, 818–819
    summary, 810–812
  elliptical, 263, 802–810
  inclination, 265, 790–794. *See also* Inclination
  low (parking), 8, 265, 785, 863–864, 869
  normal, control, 850–854
    from elements, 790–792, 814–815
    perturbations, 832–842
  period, 4, 127, 265, 782–788, 808
  perturbations, 262, 789, 831–849
  plane orientation, 790–792, 811–812, 863–865
  radius variation, 785–787
  raising, 870
  separation between satellites, 218, 230–231, 339,
    490, 494
  spectrum crowding, 10, 266–267
  transfer to geostationary (GTO), 659, 723,
    860–865, 869
Orbiter vehicle (STS):
  launch, 863, 869
  limit load, 725
  payload bay, 661–662

  vibration levels, 730–731
Orientation, 630–636, 646–661. *See also* Attitude
  control
Orthogonal:
  codes, 414, 519
  multiple access, 402
  polarization, 242, 245, 494
Orthomode coupler, 497
Oscillation, *see* Resonance
  in latitude, inclination, 488, 491, 793–795
  in longitude:
    eccentricity, 488, 491, 809–810
    inclination, 793–795
  long term, 848
  sloshing in tank, 641
  structure, 671–677
Oscillator, crystal, 435, 601
Outage, 101, 263, 771
  rf noise, 292–315
Out-band signaling, 109, 114–117
Outgassing, 542
Output multiplexer, 561, 570–571
Outside plant, 101, 104–105
Overcharging battery, 612, 625
Overdrive of TWTA, 361–363, 432, 463. *See also*
  Backoff in TWT operation; Traveling-wave
  tube
Overlap ALOHA, 536–539
Overseas traffic, 72. *See also* International
Owners, of satellite, 2
Oxygen absorption bands, 256, 686

P. (grade of service), 50, 52–54, 56
Pacific Island communications, 529
Packet:
  burst composition, 100
  loss rate, 528, 532
  routing and address, 415
  switched network, 97, 100, 102, 416
Paging, 31
PAL (video system), 82–83
PAMA (preassigned multiple access), 404, 406
Panel, 670–672, 681. *See also* Solar, array
Parabolic:
  anomaly, 818–819
  orbit, 803
  reflector, antenna, 226, 568, 573–575
Parallax correction of elevation, 826–827
Parallel redundancy, 746–752
Parallelled power amplifiers, 512, 514
Parameter:
  gravitational (GM), 781, 783
  orbit dimension, 803, 808
  unit vector, 813–814
Parametric amplifier, 273–274, 318
Parity code, 412
Parking orbit, 8, 265, 785, 863–864, 869
Particles in space, 544–545, 609, 717, 720–722
Path length, *see* Range

Path loss, 226, 254–256, 327, 338, 587
Payload, 337–354
  communications equipment, 343, 558–576
  shuttle bay, 661–662
PBS (video), 454–457
PBX or PABX (private branch exchange), 104
PCM (pulse code modulation):
  data, 582
  in FDMA, 419
  voice, 460
PCM/PSK/FDMA, 419, 421, 423
Peak deviation, 447
Peel test, solar cell, 719–720
Penetration of particles, 546, 609
Performance of telecommunications network, 101–109
Perifocus, 805
Perigee, 260, 817
  argument of, 811–813, 818–819
  motor delta-V, 859, 863–869
  motor vibration, 555–556, 661
  unit vector, 812–814
  velocity, 805, 808, 860–864
Perihelion, 603, 801
Period, orbit, 4, 127, 265, 782–788, 808
Permitted allocation, 125. *See* Allocation
Personal computer, 78
Perturbations, 262, 789, 831–849
  attitude of satellite, 642–645
  moon, 834–838, 841–842
  nonspherical earth, 260, 840–849
  sun, 832–834
  sun and moon, 838–840
PEV (pitch excited vocoder), 416
PFD (power flux density), 183–243
Phase:
  array, antenna, 295, 506
  digital transmission, 473
Phase nonlinearity, 432
Phase shift keying, 419, 452, 472–473, 484, 488.
  See also BPSK (binary PSK, 2PSK); QPSK
  (quadrature phase shift keying)
Photon energy, 604–605
Physical constants, 679–680, 682, 783–784, 801
Picture phone, 76, 82
Pilot tone, 121, 583
PIN diode attenuator, 561
Pitch, 270, 632–636
  control, *see* Attitude control
  error, 637, 660
  gravity gradient, 645
  pointing, 635–636
Pitch-excited vocoder (PEV), 416
Planar array, 278
Planar projection, 776–777
Plane change, 863–865
Plant, outside, 101, 104–105
PM nonlinearity, 432–433
PN coding, 417, 516
Pocket calculators, growth, 65

Pointing:
  earth station antenna:
    azimuth and elevation, 218–226, 825–829
    error, 340, 359, 364, 737
  satellite, control, 263, 630–631
    error, 282, 286, 364, 367, 372–375, 637, 660
    pitch, roll, and yaw, 270, 299–305, 632–636
    solar array, 543
Point-of-sale communications, 529
Point-to-point service, 86–87, 286
Poisson distribution, 53
Poisson's ratio, 664, 666
Polar antenna mount, 287
Polaris star, 649
Polarization:
  impairment, 434, 486, 586
  isolation, 246–251, 403
  reuse, 134, 242–253, 494, 559
  rotation, 244–251
Polar projection, 778, 840, 851
Poles, switch, 498
Polling, 101, 537–539
Population:
  vs. beam size, 510
  by longitude, 34, 36
  maritime, 44
  United States, 32, 38–41, 297–298, 511
  world, 71
Position vector of satellite, 810–816
Postal, telephone and telegraph authorities, 5
Potential energy *vs.* longitude, 847
Power:
  assignments in transponder, 365
  control electronics, 553, 612–618, 625
  distribution bus, 344, 597–613, 615–618, 628
  flow in power amplifier, 361, 364
  flow in satellite, 712
  flux density (PFD), 183–243
  levels in link budget, 355, 383
  limited case, TDMA, 481–482, 485
  margin, 614
  modeling, 337–354
  spacecraft budget, 614
  spectrum density, 221, 488
  subsystem equipment, 344, 605–630
  subsystem reliability, 754–755
  sum of interference, 287, 331, 374–378
  supply for transmitter, 464, 628–629
  telemetry receiver, 588
Power amplifier, 347–354, 426, 512–514, 549, 561,
  564–567. *See also* HPA; Traveling-wave tube
Preamble:
  command frame, 594, 596
  generator, 459
  TDMA frame, 479
Preassignment of channels, 8, 97, 404, 462
Precession:
  equinox, 776
  lunar orbit normal, 835–837
  north pole, 776

Precession (*Continued*)
  satellite orbit normal, 832–844
  satellite spin axis, 581, 637–641
Precipitation, 242–252, 311–312. *See also* Rain
Pressure:
  atmospheric, 541
  fuel tank, 555, 656
  from solar radiation, 642, 858–859
Pre-wired cross connection, 495
Price lists, 87
Price sensitive traffic, 83
Primary allocation, 125. *See also* Allocation
Principal axes, 638–639
Priorities, 64, 97, 125
Private line networks, 98
Probability, *see* Reliability
  call being blocked, 54
  data error, P_e, 588
  operating, 737–741
  satellite survival, 760
Processing:
  gain, 417, 519
  interacting technologies, 31
  on-board, 401, 416, 506, 510
Programs, *see* Fortran programs
Projection, map:
  hodocentric, 268–269, 297
  Mercator, 268
  polar, 778, 840, 851
Propellant, 651–657, 849–858. *See also* Fuel
Properties, of materials, 654–655, 679–680, 687
Proportional gain, control system, 657
Propulsion system, 580, 644, 651–661, 756, 867. *See also* Fuel; Thruster
Pros and cons of satellites, 10
Protons in space, 544–545, 609, 717, 720–722
Pseudorandom:
  bit stuffing, 464
  Monte Carlo method, 766–772
  noise (PN), 417, 516
PSK, 419, 452, 472–473, 484, 488. *See also* BPSK (binary PSK, 2PSK); QPSK (quadrature phase shift keying)
Psophometric weighting, 441
PTT (postal, telegraph, and telephone authorities), 5
Public Broadcasting Service, 454–457
Public switched telephone service, 17
Pulse code modulation, 419, 460, 582. *See also* PCM (pulse code modulation)
Pulsed plasma thruster, 657
Pulse width modulation (PWM) regulator, 617
Push to talk, 422
PWM audio for video service, 454
Pyrotechnic device, 556, 723

Q_d, figure of quality, 363, 446, 485
QAM/SC (quadrature amplitude modulation/suppressed carrier), 473

QASK (quadrature amplitude shift keying), 473
QPSK (quadrature phase shift keying):
  audio, 454–457
  bandwidth, 452
  bit error rate (BER), 472, 484
  demodulator, 461
  modulator, 459
  spectrum, 488
Quadrature, 473. *See also* QPSK (quadrature phase shift keying)
Quality:
  control, 558, 630, 740, 761–766
  figure of, Q_d, 363, 446, 485
  factor, Q, 274–275
  of service, 52–57

$R_{1/2}$, coding rate, 416, 472–473. *See also* Rate (coding)
Radar cross section, 597
Radian conversion to degree, 447, 816
Radiation, 552, 682
  belt, Van Allen, 547–548
  black body, 682–686
  and conduction, 712–716
  cooling and heating, 688–693
  coupling factor, 698–709
  degradation of solar cell, 544, 554, 609, 616
  from earth, 686
  heat transfer, 704
  length, 667, 716
  particle, 544–545, 609, 717, 720–722
  pattern, 276–277, 310, 589–590. *See also* Antenna
  pressure, solar, 642, 859
  of rf power, 226, 274. *See also* EIRP
  solar, 542, 603–604, 686
  stellar, 543–544
  thermal, 557–558, 682–688, 692–712
Radio:
  amateur, 528
  astronomy frequencies, 145, 164–166
  navigation satellite frequency allocations, 146–151
  regulations, 125, 128
Radius:
  of earth, 223, 271, 782–784, 825, 843, 845
  vs. eccentric anomaly, 804, 809
  of geostationary orbit, 220–223, 784
  of gyration, 631
  vs. orbit period and velocity, 785
  vs. true anomaly, 803, 809
  vector of satellite, from orbit elements, 812–815, 817
Radome, 359
Rain, 242–252, 311–312
  attenuation, 254–256, 261–262, 311, 426–428, 487
  Crane model, 256–261
  noise temperature, 297, 299, 311–312

polarization rotation, 243–249
rate, 260
regions, 256–259
Random:
    failure, 740–741, 743
    Monte Carlo simulation, 766–772
    vibration, 730
Random Multiple Access (RMA), 402–407, 515, 528–539
Range:
    delay due to, 225, 270–273, 828
    distance, 220–221, 229
    vs. elevation, nadir, central angle, 223–225, 229–230, 632–635
    intersatellite, 218–220, 230–231
    measurement, 600–602
    path loss, 226, 229, 254–255, 338, 587
    signal in satellite, 577, 579
    test on satellite, 717–718
    variation, daily, 270–273, 491–492, 830–831
Rank:
    earth station size, 37
    urban area size, 38
RARC (Regional Administrative Radio Conference), 125
Rate (coding):
    vs. bandwidth, 452
    BER vs. bit energy per noise density, 472
    one-half, $R_{1/2}$, 416, 472–473
Ratio to dB conversion, 184–185, 586–587
RBOC (Regional Bell Operating Company), 117
Reacqusition mode, attitude control, 659
Reaction control system, 262, 580, 651–661, 756. See also Attitude control
Reaction wheel, 548, 554, 630–631, 636, 650–651
Real time service, 57
Receiver, 561–564
    block diagram, 348
    figure of merit, earth station, 291
    gain switch, 444. See also Attenuator in spacecraft
    noise elements, 317, 340, 448
    output spectrum, 420
Reception, broadcasting types, 127
Rechargeable battery, 612, 619–627. See also Battery
Reciprocity law, thermal, 702
Reconditioning, battery, 612–613, 626
Record services, 76
Rectangular:
    coordinate system (x, y, z), 774–776
    panel, 664, 670–672
    spectrum, 221, 227
Redundancy, 592, 597, 736, 746–759, 761–766
References, 122, 399, 539, 772, 870
Reflector, 226, 568, 570, 573–575
Regeneration, 13, 401, 416, 479
Regional Administrative Radio Conferences (RARC), 125

Regions, ITU, 124–125
Regression, of nodes, 776, 832–844
Regulation, of bus voltage, 344, 615–618
Relay contacts, 541, 717
Relay satellite, 8
Reliability, 732–772. See also Probability
    availability, 263, 771
    battery, 626–627
    definition, 737–738
    failure density function, 739
    improvement, 759–766
    introduction, 557–558, 741
    model of Intelsat IV, 752–761
    Monte Carlo simulation, 766–772
    random failure, 742–745
    redundancy, 592, 597, 736, 746–759, 761–766
    single point of failure, 557–558, 628
    wear-out failure, 745–746
Remodulation, 238, 350, 401, 416, 418
Remote encoder, telemetry, 591
Remote telecomputing, 78
Repeater, 86, 103. See also Transponder
Replacement, of satellites, 766–771
Repositioning, satellite, 865–867
Request channel, 402–403, 405
Reservation services, 76
Residential telecommunications, 104
Residual magnetic dipole moment, 644
Resonance, 566, 645, 657, 671–677, 728. See also Oscillation
Resources, multiple access, 405
Rest frequency (radio astronomy), 164–166
Retiming, 416, 506
Retransmission for RMA, 528, 531–532
Retrieval of satellite, 869
Retrograde orbit, 791
Return to zero code (RZ), 581–582
Reuse of frequency allocations, 134, 242–253, 494, 559
Reverberation chamber, 731
Reverse frequency allocation, 559
Reverse saturation current, 606
rf:
    beacon, 577–579
    sensor, 554, 598–599, 647–648, 762
    signal leakage, 486
Right ascension, 774–776
    of ascending node, 790–793
    from **r** and **ṙ**, 818–819
    of satellite, 814–815, 822
Right-hand rule, 632
Right spherical triangle, 780
Ring:
    network, 89
    signal, telephone, 109, 114–117
RMA (Random Multiple Access), 402–407, 515, 528–539
RMS surface, antenna, 283
Robust CDMA, 515
Rocket, 865–869

Roll, 632–637
  control, *see* Attitude control
  definition, 270, 632–633
  error, 637, 660
  gravity gradient, 645
  pointing, 635–636
Rotation:
  earth, 259, 784–786, 821
  polarization, 244–251
  wheel, 548, 554, 630, 650–651
Round-Robin Reservation ALOHA, 537–539
Round trip propagation time, *see* Echo
Routing, packet, 415–416, 458
RSS of attitude control errors, 660
Rural communications, 3, 505
RZ code (return to zero), 581–582

Safety margin, structure, 667
Satellite, *see specific items*
SBPC (single burst per carrier), 414
SBPT (single burst per transponder), 415
Scanning:
  beam, 501, 504
  density, 13
  horizon sensor, 646–647
Scintillation, 252, 255, 256
Score satellite, 8
Scout launch vehicle vibration, 730–731
SCPC (single channel per carrier):
  definition, 403
  with FDMA (SPADE), 419, 421
  input power, 432
  optimization, 450
  transponder capacity, 426
Scrambling for security, 7, 410, 457, 459, 515
SDMA, 409, 490–515
  definition, 403
  forms of, 494, 495
  with satellite switch, 465, 469, 498
  time delay, 506
SDMA/BH/TDMA, 504
SDMA/FDMA, 492
SDMA/SS, block diagram, 458
SDMA/SS/CDMA (FTSS), 498
SDMA/SS/FDMA, 496, 498
SDMA/SS/TDMA, 465, 469, 479, 496
SDMA/TDMA, 498
SEASAT satellite, 628
SECAM (video system), 82–83
SECO (secondary engine cutoff), 723
Secondary allocation, 125, 134–166
Security, 7, 410, 457, 459, 515
Semimajor axis, 781
  changes in, 786–788
  earth's shadow, 796–797
  equations, 808
  from r and ṙ, 818–819
Semiminor axis, 796–797, 808
Sensor, 554, 630–631, 646–650
  infrared (horizon), 583, 636, 646–648, 660, 757

misalignment, 659
  rf, 554, 598–599, 647–648, 762
  star, 636, 649–650, 659
  sun, 582–583, 648–649, 757
Separation:
  with launch vehicle, 663, 723
  between satellites, 218, 230–231, 339, 490, 494
Series regulator, 616–617
Service:
  analog, 76
  bit rates, 532
  grade of, 50–63
  new, 11
  voice communications, 77
S glass, properties, 679–680
Shadow of earth, 794–798. *See also* Eclipse
Shaker test, 724–730
Shaped antenna beam, 281–283, 287–290, 294, 296, 371, 573–576
Shear, 664–666, 670, 678–679
Shelf, despun, 548, 550
Shell, structure, 664, 669–670
Shield, solar cell, 605, 720–722
Shielding, earth station, 275–276
Ship, 2, 7, 127. *See also* Maritime
Shock, 556, 661, 723–724
Short:
  battery cell, 760
  solar array, 618, 628
  solar cell, 610
Short-wave radio, 2
Shunt regulator, 617–618
Shuttle orbiter vehicle:
  launch, 863, 869
  limit load, 725
  payload bay, 661–662
  vibration levels, 730–731
Sideband suppressed carrier, 473
Sidelobe, 292, 435
  antenna beamwidth, 280
  interference, 219, 286–287, 435
  limits (FCC and CCIR), 276–277
  mainlobe, level to, 295
  noise, 291–292, 310, 588
  spectrum, 186, 221
Sidereal day, 786
Signal:
  cancellation, 276
  mark, 517
  polarization, *see* Polarization
  processing, 401, 416, 506, 510
  to noise ratio (S/N), 333, 383–385, 440–442
Signaling:
  TDMA, 463
  telephone, 109, 111–118, 477
  type, 12
Silicon solar cell, 561, 605–612. *See also* Solar, cell
Simplex line, 91, 95
Simulation:
  launch environment, 723–732

of satellite system (Monte Carlo), 766–772
solar cell tests, 718–722
thermal vacuum test, 716–718, 732–736
Sines, law of, 779–780
Single:
  access, 403
  carrier operation, 362
  carrier per transponder, definition of, 403
  channel per carrier, 445–451. See also SCPC (single channel per carrier)
  point of failure, 557–558, 628
Sinusoidal vibration test, 726–729
Site shielding, 275–276
Sky noise temperature, 297, 299, 313
Slant range, see Range
Slip rings, 553, 612, 614, 628, 630
Sloshing in fuel tank, 641
Slotted ALOHA (SLOHA), 531–537
Smart terminal, 78
S/N, 333, 383–385, 440–442
Snow, 299
Solar:
  absorptivity, 687
  array, 553, 602–611
    booster charging array, 612–613
    deployed, 549
    drive, 543, 593–594, 630
    mass, 344–347, 611
    shunt regulator, 617–618
    stowed, 663
  cell, 561, 605–612
    cover slide, 605, 720–722
    current vs. voltage curve (I-V), 606–608, 622
    degradation, 545, 554, 609, 616
    failure, 610–611, 618, 628
    temperature, 613, 615–616, 690
    testing, 718–720, 737
  constants, 542–543, 603, 783, 842
  day, 786
  declination, 603–604, 797–801
  earth, distance to, 543, 603, 801, 842
  eclipse, 794–798
  flaps for attitude control, 554, 643
  flare protons, 544–545, 609, 720–722
  flux, 542–543, 603
  gravitational parameter, 781, 783
  longitude, 800
  mass, 783
  noise, rf, 292–314
  orbit normal, effect on, 832–842
  outage due to rf noise, 292–315
  perturbations, 832–840
  radiation, 542, 603–604, 686
  radiation pressure, 642, 859
  reflector, thermal surface, 687
  right ascension, 801
  sensor, 582–583, 648–649, 757
  spectrum, 604, 686, 718–719, 734
  sunspot, 70, 456–457, 307, 544, 720–722

thermal effect, 543, 558, 686
torque, 642–643
Solid:
  lubricant, 542, 717
  propellant rocket, 865–869
  propellant specific impulse, 652
  state amplifier, 561
Solstice, summer and winter, 543, 603, 614, 800–801
Sound in video sync signal, 454
Space, 517, 541–548
  communications resource, 258–270
  domain (or division) multiple access, 490–514. See also SDMA
  environment, 541–548, 717
  junk, 9, 844, 848
  radiation, 544–545, 609, 717, 720–722
  research service, definition, 127
  segment, 7
  segment noise budget, 449
  shuttle, see Shuttle orbiter vehicle
  system, definition, 126
  temperature, 543–544, 682–683
  vacuum, 541, 717
Spacecraft, 541–773. See specific item
SPADE, 405
  compared to other access methods, 405, 421
  demand assignment method, 406
  threshold, 451
  transponder plan, 429, 446
Specialized common carrier (SCC), 96, 98
Specific:
  heat, 680, 688–693
  impulse, 651–652, 657, 849, 867–869
  stiffness, 679
  strength, 679
Spectrum:
  CDMA (code division multiple-access), 516
  despreading, 518
  FDMA (frequency domain multiple access), 420
  overlap, 434
  rectangular, 221, 227
  shaping, 489
  sidelobes, 186, 221
  solar radiation, 604, 686, 718–719, 734
  spread, 417, 506, 517, 520, 524–527
  spreading, 239, 411, 465
  TDMA (time division multiple access), 186, 460–461, 464–466
  television, 13, 28, 456
  thermal radiation, 686
Speech, 22
Speed of light, 180, 270, 783
Speech-predictive encoding (SPEC), 585
Spherical coordinates, 228, 774–775
Spherical trigonometry, 778–780
Spin, flat, 641
Spin infrared sensor, 646–648
Spin stabilization, 548–550, 590, 613, 637–641
Spiral nomogram, see Nomogram

Spoofing, 486
Spot beam, 251–253, 290–294, 300–305, 469–470, 511, 661
Spread spectrum, 417, 506, 517, 520. *See also* CDMA; Spectrum; SSMA (spread spectrum multiple access)
examples, 524–527
Spun shelf, 550
Spurious emission, 434
SS/FDM/TDMA, 499
SSMA (spread spectrum multiple access), 241, 506, 519
SSPA (Solid state power amplifier), 347
Stabilization, 646–661. *See also* Attitude control
Stable longitude, 845–847
Standard A earth station, 273–274
Standard Mortality Table, 739–740
Standby unit, redundant, 747–751
Star network, 88
Star tracker (sensor), 636, 646, 649–650, 659
State of charge, battery, 620, 625
State populations, 298
Static:
horizon sensor, 646
load, 667–671
test, 726
Station:
address, TDMA frame, 479
change, satellite, 865–866
class codes, ITU, 183, 217–218
Stationkeeping, 849–859
box, 262, 265, 488, 491, 787–790
east-west, 854–858
mode, 659
north-south, 850–854
use of thrusters, 651
Steel, properties, 679–680
Steerable antenna beam, 294, 573
Stefan-Boltzmann law of radiation, 682–688
Stellar radiation, 543–544
Stiffness, 679
Storage orbit, 259, 844
Store and forward, 97, 416, 501, 504, 506, 510
Strain, 663–671
Strength, 679
Stress, 663–671
Structure, 549–550, 555–557, 661–682, 723
Strut, 664, 667–669
STS, *see* Shuttle orbiter vehicle
Subband, frequency, 253
Subcarrier, FM, 453
Sublease of satellite, 2
Submarine cable, 17, 25, 72–74
Subsatellite point, 191, 220–225, 268, 632
Subscriber equipment, 88, 111
Subsystem testing, 597, 717–720, 737
Summation law, thermal, 702
Summer solstice, 543, 614, 800–801
Sun, *see* Solar
Sun-synchronous orbit, 506

Supergroup, 12, 18
Supermastergroup, 18
Superstation (TV), 81
Supervisory:
information listed, 471, 477
signaling, 109
Support subsystem module, 557
Surface:
thermal properties, 685, 687
tolerance, 283–284, 289
Survival of satellite, 760
Swept sinusoidal vibration test, 726–729
Switch (in satellite):
attenuator, 438, 444, 497, 561–562
cross-connect, 492–509
matrix, 349–350, 495–507
throw charges, 496
Switching (on ground):
centers:
AT&T, 45
international, 88, 92
in earth station, 359
office, 13
packet, network, 97, 100, 102, 416
plan, network, 90, 100–101
Switchless FDMA/SDMA, 512, 515
Symbol, telecommunications offices, 88, 92
Synchronization:
signals, packet, 412, 415
TDMA, 479–480, 485
telemetry beacon, 501
Synchronous orbit, 127, 258–260, 264–265. *See also* Geostationary orbit
Synchronous terminals, 102
Syncom satellite, 8
System:
definition, 126
margin, 471
modeling, 337–343
noise temperature, 291. *See also* Noise
quality factor, 274–275
trade-off, reliability, 761–766

TAC (Telenet access controller), 102
Tail (local loop), 101
Talk-down, disconnect, 109
Tank, fuel, 541, 555, 641, 652, 656
TASI locking tone, 116–117
TASO (Television Advisory Standards Organization), 387
TAT (submarine cable), 70, 72–74
T-carrier equipment, 104–105
TDA (tunnel diode amplifier), 318, 561, 563
TDM (time division multiplex):
down-link to VSAT, 510
multiplexer, 103, 459
scrambling, 457. *See also* Scrambling for security
spectrum, 460–461, 466
TDMA (time domain multiple access), 403, 458–489

adaptive network, 471–472
assignment methods, 462
basic characteristics, 409–411, 458–465, 476
channel assignments, 404, 462, 477
C/N and C/N$_0$, 471
earth station, 459, 493–494, 565
impairments, 484–489
operations, 463–464, 468–484
with other services, 467
shared transponder, 467
spectra nomogram, 474
spectrum, 186, 460–461, 464–466
time delay variation, 261, 488–492
types, 465–468
up-link power, 475–478
TDMA/SDMA/SS/TDMA, 515
TDM/PSK/FDMA, 421
Technology inversion, 515
Telecommunications:
    classes, 12, 14–17
    growth, 64–85, 108
    networks, 85–122
Teleconference, 76, 82
Telegraph growth, 65
Telemail, 529
Telemetry, 578–592
    antenna, 549, 591
    beacon, 577–579
    data, 581–585
    functional diagram, 574, 577
    link budget, 585–588
    list, 579–580
    subsystem reliability, 737, 753–754
Telenet access controller, (TAC), 102
Telephone:
    area code, 39, 471
    calls, see Calls, telephone
    in cities, 41
    vs. data and text processing, 31
    FDMA, 419–421
    vs. gross domestic product, 30, 35
    growth, 67, 71, 77
    lines, conditional, 5
    by longitude, 34
    off-hook, 109, 113
    office, 90–92
    signaling frequencies, 113–122
    signal to noise ratio, FM, 386
    traffic nomogram, 51
    video, 76, 82
    volume, 35, 73, 75–76
Television:
    audio handling, 453–457
    bandwidth, 386
    cable (CATV), see Cable television (CATV)
    channel assignment, 107, 145
    digital transmission, 13, 338, 388
    growth, 64, 81–85
    high definition (HDTV), 82

MAC (multiplexed analog component), 13, 18, 454
PAL and SECAM, 82–83
PBS, 454–457
quality, measurement, 387
receiver, 64, 87, 455
signal strength, 224, 233
S/N (signal-to-noise ratio), 383–385
spectrum, 13, 28, 456
telephone, video, 76, 82
TV/FM/FDMA, 451–457
two carriers per transponder, 240, 453
Telstar satellite, 8
Temperature, see Thermal
    correction, battery voltage, 622–623
    gradient, 712–716, 732–733
    noise, 307, 314, 320
    of satellite, 557–558
    of solar cell, 613, 615–616, 690
    of space, 543–544, 682–683
    telemetry of, 580–581
    transient, 688–693
Tensile strength, 677, 679, 681
Terminal:
    asynchronous, 102
    data, 78, 81
    interactive, 31
    interface processor, 102
    TDMA, 493, 494
    very small aperture (VSAT), 510
Terrestrial, see Earth; Earth station
    carriers, 4
    interference, microwave, 219–220, 252, 275–276, 593
    time delay, 271
Test, spacecraft, 716–737
Test tone, telephone, 109–112
Text processing, vs. data and audio/video, 31
Theft of service, 2
Thermal, 682–716
    balance test, 732–736
    battery requirements, 557
    black body radiation, 684–685
    conductivity, 541, 680–681, 712–716
    configuration factor, 698–706
    control, 557–558, 614, 682–716
    cycling test, solar cell, 719
    diffusivity, 680–681
    dissipation, 549, 557, 682–686
    distortion, antenna, 270
    emissivity, 686–688
    expansion, 680–681
    heat flow, 361, 364, 712
    insulation, 550, 656, 696–698
    mass, 688–693
    model, 704–712
    noise, 328–329, 441–449
    properties of spacecraft materials, 678–681
    radiation, 557–558, 682–688, 692–712
    specific heat, 680

Thermal (*Continued*)
  surface properties, 685–688
  transient, 688–693
  vacuum tests, 717–718, 732–736
Thermistor, 581, 735
Thermocouple, 735
Thin-route traffic, 271, 489
Three-axis stabilization, 636–638. *See also* Attitude
    control
  configuration, 548–549
  power, 613
Threshold E_b/N_0, 337
Throughput:
  RMA, 528–529, 531
  TDMA, 462–463
Thruster, 656–657
  control, 657–659
  fuel, 651–657, 849–858. *See also* Fuel
  function, 630–631
  inclination maneuver, 851–852
  location, 549
  misalignment, 643
  physics principles, 651–653, 849–850, 867–869
  propulsion system, 651–657
  reliability, 753, 761
Tie line, 86
TIM (terrestrial interface module), 459
Time:
  delay, satellite range, 225, 270–273, 828
  epoch, 797–800, 805
  Greenwich Mean, 799–800, 821
  guard, 415, 479
  holding, 47–48, 50–52, 73
  interval between inclination maneuvers, 853
  interval between longitude maneuvers, 857
  nutation constant, 639–640
  resonance, 566, 645, 657, 671–677, 728
  standard signal satellite, 152–160
  store and forward, 97, 416, 501, 504, 506, 510
  thermal constant, 690–693, 736
  Universal (UT), 799–800, 821
  zone, traffic, 50
Time domain (or division), 87, 403
Time domain multiple access (TDMA), 458–489.
    See also TDMA (time domain multiple access)
Time domain switch matrix, 500
Timeliness, 12, 57
Timing, via beacon, 501
TIP (terminal interface processor), 102
Titan launch vehicle, 730–731, 868
Titanium, use in spacecraft, 679–681
T1 bandwidth, 452
Tone frequencies, telephone, 109, 113–122
T1 quality video, 83
Topocentric:
  coordinate system (azimuth and elevation), 228,
      776, 825–829
  separation, 218, 230–231, 339, 490, 494
Toroidal antenna, 548, 588–590, 596
Torque, 547, 631, 642, 657

  external, 642–645
  magnetic, 547, 644
  physical concept, 630–631
  use in attitude control, 554, 657–659
Torr (or mm Hg), 542, 734. *See also* Vaccum
Touch tone frequencies, 118–120
Tracker, star, 636, 646, 649–650, 659
Tracking antenna, ground, 3, 9, 259, 264, 597–599
Tracking & Data Relay Satellite (TDRS), 499, 502
Traffic, 11–122
  digital, 412
  vs. distance, 42, 45, 83
  vs. earth station rank, 37
  economy of scale, 83–84
  estimation, 11
  growth, 9, 64–85, 101, 108
  Intelsat, 30, 36–38, 66
  vs. longitude, 36, 194
  network, 85–122
  nomogram, 51
  potential *vs.* charges, 83, 85
  sources, 28–41
  type, 12–22
  unit, erlang, 50–64, 75
  United States, 46–47, 72, 76. *See also* United
      States
  variation, daily, weekly, annual, 48–52, 108
Trajectory, 774–870. *See also* Orbit
Transducer, telemetry, 581
Transfer orbit, 659, 723, 860–865, 869
Transient, thermal, 688–693
Transmission:
  attenuation, 327, 338–340. *See also* Attenuation
  blind, 508
  delay, *see* Delay
  digital, 13, 25, 473
  line loss noise, 297, 309–312
  path loss, 226, 254–256, 327, 338, 587
  time, facsimile, 13, 22
Transmitter, *see* HPA; Power amplifier; Traveling-
    wave tube
Transponder:
  capacity, 425–431, 477, 482–484
  dual video, 240, 434, 451, 453
  filter, limitation, 484, 489
  gain switch, 444. *See also* Attenuator
  hopping, 414, 506, 507
  noise budget, 449
  power and mass, 337–354
  power flow, 361, 364
Transportable earth station, 6
Transverse moment of inertia, 641
Transverse strain, 664
Traveling-wave tube, 561, 564–567
  amplifier (TWTA), 347–354, 358–359, 383, 459,
      486
  input power, *see* Backoff in TWT operation
  multicollector, 465, 567
  noise temperature, 318
  power supply requirements, 628–629

reliability, 758, 761
saturation, 361–363, 432–433, 446, 450, 463, 565–566
Triangular waveform, 239
Triaxial earth, 262, 840–848
Trigonometry, spherical, 778–780
Trim attenuator, 444
Triple Golay code, 472
Troposphere, noise temperature, 296, 308
True anomaly, 802–804
    definition, 780–781
    equations, 809
    from **r** and **r˙**, 818–819
True equator of data, 776
Trunk:
    cable (CATV), 106
    grade of service, 52–56
    traffic, 46
TT&C system (telemetry, tracking, and command), 573–602, 753–754
TU (traffic unit), 52
Tunnel diode amplifier (TDA), 318, 561, 563
Two-wire telephone network, 90–95
TWT, 561, 564–567. *See also* Traveling-wave tube
TWTA, *see* Traveling-wave tube, amplifier (TWTA)
Type of traffic, 12–22

UC (unit call), 52
UHF (ultra high frequency), 213
Ultimate stress, 666–667
Ultraviolet, 210, 605, 719. *See also* Spectrum, solar radiation
Underestimation of facilities and traffic, 11
Undersea cable, 5–6
Underserved markets, 3
Unique code (CDMA), 87
Unique word, 412, 479
Unit call, 52
United States:
    coverage, 306, 371
    population, 32, 38–41, 297–298, 511
    traffic, 46–47, 72, 76
Universal time (UT), 799–800, 821
Unlicensed operation, 128
Unmanned earth station, 489
Unregulated bus, 615–616
Unreliability, 738, 746
Unstable longitude, 845–847
Unswitched service, 57, 86
Up-converter, 419, 459
Up-link (earth-to-space link):
    broadcasting satellite (DBS), 380
    budget calculation, 354–361
    equations (summary), 388–392
    fade, 463
    frequency allocations, 136–139
    impairment, TDMA, 486
    INTELSAT V, 379, 383
    maritime satellite, 381–382
    power control, 264

power control
    in FDMA, 402, 411, 425
    in TDMA, 464, 478
Urban areas, U.S., 38–39, 46, 297
User:
    definitions (ITU), 126
    premises, 10
    satisfaction, 97
Utilization, ALOHA, 530

V and H coordinates, AT&T, 43
Vacuum:
    chamber, 704–705, 732–734
    in space, 541, 717
Valence band, solar cell, 605–606
Valve, propulsion fuel, 656
Van Allen radiation belt, 547–548
Van de Graaff particle accelerator, 720
Vane, solar torque, 643
Variable power divider, 496
Velocity:
    circular orbit, 781
    difference, 851–866
        at apogee, 805, 808, 860–866
        equation, 787
        Hohmann transfer, 860–862
        maneuver, 853–858
        momentum, 849
    energy conservation, 804, 808, 855
    equations, 808–809
    escape, 862
    exhaust, 651, 849, 867
    of light, 180, 273, 783
    from orbit elements, 812–816
    vs. radius, 785
Verification of command, 576–577, 593
Vernal equinox, 542–543, 614, 774, 798–801
Very small aperture terminal, 11, 33, 403, 406. *See also* VSAT (very small aperture terminal)
Vestigial sideband suppressed carrier, 473
VHF (very high frequency), 213
Via station in double hop, 271–272
Vibration:
    launch environment, 555–556, 661, 723
    spacecraft structure, 673–678
    testing, 720–732
Video, *see* Television
Videotelephone, 76, 82
View from satellite, 268
Violet solar cell, 607–609
Virtual circuit, 102
Viscous damper, 639
Vis Viva equation, 804, 808, 855
Vocoder (PEV), 416
Voice:
    activation, 422, 448
    activity, 404
    circuits by longitude, 36
    digitization, 29
    PCM, 460

Voltage breakdown, 542
Voltage regulator, 344, 615–618
VOX (voice activation), 422, 448
VSAT (very small aperture terminal), 11, 33, 403, 406
    customer premises, 97
    line losses, 359
    multihop, 271
    variation between up-link and down-link, 510
    VSB/SC (vestigial sideband suppressed carrier), 473
V-slit sun sensor, 648

Wall Street Journal, 21
Walsh code, 522–523
WARC, 124–128. *See also* CCIR
Warming, 688–693
Water vapor absorption band, 164–166, 254–256, 309
Waveguide:
    attenuation, 309, 316
    radiation from, 568
Wavelength:
    vs. antenna gain, 278, 570, 572
    to frequency conversion, 180, 211, 277, 586
Wear-out failure, 741, 745–746
Weight, *see* Mass; *specific items*

Weighted antenna patterns, 288
Weighting network, audio, 438, 441
Welding, 542
Wheel, reaction or momentum, 548, 554, 630, 636–638, 650–651, 762–764
Winter solstice, 543, 603, 800–801
Wobble, 581, 638–640
Word processing, *vs.* audio/video and data, 31
World Administrative Radio Conference (WARC), 124–125
World population, 71

Xenon arc lamp, 718–719

Yaw, 270, 632–633
    control, *see* Attitude control
    error, 637, 660
    gravity gradient, 645
    sensor, 648–650
Yield stress, 666–667
Young's modulus, 664–667, 678–679

Zero gravity, 541, 656, 717, 784
Zero momentum, attitude control, 636–637, 650, 658
Zip code, 39
Ziph's law, population, 32